Quantum Information and Foundations

Quantum Information and Foundations

Special Issue Editors

Giacomo Mauro D'Ariano
Paolo Perinotti

MDPI • Basel • Beijing • Wuhan • Barcelona • Belgrade • Manchester • Tokyo • Cluj • Tianjin

Special Issue Editors
Giacomo Mauro D'Ariano
QUit Group, Department of
Physics, University of Pavia
Italy

Paolo Perinotti
QUit Group, Department of
Physics, University of Pavia
Italy

Editorial Office
MDPI
St. Alban-Anlage 66
4052 Basel, Switzerland

This is a reprint of articles from the Special Issue published online in the open access journal *Entropy* (ISSN 1099-4300) (available at: https://www.mdpi.com/journal/entropy/special_issues/quantum_information_foundations).

For citation purposes, cite each article independently as indicated on the article page online and as indicated below:

LastName, A.A.; LastName, B.B.; LastName, C.C. Article Title. *Journal Name* **Year**, *Article Number*, Page Range.

ISBN 978-3-03928-380-4 (Pbk)
ISBN 978-3-03928-381-1 (PDF)

© 2020 by the authors. Articles in this book are Open Access and distributed under the Creative Commons Attribution (CC BY) license, which allows users to download, copy and build upon published articles, as long as the author and publisher are properly credited, which ensures maximum dissemination and a wider impact of our publications.

The book as a whole is distributed by MDPI under the terms and conditions of the Creative Commons license CC BY-NC-ND.

Contents

About the Special Issue Editors ix

Giacomo Mauro D'Ariano and Paolo Perinotti
Quantum Information and Foundations
Reprinted from: *Entropy* 2020, 22, 22, doi:10.3390/e22010022 1

Alessandro Bisio, Giacomo Mauro D'Ariano, Nicola Mosco, Paolo Perinotti and Alessandro Tosini
Solutions of a Two-Particle Interacting Quantum Walk
Reprinted from: *Entropy* 2018, 20, 435, doi:10.3390/e20060435 9

Marc-Olivier Renou, Nicolas Gisin and Florian Fröwis
Robust Macroscopic Quantum Measurements in the Presence of Limited Control and Knowledge
Reprinted from: *Entropy* 2018, 20, 39, doi:10.3390/e20010039 27

Louis H. Kauffman
Iterant Algebra
Reprinted from: *Entropy* 2017, 19, 347, doi:10.3390/e19070347 41

Časlav Brukner
A No-Go Theorem for Observer-Independent Facts
Reprinted from: *Entropy* 2018, 20, 350, doi:10.3390/e20050350 71

Alexander Wilce
A Royal Road to Quantum Theory (or Thereabouts)
Reprinted from: *Entropy* 2018, 20, 227, doi:10.3390/e20040227 81

Giulio Chiribella
Agents, Subsystems, and the Conservation of Information
Reprinted from: *Entropy* 2018, 20, 358, doi:10.3390/e20050358 107

Howard Barnum, Ciarán M. Lee, Carlo Maria Scandolo and John Selby
Ruling out Higher-Order Interference from Purity Principles
Reprinted from: *Entropy* 2017, 19, 253, doi:10.3390/e19060253 161

John Selby and Bob Coecke
Leaks: Quantum, Classical, Intermediate and More
Reprinted from: *Entropy* 2017, 19, 174, doi:10.3390/e19040174 189

Alberto Barchielli, Matteo Gregoratti and Alessandro Toigo
Measurement Uncertainty Relations for Position and Momentum: Relative Entropy Formulation
Reprinted from: *Entropy* 2017, 19, 301, doi:10.3390/e19070301 213

Giovanni Amelino-Camelia
Planck-Scale Soccer-Ball Problem: A Case of Mistaken Identity
Reprinted from: *Entropy* 2017, 19, 400, doi:10.3390/e19080400 249

Mario Arnolfo Ciampini, Paolo Mataloni and Mauro Paternostro
Structure of Multipartite Entanglement in Random Cluster-Like Photonic Systems
Reprinted from: *Entropy* 2017, 19, 473, doi:10.3390/e19090473 257

Ämin Baumeler and Stefan Wolf
Non-Causal Computation
Reprinted from: *Entropy* **2017**, *19*, 326, doi:10.3390/e19070326 269

Chris Heunen
The Many Classical Faces of Quantum Structures
Reprinted from: *Entropy* **2017**, *19*, 144, doi:10.3390/e19040144 279

Philipp Andres Höhn
Quantum Theory from Rules on Information Acquisition
Reprinted from: *Entropy* **2017**, *19*, 98, doi:10.3390/e19030098 301

Catalina Curceanu, Hexi SHI, Sergio Bartalucci, Sergio Bertolucci, Massimiliano Bazzi, Carolina Berucci, Mario Bragadireanu, Michael Cargnelli, Alberto Clozza, Luca De Paolis, Sergio Di Matteo, Jean-Pierre Egger, Carlo Guaraldo, Mihail Iliescu, Johann Marton, Matthias Laubenstein, Edoardo Milotti, Marco Miliucci, Andreas Pichler, Dorel Pietreanu, Kristian Piscicchia, Alessandro Scordo, Diana Laura Sirghi, Florin Sirghi, Laura Sperandio, Oton Vazquez Doce, Eberhard Widmann and Johann Zmeskal
Test of the Pauli Exclusion Principle in the VIP-2 Underground Experiment
Reprinted from: *Entropy* **2017**, *19*, 300, doi:10.3390/e19070300 319

Kristian Piscicchia, Angelo Bassi, Catalina Curceanu, Raffaele Del Grande, Sandro Donadi, Beatrix C. Hiesmayr and Andreas Pichler
CSL Collapse Model Mapped with the Spontaneous Radiation
Reprinted from: *Entropy* **2017**, *19*, 319, doi:10.3390/e19070319 327

Robert B. Griffiths
Quantum Information: What Is It All About?
Reprinted from: *Entropy* **2017**, *19*, 645, doi:10.3390/e19120645 335

Benjamin F. Dribus
Entropic Phase Maps in Discrete Quantum Gravity
Reprinted from: *Entropy* **2017**, *19*, 322, doi:10.3390/e19070322 347

Yangyang Wang, Xiaofei Qi and Jinchuan Hou
Nonclassicality by Local Gaussian Unitary Operations for Gaussian States
Reprinted from: *Entropy* **2018**, *20*, 266, doi:10.3390/e20040266 395

Kevin Vanslette
Entropic Updating of Probabilities and Density Matrices
Reprinted from: *Entropy* **2017**, *19*, 664, doi:10.3390/e19120664 409

Andriyan Bayu Suksmono
Finding a Hadamard Matrix by Simulated Quantum Annealing
Reprinted from: *Entropy* **2018**, *20*, 141, doi:10.3390/e20020141 433

Ameneh Arjmandzadeh and Majid Yarahmadi
Quantum Genetic Learning Control of Quantum Ensembles with Hamiltonian Uncertainties
Reprinted from: *Entropy* **2017**, *19*, 376, doi:10.3390/e19080376 447

Lucas Kocia, Yifei Huang, Peter Love
Discrete Wigner Function Derivation of the Aaronson—Gottesman Tableau Algorithm
Reprinted from: *Entropy* **2017**, *19*, 353, doi:10.3390/e19070353 459

Alain Deville and Yannick Deville
Concepts and Criteria for Blind Quantum Source Separation and Blind Quantum Process Tomography
Reprinted from: *Entropy* **2017**, *19*, 311, doi:10.3390/e19070311 . . . . **477**

About the Special Issue Editors

Giacomo Mauro D'Ariano is full professor at Pavia University, where he teaches Quantum Mechanics and Foundations of Quantum Theory, and leads the group QUit. He is a fellow of the American Physical Society and of the Optical Society of America, a member of the Academy Istituto Lombardo of Scienze e Lettere, of the Center for Photonic Communication and Computing at Northwestern IL, and of FQXi. He is the (co)author of more than 350 articles in peer-reviewed physics journals. He started Quantum Information in Italy, where he created a school that spread scholars worldwide.

Paolo Perinotti is an associate professor at the Physics Department of Pavia University. He teaches "Theoretical physics and information theory" and "Statistical Mechanics". His research activity is focused on quantum information theory, quantum foundations, and quantum mechanics. In 2016 he was awarded the Birkhoff Von Neumann prize for researches in Quantum Foundations. He is a member of the Foundational Questions Institute (FQXi), and of the International Quantum Structures Association.

Editorial

Quantum Information and Foundations

Giacomo Mauro D'Ariano * and Paolo Perinotti *

QUIT Group, Dipartimento di Fisica dell'Università di Pavia, Istituto Nazionale di Fisica Nucleare, via Bassi 6, 27100 Pavia, Italy
* Correspondence: dariano@unipv.it (G.M.D.); paolo.perinotti@unipv.it (P.P.)

Received: 3 December 2019; Accepted: 10 December 2019; Published: 23 December 2019

Keywords: quantum information; quantum foundations; quantum theory and gravity

---

The new era of quantum foundations, fed by the quantum information theory experience and opened in the early 2000s by a series of memorable papers [1–3], led in a few years to a wealth of results, that can all be roughly traced back to the idea of testing quantum theory against new rivals instead of struggling in the worn-out attempt at its recomprehension within a classical imaginative world. The first remarkable construction of a toy theory for foundational purposes, in our knowledge, is represented by Ref. [4].

The study of foil theories along with their informational power lead to important progress, paralleled with an increasing understanding of the new foundational scenario [5]. Most importantly, this stream of thought is the origin of the new paradigm of the so-called *reconstructions*, which aim at singling out quantum theory in a wider scenario of possible theories of elementary physical systems [6,7]. Grant the authors an unwarranted bit of pride in stating that a clear picture of such a playground is now available thanks to the formulation of the concept of *Operational Probabilistic Theory* (OPT) [8,9]. As a result of the growing interest, we now understand quantum theory as a special kind of information theory, with postulates that regard the possibility or impossibility to carry out specific information processing tasks, instead of directly describing the mathematical structures of Hilbert spaces, operator algebras, and alike.

One of the future challenges for the informational approach to quantum foundations is then to embrace the mechanical part of the theory, besides the merely information-theoretic one, or, better, to remain on top of it.

The time demarcation represented by the year 2000 is of course artificial, just like every symbolic date, as quantum information was strictly connected to foundations since its very birth. One could not express this fact in better words than Chris Fuchs' own: *"The title of the NATO Advanced Research Workshop that gave birth to this volume was 'Decoherence and its Implications in Quantum Computation and Information Transfer' ... The life of the party was all the talks and conversations on 'Decoherence and its Implications in Quantum Foundations'."* [10]. The new approach, moreover, has some deep connections with the previous experience that can be broadly collected under the name *quantum logic*. Having said that, the turn of the century undoubtedly brought the foundations new vigour.

This special issue is meant to witness recent progress of the balanced and fertile interchange between the developments in application-oriented quantum information theory and those in foundations. As a result, the response of the authors was great, and produced a perfect blend of flavours. The subjects of the contributions can be briefly classified in three groups.

The first one can be deemed *resources*. One of the main topics in a well-organised information theory is quantification and classification of resources. It is nowadays common wisdom that *the resource for quantum computation and information is entanglement*, which is incidentally one of the main resources also for foundations. In a broader view, entanglement is one of the *nonclassical* resources allowed by quantum theory.

In the contribution, *Nonclassicality by Local Gaussian Unitary Operations for Gaussian States* [11], the authors introduce a measure of nonclassicality for Gaussian states of continuous variable systems and compare it with other measures of nonclassical correlations. The resource in this case is nonclassicality, namely, the ability to produce phenomena that are not reproducible by classical means. The proposed measure of nonclassicality is explicitly computed for a system of two bosonic modes, and estimated in the general case.

In another respect, one of the primary resources in quantum information is the ability to prepare states on demand. Methods for predicting the statistical efficiency of sources, or for sharpening our description of preparations through density matrices in the presence of partial information are then of the utmost importance.

In the paper, *Entropic Updating of Probabilities and Density Matrices* [12], the author analyses the task of reconstructing the theoretical description of a quantum state from partial experimental information. The standard relative entropy and the Umegaki entropy are derived in parallel from the same set of design criteria.

Finally, in the contribution, *Structure of Multipartite Entanglement in Random Cluster-Like Photonic Systems* [13], the authors analyse the size of multipartite entanglement in randomly generated cluster states, relating it to the density of nodes in the cluster.

A second collection of contributions regards *algorithms and protocols*. This selection witnesses progress in the ongoing challenge towards new algorithms and new tasks. In the contribution, *Finding a Hadamard Matrix by Simulated Quantum Annealing* [14], the author analyses quantum algorithms for finding a Hadamard matrix, which is itself a hard problem. The problem is reformulated in terms of energy minimisation of spin vectors connected by a complete graph, and approached via path-integral Monte-Carlo techniques. The scaling properties of the method show that the quantum algorithm outperforms its classical counterpart in solving this hard problem, providing yet another hint to quantum supremacy.

In the contribution, *Quantum Genetic Learning Control of Quantum Ensembles with Hamiltonian Uncertainties* [15], the authors propose a new method for controlling a quantum ensemble of two-level systems with uncertainties in the parameters of the Hamiltonian system. The method is based on the combination of a sample-learning control and a quantum genetic algorithm, witnessing the continuous cross-fertilisation between quantum theory and computer science.

The authors of the contribution, *Discrete Wigner Function Derivation of the Aaronson-Gottesman Tableau Algorithm* [16], present a discrete Wigner-function-based simulation algorithm for odd-d qudits that has the same time and space complexity as the Aaronson–Gottesman algorithm for qubits. The authors also discuss the differences between the Wigner function algorithm for odd-d and the Aaronson–Gottesman algorithm for qubits, conjecturing that they are due to the fact that qubits exhibit state-independent contextuality. This may provide a guide for extending the discrete Wigner function approach to qubits. Considering this result, one can easily realise how tightly quantum computation and quantum foundations are bound.

Concepts and Criteria for Blind Quantum Source Separation and Blind Quantum Process Tomography [17] discusses communication protocols for demixing a signal from the output of a communication line and establishes properties that were already used without justification in that context. The scenario considered here involves a pair of electron spins initially prepared in a pure state and then submitted to an undesired exchange coupling. The authors introduce a criterion for checking that the coupling does not produce entanglement.

In recent years, after studies that provided a fully algebraic method for analysing quantum circuits [18], it was realised that there are easy protocols challenging the circuit model, but are still amenable to a fully algebraic account [19]. Some of these protocols can be interpreted as computations that call events in a causally indefinite order, thus hinting to interesting foundational questions. In the article, *Non-Causal Computation* [20], the authors review recent results on indefinite orders and their potentiality in computation, replacing the requirement of a global ordering between gates in the computation with that of mere logical consistency.

The third collection regards *foundations*. This is the subject that encompasses all the remaining contributions, that amount to fifteen, with a very diverse span of subjects, approaches, and techniques.

One of the lessons of the quantum information theoretical approach to foundations is that very often physical concepts are easily grasped referring to the operations and processes they can undergo. In this spirit, the author of the contribution, *Agents, Subsystems, and the Conservation of Information* [21] proposes a mathematical modelling for subsystems of physical systems in the general scenario of OPTs, where subsystems are identified through a subalgebra of the full algebra of operations on the composite system they are part of. Various cases are then discussed, with a particular focus on quantum systems.

The relevance of appropriately treating subsystems of composite systems might appear somewhat technical at a superficial sight, but after giving the subject some more thought, one realizes that the notion of subsystem underlies many fundamental questions, e.g., Wigner's thought experiment popularly known as the Wigner's friend paradox. This is the subject of the contribution, *A No-Go Theorem for Observer-Independent Facts* [22], which proposes a perspective on the argument of Frauchiger and Renner [23] proving that "single-world interpretations of quantum theory cannot be self-consistent". The author derives a no-go theorem for observer-independent facts, which would be common both for Wigner and the friend. This result is claimed to undermine one of the assumptions behind the concept of "self-consistency" by the authors of Ref. [23].

The analysis of conceptual foundational questions is possible thanks to the availability of a suitable mathematical language. A continuous process of reformulation and reconsideration of the latter is an important chapter in quantum foundations, as witnessed by the contribution, *A Royal Road to Quantum Theory (or Thereabouts)* [24]. Here, the author proposes an alternate perspective for approaching the problem of reformulating the mathematical language of quantum theory from simple postulates, based on the theory of Euclidean Jordan algebras. While the paper, as declared by the author, "fails to derive quantum mechanics", it derives a more general framework that embraces the quantum along with alternate, not wildly different possible theories.

In addition, the article *Quantum Theory from Rules on Information Acquisition* [25] reviews a reconstruction of the mathematical framework of quantum theory. The starting point here is a set of rules constraining an observer's acquisition of information about physical systems. The reconstruction offers an informational explanation for entanglement, monogamy, and nonlocality, from limited accessible information and complementarity. The analysis leads to a notion of "conserved informational charges" that stems from complementarity relations that characterise the unitary group and the set of pure states.

The review *The Many Classical Faces of Quantum Structures* [26] addresses a mathematical reformulation of quantum mechanics in terms of classical mechanics. The standpoint for this approach is that interpretational problems with quantum mechanics can be phrased precisely by only talking about empirically accessible information. This review spells out the main points of the abovementioned approach in terms of the algebraic structures lying behind quantum theory.

After the reconstruction of the mathematical language of quantum theory from information theoretical postulates was completed, one of the possible developments was the attempt at a reformulation of quantum mechanics from information processing. In this respect, much progress was achieved, essentially showing that one can have a fully information-theoretic account of the basic equations at the core of relativistic quantum field theory, such as Weyl's and Dirac's [27,28], and Maxwell's [29]. The next difficult step in this direction is introducing interactions. A recent result in this direction is the study of all possible interacting cellular automata in one dimension along with a full diagonalization of their two-particle sector [30]. In the contribution, *Solutions of a two-particle interacting quantum walk* [31], the authors provide an alternative solution of the dynamics of the abovementioned class of cellular automata based on a path-sum approach.

Once again, on the exploration of the language of quantum foundations, one can read *Ruling out Higher-Order Interference from Purity Principles* [32], where the authors analyse the principles of Causality, Purity Preservation, Pure Sharpness, and Purification in the operational framework of generalised probabilistic theories, proving that these principles limit interference to second-order, namely, the interference pattern formed in a multislit experiment is a function of the interference patterns formed between pairs of slits. This behaviour is typical of quantum theory, where there are no genuinely new features resulting from considering three slits instead of two. Systems in such theories correspond to Euclidean Jordan algebras.

Another contribution that is focused on the mathematical language and its framework is *Leaks: Quantum, Classical, Intermediate and More* [33], where the authors introduce the notion of a leak for general process theories and identify quantum theory as a theory with minimal leakage, as opposed to classical theory that has maximal leakage. Leaks are processes that provide leakage of classical information, and can be introduced in most theories. These processes allow for a category theoretical account of decoherence as a mechanism for the emergence of classical theory in a quantum scenario. The authors also discuss the relation of leaks with purity of processes.

One of the main themes in the context of reconstructions and reformulations of quantum theory is to open the route to possible new post-quantum theories. The article, *Iterant Algebra* [34] moves a step beyond quantum theory, starting from a generalisation of the structure of matrix algebra, motivated by the structure of measurement for discrete processes. Iterant algebra is shown to embrace matrix and Clifford algebras, and the framework is then applied to discuss various aspects of quantum mechanics, such as the Schrödinger and Dirac equations, Majorana Fermions, and representations of the braid group.

We now move to a different chapter in foundations, where one can use the standard mathematical formalism to face questions and concepts that have interpretational issues. An example is given by *Robust Macroscopic Quantum Measurements in the Presence of Limited Control and Knowledge* [35]. The authors tackle the problem of compatibility of quantum behaviour and macroscopic measurements, focusing on the estimation of the polarization direction for a large system of spin 1/2 particles. The analysis starts from a model of von Neumann pointer measurement and shows traits of a classical measurement for an intermediate coupling strength. A relevant part of the contribution is devoted to the analysis of response of the model against relaxations of the initial assumptions, showing that the model is robust.

One of the fundamental subjects that attracted interest from the very birth of quantum mechanics is uncertainty. The study of uncertainty is still lively, and the present special issue includes one contribution that is devoted to this subject: *Measurement Uncertainty Relations for Position and Momentum: Relative Entropy Formulation* [36]. The authors analyse uncertainty as related to incompatibility of different observables, where the latter is quantified by the amount of unavoidable approximation in a joint measurement. As a quantifier of information loss, the authors consider relative entropy of a "true" probability distribution and an approximating one. Such an analysis is applied to obtain lower bound for the amount of information that is lost by replacing the distributions of the sharp position and momentum observables, as they could be obtained with two separate experiments, by the marginals of any smeared joint measurement.

The renewed interest in fundamental problems produced new approaches to the unification of quantum mechanics and the theory of gravity. Recent trends in quantum gravity are thus of high interest for the community working in foundations and, for this reason, we appreciate the value of a contribution such as *Planck-Scale Soccer-Ball Problem: A Case of Mistaken Identity* [37], which reports about reflections on the rule of composition for momenta. Over the last decade, nonlinear laws of composition of momenta were predicted by many approaches to quantum gravity. In order to dissipate concerns about such nonlinearity, the author discusses the subtle difference between the two roles that a law of momentum composition play: the first one is related to the description of space-time locality, and the second one is related to translational invariance. The contribution exhibits an example of space-time where the local structure provides a nonlinear composition of momenta and yet translational invariance is expressed by a linear law for the addition of momenta of many-particle systems.

Another contribution focused on a model aiming at a formulation of quantum gravity is *Entropic Phase Maps in Discrete Quantum Gravity* [38], where the author makes an attempt based on path summation over a space of evolutionary pathways in a history configuration space. This approach enables derivation of discrete Schrödinger-type equations, and mathematical constructions thereof are used to introduce entropic functions that obey an abstract version of the second law of thermodynamics.

One of the most remarkable consequences of the widespread interest in foundations is a flourishing of experiments aimed at testing fundamental questions, or challenging established pillars of quantum theory. A remarkable example is the Pauli exclusion principle for Fermions, that has been tested in a series of recent experiments, in an ongoing effort that is witnessed also by a contribution in the present issue, *Test of the Pauli Exclusion Principle in the VIP-2 Underground Experiment* [39]. Here, the authors report progress of the VIP-2 experiments at the Laboratori Nazionali del Gran Sasso, seeking a prohibited transition in copper atoms of a $2p$ orbit electron to the fully populated ground state, via X-ray analysis. The present limit on the probability for Pauli exclusion principle violation for electrons set by the VIP experiment is 4.7×10^{-29}. A first result from the VIP-2 experiment improves on the VIP limit, while the goal is a gain of two orders of magnitude in the long run.

A second example is the test of spontaneous collapse models, which aim at an objective solution of the measurement problem that keeps the quantum formalism untouched while tweaking its dynamical equations. In the contribution, *CSL Collapse Model Mapped with the Spontaneous Radiation* [40], new upper limits on the parameters of the Continuous Spontaneous Localization collapse models are extracted. The main idea behind the experiment is to analyse IGEX data about X-ray emission and compare them with the spectrum of the spontaneous photon emission process predicted by collapse models. This study allows for the exclusion of a broad range of the parameter space for CSL models.

Finally, we include a contribution out of line, which is more focused on interpretational issues than technical, such as *Quantum Information: What Is It All About?* [41]. In this contribution, the author answers the provocative question originally posed by John Bell, claiming that, in the consistent histories approach to quantum theory, information is meant about projectors on subspaces of the Hilbert space of a system, representing its quantum properties. The main focus is the discussion of how the single-framework rule—i.e., the rule for assigning probabilities to a projective decomposition of the identity—for consistent histories avoids contradictions and recovers both classical information theory and macroscopic physics. Room for issues is left only in the regimes without classical analogue, where a single framework is not sufficient.

As a concluding remark, we would like to thank all the authors for their contributions and declare our satisfaction in verifying the ongoing interest in fundamental problems—the only possible fuel for the science and technology of tomorrow.

Acknowledgments: We express our thanks to the authors of the above contributions, and to the journal Entropy and MDPI for their support during this work.

Conflicts of Interest: The authors declare no conflict of interest.

References

1. Hardy, L. Quantum theory from five reasonable axioms. *arxiv* **2001**, arXiv:quant-ph/0101012.
2. Fuchs, C.A. Quantum Mechanics as Quantum Information (and only a little more). *arXiv* **2002**, arXiv:quant-ph/0205039.
3. Brassard, G. Is information the key? *Nat. Phys.* **2005**, *1*, 2–4. [CrossRef]
4. Hardy, L. Disentangling nonlocality and teleportation. *arXiv* **1999**, arXiv:quant-ph/9906123.
5. Spekkens, R.W. Evidence for the epistemic view of quantum states: A toy theory. *Phys. Rev. A* **2007**, *75*, 032110. [CrossRef]

6. Dakic, B.; Brukner, C. Quantum theory and beyond: Is entanglement special? In *Deep Beauty: Understanding the Quantum World through Mathematical Innovation*; Halvorson, H., Ed.; Cambridge University Press: Cambridge, UK, 2011; pp. 365–392.
7. Masanes, L.; Müller, M.P. A derivation of quantum theory from physical requirements. *New J. Phys.* **2011**, *13*, 063001. [CrossRef]
8. Chiribella, G.; D'Ariano, G.M.; Perinotti, P. Probabilistic theories with purification. *Phys. Rev. A* **2010**, *81*, 062348. [CrossRef]
9. D'Ariano, G.M.; Chiribella, G.; Perinotti, P. *Quantum Theory from First Principles: An Informational Approach*; Cambridge University Press: Cambridge, UK, 2017.
10. Gonis, T.; Gonis, A.; Turchi, P.E. *Decoherence and Its Implications in Quantum Computation and Information Transfer*; NATO Science Series: Computer and Systems Sciences; IOS Press: Amsterdam, The Netherlands, 2001.
11. Wang, Y.; Qi, X.; Hou, J. Nonclassicality by Local Gaussian Unitary Operations for Gaussian States. *Entropy* **2018**, *20*, 266. [CrossRef]
12. Vanslette, K. Entropic Updating of Probabilities and Density Matrices. *Entropy* **2017**, *19*, 664. [CrossRef]
13. Ciampini, M.A.; Mataloni, P.; Paternostro, M. Structure of Multipartite Entanglement in Random Cluster-Like Photonic Systems. *Entropy* **2017**, *19*, 473. [CrossRef]
14. Suksmono, A.B. Finding a Hadamard Matrix by Simulated Quantum Annealing. *Entropy* **2018**, *20*, 141. [CrossRef]
15. Arjmandzadeh, A.; Yarahmadi, M. Quantum Genetic Learning Control of Quantum Ensembles with Hamiltonian Uncertainties. *Entropy* **2017**, *19*, 376. [CrossRef]
16. Kocia, L.; Huang, Y.; Love, P. Discrete Wigner Function Derivation of the Aaronson–Gottesman Tableau Algorithm. *Entropy* **2017**, *19*, 353. [CrossRef]
17. Deville, A.; Deville, Y. Concepts and Criteria for Blind Quantum Source Separation and Blind Quantum Process Tomography. *Entropy* **2017**, *19*, 311. [CrossRef]
18. Chiribella, G.; D'Ariano, G.M.; Perinotti, P. Theoretical framework for quantum networks. *Phys. Rev. A* **2009**, *80*, 022339. [CrossRef]
19. Chiribella, G.; D'Ariano, G.M.; Perinotti, P.; Valiron, B. Quantum computations without definite causal structure. *Phys. Rev. A* **2013**, *88*, 022318. [CrossRef]
20. Baumeler, Ä.; Wolf, S. Non-Causal Computation. *Entropy* **2017**, *19*, 326. [CrossRef]
21. Chiribella, G. Agents, Subsystems, and the Conservation of Information. *Entropy* **2018**, *20*, 358. [CrossRef]
22. Brukner, Č. A No-Go Theorem for Observer-Independent Facts. *Entropy* **2018**, *20*, 350. [CrossRef]
23. Frauchiger, D.; Renner, R. Quantum theory cannot consistently describe the use of itself. *Nat. Commun.* **2018**, *9*, 3711. [CrossRef]
24. Wilce, A. A Royal Road to Quantum Theory (or Thereabouts). *Entropy* **2018**, *20*, 227. [CrossRef]
25. Höhn, P.A. Quantum Theory from Rules on Information Acquisition. *Entropy* **2017**, *19*, 98. [CrossRef]
26. Heunen, C. The Many Classical Faces of Quantum Structures. *Entropy* **2017**, *19*, 144. [CrossRef]
27. Bisio, A.; D'Ariano, G.M.; Tosini, A. Dirac quantum cellular automaton in one dimension: *Zitterbewegung* and scattering from potential. *Phys. Rev. A* **2013**, *88*, 032301. [CrossRef]
28. D'Ariano, G.M.; Perinotti, P. Derivation of the Dirac equation from principles of information processing. *Phys. Rev. A* **2014**, *90*, 062106. [CrossRef]
29. Bisio, A.; D'Ariano, G.M.; Perinotti, P. Quantum Cellular Automaton Theory of Light. *arXiv* **2014**, arXiv:1407.6928.
30. Bisio, A.; D'Ariano, G.M.; Perinotti, P.; Tosini, A. Thirring quantum cellular automaton. *Phys. Rev. A* **2018**, *97*, 032132. [CrossRef]
31. Bisio, A.; D'Ariano, G.M.; Mosco, N.; Perinotti, P.; Tosini, A. Solutions of a Two-Particle Interacting Quantum Walk. *Entropy* **2018**, *20*, 435. [CrossRef]
32. Barnum, H.; Lee, C.M.; Scandolo, C.M.; Selby, J.H. Ruling out Higher-Order Interference from Purity Principles. *Entropy* **2017**, *19*, 253. [CrossRef]
33. Selby, J.; Coecke, B. Leaks: Quantum, Classical, Intermediate and More. *Entropy* **2017**, *19*, 174. [CrossRef]
34. Kauffman, L.H. Iterant Algebra. *Entropy* **2017**, *19*, 347. [CrossRef]
35. Renou, M.O.; Gisin, N.; Fröwis, F. Robust Macroscopic Quantum Measurements in the Presence of Limited Control and Knowledge. *Entropy* **2018**, *20*, 39. [CrossRef]

36. Barchielli, A.; Gregoratti, M.; Toigo, A. Measurement Uncertainty Relations for Position and Momentum: Relative Entropy Formulation. *Entropy* **2017**, *19*, 301. [CrossRef]
37. Amelino-Camelia, G. Planck-Scale Soccer-Ball Problem: A Case of Mistaken Identity. *Entropy* **2017**, *19*, 400. [CrossRef]
38. Dribus, B.F. Entropic Phase Maps in Discrete Quantum Gravity. *Entropy* **2017**, *19*, 322. [CrossRef]
39. Curceanu, C.; Shi, H.; Bartalucci, S.; Bertolucci, S.; Bazzi, M.; Berucci, C.; Bragadireanu, M.; Cargnelli, M.; Clozza, A.; De Paolis, L.; et al. Test of the Pauli Exclusion Principle in the VIP-2 Underground Experiment. *Entropy* **2017**, *19*, 300. [CrossRef]
40. Piscicchia, K.; Bassi, A.; Curceanu, C.; Grande, R.D.; Donadi, S.; Hiesmayr, B.C.; Pichler, A. CSL Collapse Model Mapped with the Spontaneous Radiation. *Entropy* **2017**, *19*, 319. [CrossRef]
41. Griffiths, R.B. Quantum Information: What Is It All About? *Entropy* **2017**, *19*, 645. [CrossRef]

© 2019 by the authors. Licensee MDPI, Basel, Switzerland. This article is an open access article distributed under the terms and conditions of the Creative Commons Attribution (CC BY) license (http://creativecommons.org/licenses/by/4.0/).

Article

Solutions of a Two-Particle Interacting Quantum Walk

Alessandro Bisio, Giacomo Mauro D'Ariano, Nicola Mosco *, Paolo Perinotti * and Alessandro Tosini

Dipartimento di Fisica dell'Università di Pavia, Istituto Nazionale di Fisica Nucleare, Pavia 27100, Italy; alessandro.bisio@unipv.it (A.B.); dariano@unipv.it (G.M.D.); alessandro.tosini@unipv.it (A.T.)
* Correspondence: nicola.mosco@unipv.it (N.M.); paolo.perinotti@unipv.it (P.P.); Tel.: +39-0382-987675 (N.M. & P.P.)

Received: 22 April 2018; Accepted: 31 May 2018; Published: 5 June 2018

Abstract: We study the solutions of an interacting Fermionic cellular automaton which is the analogue of the Thirring model with both space and time discrete. We present a derivation of the two-particle solutions of the automaton recently in the literature, which exploits the symmetries of the evolution operator. In the two-particle sector, the evolution operator is given by the sequence of two steps, the first one corresponding to a unitary interaction activated by two-particle excitation at the same site, and the second one to two independent one-dimensional Dirac quantum walks. The interaction step can be regarded as the discrete-time version of the interacting term of some Hamiltonian integrable system, such as the Hubbard or the Thirring model. The present automaton exhibits scattering solutions with nontrivial momentum transfer, jumping between different regions of the Brillouin zone that can be interpreted as Fermion-doubled particles, in stark contrast with the customary momentum-exchange of the one-dimensional Hamiltonian systems. A further difference compared to the Hamiltonian model is that there exist bound states for every value of the total momentum and of the coupling constant. Even in the special case of vanishing coupling, the walk manifests bound states, for finitely many isolated values of the total momentum. As a complement to the analytical derivations we show numerical simulations of the interacting evolution.

Keywords: quantum walks; Hubbard model; Thirring model

1. Introduction

Quantum walks (QWs) describe the evolution of one-particle quantum states on a lattice, or, more generally, on a graph. The quantum walk evolution is linear in the quantum state and the quantum aspect of the evolution occurs in the interference between the different paths available to the walker. There are two kinds of quantum walks: continuous time QWs, where the evolution operator of the system given in terms of an Hamiltonian can be applied at any time (see Farhi et al. [1]), and discrete-time QWs, where the evolution operator is applied in discrete unitary time-steps. The discrete-time model, which appeared already in the Feynman discretization of the Dirac equation [2], was later rediscovered in quantum information [3–7], and proved to be a versatile platform for various scopes. For example, QWs have been used for empowering quantum algorithms, such as database search [8,9], or graph isomorphism [10,11]. Moreover, quantum walks have been studied as a simulation tool for relativistic quantum fields [12–28], and they have been used as discrete models of spacetime [29–32].

QWs are among the most promising quantum simulators with possible realizations in a variety of physical systems, such as nuclear magnetic resonance [33,34], trapped ions [35], integrated photonics, and bulk optics [36–39].

New research perspectives are unfolding in the scenario of multi-particle interacting quantum walks where two or more walking particles are coupled via nonlinear (in the field) unitary operators. The properties of these systems are still largely unexplored. Both continuous-time [40] and discrete-time [41] quantum walks on sparse unweighted graphs are equivalent in power to the quantum circuit model. However, it is highly non-trivial to design a suitable architecture for universal quantum computation based on quantum walks. Within this perspective, a possible route has been suggested in [42] based on interacting multi-particle quantum walks with indistinguishable particles (Bosons or Fermions), proving that "almost any interaction" is universal. Among the universal interacting many-body systems are the models with coupling term of the form $\chi \delta_{x_1, x_2} \hat{n}(x_1) \hat{n}(x_2)$, with $\hat{n}(x)$ the number operator at site x. The latter two-body interaction lies at the basis of notable integrable quantum systems in one space dimension such as the Hubbard and the Thirring Hamiltonian models.

The first attempts at the analysis of interacting quantum walks were carried out in [43,44]. More recently, in [45], the authors proposed a discrete-time analogue of the Thirring model, which is indeed a Fermionic quantum cellular automaton, whose dynamics in the two-particle sector reduces to an interacting two-particle quantum walk. As for its Hamiltonian counterpart, the discrete-time interacting walk has been solved analytically in the case of two Fermions. Analogously to any Hamiltonian integrable system, also in the discrete-time case the solution is based on the Bethe Ansatz technique. However, discreteness of the evolution prevents the application of the usual Ansatz, and a new Ansatz has been introduced successfully [45].

In this paper, we present an original simplified derivation of the solution of [45], which exploits the symmetries of the interacting walk. We present the diagonalization of the evolution operator and the characterization of its spectrum. We explicitly write the two particle states corresponding to the scattering solutions of the system, having eigenvalues in the continuous spectrum of the evolution operator. We then show how the present model predicts the formation of bound states, which are eigenstates of the interacting walk corresponding to the discrete spectrum. We provide also in this case the analytic expression of such molecular states.

We comment on the phenomenological differences between the Hamiltonian model and the discrete-time one. First, we see that the set of possible scattering solutions is larger in the discrete-time case: for a fixed value total momentum, a non trivial transfer of relative momentum can occur besides the simple exchange of momentum between the two particles, differently from the Hamiltonian case. In addition, the family of bound states appearing in the discrete-time scenario is larger than the corresponding Hamiltonian one. Indeed, for any fixed value of the coupling constant, a bound state exists with any possible value of the total momentum, while, for Hamiltonian systems, bound states cannot have arbitrary total momentum.

Finally, we show that, in the set of solutions for the interacting walk, there are perfectly localized states (namely, states that lie on a finite number of lattice sites). Moreover, differently from the Hamiltonian systems, bound states exist also for null coupling constants; however, this is true only for finitely many isolated values of the total momentum. In addition to the exact analytical solution of the dynamics, we show the simulation of some significant initial states.

2. The Dirac Quantum Walk

In this section, we review the Dirac quantum cellular automaton on the line describing the free evolution of a two-component Fermionic field. The single particle Hilbert space is given by $\mathscr{H} := \mathbb{C}^2 \otimes \ell^2(\mathbb{Z})$ for which we employ the factorized basis $|a\rangle |x\rangle$, with $a \in \{\uparrow, \downarrow\}$ and $x \in \mathbb{Z}$. The Dirac automaton describes an arbitrary number of Fermions whose evolution is linear in the field:

$$\psi(x, t+1) = W\psi(x, t), \qquad \psi(x, t) = \begin{pmatrix} \psi_\uparrow(x,t) \\ \psi_\downarrow(x,t) \end{pmatrix}, \qquad (1)$$

$$[\psi_a(x), \psi_b^\dagger(x')]_+ = \delta_{a,b} \delta_{x,x'}, \qquad [\psi_a(x), \psi_b(x')]_+ = 0, \qquad (2)$$

where W is a unitary operator. In the single particle sector, the automaton can be regarded as a quantum walk on the single-particle Hilbert space $\mathscr{H}$ whose evolution unitary operator W is given by

$$W = \begin{pmatrix} \nu T_x & -i\mu \\ -i\mu & \nu T_x^\dagger \end{pmatrix}, \qquad \nu, \mu > 0, \quad \nu^2 + \mu^2 = 1, \tag{3}$$

where T_x denotes the translation operator on $\ell^2(\mathbb{Z})$, defined by $T_x |x\rangle = |x+1\rangle$.

Since the walk W is translation invariant (it commutes with the translation operator), it can be diagonalized in momentum space. In the momentum representation, defining $|p\rangle := (2\pi)^{-1/2} \sum_{x \in \mathbb{Z}} e^{-ipx} |x\rangle$, with $p \in B := (-\pi, \pi]$, the walk operator can be written as

$$W = \int_B dp\, W(p) \otimes |p\rangle\langle p|, \qquad W(p) = \begin{pmatrix} \nu e^{ip} & -i\mu \\ -i\mu & \nu e^{-ip} \end{pmatrix}, \tag{4}$$

where $|\nu|^2 + |\mu|^2 = 1$. The spectrum of the walk is given by $\{e^{-i\omega(p)}, e^{i\omega(p)}\}$, where the dispersion relation $\omega(p)$ is given by

$$\omega(p) := \mathrm{Arccos}(\nu \cos p), \tag{5}$$

where Arccos denotes the principal value of the arccosine function. The single-particle eigenstates, solving the eigenvalue problem

$$W(p) v_p^s = e^{-is\omega(p)} v_p^s, \qquad s = \pm, \tag{6}$$

can be conveniently written as

$$v_p^s = \frac{1}{|N_s|} \begin{pmatrix} -i\mu \\ g_s(p) \end{pmatrix}, \tag{7}$$

with $g_s(p) := -i(s \sin \omega(p) + \nu \sin p)$, $|N_s|^2 := \mu^2 + |g_s|^2$.

3. The Thirring Quantum Walk

In this section, we present a Fermionic cellular automaton in one spatial dimension with an on-site interaction, namely two particles interact only when they lie at the same lattice site. The linear part corresponds to the Dirac QW [17] and the interaction term is the most general number-preserving coupling in one dimension [46]. The same kind of interaction characterizes also the most studied integrable quantum systems, such as the Thirring [47] and the Hubbard [48] models.

The linear part of the automaton is given by the Dirac automaton, describing the free evolution of the particles. In order to introduce an interaction, we modify the evolution operator adding an extra unitary step of the form:

$$V_{\text{int}} := e^{i\chi n_\uparrow(x) n_\downarrow(x)}, \tag{8}$$

where $n_a(x)$, $a \in \{\uparrow, \downarrow\}$, represents the particle number at site x, namely $n_a(x) = \psi_a^\dagger(x) \psi_a(x)$, and χ is a real coupling constant. Since the interaction term preserves the total number operator, we can study the automaton for a fixed number of particles. For N interacting particles, we can describe the evolution in terms of an interacting quantum walk over $\mathscr{H}_N = \mathscr{H}^{\otimes N}$ with the free evolution given by $W_N := W^{\otimes N}$.

In this work, we focus on the two-particle sector whose solutions has been derived in [45]. As we will see, the Thirring walk features molecule states besides scattering solutions. This features is shared also by the Hadamard walk with the same on-site interaction [44].

$W_N := W^{\otimes N}$, acting on the Hilbert space $\mathcal{H}_N = \mathcal{H}^{\otimes N}$ and describing the free evolution of the particles. In order to introduce an interaction, we modify the update rule of the walk with an extra step V_{int}: $\mathcal{U}_N := W_N V_{int}$. In the present case, the term V_{int} has the form

$$V_{int} = V_N(\chi) := e^{i\chi n_\uparrow(x) n_\downarrow(x)}. \tag{9}$$

Since we focus on the solutions involving the interaction of two particles, it is convenient to write the walk in the centre of mass basis $|a_1, a_2\rangle |y\rangle |w\rangle$, with $a_1, a_2 \in \{\uparrow, \downarrow\}$, $y = x_1 - x_2$ and $w = x_1 + x_2$. Therefore, on this basis, the generic Fermionic state is $|\psi\rangle = \sum_{a_1, a_2, y, w} c(a_1, a_2, y, w) |a_1, a_2\rangle |y\rangle |w\rangle$ with $c(a_2, a_1, y, w) = -c(a_1, a_2, -y, w)$. Notice that only the pairs y, w with y and w, both even or odd, correspond to physical points in the original basis x_1, x_2.

We define the two-particle walk with both y and w in $\mathbb{Z}$, so that the linear part of walk can be written as

$$W_2 = \mu\nu \begin{pmatrix} \frac{\nu}{\mu} T_w^2 & -iT_y \otimes T_w & -iT_y^\dagger \otimes T_w & -\frac{\mu}{\nu} \\ -iT_y \otimes T_w & \frac{\nu}{\mu} T_y^2 & -\frac{\mu}{\nu} & -iT_y^\dagger \otimes T_w \\ -iT_y^\dagger \otimes T_w & -\frac{\mu}{\nu} & \frac{\nu}{\mu} T_y^{\dagger 2} & -iT_y^\dagger \otimes T_w^\dagger \\ -\frac{\mu}{\nu} & -iT_y \otimes T_w^\dagger & -iT_y^\dagger \otimes T_w^\dagger & \frac{\nu}{\mu} T_w^{\dagger 2} \end{pmatrix}, \tag{10}$$

where T_y represents the translation operator in the relative coordinate y, and T_w the translation operator in the centre of mass coordinate w, whereas the interacting term reads

$$V_2(\chi) = \begin{pmatrix} I_y \otimes I_w & 0 & 0 & 0 \\ 0 & e^{i\chi \delta_{y,0}} \otimes I_w & 0 & 0 \\ 0 & 0 & e^{i\chi \delta_{y,0}} \otimes I_w & 0 \\ 0 & 0 & 0 & I_y \otimes I_w \end{pmatrix}. \tag{11}$$

This definition gives a walk $\mathcal{U}_2 = W_2 V_2(\chi)$ that can be decomposed into two identical copies of the original walk. Indeed, defining as C the projector on the physical center of mass coordinates, one has $\mathcal{U}_2 = C\mathcal{U}_2 C + (I - C)\mathcal{U}_2 (I - C)$, where $C\mathcal{U}_2 C$ and $(I - C)\mathcal{U}_2 (I - C)$ are unitarily equivalent. We will then diagonalize the operator $\mathcal{U}_2$, reminding readers that the physical solutions will be given by projecting the eigenvectors with C.

Introducing the (half) relative momentum $k = \frac{1}{2}(p_1 - p_2)$ and the (half) total momentum $p = \frac{1}{2}(p_1 + p_2)$, the free evolution of the two particles is written in the momentum representation as

$$W_2 = \int dk dp\, W_2(p, k) \otimes |k\rangle\langle k| \otimes |p\rangle\langle p|, \tag{12}$$

where the matrix $W_2(p, k)$ is given by

$$W_2(p, k) = W(p_1) \otimes W(p_2). \tag{13}$$

Furthermore, we introduce the vectors $v_k^{sr} := v_{p+k}^s \otimes v_{p-k}^r$, with $s, r = \pm$, such that

$$W_2(p, k) v_k^{sr} = e^{-i\omega_{sr}(p,k)} v_k^{sr}, \tag{14}$$

where $\omega_{sr}(p, k) := s\omega(p + k) + r\omega(p - k)$ is the dispersion relation of the two-particle walk. Explicitly, the vectors v_k^{sr} are given by

$$v_k^{sr} = \frac{1}{|N_s(p+k)||N_r(p-k)|} \begin{pmatrix} -\mu^2 \\ -i\mu g_r(p-k) \\ -i\mu g_s(p+k) \\ g_s(p+k) g_r(p-k) \end{pmatrix}. \tag{15}$$

We focus in this work on Fermionic solutions satisfying the eigenvalue equation

$$U_2(\chi, p) |\psi\rangle = e^{-i\omega} |\psi\rangle, \qquad \omega \in \mathbb{R}, \tag{16}$$

with $|\psi(y)\rangle \in \mathbb{C}^4$. In the centre of mass basis, the antisymmetry condition reads

$$|\psi(y)\rangle = -E |\psi(-y)\rangle, \tag{17}$$

E being the exchange matrix

$$E = \begin{pmatrix} 1 & 0 & 0 & 0 \\ 0 & 0 & 1 & 0 \\ 0 & 1 & 0 & 0 \\ 0 & 0 & 0 & 1 \end{pmatrix}. \tag{18}$$

4. Symmetries of the Thirring Quantum Walk

The Thirring walk manifests some symmetries that allow for simplifying the derivation and the study of the solutions. First of all, as we already mentioned, one can show that the interaction $V(\chi)$ commutes with the total number operator. This means that one can study the walk dynamics separately for each fixed number of particles. We focus here on the two-particle walk $U_2 = W_2 V_2(\chi)$, where $W_2 = W \otimes W$ and $V_2(\chi) = e^{i\chi \delta_{y,0}(1-\delta_{a_1,a_2})}$.

Since the interacting walk U_2 commutes with the translations in the centre of mass coordinate w, the total momentum is a conserved quantity, so it is convenient to study the walk parameterized by the total momentum p. To this end, we consider the basis $|a_1, a_2\rangle |y\rangle |p\rangle$, so that, for fixed values of p, the interacting walk of two particles can be expressed in terms of a one-dimensional QW $U_2(\chi, p) = W_2(p) V(\chi)$ with a four-dimensional coin:

$$W_2(p) = \mu\nu \begin{pmatrix} \frac{\nu}{\mu} e^{i2p} & -ie^{ip} T_y & -ie^{ip} T_y^\dagger & -\frac{\mu}{\nu} \\ -ie^{ip} T_y & \frac{\nu}{\mu} T_y^2 & -\frac{\mu}{\nu} & -ie^{-ip} T_y \\ -ie^{ip} T_y^\dagger & -\frac{\mu}{\nu} & \frac{\nu}{\mu} T_y^{\dagger 2} & -ie^{-ip} T_y^\dagger \\ -\frac{\mu}{\nu} & -ie^{-ip} T_y & -ie^{-ip} T_y^\dagger & \frac{\nu}{\mu} e^{-i2p} \end{pmatrix}. \tag{19}$$

Although the range of the variable p is the interval $(-\pi, \pi]$, it is possible to show that one can restrict the study of the walk to the interval $[0, \pi/2]$. On the one hand, the two-particle walk transforms unitarily under a parity transformation in the momentum space. Starting from the single particle walk, $W(p)$ transforms under a parity transformation as

$$W(p) = \sigma_x W(-p) \sigma_x, \qquad p \in (-\pi, \pi], \tag{20}$$

so that, for the two-particle walk, we have the relation

$$W_2(-p, y) = \sigma_x \otimes \sigma_x \, E W_2(p, y) E \, \sigma_x \otimes \sigma_x. \tag{21}$$

On the other hand, a translation of π of the total momentum p entails that

$$W_2(p + \pi, y) = \sigma_z \otimes \sigma_z \, W_2(p, y) \, \sigma_z \otimes \sigma_z, \tag{22}$$

while the interaction term remains unaffected in both cases.

The Thirring walk features also another symmetry that can be exploited to simplify the derivation of the solutions. It is easy to check that the walk operator $U_2(p,\chi) = W_2(p)V(\chi)$ commutes with the projector defined by

$$P := \begin{pmatrix} P_o & 0 & 0 & 0 \\ 0 & P_e & 0 & 0 \\ 0 & 0 & P_e & 0 \\ 0 & 0 & 0 & P_o \end{pmatrix}, \qquad (23)$$

where P_e and P_o are the projectors on the even and the odd subspaces, respectively:

$$P_e = \sum_{z \in \mathbb{Z}} |2z\rangle\langle 2z|, \qquad P_o = \sum_{z \in \mathbb{Z}} |2z+1\rangle\langle 2z+1|. \qquad (24)$$

The projector P induces a splitting of the total Hilbert space $\mathcal{H}$ in two subspaces $P\mathcal{H}$ and $(I - P)\mathcal{H}$, with the interaction term acting non-trivially only in the subspace $P\mathcal{H}$. In the complementary subspace $(I - P)\mathcal{H}$, the evolution is free for Fermionic particles. This means that solutions of the free theory are also solutions of the interacting one, as opposed to the Bosonic case for which the interaction is non-trivial also in $(I - P)\mathcal{H}$.

5. Review of the Solutions

We focus in this section on the antisymmetric solutions of the Thirring walk that actually feel the interaction. From the remarks that we have made in the previous section, such solutions can only be found in the subspace $P\mathcal{H}$. Formally, we have to solve the eigenvalue equation $PU_2(\chi,p)|\psi\rangle = e^{-i\omega}|\psi\rangle$, with $|\psi\rangle \in P\mathcal{H}$. Conveniently, we write a vector $|\psi\rangle \in P\mathcal{H}$ in the form

$$|\psi\rangle = \sum_{z \in \mathbb{Z}} \begin{pmatrix} \psi^1(z) \\ 0 \\ 0 \\ \psi^4(z) \end{pmatrix} \otimes |2z+1\rangle + \sum_{z \in \mathbb{Z}} \begin{pmatrix} 0 \\ \psi^2(z) \\ \psi^3(z) \\ 0 \end{pmatrix} \otimes |2z\rangle, \qquad (25)$$

and the antisymmetry condition becomes:

$$\psi^{1,4}(-z) = -\psi^{1,4}(z-1), \qquad (26)$$
$$\psi^2(-z) = -\psi^3(z). \qquad (27)$$

The restriction of the walk to the subspace $P\mathcal{H}$ entails that the eigenvalue problem is equivalent to the following system of equations:

$$\begin{cases} e^{-i\omega}\psi^1(z) = \nu^2 e^{i2p}\psi^1(z) - i\mu\nu e^{ip}e^{i\chi\delta_{z,0}}\psi^2(z) - i\mu\nu e^{ip}e^{i\chi\delta_{z,-1}}\psi^3(z+1) - \mu^2\psi^4(z), \\ e^{-i\omega}\psi^2(z) = -i\mu\nu e^{ip}\psi^1(z-1) + \nu^2 e^{i\chi\delta_{z,1}}\psi^2(z-1) - \mu^2 e^{i\chi\delta_{z,0}}\psi^3(z) - i\mu\nu e^{-ip}\psi^4(z-1), \\ e^{-i\omega}\psi^3(z) = -i\mu\nu e^{ip}\psi^1(z) - \mu^2 e^{i\chi\delta_{z,0}}\psi^2(z) + \nu^2 e^{i\chi\delta_{z,-1}}\psi^3(z+1) - i\mu\nu e^{-ip}\psi^4(z), \\ e^{-i\omega}\psi^4(z) = -\mu^2\psi^1(z) - i\mu\nu e^{-ip}e^{i\chi\delta_{z,0}}\psi^2(z) - i\mu\nu e^{-ip}e^{i\chi\delta_{z,-1}}\psi^3(z+1) + \nu^2 e^{-i2p}\psi^4(z). \end{cases} \qquad (28)$$

The most general solution of Equation (28) for $p \notin \{0, \pi/2\}$ has two forms:

$$U_2(\chi,p)|\psi_{\pm\infty}\rangle = e^{\pm i2p}|\psi_{\pm\infty}\rangle, \qquad \psi_{\pm\infty}(z) = \begin{cases} \begin{pmatrix} \zeta_{\pm\infty} \\ \eta_{\pm\infty} \\ -\eta_{\pm\infty} \\ \zeta'_{\pm\infty} \end{pmatrix} \delta_{z,0}, & z \geq 0, \\ \text{antisymmetrized}, & z < 0, \end{cases} \qquad (29)$$

and

$$\psi(z) = \begin{cases} \sum_{s,r=\pm} \int_S dk\, g_\omega^{sr}(k) w_k^{sr}(z), & z > 0, \\ \text{antisymmetrized}, & z < 0, \end{cases} \qquad w_k^{sr}(z) := \begin{pmatrix} v_k^{sr,1} e^{-i(2z+1)k} \\ v_k^{sr,2} e^{-i(2z)k} \\ v_k^{sr,3} e^{-i(2z)k} \\ v_k^{sr,4} e^{-i(2z+1)k} \end{pmatrix}, \tag{30}$$

$$\psi(0) = \begin{pmatrix} \sum_{s,r=\pm} \int_S dk\, g_\omega^{sr}(k) v_k^{sr,1} \\ \zeta \\ -\zeta \\ \sum_{s,r=\pm} \int_S dk\, g_\omega^{sr}(k) v_k^{sr,4} \end{pmatrix},$$

with $k = k_R + ik_I$, $S := \{ k \in \mathbb{C} \mid k_R \in (-\pi, \pi] \}$, and g_ω^{sr} satisfying the condition

$$e^{-i\omega} \neq e^{-i\omega_{sr}(p,k)} \implies g_\omega^{sr}(k) = 0. \tag{31}$$

Solving Equation (28) corresponds now to find the function g_ω^{sr}. Let us now study the equation

$$e^{-i\omega_{sr}(p,k)} = e^{-i\omega}. \tag{32}$$

Since $e^{-i\omega_{sr}(p,k)}$ has to be an eigenvalue of $U_2(\chi, p)$, $\omega_{sr}(p,k)$ must be real and thus $k \in \Gamma_f$ or $k \in \Gamma_l$ with $l = 0, \pm 1, 2$, so we conveniently define the sets:

$$\Omega_f^{sr} := \left\{ e^{-i\omega_{sr}(p,k)} \mid k \in \Gamma_f \right\}, \qquad \Omega_l^{sr} := \left\{ e^{-i\omega_{sr}(p,k)} \mid k \in \Gamma_l \right\}, \tag{33}$$

$$\Gamma_f := \{ k \in S \mid k_I = 0 \}, \qquad \Gamma_l := \left\{ k \in S \mid k_R = l\frac{\pi}{2} \right\}, \qquad l = 0, \pm 1, 2. \tag{34}$$

It is easy to see that $\Omega_f^{sr} \cap \Omega_l^{sr} = \emptyset$ for all s, r and l, and the range of the function $e^{-i\omega_{sr}(p,k)}$ covers the entire unit circle except for the points $e^{\pm i2p}$. Therefore, we can discuss separately the case $e^{-i\omega} \in \Omega_f^{sr}$ and the case $e^{-i\omega} \in \Omega_l^{sr}$. A solution with $e^{-i\omega} = e^{\pm i2p}$ actually exists, corresponding to the function of Equation (29), and it will be discussed in Section 5.3.

Let us start with the case $e^{-i\omega} \in \Omega_f^{sr}$, which will lead to the characterization of the continuous spectrum of the Thirring walk $U_2(\chi, p)$ and of the scattering solutions.

5.1. Scattering Solutions

In this section, we assume $p \notin \{0, \pi/2\}$ with $e^{-i\omega} \in \Omega_f^{sr}$. This implies that $e^{-i\omega} \neq e^{\pm i2p}$: indeed, as one can notice from Figure 1, the lines $\omega = \pm 2p$ lie entirely in the gaps between the curves $\omega = \pm 2\omega(p)$ and $\omega = \pm(\pi - 2\operatorname{Arccos}(n \sin p))$. The solution is thus the one given in Equation (30). One can prove that $\Omega_f^{++} = \Omega_f^{--}$ and $\Omega_f^{+-} = \Omega_f^{-+}$. Furthermore, as one can notice from Figure 2, there are four values of the triple (s,r,k) such that $e^{-i\omega_{sr}(p,k)} = e^{-i\omega}$ for a given value of $e^{-i\omega}$: if the triple $(+,+,k)$ is a solution, so are $(+,+,\pi-k)$, $(-,-,-k)$ and $(-,-,k-\pi)$; and if $(+,-,k)$ is a solution, then also $(+,-,\pi-k)$, $(-,+,-k)$ and $(-,+,k-\pi)$ are solutions. This result greatly simplifies Equation (30). Indeed, the sum over s,r and the integral over k reduces to the sum of four terms:

$$\begin{aligned} \psi_k^{\pm,1}(z) &:= (\alpha_k^\pm v_k^{\pm,1} + \delta_k^\pm v_{k-\pi}^{\mp,1}) e^{-i(2z+1)k} - (\beta_k^\pm v_{-k}^{\pm,1} + \gamma_k^\pm v_{\pi-k}^{\mp,1}) e^{i(2z+1)k}, & z \geq 0, \\ \psi_k^{\pm,2}(z) &:= (\alpha_k^\pm v_k^{\pm,2} - \delta_k^\pm v_{k-\pi}^{\mp,2}) e^{-i2zk} - (\beta_k^\pm v_{-k}^{\pm,2} - \gamma_k^\pm v_{\pi-k}^{\mp,2}) e^{i2zk}, & z > 0, \\ \psi_k^{\pm,3}(z) &:= (\alpha_k^\pm v_k^{\pm,3} - \delta_k^\pm v_{k-\pi}^{\mp,3}) e^{-i2zk} - (\beta_k^\pm v_{-k}^{\pm,3} - \gamma_k^\pm v_{\pi-k}^{\mp,3}) e^{i2zk}, & z > 0, \\ \psi_k^{\pm,4}(z) &:= (\alpha_k^\pm v_k^{\pm,4} + \delta_k^\pm v_{k-\pi}^{\mp,4}) e^{-i(2z+1)k} - (\beta_k^\pm v_{-k}^{\pm,4} + \gamma_k^\pm v_{\pi-k}^{\mp,4}) e^{i(2z+1)k}, & z \geq 0, \\ \psi_k^{\pm,2}(0) &= -\psi_k^{\pm,3}(0) := \zeta. \end{aligned} \tag{35}$$

As we will see, the original problem can be simplified in this way to an algebraic problem with a finite set of equations. We note that the fact that the equation $e^{-i\omega_{sr}(p,k)} = e^{-i\omega}$ has a finite number of solutions is a consequence of the fact that we are considering a model in one spatial dimension. However, in analogous one-dimensional Hamiltonian models (e.g., the Hubbard model), the degeneracy of the eigenvalues is two.

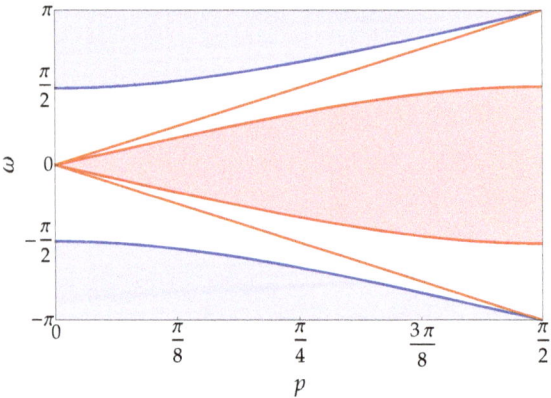

Figure 1. Continuous spectrum of the two-particle walk as a function of the total momentum $p \in [0, \pi/2]$ with mass parameter $m = 0.7$. The continuous spectrum is the same as in the free case. The solid blue curves are described by the functions $\omega = \pm 2\omega(p)$, and the red ones by $\omega = \pm(\pi - 2\operatorname{Arccos}(n \sin p))$. As one can notice, the light-red lines $\omega = \pm 2p$ lie entirely in the gaps between the solid curves, highlighting the fact that $e^{\pm i 2p}$ is not in the range of $e^{-i\omega_{sr}(p,k)}$ for $p \neq 0, \pi/2$ (see text).

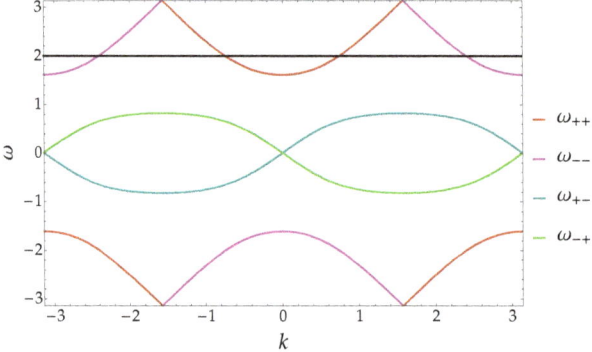

Figure 2. Spectrum of the walk for $m = 0.6$ and $p = \pi/6$ as a function of k. The colours highlight the different ranges of eigenvalues corresponding to the dispersion relation $\omega_{sr}(p,k)$. The range of $\omega_{sr}(p,k)$ is understood to be computed $\bmod(2\pi)$. One can notice that there are four values of the relative momentum k having the same value of the dispersion relation ($\omega = 2$ in the figure). This is in contrast to the Hamiltonian model for which there are only two solutions.

Let us consider for the sake of simplicity the solution of the kind $\psi_k^{+,j}(z)$, since the other one can be analysed in a similar way. Using the notation of Appendix, Equation (35) reduces to the expressions (dropping the $+$ superscript)

$$\psi_k^1(z) = a[\lambda e^{-i(2z+1)k} - \rho e^{i(2z+1)k}],$$
$$\psi_k^2(z) = \lambda b e^{-i2zk} - \rho c e^{i2zk},$$
$$\psi_k^3(z) = \lambda c e^{-i2zk} - \rho b e^{i2zk}, \qquad (36)$$
$$\psi_k^4(z) = d[\lambda e^{-i(2z+1)k} - \rho e^{i(2z+1)k}],$$
$$\lambda := \alpha_k + \delta_k, \qquad \rho := \beta_k + \gamma_k,$$
$$\psi_k^2(0) = \xi.$$

We notice that now the number of unknown parameters is further reduced to three, namely λ, ρ, and ξ. Clearly, one of the parameters can be fixed by choosing arbitrarily the normalization. From now on, we fix $\lambda = 1$ and define $T_+ := \rho$. Equation (36) has to satisfy the recurrence relations of Equation (28) for $z = 0$ and $z = 1$, while, for $z > 1$, it is automatically satisfied. For $z = 0$, Equation (28) becomes

$$e^{-i\omega}\psi_k^1(0) = \nu^2 e^{i2p}\psi_k^1(0) - i\mu\nu e^{ip}e^{i\chi}\xi - i\mu\nu e^{ip}\psi_k^3(1) - \mu^2\psi_k^4(0), \qquad (37)$$
$$e^{-i\omega}\xi = i\mu\nu e^{ip}\psi_k^1(0) - \nu^2\psi_k^3(1) - \mu^2 e^{i\chi}\xi + i\mu\nu e^{-ip}\psi_k^4(0), \qquad (38)$$
$$-e^{-i\omega}\xi = -i\mu\nu e^{ip}\psi_k^1(0) - \mu^2 e^{i\chi}\xi + \nu^2\psi_k^3(1) - i\mu\nu e^{-ip}\psi_k^4(0), \qquad (39)$$
$$e^{-i\omega}\psi_k^4(0) = -\mu^2\psi_k^1(0) - i\mu\nu e^{-ip}e^{i\chi}\xi - i\mu\nu e^{-ip}\psi_k^3(1) + \nu^2 e^{-i2p}\psi_k^4(0). \qquad (40)$$

Starting from Equation (37), we can notice that $\nu^2 e^{i2p}a - i\mu\nu e^{ip}e^{ik}b - i\mu\nu e^{ip}e^{-ik}c - \mu^2 d = e^{-i\omega}a$, where we employed the notation of Appendix A, so that we obtain $\xi = e^{-i\chi}(b - T_+c)$. We can then substitute this expression in Equation (39) and use the relations

$$-i\mu\nu e^{ip}e^{ik}a + \nu^2 e^{i2k}b - \mu^2 c - i\mu\nu e^{-ip}e^{ik}d = e^{-i\omega}b, \qquad (41)$$
$$-i\mu\nu e^{ip}e^{-ik}a - \mu^2 b + \nu^2 e^{-i2k}c - i\mu\nu e^{-ip}e^{-ik}d = e^{-i\omega}c, \qquad (42)$$

to obtain the expression

$$e^{-i\chi}(b - T_+c) = T_+b - c, \qquad (43)$$

and thus

$$T_+ = \frac{c + e^{-i\chi}b}{b + e^{-i\chi}c} = \frac{g_+(p+k) + e^{-i\chi}g_+(p-k)}{g_+(p-k) + e^{-i\chi}g_+(p+k)}. \qquad (44)$$

For these values of ξ and T_+ one can verify that Equation (28) is satisfied also for $z = 1$, thus concluding the derivation. For the solution of the kind $\psi_k^{-,j}(z)$, we can follow a similar reasoning, obtaining the analogous quantity T_-:

$$T_- := \frac{g_+(p+k) + e^{-i\chi}g_-(p-k)}{g_-(p-k) + e^{-i\chi}g_+(p+k)}. \qquad (45)$$

It is worth noticing that $T_\pm$ is of unit modulus for $k \in (-\pi, \pi]$.
The final form of the solution results in being:

$$\psi_k^{\pm,1}(z) = (v_k^{\pm,1} + v_{k-\pi}^{\mp,1})e^{-i(2z+1)k} - T_\pm(v_{-k}^{\pm,1} + v_{\pi-k}^{\mp,1})e^{i(2z+1)k},$$
$$\psi_k^{\pm,2}(z) = e^{-i\chi\delta_{z,0}}\left[(v_k^{\pm,2} - v_{k-\pi}^{\mp,2})e^{-i2zk} - T_\pm(v_{-k}^{\pm,2} - v_{\pi-k}^{\mp,2})e^{i2zk}\right], \qquad (46)$$
$$\psi_k^{\pm,3}(z) = (v_k^{\pm,3} - v_{k-\pi}^{\mp,3})e^{-i2zk} - T_\pm(v_{-k}^{\pm,3} - v_{\pi-k}^{\mp,3})e^{i2zk},$$
$$\psi_k^{\pm,4}(z) = (v_k^{\pm,4} + v_{k-\pi}^{\mp,4})e^{-i(2z+1)k} - T_\pm(v_{-k}^{\pm,4} + v_{\pi-k}^{\mp,4})e^{i(2z+1)k},$$

which in terms of the relative coordinate y can be written as

$$\psi_k^\pm(y) = \begin{cases} e^{-i\chi\delta_{z,0}\delta_{j,2}}\left[(v_k^{+\pm} + v_{k-\pi}^{-\mp})e^{-iky} - T_\pm(v_{-k}^{\pm+} + v_{\pi-k}^{\mp-})e^{iky}\right], & y \geq 0, \\ \text{antisymmetrized}, & y < 0. \end{cases} \quad (47)$$

We can interpret such a solution as a scattering of plane waves for which the coefficient $T_\pm$ plays the role of the transmission coefficient. Being the total momentum a conserved quantity, the two particles can only exchange their momenta, as expected from a theory in one dimension. Furthermore, for each value k of the relative momentum, the two particles can also acquire an additional phase of π. As the interaction is a compact perturbation of the free evolution, the continuous spectrum is the same as that of the free walk. Equation (46) provides the generalized eigenvector if $U_2(\chi, p)$ corresponding to the continuous spectrum $\sigma_c = \Omega_f^{++} \cup \Omega_f^{+-}$.

5.2. Bound States

In the previous section, we derived the solutions in the continuous spectrum, which can be interpreted as scattering plane waves in one spatial dimension. We seek now the solutions corresponding to the discrete spectrum, namely solutions with eigenvalues in any one of the sets Ω_l^{sr}. The derivation of the solution follows similar steps as for the scattering solutions. In particular, the degeneracy in k is the same: there are four solutions to the equation $e^{-i\omega_{sr}(p,k)} = e^{-i\omega}$ even in this case, as proved in [45]. Therefore, the general form of the solution in this case can be written again as in Equation (35) and, following the same reasoning, one obtains the same set of solutions as in Equation (46). At this stage, we did not impose that the solution is a proper eigenvector in the Hilbert space $\mathcal{H}$. To this end, we have to set $T_\pm = 0$ to eliminate the exponentially-divergent terms in Equation (46). As one can prove, the equation $T_\pm = 0$ has only one solution for fixed values of χ and p. More precisely, there is a unique $k \in \Gamma_0 \cup \Gamma_{-1} \cup \Gamma_1 \cup \Gamma_2$, with $k_I < 0$ and $e^{i\chi} \notin \{1, -1\}$, such that either $T_+ = 0$ or $T_- = 0$.

In other words, for each pair of values (χ, p), the walk $U_2(p)$ has one and only one eigenvector corresponding to an eigenvalue in the point spectrum. Such eigenvector can be written as

$$\begin{aligned} \psi_{\tilde{k}}^1(z) &= (v_{\tilde{k}}^{+\pm,1} + v_{\tilde{k}-\pi}^{-\mp,1})e^{-i(2z+1)\tilde{k}}, \\ \psi_{\tilde{k}}^2(z) &= e^{-i\chi\delta_{z,0}}\left[(v_{\tilde{k}}^{+\pm,2} - v_{\tilde{k}-\pi}^{-\mp,2})e^{-i2z\tilde{k}}\right], \\ \psi_{\tilde{k}}^3(z) &= (v_{\tilde{k}}^{+\pm,3} - v_{\tilde{k}-\pi}^{-\mp,3})e^{-i2z\tilde{k}}, \\ \psi_{\tilde{k}}^4(z) &= (v_{\tilde{k}}^{+\pm,4} + v_{\tilde{k}-\pi}^{-\mp,4})e^{-i(2z+1)\tilde{k}}, \end{aligned} \quad (48)$$

where $\tilde{k}$ is the solution of $T_+ = 0$ or $T_- = 0$ and $\pm$ chosen accordingly. More compactly, in the y coordinate, the solution can be written as

$$\psi_{\tilde{k}}(y) = \begin{cases} e^{-i\chi\delta_{z,0}\delta_{j,2}}\left[(v_{\tilde{k}}^{+\pm} + v_{\tilde{k}-\pi}^{-\mp})e^{-iky}\right], & y \geq 0, \\ \text{antisymmetrized}, & y < 0. \end{cases} \quad (49)$$

In Figure 3, the discrete spectrum of the interacting walk together with the continuous spectrum as a function of the total momentum p is depicted. The solid curves in the gaps between the continuous bands denote the discrete spectrum for different values of the coupling constant $\chi = 2\pi/3, 3\pi/7, -3\pi/7, -2\pi/3$. Molecule states appear also in the Hadamard walk with the same on-site interaction [44].

Referring to Figure 4, we show the evolution of two particles initially prepared in a singlet state localized at the origin. From the figure, one can appreciate the appearance of the bound state component that has non-vanishing overlapping with the initial state. The bound state, being exponentially decaying in the relative coordinate y, is localized on the diagonal of the plot, that is when the two particles lie at the same point.

In Figure 5, the probability distribution of the bound state corresponding to a choice of parameters $\chi = 0.2\pi$ and $p = 0.035\pi$ is depicted. The plot highlights the exponential decay of the tails, which is the characterizing feature of the bound state.

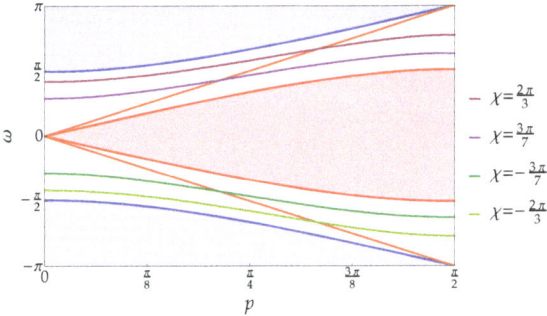

Figure 3. Complete spectrum of the two-particle Thirring walk as a function of the total momentum p with mass parameter $m = 0.7$. The continuous spectrum is as in Figure 1. The solid lines in the gaps show the point spectrum for different values of the coupling constant: from top to bottom, $\chi = 2\pi/3, 3\pi/7, -3\pi/7, -2\pi/3$. It is worth noticing that, for each pair (χ, p), there is only one value in the discrete spectrum. The light-red lines $\omega = \pm 2p$ lie entirely in the gap between the continuous bands highlighting the fact that the $e^{\pm i2p}$ is not in the range of $e^{-i\omega_{sr}(p,k)}$ for $p \neq 0, \pi/2$; for a given coupling constant χ, $e^{\pm i2p}$ is an eigenvalue for $p = \chi/2$.

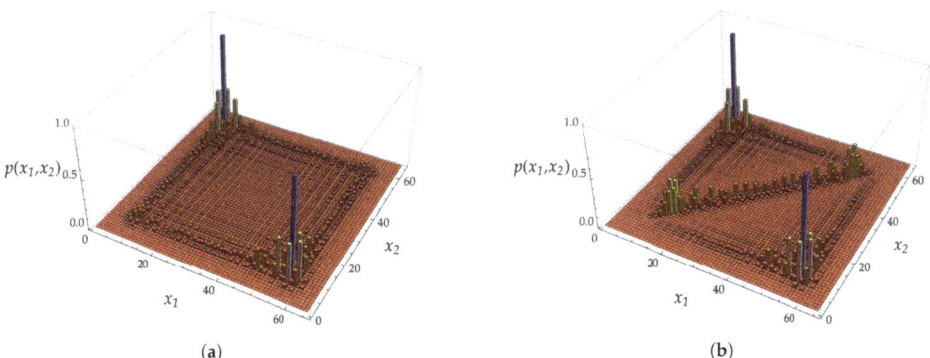

Figure 4. We show for comparison the free evolution (**a**) and the interacting one (**b**) highlighting the appearance of bound states components along the diagonal, namely when the two particles are at the same site (i.e., $x_1 = x_2$), where x_1 and x_2 denote the positions of the two particles. The plots show the probability distribution $p(x_1, x_2)$ in position space after $t = 32$ time-steps. The chosen value of the mass parameter is $m = 0.6$ and the coupling constant is $\chi = \pi/2$. The two particles are initially prepared in a singlet state located at the origin.

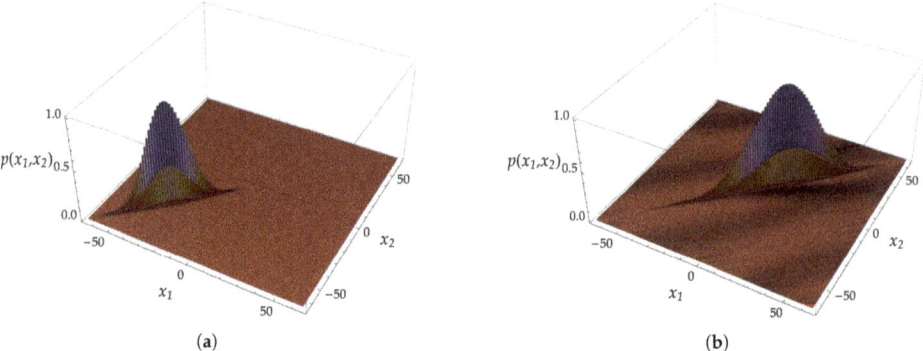

Figure 5. We show the evolution of a bound state of the two particles peaked around the value of the total momentum $p = 0.035\pi$. The mass paramater is $m = 0.6$ and the coupling constant $\chi = 0.2\pi$. In (**a**) is depicted the probability distribution of the initial state. In (**b**) is depicted the probability distribution of the evolved state after $t = 128$ time-steps. One can notice that, in the relative coordinate $x_1 - x_2$, the probability distribution remains concentrated on the diagonal, highlighting the fact that the two particles are in a bound state. The diffusion of the state happens only in the centre of a mass coordinate.

5.3. Solution for $e^{-i\omega} = e^{\pm i2p}$

Thus far, we have studied proper eigenvectors that decay exponentially as the two particles are further apart. However, the previous analysis failed to cover the particular case when $e^{-i\omega} = e^{\pm i2p}$, since the range of $e^{-i\omega_{sr}(p,k)}$ does not include the two points of the unit circle $e^{\pm i2p}$.

We now study the solutions with $e^{-i\omega} = e^{\pm i2p}$ having the form given in Equation (29). One can prove that such solutions are non-vanishing only for $z = 0$ on $P.\mathcal{H}$, namely we look for a solution of the form

$$|\psi\rangle = \begin{pmatrix} -\zeta \\ 0 \\ 0 \\ -\zeta' \end{pmatrix} \otimes |-1\rangle + \begin{pmatrix} 0 \\ \eta \\ -\eta \\ 0 \end{pmatrix} \otimes |0\rangle + \begin{pmatrix} \zeta \\ 0 \\ 0 \\ \zeta' \end{pmatrix} \otimes |1\rangle. \quad (50)$$

Subtracting the first and the last equations of (28) using (50), we obtain the following equation:

$$(e^{-i\omega} - e^{i2p})\zeta = e^{i2p}(e^{-i\omega} - e^{-i2p})\zeta'. \quad (51)$$

If both ζ and ζ' are non-zero, one can prove that a solution does not exist and thus we have to consider the two cases $\zeta = 0$ and $\zeta' = 0$ separately. Starting from $\zeta' = 0$, Equation (51) imposes that $e^{-i\omega} = e^{i2p}$, meaning that, if a solution exists in this case, it is an eigenvector corresponding to the eigenvalue e^{i2p}. From the second equation of (28), we obtain the relation

$$(1 - \mu^2 e^{i(\chi - 2p)})\eta = i\mu v e^{-ip}\zeta \quad (52)$$

and, using the first equation of (28), it turns out that a solution exists only if $e^{i\chi} = e^{i2p}$, as expected, since, otherwise, the case of Section 5.2 would have held. The other case, namely $e^{-i\omega} = e^{-i2p}$, can be studied analogously. Let us, then, denote as $|\psi_{\pm\infty}\rangle$ such proper eigenvectors with eigenvalue $e^{\pm i2p}$ for $\chi = e^{\pm i2p}$ and, choosing $\eta = \frac{\mu}{v}$ as the value for the free parameter η, we obtain the following expression for $|\psi_{\pm\infty}\rangle$:

$$|\psi_{\pm\infty}\rangle = ie^{\pm ip} \begin{pmatrix} \frac{1\pm 1}{2} \\ 0 \\ 0 \\ -\frac{-1\pm 1}{2} \end{pmatrix} \otimes |-1\rangle + \begin{pmatrix} 0 \\ \frac{\mu}{\nu} \\ -\frac{\mu}{\nu} \\ 0 \end{pmatrix} \otimes |0\rangle + ie^{\pm ip} \begin{pmatrix} -\frac{1\pm 1}{2} \\ 0 \\ 0 \\ \frac{-1\pm 1}{2} \end{pmatrix} \otimes |1\rangle. \tag{53}$$

Such solutions provide a special case of molecule states (namely, proper eigenvectors of $U_2(\chi, p)$), being localized on few sites, and differ from the previous solutions showing an exponential decay in the relative coordinate.

5.4. Solutions for $p \in \{0, \pi/2\}$

The solutions that we presented in the previous discussion do not cover the extreme values $p = 0, \pi/2$ (see [45] for a reference). Let us consider for definiteness the case $p = 0$, since the other case is obtained in a similar way. For $e^{-i\omega} \neq 1$, the previous analysis still holds. Indeed, noticing that $\omega_{\pm\pm}(0, k) = \pm 2\omega(k)$, we have $\omega(k) \in \mathbb{R}$ and $\omega(k) \neq 0$ if and only if $k \in \Gamma_f \cup \Gamma_0 \cup \Gamma_2$, whereas $\omega_{\pm\mp}(0, k) = 0$ for all $k \in \mathbb{C}$. This means that the solutions $|\psi_k^+\rangle$ of Equation (46) are actually eigenvectors of $U_2(\chi, 0)$. Thus, the spectrum is made by a continuous part, given by the arc of the unit circle containing -1 and having $e^{\pm 2i\omega(0)}$ as extremes, and a point spectrum with two points: $e^{-2i\omega(\tilde{k})}$, where $\tilde{k}$ is the solution of $T_+ = 0$ for $p = 0$, and 1. As shown in [45], 1 is a separated part of the spectrum of $U_2(\chi, 0)$ and the corresponding eigenspace is a separable Hilbert space of stationary bound states. This fact underlines an important feature of the Thirring walk not shared by analogous Hamiltonian models. It is remarkable that this behaviour occurs also for the free walk with $\chi = 0$. In Figure 6, we show the probability distribution of two states having the properties hereby discussed. It is worth noticing that all the states v_k^{+-} with $k \in (-\pi, \pi]$ are eigenvectors relative to the eigenvalue 1, and thus they generate a subspace on which the walk acts identically. We remark that this behaviour relies on the fact that the dispersion relation in one dimension is an even function of k.

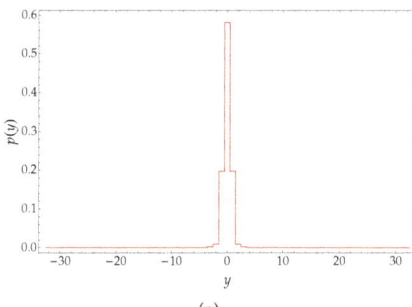
(a)

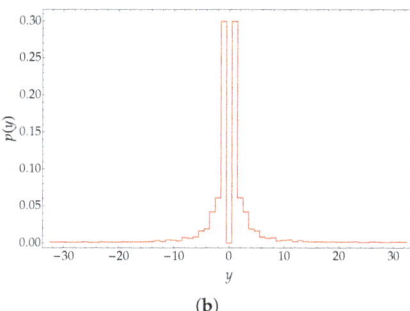
(b)

Figure 6. We show the case of two proper eigenstates for $p = 0$. In both cases the mass parameter is $m = 0.6$. (a): probability distribution in the relative coordinate y of $\int dk\, (v_k^{+-} - v_k^{-+})e^{-iyk}$. (b): probability distribution in the y-coordinate of $\int dk\, (v_k^{+-} + v_k^{-+})e^{-iyk}$.

6. Conclusions

In this work, we reviewed the Thirring quantum walk [45], providing a simplified derivation of its solutions for Fermionic particles. The simplified derivation relies on the symmetric properties of the walk evolution operator, allowing for separating the subspace of solutions affected by the interaction from the subspace where the interaction step acts trivially. The interaction term is the most general number-preserving interaction in one dimension, whereas the free evolution is provided by the Dirac QW [17].

We showed the explicit derivation of the scattering solutions (solutions for the continuous spectrum) as well as for the bound-state solutions. The Thirring walk features also localized bound states (namely, states whose support is finite on the lattice) when $e^{-i\omega} = e^{\pm i2p}$. Such solutions exist only when the coupling constant is $\chi = 2p$. Figure 4 depicts the evolution of a perfectly localized state showing the overlapping with bound state components. In Figure 5, we reported the evolution of a bound state of the two particles peaking around a certain value of the total momentum: one can appreciate that the probability distribution remains localized on the main diagonal during the evolution.

Finally, we showed that bound states exist also for a vanishing coupling constant—even though this is true only for a finite set of values of the total momentum p—which is a striking difference between the discrete model of the present work and corresponding Hamiltonian systems.

Author Contributions: G.M.D. and P.P. conceived and designed the model; A.B. and A.T. performed the calculations; N.M. reviewed the derivation exploiting the symmetries of the walk, and performed the numerical analysis.

Funding: This publication was made possible through the support of a grant from the John Templeton Foundation under the project ID# 60609 *Causal Quantum Structures*. The opinions expressed in this publication are those of the authors and do not necessarily reflect the views of the John Templeton Foundation.

Conflicts of Interest: The authors declare no conflict of interest. The founding sponsors had no role in the design of the study; in the collection, analyses, or interpretation of data; in the writing of the manuscript, and in the decision to publish the results.

Appendix A. Notation

For the single particle walk of Equation (3), the eigenstates can be written as

$$v_p^s = \frac{1}{|N_s(p)|} \begin{pmatrix} -i\mu \\ g_s(p) \end{pmatrix}, \qquad g_s(p) := -i(s\sin\omega(p) + \nu\sin p), \tag{A1}$$

with $|N_s(p)|^2 = \mu^2 + |g_s(p)|^2$. For the two-particle walk, we define $v_k^{rs} := v_{p+k}^s \otimes v_{p-k}^r$. If $s = r$, then we name the related eigenspace the *even* eigenspace; whereas, if $s \neq r$, we call the related eigenspace the *odd* eigenspace. As proven in item 3 of Lemma 1 of [45], for a given k, the degeneracy is 4 both in the even and in the odd case. Namely, if the triple $(+,+,k)$ is a solution, then also $(+,+,-k)$, $(-,-,\pi-k)$ and $(-,-,k-\pi)$ are solutions; if the triple $(+,-,k)$ is a solution, then also $(+,-,\pi-k)$ and $(-,+,k-\pi)$ are solutions.

Explicitly, for the even case, we have:

$$v_k^{++} \propto \begin{pmatrix} -\mu^2 \\ -i\mu g_+(p-k) \\ -i\mu g_+(p+k) \\ g_+(p+k)g_+(p-k) \end{pmatrix}, \qquad v_{\pi-k}^{--} \propto \begin{pmatrix} -\mu^2 \\ i\mu g_+(p+k) \\ i\mu g_+(p-k) \\ g_+(p+k)g_+(p-k) \end{pmatrix}, \tag{A2}$$

$$v_{-k}^{++} \propto \begin{pmatrix} -\mu^2 \\ -i\mu g_+(p-k) \\ -i\mu g_+(p+k) \\ g_+(p+k)g_+(p-k) \end{pmatrix}, \qquad v_{k-\pi}^{--} \propto \begin{pmatrix} -\mu^2 \\ i\mu g_+(p-k) \\ i\mu g_+(p+k) \\ g_+(p+k)g_+(p-k) \end{pmatrix}. \tag{A3}$$

Analogously for the odd case, the eigenstates are

$$v_k^{+-} \propto \begin{pmatrix} -\mu^2 \\ -i\mu g_-(p-k) \\ -i\mu g_+(p+k) \\ g_+(p+k)g_-(p-k) \end{pmatrix}, \qquad v_{\pi-k}^{+-} \propto \begin{pmatrix} -\mu^2 \\ i\mu g_+(p+k) \\ i\mu g_-(p-k) \\ g_+(p+k)g_-(p-k) \end{pmatrix}, \tag{A4}$$

$$v_{-k}^{-+} \propto \begin{pmatrix} -\mu^2 \\ -i\mu g_+(p+k) \\ -i\mu g_-(p-k) \\ g_-(p+k)g_+(p-k) \end{pmatrix}, \qquad v_{k-\pi}^{-+} \propto \begin{pmatrix} -\mu^2 \\ i\mu g_+(p+k) \\ i\mu g_-(p-k) \\ g_+(p+k)g_-(p-k) \end{pmatrix}. \tag{A5}$$

In order to simplify the derivation of the solution, we adopt the following notation:

$$v_k^{++} =: \begin{pmatrix} a \\ b \\ c \\ d \end{pmatrix}, \quad v_{-k}^{++} = \begin{pmatrix} a \\ c \\ b \\ d \end{pmatrix}, \quad v_{\pi-k}^{--} = \begin{pmatrix} a \\ -c \\ -b \\ d \end{pmatrix}, \quad v_{k-\pi}^{--} = \begin{pmatrix} a \\ -b \\ -c \\ d \end{pmatrix}, \tag{A6}$$

$$v_k^{+-} =: \begin{pmatrix} a' \\ b' \\ c' \\ d' \end{pmatrix}, \quad v_{-k}^{-+} = \begin{pmatrix} a' \\ c' \\ b' \\ d' \end{pmatrix}, \quad v_{\pi-k}^{+-} = \begin{pmatrix} a' \\ -c' \\ -b' \\ d' \end{pmatrix}, \quad v_{k-\pi}^{-+} = \begin{pmatrix} a' \\ -b' \\ -c' \\ d' \end{pmatrix}. \tag{A7}$$

References

1. Farhi, E.; Goldstone, J.; Gutmann, S. A Quantum Algorithm for the Hamiltonian NAND Tree. *Theory Comput.* **2008**, *4*, 169–190. doi:10.4086/toc.2008.v004a008. [CrossRef]
2. Feynman, R.P.; Hibbs, A.R.; Styer, D.F. *Quantum Mechanics and Path Integrals*; Volume 2, International Series in Pure and Applied Physics; McGraw-Hill: New York, NY, USA, 1965.
3. Grossing, G.; Zeilinger, A. Quantum cellular automata. *Complex Syst.* **1988**, *2*, 197–208.
4. Ambainis, A.; Bach, E.; Nayak, A.; Vishwanath, A.; Watrous, J. One-dimensional Quantum Walks. In Proceedings of the STOC '01 Thirty-Third Annual ACM Symposium on Theory of Computing, Hersonissos, Greece, 6–8 July 2001; ACM: New York, NY, USA, 2001; pp. 37–49. doi:10.1145/380752.380757. [CrossRef]
5. Reitzner, D.; Nagaj, D.; Bužek, V. Quantum Walks. *Acta Phys. Slov. Rev. Tutor.* **2011**, *61*, 603–725. [CrossRef]
6. Gross, D.; Nesme, V.; Vogts, H.; Werner, R. Index theory of one dimensional quantum walks and cellular automata. *Commun. Math. Phys.* **2012**, *310*, 419–454. [CrossRef]
7. Shikano, Y. From Discrete Time Quantum Walk to Continuous Time Quantum Walk in Limit Distribution. *J. Comput. Theor. Nanosci.* **2013**, *10*, 1558–1570. doi:10.1166/jctn.2013.3097. [CrossRef]
8. Childs, A.M.; Goldstone, J. Spatial search by quantum walk. *Phys. Rev. A* **2004**, *70*, 022314. doi:10.1103/PhysRevA.70.022314. [CrossRef]
9. Portugal, R. *Quantum Walks and Search Algorithms*; Springer Science & Business Media: Berlin, Germany, 2013.
10. Douglas, B.L.; Wang, J.B. A classical approach to the graph isomorphism problem using quantum walks. *J. Phys. A Math. Theor.* **2008**, *41*, 075303. [CrossRef]
11. Gamble, J.K.; Friesen, M.; Zhou, D.; Joynt, R.; Coppersmith, S.N. Two-particle quantum walks applied to the graph isomorphism problem. *Phys. Rev. A* **2010**, *81*, 052313. doi:10.1103/PhysRevA.81.052313. [CrossRef]
12. Bialynicki-Birula, I. Weyl, Dirac, and Maxwell equations on a lattice as unitary cellular automata. *Phys. Rev. D* **1994**, *49*, 6920. [CrossRef]
13. Meyer, D. From quantum cellular automata to quantum lattice gases. *J. Stat. Phys.* **1996**, *85*, 551–574. [CrossRef]
14. Yepez, J. Relativistic Path Integral as a Lattice-based Quantum Algorithm. *Quantum Inf. Process.* **2006**, *4*, 471–509. [CrossRef]
15. Arrighi, P.; Facchini, S. Decoupled quantum walks, models of the Klein-Gordon and wave equations. *EPL* **2013**, *104*, 60004. [CrossRef]

16. Bisio, A.; D'Ariano, G.M.; Tosini, A. Quantum field as a quantum cellular automaton: The Dirac free evolution in one dimension. *Ann. Phys.* **2015**, *354*, 244–264. doi:10.1016/j.aop.2014.12.016. [CrossRef]
17. D'Ariano, G.M.; Perinotti, P. Derivation of the Dirac equation from principles of information processing. *Phys. Rev. A* **2014**, *90*, 062106. doi:10.1103/PhysRevA.90.062106. [CrossRef]
18. D'Ariano, G.M.; Mosco, N.; Perinotti, P.; Tosini, A. Path-integral solution of the one-dimensional Dirac quantum cellular automaton. *Phys. Lett. A* **2014**, *378*, 3165–3168. doi:10.1016/j.physleta.2014.09.020. [CrossRef]
19. D'Ariano, G.M.; Mosco, N.; Perinotti, P.; Tosini, A. Discrete Feynman propagator for the Weyl quantum walk in 2 + 1 dimensions. *EPL* **2015**, *109*, 40012. doi:10.1209/0295-5075/109/40012. [CrossRef]
20. Arrighi, P.; Facchini, S.; Forets, M. Quantum walking in curved spacetime. *Quantum Inf. Process.* **2016**, *15*, 3467–3486. doi:10.1007/s11128-016-1335-7. [CrossRef]
21. Bisio, A.; D'Ariano, G.M.; Perinotti, P. Quantum cellular automaton theory of light. *Ann. Phys.* **2016**, *368*, 177–190. doi:10.1016/j.aop.2016.02.009. [CrossRef]
22. Arnault, P.; Debbasch, F. Quantum walks and discrete gauge theories. *Phys. Rev. A* **2016**, *93*, 052301. doi:10.1103/PhysRevA.93.052301. [CrossRef]
23. Bisio, A.; D'Ariano, G.M.; Erba, M.; Perinotti, P.; Tosini, A. Quantum walks with a one-dimensional coin. *Phys. Rev. A* **2016**, *93*, 062334. doi:10.1103/PhysRevA.93.062334. [CrossRef]
24. Mallick, A.; Mandal, S.; Chandrashekar, C.M. Neutrino oscillations in discrete-time quantum walk framework. *Eur. Phys. J. C* **2017**, *77*, 85. doi:10.1140/epjc/s10052-017-4636-9. [CrossRef]
25. Molfetta, G.D.; Pérez, A. Quantum walks as simulators of neutrino oscillations in a vacuum and matter. *New J. Phys.* **2016**, *18*, 103038. [CrossRef]
26. Brun, T.A.; Mlodinow, L. Discrete spacetime, quantum walks and relativistic wave equations. *Phys. Rev. A* **2018**, *97*, 042131. doi:10.1103/PhysRevA.97.042131. [CrossRef]
27. Brun, T.A.; Mlodinow, L. Detection of discrete spacetime by matter interferometry. *arXiv* **2018**, arXiv:1802.03911. [CrossRef]
28. Raynal, P. Simple derivation of the Weyl and Dirac quantum cellular automata. *Phys. Rev. A* **2017**, *95*, 062344. doi:10.1103/PhysRevA.95.062344. [CrossRef]
29. Bibeau-Delisle, A.; Bisio, A.; D'Ariano, G.M.; Perinotti, P.; Tosini, A. Doubly special relativity from quantum cellular automata. *EPL* **2015**, *109*, 50003. [CrossRef]
30. Bisio, A.; D'Ariano, G.M.; Perinotti, P. Special relativity in a discrete quantum universe. *Phys. Rev. A* **2016**, *94*, 042120. doi:10.1103/PhysRevA.94.042120. [CrossRef]
31. Bisio, A.; D'Ariano, G.M.; Perinotti, P. Quantum walks, deformed relativity and Hopf algebra symmetries. *Philos. Trans. R. Soc. Lond. A Math. Phys. Eng. Sci.* **2016**, *374*, doi:10.1098/rsta.2015.0232. [CrossRef] [PubMed]
32. Arrighi, P.; Facchini, S.; Forets, M. Discrete Lorentz covariance for quantum walks and quantum cellular automata. *New J. Phys.* **2014**, *16*, 093007. [CrossRef]
33. Du, J.; Li, H.; Xu, X.; Shi, M.; Wu, J.; Zhou, X.; Han, R. Experimental implementation of the quantum random-walk algorithm. *Phys. Rev. A* **2003**, *67*, 042316. doi:10.1103/PhysRevA.67.042316. [CrossRef]
34. Ryan, C.A.; Laforest, M.; Boileau, J.C.; Laflamme, R. Experimental implementation of a discrete-time quantum random walk on an NMR quantum-information processor. *Phys. Rev. A* **2005**, *72*, 062317. doi:10.1103/PhysRevA.72.062317. [CrossRef]
35. Xue, P.; Sanders, B.C.; Leibfried, D. Quantum Walk on a Line for a Trapped Ion. *Phys. Rev. Lett.* **2009**, *103*, 183602. doi:10.1103/PhysRevLett.103.183602. [CrossRef] [PubMed]
36. Do, B.; Stohler, M.L.; Balasubramanian, S.; Elliott, D.S.; Eash, C.; Fischbach, E.; Fischbach, M.A.; Mills, A.; Zwickl, B. Experimental realization of a quantum quincunx by use of linear optical elements. *J. Opt. Soc. Am. B* **2005**, *22*, 499–504. doi:10.1364/JOSAB.22.000499. [CrossRef]
37. Sansoni, L.; Sciarrino, F.; Vallone, G.; Mataloni, P.; Crespi, A.; Ramponi, R.; Osellame, R. Two-Particle Bosonic-Fermionic Quantum Walk via Integrated Photonics. *Phys. Rev. Lett.* **2012**, *108*, 010502. doi:10.1103/PhysRevLett.108.010502. [CrossRef] [PubMed]
38. Crespi, A.; Osellame, R.; Ramponi, R.; Giovannetti, V.; Fazio, R.; Sansoni, L.; De Nicola, F.; Sciarrino, F.; Mataloni, P. Anderson localization of entangled photons in an integrated quantum walk. *Nat. Photonics* **2013**, *7*, 322–328. [CrossRef]

39. Flamini, F.; Spagnolo, N.; Sciarrino, F. Photonic quantum information processing: A review. *arXiv* **2018**, arXiv:1803.02790. [CrossRef]
40. Childs, A.M. Universal Computation by Quantum Walk. *Phys. Rev. Lett.* **2009**, *102*, 180501. doi:10.1103/PhysRevLett.102.180501. [CrossRef] [PubMed]
41. Lovett, N.B.; Cooper, S.; Everitt, M.; Trevers, M.; Kendon, V. Universal quantum computation using the discrete-time quantum walk. *Phys. Rev. A* **2010**, *81*, 042330. doi:10.1103/PhysRevA.81.042330. [CrossRef]
42. Childs, A.M.; Gosset, D.; Webb, Z. Universal computation by multiparticle quantum walk. *Science* **2013**, *339*, 791–794. [CrossRef] [PubMed]
43. Meyer, D.A. Quantum lattice gases and their invariants. *Int. J. Mod. Phys. C* **1997**, *8*, 717–735. [CrossRef]
44. Ahlbrecht, A.; Alberti, A.; Meschede, D.; Scholz, V.B.; Werner, A.H.; Werner, R.F. Molecular binding in interacting quantum walks. *New J. Phys.* **2012**, *14*, 073050. [CrossRef]
45. Bisio, A.; D'Ariano, G.M.; Perinotti, P.; Tosini, A. Thirring quantum cellular automaton. *Phys. Rev. A* **2018**, *97*, 032132. doi:10.1103/PhysRevA.97.032132. [CrossRef]
46. Östlund, S.; Mele, E. Local canonical transformations of fermions. *Phys. Rev. B* **1991**, *44*, 12413–12416. doi:10.1103/PhysRevB.44.12413. [CrossRef]
47. Thirring, W.E. A soluble relativistic field theory. *Ann. Phys.* **1958**, *3*, 91–112. doi:10.1016/0003-4916(58)90015-0. [CrossRef]
48. Hubbard, J. Electron correlations in narrow energy bands. *Proc. R. Soc. Lond. A Math. Phys. Eng. Sci.* **1963**, *276*, 238–257. doi:10.1098/rspa.1963.0204. [CrossRef]

© 2018 by the authors. Licensee MDPI, Basel, Switzerland. This article is an open access article distributed under the terms and conditions of the Creative Commons Attribution (CC BY) license (http://creativecommons.org/licenses/by/4.0/).

Article

Robust Macroscopic Quantum Measurements in the Presence of Limited Control and Knowledge

Marc-Olivier Renou *, Nicolas Gisin and Florian Fröwis

Department of Applied Physics, University of Geneva, 1211 Geneva 4, Switzerland;
Nicolas.Gisin@unige.ch (N.G.); Florian.Froewis@unige.ch (F.F.)
* Correspondence: marcolivier.renou@unige.ch

Received: 31 October 2017; Accepted: 26 December 2017; Published: 9 January 2018

Abstract: Quantum measurements have intrinsic properties that seem incompatible with our everyday-life macroscopic measurements. Macroscopic Quantum Measurement (MQM) is a concept that aims at bridging the gap between well-understood microscopic quantum measurements and macroscopic classical measurements. In this paper, we focus on the task of the polarization direction estimation of a system of N spins 1/2 particles and investigate the model some of us proposed in Barnea et al., 2017. This model is based on a von Neumann pointer measurement, where each spin component of the system is coupled to one of the three spatial component directions of a pointer. It shows traits of a classical measurement for an intermediate coupling strength. We investigate relaxations of the assumptions on the initial knowledge about the state and on the control over the MQM. We show that the model is robust with regard to these relaxations. It performs well for thermal states and a lack of knowledge about the size of the system. Furthermore, a lack of control on the MQM can be compensated by repeated "ultra-weak" measurements.

Keywords: quantum measurement; quantum estimation; macroscopic quantum measurement

1. Introduction

In our macroscopic world, we constantly measure our environment. For instance, to find north with a compass, we perform a direction measurement by looking at the pointer. Yet, finding a quantum model for this kind of macroscopic measurement faces several problems. Many characteristics of quantum measurements seem to be incompatible with our intuitive notion of macroscopic measurements. For example, perfectly measuring two non-commuting observables is impossible in quantum mechanics, and any informative measurement has a nonvanishing invasiveness. Thus, if it exists, such a model cannot be of the standard projective kind. Although we have a good intuition of what such a measurement is, the natural characteristics it should satisfy are not obvious. Even if these characteristics can be rigorously formulated, it is not clear whether there exists a quantum model that satisfies them all.

For concreteness, quantum models for macroscopic measurements can be considered as a parameter estimation task. In this paper, we focus on the estimation of the direction of polarization of N qubits, oriented in a direction that is uniformly chosen at random. The question of the optimal way to estimate N qubit polarization is already well studied [1,2] and can be seen as part of a larger class of covariant estimation problems [3]. It is linked to covariant cloning [4] and purification of state [5]. In the limit of macroscopic systems, those optimal measurements are arbitrarily precise and potentially with low disturbance of the system [6,7]. A tradeoff between the quality of the guess and the disturbance of the state has been demonstrated [8], as well as an improvement of the guess when abstention is allowed [9]. However, these optimal measurements may not be satisfying models of our everyday-life macroscopic measurements as it is not clear how these optimal measurements could be physically implemented in a natural way. A first attempt to solve this Positive Operator Valued Measure (POVM), which is continuous, into a POVM with a finite (and small) number of elements [10,11]. However, even if this reduction exists, the resulting POVM is difficult to interpret physically, and to our best knowledge, no family of reduced POVM for every N exists.

In [12], we argue that a good model of a macroscopic measurement should be highly non-invasive, collect a large amount of information in a single shot and be described by a "fairly simple" coupling between system and observer. Measurements that fulfills these requirements are called "Macroscopic Quantum Measurements" (MQM). Invasiveness seems to be difficult to satisfy with a quantum model. Indeed, the disturbance induced on the state by a measurement is generic in quantum mechanics. This has no counterpart in classical physics, where any measurement can ideally be done without disturbance of the system. However, it is now well known that this issue can be solved by accepting quantum measurements of finite accuracy. In [13], Poulin shows the existence of a trade-off between state disturbance and measurement resolution as a function of the size of the ensemble. One macroscopic observable can behave "classically", provided we measure it with sufficiently low resolution. Yet, the question is still open for several non-commuting observables. Quantum physics allows precise measurements of only one observable among two non-commuting ones.

In this paper, we study the behavior of an MQM model for the measurement of the polarization of a large ensemble of N parallel spin 1/2 particles, which implies the measurement of the non-commuting spin operators. In this model, the measured system is first coupled to a measurement apparatus through an intuitive Hamiltonian already introduced in [14]. Then, the apparatus is measured. We extend our previous study to more general cases. In [12], it was shown that this model allows good direction estimation and low disturbance for systems of N parallel spin 1/2 particles. This system can be interpreted as the ground state of a product Hamiltonian. Here, we generalize the scenario to thermal states. We also study a different measurement procedure based on repeated weak measurements.

The paper is structured as follows: We first present a simplified technical framework that describes the measurement of a random direction for a given quantum state and observable. Considering an input state and an observable independent of the particle number and with no preferred direction, we show that the problem reduces to many sub-problems, which correspond to systems of fixed total spin j. Then, we quantitatively treat the case of the thermal state, which generalizes the N parallel spin 1/2 particle for non-zero temperature, showing that the discussed MQM is still close to the optimal measurement. In the proposed MQM, the precision of the estimated direction highly depends on the optimized coupling strength of the model. In Section 4, we follow the ideas of [13], and we show that one may relax this requirement by doing repeated "ultra-weak" measurements and a naive guess. We conclude and summarize in the last section.

2. Estimation of a Direction

In this paper, we aim to study the behavior of a specific MQM model for a direction estimation task, e.g., the estimation of the direction of a magnet or a collection of spins. Hence, we first introduce an explicit (and specific) direction estimation problem, which is presented as a game. It concerns the direction estimation of a qubit ensemble. In the following, $S_{\vec{u}} = \vec{S} \cdot \vec{u}$ represents the spin operator projected in direction $\vec{u}$, i.e., the elementary generator of rotations around $\vec{u}$. For a given state $\rho_{\vec{u}}$ of $N = 2J$ qubits, we say that $\rho_{\vec{u}}$ points in the direction $\vec{u}$ if it is positively polarized in the $\vec{u}$ direction, i.e., if $[\rho_{\vec{u}}, S_{\vec{u}}] = 0$ and $\text{Tr}(\rho_{\vec{u}} S_{\vec{u}}) > 0$. We consider the problem of polarization direction estimation from states that are all the same, but point in a direction that is chosen uniformly at random. This problem has already been widely studied [1–3,6,15]. We give here a unified framework adapted to our task.

2.1. General Framework

We consider a game with a referee, Alice, and a player, Bob. Alice and Bob agree on some initial state ρ_z. In each round of the game, Alice chooses a direction $\vec{u}$ from a uniform distribution on the unit sphere. She rotates ρ_z to $\rho_{\vec{u}} = \mathcal{R}_{\vec{u}}^\dagger \rho_z \mathcal{R}_{\vec{u}}$, where $\mathcal{R}_{\vec{u}}$ is a rotation operator, which maps $\vec{z}$ to $\vec{u}$. She sends

$\rho_{\vec{u}}$ to Bob, who measures it with some given measurement device characterized by a Positive Operator Valued Measure (POVM) Ω_r. He obtains a result r with probability $p(r|\vec{u}) = \text{Tr}(\Omega_r \rho_{\vec{u}})$, from which he deduces $\vec{v}_r$, his guess for $\vec{u}$. Bob's score is computed according to some predefined score function $g(\vec{u}, \vec{v}_r) = \vec{u} \cdot \vec{v}_r$. Given his measurement result, Bob's goal is to find the optimal estimate, i.e., the one that optimizes his mean score [16].

$$G = \int dr \int d\vec{u}\, p(r|\vec{u}) g(\vec{u}, \vec{v}_r) \tag{1}$$

For simplicity, we consider an equivalent, but simplified POVM. In our description, Bob measures the system, obtains results r and then post-processes this information to find his guess $\vec{v}_r$. We now regroup all POVM elements corresponding to the same guess and label by the guessed direction. Formally, we go from Ω_r to $O_{\vec{v}} = \int dr \Omega_r \delta(\vec{v}_r - \vec{v})$.

Some assumptions are made about ρ_z and $O_{\vec{v}}$. We suppose that ρ_z points in the z direction. Moreover, we assume that ρ_z is symmetric under the exchange of particles, which implies $[\rho_z, S^2] = 0$. Let $|\alpha, j, m\rangle$ be the basis in which S_z and S^2 are diagonal (where $j \in \{J, J-1, ...\}$ is the total spin, α the multiplicity due to particle exchange and m the spin along z). Then, ρ_z is diagonal in this basis, with coefficients independent of α, denoted as $c_m^j = \langle \alpha, j, m | \rho_z | \alpha, j, m \rangle$.

We also suppose that the measurement device does not favor any direction and treats each particle equally. Mathematically, this means that $O_{\vec{v}}$ is covariant with respect to particle exchange and rotations. Then, any POVM element is generated from one kernel O_z and the rotations $\mathcal{R}_{\vec{v}}$: $O_{\vec{v}} = \mathcal{R}_{\vec{v}}^\dagger O_z \mathcal{R}_{\vec{v}}$ (for more technical details, see [15]). With this, Equation (1) simplifies to:

$$G = \int d\vec{v} \int d\vec{u}\, p(\vec{v}|\vec{u}) g(\vec{u}, \vec{v}), \tag{2}$$

2.2. Score for Given Input State and Measurement

The following lemma is already implicitly proven in [15].

Lemma 1. *Bob's mean score is:*

$$G = \sum_j \frac{j A_j \text{Tr}\left(\rho_z^j\right)}{j+1} \text{Tr}\left(\frac{S_z}{j} \tilde{\rho}_z^j\right) \text{Tr}\left(\frac{S_z}{j} \frac{O_z^j}{2j+1}\right) \tag{3}$$

where $A_j = \binom{2J}{J-j} - \binom{2J}{J-j-1}$ is the degeneracy of the multiplicity α in a subspace of given (j, m), O_z^j is the projections of O_z over all subspaces of fixed (α, j), ρ_z^j is the projection of ρ_z over all subspaces of fixed (α, j) and $\tilde{\rho}_z^j = \frac{\rho_z^j}{\text{Tr}(\rho_z^j)}$.

Lemma 1 says that Bob cannot use any coherence between subspaces associated with different (α, j) to increase his score. In other words, the score Bob achieves is the weighted sum (where the weights are $\text{Tr}\left(\rho_z^{\alpha, j}\right)$) of the scores G^j Bob would achieve by playing with the states $\tilde{\rho}_z^j$. This property is a consequence of the assumption that no direction or particle is preferred by Bob's measurement or in the set of initial states. For self-consistency, we prove this lemma.

Proof. Bob's mean score is:

$$G = \int dr \int d\vec{u}\, p(\vec{v}|\vec{u}) g(\vec{u}, \vec{v}) = \int dv\, \text{Tr}(O_{\vec{v}} \Gamma_{\vec{v}}), \tag{4}$$

where $\Gamma_{\vec{v}} = \vec{v} \cdot \int d\vec{u}\, \rho_{\vec{u}}\, \vec{u}$. As $\rho_{\vec{u}}$ is the rotated ρ_z and $O_{\vec{v}}$ is covariant, we have:

$$G = \text{Tr}(O_z \Gamma_z). \tag{5}$$

Let $P_{\alpha,j} = \sum_m |\alpha,j,m\rangle\langle\alpha,j,m|$ be projectors, $\Gamma_z^{\alpha,j} = P_{\alpha,j}\Gamma_z P_{\alpha,j}$ and $O_z^{\alpha,j} = P_{\alpha,j} O_z P_{\alpha,j}$. Here, as ρ and O_z do not depend on the particle number, α is only a degeneracy.

As Γ_z is invariant under rotation around z and commutes with S^2, we have $\Gamma_z = \sum_{\alpha,j} \Gamma_z^{\alpha,j}$. Then, $G = \sum_{\alpha,j} \text{Tr}\left(O_z^{\alpha,j}\Gamma_z^{\alpha,j}\right) = \sum_j A_j \text{Tr}\left(O_z^j \Gamma_z^j\right)$, where O_z^j, Γ_z^j are respectively the projections of O_z, Γ_z over any spin coherent subspace of fixed α, j. Let $G^j = \text{Tr}\left(O_z^j \Gamma_z^j\right)$.

$\Gamma_z^j = \sum_m c_m^j \int d\vec{u}\, u_z \mathcal{R}_{\vec{u}}^\dagger |\alpha,j,m\rangle\langle\alpha,j,m| \mathcal{R}_{\vec{u}}$ is symmetric under rotations around z. Then, it is diagonal in the basis $|\alpha,j,m\rangle$ with fixed j,α. As $\langle\alpha,j,\mu|\int d\vec{u}\, u_z \mathcal{R}_{\vec{u}}^\dagger|\alpha,j,m\rangle\langle\alpha,j,m|\mathcal{R}_{\vec{u}}|\alpha,j,\mu\rangle = \frac{m\mu}{j(j+1)(2j+1)} = \frac{m}{j(j+1)(2j+1)}\langle\alpha,j,\mu|S_z^{\alpha,j}|\alpha,j,\mu\rangle$, we have:

$$\Gamma_z^j = \sum_m c_m^j \frac{m}{j(j+1)(2j+1)} S_z^{\alpha,j} \tag{6}$$

and:

$$G^j = \frac{1}{j(j+1)(2j+1)} \text{Tr}\left(S_z \rho_z^j\right) \text{Tr}\left(S_z O_z^j\right). \tag{7}$$

□

2.3. State Independent Optimal Measurement, Optimal State for Direction Estimation

Given the state ρ_z, the measurement that optimizes Bob's score is the set of $\left\{\Theta_{\vec{u}}^{\alpha,j}\right\}$ such that $\text{Tr}\left(S_z \Theta_z^{\alpha,j}\right)$ is maximal. The maximum is obtained when $\Theta_z^{\alpha,j}$ is proportional to a projector on the eigenspace of S_z with the maximal eigenvalue, that is for $\Theta_z^{\alpha,j} = (2j+1)|\alpha,j,\pm j\rangle\langle\alpha,j,\pm j|$. Here, the sign depends on the sign of $\text{Tr}\left(S_z \rho_z^j\right)$. In the following, we restrict ourselves to the case where the $\text{Tr}\left(S_z \rho_z^j\right)$ are all positive (this is the case for the thermal state, considered below). Then:

$$G_{\text{opt}} = \sum_j \frac{j A_j \text{Tr}\left(\rho_z^j\right)}{j+1} \text{Tr}\left(\frac{S_z}{j}\tilde{\rho}_z^j\right). \tag{8}$$

For $\rho_z = |J,J\rangle\langle J,J|$, the thermal state of temperature $T = 0$, we find $G_{\text{opt},T=0} = \frac{J}{J+1}$. Equivalently, we recover the optimal fidelity $F_{\text{opt},T=0} = \frac{1}{2}(1 + G_{\text{opt},T=0}) = \frac{N+1}{N+2}$, already found in [1]. Asymptotically, we have $G_{\text{opt},T=0} = 1 - 1/J + O(1/J^2)$. This induces a natural characterization of the optimality of an estimation procedure. Writing $G_{T=0}$ as $G_{T=0} = 1 - \epsilon_J/J$ where $\epsilon_J = J(1 - G_{T=0}) \geq 1$, we say that the procedure is asymptotically optimal if $\epsilon_J = 1 + O(1/J)$ and almost optimal if $\epsilon_J - 1$ is asymptotically not far from zero.

2.4. Optimality of a State and a Measurement for Direction Guessing

Given the input state ρ_z, we can now compare the performances of a given measurement to the optimal measurement. From Equations (3) and (8), we have, for an arbitrary measurement:

$$\Delta G \equiv G_{\text{opt}} - G = \sum_j \frac{j A_j \text{Tr}\left(\rho_z^j\right)}{j+1} \text{Tr}\left(\frac{S_z}{j}\tilde{\rho}_z^j\right) \text{Tr}\left(\frac{S_z}{j} \frac{\Theta_z^j - O_z^j}{2j+1}\right). \tag{9}$$

For every j, the three terms of the product are positive. Then, qualitatively, the measurement is nearly optimal if for each j, the product of the three is small. We give here the interpretation of each of these terms:

- A_j is the degeneracy under permutation of particles (labeled by α) and $\text{Tr}\left(\rho_z^j\right)$ the weight of ρ_z over a subspace j, α. Hence, the first term, bounded by $j/(j+1)$, only contains the total weight of ρ_z over a fixed total spin j. Hence, it is small whenever ρ has little weight in the subspace j.
- $\text{Tr}\left(\frac{S_z}{j}\tilde{\rho}_z^j\right)$ is small whenever the component of ρ_z on the subspace of total spin j, $\tilde{\rho}_z^j = P_z \rho_z P_z$, is small or not well polarized. It is bounded by one. When ρ_z^j is not well polarized, the optimality of the measurement in that subspace makes little difference. Then, this second term characterizes the quality of the component ρ_z^j for the guess of the direction.
- The last term is small when O_z^j is nearly optimal and is also bounded by one. More exactly, as O_z^j is a covariant POVM, we have $\text{Tr}\left(O_z^j\right) = 2j+1$, and all diagonal coefficients are positive. Because of S_z/j, O_z^j is (nearly) optimal when it projects (mainly) onto the subspace of S_z with the highest eigenvalue. POVMs containing other projections are sub-optimal. This effect is amplified by the operator S_z: the further away these extra projections $\propto |j,m\rangle\langle j,m|$ are from the optimal projector $\propto |j,j\rangle\langle j,j|$ (in the sense of $j-m$), the stronger the sub-optimality is. Then, the last term corresponds to the optimality of the measurement component O_z^j for the guess of the direction.

Interestingly, we see here that the state and measurement "decouple": the optimal measurement is independent of the considered state. However, if the measurement is not optimal only for subspaces where ρ_z has low weight or is not strongly polarized, it will still result in a good mean score.

2.5. Estimation from a Thermal State

We now consider the case where the game is played with a thermal state (with temperature $T = 1/\beta$) of $N = 2J$ spins:

$$\rho_z = \frac{1}{Z}\left(e^{-\beta\sigma_z/2}\right)^{\otimes N} = \frac{1}{Z}\sum_{\alpha,j,m} e^{-\beta m}|\alpha,j,m\rangle\langle\alpha,j,m|, \quad (10)$$

where $Z = (2\cosh(\beta/2))^N$ is the partition sum. ρ_z is clearly invariant under rotations around z and symmetric under particle exchange. For later purposes, we define $f_j(\beta) = Z\,\text{Tr}\left(\frac{S_z}{j}\rho_z^{\alpha,j}\right) = \left[(1+j)\sinh(j\beta) - j\sinh((1+j)\beta)\right]/(2j\sinh(\beta/2)^2)$.

Equation (3) now reads:

$$G_{T=0} = \frac{J}{J+1}\text{Tr}\left(\frac{S_z}{J}\frac{O_z}{2J+1}\right), \quad (11)$$

and for any temperature β:

$$G = \frac{1}{Z}\sum_j A_j\, f_j(\beta)\, G_{T=0}^j, \quad (12)$$

with the optimal measurement, $G_{\text{opt},T} = \frac{1}{Z}\sum_j \frac{jA_j}{j+1} f_j(\beta)$. Note that for low temperatures, this expression can be approximated with $\langle J\rangle_\beta/(\langle J\rangle_\beta + 1)$, where $\langle J\rangle_\beta$ is the mean value of the total spin operator for a thermal state.

3. A Macroscopic Quantum Measurement

3.1. The Model

In the following, we consider a model already introduced in [12,14] for polarization estimation. It is adapted from the Arthur–Kelly model, which is designed to simultaneously measure momentum and position [17–19]. The model is expressed in the von Neumann measurement formalism [20–22]. The measurement device consists of a quantum object (the pointer), which is first initialized in a well-known state and coupled to the system to be measured. At last, the pointer is measured in a projective way. The result of the measurement provides information about the state of the system. Tuning the initial state of the pointer and the strength of interaction, one can model a large range of measurements on the system, from projective measurements, which are partially informative, but destruct the state to weak measurements, which acquire little information, but do not perturb much.

More specifically, to measure the direction of $\rho_{\vec{u}}$, we use a pointer with three spatial degrees of freedom:

$$|\phi\rangle = \frac{1}{(2\pi\Delta^2)^{3/4}} \int dx\, dy\, dz\, e^{-\frac{x^2+y^2+z^2}{4\Delta^2}} |x\rangle |y\rangle |z\rangle, \quad (13)$$

where x, y, z are the coordinates of the pointer. The parameter Δ in $|\phi\rangle$ represents the width of the pointer: a small Δ corresponds to a narrow pointer and implies a strong measurement, while a large Δ gives a large pointer and a weak measurement. The interaction Hamiltonian reads:

$$H_{\text{int}} = \vec{S} \cdot \vec{p} \equiv p_x \otimes S_x + p_y \otimes S_y + p_z \otimes S_z, \quad (14)$$

where p_x, p_y, p_z are the conjugate variables of x, y, z. A longer interaction time or stronger coupling can always be renormalized by adjusting Δ. Hence, we take the two equal to one. Finally, a position measurement with outcome $\vec{r}$ is performed on the pointer. The POVM elements associated with this measurement are $O_{\vec{r}} = E_{\vec{r}} E_{\vec{r}}^\dagger$, where the Krauss operator $E_{\vec{r}}$ reads:

$$E_{\vec{r}} \propto \int d\vec{p}\, e^{i\vec{r}\cdot\vec{p}} e^{-\Delta^2 p^2} e^{-i\vec{p}\cdot\vec{S}}. \quad (15)$$

The POVM associated with this model is already covariant. Indeed, the index of each POVM element is the direction of guess (to exactly obtain the form given in Section 2.1, one has to define $O_{\vec{u}} = \int_0^\infty r^2 O_{\vec{r}}\, dr$, which is equivalent to identifying each vector with its direction). Any $O_{\vec{r}}$ is a rotation of O_z: $O_{\vec{r}} = \mathcal{R}_{\vec{r}}^\dagger O_z \mathcal{R}_{\vec{r}}$.

3.2. Behavior for Zero Temperature States

At zero temperature, it is already known that the score obtained for a game where Bob does the MQM remains close to the optimal one. In our previous study [12], we demonstrated a counter-intuitive behavior of the quality of the guess: a weaker coupling strength can achieve better results than a strong coupling; see Figure 1a. In particular, we show that for well-chosen finite coupling strength, the score of the guess is almost optimal. The optimal value of the coupling is $\Delta = \sqrt{J/4}$: it scales with the square root of the number of particles.

Additional calculations confirm this first conclusion (see Figure 1b). Exploiting the conclusion of the discussion of Section 2.4, we only considered the first diagonal coefficient of O_z, $o_J = \langle J, J | O_z | J, J \rangle$, to lower bound the performance of the POVM [23]. Numerical simulations suggest that for a coupling strength $\Delta = \sqrt{J/4}$, only considering the bound over o_J, $G_{T=0}$ develops as $G_{T=0} = 1 - \epsilon_J / J$ with $\epsilon_J = J(1 - G_J) \lesssim 19/18$ for large J. Hence, the asymptotic difference between $G_{\text{opt},T=0}$ and $G_{\text{MQM},T=0}$ is such that $J\Delta G_{T=0}$ remains bounded, in the order of 0.05.

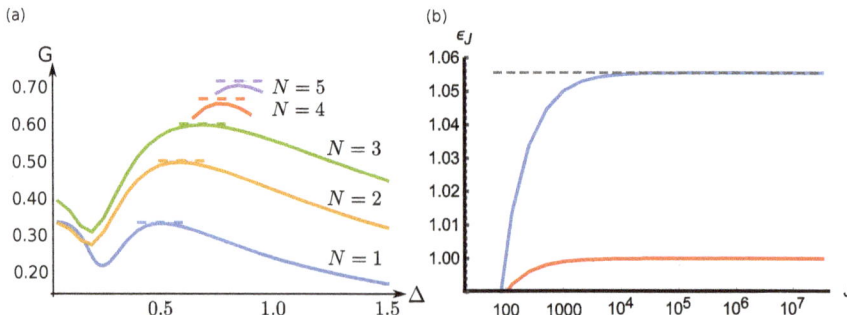

Figure 1. (a) Mean score as a function of the pointer width Δ for various $N = 2J$. The dashed lines correspond to the optimal value G_{opt}. (b) Scaling factor $\epsilon_J = J(1 - G_J)$ from the approximate lower bound on the score G (upper, blue curve) compared to the optimal scaling factor $J(1 - G_{\text{opt}})$ (lower, red curve). For large J, ϵ_J seems to go to 19/18 (dashed line). See Section 3.2 for further details.

From Equation (3) and the discussion about Equation (9), we see that, to achieve optimality, the first diagonal coefficient o_J must be maximal [24], that is equal to $2J+1$. When this is not the case, as $\text{Tr}(O_z) = 2J+1$, the difference $(2J+1) - o_J = \text{Tr}(O_z) - o_J = \sum_{m \neq J} o_m$ is distributed between the other diagonal coefficients $o_m = \langle J, m | O_z | J, m \rangle$, for $m \neq J$. The score achieved by the measurement is given by Equation (11):

$$G_{T=0} = \text{Tr}\left(\frac{S_z}{J} \frac{O_z^J}{2J+1}\right) = \frac{J}{J+1} \sum_m \frac{m}{J} \frac{o_m}{2J+1}. \tag{16}$$

Our bound only considers the coefficient o_J. However, a simple calculation shows that this is enough to deduce the strict suboptimality of the measurement. Indeed, one can derive:

$$\epsilon_J = J\left(1 - \frac{J}{J+1}\left(\frac{o_J}{2J+1} + \sum_{m \neq J} \frac{m}{J} \frac{o_m}{2J+1}\right)\right)$$

$$\geq J\left(1 - \frac{J}{J+1}\left(\frac{o_J}{2J+1} + \frac{J-1}{J}\left(1 - \frac{o_J}{2J+1}\right)\right)\right)$$

$$\geq 2 - \frac{o_J}{2J+1} + o(1),$$

where $o(1) \to 0$ when $J \to \infty$. Hence, if o_J is not asymptotically $2J+1$, ϵ_J cannot be asymptotically one.

In the following, we show that a lower bound on G for thermal states can be calculated with methods based on the $T=0$ case.

3.3. Behavior for Finite Temperature States

As it is built from the spin operators only, the measurement scheme depends only on the properties of the system with respect to the spin operators. More precisely, for a given system size $N = 2J$, we consider the basis $\{|\alpha, j, m\rangle^{(N)}\}$, and for given total spin j and permutation multiplicity α, the projector $P_{\alpha,j}^{(N)} = \sum_{\alpha,j} |\alpha, j, m\rangle\langle\alpha, j, m|^{(N)}$. Then, the projection of Equation (15) for $N = 2J$ spins onto the subspace j, α is equivalent to the projected Krauss operator for $n = 2j$ spins onto j:

$$P_{\alpha,j}^{(N)} E_{\vec{r}}^{(N)} P_{\alpha,j}^{(N)} \equiv P_j^{(n)} E_{\vec{r}}^{(n)} P_j^{(n)}, \tag{17}$$

where the equivalence $\equiv$ is interpreted as $|\alpha, j, m\rangle^{(N)} \equiv |m\rangle^{(n)}$ (there is no multiplicity for n and $j = n/2$).

For non-zero temperature, we adapt the numerical estimation model of [12]. Due to Lemma 1 and Equation (17), we can directly exploit the same model and combine the results for the different subspaces for given j. However, in this case, we are limited by the choice of the coupling strength Δ of the pointer with the system. At zero temperature, only the total spin subspace that corresponds to $j = J$ is involved. The optimal coupling strength is then $\Delta = \sqrt{J/4}$. For a non-zero temperature, all possible j appear, and the value of Δ cannot be optimized for each one. Our strategy is to choose the optimal coupling value for the equivalent total spin J_{eq} satisfying $\langle S^2 \rangle = J_{\text{eq}}(J_{\text{eq}} + 1)$, which can be deduced from $\langle S^2 \rangle = \frac{1}{4}(3J + J(2J-1)\tanh^2 \beta/2)$ (for a thermal state). Depending on the sensitivity of the MQM guessing scheme with respect to a change in the value of Δ, this method may work or not. Numeric simulations show that a change of order $O(\sqrt{J})$ perturbs the score. However, one can hope that for smaller variation, the perturbation is insignificant.

We tested the method for different values of temperature $T = 1/\beta$ corresponding to spin polarization $\langle S_z \rangle = J \tanh \beta/2$. We find again that the asymptotic difference between G_{opt} and G_{MQM} is small. More precisely, Figure 2 shows $J\Delta G_\beta$ as a function of J, for different temperature corresponding to $\langle S_z \rangle = cJ$, for various c. For each Δ, the error $J\Delta G_\beta$ seems to be bounded for large J.

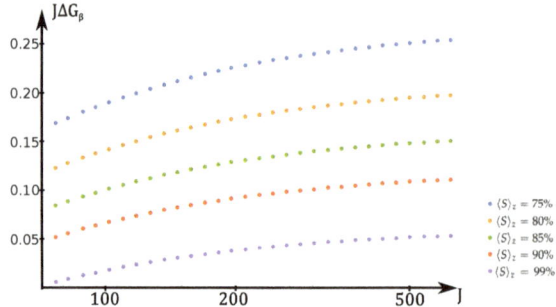

Figure 2. $J\Delta G$ (Equation (9)) as a function of J, for various β chosen such that $\langle S_z \rangle = J \tanh \beta/2$. The Macroscopic Quantum Measurement (MQM) is close to optimal even for finite temperature. See Section 3.3 for further details.

4. Estimation of a Direction through Repeated Weak Measurements

In the previous section, we considered a specific MQM and studied the mean score of the state direction for pure states, as well as for more realistic thermal states. We compared it to its optimal value, obtained with the optimal theoretical measurement. We showed that the difference remained bounded. As the model makes use of a simple Hamiltonian coupling between system and observer, it satisfies the requirements of an MQM as stated in the introduction for thermal states.

However, this model requires that three one-dimensional (1D) pointers (or equivalently one three-dimensional (3D) pointer) are coupled to the system at the very same time, to be then measured. This requirement is difficult to meet. Moreover, an optimized coupling strength between system and pointer is necessary: the pointer width has to be $\Delta = \sqrt{J/4}$ within relatively tight limits. This requires a good knowledge about the system to be measured (its size, its temperature, etc.) and fine control over the measurement. Following [13], we can overcome this problem by implementing many ultra-weak measurements. To this end, we focus on a relaxation of the measurement procedure, where we consider repeated very weak measurements (with $\Delta \gg \sqrt{J/4}$) in successive orthogonal directions on the state, which is gradually disturbed by the measurements. This idea has already been implemented experimentally [25]. The guessed state is obtained by averaging the results in each of the three directions. Note that this is not optimal, as the first measurements are more reliable than the last. However, we show in the following that this intuitive approach gives almost optimal results. For simplicity, we restrict ourselves to the case of a perfectly-polarized state, or equivalently a thermal state at a zero temperature.

4.1. The Model

We modify the game considered so far in the following way. Bob now uses a modified strategy, in which he successively repeats the same measurement potentially in different measurement basis. First, he weakly couples the state to a 1D Gaussian pointer through an interaction Hamiltonian in some direction w. The pointer state is:

$$|\phi\rangle = \frac{1}{(2\pi\Delta^2)^{1/4}} \int dw\, e^{-\frac{w^2}{4\Delta^2}} |w\rangle, \quad (18)$$

and the Hamiltonian reads:

$$H_w \propto p_w \otimes S_w, \quad (19)$$

where $w \in \{x, y, z\}$. Then, Bob measures the pointer. The post measured state is used again for the next measurement and is disturbed in each round. We first analytically derive the case where Bob only measures in one direction ($w = z$). Then, we consider the case where Bob does t measurements successively in each orthogonal direction x, y, z. He obtains results $x_1, y_1, z_1, x_2, y_2, ..., z_t$ and estimates the direction with the vector which coordinates are the average of the x_i, the y_i and the z_i.

4.2. Measurement in a Single Direction

We first study the 1D case. First, note that the optimal strategy when the measurement operators O_r are required to measure in a fixed direction z (i.e., $[O_r, S_z] = 0$) is to measure the operators S_z^j: As the O_r commutes with S_z, they can be simulated with a measurement of S_z^j. The optimum is to answer $\pm z$ depending on the sign of the result. The obtained score is then $G = \frac{J}{2J+1}$ for integer $J = N/2$ and $G = \frac{2J+1}{4(J+1)}$ otherwise.

In our model, we consider an interaction Hamiltonian H_w taken in a constant direction $w = z$. The total number of measurements is t.

The measurement results form a vector $\vec{r} = \{r_1, ..., r_t\}$. The POVM of the full measure sequence is:

$$\Omega_{\vec{r}} = \begin{bmatrix} \ddots & & \\ & F_m(\vec{r}) & \\ & & \ddots \end{bmatrix} \tag{20}$$

where:

$$F_m(\vec{r}) = \frac{1}{\left(\Delta\sqrt{2\pi}\right)^p} e^{-\frac{\|\vec{r} - m\vec{1}\|^2}{2\Delta^2}}, \tag{21}$$

where $\vec{1} = \{1, ..., 1\}$. As all measurements for each step commute, this case can be solved analytically. Note first that the ordering of the measurement results is irrelevant. From Equation (1), we find:

$$G = \frac{1}{(J+1)(2J+1)} \text{Tr}(S_z O_z)$$

$$= \frac{2}{(J+1)(2J+1)\left(\Delta\sqrt{2\pi}\right)^t} \int d\vec{r} \delta(\vec{v}_{\vec{r}} - \vec{z}) e^{-\frac{\|\vec{r}\|^2}{2\Delta^2}} \left(\sum_{m>0} m \, e^{\frac{-mt}{2\Delta^2}} \sinh\left(\frac{m \vec{r} \cdot \vec{1}}{\Delta^2}\right) \right),$$

where $\vec{v}_{\vec{r}}$ is the optimal guess. For $\vec{r}$ such that $\vec{r} \cdot \vec{1} \geq 0$, the optimal guess is clearly $\vec{v}_{\vec{r}} = \vec{z}$. By symmetry, $\vec{v}_{-\vec{r}} = -\vec{v}_{\vec{r}}$, and the optimal guess is $\vec{v}_{\vec{r}} = \text{sign}(\vec{r} \cdot \vec{1})\vec{z}$. Then:

$$G = \frac{2}{(J+1)(2J+1)} \sum_{m>0} m \, \text{erf}\left(\frac{m}{\Delta}\sqrt{\frac{t}{2}}\right) \tag{22}$$

is easily computed by integration over $\vec{r}$ and by decomposition into its parallel and orthogonal components to $\vec{1}$. We see here that the score only depends on the ratio $\frac{\sqrt{t}}{\Delta}$ and reaches the 1D strong measurement limit for $\frac{\sqrt{t}}{\Delta} \gg 1$ (see Figure 3). Here, erf is the error function. We see that $G \to 1/2$ for $J \to \infty$, which is the optimal value for optimal measurements lying on one direction.

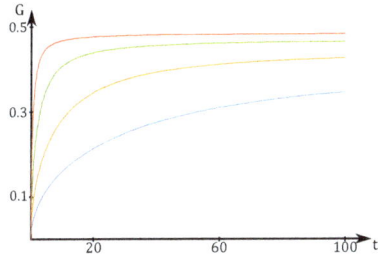

Figure 3. Score for repeated weak measurement in a single fixed direction with $\Delta = 10$ and $J = 2, 4, 8, 16$. See Section 4.2 for further details.

4.3. Ultra-Weak Measurements in Three Orthogonal Directions

We now study the relaxation of our initial MQM model. In this case, for a large number of measurement t, we could not analytically derive the mean score. We hence implemented a numerical simulation of the model. We fix the number of qubits $N = 2J$ and pointer width Δ. The vector $\vec{u}$ is drawn at random on the Bloch sphere. Then, we simulate τ successive weak measurements in directions x, y, z of the system $|\vec{u}\rangle^{\otimes N}$. For each $t \leq \tau$, we guess $\vec{u}$ from the mean of the results for x, y, z for measurements up to t.

For large Δ, our procedure can be seen as successive weak measurements of the system. Each measurement acquires a small amount of information and weakly disturbs the state. We attribute the same weight to each measurement result to find the estimated polarization. As each measurement disturbs the state, this strategy is not optimal. However, keeping the heuristic of "intuitive measurement", we consider this guessing method as being natural.

The results from the numerical simulation suggest that for a fixed number of particles $N = 2J$ and fixed pointer width Δ, the score as a function of t increases and then decreases (see Figure 4a), which is intuitive. Indeed, for few measurements, the state is weakly disturbed, and each measurement acquires only a small amount of information about the original state. Then, after a significant number of measurements, the state is strongly disturbed, and each measurement is done over a noisy state and gives no information about the initial state. Hence, there is an optimal number of measurements $t^{\max}(N, \Delta)$ that gives a maximal score $G^{\max}(N, \Delta)$. Moreover, for a fixed $N = 2J$, $G^{\max}(N, \Delta)$ increases smoothly as the measurements are weaker, i.e., as Δ increases. It reaches a limit $G^{\max}(N)$ (see Figure 4a). This suggest that for weak enough measurements, we observe the same behavior as in the 1D case. More measurements compensate a weaker interaction strength, without loss of precision. Hence, the precision of a single measurement is not important, as long as the measurement is weak enough. Moreover, in that case, we observe a plateau, which suggests that the exact value of t is not important. For $N \gg 1$, even with t far from $t^{\max}$, the mean score is close to $G^{\max}$. Interestingly, the trade-off between $t^{\max}$ and Δ found for the 1D case seems to repeat here. We numerically find that $\sqrt{t^{\max}}/\Delta$ is constant for a given $N = 2J$ (see Figure 4c) and scales as $1/\sqrt{N}$.

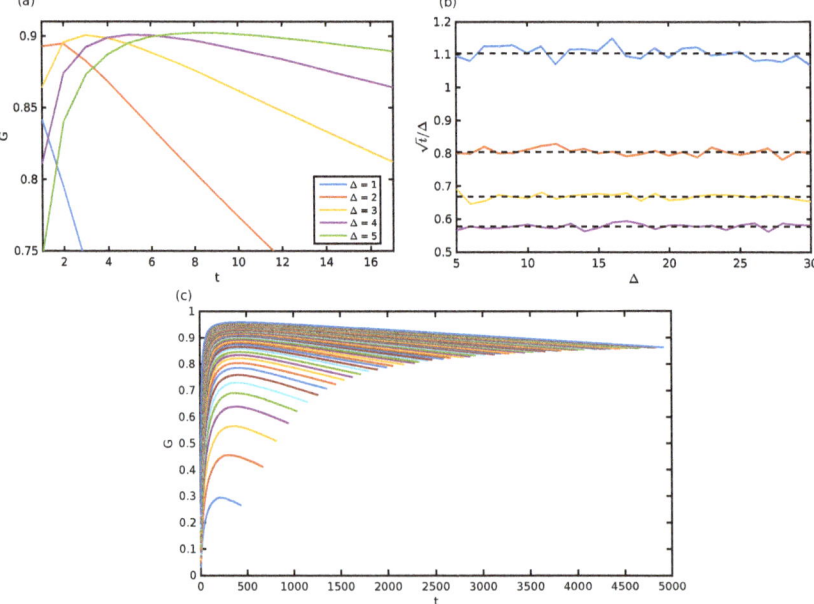

Figure 4. (a) Score as a function of the number of measurements for $N = 2J = 20$. For each Δ, there is an optimal repetition rate $t^{\max}$. The optimal score $G^{\max}$ saturates for Δ big enough. (b) Ratio $\sqrt{t^{\max}}/\Delta$ for $N = 5, 10, 15, 20$. As in the 1D case, $\sqrt{t^{\max}}/\Delta$ is constant and only depends on N. (c) Score as a function of the number of measurements t, for $N = 2J = 1.50$ and $\Delta = 8\sqrt{N}$. See Section 4.3 for further details.

Most importantly, for weak enough measurements, the obtained score is close to the optimal one, as shown in Figure 5. Numerical fluctuations prevent any precise statements about an estimation of the error, but the error is close to what was obtained with the initial measurement procedure; see Figure 5.

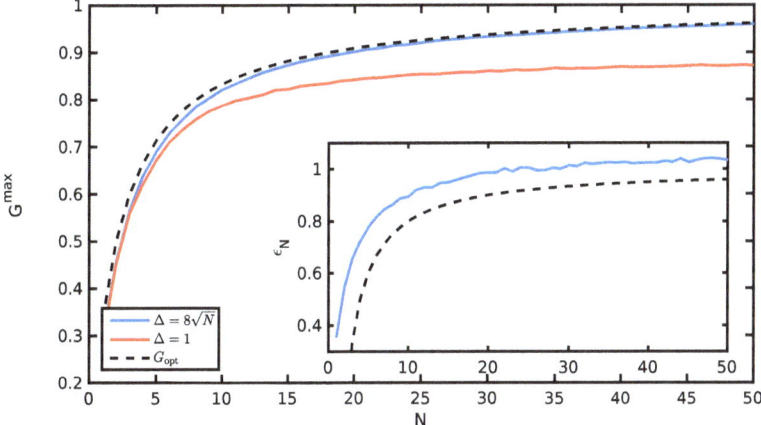

Figure 5. Mean score $G^{\max}$ as a function of N maximized over t. A too strong measurement ($\Delta = 1$) fails to achieve an optimum. A weak enough measurement ($\Delta = 8\sqrt{N}$) achieves a good score. The insert shows $\epsilon_N = \frac{N}{2}(1 - G_N)$. See Section 4.3 for further details.

5. Conclusions

In this paper, we asked the question of how to model everyday measurements of a macroscopic system within quantum mechanics. We introduced the notion of Macroscopic Quantum Measurement and argued that such a measurement should be highly non-invasive, collect a large amount of information in a single shot and be described by a "fairly simple" coupling between system and observer. We proposed a concrete model based on a pointer von Neumann measurement inspired by the Arthur–Kelly model, where a pointer is coupled to the macroscopic quantum system through a Hamiltonian and then measured. This approach applies to many situations, as long as a natural Hamiltonian for the measured system can be found.

Here, we focused on the problem of a direction estimation. The Hamiltonian naturally couples the spin of the macroscopic quantum state to the position of a pointer in three dimensions, which is then measured. This reveals information about the initial direction of the state. We extended our previous study to consider a collection of aligned spins, which exploits the non-monotonic behavior of the mean score as a function of the coupling strength. We presented more precise results. We relaxed the assumptions about the measured system, by considering a thermal state of finite temperature and showed that our initial conclusions are still valid. We also relaxed the assumptions over the measurement scheme, looking at its approximation by a repetition of ultra weak measurements in several orthogonal directions. Here again, we obtained numerical results supporting the initial conclusion. In summary, this MQM proposal tolerates several relaxations regarding lack of control or knowledge.

It is likely that these two relaxations can be unified: polarization measurement of systems with n unknown number of particle or temperature should be accessible via the repeated 1D ultra-weak measurement method. However, this claim has to be justified numerically. Further open questions include the behavior of Arthur–Kelly models in other situations where two or more non-commuting quantities have to be estimated, e.g., for position and velocity estimation.

Acknowledgments: We would like to thank Tomer Barnea for fruitful discussions. Partial financial support by ERC-AG MEC and Swiss NSF is gratefully acknowledged.

Author Contributions: Nicolas Gisin suggested the study. Marc-Olivier Renou and Florian Fröwis performed the simulations and worked out the theory. Marc-Olivier Renou wrote the paper. All authors discussed the results and implications and commented on the manuscript at all stages. All authors have read and approved the final manuscript.

Conflicts of Interest: The authors declare no conflict of interest.

References and Notes

1. Massar, S.; Popescu, S. Optimal Extraction of Information from Finite Quantum Ensembles. *Phys. Rev. Lett.* **1995**, *74*, 1259–1263.
2. Gisin, N.; Popescu, S. Spin Flips and Quantum Information for Antiparallel Spins. *Phys. Rev. Lett.* **1999**, *83*, 432–435.
3. Chiribella, G.; D'Ariano, G.M. Maximum likelihood estimation for a group of physical transformations. *Int. J. Quantum Inf.* **2006**, *4*, 453–472.
4. Scarani, V.; Iblisdir, S.; Gisin, N.; Acín, A. Quantum cloning. *Rev. Mod. Phys.* **2005**, *77*, 1225–1256.
5. Cirac, J.; Ekert, A.; Macchiavello, C. Optimal purification of single qubits. *Phys. Rev. Lett.* **1999**, *82*, 4344.
6. Bagan, E.; Monras, A.; Muñoz Tapia, R. Comprehensive analysis of quantum pure-state estimation for two-level systems. *Phys. Rev. A* **2005**, *71*, 062318.
7. Bagan, E.; Ballester, M.A.; Gill, R.D.; Monras, A.; Muñoz-Tapia, R. Optimal full estimation of qubit mixed states. *Phys. Rev. A* **2006**, *73*, 032301.
8. Sacchi, M.F. Information-disturbance tradeoff for spin coherent state e stimation. *Phys. Rev. A* **2007**, *75*, 012306.
9. Gendra, B.; Ronco-Bonvehi, E.; Calsamiglia, J.; Muñoz-Tapia, R.; Bagan, E. Optimal parameter estimation with a fixed rate of abstention. *Phys. Rev. A* **2013**, *88*, 012128.
10. Latorre, J.; Pascual, P.; Tarrach, R. Minimal optimal generalized quantum measurement. *Phys. Rev. Lett.* **1998**, *81*, 1351.
11. Chiribella, G.; D'Ariano, G.M.; Schlingemann, D.M. How Continuous Quantum Measurements in Finite Dimensions Are Actually Discrete. *Phys. Rev. Lett.* **2007**, *98*, 190403.
12. Barnea, T.; Renou, M.O.; Fröwis, F.; Gisin, N. Macroscopic quantum measurements of non-commuting observables. *Phys. Rev. A* **2017**, *96*, 012111.
13. Poulin, D. Macroscopic observables. *Phys. Rev. A* **2005**, *71*, 022102.
14. D'Ariano, G.; Presti, P.L.; Sacchi, M. A quantum measurement of the spin direction. *Phys. Lett. A* **2002**, *292*, 233–237.
15. Holevo, A. *Probabilistic and Statistical Aspects of Quantum Theory*; Edizioni della Normale: Pisa, Italy, 1982.
16. Often, the considered score is $F = \int dr \int d\vec{u}\, p(r|\vec{u}) f(\vec{u}, \vec{v}_r)$, where $f(\vec{u}, \vec{v}_r) = |\langle \vec{u}|\vec{v}_r\rangle|^2$ can be seen as the fidelity between qubits $|\vec{u}\rangle$ and $|\vec{v}_r\rangle$, where a unit vector is associated with the corresponding qubit via the Bloch sphere identification. As $F = \frac{1}{2}(1 + G)$, this is equivalent. We chose this formulation for practical reason.
17. Arthurs, E.; Kelly, J.L. On the Simultaneous Measurement of a Pair of Conjugate Observables. *Bell Syst. Tech. J.* **1965**, *44*, 725–729.
18. Pal, R.; Ghosh, S. Approximate joint measurement of qubit observables through an Arthur–Kelly model. *J. Phys. A Math. Theor.* **2011**, *44*, 485303.
19. Levine, R.Y.; Tucci, R.R. On the simultaneous measurement of spin components using spin-1/2 meters: Naimark embedding and projections. *Found. Phys.* **1989**, *19*, 175–187.
20. Von Neumann, J. *Mathematical Foundations of Quantum Mechanics*; Princeton University Press: Princeton, NJ, USA, 1955.
21. Busch, P.; Lahti, P.J.; Mittelstaedt, P. *The Quantum Theory of Measurement*; Springer: Berlin/Heidelberg, Germany, 1991; pp. 27–98.
22. Peres, A. *Quantum Theory: Concepts and Methods*; Kluwer Academic Publishers: Norwell, MA, USA, 2002.
23. This method is equivalent to the one previously used in [12]. There, $\langle z|^{\otimes N} \Omega_{\vec{r}} |z\rangle^{\otimes N}$ is lower bounded by $|\langle z|^{\otimes N} E_{\vec{r}} |\vec{r}\rangle^{\otimes N}|^2$, but as the Krauss operator is diagonal, this last term is nothing else than $o_J |\langle z|^{\otimes N} |\vec{r}\rangle^{\otimes N}|^2$.

24. We can interpret this physically. We see from Section 2.3 that the best covariant measurement is obtained from $O_z \propto |J, J\rangle\langle J, J|$. Other covariant measurements can be obtained with $O_z \propto |m, J\rangle\langle m, J|$ for $0 \leq m < J$. The coefficients o_m can be interpreted as how much each of these measurements is done. The term $|m, J\rangle\langle m, J|$ can also be thought as the physical system used to measure. When it is highly polarized ($m = J$), the measurement is efficient. However, when the polarization is low, the information gain is weak, e.g., $m = 0$, and we clearly see that all POVM elements are $\propto \mathbb{1}$.
25. Hacohen-Gourgy, S.; Martin, L.S.; Flurin, E.; Ramasesh, V.V.; Whaley, K.B.; Siddiqi, I. Quantum dynamics of simultaneously measured non-commuting observables. *Nature* **2016**, *538*, 491–494.

© 2018 by the authors. Licensee MDPI, Basel, Switzerland. This article is an open access article distributed under the terms and conditions of the Creative Commons Attribution (CC BY) license (http://creativecommons.org/licenses/by/4.0/).

Article

Iterant Algebra

Louis H. Kauffman

Department of Mathematics, Statistics and Computer Science, University of Illinois at Chicago, 851 South Morgan Street, Chicago, IL 60607-7045, USA; kauffman@uic.edu; Tel.: +1-773-363-5115

Received: 29 May 2017; Accepted: 5 July 2017; Published: 11 July 2017

Abstract: We give an exposition of iterant algebra, a generalization of matrix algebra that is motivated by the structure of measurement for discrete processes. We show how Clifford algebras and matrix algebras arise naturally from iterants, and we then use this point of view to discuss the Schrödinger and Dirac equations, Majorana Fermions, representations of the braid group and the framed braids in relation to the structure of the Standard Model for physics.

Keywords: iterant; Clifford algebra; matrix algebra; braid group; Fermion; Dirac equation

1. Introduction

This is a paper about an approach to algebra that we call *iterants*. The idea behind the definition of iterant (see Section 2) is that one is studying a periodic discrete process with an associated action of a group of permutations on the sequences of the process. The simplest such discrete system is an alternation between $+1$ and -1. We will show that this system gives rise in a natural way to the square root of minus one. This way thinking about the square root of minus one as an iterant is explained below. More generally, by starting with a discrete time series of positions, one has a non-commutativity of observations due to time-delays (the clock must tick to measure a velocity) and this non-commutativity can be encapsulated in a generalized iterant algebra as defined in Section 3 of the present paper. Iterant algebra generalizes matrix algebra and we shall see how it can be used to formulate the algebra of the framed Artin Braid Group, the Lie algebra $su(3)$ for the Standard Model for particle physics, the framed braid representations for Fermions of Sundance Bilson-Thompson, the Clifford algebra for Majorana Fermions and the structure of the Schrödinger and Dirac equations. This paper is a sequel to [1] and it uses material from that paper and extends it into the more general context of the present paper. See also [1–4] for previous work by the author about iterants. This paper also incorporates results of the author that appear in the joint paper of the author and Rukhsan Ul-Haq [5]. Our intent is to give a picture of the range of application of the basic mathematical idea of iterants and to include a description of the basic results that make them work.

This paper is organized as follows. Sections 2–4 are devoted to the mathematics of iterants. Each remaining section of the paper applies the iterant structure to a topic in mathematical physics that is of interest to the author. We hope that the reader finds the first few sections to be a readable introduction to iterants. An interested reader can then turn to the remaining sections to see how iterants can be used in specific cases. The reader should note that since applying iterants often means reformulating a topic usually written in matrix algebra in terms of iterant algebra, and the specific interest in such a formulation may be, at this time, of a formal nature. Nevertheless, the reformulation often raises many interesting questions, and these will be the subject of subsequent work.

Sections 2 and 3 are an introduction to the process algebra of iterants and how the square root of minus one arises from an alternating process. Section 4 shows how iterants give an alternative way to do matrix algebra. The section ends with the construction of the split quaternions. Section 4 considers iterants of arbitrary period (not just two) and shows, with the example of the cyclic group, how the ring of all $n \times n$ matrices can be seen as a faithful representation of an iterant algebra based on the cyclic group of order n. We then generalize this construction (Theorem 1) to arbitrary non-commutative finite groups G. Such a group has a multiplication table ($n \times n$ where n is the order of the group G). We show that, by rearranging the multiplication table so that the identity elements appear on the diagonal, we get a set of permutation matrices that represent the group faithfully as $n \times n$ matrices. This gives a faithful representation of the iterant algebra associated with the group G onto the ring of $n \times n$ matrices. As a result, we see that iterant algebra is fundamental to matrix algebra. Section 4 ends with a number of classical examples including iterant representations for quaternion algebra.

Section 5 is a discussion of the Schrödinger equation. We formulate a discrete model related to the diffusion equation by following a heuristic that would identify the square root of minus one as a controlled oscillation between plus one and minus one. The resulting discrete model has the equation (compare with [1])

$$\psi(x, t+\tau) = ((-1)^{n(t)}/2)\psi(x-\Delta, t) + (1-(-1)^{n(t)})\psi(x, t) + ((-1)^{n(t)}/2)\psi(x+\Delta, t)$$

and satisfies a discrete version of the diffusion equation with an extra coefficient of $(-1)^{n(t)}$, where $n(t)$ denotes the number of time steps τ needed to reach time t. By dividing this discrete system into its even and odd parts (the parity of $(-1)^{n(t)}$), we retrieve the Schrödinger equation, and the formalism of the complex numbers handles the parity. In the discrete model, the iterant structure appears directly.

Section 6 discusses the iterant structure of the framed Artin braid group via framed braids and discusses the basics of the Sundance Bilson-Thompson model for elementary particles. In Section 7, we apply this to a formulation of the particle model of Sundance Bilson-Thompson [6], using framed braids.

In Section 7, we give an iterant interpretation of the $su(3)$ Lie algebra for the Standard Model using [7]. The resulting formulation of the $su(3)$ Lie algebra is particularly elegant from our point of view, and we expect it to give further insight into the standard model. This iterant formulation of the $su(3)$ Lie Algebra is so concise that we show it here in the Introduction. We use the specific iterant formulas

$$T_+ = [1,0,0]A, \ T_- = [0,1,0]B,$$

$$U_+ = [0,1,0]A, \ U_- = [0,0,1]B,$$

$$V_+ = [0,0,1]A, \ V_- = [1,0,0]B,$$

$$T_3 = [1/2, -1/2, 0], \ Y = \frac{1}{\sqrt{3}}[1, 1, -2].$$

We have the permutation relations $A[x,y,z] = [y,z,x]A$ and $B = A^2 = A^{-1}$ so that $B[x,y,z] = [z,y,x]B$. This reduces the basic $su(3)$ Lie algebra to a very elementary patterning of order three cyclic operations. The details of this formulation are given in Section 7.

In Section 8 we apply this point of view on the Standard Model to obtain an embedding of the framed braid algebra for the Sundance Bilson-Thompson model into the iterant version of $su(3)$. These three sections are an account of research of the author and Rukhsan Ul-Haq in [5].

Section 9 discusses how Clifford algebras are related to the structure of Fermions. We show how the algebra of the split quaternions, the very first iterant algebra that appears in relation to the square root of minus one, is behind the structure of the operator algebra of the electron. The Clifford structure on two generators describes a pair of Majorana Fermion operators. Majorana Fermions are particles that are their own antiparticles. These Majorana Fermion operators correspond to Clifford algebra generators a and b such that $a^2 = b^2 = 1$ and $ab = -ba$. Using our iterant formulation, we can take a as the iterant corresponding to a period two oscillation, and b as the time shifting operator. The product ab is a square root of minus one in the non-commutative context of this Clifford algebra. The annihilation operator for an electron can be symbolized by $\phi = (a+ib)/2$ and the creation operator for an electron by $\phi^\dagger = (a-ib)/2$. These form the operator algebra for an electron. Note that

$$\phi^2 = (a+ib)(a+ib)/4 = (a^2 - b^2 + i(ab+ba))/4 = (0+i0)/4 = 0 = (\phi^\dagger)^2$$

and therefore

$$\phi\phi^\dagger + \phi^\dagger\phi = (\phi + \phi^\dagger)^2 = a^2 = 1.$$

The electron is seen in terms of its underlying Clifford structure in the form of a pair of Majorana Fermions. Section 9 shows how braiding is related to the Majorana Femions.

Section 10 discusses the structure of the Dirac equation and how the nilpotent and the Majorana operators arise naturally in this context. This section provides a link between our work and the work on nilpotent structures and the Dirac equation of Peter Rowlands [8]. We end this section with an expression in split quaternions for the the Majorana–Dirac equation in $(3+1)$ spacetime. The Majorana–Dirac equation can be written as follows:

$$(\partial/\partial t + \hat{\eta}\eta \partial/\partial x + \epsilon \partial/\partial y + \hat{\epsilon}\eta \partial/\partial z - \hat{\epsilon}\hat{\eta}\eta m)\psi = 0,$$

where η and ϵ are the generators of our simplest iterant algebra with $\eta^2 = \epsilon^2 = 1$ and $\eta\epsilon + \epsilon\eta = 0$. The elements $\hat{\epsilon}, \hat{\eta}$ form a commuting copy of this algebra. This use of a combination of the simplest Clifford algebra with itself is the underlying structure of the Majorana–Dirac equation.

We give a specific real solution to the Majorana–Dirac equation in our iterant/Clifford algebra formalism. Here, $\rho(x,t) = e^{(p\bullet x - Et)}$, where $p = (p_x, p_y, p_z)$ is a constant vector momentum, and x denotes the vector (x,y,z). The solution to the Majorana–Dirac equation is $\hat{\Gamma}\rho(x,t)$ as shown below:

$$\hat{\Gamma}\rho(x,t) = (-E - \hat{\eta}\eta p_x - \epsilon p_y - \hat{\epsilon}\eta p_z + \hat{\epsilon}\hat{\eta}\eta m)\rho(x,t).$$

This solution is real in the sense that its coordinates are all real valued functions once the iterant or matrix forms for the operators are made explicit. The combination of iterant and Clifford algebra language that we develop here makes the analysis of certain aspects of the Dirac equation and the Majorana–Dirac equation very clear. More work needs to be done in all these fronts.

This paper is a snapshot of a larger story. Iterant algebra is a basically simple reformulation of aspects of patterned algebra that can often illuminate correspondingly elementary topics in mathematics and physics. The present work is a beginning in the larger enterprise of understanding relationships in discrete physics and relationships between algbra and physics.

2. Iterants

An *iterant* is a sum of elements of the form

$$a\sigma = [a_1, a_2, ..., a_n]\sigma,$$

where $a = [a_1, a_2, ..., a_n]$ is a vector of elements that are scalars (usually real or complex numbers) and σ is a permutation on n letters. Vectors are added and multiplied coordinatewise (see below), and we take the following rule for multiplication of vector/permutation combinations:

$$(a\sigma)(b\tau) = (ab^\sigma)\sigma\tau,$$

where b^σ denotes the vector b with its elements permuted by the action of σ.

If a and b are vectors, then ab denotes the vector, where $(ab)_i = a_i b_i$, and $a+b$ denotes the vector where $(a+b)_i = a_i + b_i$. Then,

$$(ka)\sigma = k(a\sigma)$$

for a scalar k, and

$$(a+b)\sigma = a\sigma + b\sigma,$$

where vectors are multiplied as above and we take the usual product of the permutations. All of matrix algebra is naturally represented in the iterant framework, as we shall see in the next sections.

For example, if η is the order two permutation of two elements, then $[a, b]^\eta = [b, a]$. Thus,

$$[a, b]\eta = \eta[a, b]^\eta = \eta[b, a].$$

We define

$$i = [1, -1]\eta$$

and then

$$i^2 = [1, -1]\eta[1, -1]\eta = [1, -1][-1, 1]]\eta^2 = [1, -1][-1, 1] = [-1, -1] = -1.$$

In this way, the complex numbers arise naturally from iterants. One can interpret $[1, -1]$ as an oscillation between $+1$ and -1 and η as a temporal shift operator. Then, $i = [1, -1]\eta$ is a time sensitive element and its self-interaction has square minus one. In this way, iterants can be interpreted as a formalization of elementary discrete processes.

Note that if we let $e = [1, -1]$, then $e^2 = 1$, $\eta^2 = 1$ and $e\eta = -\eta e$. Thus, e and η generate a small Clifford algebra.

3. Iterants and Discrete Processes

The primitive idea behind an iterant is a periodic time series or "waveform"

$$\cdots abababababab \cdots.$$

We illustrate with period two. The elements of the waveform can be any mathematically or empirically well-defined objects. We can regard the ordered pairs $[a, b]$ and $[b, a]$ as abbreviations for the waveform or as two points of view about the waveform (a first or b first). We have called $[a, b]$ an *iterant*. Thinking of an iterant as a discrete process, we define a time shift operator η such that $[a, b]\eta = \eta[b, a]$ and $\eta^2 = 1$.

Discrete Calculus and the Temporal Shift Operator. If we have a discrete time series $X, X', X'', \cdots$, then it is convenient to define an operator J so that $X^t J = JX^{t+1}$, and it is this temporal shift operator that can be used to correlate discrete calculus for the time-series. For example, we can define a discrete derivative D by the equation

$$DX^t = J(X^{t+1} - X^t)/\Delta t,$$

(with time increment equal to Δt). Note then that the derivative is expressed as a commutator:

$$DX^t = J(X^{t+1} - X^t)/\Delta t = (JX^{t+1} - JX^t)\Delta t = (X^t J - JX^t)/\Delta t = [X^t, J/\Delta t],$$

where here $[R, S] = RS - SR$ is the commutator. This means that this discrete derivative satisfies the Leibniz rule for products, and it can be used for formulations of discrete physics. This use of the temporal shift operator dovetails with its use for keeping track of observation in a discrete model, where successive observations require temporal shifts. In particular, let $P = mDX^t$ and $Q = X^t$ denote momentum and position, respectively (m is mass and commutes with J, as does Δt). Then, PQ and QP do not commute and the temporal shift operator J keeps track of the fact that measuring momentum requires a tick of the clock. We can interpret PQ as first measuring Q and then measuring P, while QP represents first measuring P and then measuring Q:

$$PQ = (mDX^t)(X^t) = (mJ(X^{t+1} - X^t)/\Delta t)X^t = mJ(X^{t+1} - X^t)X^t/\Delta t,$$

$$QP = X^t(mJ(X^{t+1} - X^t)/\Delta t) = mJX^{t+1}(X^{t+1} - X^t)/\Delta t.$$

Thus,
$$[Q,P] = QP - PQ = mJ(X^{t+1} - X^t)^2/\Delta t = mJ(\Delta X)^2/\Delta t.$$

In this form of discrete physics, the commutator equation
$$[Q,P] = k,$$
where k is a constant, is satisfied by a Brownian walk with diffusion constant $(\Delta X)^2/\Delta t$. In this way, our interpretation of the square root of negative one in terms of the temporal shift operator fits into a larger context of the physics of discrete observations. In this paper, we work with periodic series and use cyclic operators such as η to keep track of the periodicity. For related discussion, see [2,3,5,9–16]. See also [17] for other uses of iterants in the context of Clifford algebras. For papers of the author about discrete physics and quantum computing see [18–28].

We have defined products and sums of iterants as follows
$$[a,b][c,d] = [ac, bd]$$
and
$$[a,b] + [c,d] = [a+c, b+d].$$

The operation of juxtapostion of waveforms is multiplication while + denotes ordinary addition of ordered pairs. These operations are natural with respect to the structural juxtaposition of iterants:

$$...abababababab...$$

$$...cdcdcdcdcdcd...$$

Structures combine at the points where they correspond. Waveforms combine at the times where they correspond. Iterants combine in juxtaposition. This theme of including the result of time in observations of a discrete system occurs at the foundation of our construction.

In the next section, we show how all matrix algebra can be formulated in terms of iterants.

4. Matrix Algebra via Iterants

Here is a direct translation of period-two iterants into 2×2 matrices. Let
$$[a,b] + [c,d]\eta = \begin{pmatrix} a & c \\ d & b \end{pmatrix},$$
where
$$[x,y] = \begin{pmatrix} x & 0 \\ 0 & y \end{pmatrix},$$
and
$$\eta = \begin{pmatrix} 0 & 1 \\ 1 & 0 \end{pmatrix}.$$

The reader will have no difficulty verifying that the usual definition of matrix multiplicaiton corresponds exactly to the iterant multiplication that we have already described. In particular,
$$[x,y][z,w] = [xy, zw]$$
and
$$[x,y] + [z,w] = [x+y, z+w]$$

are rules of matrix multiplication and addition, as are

$$[x,y]\eta = \eta[y,x].$$

Thus, matrix multiplication and addition is identical with iterant multiplication. There are many ways to motivate the rules for matrix algebra. Iterants are a natural entry into matrix structure.

The fact that the iterant expression $[a,d]1 + [b,c]\eta$ captures the whole of 2×2 matrix algebra corresponds to the fact that a two by two matrix is combinatorially the union of the identity pattern (the diagonal) and the interchange pattern (the antidiagonal) that correspond to the operators 1 and η:

$$\begin{pmatrix} * & @ \\ @ & * \end{pmatrix}.$$

In the formal diagram for a matrix shown above, we indicate the diagonal by $*$ and the anti-diagonal by @.

In the case of complex numbers, we represent

$$\begin{pmatrix} a & -b \\ b & a \end{pmatrix} = [a,a] + [-b,b]\eta = a1 + b[-1,1]\eta = a + bi.$$

In this way, we see that 2×2 matrix algebra can be seen as a hypercomplex number system based on the symmetric group S_2. In the next section, we generalize this point of view to arbitrary finite groups by generalizing Cayley's Theorem that shows that every finite group has a faithful representation as a permutation group.

The factorization of i into a product $\epsilon\eta$ of non-commuting iterant operators shows, in the iterant viewpoint, the temporal nature of i and its algebraic roots.

Note that the quaternions arise from the *split quaternions*: The split quaternions are the system

$$\{\pm 1, \pm\epsilon, \pm\eta, \pm i\}.$$

Here, $\epsilon\epsilon = 1 = \eta\eta$ while $i = \epsilon\eta$ so that $ii = -1$. The quaternions come about once we construct an extra square root of minus one that commutes with them. Call this extra root of minus one $\sqrt{-1}$. Then, the quaternions are generated by

$$I = \sqrt{-1}\epsilon, J = \epsilon\eta, K = \sqrt{-1}\eta$$

with

$$I^2 = J^2 = K^2 = IJK = -1.$$

Remark 1. *The rest of this section is an exposition of the higher period iterants and the general Theorem 1 about finite groups and iterant matrix representations. The exposition follows the corresponding exposition in our paper [1].*

4.1. Iterants of Arbirtarily High Period and General Matrix Algebras

Consider a waveform of period three.

$$\cdots abcabcabcabcabcabc \cdots$$

Here, we see three natural iterant views (depending upon whether one starts at a, b or c).

$$[a,b,c], \quad [b,c,a], \quad [c,a,b].$$

The appropriate shift operator is given by the cyclic permutation S:

$$[x, y, z]S = S[z, x, y].$$

With $T = S^2$, we have

$$[x, y, z]T = T[y, z, x]$$

and $S^3 = 1$. We obtain a closed algebra of iterants whose general element is of the form

$$[a, b, c] + [d, e, f]S + [g, h, k]S^2,$$

where $a, b, c, d, e, f, g, h, k$ are real or complex numbers. Call this algebra $\mathbb{V}ect_3(\mathbb{R})$ when the scalars are in a commutative ring with unit $\mathbb{F}$. Let $M_3(\mathbb{F})$ denote the 3×3 matrix algebra over $\mathbb{F}$. We have the:

Lemma 1. *The iterant algebra $\mathbb{V}ect_3(\mathbb{F})$ is isomorphic to the full 3×3 matrix algebra $M_3((\mathbb{F})$.*

Proof.

$$[a, b, c] + [d, e, f]S + [g, h, k]S^2$$

maps to the matrix

$$\begin{pmatrix} a & d & g \\ h & b & e \\ f & k & c \end{pmatrix},$$

preserving the algebra structure. Since any 3×3 matrix can be written uniquely in this form, it follows that $\mathbb{V}ect_3(\mathbb{F})$ is isomorphic to the full 3×3 matrix algebra $M_3(\mathbb{F})$. □

We can summarize the pattern behind this expression of 3×3 matrices by the following symbolic matrix:

$$\begin{pmatrix} 1 & S & T \\ T & 1 & S \\ S & T & 1 \end{pmatrix}.$$

Here, the letter T occupies the positions in the matrix that correspond to the permutation matrix that represents it, and the letter $T = S^2$ occupies the positions corresponding to its permutation matrix. The 1s occupy the diagonal for the corresponding identity matrix. The iterant representation corresponds to writing the 3×3 matrix as a disjoint sum of these permutation matrices such that the matrices themselves are closed under multiplication. In this case, the matrices form a permutation representation of the cyclic group of order 3, $C_3 = \{1, S, S^2\}$.

Remark 2. *Note that a permutation matrix is a matrix of zeroes and ones such that some permutation of the rows of the matrix transforms it to the identity matrix. Given an $n \times n$ permutation matrix P, we associate to it a permuation*

$$\sigma(P) : \{1, 2, \cdots, n\} \longrightarrow \{1, 2, \cdots, n\}$$

via the following formula

$$i\sigma(P) = j,$$

where j denotes the column in P where the i-th row has a 1. Note that an element of the domain of a permutation is indicated to the left of the symbol for the permutation. It is then easy to check that for permutation matrices P and Q,

$$\sigma(P)\sigma(Q) = \sigma(PQ),$$

given that we compose the permutations from left to right according to this convention.

This construction generalizes directly for iterants of any period and hence for a set of operators forming a cyclic group of any order. We shall generalize further to any finite group G. We now define $\mathbb{Vect}_n(G, \mathbb{F})$ for any finite group G.

Definition 1. *Let G be a finite group, written multiplicatively. Let $\mathbb{F}$ denote a given commutative ring with unit. Assume that G acts as a group of permutations on the set $\{1, 2, 3, \cdots, n\}$ so that given an element $g \in G$ we have (by abuse of notation)*

$$g : \{1, 2, 3, \cdots, n\} \longrightarrow \{1, 2, 3, \cdots, n\}.$$

We shall write

$$ig$$

for the image of $i \in \{1, 2, 3, \cdots, n\}$ under the permutation represented by g. The notation denotes functionality from the left. We have $(ig)h = i(gh)$ for all elements $g, h \in G$ and $i1 = i$ for all i, in order to have a representation of G as permutations. We shall call an n-tuple of elements of $\mathbb{F}$ a vector and denote it by $a = (a_1, a_2, \cdots, a_n)$. We then define an action of G on vectors over $\mathbb{F}$ by the formula

$$a^g = (a_{1g}, a_{2g}, \cdots, a_{ng}),$$

and note that $(a^g)^h = a^{gh}$ for all $g, h \in G$. Define an algebra $\mathbb{Vect}_n(G, \mathbb{F})$, the iterant algebra for G, to be the set of finite sums of formal products of vectors and group elements in the form ag with multiplication rule

$$(ag)(bh) = ab^g(gh),$$

and the understanding that $(a + b)g = ag + bg$ and for all vectors a, b and group elements g. It is understood that vectors are added coordinatewise and multiplied coordinatewise. Thus, $(a + b)_i = a_i + b_i$ and $(ab)_i = a_i b_i$.

Theorem 1. *Let G be a finite group of order n [1]. Let $\rho : G \longrightarrow S_n$ denote the right regular representation of G as permutations of n objects. List the elements of G as $G = \{g_1, \cdots, g_n\}$, and let G act on its own underlying set via the definition $g_i \rho(g) = g_i g$. Here, we describe $\rho(g)$ acting on the set of elements g_k of G. We also regard $\rho(g)$ as a mapping of the set $\{1, 2, \cdots n\}$, replacing g_k by k and $i\rho(g) = k$ where $g_i g = g_k$.*

Then, $\mathbb{Vect}_n(G, \mathbb{F})$ is isomorphic to the matrix algebra $M_n((\mathbb{F}))$. In particular, $\mathbb{Vect}_{n!}(S_n, \mathbb{F})$ is isomorphic with the matrices of size $n! \times n!$, $M_{n!}((\mathbb{F}))$.

Proof. Take the multiplication table for G to be the $n \times n$ matrix with columns and rows listed in the order $[g_1, \cdots, g_n]$. Permute the rows of this matrix so that the diagonal consists in all 1 s. Let the resulting matrix be called the *G-Table*. The *G-Table* is labeled by elements of the group. For any vector a, let $D(a)$ denote the $n \times n$ diagonal matrix whose entries in order down the diagonal are the entries of a in the order specified by a. For each group element g, let P_g denote the permutation matrix with 1 in every spot on the *G-Table* that is labeled by g and 0 in all other spots. It is now a direct verification that the mapping

$$F(\Sigma_{i=1}^n a_i g_i) = \Sigma_{i=1}^n D(a_i) P_{g_i}$$

defines an isomorphism from $\mathbb{Vect}_n(G, \mathbb{F})$ to the matrix algebra $M_n((\mathbb{F}))$. The main point to check is that $\sigma(P_g) = \rho(g)$. We now prove this fact.

In the *G-Table*, the rows correspond to $\{g_1^{-1}, g_2^{-1}, \cdots g_n^{-1}\}$ and the columns correspond to $\{g_1, g_2, \cdots g_n\}$ so that the *i-i* entry of the table is $g_i^{-1} g_i = 1$. With this, we have that, in the table, a group element g occurs in the *i*-th row at column *j* where $g_i^{-1} g_j = g$. This is equivalent to the equation $g_i g = g_j$, which, in turn, is equivalent to the statement $i\rho(g) = j$. This is exactly our functional interpretation of the action of the permutation corresponding to the matrix P_g. Thus, $\rho(g) = \sigma(P_g)$. The rest of the proof is straightforward and left to the reader. □

Example 1.

1. Consider the cyclic group of order three.

$$C_3 = \{1, S, S^2\}$$

with $S^3 = 1$. The multiplication table is

$$\begin{pmatrix} 1 & S & S^2 \\ S & S^2 & 1 \\ S^2 & 1 & S \end{pmatrix}.$$

Interchanging the second and third rows, we obtain

$$\begin{pmatrix} 1 & S & S^2 \\ S^2 & 1 & S \\ S & S^2 & 1 \end{pmatrix},$$

and this is the G-Table that we used for $\text{Vect}_3(C_3, \mathbb{F})$ prior to proving the Main Theorem. The same pattern works for abitrary cyclic groups.

2. Consider the symmetric group on six letters,

$$S_6 = \{1, R, R^2, F, RF, R^2F\},$$

where $R^3 = 1, F^2 = 1, FR = RF^2$. Then, the multiplication table is

$$\begin{pmatrix} 1 & R & R^2 & F & RF & R^2F \\ R & R^2 & 1 & RF & R^2F & F \\ R^2 & 1 & R & R^2F & F & RF \\ F & R^2F & RF & 1 & R^2 & R \\ RF & F & R^2F & R & 1 & R^2 \\ R^2F & RF & F & R^2 & R & 1 \end{pmatrix}.$$

The corresponnding G-Table is

$$\begin{pmatrix} 1 & R & R^2 & F & RF & R^2F \\ R^2 & 1 & R & R^2F & F & RF \\ R & R^2 & 1 & RF & R^2F & F \\ F & R^2F & RF & 1 & R^2 & R \\ RF & F & R^2F & R & 1 & R^2 \\ R^2F & RF & F & R^2 & R & 1 \end{pmatrix}.$$

This G-Table encodes the isomorphism of $\text{Vect}_6(S_3, \mathbb{F})$ with the full algebra of six by six matrices. Similarly, $\text{Vect}_{n!}(S_n, \mathbb{F})$ is isomorphic with the full algebra of $n! \times n!$ matrices. The permutation matrices are obtained from the G-Table by choosing a given group element and then replacing it by 1 for each appearance in the table, and replacing the other elements of the table by 0. For example, we have the permutation matrix for R given by the formula below:

$$R = \begin{pmatrix} 0 & 1 & 0 & 0 & 0 & 0 \\ 0 & 0 & 1 & 0 & 0 & 0 \\ 1 & 0 & 0 & 0 & 0 & 0 \\ 0 & 0 & 0 & 0 & 0 & 1 \\ 0 & 0 & 0 & 1 & 0 & 0 \\ 0 & 0 & 0 & 0 & 1 & 0 \end{pmatrix}.$$

3. Consider the group $G = C_2 \times C_2$, the "Klein 4-Group". Take $G = \{1, A, B, C\}$ where $A^2 = B^2 = C^2 = 1$, $AB = BA = C$. G has the multiplication table, which is also its G-Table for $\mathbb{V}\text{ect}_4(G, \mathbb{F})$:

$$\begin{pmatrix} 1 & A & B & C \\ A & 1 & C & B \\ B & C & 1 & A \\ C & B & A & 1 \end{pmatrix}.$$

Thus, we have the corresponding permutation matrices that I shall call E, A, B, C. The reader can verify that $A^2 = B^2 = C^2 = 1$, $AB = BA = C$. Let

$$\alpha = [1, -1, -1, 1], \beta = [1, 1, -1, -1], \gamma = [1, -1, 1, -1].$$

In addition, let

$$I = \alpha A, J = \beta B, K = \gamma C.$$

Then, it is easy to check that

$$I^2 = J^2 = K^2 = IJK = -1, IJ = K, JI = -K.$$

Thus, we have constructed the quaternions as iterants in relation to the Klein 4-Group. In Figure 1, we illustrate these quaternion generators with string diagrams for the permutations. The reader can check that the permuations correspond to the permutation matrices constructed for the Klein 4-Group.

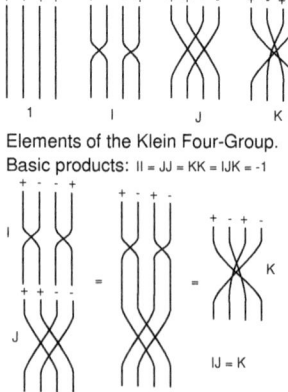

Elements of the Klein Four-Group.
Basic products: II = JJ = KK = IJK = -1

IJ = K

Product of I and J perfomed as flat framed braid multiplication.

Figure 1. Quaternions from the Klein 4-Group.

4. Since complex numbers commute with one another, we could consider iterants whose values are in the complex numbers. This is just like considering matrices whose entries are complex numbers. Thus, we shall allow a version of i that commutes with the iterant shift operator η. Let this commuting i be denoted by ι. Then, we are assuming that

$$\iota^2 = -1,$$

$$\eta\iota = \iota\eta,$$

$$\eta^2 = +1.$$

We then consider iterants of the form $[a + b\iota, c + d\iota]$ and $[a + b\iota, c + d\iota]\eta = \eta[c + d\iota, a + b\iota]$. In particular, we have $\epsilon = [1, -1]$, and $i = \epsilon\eta$ is quite distinct from ι. Note, as before, that $\epsilon\eta = -\eta\epsilon$ and that $\epsilon^2 = 1$. Now, let

$$I = \iota\epsilon,$$

$$J = \epsilon\eta,$$

$$K = \iota\eta.$$

We find the quaternions once more:

$$I^2 = \iota\epsilon\iota\epsilon = \iota\iota\epsilon\epsilon = (-1)(+1) = -1,$$

$$J^2 = \epsilon\eta\epsilon\eta = \epsilon(-\epsilon)\eta\eta = -1,$$

$$K^2 = \iota\eta\iota\eta = \iota\iota\eta\eta = -1,$$

$$IJK = \iota\epsilon\epsilon\eta\iota\eta = \iota\iota\eta\eta = \iota\iota = -1.$$

Thus,

$$I^2 = J^2 = K^2 = IJK = -1.$$

This construction shows how the structure of the quaternions comes directly from the non-commutative structure of period two iterants. The group $SU(2)$ of 2×2 unitary matrices of determinant one is isomorphic to the quaternions of length one.

5. Similarly,

$$H = [a, b] + [c + d\iota, c - d\iota]\eta = \begin{pmatrix} a & c + d\iota \\ c - d\iota & b \end{pmatrix}$$

represents a Hermitian 2×2 matrix and hence an observable for quantum processes mediated by $SU(2)$. Hermitian matrices have real eigenvalues.

If in the above Hermitian matrix form, we take $a = T + X, b = T - X, c = Y, d = Z$, then we obtain an iterant and/or matrix representation for a point in Minkowski spacetime:

$$H = [T + X, T - X] + [Y + Z\iota, Y - Z\iota]\eta = \begin{pmatrix} T + X & Y + Z\iota \\ Y - Z\iota & T - X \end{pmatrix}.$$

Note that we have the formula
$$Det(H) = T^2 - X^2 - Y^2 - Z^2.$$

It is not hard to see that the eigenvalues of H are $T \pm \sqrt{X^2 + Y^2 + Z^2}$. Thus, viewed as an observable, H can observe the time and the invariant spatial distance from the origin of the event (T, X, Y, Z). At least at this very elementary juncture, quantum mechanics and special relativity are reconciled.

6. Hamilton's Quaternions are generated by iterants, as discussed above, and we can express them purely algebraicially by writing the corresponding permutations as shown below:

$$I = [+1, -1, -1, +1]s,$$

$$J = [+1, +1, -1, -1]l,$$

$$K = [+1, -1, +1, -1]t,$$

where

$$s = (12)(34),$$

$$l = (13)(24),$$

$$t = (14)(23).$$

Here, we represent the permutations as products of transpositions (ij). The transposition (ij) interchanges i and j, leaving all other elements of $\{1, 2, ..., n\}$ fixed.

One can verify that

$$I^2 = J^2 = K^2 = IJK = -1.$$

Note that making an iterant interpretation of an entity like $I = [+1, -1, -1, +1]s$ is a conceptual departure from our original period two iterant (or cyclic period n) notion. Now, we consider iterants such as $[+1, -1, -1, +1]$ where the permutation group acts to produce other orderings of a given sequence. The iterant itself can represent a form that can be seen in any of its possible orders. These orders are subject to permutations that produce the possible views of the iterant. Algebraic structures such as the quaternions appear in the explication of such forms.

7. In all these examples, we can interpret the iterants as short hand for matrix algebra based on permutation matrices, or as indicators of discrete processes. The discrete processes become more complex in proportion to the complexity of the groups used in the construction. We began with processes of order two, then considered cyclic groups of arbitrary order, then the symmetric group S_3 in relation to 6×6 matrices, and the Klein 4-Group in relation to the quaternions. In the case of the quaternions, we know that this structure is intimately related to rotations of three- and four-dimensional space and many other geometric themes.

5. Schrödinger's Equation

In this section, we go more deeply into a treatment of Schrödinger's equation that was begun in the introduction to [1]. In that paper, we used this example for Schrödinger's equation to motivate the introduction of iterants. Here, we already have iterants, but we find that a discrete model for Schrödinger's equation instantiates an alternating pattern that is essentially of the form $\cdots + - + - + - + \cdots$, and the problem of taking the continuum limit of this discrete model leads to the complex numbers by a parity consideration. The parity consideration corresponds to our iterant construction of the square root of minus one, and so we see in this model how the iterant square root of minus one can correspond to an alternation in a discrete process while the usual square root of minus one describes the behaviour of the limit of the process.

5.1. Brownian Walks and the Diffusion Equation

Recall how the diffusion equation arises in discussing Brownian motion. We are given a Brownian process where

$$x(t + \tau) = x(t) \pm \Delta,$$

so that the time step is τ and the space step is of absolute value Δ. We regard the probability of left or right steps as equal, so that if $P(x,t)$ denotes the probability that the Brownian particle is at point x at time t, then

$$P(x, t + \tau) = P(x - \Delta, t)/2 + P(x + \Delta, t)/2.$$

From this equation for the probability, we can write a difference equation for the partial derivative of the probability with respect to time:

$$(P(x, t + \tau) - P(x, t))/\tau = (\hbar^2/2\tau)[(P(x - \Delta, t) - 2P(x, t) + P(x + \Delta))/\Delta^2].$$

The expression in brackets on the right-hand side is a discrete approximation to the second partial of $P(x,t)$ with respect to x. Thus, if the ratio $C = \Delta^2/2\tau$ remains constant as the space and time intervals approach zero, then this equation goes in the limit to the diffusion equation

$$\partial P(x,t)/\partial t = C\partial^2 P(x,t)/\partial x^2.$$

C is called the diffusion constant for the Brownian process.

5.2. An Iterant Intepretation of Schrödinger's Equation

Recall that Schrödinger's equation can be regarded as the diffusion equation with an imaginary diffusion constant. Recall how this works. The Schrödinger equation is

$$i\hbar \partial \psi / \partial t = H\psi,$$

where the Hamiltonian H is given by the equation $H = p^2/2m + V$, where $V(x,t)$ is the potential energy and $p = (\hbar/i)\partial/\partial x$ is the momentum operator. With this, we have $p^2/2m = (-\hbar^2/2m)\partial^2/\partial x^2$. Thus, with $V(x,t) = 0$, the equation becomes $i\hbar \partial \psi / \partial t = (-\hbar^2/2m)\partial^2 \psi/\partial x^2$, which simplifies to

$$\partial \psi/\partial t = (i\hbar/2m)\partial^2 \psi/\partial x^2.$$

Thus, we have arrived at the form of the diffusion equation with an imaginary constant, and it is possible to make the identification with the diffusion equation by setting

$$\hbar/m = \Delta^2/\tau,$$

where Δ denotes a space interval, and τ denotes a time interval as explained in the last section about the Brownian walk. With this, we can ask what space interval and time interval will satisfy this relationship? One answer is that this equation is satisfied when m is the Planck mass, Δ is the Planck length and τ is the Planck time. Note that $L^2/T = (\hbar/Mc)^2/(\hbar/Mc^2) = \hbar/M$. Here, $\hbar$ is Planck's constant divided by 2π. c is the speed of light. G is Newton's gravitational constant. $M = \sqrt{\hbar c/G}, L = \hbar/Mc, T = \hbar/Mc^2$.

What does all this say about the nature of the Schrödinger equation itself? Consider a discrete function $\psi(x,t)$ defined (recursively) by the following equation:

$$\psi(x, t + \tau) = (i/2)\psi(x - \Delta, t) + (1 - i)\psi(x, t) + (i/2)\psi(x + \Delta, t).$$

In other words, we are thinking here of a random "quantum walk" where the amplitude for stepping right or stepping left is proportional to i while the amplitude for not moving at all is proportional to $(1-i)$. It is then easy to see that ψ is a discretization of

$$\partial \psi / \partial t = (i\Delta^2/2\tau)\partial^2\psi/\partial x^2.$$

Just note that ψ satisfies the difference equation

$$(\psi(x,t+\tau) - \psi(x,t))/\tau = (i\Delta^2/2\tau)(\psi(x-\Delta,t) - 2\psi(x,t) + \psi(x+\Delta,t))/\Delta^2.$$

This gives a direct interpretation of the solution to the Schrödinger equation as a limit of a sum over generalized Brownian paths with complex amplitudes.

Replacing i by An Iterant. Now, however, suppose that we replace i by $(-1)^{n(t)}$ at time step $t = n(t)\tau$ where $n(t)$ is a non-negative integer. Instead of writing

$$\psi(x,t+\tau) = (i/2)\psi(x-\Delta,t) + (1-i)\psi(x,t) + (i/2)\psi(x+\Delta,t),$$

we will write

$$\psi(x,t+\tau) = ((-1)^{n(t)}/2)\psi(x-\Delta,t) + (1-(-1)^{n(t)})\psi(x,t) + ((-1)^{n(t)}/2)\psi(x+\Delta,t).$$

Then, we will find that

$$(\psi(x,t+\tau) - \psi(x,t))/\tau = (-1)^{n(t)}(\Delta^2/2\tau)(\psi(x-\Delta,t) - 2\psi(x,t) + \psi(x+\Delta,t))/\Delta^2,$$

so that the diffusion equation seems to have been replaced with an equation of the form

$$\partial \psi / \partial t = \pm\kappa \partial^2 \psi/\partial x^2.$$

We wish to consider the continuum limit. However, there is no meaning to

$$(-1)^{n(t)}$$

in the realm of continuous time. In the discrete world, the wave function ψ divides into ψ_e and ψ_o where the (discrete) time, $n(t)$, is either even or odd. We write

$$\partial_t \psi_e = \kappa \partial_x^2 \psi_o,$$

$$\partial_t \psi_o = -\kappa \partial_x^2 \psi_e,$$

and take the continuum limit of ψ_e and ψ_o separately.

In fact, we can interpret the $\{\pm\}$ as the complex number i. We write

$$\psi = \psi_e + i\psi_o,$$

so that

$$i\partial_t \psi = i\partial_t(\psi_e + i\psi_o) = i\partial_t\psi_e - \partial_t\psi_o$$
$$= i\kappa\partial_x^2\psi_o + \kappa\partial_x^2\psi_e = \kappa\partial_x^2(\psi_e + i\psi_o)$$
$$= \kappa\partial_x^2\psi.$$

Thus,

$$i\partial \psi/\partial t = \kappa \partial^2 \psi/\partial x^2.$$

This the Schrödinger equation. Instead of the simple diffusion equation, we have a mutual dependency where the temporal variation of ψ_e is mediated by the spatial variation of ψ_o and vice-versa:

$$\psi = \psi_e + i\psi_o,$$

$$\partial_t \psi_e = \kappa \partial_x^2 \psi_o,$$

$$\partial_t \psi_o = -\kappa \partial_x^2 \psi_e,$$

$$i\partial \psi / \partial t = \kappa \partial^2 \psi / \partial x^2.$$

Note that in terms of the iterant interpretation, the pair $[\psi_e, \psi_o]$ is an abbreviation of the temporal series $\cdots \psi_t, \psi_{t+\tau}, \psi_{t+2\tau}, \cdots$ that represents the discrete process $\psi_{t+\tau}(x) = ((-1)^{n(t)}/2)\psi_t(x - \Delta) + (1 - (-1)^{n(t)})\psi_t(x) + ((-1)^{n(t)}/2)\psi_t(x + \Delta)$ Here, the process itself is not periodic, but the underlying alternation of the parity of $(-1)^{n(t)}$ gives the iterant stucture that allows the use of i as a combination of shift and permutation.

Remark 3. *The discrete recursion at the beginning of this section can be implemented to approximate solutions to the Schrödinger equation. This will be the subject of another paper. The main point of this section is that a discrete version of the Schrödinger equation actually uses the temporal iterant interpretation of the square root of minus one, so that one can think of this oscillation as part of a discrete process in back of the Schrödinger evolution. This reformulation of basic quantum mechanics deserves further study.*

6. The Framed Braid Group and the Sundance Bilson-Thompson Model for Elementary Particles

The reader should recall that the symmetric group S_n has presentation

$$S_n = (T_1, \cdots T_{n-1} | T_i^2 = 1, T_i T_{i+1} T_i = T_{i+1} T_i T_{i+1}, T_i T_j = T_j T_i; |i - j| > 1).$$

The Artin Braid Group B_n is a relative of the symmetric group that is obtained by removing the condition that each generator has a square equal to the identity:

$$B_n = (\sigma_1, \cdots \sigma_{n-1} | \sigma_i \sigma_{i+1} \sigma_i = \sigma_{i+1} \sigma_i \sigma_{i+1}, \sigma_i \sigma_j = \sigma_j \sigma_i; |i - j| > 1).$$

In Figure 2, we illustrate the the generators $\sigma_1, \sigma_2, \sigma_3$ of the 4-strand braid group and we show the topological nature of the relation $\sigma_1 \sigma_2 \sigma_1 = \sigma_2 \sigma_1 \sigma_2$ and the commuting relation $\sigma_1 \sigma_3 = \sigma_3 \sigma_1$. Topological braids are represented as collections of always descending strings, starting from a row of points and ending at another row of points. The strings are embedded in three-dimensional space and can wind around one another. The elementary braid generators σ_i correspond to the i-th strand interchanging with the $(i + 1)$-th strand. Two braids are multiplied by attaching the bottom endpoiints of one braid to the top endpoints of the other braid to form a new braid.

There is a fundamental homomorphism

$$\pi : B_n \longrightarrow S_n$$

defined on generators by

$$\pi(\sigma_i) = T_i$$

in the language of the presentations above. In terms of the diagrams in Figure 2, a braid diagram is a permutation diagram if one forgets about its weaving structure of over and under strands at a crossing.

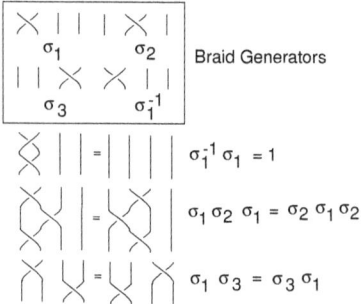

Figure 2. Braid generators.

We now turn to a generalization of the braid group, the *framed braid group*. In this generalization, we associate elements of the form t^a to the top of each braid strand. For these purposes, it is useful to take t as an algebraic variable and a as an integer. To interpret this framing, geometrically replace each braid strand by a ribbon and interpret t^a as a $2\pi a$ twist in the ribbon. In Figure 3, we illustrate how to multiply two framed braids. In our formalism, the braids A and B in this figure are given by the formulas

$$A = [t^a, t^b, t^c]\sigma_1\sigma_2\sigma_3,$$
$$B = [t^d, t^e, t^f]\sigma_2\sigma_3,$$

in the framed braid group on three strands, denoted FB_3. As the Figure 3 illustrates, we have the basic formula

$$v\sigma = \sigma v^{\pi(\sigma)},$$

where v is a vector of the form $v = [t^a, t^b, t^c]$ (for $n = 3$) and $v^{\pi(\sigma)}$ denotes the action of the permutation associated with the braid σ on the vector v. In the figure, the permutation is accomplished by sliding the algebra along the strings of the braid.

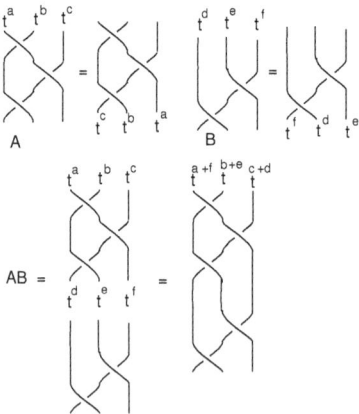

Figure 3. Framed braids.

We can form an algebra $Alg[FB_n]$ by taking formal sums of framed braids of the form $\sum c_k v_k G_k$, where c_k is a scalar, v_k is a framing vector and G_k is an element of the Artin Braid group B_n. Since braids act on framing vectors by permutations, this algebra is a generalization of the iterant algebras we have defined so far. The algebra of framed braids uses an action of the braid group based on its representation to the symmetric group. Furthemore, the representation $\pi : B_n \longrightarrow S_n$ induces a map of algebras

$$\hat{\pi} : Alg[FB_n] \longrightarrow Alg[FS_n],$$

where we recognize $Alg[FS_n]$ as exactly an iterant algebra based in S_n.

In [6], Sundance Bilson-Thompson represents Fermions as framed braids. See Figure 4 for his diagrammatic representations. In this theory, each Fermion is associated with a framed braid. Thus, from the figure, we see that the positron and the electron are given by the framed braids

$$e^+ = [t,t,t]\sigma_1\sigma_2^{-1},$$

and

$$e^- = \sigma_2\sigma_1^{-1}[t^{-1},t^{-1},t^{-1}].$$

Here, we use $[t^a, t^b, t^c]$ for the framing numbers (a,b,c). Products of framed braids correspond to particle interactions. Note that $e^+e^- = [1,1,1] = \gamma$ so that the electron and the positron are inverses in this algebra. In Figure 5 are illustrated the representations of bosons, including γ, a photon and the identity element in this algebra. Other relations in the algebra correspond to particle interactions. For example, in Figure 6 the muon decay is illustrated:

$$\mu \to \nu_\mu + W_- \to \nu_\mu + \bar{\nu}_e + e^-.$$

The reader can see the definitions of the different parts of this decay sequence from the three figures we have just mentioned. Note that strictly speaking the muon decay is a multiplicative identity in the braid algebra:

$$\mu = \nu_\mu W_- = \nu_\mu \bar{\nu}_e e^-.$$

Particle interactions in this model are mediated by factorizations in the non-commutative algebra of the framed braids.

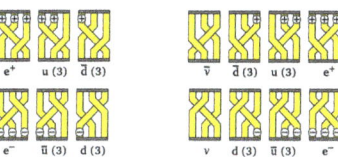

Figure 4. Sundance Bilson-Thompson Framed Braid Fermions ("(3)" under the labels for the up and down quarks and antiquarks represent the fact that there are three permutations of charge placement giving the three colours).

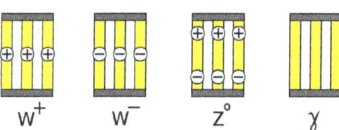

Figure 5. Bosons.

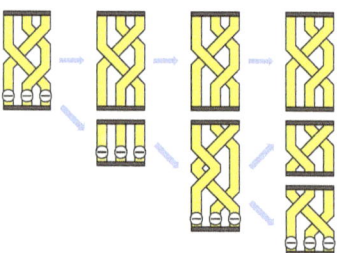

Figure 6. Representation of $\mu \to \nu_\mu + W_- \to \nu_\mu + \bar{\nu}_e + e^-$.

By using the representation $\hat{\pi} : Alg[FB_3] \longrightarrow Alg[FS_3]$, we can image the structure of Bilson-Thompson's framed braids in the the iterant algebra corresponding to the symmetric group. However, we propose to change this map so that we have a non-trivial representation of the Artin braid group. This can be accomplished by defining

$$\rho : Alg[FB_3] \longrightarrow Alg[FS_3],$$

where

$$\rho(\sigma_k) = [t, t, t] T_k$$

and

$$\rho(\sigma_k^{-1}) = [t^{-1}, t^{-1}, t^{-1}] T_k$$

for $k = 1, 2$. The reader will find that we have now represented the braid group in the iterant algebra $Alg[FS_3]$ and extended the representation to the framed braid group algebra. Thus, *the Sundance Bilson-Thompson representation of elementary particles as framed braids is represented inside the iterant algebra for the symmetric group on three letters.* In Section 10, we carry this further and place the representation inside the Lie Algebra $su(3)$.

7. Iterants and the Standard Model

In this section, we shall give an iterant interpretation for the Lie algebra of the special unitary group $SU(3)$. The Lie algebra in question is denoted as $su(3)$ and is often described by a matrix basis. The Lie algebra $su(3)$ is generated by the following eight Gell Man Matrices [29]:

$$\lambda_1 = \begin{pmatrix} 0 & 1 & 0 \\ 1 & 0 & 0 \\ 0 & 0 & 0 \end{pmatrix}, \lambda_2 = \begin{pmatrix} 0 & -i & 0 \\ i & 0 & 0 \\ 0 & 0 & 0 \end{pmatrix}, \lambda_3 = \begin{pmatrix} 1 & 0 & 0 \\ 0 & -1 & 0 \\ 0 & 0 & 0 \end{pmatrix},$$

$$\lambda_4 = \begin{pmatrix} 0 & 0 & 1 \\ 0 & 0 & 0 \\ 1 & 0 & 0 \end{pmatrix}, \lambda_5 = \begin{pmatrix} 0 & 0 & i \\ 0 & 0 & 0 \\ -i & 0 & 0 \end{pmatrix}, \lambda_6 = \begin{pmatrix} 0 & 0 & 0 \\ 0 & 0 & 1 \\ 0 & 1 & 0 \end{pmatrix},$$

$$\lambda_7 = \begin{pmatrix} 0 & 0 & 0 \\ 0 & 0 & -i \\ 0 & i & 0 \end{pmatrix}, \lambda_8 = \frac{1}{\sqrt{3}} \begin{pmatrix} 1 & 0 & 0 \\ 0 & 1 & 0 \\ 0 & 0 & -2 \end{pmatrix}.$$

The group $SU(3)$ consists of the matrices $U(\epsilon_1, \cdots, \epsilon_8) = e^{i \sum_a \epsilon_a \lambda_a}$, where $\epsilon_1, \cdots, \epsilon_8$ are real numbers and a ranges from 1 to 8. The Gell Man matrices satisfy the following relations:

$$tr(\lambda_a \lambda_b) = 2\delta_{ab},$$

$$[\lambda_a/2, \lambda_b/2] = if_{abc}\lambda_c/2.$$

Here, we use the summation convention summing over repeated indices, and tr denotes standard matrix trace, $[A, B] = AB - BA$ is the matrix commutator and δ_{ab} is the Kronecker delta, equal to 1 when $a = b$, and equal to 0, otherwise. The structure coefficients f_{abc} take the following non-zero values:

$$f_{123} = 1, f_{147} = 1/2, f_{156} = -1/2, f_{246} = 1/2, f_{257} = 1/2,$$

$$f_{345} = 1/2, f_{367} = -1/2, f_{458} = \sqrt{3}/2, f_{678} = \sqrt{3}/2.$$

We now give an iterant representation for these matrices that is based on the pattern

$$\begin{pmatrix} 1 & A & B \\ B & 1 & A \\ A & B & 1 \end{pmatrix}$$

as described in the previous section. That is, we use the cyclic group of order three to represent all 3×3 matrices at iterants based on the permutation matrices

$$A = \begin{pmatrix} 0 & 1 & 0 \\ 0 & 0 & 1 \\ 1 & 0 & 0 \end{pmatrix}, B = \begin{pmatrix} 0 & 0 & 1 \\ 1 & 0 & 0 \\ 0 & 1 & 0 \end{pmatrix}.$$

Recalling that $[a, b, c]$ as an iterant, denotes a diagonal matrix

$$[a, b, c] = \begin{pmatrix} a & 0 & 0 \\ 0 & b & 0 \\ 0 & 0 & c \end{pmatrix},$$

the reader will have no difficulty verifying the following formulas for the Gell Mann Matrices in the iterant format:

$$\lambda_1 = [1, 0, 0]A + [0, 1, 0]B,$$

$$\lambda_2 = [-i, 0, 0]A + [0, i, 0]B,$$

$$\lambda_3 = [1, -1, 0],$$

$$\lambda_4 = [1, 0, 0]B + [0, 0, 1]A,$$

$$\lambda_5 = [i, 0, 0]B + [0, 0, -i]A,$$

$$\lambda_6 = [0, 1, 0]A + [0, 0, 1]B,$$

$$\lambda_7 = [0, -i, 0]A + [0, 0, i]B,$$

$$\lambda_8 = \frac{1}{\sqrt{3}}[1, 1, -2].$$

Letting $F_a = \lambda_a/2$, we can now rewrite the Lie algebra into simple iterants of the form $[a, b, c]G$ where G is a cyclic group element. Compare with [7]. Let

$$T_\pm = F_1 \pm iF_2,$$

$$U_\pm = F_6 \pm iF_7,$$

$$V_\pm = F_4 \pm iF_5,$$

$$T_3 = F_3,$$

$$Y = \frac{2}{\sqrt{3}} F_8.$$

Iterant Formulation of the $su(3)$ Lie Algebra. We now have the specific iterant formulas

$$T_+ = [1,0,0]A,$$
$$T_- = [0,1,0]B,$$
$$U_+ = [0,1,0]A,$$
$$U_- = [0,0,1]B,$$
$$V_+ = [0,0,1]A,$$
$$V_- = [1,0,0]B,$$
$$T_3 = [1/2, -1/2, 0],$$
$$Y = \frac{1}{\sqrt{3}}[1,1,-2].$$

We have that $A[x,y,z] = [y,z,x]A$ and $B = A^2 = A^{-1}$ so that $B[x,y,z] = [z,y,x]B$. We have reduced the basic $su(3)$ Lie algebra to a very elementary patterning of order three cyclic operations. In a subsequent paper, we will use this point to view to examine the irreducible representations of this algebra and to illuminate the Standard Model's Eightfold Way.

8. Iterants, Braiding and the Sundance Bilson-Thompson Model for Fermions

In the last section, we based our iterant representations on the following patterns and matrices. The pattern,

$$\begin{pmatrix} 1 & A & B \\ B & 1 & A \\ A & B & 1 \end{pmatrix},$$

uses the cyclic group of order three to represent all 3×3 matrices at iterants based on the permutation matrices

$$A = \begin{pmatrix} 0 & 1 & 0 \\ 0 & 0 & 1 \\ 1 & 0 & 0 \end{pmatrix}, B = \begin{pmatrix} 0 & 0 & 1 \\ 1 & 0 & 0 \\ 0 & 1 & 0 \end{pmatrix}.$$

Recalling that $[a,b,c]$ as an iterant denotes a diagonal matrix

$$[a,b,c] = \begin{pmatrix} a & 0 & 0 \\ 0 & b & 0 \\ 0 & 0 & c \end{pmatrix}.$$

In fact, there are six 3×3 permuation matrices: $\{I, A, B, P, Q, R\}$, where

$$P = \begin{pmatrix} 0 & 1 & 0 \\ 1 & 0 & 0 \\ 0 & 0 & 1 \end{pmatrix}, Q = \begin{pmatrix} 1 & 0 & 1 \\ 0 & 0 & 1 \\ 0 & 1 & 0 \end{pmatrix}, R = \begin{pmatrix} 0 & 0 & 1 \\ 0 & 1 & 0 \\ 1 & 0 & 0 \end{pmatrix}.$$

We then have $A = QP, B = PQ, R = PQP = QPQ$. The two transpositions P and Q generate the entire group of permuatations S_3. It is usual to think of the order-three transformations A and B as expressed in terms of these transpositons, but we can also use the iterant structure of the 3×3 matrices to express P, Q and R in terms of A and B. The result is as follows:

$$P = [0,0,1] + [1,0,0]A + [0,1,0]B,$$
$$Q = [1,0,0] + [0,1,0]A + [0,0,1]B,$$
$$R = [0,1,0] + [0,0,1]A + [1,0,0]B.$$

Recall from the previous section that we have the iterant generators for the $su(3)$ Lie algebra:

$$T_+ = [1,0,0]A,$$
$$T_- = [0,1,0]B,$$
$$U_+ = [0,1,0]A,$$
$$U_- = [0,0,1]B,$$
$$V_+ = [0,0,1]A,$$
$$V_- = [1,0,0]B.$$

Thus, we can express these transpositions P and Q in the iterant form of the Lie algebra as

$$P = [0,0,1] + T_+ + T_-,$$
$$Q = [1,0,0] + U_+ + U_-,$$
$$R = [0,1,0] + V_+ + V_-.$$

The basic permutations receive elegant expressions in the iterant Lie algebra.

Now that we have basic permutations in the Lie algebra, we can take the map from Section 7

$$\rho : Alg[FB_3] \longrightarrow Alg[FS_3]$$

with

$$\rho(\sigma_k) = [t,t,t]T_k$$

and

$$\rho(\sigma_k^{-1}) = [t^{-1}, t^{-1}, t^{-1}]T_k$$

for $k = 1, 2$ and send T_1 to P and T_2 to Q. Then, we have

$$\rho(\sigma_1) = [t,t,t]P$$

and

$$\rho(\sigma_1^{-1}) = [t^{-1}, t^{-1}, t^{-1}]P$$

and

$$\rho(\sigma_2) = [t,t,t]Q$$

and

$$\rho(\sigma_1^{-1}) = [t^{-1}, t^{-1}, t^{-1}]Q.$$

By choosing $t \neq 1$ on the unit circle in the complex plane, we obtain representations of the Sundance Bilson-Thompson constructions of Fermions via framed braids *inside* the $su(3)$ Lie algebra. This brings the Bilson-Thompson formalism in direct contact with the Standard Model via our iterant representations. We shall return to these relationships in a sequel to the present paper.

9. Clifford Algebra, Majorana Fermions and Braiding

This section is based on our paper [1]. We show how the very simple Clifford algebra(s) that come from iterants figure in studying Fermions and Majorana Fermions. This section also provides the background for the next section on the Dirac equation. The original paper by Ettore Majorana [30] led to the notion of Clifford algebraic Majorana operators that we discuss in this section. In the next section on the Dirac equation, we show how this Clifford algebra is related to Majorana's original equation. A key relationship between the physics of the Quantum Hall effect and the kind of braiding representations considered here originates with the paper of Moore and Read [31]. See also [28] where we look at the combinatorial topology behind the braid group representations of Moore and Read.

Recall Fermion algebra. One has Fermion annihiliation operators ψ and their conjugate creation operators $\psi^\dagger$. One has $\psi^2 = 0 = (\psi^\dagger)^2$. There is a fundamental commutation relation

$$\psi \psi^\dagger + \psi^\dagger \psi = 1.$$

If you have more than one of them, say ψ and ϕ, then they anti-commute:

$$\psi \phi = -\phi \psi.$$

Majorana Fermion operators c satisfy $c^\dagger = c$ so that the corresponding particles are their own anti-particles. A group of researchers [32] claims, at this writing, to have found Majorana Fermions in edge effects in nano-wires.

Majorana operators are related to standard Fermions as follows: the algebra for Majoranas is $c = c^\dagger$ and $cc' = -c'c$ if c and c' are distinct Majorana Fermions with $c^2 = 1$ and $c'^2 = 1$. One can make a standard Fermion operator from two Majorana operators via

$$\psi = (c + ic')/2,$$

$$\psi^\dagger = (c - ic')/2.$$

Similarly, one can mathematically make two Majoranas from any single Fermion. If one takes a set of Majoranas

$$\{c_1, c_2, c_3, \cdots, c_n\},$$

then there are natural braiding operators that act on the vector space with these c_k as the basis. The operators are mediated by algebra elements that *themselves* satisfy braiding relations

$$\tau_k = (1 + c_{k+1}c_k)/\sqrt{2},$$

$$\tau_k^{-1} = (1 - c_{k+1}c_k)/\sqrt{2}.$$

The Ivanov [33] braiding operators are

$$T_k : Span\{c_1, c_2, \cdots, c_n\} \longrightarrow Span\{c_1, c_2, \cdots, c_n\}$$

via

$$T_k(x) = \tau_k x \tau_k^{-1}.$$

The braiding is simply:

$$T_k(c_k) = c_{k+1},$$

$$T_k(c_{k+1}) = -c_k,$$

and T_k is the identity otherwise. We have then a unitary representaton of the Artin braid group. See Figure 7 for a depiction of the braiding of Majorana Fermions in relation to the topology of a belt that connects them. In quantum mechanics, we must represent rotations of three-dimensional space as unitary transformations. This relationship between rotations and unitary transformations is encoded in the topology of the belt. See [34] for more about this topological view of the physics of Fermions. In the figure, we see that the strictly topological belt does not know which of the two Fermions will individually acquire a phase change, but the Ivanov algebra above makes this decision. More understanding is needed in this area of subtle topological structure of Fermions.

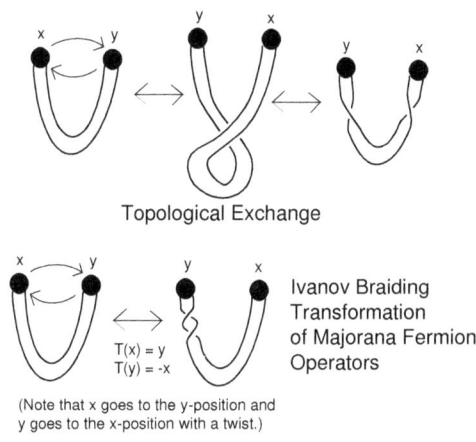

Figure 7. Braiding action on a pair of fermions.

Recall that, in discussing the inception of iterants, we introduce a *temporal shift operator* η such that

$$[a,b]\eta = \eta[b,a]$$

and

$$\eta\eta = 1$$

for any iterant $[a,b]$. In this way, we have a Clifford algebra generated by $e = [1,-1]$ and η. We can take e and η as Majorana Fermion operators and construct Fermion operators

$$\psi = (e + i\eta)/2,$$

$$\psi^\dagger = (e - i\eta)/2.$$

Here, i is an extra square root of minus one that commutes with the operators e and η. We arrive at fermions in a few short steps from the origin of the iterants. Algebraically, we have controlled the period two oscillation e so that it satisfies the fermion algebra. From the point of view taken in this paper, it is worth examining if this discrete process view of fermion algebra and Majorana operator algebra can shed light on the many properties in this domain. In particular, I would like to see if there is insight into the braiding of Majorana Fermion operators to be gained from the iterant viewpoint.

10. The Dirac Equation and Majorana Fermions

This section goes beyond our paper [1]. We expand on the relationship of a nilpotent formulation of the Dirac equation and an iterant formulation. We first construct the Dirac equation. The algebra underlying this equation has the same properties as the creation and annihilation algebra for Majorana Fermion operators, so it is by way of this algebra that we will come to the Dirac equation.

If the speed of light is equal to 1 (by convention), then energy E, momentum p and mass m are related by the (Einstein) equation
$$E^2 = p^2 + m^2.$$
Dirac constructed his equation by finding an algebraic square root of $p^2 + m^2$. A corresponding linear operator for E can then take the role of the Hamiltonian in the Schrödinger equation. We first assume that p is a scalar (using one dimension of space and one dimension of time). Let $E = \alpha p + \beta m$, where α and β are elements of a non-commutative, associative algebra. Then,
$$E^2 = \alpha^2 p^2 + \beta^2 m^2 + pm(\alpha\beta + \beta\alpha).$$
Hence, $E^2 = p^2 + m^2$ if $\alpha^2 = \beta^2 = 1$ and $\alpha\beta + \beta\alpha = 0$. We can use the iterant algebra generated by e and η with $\alpha = e$ and $\beta = \eta$. Recall that the quantum operator for momentum is $\hat{p} = -i\partial/\partial x$ and the operator for energy is $\hat{E} = i\partial/\partial t$. The Dirac equation is
$$\hat{E}\psi = \alpha\hat{p}\psi + \beta m\psi.$$
This becomes the explicit equation:
$$i\partial\psi/\partial t = -i\alpha\partial\psi/\partial x + \beta m\psi.$$
Let
$$\mathcal{O} = i\partial/\partial t + i\alpha\partial/\partial x - \beta m$$
so that the Dirac equation takes the form
$$\mathcal{O}\psi(x,t) = 0.$$

A Plane Wave Solution to the Dirac Equation. Note that
$$\mathcal{O}e^{i(px-Et)} = (E - \alpha p - \beta m)e^{i(px-Et)}$$
and note also that
$$(E + \alpha p + \beta m)(E - \alpha p - \beta m) = E^2 - p^2 - m^2 = 0.$$
Thus, it follows that
$$\phi = (E + \alpha p + \beta m)e^{i(px-Et)}$$
is a solution of the Dirac equation.

Now let $\Delta = (E - \alpha p - \beta m)$ and let
$$U = \Delta\beta\alpha = (E - \alpha p - \beta m)\beta\alpha = \beta\alpha E + \beta p - \alpha m.$$
Then,
$$U^2 = -E^2 + p^2 + m^2 = 0.$$
The nilpotent element U leads to the same plane wave solution to the Dirac equation as follows. We have shown that
$$\mathcal{O}\psi = \Delta\psi$$
for $\psi = e^{i(px-Et)}$. It then follows that
$$\mathcal{O}(\beta\alpha\Delta\beta\alpha\psi) = \Delta\beta\alpha\Delta\beta\alpha\psi = U^2\psi = 0,$$

from which it follows that
$$\psi = \beta\alpha U e^{i(px-Et)}$$
is a plane wave solution to the Dirac equation.

We can multiply the operator $\mathcal{O}$ by $\beta\alpha$ on the right, obtaining the operator
$$\mathcal{D} = \mathcal{O}\beta\alpha = i\beta\alpha\partial/\partial t + i\beta\partial/\partial x - \alpha m,$$
and the equivalent Dirac equation
$$\mathcal{D}\psi = 0.$$
For ψ above, we have $\mathcal{D}(Ue^{i(px-Et)}) = U^2 e^{i(px-Et)} = 0$. This beautiful observation that the Dirac operator can be modified so that one can directly construct nilpotent solutions to the Dirac equation was first made by Peter Rowlands [8] in the context of doubled quaternion algebra. Here we have shown how Rowland's work fits into the Clifford algebra and iterant approach to the Dirac equation. Such solutions can be articulated into specific vector solutions by using either an iterant or matrix representation of the algebra.

10.1. U and $U^\dagger$ as Creation and Annihilation Operators

The Clifford algebra element U can be regarded (in the context of this rewrite of the Dirac equation) as a creation operator for a Fermion.

If, reversing time, we let
$$\tilde{\psi} = e^{i(px+Et)},$$
then
$$\mathcal{D}\tilde{\psi} = (-\beta\alpha E + \beta p - \alpha m)\psi = U^\dagger\tilde{\psi},$$
giving a definition of $U^\dagger$ for the anti-particle for $U\psi$.
$$U = \beta\alpha E + \beta p - \alpha m$$
and
$$U^\dagger = -\beta\alpha E + \beta p - \alpha m.$$
Note that here we have
$$(U + U^\dagger)^2 = (2\beta p + \alpha m)^2 = 4(p^2 + m^2) = 4E^2,$$
and
$$(U - U^\dagger)^2 = -(2\beta\alpha E)^2 = -4E^2.$$
$$U^2 = (U^\dagger)^2 = 0,$$
and
$$UU^\dagger + U^\dagger U = 4E^2.$$

The Fermion operator algebra emerges from these plane wave solutions to the Dirac equation. The decomposition of U and $U^\dagger$ into the corresponding Majorana Fermion operators corresponds to the decomposition of the energy into momentum and mass: $E^2 = p^2 + m^2$. Normalizing by dividing by $2E$, we have
$$A = (\beta p + \alpha m)/E$$
and
$$B = i\beta\alpha,$$

so that
$$A^2 = B^2 = 1$$
and
$$AB + BA = 0.$$
Then,
$$U = (A + Bi)E$$
and
$$U^\dagger = (A - Bi)E,$$
showing how the Fermion operators are expressed in terms of the simpler Clifford algebra of Majorana operators (split quaternions once again). We can take $A = e$ and $B = \eta$ and regard these Fermion annihilation and creation operators in the simplest iterant framework.

10.2. Iterant Formulation of the Dirac Equation

Note that the solutions to the Dirac equation that we have written are expressed using abstract algebra. To write explicit solutions using this algebraic approach, we can write
$$\mathcal{O} = \hat{E} - \alpha \hat{P} - \beta m,$$
where $\hat{E}$ is the energy operator and $\hat{p}$ is the momentum operator. Then, a solution
$$\phi = A + \alpha B + \beta C + \alpha \beta D$$
of the Dirac equation consists in a quadruple of complex functions of (x, t) such that
$$\mathcal{O}\phi = 0.$$
We can regard $[A, B, C, D] = \phi = A + \alpha B + \beta C + \alpha \beta D$ as an iterant that is acted upon by α and β. We see that (by multiplying on the left)
$$[A, B, C, D]^\alpha = [B, A, D, C]$$
and
$$[A, B, C, D]^\beta = [C, -D, A, -B].$$
Thus, the structure corresponds to the action of the split quaternions as a signed Klein 4-group. The equation $\mathcal{O}\phi = 0$ becomes four operator equations involving these signed permutations:
$$\mathcal{O}\phi = (\hat{E} - \alpha \hat{p} - \beta m)(A + \alpha B + \beta C + \alpha \beta D) =$$
$$\hat{E}A + \alpha \hat{E}B + \beta \hat{E}C + \alpha \beta \hat{E}D$$
$$-\alpha \hat{p}A - \hat{p}B - \alpha \beta \hat{p}C - \beta \hat{p}D$$
$$-\beta mA + \alpha \beta mB - mC + \alpha mD$$
$$= (\hat{E}A - \hat{p}B - mC) + \alpha(\hat{E}B - \hat{p}A + mD) + \beta(\hat{E}C - \hat{p}D - mA) + \alpha\beta(\hat{E}D - \hat{p}C + mB).$$
Thus, $\mathcal{O}\phi = 0$ is equivalent to the set of equations
$$\hat{E}A = \hat{p}B + mC,$$
$$\hat{E}B = \hat{p}A - mD,$$

$$\hat{E}C = \hat{p}D + mA,$$
$$\hat{E}D = \hat{p}C - mB.$$

This, in turn, can be written in iterant form as

$$\hat{E}[A, B, C, D] = \hat{p}[B, A, D, C] + m[C, -D, A, -B] = \hat{p}[A, B, C, D]^\alpha + m[A, B, C, D]^\beta.$$

The plane wave solution $\phi = (E + \alpha p + \beta m)e^{i(px-Et)}\mathbf{k}$ corresponds, in this iterant formalism, to $\phi = [E, p, m, 0]e^{i(px-Et)}$.

In this way, we can think of a solution to the Dirac equation as an iterant composed of four complex valued functions taken in order with the given action of the split quaternions as described above. This can then be reformulated as single recursive system, as we did for the Schrödinger equation in the introduction. The analogs for the way the recursion acts on the time steps of the recursion are given by the action of the split quaternions rather than the action of the complex numbers ($[a, b]^i = [-b, a]$). The idea remains the same, and the matrix representations for the Dirac algebra arise naturally from the algebra itself.

10.3. Writing in the Full Dirac Algebra

This section closely follows our paper [1] and is expanded for the discussion at the end. The aim is to write the Dirac equation for three dimensions of space and one dimension of time, and then to write a version of the Majorana–Dirac Equation (that can have real solutions) in terms of a doubled split quaternion algebra, expressed in iterant language. This provides an alternative to working with modifications of the 4 × 4 Dirac matrices. We formulate it to illustrate again the iterant concept and to raise the question of finding other matrix representations for equations of Majorana type.

We have written the Dirac equation so far in one dimension of space and one dimension of time. In order to write in three spatial dimensions, we take an independent Clifford algebra generated by $\sigma_1, \sigma_2, \sigma_3$ with $\sigma_i^2 = 1$ for $i = 1, 2, 3$ and $\sigma_i\sigma_j = -\sigma_j\sigma_i$ for $i \neq j$. Assume that α and β generate an independent Clifford algebra that commutes with the algebra of the σ_i. Replace the scalar momentum p by a 3-vector momentum $p = (p_1, p_2, p_3)$ and let $p \bullet \sigma = p_1\sigma_1 + p_2\sigma_2 + p_3\sigma_3$. Replace $\partial/\partial x$ with $\nabla = (\partial/\partial x_1, \partial/\partial x_2, \partial/\partial x_2)$ and $\partial p/\partial x$ with $\nabla \bullet p$.

The Dirac equation is then written

$$i\partial\psi/\partial t = -i\alpha\nabla \bullet \sigma\psi + \beta m\psi.$$

The Dirac operator is

$$\mathcal{O} = i\partial/\partial t + i\alpha\nabla \bullet \sigma - \beta m.$$

Using the Dirac operator, the Dirac equation is is

$$\mathcal{O}\psi(x, t) = 0.$$

Let

$$\psi(x, t) = e^{i(p\bullet x - Et)}$$

and construct solutions by first applying the Dirac operator to this ψ. The modified Dirac operator is

$$\mathcal{D} = i\beta\alpha\partial/\partial t + \beta\nabla \bullet \sigma - \alpha m.$$

We have that

$$\mathcal{D}\psi = U\psi,$$

where $U = \beta\alpha E + \beta p \bullet \sigma - \alpha m$. Here, $U^2 = 0$ and $U\psi$ is a solution to the modified Dirac Equation. We can use the Fermion operators as creation and annihilation operators, and locate the corresponding Majorana Fermion operators. We leave these details to the reader.

10.4. Majorana Fermions in the Sense of Majorana

We end with a brief discussion making Dirac algebra distinct from the one generated by $\alpha, \beta, \sigma_1, \sigma_2, \sigma_3$ to obtain an equation that can have real solutions. This was the strategy that Majorana [30] followed to construct his Majorana Fermions. A real equation can have solutions that are invariant under complex conjugation and so can correspond to particles that are their own anti-particles. We will describe this Majorana algebra in terms of the split quaternions ϵ and η. For convenience, we use the matrix representation given below. The reader of this paper can substitute the corresponding iterants:

$$\epsilon = \begin{pmatrix} -1 & 0 \\ 0 & 1 \end{pmatrix}, \eta = \begin{pmatrix} 0 & 1 \\ 1 & 0 \end{pmatrix}.$$

Let $\hat{\epsilon}$ and $\hat{\eta}$ generate another, independent algebra of split quaternions, commuting with the first algebra generated by ϵ and η. Then, a totally real Majorana Dirac equation can be written as follows:

$$(\partial/\partial t + \hat{\eta}\eta\partial/\partial x + \epsilon\partial/\partial y + \hat{\epsilon}\eta\partial/\partial z - \hat{\epsilon}\hat{\eta}\eta m)\psi = 0.$$

To see that this is a correct Dirac equation, note that

$$\hat{E} = \alpha_x \hat{p}_x + \alpha_y \hat{p}_y + \alpha_z \hat{p}_z + \beta m$$

(Here, the "hats" denote the quantum differential operators corresponding to the energy and momentum.) will satisfy

$$\hat{E}^2 = \hat{p}_x^2 + \hat{p}_y^2 + \hat{p}_z^2 + m^2$$

if the algebra generated by $\alpha_x, \alpha_y, \alpha_z, \beta$ has each generator of square one and each distinct pair of generators anti-commuting. From there, we obtain the general Dirac equation by replacing $\hat{E}$ by $i\partial/\partial t$, and $\hat{p}_x$ with $-i\partial/\partial x$ (and same for y, z):

$$(i\partial/\partial t + i\alpha_x \partial/\partial x + i\alpha_y \partial/\partial y + i\alpha_z \partial/\partial z - \beta m)\psi = 0.$$

This is equivalent to

$$(\partial/\partial t + \alpha_x \partial/\partial x + \alpha_y \partial/\partial y + \alpha_z \partial/\partial z + i\beta m)\psi = 0.$$

Thus, here we take

$$\alpha_x = \hat{\eta}\eta, \alpha_y = \epsilon, \alpha_z = \hat{\epsilon}\eta, \beta = i\hat{\epsilon}\hat{\eta}\eta,$$

and observe that these elements satisfy the requirements for the Dirac algebra. Since the algebra appearing in the Majorana–Dirac operator is constructed entirely from two commuting copies of the split quaternions, there is no appearance of the complex numbers, and when written out in 2×2 matrices, we obtain coupled real differential equations to be solved.

A solution to the Majorana–Dirac Equation. Let $\rho(x, t) = e^{(p \bullet x - Et)}$. Note that ρ is a a real-valued function. Let

$$\mathcal{MO} = (\partial/\partial t + \hat{\eta}\eta\partial/\partial x + \epsilon\partial/\partial y + \hat{\epsilon}\eta\partial/\partial z - \hat{\epsilon}\hat{\eta}\eta m).$$

This is the Majorana–Dirac operator, as we have explained above. Then, we have the equation

$$\mathcal{MO}\rho(x, t) = (-E + \hat{\eta}\eta p_x + \epsilon p_y + \hat{\epsilon}\eta p_z - \hat{\epsilon}\hat{\eta}\eta m)\rho(x, t).$$

Let
$$\Gamma = -E + \hat{\eta}\eta p_x + \epsilon p_y + \hat{\epsilon}\eta p_z - \hat{\epsilon}\hat{\eta}\eta m,$$
and
$$\hat{\Gamma} = -E - \hat{\eta}\eta p_x - \epsilon p_y - \hat{\epsilon}\eta p_z + \hat{\epsilon}\hat{\eta}\eta m.$$
Then, we have
$$\hat{\Gamma}\Gamma = 0,$$
since all algebraic coefficients square to minus one, and anti-commute. Therefore,
$$\mathcal{MO}(\hat{\Gamma}\rho(x,t)) = \hat{\Gamma}\Gamma\rho(x,t) = 0.$$
Thus,
$$\hat{\Gamma}\rho(x,t) = (-E - \hat{\eta}\eta p_x - \epsilon p_y - \hat{\epsilon}\eta p_z + \hat{\epsilon}\hat{\eta}\eta m)\rho(x,t)$$
is a solution to the Majorana–Dirac equation. When this solution is written out into its components, it is an entirely real valued solution since the components of the matrices representing the algebra are all real numbers. Recall from the earlier part of this section that we were able to reformulate solutions of this kind for the usual Dirac equation in terms of the nilpotent formalism with the algebraic element U with $U^2 = 0$. Here, we can produce real solutions to the Majorana–Dirac equation, but it does not seem possible to put them in the nilpotent formalism. This is surely a reflection of the fact that these solutions are not Fermions in the usual sense. On the other hand, one can regard the solution $\hat{\Gamma}\rho(x,t)$ in relation to the algebra element $\hat{\Gamma}$, and this algebra element is a combination of Majorana Fermion operators $\{\hat{\eta}\eta, \epsilon, \hat{\epsilon}\eta, \hat{\epsilon}\hat{\eta}\eta\}$ in the sense of Clifford algebra or iterant operators that we have used earlier in this paper. Thus, we see that there is at least the beginning of a relationship between the modern use of the Majorana Fermion operators and the original intents of Ettore Majorana to find real solutions to the Dirac equation.

We would like to know if there are other ways to produce such real Dirac equations, and particularly if there are ways to accomplish this aim that do not algebraically entangle the two copies of the split quaternions as our construction (and Majorana's original construction) seems to require.

Acknowledgments: It gives the author great pleasure to thank G. Spencer-Brown, James Flagg, Alex Comfort, David Finkelstein, Pierre Noyes, Peter Rowlands, Sam Lomonaco, Bernd Schmeikal and Rukhsan Ul-Haq for conversations related to the considerations in this paper.

Conflicts of Interest: The author declares no conflict of interest.

References

1. Kauffman, L.H. Iterants, Fermions and Majorana Operators. In *Unified Field Mechanics—Natural Science Beyond the Veil of Spacetime*; Amoroso, R., Kauffman, L.H., Rowlands, P., Eds.; World Scientific Pub. Co.: Singapore, 2015; pp. 1–32.
2. Kauffman, L.H. Knot Logic. In *Knots and Applications*; Kauffman, L., Ed.; World Scientific Pub. Co.: Singapore, 1994; pp. 1–110.
3. Kauffman, L.H. Knot logic and topological quantum computing with Majorana fermions. In *Logic and Algebraic Structures in Quantum Computing and Information*; Lecture Notes in Logic; Chubb, J., Eskandarian, A., Harizanov, V., Eds.; Cambridge University Press: Cambridge, UK, 2016; 124p.
4. Kauffman, L.H.; Lomonaco, S.J. Braiding, Majorana Fermions and Topological Quantum Computing, (to appear in Special Issue of QIP on Topological Quantum Computing). In Proceedings of the 2nd International Conference and Exhibition on Mesoscopic and Condensed Matter Physics, Chicago, IL, USA, 26–28 October 2016.
5. Ul Haq, R.; Kauffman, L.H. Iterants, Idempotents and Clifford algebra in Quantum Theory. *arXiv* **2017**, arXiv:1705.06600.
6. Bilson-Thompson, S.O. A topological model of composite fermions. *arXiv* **2006**, arXiv:hep-ph/0503213.
7. Gasiorowicz, S. *Elementary Particle Physics*; Wiley: New York, NY, USA, 1966.

8. Rowlands, P. *Zero to Infinity: The Foundations of Physics*; Series on Knots and Everything, Volume 41; World Scientific Publishing Co.: Singapore, 2007.
9. Spencer-Brown, G. *Laws of Form*; George Allen and Unwin Ltd.: London, UK, 1969.
10. Kauffman, L. Sign and Space. In *Religious Experience and Scientific Paradigms, Proceedings of the 1982 IASWR Conference*; Institute of Advanced Study of World Religions: Stony Brook, NY, USA, 1985; pp. 118–164.
11. Kauffman, L. Self-reference and recursive forms. *J. Soc. Biol. Struct.* **1987**, *10*, 53–72.
12. Kauffman, L. Special relativity and a calculus of distinctions. In Proceedings of the 9th Annual International Meeting of ANPA, Cambridge, UK, 23–28 September 1987; pp. 290–311.
13. Kauffman, L. Imaginary values in mathematical logic. In Proceedings of the Seventeenth International Conference on Multiple Valued Logic, Boston, MA, USA, 26–28 May 1987; pp. 282–289.
14. Kauffman, L.H. Biologic. *AMS Contemp. Math. Ser.* **2002**, *304*, 313–340.
15. Kauffman, L.H. *Temperley-Lieb Recoupling Theory and Invariants of Three-Manifolds (Annals Studies-114)*; Princeton University Press: Princeton, NJ, USA, 1994.
16. Kauffman, L.H. Time imaginary value, paradox sign and space. In *Computing Anticipatory Systems, Proceedings of the AIP Conference CASYS—Fifth International Conference, Liege, Belgium, 13–18 August 2001*; Dubois, D., Ed.; AIP Conference Publishing: Melville, NY, USA, 2002; Volume 627.
17. Schmiekal, B. *Decay of Motion: The Anti-Physics of SpaceTime*; Nova Publishers, Inc.: Hauppauge, NY, USA, 2014.
18. Kauffman, L.H.; Noyes, H.P. Discrete Physics and the Derivation of Electromagnetism from the formalism of Quantum Mechanics. *Proc. R. Soc. Lond. A* **1996**, *452*, 81–95.
19. Kauffman, L.H.; Noyes, H.P. Discrete Physics and the Dirac Equation. *Phys. Lett. A* **1996**, *218*, 139–146.
20. Kauffman, L.H. Noncommutativity and discrete physics. *Phys. D Nonlinear Phenom.* **1998**, *120*, 125–138.
21. Kauffman, L.H. Space and time in discrete physics. *Int. J. Gen. Syst.* **1998**, *27*, 241–273.
22. Kauffman, L.H. A non-commutative approach to discrete physics. In *Aspects II: Proceedings of ANPA 20*; ANPA: Stanford, CA, USA, 1999; pp. 215–238.
23. Kauffman, L.H. Non-commutative calculus and discrete physics. In *Boundaries: Scientific Aspects of ANPA 24*; ANPA: Stanford, CA, USA, 2003; pp. 73–128.
24. Kauffman, L.H. Non-commutative worlds. *New J. Phys.* **2004**, *6*, 73.
25. Kauffman, L.H. Non-commutative worlds and classical constraints. In *Scientific Essays in Honor of Pierre Noyes on the Occasion of His 90-th Birthday*; Amson, J., Kaufman, L.H., Eds.; World Scientific Pub. Co.: Singapore, 2013; pp. 169–210.
26. Kauffman, L.H. Differential geometry in non-commutative worlds. In *Quantum Gravity: Mathematical Models and Experimental Bounds*; Fauser, B., Tolksdorf, J., Zeidler, E., Eds.; Birkhauser: Basel, Switzerland, 2007; pp. 61–75.
27. Kauffman, L.H. Knot Logic and Topological Quantum Computing with Majorana Fermions. *arXiv* **2013**, arXiv:1301.6214.
28. Kauffman, L.H.; Lomonaco, S.J., Jr. q-deformed spin networks, knot polynomials and anyonic topological quantum computation. *J. Knot Theory Ramif.* **2007**, *16*, 267–332.
29. Cheng, T.P.; Lee, L.F. *Gauge Theory of Elementary Particles*; Clarendon Press: Oxford, UK, 1988.
30. Majorana, E. A symmetric theory of electrons and positrons. *I Nuovo Cimento* **1937**, *14*, 171–184.
31. Moore, G.; Read, N. Noabelions in the fractional quantum Hall effect. *Nucl. Phys. B* **1991**, *360*, 362–396.
32. Mourik, V.; Zuo, K.; Frolov, S.M.; Plissard, S.R.; Bakkers, E.P.A.M.; Kouwenhuven, L.P. Signatures of Majorana fermions in hybred superconductor-semiconductor devices. *Science* **2012**, *336*, 1003–1007.
33. Ivanov, D.A. Non-abelian statistics of half-quantum vortices in p-wave superconductors. *Phys. Rev. Lett.* **2001**, *86*, 268, doi:10.1103/PhysRevLett.86.268.
34. Kauffman, L.H. *Knots and Physics*; World Scientific Pub., Co.: Singapore, 2012.

© 2017 by the author. Licensee MDPI, Basel, Switzerland. This article is an open access article distributed under the terms and conditions of the Creative Commons Attribution (CC BY) license (http://creativecommons.org/licenses/by/4.0/).

Article
A No-Go Theorem for Observer-Independent Facts

Časlav Brukner [1,2]

[1] Vienna Center for Quantum Science and Technology (VCQ), Faculty of Physics, University of Vienna, Boltzmanngasse 5, A-1090 Vienna, Austria; caslav.brukner@univie.ac.at
[2] Institute of Quantum Optics and Quantum Information (IQOQI), Austrian Academy of Sciences, Boltzmanngasse 3, A-1090 Vienna, Austria

Received: 5 April 2018; Accepted: 2 May 2018; Published: 8 May 2018

Abstract: In his famous thought experiment, Wigner assigns an entangled state to the composite quantum system made up of Wigner's friend and her observed system. While the two of them have different accounts of the process, each Wigner and his friend can in principle verify his/her respective state assignments by performing an appropriate measurement. As manifested through a click in a detector or a specific position of the pointer, the outcomes of these measurements can be regarded as reflecting directly observable "facts". Reviewing arXiv:1507.05255, I will derive a no-go theorem for observer-independent facts, which would be common both for Wigner and the friend. I will then analyze this result in the context of a newly-derived theorem arXiv:1604.07422, where Frauchiger and Renner prove that "single-world interpretations of quantum theory cannot be self-consistent". It is argued that "self-consistency" has the same implications as the assumption that observational statements of different observers can be compared in a single (and hence an observer-independent) theoretical framework. The latter, however, may not be possible, if the statements are to be understood as relational in the sense that their determinacy is relative to an observer.

Keywords: Wigner-friend experiment; no-go theorem; quantum foundations; interpretations of quantum mechanics

1. Introduction

One of the most debated situations concerning the quantum measurement problem is described in the thought experiment of the so-called "Wigner's friend". The experiment involves a quantum system and an observer (Wigner's friend) who performs measurements on this system in a sealed laboratory. A "super-observer" (Wigner) is placed outside the laboratory. While for the friend, the measurement outcome is reflected in a property of the device recording it (e.g., in the form of a click in a photo-detector or a certain position of a pointer device), Wigner can describe the process unitarily on the basis of the information that is in principle available to him. At the end of the process, the friend projects the state of the system corresponding to the observed outcome, whereas Wigner assigns a specific entangled state to the system and the friend, which he can verify performing a further experiment. When Wigner's friend observes an outcome, does the state collapse for Wigner as well? If not, how can we reconcile their different accounts of the process?

The thought experiment of Wigner's friend has great conceptual value, as it challenges different approaches to understanding quantum theory. In his original work [1], Wigner designed the experiment to support his view that consciousness is necessary to complete the quantum measurement process. According to the many-worlds interpretation [2], there are many copies of Wigner's friend in different "worlds". Each copy observes one outcome, a different one in each world. According to the Copenhagen, relational [3] and quantum Bayesian [4] interpretations, the state is defined only relative to the observer; relative to the friend, the state is projected, while relative to Wigner, it is in a superposition. Either way, supporters of any of these interpretations will arrive at the same predictions in Wigner's verifying experiment. In contrast, objective collapse theories [5–7] predict that the quantum state collapses when a superposed system reaches a certain threshold of mass, size, complexity, etc., such that it becomes impossible to even prepare the entangled state of Wigner's friend and the system. Consequently, Wigner's state assignment can statistically be disproved repeating the verifying experiment.

The descriptions of "what is happening inside the lab" as given by Wigner and Wigner's friend respectively will differ. This difference need not pose a consistency problem for quantum theory, for example if one takes the view that the theory gives the physical description relative to the observer and her/his measuring apparatus in agreement with [3]. As long as the two observers do not exchange the information about their outcomes, they will remain separated from each other, each holding a different description of the systems with respect to their individual experimental arrangements. If they do compare their predictions, they will agree. For example, should the friend communicate her result to Wigner, this would collapse the state he assigns to the friend and the system. This suggests that there should be no tension in accepting that, relative to their experimental arrangements, Wigner's friend in her measurement, as well as Wigner in his verifying measurement, each obtains a respective measurement outcome. Since these outcomes are usually manifested as clicks in detectors or definite positions of a pointer, they can be considered as directly accessible "facts". Quite naturally, the question arises: Can the facts as observed by Wigner and by Wigner's friend be jointly considered as objective properties of the world, in which case we might call them "facts of the world"? What we mean with this question is whether there exists any theory, potentially different from quantum theory, where a joint probability may be assigned for Wigner's outcome and for that of his friend.

Reviewing the results of [8], I will derive a Bell-type no-go theorem for observer-independent facts, showing that there can be no theory in which Wigner's and Wigner's friend' facts can jointly be considered as (local) objective properties. More precisely, I will show that the assumptions of "locality", "freedom of choice" and "universality of quantum theory" (the latter in the sense that there are no constraints of the system to which the theory can be applied) are incompatible with the assumption of observer-independent facts, i.e., under the assumptions one cannot define joint probabilities for Wigner's outcome and for that of Wigner's friend. This might indicate that in quantum theory, we can only define facts relative to an observation and an observer. I will then analyze the relation of these results to the theorem developed by Frauchinger and Renner [9], which proves that "single-world interpretations of quantum theory cannot be self-consistent". In particular, I will argue that the implications of their "self-consistency" requirement are equivalent to those of a theoretical framework in which the truth values of the observational statements by Wigner and Wigner's friend can be jointly assigned and then whether they are consistent or not verified. However, in the view of the no-go theorem, this in general need not be possible in a physical theory; the theory may operate only with facts relative to the observer.

It should be emphasized that the no-go theorem applies to "facts" understood as "immediate experiences of observers"; it may refer to what various interpretations of quantum mechanics assume to be "real" (e.g., the wave function of the Universe, Bohmian's trajectories, etc.) only to the extent to which these "realities" give rise to directly observable facts in terms of detector clicks or pointer positions.

2. Deutsch's Version of Wigner's Friend Experiment

The standard description of the Wigner-friend thought experiment involves a quantum two-level system (System 1, e.g., a spin-1/2 particle), which can give rise to two outcomes upon measurement (e.g., two opposite directions when passing through a Stern–Gerlach apparatus). The outcomes are recorded by a measurement apparatus and eventually in the friend's memory (System 2). Now, Wigner is placed outside the isolated laboratory in which the experiment takes place and can perform a quantum measurement on the overall system (spin-1/2 particle + friend's laboratory). Take it that all experiments are carried out a sufficient number of times to collect statistics.

For concreteness, suppose that a measurement of spin along z is performed on a particle initially prepared in state $|x+\rangle_S = \frac{1}{\sqrt{2}}(|z+\rangle_S + |z-\rangle_S)$, where subscript S refers to the spin. After the measurement is completed, the measurement apparatus is found in one of many perceptively different macroscopic configurations, like different positions of a pointer along a scale. If the apparatus pointer is found in a specific position along the scale, the friend can say that the observable spin z has the value "up" or "down". Note that for the present argument, we need not make any assumption about how the friend formally describes the spin and the apparatus, which measurement formalism she uses or even if she uses quantum theory for it. All that is needed is the assumption that the friend perceives a definite outcome.

Wigner uses quantum theory to describe the friend's measurement. From his perspective, the measurement is described by a unitary transformation. The different possible spin states $|z+\rangle_S$ and $|z-\rangle_S$ are supposed to get entangled to the perceptively different macroscopic configurations of the apparatus and the parts of the laboratory including the friend's memory. The states of different macroscopic configurations are represented by orthogonal states $|F_{z+}\rangle_F$ and $|F_{z-}\rangle_F$, respectively. We assume that the state of the composite system "spin + friend's laboratory" is given by:

$$|\Phi\rangle_{SF} = \frac{1}{\sqrt{2}}(|z+\rangle_S|F_{z+}\rangle_F + |z-\rangle_S|F_{z-}\rangle_F), \qquad (1)$$

where the particular phase (here "+") between the two amplitudes in Equation (1) is specified by the measurement interaction in control of Wigner (note that if Wigner did not know this phase due to the lack of control of it, he would describe the "spin + friend's laboratory" in an incoherent mixture of the two possibilities). Wigner can verify his state assignment (1), for example by performing a Bell state measurement in the basis: $|\Phi^{\pm}\rangle_{SF} = \frac{1}{\sqrt{2}}(|z+\rangle_S|F_{z+}\rangle_F \pm |z-\rangle_S|F_{z-}\rangle_F)$ and $|\Psi^{\pm}\rangle_{SF} = \frac{1}{\sqrt{2}}(|z+\rangle_S|F_{z-}\rangle_F \pm |z-\rangle_S|F_{z+}\rangle_F)$.

The fact that the friend and Wigner have different accounts of the friend's measurement process is at the heart of the discussion surrounding the Wigner-friend thought experiment. Still, the difference need not give rise to any inconsistency in practicing quantum theory, since the two descriptions belong to two different observers, who remain separated in making predictions for their respective systems. The novelty of Deutsch's proposal [10] lies in the possibility for Wigner to acquire direct knowledge on whether the friend has observed a definite outcome upon her measurement or not without revealing what outcome she has observed. The friend could open the laboratory in a manner that allowed communication (e.g., a specific message written on a piece of paper) to be passed outside to Wigner, keeping all other degrees of freedom fully isolated, as illustrated in Figure 1. Obviously, it is of central importance that the message does not contain any information concerning the specific observed outcome (which would destroy the coherence of state (1)), but merely an indication of the kind: "I have observed a definite outcome" or "I have not observed a definite outcome". If the message is encoded in the state of system M, the overall state is:

$$|\Phi\rangle_{SFM} = \frac{1}{\sqrt{2}}(|z+\rangle_S|F_{z+}\rangle_F + |z-\rangle_S|F_{z-}\rangle_F)|\text{"I have observed a definite outcome"}\rangle_M, \qquad (2)$$

since the state of the message is factorized out from the total state (I leave the option for the message "I have not observed a definite outcome" out, as it conflicts with our experience of the situation that we refer to as measurement and it also can be used to violate the bound on quantum state discrimination [8]).

If we assume the universality of quantum theory in the sense that it can be applied at any scale, including the apparatus, the entire laboratory and even the observer's memory, we conclude that the message will indicate that the friend perceives a definite outcome and yet Wigner will confirm his state assignment (1). This should be contrasted to the "collapse models" by Ghirardi, Rimini and Weber [5] or by Diosi [6] and Penrose [7], which predict a breakdown of the quantum-mechanical laws at some scale. In the presence of such a collapse, the prediction based on Wigner's state assignment will statistically deviate from the result obtained in the verification test.

Figure 1. Deutsch's version of the Wigner-friend thought experiment. An observer (Wigner's friend) performs a Stern–Gerlach experiment on a spin 1/2 particle in a sealed laboratory. The outcome, either "spin up" or "spin down", is recorded in the friend's laboratory, including her memory. A super-observer (Wigner) describes the entire experiment as a unitary transformation resulting in an encompassing entangled state between the system and the friend's laboratory. The friend is allowed to communicate a message, which only reports whether she sees a definite outcome or not, without in any way revealing the actual outcome she observes.

3. The No-Go Theorem

We have seen that Wigner not only perceives his own facts, he is also able to obtain a direct evidence for the existence of the friend's facts (although without knowing which specific outcome has been realized in the laboratory). This strongly suggests that Wigner's and Wigner's friend's facts coexist. We pose the question: Is there a theoretical framework, potentially going beyond quantum theory, in which one can account for observer-independent facts, ones that hence can be called "facts of the world"? In such a framework, one could assign jointly truth values to both the observational statement A_1: "The pointer of Wigner's friend's apparatus points to result z+" and A_2: "The pointer of Wigner's apparatus points to result Φ".

One important remark: Whenever Wigner performs his measurement, he can inform the friend about the outcome he observed. Hence, Wigner's friend can learn Wigner's outcome in addition to the outcome she herself observed directly. In this way, Wigner's friend can know the truth values of both statements A_1 and A_2. The assumption of "observer-independent facts" is a stronger condition: we require an assignment of truth values to statements A_1 and A_2 independently of which measurement Wigner performs. Wigner can either perform his verifying experiment or he can perform Wigner's friend's measurement (for example, by opening the lab, or learning it from the friend). In either experiment, the observed outcome (e.g., "Φ" and "z+", respectively) is required to reveal the assigned truth value for A_1 or A_2. We formalize the requirement of "observer-independent facts" in the following assumption.

Postulate 1. *("Observer-independent facts") The truth values of the propositions A_i of all observers form a Boolean algebra $\mathcal{A}$. Moreover, the algebra is equipped with a (countably additive) positive measure $p(A) \geq 0$ for all statements $A \in \mathcal{A}$, which is the probability for the statement to be true.*

In the proof, we will only use the conjunction of propositions of different observers, which is a weaker requirement. Furthermore, we use a countably additive measure since we are dealing with only a countable (in fact only a finite) set of elements. In Boolean algebra, one can build the conjunction, the disjunction and the negation of the statements. A typical example of a Boolean algebra is set theory. The operations are identified with the set theoretic intersection, union and complement, respectively. This is significant in the context of classical physics, where the propositions can be represented by subsets of a phase space. In the present context, one can jointly assign truth values "true" or "false" to statements A_1 and A_2 about observations made by Wigner's friend and Wigner, respectively. Moreover, one can build the conjunction $A_1 \cap A_2$ and assign joint probability $p(A_1 = \pm 1, A_2 = \pm 1)$, where A_1 is observed by the friend and A_2 by Wigner (and where truth value "true" corresponds to a value of one and "false" to -1). Note that since observables corresponding to A_1 and A_2 do not commute with each other, this amounts to introducing "hidden variables", for which we now formulate a Bell's theorem [11].

Theorem 1. (No-go theorem for "observer-independent facts") *The following statements are incompatible (i.e., lead to a contradiction)*

1. *"Universal validity of quantum theory". Quantum predictions hold at any scale, even if the measured system contains objects as large as an "observer" (including her laboratory, memory etc.).*
2. *"Locality". The choice of the measurement settings of one observer has no influence on the outcomes of the other distant observer(s).*
3. *"Freedom of choice". The choice of measurement settings is statistically independent from the rest of the experiment.*
4. *"Observer-independent facts". One can jointly assign truth values to the propositions about observed outcomes ("facts") of different observers (as specified in the postulate above).*

Before going to the proof, I make two comments. Firstly, we use word "universal" in assumption 1 in the sence of Peres [12]: "There is nothing in quantum theory making it applicable to three atoms and inapplicable to 10^{23} ... Even if quantum theory is universal, it is not closed. A distinction must be made between endophysical systems—those which are described by the theory—and exophysical ones, which lie outside the domain of the theory (for example, the telescopes and photographic plates used by astronomers for verifying the laws of celestial mechanics). While quantum theory can in principle describe anything, a quantum description cannot include everything. In every physical situation something must remain unanalyzed. This is not a flaw of quantum theory, but a logical necessity ...".

Secondly, the theorem can be derived by replacing assumptions 2, 3 and 4 with a single assumption of Bell's "local causality". The latter already implies the existence of (local) probabilities for "joint facts" for Wigner and Wigner's friend [13], which is the subject of the present no-go theorem. The reason for working with the present choice of assumptions is that the relevance of the theorem for the propositions different observers make about their respective outcome becomes apparent.

Proof. With reference to Figure 2, consider a pair of super-observers (Alice and Bob) who can carry out experiments on two systems that include a laboratory for each system, in each of which an observer (Charlie and Debbie, respectively) performs a measurement on a spin-1/2 particle. We consider a Bell inequality test and assume that Alice chooses between two measurement settings A_1 and A_2, and similarly, Bob chooses between B_1 and B_2. The settings A_1 and A_2 correspond to the observational statements Charlie and Alice can make about their respective outcomes, respectively. Similarly, the settings B_1 and B_2 correspond to observational statements of Debbie and Bob, respectively. Assumptions (2), (3) and (4) together account for the existence of local hidden variables that predefine the values for A_1, A_2, B_1 and B_2 to be $+1$ or -1. Moreover, the assumptions imply the existence of the joint probability $p(A_1, A_2, B_1, B_2)$ whose marginals satisfy the Clauser–Horne–Shimony–Holt inequality (CHSH): $S = \langle A_1 B_1 \rangle + \langle A_1 B_2 \rangle + \langle A_2 B_1 \rangle - \langle A_2 B_2 \rangle \leq 2$. Here, for example, $\langle A_1 B_1 \rangle = \sum_{A_1, B_1 = -1,1} A_1 B_1 p(A_1, B_1)$ and $p(A_1, B_1) = \sum_{A_2, B_2 = -1,1} p(A_1, A_2, B_1, B_2)$ and similarly for other cases.

Suppose that Charlie and Debbie initially share an entangled state of two respective spin-1/2 particles S_1 and S_2 in a state:

$$|\psi\rangle_{S_1 S_2} = -\sin\frac{\theta}{2} |\phi^+\rangle_{S_1 S_2} + \cos\frac{\theta}{2} |\psi^-\rangle_{S_1 S_2}, \tag{3}$$

where $|\phi^+\rangle_{S_1 S_2} = \frac{1}{\sqrt{2}}(|z+\rangle_{S_1}|z+\rangle_{S_2} + |z-\rangle_{S_1}|z-\rangle_{S_2})$ and $|\psi^-\rangle_{S_1 S_2} = \frac{1}{\sqrt{2}}(|z+\rangle_{S_1}|z-\rangle_{S_2} - |z-\rangle_{S_1}|z+\rangle_{S_2})$, and the first spin is in possession of Charlie and the second of Debbie. The state can be obtained by applying rotation $(\mathbb{1} \otimes e^{-\frac{i}{2}\theta\sigma_y})|\psi^-\rangle_{S_1 S_2}$ to the singlet state $|\psi^-\rangle_{S_1 S_2} = \frac{1}{\sqrt{2}}(|z+\rangle_{S_1}|z-\rangle_{S_2} - |z-\rangle_{S_1}|z+\rangle_{S_2})$, where θ is the angle of rotation of Debbie's spin around the y-axis and σ_y is a Pauli matrix. This particular choice of the state enables all measured observables to be either of the Wigner's friend type, or of the Wigner type.

For Alice and Bob, the overall state of the spins together with Charlie's and Debbie's laboratories is initially:

$$|\Psi_0\rangle = |\psi\rangle_{S_1 S_2}|0\rangle_C|0\rangle_D, \qquad (4)$$

in agreement with Assumption 1. The state $|0\rangle_C|0\rangle_D$ of the two observers does not require further characterization, except for the description of observers capable of completing a measurement.

Now, Charlie and Debbie each perform a measurement of the respective spin along the z direction. This measurement procedure is described as a unitary transformation from the point of view of Alice and Bob. We assume that after Charlie and Debbie complete their measurement, the overall state becomes:

$$|\check{\Psi}\rangle = -\sin\frac{\theta}{2}|\Phi^+\rangle + \cos\frac{\theta}{2}|\Psi^-\rangle, \qquad (5)$$

where:

$$|\Phi^+\rangle = \frac{1}{\sqrt{2}}(|A_{up}|B_{up}\rangle + |A_{down}\rangle|B_{down}\rangle), \qquad (6)$$

$$|\Psi^-\rangle = \frac{1}{\sqrt{2}}(|A_{up}\rangle|B_{down}\rangle - |A_{down}\rangle|B_{up}\rangle) \qquad (7)$$

and:

$$|A_{up}\rangle = |z+\rangle_{S_1}|C_{z+}\rangle_C, \qquad |B_{up}\rangle = |z+\rangle_{S_2}|D_{z+}\rangle_D, \qquad (8)$$
$$|A_{down}\rangle = |z-\rangle_{S_1}|C_{z-}\rangle_C, \qquad |B_{down}\rangle = |z-\rangle_{S_2}|D_{z-}\rangle_D. \qquad (9)$$

We take now $\theta = \pi/4$ and define two sets of (binary) observables, which play the same role of spin (Pauli) operators along the z and x axis, respectively: $A_z = |A_{up}\rangle\langle A_{up}| - |A_{down}\rangle\langle A_{down}|$ and $A_x = |A_{up}\rangle\langle A_{down}| + |A_{down}\rangle\langle A_{up}|$ for Alice and similarly B_z and B_x for Bob. In the Bell experiment, Alice chooses between $A_1 = A_z$ and $A_2 = A_x$, whereas Bob chooses between $B_1 = B_z$ and $B_2 = B_x$. Note that Alice and Bob each choose between the friend's (A_1 and B_1) and Wigner's (A_2 and B_2) type of measurement. The Bell test with these measurement settings and state (5) results in $S_Q = 2\sqrt{2}$. The violation of the inequality implies that the conjunction of the assumptions (1–4) used to derive it is untenable. □

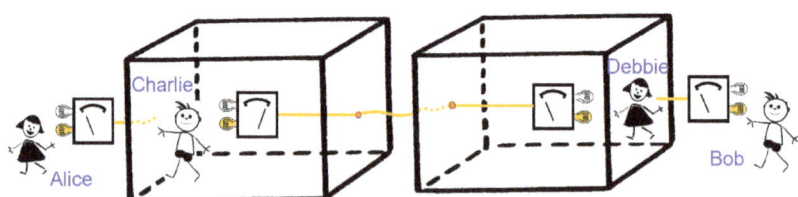

Figure 2. A Bell experiment on two entangled observers in a Wigner-friend scenario. The super-observers Alice and Bob perform their respective measurements on laboratories containing the observers Charlie and Debbie, who both perform a Stern–Gerlach measurement on their respective spin-1/2 particles.

In Appendix A, we present a Greenberger–Horne–Zeilinger type of the theorem with three Wigners and three friends. There, the discrepancy between quantum theory and the theories respecting (2–4) is no more of a probabilistic, but of a deterministic nature.

We conclude that Wigner, even as he has clear evidence for the occurrence of a definite outcome in the friend's laboratory, cannot assume any specific value for the outcome to coexist together with the directly observed value of his outcome, given that all other assumptions are respected. Moreover, there is no theoretical framework where one can assign jointly the truth values to observational propositions of different observers (they cannot build a single Boolean algebra) under these assumptions. A possible consequence of the result is that there cannot be facts of the world per se, but only relative to an observer, in agreement with Rovelli's relative-state interpretation [3], quantum Bayesianism (already in 1996, in the "Replies to Referee 4" of [14], Fuchs drew a distinction between "facts for the agent" and "facts for everybody") [4], as well as the (neo)-Copenhagen interpretation [8]. It is interesting to note that a similar view was expressed by Jammer as early as in 1974 [15], when he wrote that "the description of the state of a system, rather than being restricted to the particle (or systems of particles) under observation, expresses a relation between the particle and all the measurement devices involved." Other possible interpretations of the violation of Bell's inequalities include violations of Assumption 1 in collapse models [5–7], of Assumption 2 in non-local hidden variable models such as de Broglie–Bohm theory [16] or of Assumption 3 in superdeterministic theories [17]. The proper account of the result in the many-worlds interpretation should be found in the interpretation's account of Bell's inequality violation [18,19] and points again to observer-dependent facts as they depend on the branch of the many worlds.

4. Relation to the Paper by Frauchiger and Renner, arXiv: 1604.07422

Building upon works by Deutsch [10], Hardy [20,21] and [8] reviewed above, Frauchiger and Rennen [9] proposed an "extended Wigner-friend thought experiment", from which they concluded that "single-world interpretations of quantum theory cannot be self-consistent". The implications of their argument have been discussed since then [4,22–24].

The claim of [9] is based on an incompatibility proof stating that there cannot exist a physical theory Tthat would fulfill the following three properties (informal versions; see [9] for details):

(QT) "Compliance with quantum theory: T forbids all measurement results that are forbidden by standard quantum theory (and this condition holds even if the measured system is large enough to contain itself an experimenter)."
(SW) "Single-world: T rules out the occurrence of more than one single outcome if an experimenter measures a system once."
(SC) "Self-consistency: T's statements about measurement outcomes are logically consistent (even if they are obtained by considering the perspectives of different experimenters)."

Property (QT) is essentially a weaker version of our Assumption 1 where it is sufficient to require the validity of quantum theory for results with vanishing probability (as the argument is possibilistic, not probabilistic). An example of a theory-violating property (SW) is the many-worlds interpretation of quantum theory.

The argument combines a set of statements that involves different observers F_1, F_2, A and Wand can be drawn on the basis of theory T:

S_1 If F_1 sees $r = t$, then W sees $w \neq ok$.
S_2 If F_2 sees $z = +$, then F_1 sees $r = t$.
S_3 If A sees $x = ok$, then F_2 sees $z = +$.
S_4 W sees $w = ok$ and is told by A that $x = ok$.

The specific type of quantum state, measurements and outcomes involved in the argument is not relevant for further discussion and will be omitted here.

Property (SC) is crucial in a step of the proof, where one combines "nested" statements (S_1–S_4) [25]. In the first step, the self-consistency property (SC) implies the following:

$$S_a \cap S_b \implies S_c \qquad (10)$$

where $\cap$ denotes logical "and" and the statements are of the type:

S_a Observer W assigns the truth value "true" to the statement: "A sees $x = ok$";
S_b Observer A assigns the truth value "true" to the statement: "If $x = ok$, then F_2 sees $z = +$";
S_c Observer W assigns the truth value "true" to the statement: "A concludes that F_2 sees $z = +$".

By repeating reasoning (10) in an iterative way, starting from statement S_4–S_1, one arrives at a new statement:

T   Observer W concludes that A concludes that F_2 concludes that F_1 concludes that $w \neq ok$.

It is important to note that this statement refers to W's conclusion about what other observers conclude when they apply T conditional on the outcomes they observe. It is not a statement about his directly observed outcome.

In the second step, the self-consistency property (SC) is used to arrive at an implication of the following type:

$$T \implies S. \qquad (11)$$

where the implied statement is:

S   Observer W concludes that $w \neq ok$,

which stands in logical contradiction with W's directly observed outcome $w = ok$.

The second step is non-trivial. It enables promoting others' knowledge based on their observations to ones' own knowledge and then to put this "promoted knowledge" in logical comparison with ones' own knowledge gained through direct observation. Through implication (11), the self-consistency property (SC) enables observational statements of other observers (A, F_2 and F_1) to be logically compared with ones (W) own. This has the same predictive power as a theoretical framework in which the truth values of statements of different observers can jointly be assigned and compared. To see this, denote statements S_i, $i = 1, 2, 3$ as implications S_1: (P $\implies$ Q), S_2: (Q $\implies$ R) and S_3: (R $\implies$ S), where P: "A sees $x = ok$", Q: "F_2 sees $z = +$", R: "F_1 sees $r = t$" and S: "W sees $w \neq ok$". Then, "collapsing" others' knowledge into W's knowledge via Equation (11) is equivalent in its implications to considering all the statements as belonging to a single Boolean algebra (i.e., they are now all propositions of observer W, who can apply logical operations on them) for which one can use the transitivity of implication to arrive at [P $\cap$ (P $\implies$ Q) $\cap$ (Q $\implies$ R) $\cap$ (R $\implies$ S)] $\implies$ S. Statement S is again in logical contradiction to W's directly observed outcome $w = ok$.

We have seen that the existence of a single Boolean algebra for truth values for observational statements of different observers is incompatible with the assumptions of "locality", "freedom of choices" and the predictions of quantum theory, which does not impose any constraints on the objects to which it is applied. This might be interpreted as an indication that the strong conclusions implied by the theorem of [9] rely on a too restrictive requirement of property (SC) on a physical theory. The requirement needs not only be fulfilled in quantum theory, but in other physical theories, as well. An example was provided by Sudbery [23]: In the special theory of relativity, due to time dilation, every inertial observer can claim that her/his clock ticks slower than that of a moving partner. This apparent contradiction in predictions of different observers is resolved when one realizes that the statements only have meaning with respect to the specific, observer-dependent measurement procedures that define "simultaneity". Similarly, the states referring to outcomes of different observers in a Wigner-friend type of experiment cannot be defined without referring to the specific experimental arrangements of the observers, in agreement with Bohr's idea of contextuality as formulated by him in 1963 [26]: "the unambiguous account of proper quantum phenomena must, in principle, include a description of all relevant features of experimental arrangement."

I conclude with a remark that the theorem by Frauchiger and Renner has deep conceptual value, as it points to the necessity to differentiate between ones' knowledge about direct observations and ones' knowledge about others' knowledge that is compatible with physical theories. It is likely that understanding this difference will be an important ingredient in further development of the method of Bayesian inference in situations as in the Wigner-friend experiment.

Funding: I acknowledge the support of the Austrian Science Fund (FWF) through the project I-2526-N27. This research was funded by [John Templeton Foundation] grant number [60609]. The opinions expressed in this publication are those of the authors and do not necessarily reflect the views of the John Templeton Foundation.

Acknowledgments: I acknowledge helpful discussions with Mateus Araújo, Veronika Baumann, Adán Cabello, Giulio Chiribella, Christopher Fuchs, Borivoje Dakić, Philipp Höhn, Nikola Paunković, Lídia del Rio, Rüdiger Schack and Stefan Wolf. I would like to especially acknowledge the fruitful discussions with Renato Renner and thank him for providing notes summarizing that discussion.

Conflicts of Interest: The author declares no conflict of interest.

Appendix A

The Bell theorem from the main text can be extended to a Greenberger–Horne–Zeilinger (GHZ) version [27] with three friends and three Wigners. Since the incompatibility of Assumptions 1–4 is not of a probabilistic, but rather of a deterministic nature, this version of the theorem completely bypasses any use of the notion of probability, similarly to the version by Frauchiger and Renner [9]. The experiment was independently introduced in [28], where it was argued that it suggests a violation of Lorentz symmetry.

Consider three spatially-separated observers (Wigners), Alice, Bob and Cleve. They each perform a measurement on a subsystem of a tripartite system. Each of the subsystems includes a further observer, Debbie, Eric and Fiona (Wigner's friends), who perform a Stern–Gerlach measurement of spin along x of their respective spin-1/2 particles. Alice measures Debbie and her spin particle; Bob measures Eric and his spin particle; and finally, Cleve measures Fiona and her spin particle. We consider a GHZ test where Alice chooses between two measurement settings: A_1 and A_2, Bob between B_1 and B_2 and Cleve between C_1 and C_2. Assumptions 2, 3 and 4 imply that A_1, A_2, B_1, B_2, C_1 and C_2 have predefined values of $+1$ or -1.

Define $\hat{A}_x = |A_{up}\rangle\langle A_{up}| - |A_{down}\rangle\langle A_{down}|$ and $\hat{A}_y = i(|A_{up}\rangle\langle A_{down}| - |A_{down}\rangle\langle A_{up}|)$ for Alice and similarly $\hat{B}_x$ and $\hat{B}_y$ for Bob and $\hat{C}_x$ and $\hat{C}_y$ for Cleve, where:

$$|A_{up}\rangle = |x+\rangle_{A1}|D_{x+}\rangle_{A2}, \quad |B_{up}\rangle = |x+\rangle_{B1}|E_{x+}\rangle_{B2}, \quad |C_{up}\rangle = |x+\rangle_{C1}|F_{x+}\rangle_{C2}, \quad \text{(A1)}$$

$$|A_{down}\rangle = |x-\rangle_{A1}|D_{x-}\rangle_{A2}, \quad |B_{down}\rangle = |x-\rangle_{B1}|E_{x-}\rangle_{B2}, \quad |C_{down}\rangle = |x-\rangle_{C1}|F_{x-}\rangle_{C2}. \quad \text{(A2)}$$

In the GHZ test, we choose $\hat{A}_1 = \hat{A}_x$, $\hat{A}_2 = \hat{A}_y$ for Alice and similarly for Bob and Cleve. Assume that Alice, Bob and Cleve perform these measurements on a shared GHZ state:

$$|\Psi_{GHZ}\rangle_{ABC} = \frac{1}{\sqrt{2}}(|A+\rangle|B+\rangle|C+\rangle - |A-\rangle|B-\rangle|C-\rangle), \quad \text{(A3)}$$

where due to Assumption 1, we presume that such a state can be prepared and $|A\pm\rangle = \frac{1}{\sqrt{2}}(|A_{up}\rangle \pm |A_{down}\rangle)$, $|B\pm\rangle = \frac{1}{\sqrt{2}}(|B_{up}\rangle \pm |B_{down}\rangle)$ and $|C\pm\rangle = \frac{1}{\sqrt{2}}(|C_{up}\rangle \pm |C_{down}\rangle)$.

In order to reproduce perfect correlations in the GHZ state, the predefined values need to satisfy $A_x B_y C_y = A_y B_x C_y = A_y B_y C_x = 1$. These equations imply then that $A_x B_x C_x = 1$; however, one finds the opposite result in quantum mechanics: $\hat{A}_x \hat{B}_x \hat{C}_x |\Psi_{GHZ}\rangle_{ABC} = -|\Psi_{GHZ}\rangle_{ABC}$.

References

1. Wigner, E.P. Remarks on the mind-body question. In *The Scientist Speculates*; Good, I.J., Ed.; Heinemann: London, UK, 1961.
2. Everett, H. "Relative State" Formulation of Quantum Mechanics. *Rev. Mod. Phys.* **1957**, *29*, 454–462. [CrossRef]
3. Rovelli, C. Relational quantum mechanics. *Int. J. Theor. Phys.* **1996**, *35*, 1637–1678. [CrossRef]
4. Fuchs, C.A. Notwithstanding Bohr, the Reasons for QBism. *Mind Matter* **2017**, *15*, 245–300.
5. Ghirardi, G.C.; Rimini, A.; Weber, T. Unified dynamics for microscopic and macroscopic systems. *Phys. Rev. D* **1986**, *34*, 470. [CrossRef]

6. Diosi, L. Models for universal reduction of macroscopic quantum fluctuations. *Phys. Rev. A* **1989**, *40*, 1165. [CrossRef]
7. Penrose, R. On gravity's role in quantum state reduction. *Gen. Relat. Gravit.* **1996**, *28*, 581–600. [CrossRef]
8. Brukner, Č. On the quantum measurement problem. In *Quantum [Un]speakables II*; Bertlmann, R., Zeilinger, A., Eds.; The Frontiers Collection; Springer: New York, NY, USA, 2016. [CrossRef]
9. Frauchiger, D.; Renner, R. Single-world interpretations of quantum theory cannot be self-consistent. *arXiv* **2016**, arXiv:1604.07422. [CrossRef]
10. Deutsch, D. Quantum theory as a universal physical theory. *Int. J. Theor. Phys.* **1985**, *24*, 1–41. [CrossRef]
11. Bell, J.S. *Speakable and Unspeakable in Quantum Mechanics*; Collected Papers on Quantum Philosophy; Cambridge University Press: Cambridge, MA, USA, 2004. [CrossRef]
12. Peres, A. *Quantum Theory: Concepts and Methods*; Springer: New York, NY, USA, 1995; p. 173. [CrossRef]
13. Zukowski, M.; Brukner, Č. Quantum non-locality—It ain't necessarily so ... *J. Phys. A Math. Theor.* **2014**, *47*, 424009. [CrossRef]
14. Fuchs, C.A.; Schlosshauer, M.; Stacey, B.C. My Struggles with the Block Universe. *arXiv* **2015**, arXiv:1405.2390. [CrossRef]
15. Jammer, M. *The Philosophy of Quantum Merchanics: The Interpretations of QM in Historical Perspective*; John Wiley and Sons: Hoboken, NJ, USA, 1974; pp. 197–198.
16. Bohm, D. A Suggested Interpretation of the Quantum Theory in Terms of "Hidden" Variables, I and II. *Phys. Rev.* **1952**, *85*, 166–193. [CrossRef]
17. Hooft, G 't. Free Will in the Theory of Everything. *arXiv* **2017**, arXiv:1709.02874. [CrossRef]
18. Brown, H.R.; Timpson, C.G. Bell on Bell's theorem: The changing face of nonlocality. In *Quantum Nonlocality and Reality: 50 Years of Bell's Theorem*; Bell, M., Gao, S., Eds.; Cambridge University Press: Cambridge, MA, USA, 2016.
19. Araújo, M. Understanding Bell's Theorem Part 3: The Many-Worlds Version. Blog: More Quantum. Available online: http://mateusaraujo.info/2016/08/02/understanding-bells-theorem-part-3-the-many-worlds-version/ (accessed on 2 August 2016).
20. Hardy, L. Quantum mechanics, local realistic theories, and Lorentz-invariant realistic theories. *Phys. Rev. Lett.* **1992**, *68*, 2981. [CrossRef] [PubMed]
21. Hardy, L. Nonlocality for two particles without inequalities for almost all entangled states. *Phys. Rev. Lett.* **1993**, *71*, 1665. [CrossRef] [PubMed]
22. Baumann, V.; Hansen, A.; Wolf, S. The measurement problem is the measurement problem is the measurement problem. *arXiv* **2016**, arXiv:1611.01111 . [CrossRef]
23. Sudbery, A. Single-World Theory of the Extended Wigner's Friend Experiment. *Found. Phys.* **2017**, *47*, 658–669. [CrossRef]
24. Bub, J. Why Bohr was (Mostly) Right. *arXiv* **2017**, arXiv:1711.01604. [CrossRef]
25. Brukner, Č. (University of Vienna, Austria; Austrian Academy of Sciences, Austria); Renner, R. (Institute for Theoretical Physics, ETH Zürich, Switzerland). Personal communication, 2017.
26. Bohr, N. Quantum Physics and Philosophy: Causality and Complementarity. In *Philosophy in Mid-Century: A Survey*; Klibansky, R., Ed.; La Nuova Italia Editrice: Florence, Italy, 1963.
27. Greenberger, D.M.; Horne, M.A.; Shimony, A.; Zeilinger, A. Going beyond Bell's Theorem.*Am. J. Phys.* **1990**, *58*, 1131–1143. [CrossRef]
28. Leegwater, G. When GHZ Meet Wigner's Friend. Erasmus University Rotterdam, Rotterdam, The Netherlands. Unpublished manuscript, 2017.

© 2018 by the author. Licensee MDPI, Basel, Switzerland. This article is an open access article distributed under the terms and conditions of the Creative Commons Attribution (CC BY) license (http://creativecommons.org/licenses/by/4.0/).

Article
A Royal Road to Quantum Theory (or Thereabouts)

Alexander Wilce

Department of Mathematics, Susquehanna University, Selinsgrove, PA 17870, USA; wilce@susqu.edu

Received: 15 January 2018; Accepted: 19 March 2018; Published: 26 March 2018

Abstract: This paper fails to derive quantum mechanics from a few simple postulates. However, it gets very close, and does so without much exertion. More precisely, I obtain a representation of finite-dimensional probabilistic systems in terms of Euclidean Jordan algebras, in a strikingly easy way, from simple assumptions. This provides a framework within which real, complex and quaternionic QM can play happily together and allows some (but not too much) room for more exotic alternatives. (This is a leisurely summary, based on recent lectures, of material from the papers arXiv:1206:2897 and arXiv:1507.06278, the latter joint work with Howard Barnum and Matthew Graydon. Some further ideas are also explored, developing the connection between conjugate systems and the possibility of forming stable measurement records and making connections between this approach and the categorical approach to quantum theory.)

Keywords: reconstruction of quantum mechanics; conjugate systems; Jordan algebras

1. Introduction and Overview

Whatever else it may be, Quantum mechanics (QM) is a machine for making probabilistic predictions about the results of measurements. To this extent, QM is, at least in part, about information. Over the last decade or so, it has become clear that the formal apparatus of quantum theory, at least in finite dimensions, can be recovered from constraints on how physical systems store and process information. To this extent, finite-dimensional QM is just about information.

The broad idea of regarding QM in this way, and of attempting to derive its mathematical structure from simple operational or probabilistic axioms, is not new. Efforts in this direction go back at least to the work of von Neumann [1], and include also attempts by Schwinger [2], Mackey [3], Ludwig [4], Piron [5], and many others. However, the consensus is that these were not entirely successful: partly because the results they achieved (e.g., Piron's well-known representation theorem) did not rule out certain rather exotic alternatives to QM, but mostly because the axioms deployed seem, in retrospect, to lack sufficient physical or operational motivation.

More recently, with inspiration from quantum information theory, attention has focused on finite-dimensional systems, where the going is a bit easier. Just as importantly, quantum information theory prompts us to treat properties of composite systems as fundamental, where earlier work focused largely on systems in isolation (a recent exception to this trend is the paper [6] of Barnum, Müller and Ududec). These shifts of emphasis are illustrated by the work of Hardy [7], who presented five simple, broadly information-theoretic postulates governing the states and measurements associated with a physical system, determining a very restricted set of possible theories, parametrized by a positive integer r, with finite-dimensional quantum and classical probability theory corresponding to $r = 1$ and $r = 2$. Following this lead, several papers, notably [8–10], have derived finite-dimensional QM from various packages of axioms governing the information-carrying and information-processing capacity of finite-dimensional systems.

Problems with existing approaches. These recent reconstructive efforts suffer from two related problems. First, they make use of assumptions that seem too strong. Secondly, in trying to derive exactly complex, finite-dimensional quantum theory, they derive too much.

- All of the cited papers assume local tomography. This is the doctrine that the state of a bipartite composite system is entirely determined by the joint probabilities it assigns to outcomes of measurements on the two subsystems. This rules out both real and quaternionic QM, both of which are legitimate quantum theories [11].
- These papers also all make some version of a uniformity assumption: that all systems having the same information-carrying capacity are isomorphic, or that all systems are composed, in a uniform way, from "bits" of a uniform type. Here, "information carrying capacity" means essentially the maximum number of states that can be distinguished from one another with probability one by a single measurement. A bit is a system for which this number is two. This rules out systems involving superselection rules, i.e., those that admit both real and classical degrees of freedom (for example, the quantum system corresponding to $M_2(\mathbb{C}) \oplus M_2(\mathbb{C})$, corresponding to a classical choice between one of two qubits, has the same information-carrying capacity as a single, four-level quantum system). More seriously, it rules out any theory that includes, e.g., real and complex, or real and quaternionic systems, as the state spaces of the bits of these theories have different dimensions. As I will discuss below, one can indeed construct mathematically-reasonable theories that embrace finite-dimensional quantum systems of all three types.
- Another shortcoming, not related to the exclusion of real and quaternionic QM, is the technical assumption (explicit in [10] for bits) that all positive affine functionals on the state space taking values between zero and one correspond to physically-accessible "effects", i.e., possible measurement results. From an operational point of view, this principle (called the "no-restriction hypothesis" in [12]) seems to call for further motivation.

Another approach. In these notes, I am going to describe an alternative approach that avoids these difficulties. This begins by associating with every physical system a convex set of states and a distinguished set of basic measurements (or experiments) that can be made on the system. We then isolate two striking features shared by classical and quantum probabilistic systems. The first is the possibility of finding a joint state that perfectly correlates a system A with an isomorphic system $\overline{A}$ (call it a conjugate system) in the sense that every basic measurement on A is perfectly correlated with the corresponding measurement on $\overline{A}$. In finite-dimensional QM, where A is represented by a finite-dimensional Hilbert space $\mathcal{H}$, $\overline{A}$, corresponds to the conjugate Hilbert space $\overline{\mathcal{H}}$, and the perfectly-correlating state is the maximally-entangled "EPR" state on $\mathcal{H} \otimes \overline{\mathcal{H}}$.

The second feature is the existence of what I call filters associated with each basic measurement. These are processes that independently attenuate the "response" of each outcome of the measurement by some specified factor. Such a process will generally not preserve the normalization of states, but up to a constant factor, in both classical and quantum theory, one can prepare any desired state by applying a suitable filter to the maximally-mixed state. Moreover, when the target state is not singular (that is, when it does not assign probability zero to any nonzero measurement outcome), one can reverse the filtering process, in the sense that it can be undone by another process with positive probability.

The upshot is that all probabilistic systems having conjugates and a sufficiently lavish supply of (probabilistically) reversible filters can be represented by formally real Jordan algebras, a class of structures that includes real, complex and quaternionic quantum systems, and just two further well-studied additional possibilities, which I will review below.

In addition to leaving room for real and quaternionic quantum mechanics (which I take to be a virtue), this approach has another advantage: it is much easier! The assumptions involved are few and easily stated, and the proof of the main technical result (Lemma 1 in Section 4) is short and straightforward. By contrast, the mathematical developments in the papers listed above are significantly more difficult and ultimately lean on the (even more difficult) classification of compact groups acting on spheres. My approach, too, leans on a received result, but one that is relatively accessible. This is the Koecher–Vinberg theorem, which characterizes formally real, or Euclidean, Jordan algebras in terms of ordered real vector spaces with homogeneous, self-dual cones. A short and non-taxing proof of this classical result can be found in [13].

These ideas were developed in [14–16] and especially [17], of which this paper is, to an extent, a summary. However, the presentation here is slightly different, and some additional ideas are also explored. In particular, I have spelled out in more detail the connection between conjugate systems and measurement records, only alluded to in the earlier paper. I also link this approach to the categorical approach to quantum theory due to Abramsky, Coecke and others [18], along the way briefly discussing recent work with Howard Barnum and Matthew Graydon [19] on the construction of probabilistic theories in which real, complex and quaternionic quantum systems coexist. Finally, Appendix B presents a uniqueness result for spectral decompositions of states, which may find further application.

A bit of background. At this point, I had better pause to explain some terms. A *Jordan algebra* is a real commutative algebra (a real vector space **E** with a commutative bilinear multiplication $a, b \mapsto a \bullet b$) having a multiplicative unit u and satisfying the Jordan identity: $a^2 \bullet (a \bullet b) = a \bullet (a^2 \bullet b)$, for all $a, b, c \in \mathbf{E}$, where $a^2 = a \bullet a$. A Jordan algebra is *formally real* if sums of squares of nonzero elements are always nonzero. The basic, and motivating, example is the space $\mathcal{L}_{\text{sa}}(\mathcal{H})$ of self-adjoint operators on a complex Hilbert space, with the Jordan product given by $a \bullet b = \frac{1}{2}(ab + ba)$. Note that here, $a \bullet a = aa$, so the notation a^2 is unambiguous. To see that $\mathcal{L}_{\text{sa}}(\mathcal{H})$ is formally real, just note that a^2 is always a positive operator.

If $\mathcal{H}$ is finite dimensional, $\mathcal{L}_{\text{sa}}(\mathcal{H})$ carries a natural inner product, namely $\langle a, b \rangle = \text{Tr}(ab)$. This plays well with the Jordan product: $\langle a \bullet b, c \rangle = \langle b, a \bullet c \rangle$ for all $a, b, c \in \mathcal{L}_{\text{sa}}(\mathcal{H})$. More generally, a finite-dimensional Jordan algebra equipped with an inner product having this property is said to be *Euclidean*. For finite-dimensional Jordan algebras, being formally real and being Euclidean are equivalent [13]. In what follows, I will abbreviate "Euclidean Jordan algebra" to EJA.

Jordan algebras were originally proposed, with what now looks like slightly thin motivation, by P. Jordan [20]: if a and b are quantum-mechanical observables, represented by $a, b \in \mathcal{L}_{\text{sa}}(\mathcal{H})$, then while $a + b$ is again self-adjoint, ab and ba are not, unless a and b commute; however, their average, $a \bullet b$, is self-adjoint and, thus, represents another observable. Almost immediately, Jordan, von Neumann and Wigner showed [21] that all formally real Jordan algebras are direct sums of simple such algebras, with the latter falling into just five classes, parametrized by positive integers n: the self-adjoint parts, $M_n(\mathbb{F})_{\text{sa}}$, of matrix algebras $M_n(\mathbb{F})$, where $\mathbb{F} = \mathbb{R}, \mathbb{C}$ or $\mathbb{H}$ (the quaternions) or, for $n = 3$, over $\mathbb{O}$ (the octonions); and also what are called spin factors V_n (closely related to Clifford algebras). There is some overlap: $V_2 \simeq M_2(\mathbb{R})$, $V_3 \simeq M_2(\mathbb{C})$ and $V_5 \simeq M_2(\mathbb{H})$. In all but one case, one can show that a simple Jordan algebra is a Jordan subalgebra of $M_n(\mathbb{C})$ for suitable n. The exceptional Jordan algebra, $M_3(\mathbb{O})_{\text{sa}}$, admits no such representation.

Besides this classification theorem, there is only one other important fact about Euclidean Jordan algebras that is needed for what follows. This is the Koecher–Vinberg (KV) theorem alluded to above. Recall that an ordered vector space is a real vector space, call it **E**, spanned by a distinguished convex cone $\mathbf{E}_+$ having its vertex at the origin. Such a cone induces a translation-invariant partial order on **E**, namely $a \leq b$ iff $b - a \in \mathbf{E}_+$. As an example, the space $\mathcal{L}_{\text{sa}}(\mathcal{H})$ is ordered by the cone of positive operators. More generally, any EJA is an ordered vector space, with positive cone $\mathbf{E}_+ := \{a^2 | a \in A\}$. This cone has two special features: first, it is *homogeneous*, i.e., for any points a, b in the interior of $\mathbf{E}_+$, there exists an automorphism of the cone (a linear isomorphism $\mathbf{E} \to \mathbf{E}$, taking $\mathbf{E}_+$ onto itself) that maps a to b. In other words, the group of automorphisms of the cone acts transitively on the cone's interior. The other special property is that $\mathbf{E}_+$ is *self-dual*. This means that **E** carries an inner product (in fact, the given one making **E** Euclidean) such that $a \in \mathbf{E}_+$ iff $\langle a, b \rangle \geq 0$ for all $b \in \mathbf{E}_+$.

An *order unit* in an ordered vector space **E** is an element $u \in \mathbf{E}_+$ such that, for all $a \in \mathbf{E}$, there exists some $n \in \mathbb{N}$ with $a \leq nu$. In finite dimensions, this is equivalent to u's belonging to the interior of the cone $\mathbf{E}_+$ [22]. In the following, by a Euclidean order unit space, I mean an ordered vector space **E** equipped with an inner product $\langle , \rangle$ with $\langle a, b \rangle \geq 0$ for all $a, b \in \mathbf{E}_+$, and a distinguished order-unit u. I will say that such a space **E** is HSD iff $\mathbf{E}_+$ is homogeneous, and also self-dual with respect to the given inner product.

Theorem 1 (Koecher 1958; Vinberg 1961). *Let **E** be a finite-dimensional euclidean order-unit space. If **E** is HSD, then there exists a unique product • with respect to which **E** (with its given inner product) is a euclidean Jordan algebra, u is the Jordan unit, and $\mathbf{E}_+$ is the cone of squares.*

It seems, then, that if we can motivate a representation of physical systems in terms of HSD order-unit spaces, we will have "reconstructed" what with a little license we might call finite-dimensional Jordan-quantum mechanics. In view of the classification theorem glossed above, this gets us into the neighborhood of orthodox QM, but still leaves open the possibility of taking real and quaternionic quantum systems seriously. (It also leaves the door open to two possibly unwanted guests, namely spin factors and the exceptional Jordan algebra. I will discuss below some constraints that at least bar the latter.)

Some notational conventions. My notation is mostly consistent with the following conventions (more standard in the mathematics than the physics literature, but in places slightly excentric relative to either). Capital Roman letters A, B, C serve as labels for systems. $M_n(\mathbb{F})$ stands for the set of $n \times n$ matrices over $\mathbb{F} = \mathbb{R}$ or $\mathbb{H}$; $M_n(\mathbb{F})_{\text{sa}}$ is the set of self-adjoint such matrices. Vectors in a Hilbert space $\mathcal{H}$ are denoted by little Roman letters x, y, z from the end of the alphabet. Operators on $\mathcal{H}$ will usually be denoted by little Roman letters $a, b, c, ...$ from the beginning of the alphabet. Roman letters t, s typically stand for real numbers. The space of all linear operators on $\mathcal{H}$ is denoted $\mathcal{L}(\mathcal{H})$; as already indicated above, $\mathcal{L}_{\text{sa}}(\mathcal{H})$ is the (real) vector space of self-adjoint operators on $\mathcal{H}$.

As above, the conjugate Hilbert space is denoted $\overline{\mathcal{H}}$. I will write $\overline{x}$ for the vectors in $\overline{\mathcal{H}}$ corresponding to $x \in \mathcal{H}$. From a certain point of view, this is the same vector; the bar serves to remind us that $\overline{cx} = \overline{c}\,\overline{x}$ for scalars $c \in \mathbb{C}$. Alternatively, one can regard $\overline{\mathcal{H}}$ as the space of "bra" vectors $\langle x|$ corresponding to the "kets" $|x\rangle$ in $\mathcal{H}$, i.e., as the dual space of $\mathcal{H}$.

The inner product of $x, y \in \mathcal{H}$ is written as $\langle x, y \rangle$ and is linear in the first argument (if you like: $\langle x, y \rangle = \langle y|x \rangle$ in Dirac notation). The inner product on $\overline{\mathcal{H}}$ is then $\langle \overline{x}, \overline{y} \rangle = \langle y, x \rangle$. The rank-one projection operator associated with a unit vector $x \in \mathcal{H}$ is p_x. Thus, $p_x(y) = \langle y, x \rangle x$. I denote functionals on $\mathcal{L}_{\text{sa}}(\mathcal{H})$ by little Greek letters, e.g., α, β..., and operators on $\mathcal{L}_{\text{sa}}(\mathcal{H})$ by capital Greek letters, e.g., Φ. Two exceptions to this scheme: a generic density operator on $\mathcal{H}$ is denoted by the capital Roman letter W, and a certain special unit vector in $\mathcal{H} \otimes \overline{\mathcal{H}}$ is denoted by the capital Greek letter Ψ. With luck, context will help keep things straight.

2. Homogeneity and Self-Duality in Quantum Theory

Why should a probabilistic physical system be represented by a Euclidean order-unit space that is either homogeneous or self-dual? One place to start hunting for an answer might be to look at standard quantum probability theory, to see if we can isolate, in operational or probabilistic terms, what makes this self-dual and homogeneous.

Correlation and self-duality. Let $\mathcal{H}$ be a finite-dimensional complex Hilbert space, representing some finite-dimensional quantum system. The system's states are represented by density operators, i.e., positive trace-one operators $W \in \mathcal{L}_{\text{sa}}(\mathcal{H})$; possible measurement-outcomes are represented by effects, i.e., positive operators $a \in \mathcal{L}_{\text{sa}}(\mathcal{H})$ with $a \leq 1$. The Born rule specifies the probability of observing effect a in state W as $\mathrm{Tr}(Wa)$. If W is a pure state, i.e., $W = p_v$ where v is a unit vector in $\mathcal{H}$, then $\mathrm{Tr}(Wa) = \langle av, v \rangle$; by the same token, if $a = p_x$, then $\mathrm{Tr}(Wa) = \langle Wx, x \rangle$.

For $a, b \in \mathcal{L}_{sa}(\mathcal{H})$, let $\langle a, b \rangle := \mathrm{Tr}(ab)$. This is an inner product. By the spectral theorem, $\mathrm{Tr}(ab) \geq 0$ for all $b \in \mathcal{L}_h(\mathcal{H})_+$ iff $\mathrm{Tr}(ap_x) \geq 0$ for all unit vectors x. However, $\mathrm{Tr}(ap_x) = \langle ax, x \rangle$. So $\mathrm{Tr}(ab) \geq 0$ for all $b \in \mathcal{L}_h(\mathcal{H})_+$ iff $a \in \mathcal{L}_h(\mathcal{H})_+$, i.e., the trace inner product is self-dualizing. However, this now leaves us with the following:

Question: *What does the trace inner product represent, oprationally or probabilistically?*

Let $\overline{\mathcal{H}}$ be the conjugate Hilbert space to $\mathcal{H}$. Suppose $\mathcal{H}$ has dimension n. Any unit vector Ψ in $\mathcal{H} \otimes \overline{\mathcal{H}}$ gives rise to a joint probability assignment to effects a on $\mathcal{H}$ and $\overline{b}$ on $\overline{\mathcal{H}}$, namely $\langle (a \otimes \overline{b}) \Psi, \Psi \rangle$. Consider the EPR state for $\mathcal{H} \otimes \overline{\mathcal{H}}$ defined by the unit vector:

$$\Psi = \frac{1}{\sqrt{n}} \sum_{x \in E} x \otimes \overline{x} \in \mathcal{H} \otimes \overline{\mathcal{H}},$$

where E is any orthonormal basis for $\mathcal{H}$. A straightforward computation shows that the joint probability of observing a and b in the state Ψ is:

$$\langle (a \otimes \overline{b}) \Psi, \Psi \rangle = \tfrac{1}{n} \mathrm{Tr}(ab).$$

In other words, the normalized trace inner product just is the joint probability function determined by the pure state vector Ψ!

As a consequence, the state represented by Ψ has a very strong correlational property: if x, y are two orthogonal unit vectors with corresponding rank-one projections p_x and p_y, we have $p_x p_y = 0$, so $\langle (p_x \otimes \overline{p_y}) \Psi, \Psi \rangle = 0$. On the other hand, $\langle (p_x \otimes \overline{p_x}) \Psi, \Psi \rangle = \tfrac{1}{n} \mathrm{Tr}(p_x) = \tfrac{1}{n}$. Hence, Ψ perfectly, and uniformly, correlates every basic measurement (orthonormal basis) of $\mathcal{H}$ with its counterpart in $\overline{\mathcal{H}}$.

Filters and homogeneity. Next, let us see why the cone $\mathcal{L}_h(\mathcal{H})_+$ is homogeneous. Recall that this means that any state in the interior of the cone (here, any non-singular density operator) can be obtained from any other by an automorphism of the cone. However, in fact, something better is true: this order-automorphism can be chosen to represent a probabilistically-reversible physical process, i.e., an invertible CP mapping with a CP inverse.

To see how this works, suppose W is a positive operator on $\mathcal{H}$. Consider the pure CP mapping $\Phi_W : \mathcal{L}_{sa}(\mathcal{H}) \to \mathcal{L}_{sa}(\mathcal{H})$ given by:

$$\Phi_W(a) = W^{1/2} a W^{1/2}.$$

Then, $\Phi_W(1) = W$. If W is nonsingular, so is $W^{1/2}$, so Φ_W is invertible, with inverse $\Phi_W^{-1} = \Phi_{W^{-1}}$, again a pure CP mapping. Now, given another nonsingular density operator M, we can get from W to M by applying $\Phi_M \circ \Phi_{W^{-1}}$.

All well and good, but we are still left with the following:

Question: *What does the mapping Φ_W represent, physically?*

To answer this, suppose W is a density operator, with spectral expansion $W = \sum_{x \in E} t_x p_x$. Here, E is an orthonormal basis for $\mathcal{H}$ diagonalizing W, and t_x is the eigenvalue of W corresponding to $x \in E$. Then, for each vector $x \in E$,

$$\Phi_W(p_x) = t_x p_x$$

where p_x is the projection operator associated with x. We can understand this to mean that Φ_W acts as a filter on the test E: the response of each outcome $x \in E$ is attenuated by a factor $0 \leq t_x \leq 1$ (my usage here is slightly non-standard, in that I allow filters that "pass" the system with a probability strictly between zero and one). Thus, if M is another density operator on $\mathcal{H}$, representing some state of the corresponding system, then the probability of obtaining outcome x after preparing the system in state M and applying the process Φ is t_x times the probability of x in state M. In detail: suppose p_x is the rank-one projection operator associated with x, and note that $W^{1/2}p_x = p_x W^{1/2} = t_x^{1/2} p_x$. Thus,

$$\mathrm{Tr}(\Phi_W(M)p_x) = \mathrm{Tr}(W^{1/2}MW^{1/2}p_x) = \mathrm{Tr}(W^{1/2}Mt_x^{1/2}p_x) = \mathrm{Tr}(t_x^{1/2}p_x W^{1/2}M)$$
$$= \mathrm{Tr}(t_x p_x M) = t_x \mathrm{Tr}(Mp_x).$$

If we think of the basis E as representing a set of alternative channels plus detectors, as in the figure below, we can add a classical filter attenuating the response of one of the detectors (say, x) by a fraction t_x. What the computation above tells us is that we can achieve the same result by applying a suitable CP map to the system's state. Moreover, this can be done independently for each outcome of E. In Figure 1, this is illustrated for a three-level quantum system: $E = \{x, y, z\}$ is an orthonormal basis, representing three possible outcomes of a Stern–Gerlach-like experiment; the filter Φ acts on the system's state in such a way that the probability of outcome x is attenuated by a factor of $t_x = 1/2$, while outcomes y and z are unaffected. Returning to the general situation, if we apply a filter Φ_W to the maximally-mixed state $\frac{1}{n}\mathbf{1}$, we obtain $\frac{1}{n}W$. Thus, we can prepare W, up to normalization, by applying the filter Φ_W to the maximally mixed state.

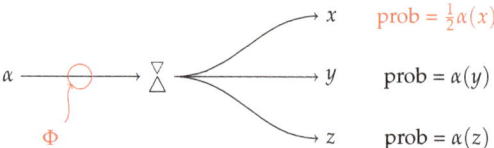

Figure 1. Φ attenuates x's sensitivity by $1/2$.

Filters are symmetric. Here is a final observation, linking these last two: the filter Φ_W is symmetric with respect to the uniformly-correlating "EPR" state Ψ, in the sense that:

$$\langle (\Phi_W(a) \otimes \overline{b})\Psi, \Psi \rangle = \langle (a \otimes \overline{\Phi}_W(b))\Psi, \Psi \rangle$$

for all effects $a, b \in \mathcal{L}_{\mathrm{sa}}(\mathcal{H})_+$. Remarkably, this is all that is needed to recover the Jordan structure of finite-dimensional quantum theory: the existence of a conjugate system, with a uniformly-correlating joint state, plus the possibility of preparing non-singular states by means of filters that are symmetric with respect to this state, and doing so reversibly when the state is nonsingular.

In a very rough outline, the argument is that states preparable (up to normalization) by symmetric filters have spectral decompositions, and the existence of spectral decompositions makes the uniformly-correlating joint state a self-dualizing inner product. However, to spell this out in a precise way, I need a general mathematical framework for discussing states, effects and processes in abstraction from quantum theory. The next section reviews the necessary apparatus.

3. General Probabilistic Theories

A characteristic feature of quantum mechanics is the existence of incompatible, or non-comeasurable, observables. This suggests the following simple, but very fruitful, notion:

Definition 1. *A test space is a collection $\mathcal{M}$ of non-empty sets $E, F, ...$, each representing the outcome-set of some measurement, experiment, or test. At the outset, one makes no special assumptions about the combinatorial structure of $\mathcal{M}$. In particular, distinct tests are permitted to overlap. Let $X := \bigcup \mathcal{M}$ denote the set of all* outcomes of all tests in $\mathcal{M}$: a probability weight on $\mathcal{M}$ is a function $\alpha : X \to [0,1]$ such that $\sum_{x \in E} \alpha(x) = 1$ for every $E \in \mathcal{M}$.

Test spaces were introduced and studied by D. J. Foulis and C. H. Randall in a long series of papers beginning around 1970. The original term for a test was an operation, which has the advantage of signaling that the concept has wider applicability than simply reading a number off a meter: anything an agent can do that leads to a well-defined, exhaustive set of mutually-exclusive outcomes defines an operation. Accordingly, test spaces were originally called "manuals of operations".

It can happen that a test space admits no probability weights at all. However, to serve as a model of a real family of experiments associated with an actual physical system, a test space should obviously carry a lavish supply of such weights. One might want to single out some of these as describing physically (or otherwise) possible states of the system. This suggests the following:

Definition 2. *A probabilistic model is a pair $A = (\mathcal{M}, \Omega)$, where $\mathcal{M}$ is a test space and Ω is some designated convex set of probability weights, called the states of the model.*

The definition is deliberately spare. Nothing prohibits us from adding further structure (a group of symmetries, say, or a topology on the space of outcomes). However, no such additional structure is needed for the results I will discuss below. I will write $\mathcal{M}(A), X(A)$ and $\Omega(A)$ for the test space, associated outcome space and state space of a model A. The convexity assumption on $\Omega(A)$ is intended to capture the possibility of forming mixtures of states. To allow the modest idealization of taking outcome-wise limits of states to be states, I will also assume that $\Omega(A)$ is closed as a subset of $[0,1]^{X(A)}$ (in its product topology). This makes $\Omega(A)$ compact and, so, guarantees the existence of pure states, that is, extreme points of $\Omega(A)$. If $\Omega(A)$ is the set of all probability weights on $\mathcal{M}(A)$, I will say that A has a full state space.

Two bits. Here is a simple, but instructive illustration of these notions. Consider a test space $\mathcal{M} = \{\{x, x'\}, \{y, y'\}\}$. Here, we have two tests, each with two outcomes. We are permitted to perform either test, but not both at once. A probability weight is determined by the values it assigns to x and to y, and since the sets $\{x, x'\}$ and $\{y, y'\}$ are disjoint, these values are independent. Thus, geometrically, the space of all probability weights is the unit square in $\mathbb{R}^2$ (Figure 2a, below). To construct a probabilistic model, we can choose any closed, convex subset of the square for Ω. For instance, we might let Ω be the convex hull of the four probability weights $\delta_x, \delta_{x'}, \delta_y$ and $\delta_{y'}$ corresponding to the midpoints of the four sides of the square, as in Figure 2b, that is,

$$\delta_x(x) = 1, \; \delta_x(x') = 0, \; \delta_x(y) = \delta_x(y') = 1/2,$$

$$\delta_{x'}(x) = 0, \; \delta_{x'}(x') = 1, \; \delta_{x'}(y) = \delta_{x'}(y') = 1/2,$$

and similarly for δ_y and $\delta_{y'}$.

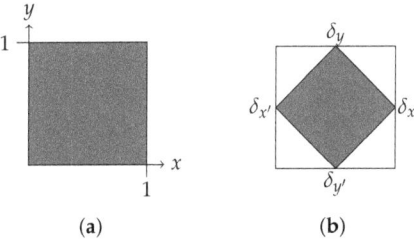

Figure 2. The state spaces of two bits. (**a**) The square bit; (**b**) The diamond bit.

The model of Figure 2a, in which we take Ω to be the entire set of probability weights on $\mathcal{M} = \{\{x, x'\}, \{y, y'\}\}$, is sometimes called the *square bit*. I will call the model of Figure 2b the *diamond bit*.

Classical, quantum and Jordan models. If E is a finite set, the corresponding *classical model* is $A(E) = (\{E\}, \Delta(E))$ where $\Delta(E)$ is the simplex of probability weights on E. If $\mathcal{H}$ is a finite-dimensional complex Hilbert space, let $\mathcal{M}(\mathcal{H})$ denote the set of orthonormal bases of $\mathcal{H}$: then $X = \bigcup \mathcal{M}(\mathcal{H})$ is the unit sphere of $\mathcal{H}$, and any density operator W on $\mathcal{H}$ defines a probability weight α_W, given by $\alpha_W(x) = \langle Wx, x \rangle$ for all $x \in X$. Letting $\Omega(\mathcal{H})$ denote the set of states of this form, we obtain the *quantum model*, $A(\mathcal{H}) = (\mathcal{M}(\mathcal{H}), \Omega(\mathcal{H}))$, associated with $\mathcal{H}$ (Gleason's theorem tells us that $A(\mathcal{H})$ has a full state space for $\dim(\mathcal{H}) > 2$, but we will not need this fact).

More generally, every Euclidean Jordan algebra $\mathbf{E}$ gives rise to a probabilistic model as follows. A minimal or primitive idempotent of $\mathbf{E}$ is an element $p \in \mathbf{E}$ with $p^2 = p$ and, for $q = q^2 < p$, $q = 0$. A Jordan frame is a maximal pairwise orthogonal set of primitive idempotents. Let $X(\mathbf{E})$ be the set of primitive idempotents; let $\mathcal{M}(\mathbf{E})$ be the set of Jordan frames; and let $\Omega(\mathbf{E})$ be the set of probability weights of the form $\alpha(p) = \langle a, p \rangle$ where $a \in \mathbf{E}_+$ with $\langle a, u \rangle = 1$. These data define the *Jordan model* $A(\mathbf{E})$ associated with $\mathbf{E}$. In the case where $\mathbf{E} = \mathcal{L}_h(\mathcal{H})$ for a finite-dimensional Hilbert space $\mathcal{H}$, this almost gives us back the quantum model $A(\mathcal{H})$: the difference is that we replace unit vectors by their associated projection operators, thus conflating outcomes that differ only by a phase.

Sharp models. Jordan models enjoy many special features that the generic probabilistic model lacks. I want to take a moment to discuss one such feature, which will be important below.

Definition 3. *A model A is unital iff, for every outcome $x \in X(A)$, there exists a state $\alpha \in \Omega(A)$ with $\alpha(x) = 1$, and sharp if this state is unique (from which it follows easily that it must be pure). If A is sharp, I will write δ_x for the unique state making $x \in X(A)$ certain.*

If A is sharp, then there is a sense in which each test $E \in \mathcal{M}(A)$ is maximally informative: if we are certain which outcome $x \in E$ will occur, then we know the system's state exactly, as there is only one state in which x has probability 1.

Classical and quantum models are obviously sharp. More generally, every Jordan model is sharp. To see this, note first that every state α on a Euclidean Jordan algebra $\mathbf{E}$ has the form $\alpha(x) = \langle a, x \rangle$ where $a \in \mathbf{E}_+$ with $\langle a, u \rangle = 1$ and where $\langle , \rangle$ is the given inner product on $\mathbf{E}$, normalized so that $\|x\| = 1$ for all primitive idempotents (equivalently, so that $\|u\| = n$, the rank of $\mathbf{E}$). The spectral theorem for EJAs [13] shows that $a = \sum_{p \in E} t_p p$ where E is a Jordan frame and the coefficients t_p are non-negative and sum to one (since $\langle a, u \rangle = 1$). If $\langle a, x \rangle = 1$, then $\sum_{p \in E} t_p \langle p, x \rangle = 1$ implies that, for every $p \in E$ with $t_p > 0$, $\langle p, x \rangle = 1$. However, $\|p\| = \|x\| = 1$, so this implies that $\langle p, x \rangle = \|p\| \|x\|$, which in turn implies that $p = x$.

In general, a probabilistic model need not even be unital, much less sharp. On the other hand, given a unital model A, it is often possible to construct a sharp model by suitably restricting the state space. This is illustrated in Figure 2b above: the full state space of the square bit is unital, but far from sharp; however, by restricting the state space to the convex hull of the barycenters of the faces, we obtain a sharp model. This is possible whenever A is unital and carries a group of symmetries acting transitively on the outcome-set $X(A)$. For details, see Appendix A. The point here is that sharpness is not, by itself, a very stringent condition: since we should expect to find highly symmetric, unital models represented abundantly "in nature", we can also expect to encounter an abundance of systems represented by sharp models.

The spaces $\mathbf{V}(A)$, $\mathbf{V}^*(A)$. Any probabilistic model gives rise to a pair of ordered vector spaces in a canonical way. These will be essential in the development below, so I am going to go into a bit of detail here.

Definition 4. *Let A be any probabilistic model. Let $\mathbf{V}(A)$ be the span of the state space $\Omega(A)$ in $\mathbb{R}^{X(A)}$, ordered by the cone $\mathbf{V}(A)_+$ consisting of non-negative multiples of states, i.e.,*

$$\mathbf{V}(A)_+ = \{t\alpha | \alpha \in \Omega(A), t \geq 0\}.$$

Call the model A finite-dimensional iff $\mathbf{V}(A)$ is finite-dimensional. From now on, I assume that all models are finite-dimensional.

Let $\mathbf{V}^*(A)$ denote the dual space of $\mathbf{V}(A)$, ordered by the dual cone of positive linear functionals, i.e., functionals f with $f(\alpha) \geq 0$ for all $\alpha \in \mathbf{V}(A)_+$. Any measurement-outcome $x \in X(A)$ yields an evaluation functional $\hat{x} \in \mathbf{V}^*(A)$, given by $\hat{x}(\alpha) = \alpha(x)$ for all $\alpha \in \mathbf{V}(A)$. More generally, an effect is a positive linear functional $f \in \mathbf{V}^*(A)$ with $0 \leq f(\alpha) \leq 1$ for every state $\alpha \in \Omega(A)$. The functionals $\hat{x}$ are effects. One can understand an arbitrary effect a to represent a mathematically possible measurement outcome, having probability $a(\alpha)$ in state α. I stress the adjective mathematically because, a priori, there is no guarantee that every effect will correspond to a physically-realizable measurement outcome. In fact, at this stage, I make no assumption at all about what, apart from the tests $E \in \mathcal{M}(A)$, is or is not physically realizable. (Later, it will follow from further assumptions that every element of $\mathbf{V}^*(A)$ represents a random variable associated with some $E \in \mathcal{M}(A)$ and is, therefore, operationally meaningful. However, this will be a theorem, not an assumption.)

The *unit effect* is the functional $u_A := \sum_{x \in E} \hat{x}$, where E is any element of $\mathcal{M}(A)$. This takes the constant value of one on $\Omega(A)$, and, thus, represents a trivial measurement outcome that occurs with probability one in every state. This is an order unit for $\mathbf{V}^*(A)$ (to see this, let $a \in \mathbf{V}(A)^*$, and let N be the maximum value of $|a(\alpha)|$ for $\alpha \in \Omega(A)$, remembering that the latter is compact: then $a \leq Nu$).

For both classical and quantum models, the ordered vector spaces $\mathbf{V}^*(A)$ and $\mathbf{V}(A)$ are naturally isomorphic. If $A(E)$ is the classical model associated with a finite set E, both are isomorphic to the space $\mathbb{R}^E$ of all real-valued functions on E, ordered pointwise. If $A = A(\mathcal{H})$ is the quantum model associated with a finite-dimensional Hilbert space $\mathcal{H}$, $\mathbf{V}(A)$ and $\mathbf{V}^*(A)$ are both naturally isomorphic to the space $\mathcal{L}_h(\mathcal{H})$ of Hermitian operators on $\mathcal{H}$, ordered by its usual cone of positive semi-definite operators. More generally, if $\mathbf{E}$ is a Euclidean Jordan algebra and $A = A(\mathbf{E})$ is the corresponding Jordan model, then $\mathbf{V}(A) \simeq \mathbf{E} \simeq \mathbf{V}^*(A)$, with $\mathbf{E}$ ordered as usual, i.e., by its cone of squares. The first of these isomorphisms is due to the definition of the model $A(\mathbf{E})$ and the second to $\mathbf{E}$'s self-duality.

The space $\mathbf{E}(A)$. It is going to be technically useful to introduce a third ordered vector space, which I will denote by $\mathbf{E}(A)$. This is the span of the evaluation-effects $\hat{x}$, associated with measurement outcomes $x \in X(A)$, in $\mathbf{V}^*(A)$, ordered by the cone:

$$\mathbf{E}(A)_+ := \left\{ \sum_i t_i \hat{x}_i \,\middle|\, t_i \geq 0 \right\}.$$

That is, $\mathbf{E}(A)_+$ is the set of linear combinations of effects $\hat{x}$ having non-negative coefficients. It is important to note that this is, in general, a proper sub-cone of $\mathbf{V}(A)^*_+$. To see this, we can revisit the example of the "diamond bit" of Figure 2b. Letting x and y be the outcomes corresponding to the right face and the top face of the larger (full) state space pictured below in Figure 3a, consider the functional $f := \hat{x} + \hat{y} - \frac{1}{2}u$. This takes positive values on the smaller state space of the diamond bit, but is negative on, for example, the state γ corresponding to the lower-left corner of the full state space (see Figure 3b). Thus, $f \in \mathbf{V}(A)_+$, but $f \notin \mathbf{E}(A)_+$.

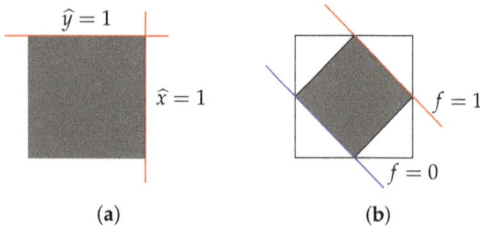

Figure 3. (a) Two outcome-effects for the square bit; (b) An effect for the diamond bit not positive on the square bit.

Since we are working in finite dimensions, the outcome-effects $\hat{x}$ span $\mathbf{V}^*(A)$. Thus, as vector spaces, $\mathbf{E}(A)$ and $\mathbf{V}^*(A)$ are the same. However, as the diamond bit illustrates, they can have quite different positive cones and, thus, need not be isomorphic as ordered vector spaces.

Processes and subnormalized states. A *subnormalized state* of a model A is an element α of $\mathbf{V}(A)_+$ with $u(\alpha) < 1$. These can be understood as states that allow a nonzero probability $1 - u(\alpha)$ of some generic "failure" event, (e.g., the destruction of the system), represented by the zero functional in $\mathbf{V}^*(A)$.

More generally, we may wish to regard two systems, represented by models A and B, as the input to and output from some process, whether dynamical or purely information-theoretic, that has some probability to destroy the system or otherwise "fail". Since such a process should preserve probabilistic mixtures, it should be represented mathematically by an affine mapping $T : \Omega(A) \to \mathbf{V}(B)_+$, taking each normalized state α of A to a possibly sub-normalized state $T(\alpha)$ of B. One can show that such a mapping extends uniquely to a positive linear mapping:

$$T : \mathbf{V}(A) \to \mathbf{V}(B),$$

so from now on, this is how I represent processes.

Even if a process T has a nonzero probability of failure, it may be possible to reverse its effect with nonzero probability.

Definition 5. *A process $T : A \to B$ is probabilistically reversible iff there exists a process S such that, for all $\alpha \in \Omega(A)$, $(S \circ T)(\alpha) = p\alpha$, where $p \in (0, 1]$.*

This means that there is a probability $1 - p$ of the composite process $S \circ T$ failing, but a probability p that it will leave the system in its initial state (note that, since $S \circ T$ is linear, p must be constant); where T preserves normalization, so that $T(\Omega(A)) \subseteq \Omega(B)$, S can also be taken to be normalization-preserving and will undo the result of T with probability one. This is the more usual meaning of "reversible" in the literature.

Given a process $T : \mathbf{V}(A) \to \mathbf{V}(B)$, there is a dual mapping $T^* : \mathbf{V}^*(B) \to \mathbf{V}^*(A)$, also positive, given by $T^*(b)(\alpha) = b(T(\alpha))$ for all $b \in \mathbf{V}^*(B)$ and $\alpha \in \mathbf{V}(A)$. The assumption that T takes normalized states to subnormalized states is equivalent to the requirement that $T^*(u_B) \leq u_A$, that is that T^* maps effects to effects.

Remark 1. *Since we are attaching no special physical interpretation to the cone $\mathbf{E}_+(A)$, we do not require a physical process $T : \mathbf{V}(A) \to \mathbf{V}(B)$ to have a dual process T^* that maps $\mathbf{E}_+(B)$ to $\mathbf{E}_+(A)$. That is, we do not require T^* to be positive as a mapping $\mathbf{E}(B) \to \mathbf{E}(A)$.*

Joint probabilities and joint states. If $\mathcal{M}_1$ and $\mathcal{M}_2$ are two test spaces, with outcome-spaces X_1 and X_2, we can construct a space of product tests (note here the savage abuse of notation: $\mathcal{M}_1 \times \mathcal{M}_2$ is not the Cartesian product of $\mathcal{M}_1$ and $\mathcal{M}_2$):

$$\mathcal{M}_1 \times \mathcal{M}_2 = \{ E \times F \mid E \in \mathcal{M}_1, F \in \mathcal{M}_2 \}$$

This models a situation in which tests from $\mathcal{M}_1$ and from $\mathcal{M}_2$ can be performed separately, and the results collated. Note that the outcome-space for $\mathcal{M}_1 \times \mathcal{M}_2$ is $X_1 \times X_2$. A joint probability weight on $\mathcal{M}_1$ and $\mathcal{M}_2$ is just a probability weight on $\mathcal{M}_1 \times \mathcal{M}_2$, that is a function $\omega : X_1 \times X_2 \rightarrow [0,1]$ such that $\sum_{(x,y) \in E \times F} \omega(x,y) = 1$ for all tests $E \in \mathcal{M}_1$ and $F \in \mathcal{M}_2$. One says that ω is non-signaling iff the marginal (or reduced) probability weights ω_1 and ω_2, given by:

$$\omega_1(x) = \sum_{y \in F} \omega(x,y) \text{ and } \omega_2(y) = \sum_{x \in E} \omega(x,y)$$

are well-defined, i.e., independent of the choice of the tests E and F, respectively. One can understand this to mean that the choice of which test to measure on $\mathcal{M}_1$ has no observable, i.e., no statistical, influence on the outcome of tests made of $\mathcal{M}_2$, and vice versa. In this case, one also has well-defined conditional probability weights:

$$\omega_{2|x}(y) := \omega(x,y)/\omega_1(x) \text{ and } \omega_{1|y} := \omega(x,y)/\omega_2(y)$$

(with, say, $\omega_{2|x} = 0$ if $\omega_1(x) = 0$, and similarly for $\omega_{1|y}$). This gives us the following bipartite version of the law of total probability [23]: for any choice: of $E \in \mathcal{M}_1$ or $F \in \mathcal{M}_2$,

$$\omega_2 = \sum_{x \in E} \omega_1(x)\omega_{2|x} \text{ and } \omega_1 = \sum_{y \in F} \omega_2(y)\omega_{1|y}. \tag{1}$$

Definition 6. *A joint state on a pair of probabilistic models A and B is a non-signaling joint probability weight ω on $\mathcal{M}(A) \times \mathcal{M}(B)$ such that, for every $x \in X(A)$ and every $y \in X(B)$, the conditional probability weights $\omega_{2|x}$ and $\omega_{1|y}$ belong to $\Omega(A)$ and $\Omega(B)$, respectively. It follows from (1) that the marginal weights ω_1 and ω_2 are also states of A and B, respectively.*

This naturally suggests that one should define, for models A and B, a composite model AB, the states of which would be precisely the joint states on A and B. If one takes $\mathcal{M}(AB) = \mathcal{M}(A) \times \mathcal{M}(B)$, this is essentially the "maximal tensor product" of A and B [24]. However, this does not coincide with the usual composite of quantum-mechanical systems. In Section 6, I will discuss composite systems in more detail. Meanwhile, for the main results of this paper, the idea of a joint state is sufficient.

For a simple example of a joint state that is neither classical, nor quantum, let B denote the "square bit" model discussed above. That is, $B = (\mathcal{B}, \Omega)$ where e $\mathcal{B} = \{\{x,x'\},\{y,y'\}\}$ is a test space with two non-overlapping, two-outcome tests, and Ω is the set of all probability weights thereon, amounting to the unit square in $\mathbb{R}^2$. The joint state on $\mathcal{B} \times \mathcal{B}$ given by Table 1 (a variant of the "non-signaling box" of Popescu and Rohrlich [25]) is clearly non-signaling. Notice that it also establishes a perfect, uniform correlation between the outcomes of any test on the first system and its counterpart on the second.

Table 1. A joint state for two square bits.

	x	x'	y	y'
x	1/2	0	1/2	0
x'	0	1/2	0	1/2
y	0	1/2	1/2	0
y'	1/2	0	0	1/2

Conditioning maps. If ω is a joint state on A and B, define the associated *conditioning maps* $\hat{\omega} : X(A) \to \mathbf{V}(B)$ and $\hat{\omega}^* : X(B) \to \mathbf{V}(A)$ by:

$$\hat{\omega}(x)(y) = \omega(x,y) = \hat{\omega}^*(y)(x)$$

for all $x \in X(A)$ and $y \in X(B)$. Note that $\hat{\omega}(x) = \omega_1(x)\omega_{2|x}$ for every $x \in X(A)$, i.e., $\hat{\omega}(x)$ can be understood as the un-normalized conditional state of B given the outcome x on A. Similarly, $\hat{\omega}^*(y)$ is the unnormalized conditional state of A given outcome y on B.

The conditioning map $\hat{\omega}$ extends uniquely to a positive linear mapping $\mathbf{E}(A) \to \mathbf{V}(B)$, which I also denote by $\hat{\omega}$, such that $\hat{\omega}(\hat{x}) = \hat{\omega}(x)$ for all outcomes $x \in X(A)$. To see this, consider the linear mapping $T : \mathbf{V}^*(A) \to \mathbb{R}^{X(B)}$ defined, for $f \in \mathbf{V}^*(A)$, by $T(f)(y) = f(\hat{\omega}^*(y))$ for all $y \in X(B)$. If $f = \hat{x}$, we have $T(\hat{x}) = \omega_1(x)\omega_{2|x} \in \mathbf{V}(B)_+$, whence, for all $y \in X(B)$, $T(\hat{x})(y) = \omega(x,y) = \hat{\omega}(x)(y)$. Since the evaluation functionals $\hat{x}$ span $\mathbf{E}(A)$, the range of T lies in $\mathbf{V}(B)$, and moreover, T is positive on the cone $\mathbf{E}(A)_+$. Hence, as advertised, T defines a positive linear mapping $\mathbf{E}(B) \to \mathbf{V}(A)$, extending $\hat{\omega}$. In the same way, $\hat{\omega}^*$ defines a positive linear mapping $\hat{\omega}^* : \mathbf{E}(B) \to \mathbf{V}(A)$.

An immediate and important corollary is that any joint state ω on A and B defines a bilinear form, which by abuse of notation I also call ω, on $\mathbf{E}(A) \times \mathbf{E}(B)$, given by $\omega(a,b) := \hat{\omega}(a)(b)$ for all $a, b \in \mathbf{E}(A)$. Note that $\omega(\hat{x}, \hat{y}) = \omega(x,y)$ for all $x \in X(A), y \in X(B)$ and also that the bilinear form ω is positive, in the sense that $\omega(a,b) \geq 0$ for all $a \in \mathbf{E}(A)_+$ and all $b \in \mathbf{E}(B)_+$.

4. Conjugates and Filters

We are now in a position to abstract the two features of QM discussed earlier. Call a test space $(X, \mathcal{M})$ *uniform* iff all tests $E \in \mathcal{M}$ have the same size, which we then call the *rank* of the test space. The test spaces associated with quantum models are uniform, and it is quite easy to generate many other examples (see Appendix A).

A uniform test space of rank n always admits at least one probability weight, namely the maximally-mixed probability weight $\rho(x) = 1/n$ for all $x \in X$. I will say that a probabilistic model A is uniform if the test space $\mathcal{M}(A)$ is uniform and the maximally-mixed state ρ belongs to $\Omega(A)$.

By an *isomorphism* $\gamma : A \to B$ from a probabilistic model A to a probabilistic model B, I mean the obvious thing: a bijection $\gamma : X(A) \to X(B)$ taking $\mathcal{M}(A)$ onto $\mathcal{M}(\overline{A})$, and such that $\beta \mapsto \beta \circ \gamma$ maps $\Omega(\overline{A})$ onto $\Omega(A)$.

Definition 7. *Let A be uniform probabilistic model with tests of size n. A conjugate for A is a model $\overline{A}$, plus a chosen isomorphism $\gamma_A : A \simeq \overline{A}$ and a joint state η_A on A and $\overline{A}$ such that for all $x, y \in X(A)$,*

(a) $\eta_A(x, \overline{x}) = 1/n$
(b) $\eta_A(x, \overline{y}) = \eta(y, \overline{x})$

where $\overline{x} := \gamma_A(x)$.

This corresponds to what is called a "weak conjugate" in [17]. Note that if $E \in \mathcal{M}(A)$, we have $\sum_{x,y \in E \times E} \eta_A(x, \overline{y}) = 1$ and $|E| = n$. Hence, $\eta_A(x, \overline{y}) = 0$ for $x, y \in E$ with $x \neq y$. Thus, η_A establishes a perfect, uniform correlation between any test $E \in \mathcal{M}(A)$ and its counterpart, $\overline{E} := \{\overline{x} | x \in E\}$, in $\mathcal{M}(\overline{A})$.

The symmetry condition (b) is pretty harmless. If η is a joint state on A and $\overline{A}$ satisfying (a), then so is $\eta^t(x, \overline{y}) := \eta(y, \overline{x})$; thus, $\frac{1}{2}(\eta + \eta^t)$ satisfies both (a) and (b). In fact, if A is sharp, (b) is automatic: if η satisfies (a), then the conditional state $(\eta_A)_{1|\overline{x}}$ assigns probability one to the outcome x. If A is sharp, this implies that $\eta_{1|\overline{x}} = \delta_x$ is uniquely defined, whence $\eta(x, \overline{y}) = n\delta_y(x)$ is also uniquely defined. In other words, for a sharp model A and a given isomorphism $\gamma : A \simeq \overline{A}$, there exists at most one joint state η satisfying (a); whence, in particular, $\eta = \eta^t$.

If $A = A(\mathcal{H})$ is the quantum-mechanical model associated with an n-dimensional Hilbert space $\mathcal{H}$, then we can take $\overline{A} = A(\overline{\mathcal{H}})$ and define $\eta_A(x,\overline{y}) = |\langle \Psi, x \otimes \overline{y}\rangle|^2$, where Ψ is the EPR state on $\mathcal{H} \otimes \overline{\mathcal{H}}$, as discussed in Section 3.

So much for conjugates. We generalize the filters associated with pure CP mappings as follows:

Definition 8. *A filter associated with a test $E \in \mathcal{M}(A)$ is a positive linear mapping $\Phi : \mathbf{V}(A) \to \mathbf{V}(A)$ such that for every outcome $x \in E$, there is some coefficient $t_x \in [0,1]$ with $\Phi(\alpha)(x) = t_x \alpha(x)$ for every state $\alpha \in \Omega(A)$.*

Equivalently, Φ is a filter iff the dual process $\Phi^* : \mathbf{V}^*(A) \to \mathbf{V}^*(A)$ satisfies $\Phi^*(\hat{x}) = t_x \hat{x}$ for each $x \in E$. Just as in the quantum-mechanical case, a filter independently attenuates the "sensitivity" of the outcomes $x \in E$. (The extreme case is one in which the coefficient t_x corresponding to a particular outcome is one, and the other coefficients are all zero. In that case, all outcomes other than x are, so to say, blocked by the filter. Conversely, given such an "all or nothing" filter Φ_x for each $x \in E$, we can construct an arbitrary filter with coefficients t_x by setting $\Phi = \sum_{x \in E} t_x \Phi_x$.)

Call a filter Φ *reversible* iff Φ is an order-automorphism of $\mathbf{V}(A)$; that is, iff it is probabilistically reversible as a process. Evidently, this requires that all the coefficients t_x be nonzero. We will eventually see that the existence of a conjugate, plus the preparability of arbitrary nonsingular states by symmetric reversible filters, will be enough to force A to be a Jordan model. Most of the work is done by the easy Lemma 1, below. First, some terminology.

Definition 9. *Suppose $\Delta = \{\delta_x | x \in X(A)\}$ is a family of states indexed by outcomes $x \in X(A)$ and such that $\delta_x(x) = 1$. Say that a state α is spectral with respect to Δ iff there exists a test $E \in \mathcal{M}(A)$ such that $\alpha = \sum_{x \in E} \alpha(x) \delta_x$. Say that the model A itself is spectral with respect to Δ if every state of A is spectral with respect to Δ.*

If A has a conjugate $\overline{A}$, then the bijection $\gamma_A : X(A) \to X(\overline{A})$ extends to an order-isomorphism $\mathbf{E}(A) \simeq \mathbf{E}(\overline{A})$. It follows that every non-signaling joint probability weight ω on A and $\overline{A}$ defines a bilinear form $a, b \mapsto \omega(a, \overline{b})$ on $\mathbf{E}(A)$.

The following is essentially proven in [17], but the presentation here is somewhat different.

Lemma 1. *Let A have a conjugate $(\overline{A}, \eta_A)$. Suppose A is spectral with respect to the states $\delta_x := \eta_{1|\overline{x}}$, $x \in X(A)$. Then:*
$$\langle a, b \rangle := n \eta_A(a, \overline{b}),$$
where n is the rank of A, defines a self-dualizing inner product on $\mathbf{E}(A)$, with respect to which $\mathbf{V}(A)_+ \simeq \mathbf{E}(A)_+$. Moreover, A is sharp, and $\mathbf{E}(A)_+ = \mathbf{V}^*(A)_+$.

Proof. That $\langle \, , \, \rangle$ is symmetric and bilinear follows from η_A's being symmetric and non-signaling. Note that $\langle \hat{x}, \hat{x} \rangle = 1$ for every $x \in X(A)$ and $\langle \hat{x}, \hat{y} \rangle = 0$ for any distinct $x, y \in X(A)$ lying in a common test. We need to show that $\langle \, , \, \rangle$ is positive-definite. Since $\hat{\overline{A}} \simeq A$ and the latter is spectral, so is the former. It follows that $\hat{\eta}$ takes $\mathbf{E}(A)_+$ onto $\mathbf{V}(\overline{A})_+$ and, hence, is an order-isomorphism. From this, it follows that every $a \in \mathbf{E}(A)_+$ has a "spectral" decomposition of the form $\sum_{x \in E} t_x x$ for some coefficients $t_x \geq 0$ and some test $E \in \mathcal{M}(A)$. In fact, any $a \in \mathbf{E}(A)$, positive or otherwise, has such a decomposition (albeit with possibly negative coefficients). If $a \in \mathbf{E}(A)$ is arbitrary, with $a = a_1 - a_2$ for some $a_1, a_2 \in \mathbf{E}(A)_+$, we can find $N \geq 0$ with $a_2 \leq Nu$. Thus, $b := a + Nu = a_1 + (Nu - a_2) \geq 0$, and so, $b := \sum_{x \in E} t_x x$ for some $E \in \mathbf{A}$, and hence, $a = b - Nu = \sum_{x \in E} t_x x - N(\sum_{x \in E} x) = \sum_{x \in E} (t_x - N) x$.

Now, let $a \in \mathbf{E}(A)$. Decomposing $a = \sum_{x \in E} t_x x$ for some test E and some coefficients t_x, we have:
$$\langle a, a \rangle = \sum_{x, y \in E \times E} t_x t_y \langle \hat{x}, \hat{y} \rangle = \sum_{x \in E} t_x^2 \geq 0.$$

This is zero only where all coefficients t_x are zero, i.e., only for $a = 0$. Therefore, $\langle \, , \, \rangle$ is an inner product, as claimed.

We need to show that $\langle \, , \, \rangle$ is self-dualizing. Clearly $\langle a, b \rangle = n\eta_A(a, \overline{b}) \geq 0$ for all $a, b \in \mathbf{E}(A)_+$. Suppose $a \in \mathbf{E}(A)$ is such that $\langle a, b \rangle \geq 0$ for all $b \in \mathbf{E}(A)_+$. Then, $\langle a, \widehat{y} \rangle \geq 0$ for all $y \in X$. Now, $a = \sum_{x \in E} t_x \widehat{x}$ for some test E; thus, for all $y \in E$, we have $\langle a, \widehat{y} \rangle = t_y \geq 0$, whence, $a \in \mathbf{E}(A)_+$.

Next, we want to show that $\mathbf{E}(A)_+ = \mathbf{V}(A)^*_+$. Since $\widehat{\eta} : \mathbf{E}(A) \to \mathbf{V}(\overline{A})$ is an order-isomorphism, for every $\alpha \in \mathbf{V}(A)$, there exists a unique $a \in \mathbf{E}(A)$ with $\widehat{\eta}(a) = \frac{1}{n}\overline{\alpha}$. In particular,

$$\langle a, x \rangle = n\eta_A(a, \overline{x}) = \overline{\alpha}(\overline{x}) = \alpha(x).$$

It follows that if $b \in \mathbf{E}(A) = \mathbf{V}^*(A)$,

$$b(\alpha) = \overline{b}(\overline{\alpha}) = \overline{b}n\widehat{\eta}_A(a) = n\eta(a, \overline{b}) = \langle a, b \rangle.$$

Since every $a \in \mathbf{E}(A)_+$ has the form $a = \widehat{\eta}^{-1}(\frac{1}{n}\overline{\alpha})$ for some $\alpha \in \mathbf{V}(A)_+$, if $b \in \mathbf{V}^*(A)_+$, we have $\langle a, b \rangle \geq 0$ for all $a \in \mathbf{E}(A)_+$, whence, by the self-duality of the latter cone, $b \in \mathbf{E}(A)_+$. Thus, $\mathbf{V}^*(A) = \mathbf{E}(A)_+$.

Finally, let us see that A is sharp. If $\alpha \in \Omega(A)$, let a be the unique element of $\mathbf{E}(A)_+$ with $\langle a, x \rangle = \alpha(x)$. In particular, $\langle a, u \rangle = 1$. If a has spectral decomposition $a = \sum_{x \in E} t_x \widehat{x}$, where $E \in \mathcal{M}(A)$, then for all $x \in E$, $\langle a, x \rangle = t_x$; hence, $\sum_{x \in E} t_x = \sum_{x \in E} \langle a, x \rangle = \langle a, u \rangle = 1$. Thus, $\|a\|^2 = \sum_{x \in E} t_x^2 \leq 1$, whence, $\|a\| \leq 1$. Now, suppose $\alpha(x) = 1$ for some $x \in X(A)$: then, $1 = \langle a, x \rangle \leq \|a\| \|x\|$; as $\|x\| = 1$, we have $\|a\| = 1$. However, now $\langle a, \widehat{x} \rangle = \|a\| \|\widehat{x}\|$, whence, $a = \widehat{x}$. Hence, there is only one weight α with $\alpha(x) = 1$, namely, $\alpha = \langle x, \cdot \rangle$, so A is sharp. □

If A is sharp, then we say that A is *spectral* iff it is spectral with respect to the pure states δ_x defined by $\delta_x(x) = 1$. If A is sharp and has a conjugate $\overline{A}$, then, as noted earlier, the state $\eta_{1|\overline{x}}$ is exactly δ_x, so the spectrality assumption in Lemma 1 is fulfilled if we simply say that A is spectral. Hence, a sharp, spectral model with a conjugate is self-dual.

For the simplest systems, this is already enough to secure the desired representation in terms of a Euclidean Jordan algebra.

Definition 10. *Call A a bit iff it has rank two (that is, all tests have two outcomes) and if every state $\alpha \in \Omega(A)$ can be expressed as a mixture of two sharply distinguishable states; that is, $\alpha = t\delta_x + (1-t)\delta_y$ for some $t \in [0, 1]$ and states δ_x and δ_y with $\delta_x(x) = 1$ and $\delta_y(y) = 1$ for some test $\{x, y\}$.*

Corollary 1. *If A is a sharp bit, then $\Omega(A)$ is a ball of some finite dimension d.*

The proof is given in Appendix C. If d is $2, 3$ or 5, we have a real, complex or quaternionic bit. For $d = 4$ or $d \geq 6$, we have a non-quantum spin factor.

For systems of higher rank (higher "information capacity"), we need to assume a bit more. Suppose A satisfies the hypotheses of Lemma 1. Appealing to the Koecher–Vinberg theorem, we see that if $\mathbf{V}(A)$ and, hence, $\mathbf{V}^*(A)$ are also homogeneous, then $\mathbf{V}^*(A)$ carries a canonical Jordan structure. In fact, we can say something a little stronger.

Theorem 2. *Let A be spectral with respect to a conjugate system $\overline{A}$. If $\mathbf{V}(A)$ is homogeneous, then there exists a canonical Jordan product on $\mathbf{E}(A)$ with respect to which u_A is the Jordan unit. Moreover, with respect to this product, $X(A)$ is exactly the set of primitive idempotents, and $\mathcal{M}(A)$ is exactly the set of Jordan frames.*

The first part is almost immediate from the Koecher–Vinberg theorem, together with Lemma 1. The KV theorem gives us an isomorphism between the ordered vector spaces $\mathbf{V}(A)$ and $\mathbf{E}(A)$, so if one is homogeneous, so is the other. Since $\mathbf{E}(A)$ is also self-dual by Lemma 1, the KV theorem yields the requisite unique Euclidean Jordan structure having u as the Jordan unit. One can then show without much trouble that every outcome $x \in X(A)$ is a primitive idempotent of $\mathbf{E}(A)$ with respect to this Jordan structure and that every test is a Jordan frame. The remaining claims (that every minimal idempotent belongs to $X(A)$ and every Jordan frame, to $\mathcal{M}(A)$) take a little bit more work. I will not reproduce the proof here; the details (which are not especially difficult, but depend on some facts concerning Euclidean Jordan algebras) can be found in [17].

The homogeneity of $\mathbf{V}(A)$ can be understood as a preparability assumption: it is equivalent to saying that every state in the interior of $\Omega(A)$ can be obtained, up to normalization, from the maximally-mixed state by a reversible process. That is, if $\alpha \in \Omega(A)$, there is some such process ϕ such that $\phi(\rho) = p\alpha$ where $0 < p \leq 1$. One can think of the coefficient p as the probability that the process ϕ will yield a nonzero result (more dramatically: will not destroy the system). Thus, if we prepare an ensemble of identical copies of the system in the maximally-mixed state ρ and subject them all to the process ϕ, the fraction that survives will be about p, and these will all be in state α.

In fact, if the hypotheses of Lemma 1 hold, the homogeneity of $\mathbf{E}(A)$ follows directly from the mere existence of reversible filters with arbitrary non-zero coefficients. To see this, suppose $a \in \mathbf{E}(A)_+$ has a spectral decomposition $\sum_{x \in E} t_x \hat{x}$ for some $E \in \mathcal{M}(A)$, with $t_x > 0$ for all x when a belongs to the interior of $\mathbf{E}(A)_+$. Now, if we can find a reversible filter for E with $\Phi(x) = t_x \hat{x}$ for all $x \in E$, then applying this to the order-unit $u = \sum_{x \in E} \hat{x}$ yields a. Thus, $\mathbf{V}^*(A)$ is homogeneous.

Two paths to spectrality. Some axiomatic treatments of quantum theory have taken one or another form of spectrality as an axiom [6,26]. If one is content to do this, then Lemma 1 above provides a very direct route to the Jordan structure of quantum theory. However, spectrality can actually be derived from assumptions that, on their face, seem a good deal weaker, or anyway more transparent (a different path to spectrality is charted in a recent paper [27] by G. Chiribella and C. M. Scandolo).

I will call a joint state on models A and B *correlating* iff it sets up a perfect correlation between some pair of tests $E \in \mathcal{M}(A)$ and $F \in \mathcal{M}(B)$. More exactly:

Definition 11. *A joint state ω on probabilistic models A and B correlates a test $E \in \mathcal{M}(A)$ with a test $F \in \mathcal{M}(B)$ iff there exist subsets $E_0 \subseteq E$ and $F_0 \subseteq F$, and a bijection $f : E_0 \to F_0$ such that $\omega(x, y) = 0$ for $(x, y) \in E \times F$ unless $y = f(x)$. In this case, say that ω correlates E with F along f. A joint state on A and B is correlating iff it correlates some pair of tests $E \in \mathcal{M}(A), F \in \mathcal{M}(B)$.*

Note that ω correlates E with F along f iff $\omega(x, f(x)) = \omega_1(x) = \omega_2(f(x))$, which, in turn, is equivalent to saying that $\omega_{2|x}(f(x)) = 1$ for $\omega_1(x) \neq 0$.

Lemma 2. *Suppose A is sharp and that every state α of A arises as the marginal of a correlating joint state between A and some model B. Then, A is spectral.*

Proof. Suppose $\alpha = \omega_1$, where ω is a joint state correlating a test $E \in \mathcal{M}(A)$ with a test $F \in \mathcal{M}(B)$, say along a bijection $f : E_0 \to F_0$, where $E_0 \subseteq E$ and $F_0 \subseteq F$. Then, for any $x \in E$ with $\alpha(x) \neq 0$, $\omega_{1|f(x)}(x) = 1$, whence, as A is sharp, $\omega_{1|f(x)} = \delta_x$, the unique state making x certain. It follows from the law of total probability that $\alpha = \sum_{x \in E} \alpha(x) \delta_x$. □

In principle, the model B can vary with the state α. Lemma 2 suggests the following language:

Definition 12. *A model A satisfies the correlation condition iff every state $\alpha \in \Omega(A)$ is the marginal of some correlating joint state of A and some model B.*

This has something of the same flavor as the *purification postulate* of [8], which requires that all states of a given system arise as marginals of a pure state on a larger, composite system, unique up to symmetries on the purifying system. However, note that we do not require the correlating joint state to be either pure (which, in classical probability theory, it will not be) or unique.

If A is sharp and satisfies the correlation condition, then every state of A is spectral. If, in addition, A has a conjugate, then for every $x \in X(A)$, we have $\eta_{1|\bar{x}} = \delta_x$. In this case, A is spectral with respect to the family of states $\eta_{1|\bar{x}}$, and the hypotheses of Lemma 1 are satisfied.

Here is another, superficially quite different, way of arriving at spectrality. Suppose A has a conjugate, $\overline{A}$. Call a transformation Φ symmetric with respect to η_A iff, for all $x, y \in X(A)$,

$$\eta_A(\Phi^* x, \bar{y}) = \eta_A(x, \overline{\Phi}^* y).$$

Say that a state α is preparable by a filter Φ iff $\alpha = \Phi(\rho)$, where ρ is the maximally-mixed state.

Lemma 3. *Let A have a conjugate, $\overline{A}$, and suppose every state of A is preparable by a symmetric filter. Then, A is spectral.*

Proof. Let $\alpha = \Phi(\rho)$ where Φ is a filter on a test $E \in \mathcal{M}(A)$, say $\Phi(x) = t_x x$ for all $x \in E$. Then:

$$\alpha = \Phi(\hat{\eta}^*(\bar{u})) = \eta(\Phi^*(\cdot), \bar{u}) = \eta(\,\cdot\,, \overline{\Phi}^*(\bar{u})) = \sum_{x \in E} \eta(\,\cdot\,, t_x \bar{x}) = \sum_{x \in E} t_x \tfrac{1}{n} \delta_x.$$

□

Thus, the hypotheses of either Corollary 2 or Lemma 3 will supply the needed spectral assumption that makes Lemma 1 work (in fact, it is not hard to see that these hypotheses are actually equivalent, an exercise I leave for the reader).

To obtain a Jordan model, we still need homogeneity. This is obviously implied by the preparability condition in Lemma 3, provided the preparing filters Φ can be taken to be reversible whenever the state to be prepared is non-singular. On the other hand, as noted above, in the presence of spectrality, it is enough to have arbitrary reversible filters, as these allow one to prepare the spectral decompositions of arbitrary non-singular states. Thus, conditions (a) and (b) below both imply that A is a Jordan model. Conversely, one can show that any Jordan model satisfies both (a) and (b), closing the loop [17]:

Theorem 3. *The following are equivalent:*

(a) *A has a conjugate, and every non-singular state can be prepared by a reversible symmetric filter;*
(b) *A is sharp, has a conjugate, satisfies the correlation condition and has arbitrary reversible filters;*
(c) *A is a Jordan model.*

5. Measurement, Memory and Correlation

Of the spectrality-underwriting conditions given in Lemmas 2 and 3, the one that seems less transparent (to me, anyway) is the correlation condition, i.e., that every state arises as the marginal of a correlating bipartite state. While surely less ad hoc than spectrality, this still calls for further explanation. Suppose we hope to implement a measurement of a test $E \in \mathcal{M}(A)$ dynamically. This would involve bringing up an ancilla system B (also uniform, suppose; and which we can suppose, by suitable coarse-graining, if necessary, to have tests of the same cardinality as A's) in some "ready" state β_o. We would then subject the combined system AB to some physical process, at the end of which, AB is in some final joint state ω, and B is (somehow!) in one of a set of record states, β_x, each corresponding to an outcome $x \in X(A)$. (This way of putting things takes us close to the usual formulation of the quantum-mechanical "measurement problem", which I certainly do not propose to discuss here. The point is only that, if any dynamical process, describable within the theory, can account for measurement results, it should be consistent with this description.)

We would like to insist that:

(a) The states β_x are distinguishable, or readable, by some test $F \in \mathcal{M}(B)$. This means that for each $x \in E$, there is a unique $y \in F$ such that $\beta_x(y) = 1$. Note that this sets up an injection $f : E \to F$.

(b) The record states must be accurate, in the sense that if we were to measure E on A, and secure $x \in E$, the record state β_x should coincide with the conditional state $\omega_{2|x}$ (if this is not the case, then a measurement of A cannot correctly calibrate the system B as a measuring device for E).

It follows from (a) and (b) that, for $x \in E$ and $y \neq f(x) \in F$,

$$\omega(x,y) = \omega_1(x)\omega_{2|x}(y) = \omega_1(x)\beta_x(y) = 0.$$

In other words, ω must correlate E with F, along the bijection $f : E \to F_o \subseteq F$. If the measurement process leaves α undisturbed, in the sense that $\omega_1 = \alpha$, then α dilates to a correlating state. This suggests the following *non-disturbance principle*: every state can be measured, by some test $E \in \mathcal{M}(A)$, without disturbance. Lemma 2 then tells us that if A is sharp and satisfies the non-disturbance principle, every state of A is spectral.

Here is a slightly different, but possibly more compelling, version of this story. Suppose we can perform a test E on A directly (setting aside, that is, any issue of whether or not this can be achieved through some dynamical process): this will result in an outcome x occurring. To do anything with this, we need to record its having occurred. This means we need a storage medium, B and a family of states β_x, one for each $x \in E$, such that if, on performing the test E, we obtain x, then B will be in state β_x. Moreover, these record states need to be readable at a later time, i.e., distinguishable by a later measurement on B. To arrange this, we need A and B to be in a joint state, associated with a joint probability weight ω, such that $\omega_1 = \alpha$ (because we want to have prepared A in the state α) and $\beta_x = \omega_{2|x}$ for every $x \in E$. We then measure E on A; upon our obtaining outcome $x \in E$, B is in the state β_x. Since the ensemble of states β_x is readable by some $F \in \mathcal{M}(B)$ with $|F| \geq |E|$, we have correlation, and α must also be spectral.

Of course, these desiderata cannot always be satisfied. What is true, in QM, is that for every choice of state α, there will exist some test that is recordable in that state, in the foregoing sense. If we promote this to the general principle, we again see that every state is the marginal of a correlating state, and hence spectral, if A is sharp.

6. Composites and Categories

Thus far, we have been referring to the correlator η_A as a joint state, but dodging the question: *state of what?* Mathematically, nothing much hangs on this question: it is sufficient to regard η_A as a bipartite probability assignment on A and $\overline{A}$. However, it would surely be more satisfactory to be able to treat it as an actual physical state of some composite system $A\overline{A}$. How should this be chosen? As mentioned above, one possibility is to take $A\overline{A}$ to be the maximal tensor product of the models A and $\overline{A}$ [24]. By definition, this has for its states all non-signaling probability assignments with conditional states belonging to A and $\overline{A}$. However, we might want composite systems, in particular $A\overline{A}$, to satisfy the same conditions we are imposing on A and $\overline{A}$, i.e., to be a Jordan model. If so, we need to work somewhat harder: the maximal tensor product will be self-dual only if A is classical.

In order to be more precise about all this, the first step is to decide what ought to count as a composite of two probabilistic models. If we mean to capture the idea of two physical systems that can be acted upon separately, but which cannot influence one another in any observable way (e.g., two spacelike-separated systems), the following seems to capture the minimal requirements:

Definition 13. *A non-signaling composite of models A and B is a model AB, together with a mapping $\pi : X(A) \times X(B) \to \mathbf{V}^*(AB)_+$ such that:*

$$\sum_{x \in E, y \in F} \pi(x,y) = u_{AB}$$

and, for $\omega \in \Omega(AB)$, $\omega \circ \pi$ is a joint state on A and B, as defined in Section 2.

The idea here, expressed in Alice-and-Bob language (Alice controlling system A, Bob controlling system B), is that $\pi(x,y)$ is an effect of the composite system AB, corresponding to x being observed by Alice and y, by Bob. In many cases, $\pi(x,y)$ will actually be an outcome in $X(AB)$. Indeed, we usually have $\pi : X(A) \times X(B) \to X(AB)$ injective, and for $E \in \mathcal{M}(A), F \in \mathcal{M}(B)$, $\pi(E \times F) = \{\pi(x,y)|x \in E, y \in F\}$ a test in $\mathcal{M}(AB)$. The rank of AB will then be the product of the ranks of A and B. Accordingly, let us call a non-signaling composite with these these properties *multiplicative*. Composites in real and complex quantum mechanics are multiplicative; in quaternionic quantum mechanics, with the most plausible definition of tensor product, they are not [28].

Therefore, the question becomes: can one construct, for Jordan models A and B, a non-signaling composite AB that is also a Jordan model? At present, and in this generality, this question seems to be open, but some progress is made in [28]: if neither A, nor B contain the exceptional Jordan algebra as a summand, such a composite can indeed be constructed, and in multiple ways. Moreover, under a considerably more restrictive definition of "Jordan composite", no Jordan composite AB can exist if either factor has an exceptional summand.

Categories of Self-Dual Probabilistic Models. It is natural to interpret a physical theory as a category, in which objects represent physical systems and morphisms represent physical processes having these systems (or their states) as inputs and outputs. In order to discuss composite systems, this should be a symmetric monoidal category. That is, for every pair of objects A, B, there should be an object $A \otimes B$, and for every pair of morphisms $f : A \to A'$ and $g : B \to B'$, there should be a morphism $f \otimes g : A \otimes B \to A' \otimes B'$, representing the two processes f and g occurring "in parallel". One requires that $\otimes$ be associative and commutative, and have a unit object I, in the sense that there exist canonical isomorphisms $\alpha_{A,B;C} : A \otimes (B \otimes C) \simeq (A \otimes B) \otimes C$, $\sigma_{A,B} : A \otimes B \simeq B \otimes A$, $\lambda_A : I \otimes A \simeq A$ and $\rho_A : A \otimes I \to A/$ These must satisfy various "naturality conditions", guaranteeing that they interact correctly; see [29] for details. One also requires that $\otimes$ be bifunctorial, meaning that $\text{id}_A \otimes \text{id}_B = \text{id}_{A \otimes B}$, and if $f : A \to A', f' : A' \to A'', g : B \to B'$ and $g' : B' \to B''$, then:

$$(f' \otimes g') \circ (f \otimes g) = (f' \circ f) \otimes (g' \circ g).$$

By a *probabilistic theory*, I mean a category of probabilistic models and processes; that is, objects of $\mathcal{C}$ are models, and a morphism $A \to B$, where $A, B \in \mathcal{C}$, is a process $\mathbf{V}(A) \to \mathbf{V}(B)$. A *monoidal probabilistic theory* is such a category, $\mathcal{C}$, carrying a symmetric monoidal structure $A, B \mapsto AB$, where AB is a non-signaling composite in the sense of the definition above. I also assume that the monoidal unit, I, is the trivial Model 1 with $\mathbf{V}(1) = \mathbb{R}$, and that, for all $A \in \mathcal{C}$,

(a) $\alpha \in \Omega(A)$ iff the mapping $\alpha : \mathbb{R} \to \mathbf{V}(A)$ given by $\alpha(1) = \alpha$ belongs to $\mathcal{C}(I, A)$;
(b) The evaluation functional $\hat{x}$ belongs to $\mathcal{C}(A, I)$ for all outcomes $x \in X(A)$.

Call $\mathcal{C}$ *locally tomographic* iff AB is a locally tomographic composite for all $A, B \in \mathcal{C}$. Much of the qualitative content of (finite-dimensional) quantum information theory can be formulated in purely categorical terms [11,18,30]. In particular, in the work of Abramsky and Coecke [18], it is shown that a range of quantum phenomena, notably gate teleportation, is available in any *dagger-compact* category. For a review of this notion, as well as a proof of the following result, see Appendix D:

Theorem 4. *Let $\mathcal{C}$ be a locally-tomographic monoidal probabilistic theory, in which every object $A \in \mathcal{C}$ is sharp, spectral and has a conjugate $\overline{A} \in \mathcal{C}$, with $\eta_A \in \Omega(A\overline{A})$. Assume also that, for all $A, B \in \mathcal{C}$,*

(i) $\overline{\overline{A}} = A$, *with* $\eta_{\overline{A}}(\overline{a}, b) = \eta_A(a, \overline{b})$;
(ii) *If* $\phi \in \mathcal{C}(A, B)$, *then* $\overline{\phi} \in \mathcal{C}(\overline{A}, \overline{B})$.

Then, $\mathcal{C}$ has a canonical dagger-compact structure, in which $\overline{A}$ is the dual of A with $\eta_A : \mathbb{R} \to \mathbf{V}(A\overline{A})$ as the co-unit.

Jordan composites. The local tomography assumption in Theorem 4 is a strong constraint. As is well known, the standard composite of two real quantum systems is not locally tomographic, yet the category of finite-dimensional real mixed-state quantum systems is certainly dagger-compact and satisfies the other assumptions of Theorem 4, so local tomography is definitely not a necessary condition for dagger-compactness.

This raises some questions. One is whether local tomography can simply be dropped in the statement of Theorem 4. At any rate, at present, I do not know of any non-dagger-compact monoidal probabilistic theory satisfying the other assumptions.

Another question is whether there exist examples other than real QM of non-locally-tomographic, but still dagger-compact, monoidal probabilistic theories satisfying the assumptions of Theorem 2. The answer to this is yes. Without going into detail, the main result of [28] is that one can construct a dagger-compact category in which the objects are Hermitian parts of finite-dimensional real, complex and quaternionic matrix algebras, that is the Euclidean Jordan algebras corresponding to finite-dimensional real, complex or quaternionic quantum-mechanical systems, and morphisms are certain completely positive mappings between enveloping complex ∗-algebras for these Jordan algebras. The monoidal structure gives *almost* the expected results: the composite of two real quantum systems is the real system corresponding to the usual (real) quantum-mechanical composite of the two components (and, in particular, is not locally tomographic). The composite of two quaternionic systems is a real system (see [11] for an account of why this is just what one wants). The composite of a real and a complex, or a quaternionic and a complex, system is again complex. The one surprise is that the composite of two standard complex quantum systems, in this category, is not the usual thing, but rather, comes with an extra superselection rule. This functions to make time-reversal a legitimate physical operation on complex systems, as it is for real and quaternionic systems. This is part of the price one pays for the dagger-compactness of this category.

7. Conclusions

As promised, we have here an easy derivation of something close to orthodox, finite-dimensional QM, from operationally or probabilistically transparent assumptions. As discussed earlier, this approach offers, in addition to its relative simplicity, greater latitude than the locally-tomographic axiomatic reconstructions of [7–10], putting us in the slightly less constrained realm of formally real Jordan algebras. This allows for real and quaternionic quantum systems, superselection rules and even theories, such as the ones discussed in Section 6, in which real, complex and quaternionic quantum systems coexist and interact.

There remains some mystery as to the proper interpretation of the conjugate system $\overline{A}$. Operationally, the situation is clear enough: if we understand A as controlled by Alice and $\overline{A}$, by Bob, then if Alice and Bob share the state η_A, then they will always obtain the same result, as long as they perform the same test. However, what does it mean physically that this should be possible (in a situation in which Alice and Bob are still able to choose their tests independently)? In fact, there is little consensus (that I can find, anyway) among physicists as to the proper interpretation of the conjugate of the Hilbert space representing a given quantum-mechanical system. One popular idea is that the conjugate is a time-reversed version of the given system; but why, then, should we expect to find a state that perfectly correlates the two? At any rate, finding a clear physical interpretation of conjugate systems, even (or especially!) in orthodox quantum mechanics, seems to me an urgently important problem.

I would like to close with another problem, this one of mainly mathematical interest. The hypotheses of Theorem 2 yield a good deal more structure than just a homogeneous, self-dual cone. In particular, we have a distinguished set $\mathcal{M}(A)$ of orthonormal observables in $\mathbf{V}^*(A)$, with respect to which every effect has a spectral decomposition. Moreover, with a bit of work, one can show that this decomposition is essentially unique. More exactly, if $a = \sum_i t_i p_i$ where the coefficients t_i are all distinct and the effects $p_1, ..., p_k$ are associated with a coarse-graining of a test $E \in \mathcal{M}(A)$, then both the coefficients and the effects are uniquely determined. The details are in Appendix B. Using this, we have a functional calculus on $\mathbf{V}^*(A)$, i.e., for any real-valued function f of a real variable and any effect a with spectral decomposition $\sum_i t_i p_i$ as above, we can define $f(a) = \sum_i f(t_i) p_i$. This gives us a unique candidate for the Jordan product of effects a and b, namely,

$$a \bullet b = \tfrac{1}{2}((a+b)^2 - a^2 - b^2)).$$

We know from Theorem 2 (and thus, ultimately, from the KV theorem) that this is bilinear. The challenge is to show this without appealing to the KV theorem (the fact that the state spaces of "bits" are always balls, as shown in Appendix C, is perhaps relevant here).

Acknowledgments: This paper is partly based on talks given in workshops and seminars in Amsterdam, Oxford, in 2014 and 2015, and was largely written while the author was a guest of the Quantum Group at the Oxford Computing Laboratory, supported by a grant (FQXi-RFP3-1348) from the FQXifoundation. I would like to thank Sonja Smets (in Amsterdam) and Bob Coecke (in Oxford) for their hospitality on these occasions. I also wish to thank Carlo Maria Scandolo for his careful reading of, and useful comments on, two earlier drafts of this paper.

Conflicts of Interest: The author declares no conflict of interest.

Appendix A. Models with Symmetry

Recall that a probabilistic model A is sharp iff, for every measurement outcome $x \in X(A)$, there exists a unique state $\delta_x \in \Omega(A)$ with $\delta_x(x) = 1$. While this is clearly a very strong condition, it is not an unreasonable one. In fact, given the test space $\mathcal{M}(A)$, we can often choose the state space $\Omega(A)$ in such a way as to guarantee that A is sharp. In particular, this is the case when $\mathcal{M}(A)$ enjoys enough symmetry.

Definition A1. *Let G be a group. A G-test space is a test space $(X, \mathcal{M})$ where X is a G-space, that is, where X comes equipped with a preferred G-action $G \times X \to X$, $(g, x) \mapsto gx$, such that $gE \in \mathcal{M}$ for all $E \in \mathcal{M}$. A G-model is a probabilistic model A such that (i) $\mathcal{M}(A)$ is a G-test space and (ii) $\Omega(A)$ is invariant under the action of G on probability weights given by $\alpha \mapsto g\alpha := \alpha \circ g^{-1}$ for $g \in G$.*

Lemma A1. *Let A be a finite-dimensional G-model, and suppose G acts transitively on the outcome space $X(A)$. Suppose also that A is unital, i.e., for every $x \in X(A)$, there exists at least one state α with $\alpha(x) = 1$. Then, there exists a G-invariant convex subset $\Delta \subseteq \Omega(A)$ such that $A' = (\mathcal{M}(A), \Delta)$ is a sharp G-model.*

Proof. For each $x \in X(A)$, let F_x denote the face of $\Omega(A)$ consisting of states α with $\alpha(x) = 1$. Let β_x be the barycenter of F_x. It is easy to check that $F_{gx} = gF_x$ for every $g \in G$. Thus, $g\beta_x = \beta_{gx}$, i.e., the set of barycenters β_x is an orbit. Let Δ be the convex hull of these barycenters. Then, Δ is invariant under G. If $\alpha \in \Delta$ with $\alpha(x) = 1$, then $\alpha \in F_x \cap \Delta = \{\beta_x\}$, so $(\mathcal{M}(A), \Delta)$ is sharp. □

Appendix B. Uniqueness of Spectral Decompositions

Let A be a model satisfying the conditions of Lemma 1. In particular, every $a \in \mathbf{E}(A) = \mathbf{V}^*(A)$ has a spectral representation $a = \sum_{x \in E} t_x \hat{x}$ for some test $E \in \mathcal{M}(A)$. In general, this expansion is highly non-unique. For instance, the unit u_A can be expanded as $\sum_{x \in E} \hat{x}$ for any test $E \in \mathcal{M}(A)$. The aim in this Appendix is to obtain a form of spectral expansion for effects that is unique.

Call a subset of a test an *event*. That is, $D \subseteq X(A)$ is an event iff there exists a test $E \in \mathcal{M}(A)$ with $D \subseteq E$. The probability of an effect D in a state α is $\alpha(D) = \sum_{x \in E} \alpha(x)$. Thus, any event gives rise to an effect, $\hat{D}$, given by $\hat{D}(\alpha) = \alpha(D)$. Evidently,

$$\hat{D} := \sum_{x \in E} \hat{x}.$$

A test is a maximal event, and for any test $E \in \mathcal{M}(A)$, $\hat{D} = u$.

Definition A2. *An effect $p \in \mathbf{V}^*(A)$ is sharp iff it has the form $p = \hat{D}$ for some event D. A set of sharp effects $p_1, ..., p_n \in \mathbf{V}^*(A)$ is jointly orthogonal with respect to $\mathcal{M}(A)$ iff there exists a test $E \in \mathcal{M}(A)$ and pairwise disjoint events $D_1, ..., D_n \subseteq E$ with $p_i = \hat{D}_i$ for $i = 1, ..., n$.*

Given an arbitrary element $a \in \mathbf{V}^*(A)$ with spectral decomposition $a = \sum_{x \in E} t_x \hat{x}$, we can isolate distinct values $t_0 > t_1 > ... > t_k$ of the coefficients t_x. Letting $E_i = \{x \in E | t_x = t_i\}$ and setting $p_i = p(E_i) = \sum_{x \in E_i} \hat{x}$, we have $a = \sum_i t_i p_i$, with $p_1, ..., p_n$ jointly orthogonal. Suppose there is another such decomposition, say $a = \sum_j s_j q_j$, with $q_j = \hat{F}_j = \sum_{y \in F_j} \hat{y}$, where $F_1, ..., F_l \subseteq F \in \mathcal{M}(A)$ are pairwise disjoint, and again, with the coefficients in descending order, say $s_0 > s_1 > \cdots > s_l$.

Lemma A2. *In the situation described above, $t_0 = s_0$ and $p_0 = q_0$.*

Proof. Normalize the inner product on $\mathbf{E}(A)$ so that $\|x\| = 1$ for all outcomes x. Then, for any sharp effect $p = \hat{D}$, D an event, we have $\|D\|^2 = |D|$, the cardinality of D. Choosing any outcome $x_0 \in E_0$, set $\alpha = |x_0\rangle$, i.e., $\alpha(\hat{x}) = \langle \hat{x}, \hat{x}_0 \rangle$ for all $x \in X(A)$. Then, $\alpha \in \Omega(A)$, $\alpha(p_0) = 1$ and $\alpha(p_i) = 0$ for $i > 0$. Thus,

$$t_0 = \alpha(a) = \sum_j s_j \alpha(q_j).$$

Since the coefficients $\alpha(q_j)$ are sub-convex, the right-hand side is no larger than the largest of the values s_j, namely, s_0. Thus, $t_0 \leq s_0$. The same argument, with the roles of the two decompositions reversed, shows that $s_0 \leq t_0$. Thus, $s_0 = t_0$.

Now again, let $x \in E_0$: then,

$$\langle \hat{x}, p_0 \rangle = \sum_{y \in E_0} \langle \hat{x}, \hat{y} \rangle = \langle x, x \rangle = 1,$$

whence, $\langle \hat{x}, a \rangle = t_0$. However, we then have (using the fact that $s_0 = t_0$):

$$t_0 = \langle \hat{x}, a \rangle = \left\langle \hat{x}, t_0 q_0 + \sum_{j=1}^l s_j q_j \right\rangle = t_0 \langle \hat{x}, q_0 \rangle + \sum_{j=1}^l s_j \langle \hat{x}, q_j \rangle.$$

Since $\sum_{j=0}^l \langle \hat{x}, q_j \rangle \leq \langle \hat{x}, u \rangle = \leq 1$, the sum in the last expression above is a sub-convex combination of the distinct values $s_0 > \cdots > s_l$. This can equal $t_0 = s_0$, the maximum of these values, only if $\langle \hat{x}, q_0 \rangle = 1$ and $\langle \hat{x}, q_j \rangle = 0$ for the remaining q_j. It follows that $\langle p_0, q_0 \rangle = \sum_{x \in E_0} \langle \hat{x}, q_0 \rangle = |E_0| = \|p_0\|^2$. The same argument, with p's and q's interchanged, shows that $\langle p_0, q_0 \rangle = \|q_0\|^2$. Hence, $\|p_0\| = \|q_0\|$, and $\langle p_0, q_0 \rangle = \|p_0\|^2 = \|p_0\| \|q_0\|$, whence, $p_0 = q_0$. □

Proposition A1. *Every $a \in \mathbf{V}^*(A)$ has a unique expansion of the form $a = \sum_{i=0}^k t_i p_i$ where $t_0 > t_1 > ... > t_k$ are non-zero coefficients and $p_1, ..., p_n$ are jointly orthogonal sharp effects.*

Proof. Suppose $a = \sum_{i=1}^k t_i p_i$, as above, and also $a = \sum_{j=1}^l s_j q_j$, with $s_0 > \cdots > s_l > 0$ and q_j pairwise orthogonal sharp effects. We shall show that $k = l$, and that $t_i = s_i$ and $p_i = q_i$ for each $i = 1, ..., k$. Lemma A2 tells us that $t_0 = s_0$ and $p_0 = s_0$. Hence,

$$\sum_{i=1}^k t_i p_i = a - t_0 p_0 = a - s_0 q_0 = \sum_{j=1}^l s_j q_j.$$

Applying Lemma A2 recursively, we find that $t_i = s_i$ and $p_i = q_i$ for $i = 1, ..., \min(k, l)$. If $k \neq l$, say $k < l$; we then have:

$$t_k p_k = s_k q_k + \sum_{j=k+1}^l s_j q_j = t_k p_k + \sum_{j=k+1}^l s_j q_j$$

whence, $\sum_{j=k+1}^{l} s_j q_j = 0$, which is impossible since all q_j are sharp and the coefficients s_j are strictly positive. Hence, $l = k$, and the proof is complete. □

Appendix C. Bits Are Balls

In most other reconstructions of QM [8–10], the first step is to show that the state space of a bit, that is, a system in which every state is the mixture of two sharply-distinguishable pure states, is a ball. In our approach, this fact is an easy consequence of Lemma 1. In our framework, we will define a *bit* to be a sharp, uniform model A with rank two, in which every state has the form $t\delta_x + (1-t)\delta_{x'}$, where $\{x, x'\} \in \mathcal{M}(A)$. Note that this implies that A is spectral.

Lemma A3. *Let A be a bit with conjugate $\overline{A}$. Then, $\Omega(A)$ is a Euclidean ball, the extreme points of which are the states δ_x, $x \in X(A)$.*

Proof. By Lemma 1, $\mathbf{E}(A)$ carries a self-dualizing inner product such that $\langle \hat{x}, \hat{y} \rangle = 0$ for $\{x, y\} \in \mathcal{M}(A)$, and which we can normalize so that $\|\hat{x}\| = 1$ for each outcome $x \in X(A)$, so that $\langle u, \hat{x} \rangle = \langle \hat{x}, \hat{x} \rangle = 1$ and $\|u\|^2 = 2$. Every state $\alpha \in \Omega(A)$ corresponds to a unique vector $a \in \mathbf{E}(A)_+$ with $\langle a, u \rangle = 1$, where $\alpha(x) = \langle a, \hat{x} \rangle$ for all $x \in X(A)$; conversely, every vector $a \in \mathbf{E}(A)_+$ with $\langle a, u \rangle = 1$ corresponds in this way to a state. In particular, the state δ_x corresponds to the unit vectors $\hat{x}$, and the maximally-mixed state corresponds to the vector $\frac{1}{n}u$. To simplify the notation, let us agree for the moment to write ρ for this vector. Thus, $\langle \rho, \hat{x} \rangle = \frac{1}{2}$, $\|\rho\|^2 = \frac{1}{4}\langle u, u \rangle = \frac{1}{2}$, and hence,

$$\|\rho - \hat{x}\|^2 = \|\rho\|^2 - 2\langle \rho, \hat{x} \rangle + \|\hat{x}\|^2 = \frac{1}{2}.$$

Thus, $\widehat{X}(A) := \{\hat{x} \mid x \in X(A)\}$ lies on the sphere of radius $1/\sqrt{2}$ about the state ρ. I now claim that any $a \in \mathbf{E}(A)$ with $\langle a, u \rangle = 1$ (in effect, any state) such that $\|\rho - a\| \leq 1/\sqrt{2}$ belongs to the positive cone $\mathbf{E}(A)_+$. To see this, use spectrality to decompose a as $s\hat{x} + t\hat{y}$ where $\{x, y\} \in \mathcal{M}(A)$ and $s, t \in \mathbb{R}$. Consider now the two-dimensional subspace $\mathbf{E}_{x,y}$ spanned by $\hat{x}$ and $\hat{y}$. With respect to the inner product inherited from $\mathbf{E}$, we can regard this as a two-dimensional Euclidean space, in which a is represented by the Cartesian coordinate pair (s, t). Expanding ρ as $\rho = \frac{1}{2}(\hat{x} + \hat{y})$, we see that $\rho \in \mathbf{E}_{x,y}$ with coordinates $(1/2, 1/2)$. The point (t, s) lies, therefore, in the disk of radius $1/\sqrt{2}$ centered at $(1/2, 1/2)$ in $\mathbf{E}_{x,y}$. Moreover, as $\langle a, u \rangle = 1$, we see that $s + t = 1$, i.e., (s, t) lies on the line of slope -1 through $(1/2, 1/2)$. This puts (s, t) in the positive quadrant of this plane, i.e., $s \geq 0$ and $t \geq 0$. However, then $a \in \mathbf{E}(A)_+$, as claimed. □

It follows that, for rank-two models, we do not even need to invoke homogeneity: they all correspond to spin factors. Letting d denote the dimension of the state space (that is, $d = \dim(\mathbf{E}) - 1$), we see that if $d = 1$, we have the classical bit; $d = 2$ gives the real quantum-mechanical bit, $d = 3$ gives the familiar Bloch sphere, i.e., the usual qubit of complex QM; while $d = 5$ corresponds to the quaternionic unit sphere, giving us the quaternionic bit. The generalized bits with $d = 4$ and $d \geq 6$ are more exotic "post-quantum" possibilities.

Appendix D. Locally-Tomographic and Dagger-Compactness

A *dagger* on a category $\mathcal{C}$ is a contravariant functor $\dagger : \mathcal{C} \to \mathcal{C}$ that is the identity on objects and satisfies $\dagger \circ \dagger = \mathrm{id}_{\mathcal{C}}$. That is, if $A \xrightarrow{f} B$ is a morphism in $\mathcal{C}$, then $A \xleftarrow{f^\dagger} B$, with $f^{\dagger\dagger} = f$ and $(f \circ g)^\dagger = g^\dagger \circ f^\dagger$ whenever $f \circ g$ is defined. An isomorphism $f : A \simeq B$ in $\mathcal{C}$ is then said to be *unitary* iff $f^\dagger = f^{-1}$. One says that $\mathcal{C}$ is $\dagger$-*monoidal* iff $\mathcal{C}$ is equipped with a symmetric monoidal structure $\otimes$ such that $(f \otimes g)^\dagger = f^\dagger \otimes g^\dagger$, and such that the canonical isomorphisms $\alpha_{A,B,C}$, $\sigma_{A,B}$, λ_A and ρ_A are all unitary.

A *dual* for an object A in a symmetric monoidal category $\mathcal{C}$ is a structure (A', η, ϵ) where $A' \in \mathcal{C}$ and $\eta : I \to A \otimes A'$ and $\epsilon : A' \otimes A \to I$, such that:

$$(\text{id}_A \otimes \epsilon) \circ (\eta \otimes \text{id}_A) = \text{id}_A \text{ and } (\epsilon \otimes \text{id}_{A'}) \circ (\text{id}_{A'} \otimes \eta) = \text{id}_{A'}$$

up to the natural associator and unit isomorphisms. If $\mathcal{C}$ is †-monoidal and $\epsilon = \sigma_{A,A'} \circ \eta_{A'}^\dagger$, then (A', η, ϵ) is a *dagger-dual*. A category in which every object A has a specified dual (A', η_A, ϵ_A) is compact closed, and a dagger-monoidal category in which every object has a given dagger-dual is dagger-compact. See [18,30] for details.

An important example of all this is the category **FdHilb**$_\mathbb{R}$ of finite-dimensional real Hilbert spaces and linear mappings. If $\mathcal{H}$ and $\mathcal{K}$ are two such spaces and $\phi : \mathcal{H} \to \mathcal{K}$, let $\phi^\dagger$ be the usual adjoint of ϕ with respect to the given inner products. Letting $\mathcal{H} \otimes \mathcal{K}$ be the usual tensor product of $\mathcal{H}$ and $\mathcal{K}$ (in particular, with $\langle x \otimes y, u \otimes v\rangle = \langle x, u \rangle \langle y, v \rangle$ for $x, u \in \mathcal{H}$ and $y, v \in \mathcal{K}$), **FdHilb**$_\mathbb{R}$ is a dagger-monoidal category with $\mathbb{R}$ as the monoidal unit.

Since any $\mathcal{H} \in $ **FdHilb**$_\mathbb{R}$ is canonically isomorphic to its dual space, we have also a canonical isomorphism $\mathcal{H} \otimes \mathcal{H} \simeq \mathcal{H}^* \otimes \mathcal{H} = \mathcal{L}(\mathcal{H})$ and a canonical trace functional $\text{Tr}_\mathcal{H} : \mathcal{H} \otimes \mathcal{H} \to \mathbb{R}$, uniquely defined by $\text{Tr}_\mathcal{H}(x \otimes y) = \langle x, y \rangle$ for all $x, y \in \mathcal{H}$. Taking $\mathcal{H}' = \mathcal{H}$, let $\eta_\mathcal{H} \in \mathcal{H} \otimes \mathcal{H}$ be given by $\eta_\mathcal{H} = \sum_i x_i \otimes x_i$, where the sum is taken over any orthonormal basis $\{x_i\}$ for $\mathcal{H}$; then, for any $a \in \mathcal{H} \otimes \mathcal{H}$, $\langle \eta_A, a \rangle = \text{Tr}(a)$. It is routine to show that $\text{Tr}_\mathcal{H} = \sigma_{\mathcal{H},\mathcal{H}} \circ \eta_\mathcal{H}^\dagger$, so that $\eta_\mathcal{H}$ and $\text{Tr}_\mathcal{H}$ make $\mathcal{H}$ its own dagger-dual.

In any compact closed symmetric monoidal category $\mathcal{C}$, every morphism $\phi : A \to B$ yields a dual morphism $\phi' : B' \to A'$ defined by:

$$\phi' = (\text{id}_{A'} \otimes \epsilon_B) \circ (\text{id}_{A'} \otimes f \otimes \text{id}_{B'}) \circ (\eta_A \otimes \text{id}_{B'}).$$

(again, suppressing associators and left and right units). For $\phi : \mathcal{H} \to \mathcal{K}$ in **FdHilb**$_\mathbb{R}$, one has, for any $v \in A$,

$$\phi'(v) = \sum_{x \in M} \langle v, f(x) \rangle x = \sum_{x \in M} \langle f^\dagger(v), x \rangle x = f^\dagger(v),$$

i.e., $\phi' = \phi^\dagger$.

Now, let $\mathcal{C}$ be a monoidal probabilistic theory; that is, a category of probabilistic models and processes, with a symmetric monoidal structure $A, B \mapsto AB$, where AB is a (non-signaling) composite in the sense discussed in Section 6. Let $\mathbb{C}$ is multiplicative, so that for $A, B \in \mathcal{C}$, we have $\pi_{AB} : X(A) \times X(B) \to X(AB)$. Henceforward, I will write $x \otimes y$ for $\pi(x, y)$ where $x \in X(A)$ and $y \in X(B)$. I will further assume that $\mathcal{C}$'s tensor unit is $I = \mathbb{R}$, and that:

(a) Every $A \in \mathcal{C}$ has a conjugate, $\overline{A} \in \mathcal{C}$, with $\overline{\overline{A}} = A$;
(b) For all $A, B \in \mathcal{C}$ and $\phi \in \mathcal{C}(A, B)$, $\overline{\phi} \in \mathcal{C}(\overline{A}, \overline{B})$;
(c) $\overline{\overline{A}} = A$, with $\eta_{\overline{A}}(\overline{a}, b) := \eta_A(a, \overline{b})$.

Remark A1. *(1) The chosen conjugate $\overline{A}$ for $A \in \mathcal{C}$ required by Condition (a) is equipped with a canonical isomorphism $\gamma_A : A \simeq \overline{A}$, with $\overline{x} = \gamma(x)$ for every $x \in X(A)$. As discussed in Section 4, this extends to an order-isomorphism $\mathbf{E}(A) \simeq \mathbf{E}(\overline{A})$, which we again write as $\gamma_A(a) = \overline{a}$ for $a \in \mathbf{E}(A)$. Notice, however, that γ_A is not assumed to be a morphism in $\mathcal{C}$.*

(2) In spite of this, Condition (b) requires that $\overline{\phi} = \gamma_B \circ \phi \circ \gamma_A^{-1}$ does belong to $\mathcal{C}(\overline{A}, \overline{B})$ for $\phi \in \mathcal{C}(A, B)$. Notice here that $\phi \mapsto \overline{\phi}$ is functorial.

(3) The second part of Condition (c) is redundant if every model A in $\mathbb{C}$ is sharp (since in this case, there is at most one correlator between $\overline{A}$ and A). Notice, too, that Condition (c) implies that:

$$\langle \overline{x}, \overline{y} \rangle = \eta_{\overline{A}}(\overline{x}, y) = \eta_A(x, \overline{y}) = \langle \hat{x}, \hat{y} \rangle$$

for all $x, y \in \mathbf{E}(A)$.

We are now ready to prove Theorem 4. We continue to assume that $\mathcal{C}$ is a locally-tomographic, multiplicative monoidal probabilistic theory, satisfying Conditions (a), (b) and (c) above. We wish to show that if every $A \in \mathcal{C}$ is sharp and spectral, then $\mathcal{C}$ has a canonical dagger, with respect to which it is dagger-compact.

Before proceeding, it will be convenient to dualize our representation of morphisms, so that $\phi \in \mathcal{C}(A, B)$ means that ϕ is a positive linear mapping $\mathbf{E}(B) \to \mathbf{E}(A)$ (thus, our co-unit $\eta \in \mathcal{C}(I, A \otimes A')$ becomes a positive linear mapping $\eta_A : \mathbf{E}(A \otimes A') \to \mathbb{R}$, and similarly, a unit $\epsilon_A \in \mathcal{C}(A' \otimes A, I)$ becomes a positive linear mapping $\mathbb{R} \to \mathbf{E}(A' \otimes A)$, i.e, an element of $\mathbf{E}(A \otimes A')$). By Lemma 1, for every $A \in \mathcal{C}$, the space $\mathbf{E}(A)$ carries a canonical self-dualizing inner product $\langle\,,\,\rangle_A$, with respect to which $\mathbf{E}(A) \simeq \mathbf{V}(A)$.

Lemma A4. *For all models $A, B \in \mathcal{C}$, the inner product on $\mathbf{E}(AB)$ factors, in the sense that if $u, x \in \mathbf{E}(A)$ and $v, y \in \mathbf{E}(B)$, then $\langle u \otimes v, x \otimes y \rangle = \langle u, x \rangle \langle v, y \rangle$.*

Proof. This follows from the sharpness of A, B and AB. For $u \in X(A)$, $v \in X(B)$, let δ_u, δ_v and $\delta_{u \otimes v}$ denote the unique states of A, B and AB such that $\delta_u(u) = \delta_v(v) = \delta_{u \otimes v}(u \otimes v) = 1$. Since $(\delta_u \otimes \delta_v)(u \otimes v)$ is also one, we conclude that $\delta_{u \otimes v} = \delta_u \otimes \delta_v$. However, we also have $\delta_u(x) = n \langle \hat{u}, \hat{x} \rangle$, $\delta_v(y) = m \langle \hat{v}, \hat{y} \rangle$ and $\delta_{u \otimes v}(x \otimes y) = nm \langle \hat{u} \otimes \hat{v}, \hat{x} \otimes \hat{y} \rangle$, where n, m and nm are the ranks, respectively, of A, B and $A \otimes B$. This establishes the claim. □

It follows that $\mathcal{C}$ is a monoidal subcategory of **FdHilb**$_{\mathbb{R}}$. In effect, we are going to show that $\mathcal{C}$ inherits a dagger-compact structure from **FdHilb**$_{\mathbb{R}}$, with the minor twist that we will take $\overline{A}$, rather than A, as the dual for $A \in \mathcal{C}$. We define the dagger of $\phi \in \mathcal{C}(A, B)$ to be the Hermitian adjoint of $\phi : \mathbf{E}(A) \to \mathbf{E}(B)$ with respect to the canonical inner products on $\mathbf{E}(A)$ and $\mathbf{E}(B)$. At this point, it is not obvious that $\phi^\dagger$ belongs to $\mathcal{C}$. In order to show that it does, we first need to show that $\mathcal{C}$ is compact closed. To define the unit, let $e_A \in \mathbf{E}(\overline{A}) \otimes \mathbf{E}(\overline{A}) = \mathbf{E}(\overline{A}A)$ (note the use of local tomography here) to be the vector with $\langle e_A, \cdot \rangle = \eta_A$, i.e., for all $a, b \in \mathbf{E}(A)$,

$$\langle e_A, \overline{a} \otimes b \rangle = \eta_A(a \otimes \overline{b}) = \langle a, b \rangle.$$

Since $\mathbf{E}(A\overline{A})$ is self-dual, $e_A \in \mathbf{E}(A\overline{A})_+$.

Lemma A5. *With η_A and e_A defined as above, $\overline{A}$ is a dual for A for every $A \in \mathcal{C}$. In particular, $\mathcal{C}$ is compact closed.*

Proof. Choose an orthonormal basis $M \subseteq \mathbf{E}(A)$. Local tomography and Lemma A4 tell us that $\overline{M} \otimes M = \{\overline{a} \otimes a | a \in M\}$ is then an orthonormal basis for $\mathbf{E}(\overline{A}A)$ (note here that $a, b \in M$ are not necessarily even positive, let alone in $X(A)$). If we expand e_A with respect to this basis, we have:

$$e_A = \sum_{a,b \in M} \langle e_A, \overline{a} \otimes b \rangle \overline{a} \otimes b$$

Since the basis is orthonormal, we have:

$$\langle e_A, \overline{a} \otimes a \rangle = \langle a, a \rangle = \|a\|^2 = 1$$

and for $a \neq b$, both in M,

$$\langle e_A, \overline{a} \otimes b \rangle = \langle a, b \rangle = 0$$

Hence, $e_A = \sum_{a \in M} \bar{a} \otimes a$. Regarding e_A as a morphism $I \to \bar{A} \otimes A$, we now have, for any $v \in \mathbf{E}(A)$,

$$(\eta_A \otimes \mathrm{id}_A) \circ (\mathrm{id}_A \otimes e_A)(v) = (\eta_A \otimes \mathrm{id}_A) \left(\sum_{x \in M} v \otimes \bar{a} \otimes a \right)$$
$$= \sum_{x \in M} \eta_A(v \otimes \bar{a})a$$
$$= \sum_{x \in M} \langle v, a \rangle a = v.$$

Similarly, for $\bar{v} \in \bar{A}$,

$$(\mathrm{id}_{\bar{A}} \otimes \eta_{\bar{A}}) \circ (e_{\bar{A}} \otimes \mathrm{id}_{\bar{A}})(\bar{v}) = (\mathrm{id}_{\bar{A}} \otimes \eta_{\bar{A}}) \left(\sum_{a \in M} \bar{a} \otimes a \otimes \bar{v} \right)$$
$$= \sum_{x \in M} \bar{a} \eta_A(a, \bar{v}) = \sum_{a \in M} \langle a, v \rangle \bar{a}$$
$$= \sum_{a \in M} \langle \bar{v}, \bar{a} \rangle \bar{a} = \bar{v}.$$

□

Lemma A6. *If $\phi : \mathbf{E}(A) \to \mathbf{E}(B)$ belongs to $\mathcal{C}$, then so does $\phi^\dagger : \mathbf{E}(B) \to \mathbf{E}(A)$.*

Proof. Using the compact structure on $\mathcal{C}$ defined above, if $\phi : A \to B$, we construct the dual of $\bar{\phi}$,

$$\bar{\phi}' := (\eta_B \otimes \mathrm{id}_A) \circ (\mathrm{id}_B \otimes \bar{\phi} \otimes \mathrm{id}_A) \circ (\mathrm{id}_B \otimes e_A) : \mathbf{E}(B) \to \mathbf{E}(A).$$

Applying this mapping to $b \in \mathbf{E}(B)$, we have:

$$b \mapsto (\eta_B \otimes \mathrm{id}_A) \left(\sum_{a \in M} b \otimes \bar{\phi}(\bar{a}) \otimes a \right) = \sum_{a \in M} \eta_B(b, \bar{\phi}(\bar{a}))a.$$
$$= \sum_{a \in M} \langle b, \phi(a) \rangle a$$
$$= \sum_{a \in M} \langle \phi^\dagger(b), a \rangle a = \phi^\dagger(b).$$

Thus, $\phi^\dagger = \bar{\phi}'$, which is evidently a morphism in $\mathcal{C}$.

Thus, $\mathcal{C}$ is a dagger-, as well as a monoidal, sub-category of **FdHilb**$_\mathbb{R}$. Hence, the associator, swap and left- and right-unit morphisms associated with an object $A \in \mathcal{C}$ are all unitary (since they are unitary in **FdHilb**$_\mathbb{R}$), whence $\mathcal{C}$ is dagger-monoidal. To complete the proof of Theorem 4, we need to check that $\eta_A = e_A^\dagger \circ \sigma_{A,\bar{A}} : \mathbf{E}(A\bar{A}) \to \mathbb{R}$. In view of our local tomography assumption, it is enough to check this on pure tensors, where a routine computation gives us $e_A^\dagger(\sigma_{A,\bar{A}}(a \otimes \bar{b})) = \langle e_A^\dagger(\bar{b} \otimes a), 1 \rangle_1 = \langle \bar{b} \otimes a, e_A \rangle_{\bar{A}A} = \langle a, b \rangle = \eta_A(a \otimes \bar{b})$. □

Remark A2. *Given that $\mathcal{C}$ is compact closed, with $\bar{A}$ the dual of A, the functoriality of $\phi \mapsto \bar{\phi}$ makes $\mathcal{C}$ strongly compact closed, in the sense of [18]. This is equivalent to dagger-compactness.*

References

1. Von Neumann, J. *Mathematical Foundations of Quantum Mechanics*; Princeton University Press: Princeton, NJ, USA, 1955.
2. Schwinger, J. The algebra of microscopic measurement. *Proc. Natl. Acad. Sci. USA* **1959**, *45*, 1542–1553.

3. Mackey, G.W. *Mathematical Foundations of Quantum Mechanics*; Dover Publications, Inc.: Mineola, NY, USA, 2004.
4. Ludwig, G. *Foundations of Quantum Mechanics I*; Springer: New York, NY, USA, 1983.
5. Piron, C. *Mathematical Foundations of Quantum Mechanics*; Academic Press: Cambridge, MA, USA, 1978.
6. Barnum, H.; Müller, M.; Ududec, C. Higher-order interference and single-system postulates characterizing quantum theory. *New J. Phys.* **2014**, *16*, 123029.
7. Hardy, L. Quantm theory from five reasonable axioms. *arXiv* **2001**, arXiv:quant-ph/0101012.
8. Chiribella, G.; D'Ariano, M.; Perinotti, P. Informational derivation of quantum theory. *Phys. Rev. A* **2011**, *84*, 012311.
9. Dakic, B.; Brukner, C. Quantum theory and beyond: Is entanglement special? *arXiv* **2009**, arXiv:0911.0695.
10. Masanes, L.; Müller, M. A derivation of quantum theory from physical requirements. *New J. Phys.* **2011**, *13*, 063001.
11. Baez, J. Division algebras and quantum theory. *Found. Phys.* **2012**, *42*, 819–855.
12. Janotta, P.; Lal, R. Generalized probabilistic theories without the no-restriction hypothesis. *Phys. Rev. A* **2013**, *87*, 052131.
13. Faraut, J.; Koranyi, A. *Analysis on Symmetric Cones*; Oxford University Press: London, UK, 1994.
14. Wilce, A. 4.5 axioms for finite-dimensional quantum probability. In *Probability in Physics*; Ben-Menahem, Y., Hemmo, M., Eds.; Springer: New York, NY, USA, 2012.
15. Wilce, A. Symmetry and composition in probabilistic theories. *Electron. Notes Theor. Comput. Sci.* **2011**, *270*, 191–207.
16. Wilce, A. Symmetry, self-duality and the Jordan structure of finite-dimensional quantum mechanics. *arXiv* **2011**, arxiv:1110.6607.
17. Wilce, A. Conjugates, Filters and Quantum Mechanics. *arXiv* **2012**, arxiv.org/pdf/1206.2897.
18. Abramsky, S.; Coecke, B. Abstract Physical Traces. *Theor. Appl. Categories* **2005**, *14*, 111–124.
19. Barnum, H.; Graydon, M.A.; Wilce, A. Some nearly quantum theories. *arXiv* **2015**, arXiv:1507.06278.
20. Jordan, P. Über ein Klasse nichtassoziativer hypercomplexe algebren. *Nachr. Akad. Wiss. Göttingen Math. Phys. Kl. I.* **1933**, *33*, 569–575. (In German)
21. Von Neumann, J. On an algebraic generalization of the quantum mechanical formalism (Part I). *Ann. Math.* **1936**, *1*, 415–484.
22. Aliprantis, C.D.; Toukey, R. *Cones and Duality*; American Mathematical Society: Providence, RI, USA, 2007.
23. Foulis, D.J.; Randall, C.H. Empirical logic and tensor products. In *Interpretations and Foundations of Quantum Theory*; Neumann, H., Ed.; Bibliographisches Inst.: Mannheim, Germany, 1981.
24. Barnum, H.; Wilce, A. Post-classical probability theory. In *Quantum Theory: Informational Foundations and Foils*; Chiribella, G., Spekkens, R., Eds.; Springer: Dordrecht, The Netherlands, 2016.
25. Popescu, S.; Rohrlich, D. Nonlocality as an axiom. *Found. Phys.* **1994**, *24*, 379–385.
26. Gunson, J. On the algebraic structure of quantum mechanics. *Commun. Math. Phys.* **1967**, *6*, 262–285.
27. Chribella, G.; Scandolo, C.M. Operational axioms for state diagonalization. *arXiv* **2015**, arXiv:1506:00380.
28. Barnum, H.; Graydon, M.; Wilce, A. Composites and categories of Euclidean Jordan algebras. *arXiv* **2016**, arXiv:1606.09331.
29. Mac Lane, S. *Categories for the Working Mathematician*; Springer: New York, NY, USA, 1978.
30. Selinger, P. Dagger compact closed categories and completely positive maps. *Electron. Notes Theor. Comput. Sci.* **2007**, *170*, 139–163.

© 2018 by the author. Licensee MDPI, Basel, Switzerland. This article is an open access article distributed under the terms and conditions of the Creative Commons Attribution (CC BY) license (http://creativecommons.org/licenses/by/4.0/).

Article

Agents, Subsystems, and the Conservation of Information

Giulio Chiribella [1,2,3,4]

[1] Department of Computer Science, University of Oxford, Parks Road, Oxford OX1 3QD, UK; giulio.chiribella@cs.ox.ac.uk
[2] Canadian Institute for Advanced Research, CIFAR Program in Quantum Information Science, 661 University Ave, Toronto, ON M5G 1M1, Canada
[3] Department of Computer Science, The University of Hong Kong, Pokfulam Road, Hong Kong, China
[4] HKU Shenzhen Institute of Research and Innovation, Yuexing 2nd Rd Nanshan, Shenzhen 518057, China

Received: 12 March 2018; Accepted: 5 May 2018; Published: 10 May 2018

Abstract: Dividing the world into subsystems is an important component of the scientific method. The choice of subsystems, however, is not defined *a priori*. Typically, it is dictated by experimental capabilities, which may be different for different agents. Here, we propose a way to define subsystems in general physical theories, including theories beyond quantum and classical mechanics. Our construction associates every agent A with a subsystem S_A, equipped with its set of states and its set of transformations. In quantum theory, this construction accommodates the notion of subsystems as factors of a tensor product, as well as the notion of subsystems associated with a subalgebra of operators. Classical systems can be interpreted as subsystems of quantum systems in different ways, by applying our construction to agents who have access to different sets of operations, including multiphase covariant channels and certain sets of free operations arising in the resource theory of quantum coherence. After illustrating the basic definitions, we restrict our attention to closed systems, that is, systems where all physical transformations act invertibly and where all states can be generated from a fixed initial state. For closed systems, we show that all the states of all subsystems admit a canonical purification. This result extends the purification principle to a broader setting, in which coherent superpositions can be interpreted as purifications of incoherent mixtures.

Keywords: subsystem; agent; conservation of information; purification; group representations; commuting subalgebras

1. Introduction

The composition of systems and operations is a fundamental primitive in our modelling of the world. It has been investigated in depth in quantum information theory [1,2], and in the foundations of quantum mechanics, where composition has played a key role from the early days of Einstein–Podolski–Rosen [3] and Schroedinger [4]. At the level of frameworks, the most recent developments are the compositional frameworks of general probabilistic theories [5–15] and categorical quantum mechanics [16–20].

The mathematical structure underpinning most compositional approaches is the structure of monoidal category [18,21]. Informally, a monoidal category describes circuits, in which wires represent systems and boxes represent operations, as in the following diagram:

$$\begin{array}{c}\underline{A}\underline{A}\\ \underline{B}\boxed{U}\underline{B}\underline{B}\\ \underline{C}\boxed{V}\underline{C}\end{array} \qquad (1)$$

The composition of systems is described by a binary operation denoted by $\otimes$, and referred to as the "tensor product" (note that $\otimes$ is not necessarily a tensor product of vector spaces). The system $A \otimes B$ is interpreted as the composite system made of subsystems A and B. Larger systems are built in a bottom-up fashion, by combining subsystems together. For example, a quantum system of dimension $d = 2^n$ can arise from the composition of n single qubits.

In some situations, having a rigid decomposition into subsystems is neither the most convenient nor the most natural approach. For example, in algebraic quantum field theory [22], it is natural to start from a single system—the field—and then to identify subsystems, e.g., spatial or temporal modes. The construction of the subsystems is rather flexible, as there is no privileged decomposition of the field into modes. Another example of flexible decomposition into subsystems arises in quantum information, where it is crucial to identify degrees of freedom that can be treated as "qubits". Viola, Knill, and Laflamme [23] and Zanardi, Lidar, and Lloyd [24] proposed that the partition of a system into subsystems should depend on which operations are experimentally accessible. This flexible definition of subsystem has been exploited in quantum error correction, where decoherence free subsystems are used to construct logical qubits that are untouched by noise [25–30]. The logical qubits are described by "virtual subsystems" of the total Hilbert space [31], and in general such subsystems are spread over many physical qubits. In all these examples, the subsystems are constructed through an algebraic procedure, whereby the subsystems are associated with algebras of observables [32]. However, the notion of "algebra of observables" is less appealing in the context of general physical theories, because the multiplication of two observables may not be defined. For example, in the framework of general probabilistic theories [5–15], observables represent measurement procedures, and there is no notion of "multiplication of two measurement procedures".

In this paper, we propose a construction of subsystems that can be applied to general physical theories, even in scenarios where observables and measurements are not included in the framework. The core of our construction is to associate subsystems to sets of *operations*, rather than observables. To fix ideas, it is helpful to think that the operations can be performed by some *agent*. Given a set of operations, the construction extracts the degrees of freedom that are acted upon *only* by those operations, identifying a "private space" that only the agent can access. Such a private space then becomes the subsystem, equipped with its own set of states and its own set of operations. This construction is closely related to an approach proposed by Krämer and del Rio, in which the states of a subsystem are identified with equivalence classes of states of the global system [33]. In this paper, we extend the equivalence relation to transformations, providing a complete description of the subsystems. We illustrate the construction in a several examples, including

1. quantum subsystems associated with the tensor product of two Hilbert spaces,
2. subsystems associated with an subalgebra of self-adjoint operators on a given Hilbert space,
3. classical systems of quantum systems,
4. subsystems associated with the action of a group representation on a given Hilbert space.

The example of the classical systems has interesting implications for the resource theory of coherence [34–41]. Our construction implies that different types of agents, corresponding to different choices of free operations, are associated with the same subsystem, namely the largest classical subsystem of a given quantum system. Specifically, classical systems arise from strictly incoherent operations [41], physically incoherent operations [38,39], phase covariant operations [38–40], and multiphase covariant operations (to the best of our knowledge, multiphase covariant operations have not been considered so far in the resource theory of coherence). Notably, we do not obtain classical subsystems from the maximally incoherent operations [34] and from the incoherent operations [35,36], which are the first two sets of free operations proposed in the resource theory of coherence. For these two types of operations, we find that the associated subsystem is the whole quantum system.

After examining the above examples, we explore the general features of our construction. An interesting feature is that certain properties, such as the impossibility of instantaneous signalling between two distinct subsystems, arise *by fiat*, rather then being postulated as physical requirements. This fact is potentially useful for the project of finding new axiomatizations of quantum theory [42–48] because it suggests that some of the axioms assumed in the usual (compositional) framework may turn out to be consequences of the very definition of subsystem. Leveraging on this fact, one could hope to find axiomatizations with a smaller number of axioms that pinpoint exactly the distinctive features of quantum theory. In addition, our construction suggests a *desideratum* that every truly fundamental axiom should arguably satisfy: *an axiom for quantum theory should hold for all possible subsystems of quantum systems*. We call this requirement Consistency Across Subsystems. If one accepts our broad definition of subsystems, then Consistency Across Subsystems is a very non-trivial requirement, which is not easily satisfied. For example, the Subspace Axiom [5], stating that all systems with the same number of distinguishable states are equivalent, does not satisfy Consistency Across Subsystems because classical subsystems are not equivalent to the corresponding quantum systems, even if they have the same number of distinguishable states.

In general, proving that Consistence Across Subsystems is satisfied may require great effort. Rather than inspecting the existing axioms and checking whether or not they are consistent across subsystems, one can try to formulate the axioms in a way that guarantees the validity of this property. We illustrate this idea in the case of the Purification Principle [8,12,13,15,49–51], which is the key ingredient in the quantum axiomatization of Refs. [13,15,42] and plays a central role in the axiomatic foundation of quantum thermodynamics [52–54] and quantum information protocols [8,15,55–57]. Specifically, we show that the Purification Principle holds for *closed systems*, defined as systems where all transformations are invertible, and where every state can be generated from a fixed initial state by the action of a suitable transformation. Closed systems satisfy the Conservation of Information [58], i.e., the requirement that physical dynamics should send distinct states to distinct states. Moreover, the states of the closed systems can be interpreted as "pure". In this setting, the general notion of subsystem captures the idea of purification, and extends it to a broader setting, allowing us to regard coherent superpositions as the "purifications" of classical probability distributions.

The paper is structured as follows. In Section 2, we outline related works. In Section 3, we present the main framework and the construction of subsystems. The framework is illustrated with five concrete examples in Section 4. In Section 5, we discuss the key structures arising from our construction, such as the notion of partial trace and the validity of the no-signalling property. In Section 6, we identify two requirements, concerning the existence of agents with non-overlapping sets of operations, and the ability to generate all states from a given initial state. We also highlight the relation between the second requirement and the notion of causality. We then move to systems satisfying the Conservation of Information (Section 7) and we formalize an abstract notion of closed systems (Section 8). For such systems, we provide a dynamical notion of pure states, and we prove that every subsystem satisfies the Purification Principle (Section 9). A macro-example, dealing with group representations in quantum theory is provided in Section 10. Finally, the conclusions are drawn in Section 11.

2. Related Works

In quantum theory, the canonical route to the definition of subsystems is to consider commuting algebras of observables, associated with independent subsystems. The idea of defining independence in terms of commutation has a long tradition in quantum field theory and, more recently, quantum information theory. In algebraic quantum field theory [22], the local subsystems associated with causally disconnected regions of spacetime are described by commuting C*-algebras. A closely related approach is to associate quantum systems to von Neumann algebras, which can be characterized as double commutants [59]. In quantum error correction, decoherence free subsystems are associated with the commutant of the noise operators [28,29,31]. In this context, Viola, Knill, and Laflamme [23] and Zanardi, Lidar, and Lloyd [24] made the point that subsystems should be defined operationally, in terms of the experimentally accessible operations. The canonical approach of associating subsystems to subalgebras was further generalized by Barnum, Knill, Ortiz, and Viola [60,61], who proposed the notion of generalized entanglement, i.e., entanglement relative to a subspace of operators. Later, Barnum, Ortiz, Somma, and Viola explored this notion in the context of general probabilistic theories [62].

The above works provided a concrete model of subsystems that inspired the present work. An important difference, however, is that here we will not use the notions of observable and expectation value. In fact, we will not use any probabilistic notion, making our construction usable also in frameworks where no notion of measurement is present. This makes the construction appealingly simple, although the flip side is that more work will have to be done in order to recover the probabilistic features that are built-in in other frameworks.

More recently, del Rio, Krämer, and Renner [63] proposed a general framework for representing the knowledge of agents in general theories (see also the Ph.D. theses of del Rio [64] and Krämer [65]). Krämer and del Rio further developed the framework to address a number of questions related to locality, associating agents to monoids of operations, and introducing a relation, called *convergence through a monoid*, among states of a global system [33]. Here, we will extend this relation to transformations, and we will propose a general definition of subsystem, equipped with its set of states and its set of transformations.

Another related work is the work of Brassard and Raymond-Robichaud on no-signalling and local realism [66]. There, the authors adopt an equivalence relation on transformations, stating that two transformations are equivalent iff they can be transformed into one another through composition with a local reversible transformation. Such a relation is related to the equivalence relation on transformations considered in this paper, in the case of systems satisfying the Conservation of Information. It is interesting to observe that, notwithstanding the different scopes of Ref. [66] and this paper, the Conservation of Information plays an important role in both. Ref. [66], along with discussions with Gilles Brassard during QIP 2017 in Seattle, provided inspiration for the present paper.

3. Constructing Subsystems

Here, we outline the basic definitions and the construction of subsystems.

3.1. A Pre-Operational Framework

Our starting point is to consider a single system S, with a given set of states and a given set of transformations. One could think S to be the whole universe, or, more modestly, our "universe of discourse", representing the fragment of the world of which we have made a mathematical model. We denote by $\mathsf{St}(S)$ the set of states of the system (sometimes called the "state space"), and by $\mathsf{Transf}(S)$ be the set of transformations the system can undergo. We assume that $\mathsf{Transf}(S)$ is equipped with a composition operation $\circ$, which maps a pair of transformations $\mathcal{A}$ and $\mathcal{B}$ into the transformation $\mathcal{B} \circ \mathcal{A}$. The transformation $\mathcal{B} \circ \mathcal{A}$ is interpreted as the transformation occurring when $\mathcal{B}$ happens right before $\mathcal{A}$. We also assume that there exists an identity operation $\mathcal{I}_S$, satisfying the condition $\mathcal{A} \circ \mathcal{I}_S = \mathcal{I}_S \circ \mathcal{A} = \mathcal{A}$ for every transformation $\mathcal{A} \in \mathsf{Transf}(A)$. In short, we assume that the physical transformations form a *monoid*.

We do not assume any structure on the state space $\mathsf{St}(S)$: in particular, we do not assume that $\mathsf{St}(S)$ is convex. We do assume, however, is that there is an action of the monoid $\mathsf{Transf}(S)$ on the set $\mathsf{St}(S)$: given an input state $\psi \in \mathsf{St}(S)$ and a transformation $\mathcal{T} \in \mathsf{Transf}(S)$, the action of the transformation produces the output state $\mathcal{T}\psi \in \mathsf{St}(S)$.

Example 1 (Closed quantum systems). *Let us illustrate the basic framework with a textbook example, involving a closed quantum system evolving under unitary dynamics. Here, S is a quantum system of dimension*

d, and the state space $\mathsf{St}(S)$ is the set of pure quantum states, represented as rays on the complex vector space $\mathbb{C}^d$, or equivalently, as rank-one projectors. With this choice, we have

$$\mathsf{St}(S) = \left\{ |\psi\rangle\langle\psi| : \quad |\psi\rangle \in \mathbb{C}^d, \quad \langle\psi|\psi\rangle = 1 \right\}. \tag{2}$$

The physical transformations are represented by unitary channels, i.e., by maps of the form $|\psi\rangle\langle\psi| \mapsto U|\psi\rangle\langle\psi|U^\dagger$, where $U \in M_d(\mathbb{C})$ is a unitary d-by-d matrix over the complex field. In short, we have

$$\mathsf{Transf}(S) = \left\{ U \cdot U^\dagger : \quad U \in M_d(\mathbb{C}), \quad U^\dagger U = U U^\dagger = I \right\}, \tag{3}$$

where I is the d-by-d identity matrix. The physical transformations form a monoid, with the composition operation induced by the matrix multiplication $(U \cdot U^\dagger) \circ (V \cdot V^\dagger) := (UV) \cdot (UV)^\dagger$.

Example 2 (Open quantum systems). *Generally, a quantum system can be in a mixed state and can undergo an irreversible evolution. To account for this scenario, we must take the state space $\mathsf{St}(S)$ to be the set of all density matrices. For a system of dimension d, this means that the state space is*

$$\mathsf{St}(S) = \left\{ \rho \in M_d(\mathbb{C}) : \quad \rho \geq 0 \quad \mathsf{Tr}[\rho] = 1 \right\}, \tag{4}$$

where $\mathsf{Tr}[\rho] = \sum_{n=1}^{d} \langle n|\rho|n\rangle$ denotes the matrix trace, and $\rho \geq 0$ means that the matrix ρ is positive semidefinite. $\mathsf{Transf}(S)$ is the set of all quantum channels [67], i.e., the set of all linear, completely positive, and trace-preserving maps from $M_d(\mathbb{C})$ to itself. The action of the quantum channel $\mathcal{T}$ on a generic state ρ can be specified through the Kraus representation [68]

$$\mathcal{T}(\rho) = \sum_{i=1}^{r} T_i \rho T_i^\dagger, \tag{5}$$

where $\{T_i\}_{i=1}^{r} \subseteq M_d(\mathbb{C})$ is a set of matrices satisfying the condition $\sum_{i=1}^{r} T_i^\dagger T_i = I$. The composition of two transformations $\mathcal{T}$ and $\mathcal{S}$ is given by the composition of the corresponding linear maps.

Note that, at this stage, there is no notion of measurement in the framework. The sets $\mathsf{St}(S)$ and $\mathsf{Transf}(S)$ are meant as a model of system S irrespectively of anybody's ability to measure it, or even to operate on it. For this reason, we call this layer of the framework *pre-operational*. One can think of the pre-operational framework as the arena in which agents will act. Of course, the physical description of such an arena might have been suggested by experiments done earlier on by other agents, but this fact is inessential for the scope of our paper.

3.2. Agents

Let us introduce agents into the picture. In our framework, an agent A is identified a set of transformations, denoted as $\mathsf{Act}(A;S)$ and interpreted as the possible actions of A on S. Since the actions must be allowed physical processes, the inclusion $\mathsf{Act}(A;S) \subseteq \mathsf{Transf}(S)$ must hold. It is natural, but not strictly necessary, to assume that the concatenation of two actions is a valid action, and that the identity transformation is a valid action. When these assumptions are made, $\mathsf{Act}(A;S)$ is a monoid. Still, the construction presented in the following will hold not only for monoids, but also for generic sets $\mathsf{Act}(A;S)$. Hence, we adopt the following minimal definition:

Definition 1 (Agents). *An agent A is identified by a subset $\mathsf{Act}(A;S) \subseteq \mathsf{Transf}(S)$.*

Note that this definition captures only one aspect of agency. Other aspects—such as the ability to gather information, make decisions, and interact with other agents—are important too, but not necessary for the scope of this paper.

We also stress that the interpretation of the subset $\text{Act}(A;S) \subseteq \text{Transf}(S)$ as the set of *actions of an agent* is not strictly necessary for the validity of our results. Nevertheless, the notion of "agent" here is useful because it helps explaining the rationale of our construction. The role of the agent is somehow similar to the role of a "probe charge" in classical electromagnetism. The probe charge need not exist in reality, but helps—as a conceptual tool—to give operational meaning to the magnitude and direction of the electric field.

In general, the set of actions available to agent A may be smaller than the set of all physical transformations on S. In addition, there may be other agents that act on system S independently of agent A. We define the independence of actions in the following way:

Definition 2. *Agents A and B act independently if the order in which they act is irrelevant, namely*

$$\mathcal{A} \circ \mathcal{B} = \mathcal{B} \circ \mathcal{A}, \qquad \forall \mathcal{A} \in \text{Act}(A;S), \mathcal{B} \in \text{Act}(B;S). \tag{6}$$

In a very primitive sense, the above relation expresses the fact that A and B act on "different degrees of freedom" of the system.

Remark 1 (Commutation of transformations vs. commutation of observables)**.** *Commutation conditions similar to Equation (6) are of fundamental importance in quantum field theory, where they are known under the names of "Einstein causality" [69] and "Microcausality" [70]. However, the similarity should not mislead the reader. The field theoretic conditions are expressed in terms of operator algebras. The condition is that the operators associated with independent systems commute. For example, a system localized in a certain region could be associated with the operator algebra* A*, and another system localized in another region could be associated with the operator algebra* B*. In this situation, the commutation condition reads*

$$CD = DC \qquad \forall C \in \mathsf{A}, \quad \forall D \in \mathsf{B}. \tag{7}$$

In contrast, Equation (6) is a condition on the **transformations**, *and not on the observables, which are not even described by our framework. In quantum theory, Equation (6) is a condition on the completely positive maps, and not to the elements of the algebras* A *and* B*. In Section 4, we will bridge the gap between our framework and the usual algebraic framework, focussing on the scenario where* A *and* B *are finite dimensional von Neumann algebras.*

3.3. Adversaries and Degradation

From the point of view of agent A, it is important to identify the degrees of freedom that no other agent B can affect. In an adversarial setting, agent B can be viewed as an adversary that tries to control as much of the system as possible.

Definition 3 (Adversary)**.** *Let A be an agent and let* $\text{Act}(A;S)$ *be her set of operations. An* **adversary** *of A is an agent B that acts independently of A, i.e., an agent B whose set of actions satisfies*

$$\text{Act}(B;S) \subseteq \text{Act}(A;S)' := \left\{ \mathcal{B} \in \text{Transf}(S) : \quad \mathcal{B} \circ \mathcal{A} = \mathcal{A} \circ \mathcal{B}, \forall \mathcal{A} \in \text{Act}(A;S) \right\}. \tag{8}$$

Like the agent, the adversary is a conceptual tool, which will be used to illustrate our notion of subsystem. The adversary need not be a real physical entity, localized outside the agent's laboratory, and trying to counteract the agent's actions. Mathematically, the adversary is just a subset of the commutant of $\text{Act}(A;S)$. The interpretation of B as an "adversary" is a way to "give life to to the mathematics", and to illustrate the rationale of our construction.

When B is interpreted as an adversary, we can think of his actions as a "degradation", which compromises states and transformations. We denote the degradation relation as $\succeq_B$, and write

$$\phi \succeq_B \psi \quad \text{iff} \quad \exists \mathcal{B} \in \mathsf{Act}(B;S) : \psi = \mathcal{B}\phi, \tag{9}$$

$$\mathcal{S} \succeq_B \mathcal{T} \quad \text{iff} \quad \exists \mathcal{B}_1, \mathcal{B}_2 \in \mathsf{Act}(B;S) : \mathcal{T} = \mathcal{B}_1 \circ \mathcal{S} \circ \mathcal{B}_2 \tag{10}$$

for $\phi, \psi \in \mathsf{St}(S)$ or $\mathcal{S}, \mathcal{T} \in \mathsf{Transf}(S)$.

The states that can be obtained by degrading ψ will be denoted as

$$\mathsf{Deg}_B(\psi) := \left\{ \mathcal{B}\psi : \quad \mathcal{B} \in \mathsf{Act}(B;S) \right\}. \tag{11}$$

The transformations that can be obtained by degrading $\mathcal{T}$ will be denoted as

$$\mathsf{Deg}_B(\mathcal{T}) := \left\{ \mathcal{B}_1 \circ \mathcal{T} \circ \mathcal{B}_2 : \quad \mathcal{B}_1, \mathcal{B}_2 \in \mathsf{Act}(B;S) \right\}. \tag{12}$$

The more operations B can perform, the more powerful B will be as an adversary. The most powerful adversary compatible with the independence condition (6) is the adversary that can implement all transformations in the commutant of $\mathsf{Act}(A;S)$:

Definition 4. *The* maximal adversary *of agent A is the agent A' that can perform the actions* $\mathsf{Act}(A';S) := \mathsf{Act}(A;S)'$.

Note that the actions of the maximal adversary are automatically a monoid, even if the set $\mathsf{Act}(A;S)$ is not. Indeed,

- the identity map $\mathcal{I}_S$ commutes with all operations in $\mathsf{Act}(A;S)$, and
- if $\mathcal{B}$ and $\mathcal{B}'$ commute with every operation in $\mathsf{Act}(A;S)$, then also their composition $\mathcal{B} \circ \mathcal{B}'$ will commute with all the operations in $\mathsf{Act}(A;S)$.

In the following, we will use the maximal adversary to define the subsystem associated with agent A.

3.4. The States of the Subsystem

Given an agent A, we think of the subsystem S_A to be the collection of all degrees of freedom that are unaffected by the action of the maximal adversary A'. Consistently with this intuitive picture, we partition the states of S into disjoint subsets, with the interpretation that two states are in the same subset if and only if they correspond to the same state of subsystem S_A.

We denote by Λ_ψ the subset of $\mathsf{St}(S)$ containing the state ψ. To construct the state space of the subsystem, we adopt the following rule:

Rule 1. *If the state ψ is obtained from the state ϕ through degradation, i.e., if $\psi \in \mathsf{Deg}_{A'}(\phi)$, then ψ and ϕ must correspond to the same state of subsystem S_A, i.e., one must have $\Lambda_\psi = \Lambda_\phi$.*

Rule 1 imposes that all states in the set $\mathsf{Deg}_{A'}(\psi)$ must be contained in the set Λ_ψ. Furthermore, we have the following fact:

Proposition 1. *If the sets $\mathsf{Deg}_{A'}(\phi)$ and $\mathsf{Deg}_{A'}(\psi)$ have non-trivial intersection, then $\Lambda_\phi = \Lambda_\psi$.*

Proof. By Rule 1, every element of $\mathsf{Deg}_{A'}(\phi)$ is contained in Λ_ϕ. Similarly, every element of $\mathsf{Deg}_{A'}(\psi)$ is contained in Λ_ψ. Hence, if $\mathsf{Deg}_{A'}(\phi)$ and $\mathsf{Deg}_{A'}(\psi)$ have non-trivial intersection, then also Λ_ϕ and Λ_ψ have non-trivial intersection. Since the sets Λ_ϕ and Λ_ψ belong to a disjoint partition, we conclude that $\Lambda_\phi = \Lambda_\psi$. □

Generalizing the above argument, it is clear that two states ϕ and ψ must be in the same subset $\Lambda_\phi = \Lambda_\psi$ if there exists a finite sequence $(\psi_1, \psi_2, \ldots, \psi_n) \subseteq \mathsf{St}(S)$ such that

$$\psi_1 = \phi, \qquad \psi_n = \psi, \qquad \text{and} \qquad \mathsf{Deg}_{A'}(\psi_i) \cap \mathsf{Deg}_{A'}(\psi_{i+1}) \neq \emptyset \quad \forall i \in \{1, 2, \ldots, n-1\}. \tag{13}$$

When this is the case, we write $\phi \simeq_A \psi$. Note that the relation $\phi \simeq_A \psi$ is an equivalence relation. When the relation $\phi \simeq_A \psi$ holds, we say that ϕ and ψ are *equivalent for agent A*. We denote the equivalence class of the state ψ by $[\psi]_A$.

By Rule 1, the whole equivalence class $[\psi]_A$ must be contained in the set Λ_ψ, meaning that all states in the equivalence class must correspond to the same state of subsystem S_A. Since we are not constrained by any other condition, we make the minimal choice

$$\Lambda_\psi := [\psi]_A. \tag{14}$$

In summary, the state space of system S_A is

$$\mathsf{St}(S_A) := \left\{ [\psi]_A : \psi \in \mathsf{St}(S) \right\}. \tag{15}$$

3.5. The Transformations of a Subsystem

The transformations of system S_A can also be constructed through equivalence classes. Before taking equivalence classes, however, we need a candidate set of transformations that can be interpreted as acting exclusively on subsystem S_A. The largest candidate set is the set of all transformations that commute with the actions of the maximal adversary A', namely

$$\mathsf{Act}(A'; S)' = \mathsf{Act}(A; S)''. \tag{16}$$

In general, $\mathsf{Act}(A; S)''$ could be larger than $\mathsf{Act}(A; S)$, in agreement with the fact the set of physical transformations of system S_A could be larger than the set of operations that agent A can perform. For example, agent A could have access only to noisy operations, while another, more technologically advanced agent could perform more accurate operations on the same subsystem.

For two transformations $\mathcal{S}$ and $\mathcal{T}$ in $\mathsf{Act}(A; S)''$, the degradation relation $\succeq_{A'}$ takes the simple form

$$\mathcal{S} \succeq_{A'} \mathcal{T} \qquad \text{iff} \qquad \mathcal{T} = \mathcal{B} \circ \mathcal{S} \quad \text{for some } \mathcal{B} \in \mathsf{Act}(A'; S). \tag{17}$$

As we did for the set of states, we now partition the set $\mathsf{Act}(A; S)''$ into disjoint subsets, with the interpretation that two transformations act in the same way on the subsystem S_A if and only if they belong to the same subset.

Let us denote by $\Theta_{\mathcal{A}}$ the subset containing the transformation $\mathcal{A}$. To find the appropriate partition of $\mathsf{Act}(A; S)''$ into disjoint subsets, we adopt the following rule:

Rule 2. *If the transformation $\mathcal{T} \in \mathsf{Act}(A; S)''$ is obtained from the transformation $\mathcal{S} \in \mathsf{Act}(A; S)''$ through degradation, i.e., if $\mathcal{T} \in \mathsf{Deg}_{A'}(\mathcal{S})$, then $\mathcal{T}$ and $\mathcal{S}$ must act in the same way on the subsystem S_A, i.e., they must satisfy $\Theta_{\mathcal{T}} = \Theta_{\mathcal{S}}$.*

Intuitively, the motivation for the above rule is that system S_A is *defined* as the system that is not affected by the action of the adversary.

Rule 2 implies that all transformations in $\mathsf{Deg}_{A'}(\mathcal{T})$ must be contained in $\Theta_{\mathcal{T}}$. Moreover, we have the following:

Proposition 2. *If the sets $\mathsf{Deg}_{A'}(\mathcal{S})$ and $\mathsf{Deg}_{A'}(\mathcal{T})$ have non-trivial intersection, then $\Theta_{\mathcal{S}} = \Theta_{\mathcal{T}}$.*

Proof. By Rule 2, every element of $\mathrm{Deg}_{A'}(\mathcal{S})$ is contained in $\Theta_\mathcal{S}$. Similarly, every element of $\mathrm{Deg}_{A'}(\mathcal{T})$ is contained in $\Theta_\mathcal{T}$. Hence, if $\mathrm{Deg}_{A'}(\mathcal{S})$ and $\mathrm{Deg}_{A'}(\mathcal{T})$ have non-trivial intersection, then also $\Theta_\mathcal{S}$ and $\Theta_\mathcal{T}$ have non-trivial intersection. Since the sets $\Lambda_\mathcal{S}$ and $\Lambda_\mathcal{T}$ belong to a disjoint partition, we conclude that $\Lambda_\mathcal{S} = \Lambda_\mathcal{T}$. □

Using the above proposition, we obtain that the equality $\Theta_\mathcal{T} = \Theta_\mathcal{S}$ holds whenever there exists a finite sequence $(\mathcal{A}_1, \mathcal{A}_2, \ldots, \mathcal{A}_n) \subseteq \mathrm{Act}(A;S)''$ such that

$$\mathcal{A}_1 = \mathcal{S}, \quad \mathcal{A}_n = \mathcal{T}, \quad \text{and} \quad \mathrm{Deg}_{A'}(\mathcal{A}_i) \cap \mathrm{Deg}_{A'}(\mathcal{A}_{i+1}) \neq \emptyset \quad \forall i \in \{1, 2, \ldots, n-1\}. \tag{18}$$

When the above relation is satisfied, we write $\mathcal{S} \simeq_A \mathcal{T}$ and we say that $\mathcal{S}$ and $\mathcal{T}$ are equivalent for agent A. It is immediate to check that $\simeq_A$ is an equivalence relation. We denote the equivalence class of the transformation $\mathcal{T} \in \mathrm{Act}(A;S)''$ as $[\mathcal{T}]_A$.

By Rule 2, all the elements of $[\mathcal{T}]_A$ must be contained in the set $\Theta_\mathcal{T}$, i.e., they should correspond to the same transformation on S_A. Again, we make the minimal choice: we stipulate that the set $\Theta_\mathcal{T}$ coincides exactly with the equivalence class $[\mathcal{T}]_A$. Hence, the transformations of subsystem S_A are

$$\mathrm{Transf}(S_A) := \left\{ [\mathcal{T}]_A : \mathcal{T} \in \mathrm{Act}(A;S)'' \right\}. \tag{19}$$

The composition of two transformations $[\mathcal{T}_1]_A$ and $[\mathcal{T}_2]_A$ is defined in the obvious way, namely

$$[\mathcal{T}_1]_A \circ [\mathcal{T}_2]_A := [\mathcal{T}_1 \circ \mathcal{T}_2]_A. \tag{20}$$

Similarly, the action of the transformations on the states is defined as

$$[\mathcal{T}]_A [\psi]_A := [\mathcal{T}\psi]_A. \tag{21}$$

In Appendix A, we show that definitions (20) and (21) are well-posed, in the sense that their right-hand sides are independent of the choice of representatives within the equivalence classes.

Remark 1. *It is important not to confuse the transformation $\mathcal{T} \in \mathrm{Act}(A;S)''$ with the equivalence class $[\mathcal{T}]_A$: the former is a transformation on the whole system S, while the latter is a transformation only on subsystem S_A. To keep track of the distinction, we define the restriction of the transformation $\mathcal{T} \in \mathrm{Act}(A;S)''$ to the subsystem S_A via the map*

$$\pi_A(\mathcal{T}) := [\mathcal{T}]_A. \tag{22}$$

Proposition 3. *The restriction map $\pi_A : \mathrm{Act}(A;S)'' \to \mathrm{Transf}(S_A)$ is a monoid homomorphism, namely $\pi_A(\mathcal{I}_S) = \mathcal{I}_{S_A}$ and $\pi_A(\mathcal{S} \circ \mathcal{T}) = \pi_A(\mathcal{S}) \circ \pi_A(\mathcal{T})$ for every pair of transformations $\mathcal{S}, \mathcal{T} \in \mathrm{Act}(A;S)''$.*

Proof. Immediate from the definition (20). □

4. Examples of Agents, Adversaries, and Subsystems

In this section, we illustrate the construction of subsystems in five concrete examples.

4.1. Tensor Product of Two Quantum Systems

Let us start from the obvious example, which will serve as a sanity check for the soundness of our construction. Let S be a quantum system with Hilbert space $\mathcal{H}_S = \mathcal{H}_A \otimes \mathcal{H}_B$. The states of S are all the density operators on the Hilbert space $\mathcal{H}_S$. The space of all linear operators from $\mathcal{H}_S$ to itself will be denoted as $\mathrm{Lin}(\mathcal{H}_S)$, so that

$$\mathrm{St}(S) = \left\{ \rho \in \mathrm{Lin}(\mathcal{H}_S) : \quad \rho \geq 0, \quad \mathrm{Tr}[\rho] = 1 \right\}. \tag{23}$$

The transformations are all the quantum channels (linear, completely positive, and trace-preserving linear maps) from $\mathrm{Lin}(\mathcal{H}_S)$ to itself. We will denote the set of all channels on system S as $\mathrm{Chan}(S)$. Similarly, we will use the notation $\mathrm{Lin}(\mathcal{H}_A)$ [$\mathrm{Lin}(\mathcal{H}_B)$] for the spaces of linear operators from $\mathcal{H}_A$ [$\mathcal{H}_B$] to itself, and the notation $\mathrm{Chan}(A)$ [$\mathrm{Chan}(B)$] for the quantum channels from $\mathrm{Lin}(\mathcal{H}_A)$ [$\mathrm{Lin}(\mathcal{H}_B)$] to itself.

We can now define an agent A whose actions are all quantum channels acting locally on system A, namely

$$\mathrm{Act}(A; S) := \left\{ \mathcal{A} \otimes \mathcal{I}_B : \quad \mathcal{A} \in \mathrm{Chan}(A) \right\}, \tag{24}$$

where $\mathcal{I}_B$ denotes the identity map on $\mathrm{Lin}(\mathcal{H}_B)$. It is relatively easy to see that the commutant of $\mathrm{Act}(A; S)$ is

$$\mathrm{Act}(A; S)' = \left\{ \mathcal{I}_A \otimes \mathcal{B} : \quad \mathcal{B} \in \mathrm{Chan}(B) \right\} \tag{25}$$

(see Appendix B for the proof). Hence, the maximal adversary of agent A is the adversary $A' = B$ that has full control on the Hilbert space $\mathcal{H}_B$. Note also that one has $\mathrm{Act}(A; S)'' = \mathrm{Act}(A; S)$.

Now, the following fact holds:

Proposition 4. *Two states $\rho, \sigma \in \mathrm{St}(S)$ are equivalent for agent A if and only if $\mathrm{Tr}_B[\rho] = \mathrm{Tr}_B[\sigma]$, where Tr_B denotes the partial trace over the Hilbert space $\mathcal{H}_B$.*

Proof. Suppose that the equivalence $\rho \simeq_A \sigma$ holds. By definition, this means that there exists a finite sequence $(\rho_1, \rho_2, \ldots, \rho_n)$ such that

$$\rho_1 = \rho, \quad \rho_n = \sigma, \quad \text{and} \quad \mathrm{Deg}_B(\rho_i) \cap \mathrm{Deg}_B(\rho_{i+1}) \neq \emptyset \quad \forall i \in \{1, 2, \ldots, n-1\}. \tag{26}$$

In turn, the condition of non-trivial intersection implies that, for every $i \in \{1, 2, \ldots, n-1\}$, one has

$$(\mathcal{I}_A \otimes \mathcal{B}_i)(\rho_i) = (\mathcal{I}_A \otimes \tilde{\mathcal{B}}_i)(\rho_{i+1}), \tag{27}$$

where $\mathcal{B}_i$ and $\tilde{\mathcal{B}}_i$ are two quantum channels in $\mathrm{Chan}(B)$. Since $\mathcal{B}_i$ and $\tilde{\mathcal{B}}_i$ are trace-preserving, Equation (27) implies $\mathrm{Tr}_B[\rho_i] = \mathrm{Tr}_B[\rho_{i+1}]$, as one can see by taking the partial trace on $\mathcal{H}_B$ on both sides. In conclusion, we obtained the equality $\mathrm{Tr}_B[\rho] \equiv \mathrm{Tr}_B[\rho_1] = \mathrm{Tr}_B[\rho_2] = \cdots = \mathrm{Tr}_B[\rho_n] \equiv \mathrm{Tr}_B[\sigma]$.

Conversely, suppose that the condition $\mathrm{Tr}_B[\rho] = \mathrm{Tr}_B[\sigma]$ holds. Then, one has

$$(\mathcal{I}_A \otimes \mathcal{B}_0)(\rho) = (\mathcal{I}_A \otimes \mathcal{B}_0)(\sigma), \tag{28}$$

where $\mathcal{B}_0 \in \mathrm{Chan}(B)$ is the erasure channel defined as $\mathcal{B}_0(\cdot) = \beta_0 \, \mathrm{Tr}_B[\cdot]$, β_0 being a fixed (but otherwise arbitrary) density matrix in $\mathrm{Lin}(\mathcal{H}_B)$. Since $\mathcal{I}_A \otimes \mathcal{B}_0$ is an element of $\mathrm{Act}(B; S)$, Equation (28) shows that the intersection between $\mathrm{Deg}_B(\rho)$ and $\mathrm{Deg}_B(\sigma)$ is non-empty. Hence, ρ and σ correspond to the same state of system S_A. □

We have seen that two global states $\rho, \sigma \in \mathrm{St}(S)$ are equivalent for agent A if and only if they have the same partial trace over B. Hence, the state space of the subsystem S_A is

$$\mathrm{St}(S_A) = \left\{ \mathrm{Tr}_B[\rho] : \quad \rho \in \mathrm{St}(S) \right\}, \tag{29}$$

consistently with the standard prescription of quantum mechanics.

Now, let us consider the transformations. It is not hard to show that two transformations $\mathcal{T}, \mathcal{S} \in \text{Act}(A;S)''$ are equivalent if and only if $\text{Tr}_B \circ \mathcal{T} = \text{Tr}_B \circ \mathcal{S}$ (see Appendix B for the details). Recalling that the transformations in $\text{Act}(A;S)''$ are of the form $\mathcal{A} \otimes \mathcal{I}_B$, for some $\mathcal{A} \in \text{Chan}(A)$, we obtain that the set of transformations of S_A is

$$\text{Transf}(S_A) = \text{Chan}(A). \tag{30}$$

In summary, our construction correctly identifies the quantum subsystem associated with the Hilbert space $\mathcal{H}_A$, with the right set of states and the right set of physical transformations.

4.2. Subsystems Associated with Finite Dimensional Von Neumann algebras

In this example, we show that our notion of subsystem encompasses the traditional notion of subsystem based on an algebra of observables. For simplicity, we restrict our attention to a quantum system S with finite dimensional Hilbert space $\mathcal{H}_S \simeq \mathbb{C}^d$, $d < \infty$. With this choice, the state space $\text{St}(S)$ is the set of all density matrices in $M_d(\mathbb{C})$ and the transformation monoid $\text{Transf}(S)$ is the set of all quantum channels (linear, completely positive, trace-preserving maps) from $M_d(\mathbb{C})$ to itself.

We now define an agent A associated with a von Neumann algebra $\mathsf{A} \subseteq M_d(\mathbb{C})$. In the finite dimensional setting, a von Neumann algebra is just a matrix algebra that contains the identity operator and is closed under the matrix adjoint. Every such algebra can be decomposed in a block diagonal form. Explicitly, one can decompose the Hilbert space $\mathcal{H}_S$ as

$$\mathcal{H}_S = \bigoplus_k \left(\mathcal{H}_{A_k} \otimes \mathcal{H}_{B_k} \right), \tag{31}$$

for appropriate Hilbert spaces $\mathcal{H}_{A_k}$ and $\mathcal{H}_{B_k}$. Relative to this decomposition, the elements of the algebra A are characterized as

$$C \in \mathsf{A} \iff C = \bigoplus_k \left(C_k \otimes I_{B_k} \right), \tag{32}$$

where C_k is an operator in $\text{Lin}(\mathcal{H}_{A_k})$, and I_{B_k} is the identity on $\mathcal{H}_{B_k}$. The elements of the commutant algebra A' are characterized as

$$D \in \mathsf{A}' \iff D = \bigoplus_k \left(I_{A_k} \otimes D_k \right), \tag{33}$$

where I_{A_k} is the identity on $\mathcal{H}_{A_k}$ and D_k is an operator in $\text{Lin}(\mathcal{H}_{B_k})$.

We grant agent A the ability to implement all quantum channels with Kraus operators in the algebra A, i.e., all quantum channels in the set

$$\text{Chan}(\mathsf{A}) := \left\{ \mathcal{C} \in \text{Chan}(S) : \mathcal{C}(\cdot) = \sum_{i=1}^r C_i \cdot C_i^\dagger, \ C_i \in \mathsf{A} \ \forall i \in \{1, \ldots, r\} \right\}. \tag{34}$$

The maximal adversary of agent A is the agent B who can implement all the quantum channels that commute with the channels in $\text{Chan}(\mathsf{A})$, namely

$$\text{Act}(B;S) = \text{Chan}(\mathsf{A})'. \tag{35}$$

In Appendix C, we prove that $\text{Chan}(\mathsf{A})'$ coincides with the set of quantum channels with Kraus operators in the commutant of the algebra A: in formula,

$$\text{Chan}(\mathsf{A})' = \text{Chan}(\mathsf{A}'). \tag{36}$$

As in the previous example, the states of subsystem S_A can be characterized as "partial traces" of the states in S, provided that one adopts the right definition of "partial trace". Denoting the commutant of the algebra A by $B := A'$, one can define the "partial trace over the algebra B" as the channel $\mathrm{Tr}_B : \mathrm{Lin}(\mathcal{H}_S) \to \bigoplus_k \mathrm{Lin}(\mathcal{H}_{A_k})$ specified by the relation

$$\mathrm{Tr}_B(\rho) := \bigoplus_k \mathrm{Tr}_{B_k}\left[\Pi_k \rho \Pi_k\right], \qquad (37)$$

where Π_k is the projector on the subspace $\mathcal{H}_{A_k} \otimes \mathcal{H}_{B_k} \subseteq \mathcal{H}_S$, and Tr_{B_k} denotes the partial trace over the space $\mathcal{H}_{B_k}$. With definition (37), is not hard to see that two states are equivalent for A if and only if they have the same partial trace over B:

Proposition 5. *Two states $\rho, \sigma \in \mathrm{St}(S)$ are equivalent for A if and only if $\mathrm{Tr}_B[\rho] = \mathrm{Tr}_B[\sigma]$.*

The proof is provided in Appendix C. In summary, the states of system $\mathrm{St}(S_A)$ are obtained from the states of S via partial trace over B, namely

$$\mathrm{St}(S_A) = \left\{ \mathrm{Tr}_B(\rho) : \rho \in \mathrm{St}(S) \right\}. \qquad (38)$$

Our construction is consistent with the standard algebraic construction, where the states of system S_A are defined as restrictions of the global states to the subalgebra A: indeed, for every element $C \in A$, we have the relation

$$\begin{aligned}
\mathrm{Tr}[C\rho] &= \mathrm{Tr}\left[\left(\bigoplus_k C_k \otimes I_{B_k}\right)\rho\right] \\
&= \sum_k \mathrm{Tr}[(C_k \otimes I_{B_k})\Pi_k \rho \Pi_k] \\
&= \sum_k \mathrm{Tr}\left\{C_k \, \mathrm{Tr}_{B_k}[\Pi_k \rho \Pi_k]\right\} \\
&= \mathrm{Tr}\left\{\check{C} \, \mathrm{Tr}_B[\rho]\right\}, \qquad \check{C} := \bigoplus_k C_k,
\end{aligned} \qquad (39)$$

meaning that the restriction of the state ρ to the subalgebra A is in one-to-one correspondence with the state $\mathrm{Tr}_B[\rho]$.

Alternatively, the states of subsystem S_A can be characterized as density matrices of the block diagonal form

$$\sigma = \bigoplus_k p_k \sigma_k, \qquad (40)$$

where (p_k) is a probability distribution, and each σ_k is a density matrix in $\mathrm{Lin}(\mathcal{H}_{A_k})$. In Appendix C, we characterize the transformations of the subsystem S_A as quantum channels $\mathcal{A}$ of the form

$$\mathcal{A} = \bigoplus_k \mathcal{A}_k, \qquad (41)$$

where $\mathcal{A}_k : \mathrm{Lin}(\mathcal{H}_{A_k}) \to \mathrm{Lin}(\mathcal{H}_{A_k})$ is a linear, completely positive, and trace-preserving map. In summary, the subsystem S_A is a direct sum of quantum systems.

4.3. Coherent Superpositions vs. Incoherent Mixtures in Closed-System Quantum Theory

We now analyze an example involving only pure states and reversible transformations. Let S be a single quantum system with Hilbert space $\mathcal{H}_S = \mathbb{C}^d$, $d < \infty$, equipped with a distinguished orthonormal basis $\{|n\rangle\}_{n=1}^d$. As the state space, we consider the set of pure quantum states: in formula,

$$\mathsf{St}(S) = \left\{ |\psi\rangle\langle\psi| \ : \ |\psi\rangle \in \mathbb{C}^d, \ \langle\psi|\psi\rangle = 1 \right\}. \tag{42}$$

As the set of transformations, we consider the set of all unitary channels: in formula,

$$\mathsf{Transf}(S) = \left\{ U \cdot U^\dagger \ : \ U \in M_d(\mathbb{C}), \ U^\dagger U = U^\dagger U = I \right\}. \tag{43}$$

To agent A, we grant the ability to implement all unitary channels corresponding to diagonal unitary matrices, i.e., matrices of the form

$$U_\theta = \sum_k e^{i\theta_k} |k\rangle\langle k|, \qquad \theta = (\theta_1, \ldots, \theta_d) \in [0, 2\pi)^{\times d}, \tag{44}$$

where each phase θ_k can vary independently of the other phases. In formula, the set of actions of agent A is

$$\mathsf{Act}(A; S) = \left\{ U_\theta \cdot U_\theta^\dagger \ : \ U_\theta \in \mathsf{Lin}(\mathcal{H}_S), \ U_\theta \text{ as in Equation } (44) \right\}. \tag{45}$$

The peculiarity of this example is that the actions of the maximal adversary A' are exactly the same as the actions of A. It is immediate to see that $\mathsf{Act}(A; S)$ is included in $\mathsf{Act}(A'; S)$ because all operations of agent A commute. With a bit of extra work, one can see that, in fact, $\mathsf{Act}(A; S)$ and $\mathsf{Act}(A'; S)$ coincide.

Let us look at the subsystem associated with agent A. The equivalence relation among states takes a simple form:

Proposition 6. *Two pure states with unit vectors $|\phi\rangle, |\psi\rangle \in \mathcal{H}_S$ are equivalent for A if and only if $|\psi\rangle = U|\phi\rangle$ for some diagonal unitary matrix U.*

Proof. Suppose that there exists a finite sequence $(|\psi_1\rangle, |\psi_2\rangle, \ldots, |\psi_n\rangle)$ such that

$$|\psi_1\rangle = |\phi\rangle, \quad |\psi_n\rangle = |\psi\rangle, \quad \text{and} \quad \mathsf{Deg}_{A'}(|\psi_i\rangle\langle\psi_i|) \cap \mathsf{Deg}_{A'}(|\psi_{i+1}\rangle\langle\psi_{i+1}|) \neq \emptyset \quad \forall i \in \{1, 2, \ldots, n-1\}.$$

This means that, for every $i \in \{1, \ldots, n-1\}$, there exist two diagonal unitary matrices U_i and $\widetilde{U}_i$ such that $U_i|\psi_i\rangle = \widetilde{U}_i|\psi_{i+1}\rangle$, or equivalently,

$$|\psi_{i+1}\rangle = \widetilde{U}_i^\dagger U_i |\psi_i\rangle. \tag{46}$$

Using the above relation for all values of i, we obtain $|\psi\rangle = U|\phi\rangle$ with $U := \widetilde{U}_{n-1}^\dagger U_{n-1} \cdots \widetilde{U}_2^\dagger U_2 \widetilde{U}_1^\dagger U_1$.

Conversely, suppose that the condition $|\psi\rangle = U|\phi\rangle$ holds for some diagonal unitary matrix U. Then, the intersection $\mathsf{Deg}_{A'}(|\phi\rangle\langle\phi|) \cap \mathsf{Deg}_{A'}(|\psi\rangle\langle\psi|)$ is non-empty, which implies that $|\phi\rangle\langle\phi|$ and $|\psi\rangle\langle\psi|$ are in the same equivalence class. □

Using Proposition 6, it is immediate to see that the equivalence class $[|\psi\rangle\langle\psi|]_{A'}$ is uniquely identified by the diagonal density matrix $\rho = \sum_k |\psi_k|^2 |k\rangle\langle k|$. Hence, the state space of system S_A is the set of diagonal density matrices

$$\mathsf{St}(S_A) = \left\{ \rho = \sum_k p_k |k\rangle\langle k| \,:\, p_k \geq 0 \,\forall k, \, \sum_k p_k = 1 \right\}. \tag{47}$$

The set of transformations of system S_A is trivial because the actions of A coincide with the actions of the adversary A', and therefore they are all in the equivalence class of the identity transformation. In formula, one has

$$\mathsf{Transf}(S_A) = \left\{ \mathcal{I}_{S_A} \right\}. \tag{48}$$

4.4. Classical Subsystems in Open-System Quantum Theory

This example is of the same flavour as the previous one but is more elaborate and more interesting. Again, we consider a quantum system S with Hilbert space $\mathcal{H} = \mathbb{C}^d$. Now, we take $\mathsf{St}(S)$ to be the whole set of density matrices in $M_d(\mathbb{C})$ and $\mathsf{Transf}(S)$ to be the whole set of quantum channels from $M_d(\mathbb{C})$ to itself.

We grant to agent A the ability to perform every multiphase covariant channel, that is, every quantum channel $\mathcal{M}$ satisfying the condition

$$\mathcal{U}_\theta \circ \mathcal{M} = \mathcal{M} \circ \mathcal{U}_\theta \qquad \forall \boldsymbol{\theta} = (\theta_1, \theta_2, \ldots, \theta_d) \in [0, 2\pi)^{\times d}, \tag{49}$$

where $\mathcal{U}_\theta = U_\theta \cdot U_\theta^\dagger$ is the unitary channel corresponding to the diagonal unitary $U_\theta = \sum_k e^{i\theta_k} |k\rangle\langle k|$. Physically, we can interpret the restriction to multiphase covariant channels as the lack of a reference for the definition of the phases in the basis $\{|k\rangle, k = 1, \ldots, d\}$.

It turns out that the maximal adversary of agent A is the agent A' that can perform every *basis-preserving channel* $\mathcal{B}$, that is, every channel satisfying the condition

$$\mathcal{B}(|k\rangle\langle k|) = |k\rangle\langle k| \qquad \forall k \in \{1, \ldots, d\}. \tag{50}$$

Indeed, we have the following:

Theorem 1. *The monoid of multiphase covariant channels and the monoid of basis-preserving channels are the commutant of one another.*

The proof, presented in Appendix D.1, is based on the characterization of the basis-preserving channels provided in [71,72].

We now show that states of system S_A can be characterized as classical probability distributions.

Proposition 7. *For every pair of states $\rho, \sigma \in \mathsf{St}(S)$, the following are equivalent:*

1. *ρ and σ are equivalent for agent A,*
2. *$\mathcal{D}(\rho) = \mathcal{D}(\sigma)$, where $\mathcal{D}$ is the completely dephasing channel $\mathcal{D}(\cdot) := \sum_k |k\rangle\langle k| \cdot |k\rangle\langle k|$.*

Proof. Suppose that Condition 1 holds, meaning that there exists a sequence $(\rho_1, \rho_2, \ldots, \rho_n)$ such that

$$\rho_1 = \rho, \qquad \rho_n = \sigma, \qquad \forall i \in \{1, \ldots, n-1\} \,\exists \mathcal{B}_i, \widetilde{\mathcal{B}}_i \in \mathsf{Act}(B; S) \,:\, \mathcal{B}_i(\rho_i) = \widetilde{\mathcal{B}}_i(\rho_{i+1}), \tag{51}$$

where $\mathcal{B}_i$ and $\widetilde{\mathcal{B}}_i$ are basis-preserving channels. The above equation implies

$$\langle k| \mathcal{B}_i(\rho_i) |k\rangle = \langle k| \widetilde{\mathcal{B}}_i(\rho_{i+1}) |k\rangle. \tag{52}$$

Now, the relation $\langle k|\mathcal{B}(\rho)|k\rangle = \langle k|\rho|k\rangle$ is valid for every basis-preserving channel $\mathcal{B}$ and for every state ρ [71]. Applying this relation on both sides of Equation (52), we obtain the condition

$$\langle k|\rho_i|k\rangle = \langle k|\rho_{i+1}|k\rangle, \tag{53}$$

valid for every $k \in \{1,\ldots,d\}$. Hence, all the density matrices $(\rho_1, \rho_2, \ldots, \rho_n)$ must have the same diagonal entries, and, in particular, Condition 2 must hold.

Conversely, suppose that Condition 2 holds. Since the dephasing channel $\mathcal{D}$ is obviously basis-preserving, we obtained the condition $\mathrm{Deg}_{A'}(\rho) \cap \mathrm{Deg}_{A'}(\sigma) \neq \emptyset$, which implies that ρ and σ are equivalent for agent A. In conclusion, Condition 1 holds. □

Proposition 7 guarantees that the states of system S_A is in one-to-one correspondence with diagonal density matrices, and therefore, with classical probability distributions: in formula,

$$\mathrm{St}(S_A) = \left\{ (p_k)_{k=1}^d : \ p_k \geq 0 \, \forall k, \ \sum_k p_k = 1 \right\}. \tag{54}$$

The transformations of system S_A can be characterized as *transition matrices*, namely

$$\mathrm{Transf}(S_A) = \left\{ [P_{jk}]_{j\leq d, k\leq d} : \ P_{jk} \geq 0 \, \forall j, k \in \{1,\ldots,d\}, \ \sum_j P_{jk} = 1 \, \forall k \in \{1,\ldots,d\} \right\}. \tag{55}$$

The proof of Equation (55) is provided in Appendix D.2.

In summary, agent A has control on a classical system, whose states are probability distributions, and whose transformations are classical transition matrices.

4.5. Classical Systems From Free Operations in the Resource Theory of Coherence

In the previous example, we have seen that classical systems arise from agents who have access to the monoid of multiphase covariant channels. In fact, classical systems can arise in many other ways, corresponding to agents who have access to different monoids of operations. In particular, we find that several types of free operations in the resource theory of coherence [34–41] identify classical systems. Specifically, consider the monoids of

1. *Strictly incoherent operations* [41], i.e., quantum channels $\mathcal{T}$ with the property that, for every Kraus operator T_i, the map $\mathcal{T}_i(\cdot) = T_i \cdot T_i$ satisfies the condition $\mathcal{D} \circ \mathcal{T}_i = \mathcal{T}_i \circ \mathcal{D}$, where $\mathcal{D}$ is the completely dephasing channel.
2. *Dephasing covariant operations* [38–40], i.e., quantum channels $\mathcal{T}$ satisfying the condition $\mathcal{D} \circ \mathcal{T} = \mathcal{T} \circ \mathcal{D}$.
3. *Phase covariant channels* [40], i.e., quantum channels $\mathcal{T}$ satisfying the condition $\mathcal{T} \circ \mathcal{U}_\varphi = \mathcal{U}_\varphi \circ \mathcal{T}$, $\forall \varphi \in [0, 2\pi)$, where $\mathcal{U}_\varphi$ is the unitary channel associated with the unitary matrix $U_\varphi = \sum_k e^{ik\varphi}|k\rangle\langle k|$.
4. *Physically incoherent operations* [38,39], i.e., quantum channels that are convex combinations of channels $\mathcal{T}$ admitting a Kraus representation where each Kraus operator T_i is of the form

$$T_i = U_{\pi_i} U_{\theta_i} P_i, \tag{56}$$

where U_{π_i} is a unitary that permutes the elements of the computational basis, U_{θ_i} is a diagonal unitary, and P_i is a projector on a subspace spanned by a subset of vectors in the computational basis.

For each of the monoids 1–4, our construction yields the classical subsystem consisting of diagonal density matrices. The transformations of the subsystem are just the classical channels. The proof is presented in Appendix E.1.

Notably, other choices of free operations, such as the *maximally incoherent operations* [34] and the *incoherent operations* [35], do *not* identify classical subsystems. The maximally incoherent operations are the quantum channels $\mathcal{T}$ that map diagonal density matrices to diagonal density matrices, namely $\mathcal{T} \circ \mathcal{D} = \mathcal{D} \circ \mathcal{T} \circ \mathcal{D}$, where $\mathcal{D}$ is the completely dephasing channel. The incoherent operations are the quantum channels $\mathcal{T}$ with the property that, for every Kraus operator T_i, the map $\mathcal{T}_i(\cdot) = T_i \cdot T_i$ sends diagonal matrices to diagonal matrices, namely $\mathcal{T}_i \circ \mathcal{D} = \mathcal{D} \circ \mathcal{T}_i \circ \mathcal{D}$.

In Appendix E.2, we show that incoherent and maximally incoherent operations do not identify classical subsystems: the subsystem associated with these operations is the whole quantum system. This result can be understood from the analogy between these operations and non-entangling operations in the resource theory of entanglement [38,39]. Non-entangling operations do not generate entanglement, but nevertheless they cannot (in general) be implemented with local operations and classical communication. Similarly, incoherent and maximally incoherent operations do not generate coherence, but they cannot (in general) be implemented with incoherent states and coherence non-generating unitary gates. An agent that performs these operations must have access to more degrees of freedom than just a classical subsystem.

At the mathematical level, the problem is that the incoherent and maximally incoherent operations do not necessarily commute with the dephasing channel $\mathcal{D}$. In our construction, commutation with the dephasing channel is essential for retrieving classical subsystems. In general, we have the following theorem:

Theorem 2. *Every set of operations that*

1. *contains the set of classical channels, and*
2. *commutes with the dephasing channel*

identifies a d-dimensional classical subsystem of the original d-dimensional quantum system.

The proof is provided in Appendix E.1.

5. Key Structures: Partial Trace and No Signalling

In this section, we go back to the general construction of subsystems, and we analyse the main structures arising from it. First, we observe that the definition of subsystem guarantees *by fiat* the validity of the no-signalling principle, stating that operations performed on one subsystem cannot affect the state of an independent subsystem. Then, we show that our construction of subsystems allows one to build a category.

5.1. The Partial Trace and the No Signalling Property

We defined the states of system S_A as equivalence classes. In more physical terms, we can regard the map $\psi \mapsto [\psi]_A$ as an operation of discarding, which takes system S and throws away the degrees of freedom reachable by the maximal adversary A'. In our adversarial picture, "throwing away some degrees of freedom" means leaving them under the control of the adversary, and considering only the part of the system that remains under the control of the agent.

Definition 5. *The partial trace over A' is the function* $\mathrm{Tr}_{A'} : \mathsf{St}(S) \to \mathsf{St}(S_A)$*, defined by* $\mathrm{Tr}_{A'}(\psi) = [\psi]_A$ *for a generic* $\psi \in \mathsf{St}(S)$*.*

The reason for the notation $\mathrm{Tr}_{A'}$ is that in quantum theory the operation $\mathrm{Tr}_{A'}$ coincides with the partial trace of matrices, as shown in the example of Section 4.1. For subsystems associated with von Neumann algebras, the partial trace is the "partial trace over the algebra" defined in Section 4.2. For subsystems associated with multiphase covariant channels or dephasing covariant operations, the partial trace is the completely dephasing channel, which "traces out" the off-diagonal elements of the density matrix.

With the partial trace notation, the states of system S_A can be succinctly written as

$$\mathsf{St}(S_A) = \left\{ \rho = \mathrm{Tr}_{A'}(\psi) : \quad \psi \in \mathsf{St}(S) \right\}. \tag{57}$$

Denoting $B := A'$, we have the important relation

$$\mathrm{Tr}_B \circ \mathcal{B} = \mathrm{Tr}_B \qquad \forall \mathcal{B} \in \mathsf{Act}(B; S). \tag{58}$$

Equation (58) can be regarded as the *no signalling property*: the actions of agent B cannot lead to any change on the system of agent A. Of course, here the no signalling property holds *by fiat*, precisely because of the way the subsystems are defined!

The construction of subsystems has the merit to clarify the status of the no-signalling principle. No-signalling is often associated with space-like separation, and is heuristically justified through the idea that physical influences should propagate within the light cones. However, locality is only *a sufficient condition* for the no signalling property. Spatial separation implies no signalling, but the converse is not necessarily true: every pair of distinct quantum systems satisfies the no-signalling condition, even if the two systems are spatially contiguous. In fact, the no-signalling condition holds even for virtual subsystems of a *single*, spatially localized system. Think for example of a quantum particle localized in the xy plane. The particle can be regarded as a composite system, made of two virtual subsystems: a particle localized on the x-axis, and another particle localized on the y-axis. The no-signalling property holds for these two subsystems, even if they are not separated in space. As Equation (58) suggests, the validity of the no-signalling property has more to do with the way subsystems are constructed, rather than the way the subsystems are distributed in space.

5.2. A Baby Category

Our construction of subsystems defines a category, consisting of three objects, S, S_A, and S_B, where S_B is the subsystem associated with the agent $B = A'$. The sets $\mathsf{Transf}(S)$, $\mathsf{Transf}(S_A)$, and $\mathsf{Transf}(S_B)$ are the endomorphisms from S to S, S_A to S_A, and S_B to S_B, respectively. The morphisms from S to S_A and from S to S_B are defined as

$$\mathsf{Transf}(S \to S_A) = \left\{ \mathrm{Tr}_B \circ \mathcal{T} : \quad \mathcal{T} \in \mathsf{Transf}(S) \right\} \tag{59}$$

and

$$\mathsf{Transf}(S \to S_B) = \left\{ \mathrm{Tr}_A \circ \mathcal{T} : \quad \mathcal{T} \in \mathsf{Transf}(S) \right\}, \tag{60}$$

respectively.

Morphisms from S_A to S, from S_B to S, from S_A to S_B, or from S_B to S_A, are not naturally defined. In Appendix F, we provide a mathematical construction that enlarges the sets of transformations, making all sets non-empty. Such a construction allows us to reproduce a categorical structure known as a *splitting of idempotents* [73,74]

6. Non-Overlapping Agents, Causality, and the Initialization Requirement

In the previous sections, we developed a general framework, applicable to arbitrary physical systems. In this section, we identify some desirable properties that the global systems may enjoy.

6.1. Dual Pairs of Agents

So far, we have taken the perspective of agent A. Let us now take the perspective of the maximal adversary A'. We consider A' as the agent, and denote his maximal adversary as A''. By definition, A'' can perform every action in the commutant of $\mathsf{Act}(A'; S)$, namely

$$\text{Act}(A''; S) = \text{Act}(A'; S)' = \text{Act}(A; S)''. \tag{61}$$

Obviously, the set of actions allowed to agent A'' includes the set of actions allowed to agent A. At this point, one could continue the construction and consider the maximal adversary of agent A''. However, no new agent would appear at this point: the maximal adversary of agent A'' is agent A' again. When two agents have this property, we call them a *dual pair*:

Definition 6. *Two agents A and B form a dual pair iff* $\text{Act}(A; S) = \text{Act}(B; S)'$ *and* $\text{Act}(B; S) = \text{Act}(A; S)'$.

All the examples in Section 4 are examples of dual pairs of agents.

It is easy to see that an agent A is part of a dual pair if and only if the set $\text{Act}(A; S)$ coincides with its double commutant $\text{Act}(A; S)''$.

6.2. Non-Overlapping Agents

Suppose that agents A and B form a dual pair. In general, the actions in $\text{Act}(A; S)$ may have a non-trivial intersection with the actions in $\text{Act}(B; S)$. This situation does indeed happen, as we have seen in Sections 4.3 and 4.4. Still, it is important to examine the special case where the actions of A and B have only trivial intersection, corresponding to the identity action $\mathcal{I}_S$. When this is the case, we say that the agents A and B are *non-overlapping*:

Definition 7. *Two agents A and B are non-overlapping iff* $\text{Act}(A; S) \cap \text{Act}(B; S) \subseteq \{\mathcal{I}_S\}$.

Dual pairs of non-overlapping agents are characterized by the fact that the sets of actions have trivial center:

Proposition 8. *Let A and B be a dual pair of agents. Then, the following are equivalent:*

1. *A and B are non-overlapping,*
2. *$\text{Act}(A; S)$ has trivial center,*
3. *$\text{Act}(B; S)$ has trivial center.*

Proof. Since agents A and B are dual to each other, we have $\text{Act}(B; S) = \text{Act}(A; S)'$ and $\text{Act}(A; S) = \text{Act}(B; S)'$. Hence, the intersection $\text{Act}(A; S) \cap \text{Act}(B; S)$ coincides with the center of $\text{Act}(A; S)$, and with the center of $\text{Act}(B; S)$. The non-overlap condition holds if and only if the center is trivial. □

Note that the existence of non-overlapping dual pairs is a condition on the transformations of the whole system S:

Proposition 9. *The following are equivalent:*

1. *system S admits a dual pair of non-overlapping agents,*
2. *the monoid $\text{Transf}(S)$ has trivial center.*

Proof. Assume that Condition 1 holds for a pair of agents A and B. Let $C(S)$ be the center of $\text{Transf}(S)$. By definition, $C(S)$ is contained into $\text{Act}(B; S)$ because $\text{Act}(B; S)$ contains all the transformations that commute with those in $\text{Act}(A; S)$. Moreover, the elements of $C(S)$ commute with all elements of $\text{Act}(B; S)$, and therefore they are in the center of $\text{Act}(B; S)$. Since A and B are a non-overlapping dual pair, the center of $\text{Act}(B; S)$ must be trivial (Proposition 8), and therefore $C(S)$ must be trivial. Hence, Condition 2 holds.

Conversely, suppose that Condition 2 holds. In that case, it is enough to take A to be the *maximal agent*, i.e., the agent $A_{\max}$ with $\text{Act}(A_{\max}; S) = \text{Transf}(S)$. Then, the maximal adversary of $A_{\max}$ is the agent $B = A'_{\max}$ with $\text{Act}(B; S) = \text{Act}(A_{\max}; S)' = C(S) = \{\mathcal{I}_S\}$. By definition, the two agents form a non-overlapping dual pair. Hence, Condition 1 holds. □

The existence of dual pairs of non-overlapping agents is a desirable property, which may be used to characterize "good systems":

Definition 8 (Non-Overlapping Agents). *We say that system S satisfies the* Non-Overlapping Agents Requirement *if there exists at least one dual pair of non-overlapping agents acting on S.*

The Non-Overlapping Agents Requirement guarantees that the total system S can be regarded as a subsystem: if A_{max} is the *maximal agent* (i.e., the agent who has access to all transformations on S), then the subsystem $S_{A_{max}}$ is the whole system S. A more formal statement of this fact is provided in Appendix G.

6.3. Causality

The Non-Overlapping Agents Requirement guarantees that the subsystem associated with a maximal agent (i.e., an agent who has access to all possible transformations) is the whole system S. On the other hand, it is natural to expect that a minimal agent, who has no access to any transformation, should be associated with the trivial system, i.e., the system with a single state and a single transformation. The fact that the minimal agent is associated with the trivial system is important because it equivalent to a property of causality [8,13,75,76]: indeed, we have the following

Proposition 10. *Let A_{min} be the minimal agent and let A_{max} be its maximal adversary, coinciding with the maximal agent. Then, the following conditions are equivalent*

1. $S_{A_{min}}$ *is the trivial system,*
2. *one has* $\text{Tr}_{A_{max}}[\rho] = \text{Tr}_{A_{max}}[\sigma]$ *for every pair of states* $\rho, \sigma \in \text{St}(S)$.

Proof. $1 \Rightarrow 2$: By definition, the state space of $S_{A_{min}}$ consists of states of the form $\text{Tr}_{A_{max}}[\rho]$, $\rho \in \text{St}(S)$. Hence, the state space contains only one state if and only if Condition 2 holds. $2 \Rightarrow 1$: Condition 2 implies that every two states of system S are equivalent for agent A_{max}. The fact that $S_{A_{min}}$ has only one transformation is true by definition: since the adversary of A_{min} is the maximal agent, one has $\mathcal{T} \in \text{Deg}_{A_{max}}(\mathcal{I}_S)$ for every transformation $\mathcal{T} \in \text{Transf}(S)$. Hence, every transformation is in the equivalence class of the identity. □

With a little abuse of notation, we may denote the trace over A_{max} as Tr_S because A_{max} has access to all transformations on system S. With this notation, the causality condition reads

$$\text{Tr}_S[\rho] = \text{Tr}_S[\sigma] \qquad \forall \rho, \sigma \in \text{St}(S). \tag{62}$$

It is interesting to note that, unlike no signalling, causality does not necessarily hold in the framework of this paper. This is because the trace Tr_S is defined as the quotient with respect to all possible transformations, and having a single equivalence class is a non-trivial property. One possibility is to demand the validity of this property, and to call a system *proper*, only if it satisfies the causality condition (62). In the following subsection, we will see a requirement that guarantees the validity of the causality condition.

6.4. The Initialization Requirement

The ability to prepare states from a fixed initial state is important in the circuit model of quantum computation, where qubits are initialized to the state $|0\rangle$, and more general states are generated by applying quantum gates. More broadly, the ability to initialize the system in a given state and to generate other states from it is important for applications in quantum control and adiabatic quantum computing. Motivated by these considerations, we formulate the following definition:

Definition 9. *A system S satisfies the* Initialization Requirement *if there exists a state $\psi_0 \in \text{St}(S)$ from which any other state can be generated, meaning that, for every other state $\psi \in \text{St}(S)$, there exists a transformation $\mathcal{T} \in \text{Transf}(S)$ such that $\psi = \mathcal{T}\psi_0$. When this is the case, the state ψ_0 is called* cyclic.

The Initialization Requirement is satisfied in quantum theory, both at the pure state level and at the mixed state level. At the pure state level, every unit vector $|\psi\rangle \in \mathcal{H}_S$ can be generated from a fixed unit vector $|\psi_0\rangle \in \mathcal{H}_S$ via a unitary transformation U. At the mixed state level, every density matrix ρ can be generated from a fixed density matrix ρ_0 via the erasure channel $\mathcal{C}_\rho(\cdot) = \rho \, \text{Tr}[\cdot]$. By the same argument, the initialization requirement is also satisfied when S is a system in an operational-probabilistic theory [8,10–13] and when S is a system in a causal process theory [75,76].

The Initialization Requirement guarantees that minimal agents are associated with trivial systems:

Proposition 11. *Let S be a system satisfying the Initialization Requirement, and let $A_{\min}$ be the minimal agent, i.e., the agent that can only perform the identity transformation. Then, the subsystem $S_{A_{\min}}$ is trivial: $\text{St}\left(S_{A_{\min}}\right)$ contains only one state and $\text{Transf}\left(S_{A_{\min}}\right)$ contains only one transformation.*

Proof. By definition, the maximal adversary of $A_{\min}$ is the maximal agent $A_{\max}$, who has access to all physical transformations. Then, every transformation is in the equivalence class of the identity transformation, meaning that system $S_{A_{\min}}$ has a single transformation. Now, let ψ_0 be the cyclic state. By the Initialization Requirement, the set $\text{Deg}_{A_{\max}}(\psi_0)$ is the whole state space $\text{St}(S)$. Hence, every state is equivalent to the state ψ_0. In other words, $\text{St}\left(S_{A_{\min}}\right)$ contains only one state. □

The Initialization Requirement guarantees the validity of causality, thanks to Proposition 10. In addition, the Initialization Requirement is important independently of the causality property. For example, we will use it to formulate an abstract notion of *closed system*.

7. The Conservation of Information

In this section, we consider systems where all transformations are *invertible*. In such systems, every transformation can be thought as the result of some deterministic dynamical law. The different transformations in $\text{Transf}(S)$ can be interpreted as different dynamics, associated with different values of physical parameters, such as coupling constants or external control parameters.

7.1. Logically Invertible vs. Physically Invertible

Definition 10. *A transformation $\mathcal{T} \in \text{Transf}(S)$ is* logically invertible *iff the map*

$$\hat{\mathcal{T}}: \quad \text{St}(S) \to \text{St}(S), \quad \psi \mapsto \mathcal{T}\psi \tag{63}$$

is injective.

Logically invertible transformations can be interpreted as evolutions of the system that preserve the distictness of states. At the fundamental level, one may require that all physical evolutions be logically invertible, a requirement that is sometimes called the *Conservation of Information* [58]. In the following, we will explore the consequences of such requirement:

Definition 11 (Logical Conservation of Information)**.** *System S satisfies the Logical Conservation of Information if all transformations in $\text{Transf}(S)$ are logically invertible.*

The requirement is well-posed because the invertible transformations form a monoid. Indeed, the identity transformation is logically invertible, and that the composition of two logically invertible transformations is logically invertible.

A special case of logical invertibility is physical invertibility, defined as follows:

Definition 12. *A transformation* $\mathcal{T} \in \mathsf{Transf}(S)$ *is physically invertible iff there exists another transformation* $\mathcal{T}' \in \mathsf{Transf}(S)$ *such that* $\mathcal{T}' \circ \mathcal{T} = \mathcal{I}_S$.

Physical invertibility is more than injectivity: not only should the map $\mathcal{T}$ be injective on the state space, but also its inverse should be a physical transformation. In light of this observation, we state a stronger version of the Conservation of Information, requiring physical invertibility:

Definition 13 (Physical Conservation of Information)**.** *System S satisfies the Physical Conservation of Information if all transformations in* $\mathsf{Transf}(S)$ *are physically invertible.*

The difference between Logical and Physical Conservation of Information is highlighted by the following example:

Example 3 (Conservation of Information in closed-system quantum theory)**.** *Let S be a closed quantum system described by a separable, infinite-dimensional Hilbert space* $\mathcal{H}_S$, *and let* $\mathsf{St}(S)$ *be the set of pure states, represented as rank-one density matrices*

$$\mathsf{St}(S) = \left\{ |\psi\rangle\langle\psi| \ : \ |\psi\rangle \in \mathcal{H}_S, \ \langle\psi|\psi\rangle = 1 \right\}. \tag{64}$$

One possible choice of transformations is the monoid of isometric channels

$$\mathsf{Transf}(S) = \left\{ V \cdot V^\dagger \ : \ V \in \mathsf{Lin}(S), \ V^\dagger V = I \right\}. \tag{65}$$

This choice of transformations satisfies the Logical Conservation of Information, but violates the Physical Conservation of Information because in general the map $V^\dagger \cdot V$ *fails to be trace-preserving, and therefore fails to be an isometric channel. For example, consider the shift operator*

$$V = \sum_{n=0}^{\infty} |n+1\rangle\langle n|. \tag{66}$$

The operator V is an isometry but its left-inverse $V^\dagger$ *is not an isometry. As a result, the channel* $V^\dagger \cdot V$ *is not an allowed physical transformation according to Equation (65).*

An alternative choice of physical transformations is the set of unitary channels

$$\mathsf{Transf}(S) = \left\{ V \cdot V^\dagger \ : \ V \in \mathsf{Lin}(S), \ V^\dagger V = V V^\dagger = I \right\}. \tag{67}$$

With this choice, the Physical Conservation of Information is satisfied: every physical transformation is invertible and the inverse is a physical transformation.

7.2. Systems Satisfying the Physical Conservation of Information

In a system satisfying the Physical Conservation of Information, the transformations are not only physically invertible, but also physically *reversible*, in the following sense:

Definition 14. *A transformation* $\mathcal{T} \in \mathsf{Transf}(S)$ *is physically reversible iff there exists another transformation* $\mathcal{T}' \in \mathsf{Transf}(S)$ *such that* $\mathcal{T}' \circ \mathcal{T} = \mathcal{T} \circ \mathcal{T}' = \mathcal{I}_S$.

With the above definition, we have the following:

Proposition 12. *If system S satisfies the Physical Conservation of Information, then every physical transformation is physically reversible. The monoid* $\mathsf{Transf}(S)$ *is a group, hereafter denoted as* $\mathsf{G}(S)$.

Proof. Since $\mathcal{T}$ is physically invertible, there exists a transformation $\mathcal{T}'$ such that $\mathcal{T}' \circ \mathcal{T} = \mathcal{I}_S$. Since the Physical Conservation of Information holds, $\mathcal{T}'$ must be physically invertible, meaning that there exists a transformation $\mathcal{T}''$ such that $\mathcal{T}'' \circ \mathcal{T}' = \mathcal{I}_S$. Hence, we have

$$\mathcal{T}'' = \mathcal{T}'' \circ (\mathcal{T}' \circ \mathcal{T}) = (\mathcal{T}'' \circ \mathcal{T}') \circ \mathcal{T} = \mathcal{T}. \tag{68}$$

Since $\mathcal{T}'' = \mathcal{T}$, the invertibility condition $\mathcal{T}'' \circ \mathcal{T}' = \mathcal{I}_S$ becomes $\mathcal{T} \circ \mathcal{T}' = \mathcal{I}_S$. Hence, $\mathcal{T}$ is reversible and Transf(S) is a group. □

7.3. Subsystems of Systems Satisfying the Physical Conservation of Information

Imagine that an agent A acts on a system S satisfying the Physical Conservation of Information. We assume that the actions of agent A form a subgroup of $G(S)$, denoted as G_A. The maximal adversary of A is the adversary $B = A'$, who has access to all transformations in the set

$$G_B := G'_A = \left\{ \mathcal{U}_B \in G(S) : \ \mathcal{U}_B \circ \mathcal{U}_A = \mathcal{U}_A \circ \mathcal{U}_B, \ \forall \mathcal{U}_A \in G(A) \right\}. \tag{69}$$

It is immediate to see that the set G_B is a group. We call it the *adversarial group*.

The equivalence relations used to define subsystems can be greatly simplified. Indeed, it is easy to see that two states $\psi, \psi' \in \mathrm{St}(S)$ are equivalent for A if and only if there exists a transformation $\mathcal{U}_B \in G_B$ such that

$$\psi' = \mathcal{U}_B \psi. \tag{70}$$

Hence, the states of the subsystem S_A are orbits of the group G_B: for every $\psi \in \mathrm{St}(S)$, we have

$$\mathrm{Tr}_B[\psi] := \left\{ \mathcal{U}_B \psi : \ \mathcal{U}_B \in G_B \right\}. \tag{71}$$

Similarly, the degradation of a transformation $\mathcal{U} \in G(S)$ yields the orbit

$$\mathrm{Deg}_B(\mathcal{U}) = \left\{ \mathcal{U}_{B,1} \circ \mathcal{U} \circ \mathcal{U}_{B,2} : \quad \mathcal{U}_{B,1}, \mathcal{U}_{B,2} \in G_B \right\}. \tag{72}$$

It is easy to show that the transformations of the subsystem S_A are the orbits of the group G_B:

$$\mathrm{Transf}(S_A) = \left\{ \pi_A(\mathcal{U}) : \ \mathcal{U} \in G''_A \right\}, \qquad \pi_A(\mathcal{U}) := \left\{ \mathcal{U}_B \circ \mathcal{U} : \ \mathcal{U}_B \in G_B \right\}. \tag{73}$$

8. Closed Systems

Here, we define an abstract notion of "closed systems", which captures the essential features of what is traditionally called a closed system in quantum theory. Intuitively, the idea is that all the states of the closed system are "pure" and all the evolutions are reversible.

An obvious problem in defining closed system is that our framework does not include a notion of "pure state". To circumvent the problem, we define the closed systems in the following way:

Definition 15. *System S is closed iff it satisfies the Logical Conservation of Information and the Initialiation Requirement, that is, iff*

1. *every transformation is logically invertible,*
2. *there exists a state $\psi_0 \in \mathrm{St}(S)$ such that, for every other state $\psi \in \mathrm{St}(S)$, one has $\psi = \mathcal{V}\psi_0$ for some suitable transformation $\mathcal{V} \in \mathrm{Transf}(S)$.*

For a closed system, we nominally say that all the states in $\mathrm{St}(S)$ are "pure", or, more precisely, "dynamically pure". This definition is generally different from the usual definition of pure states as extreme points of convex sets, or from the compositional definition of pure states as states with only product extensions [77]. First of all, dynamically pure states are *not a subset* of the state space: provided that the right conditions are met, they are *all* the states. Other differences between the usual notion of pure states and the notion of dynamically pure states are highlighted by the following example:

Example 4. *Let S be a system in which all states are of the form $U\rho_0 U^\dagger$, where U is a generic 2-by-2 unitary matrix, and $\rho_0 \in M_2(\mathbb{C})$ is a fixed 2-by-2 density matrix. For the transformations, we allow all unitary channels $U \cdot U^\dagger$. By construction, system S satisfies the initialization Requirement, as one can generate every state from the initial state ρ_0. Moreover, all the transformations of system S are unitary and therefore the Conservation of Information is satisfied, both at the physical and the logical level. Therefore, the states of system S are dynamically pure. Of course, the states $U\rho_0 U^\dagger$ need not be extreme points of the convex set of all density matrices, i.e., they need not be rank-one projectors. They are so only when the cyclic state ρ_0 is rank-one.*

On the other hand, consider a similar example, where

- *system S is a qubit,*
- *the states are pure states, of the form $|\psi\rangle\langle\psi|$ for a generic unit vector $|\psi\rangle \in \mathbb{C}^2$,*
- *the transformations are unitary channels $V \cdot V^\dagger$, where the unitary matrix V has real entries.*

Using the Bloch sphere picture, the physical transformations are rotations around the y axis. Clearly, the Initialization Requirement is not satisfied because there is no way to generate arbitrary points on the sphere using only rotations around the y-axis. In this case, the states of S are pure in the convex set sense, but not dynamically pure.

For closed systems satisfying the Physical Conservation of Information, every pair of pure states are interconvertible:

Proposition 13 (Transitive action on the pure states). *If system S is closed and satisfies the Physical Conservation of Information, then, for every pair of states $\psi, \psi' \in \mathsf{St}(S)$, there exists a reversible transformation $\mathcal{U} \in \mathsf{G}(S)$ such that $\psi' = \mathcal{U}\psi$.*

Proof. By the Initialization Requirement, one has $\psi = \mathcal{V}\psi_0$ and $\psi' = \mathcal{V}'\psi_0$ for suitable $\mathcal{V}, \mathcal{V}' \in \mathsf{Transf}(S)$. By the Physical Conservation of Information, all the tranformations in $\mathsf{Transf}(S)$ are physically reversible. Hence, $\psi' = \mathcal{V}' \circ \mathcal{V}^{-1}\psi = \mathcal{U}\psi$, having defined $\mathcal{U} = \mathcal{V}' \circ \mathcal{V}^{-1}$. □

The requirement that all pure states be connected by reversible transformations has featured in many axiomatizations of quantum theory, either directly [5,44–46], or indirectly as a special case of other axioms [42,48]. Comparing our framework with the framework of general probabilistic theories, we can see that the dynamical definition of pure states refers to a rather specific situation, in which all pure states are connected, either to each other (in the case of physical reversibility) or with to a fixed cyclic state (in the case of logical reversibility).

9. Purification

Here, we show that closed systems satisfying the Physical Conservation of Information also satisfy the purification property [8,12,13,15,49–51], namely the property that every mixed state can be modelled as a pure state of a larger system in a canonical way. Under a certain regularity assumption, the same holds for closed systems satisfying only the Logical Conservation of Information.

9.1. Purification in Systems Satisfying the Physical Conservation of Information

Proposition 14 (Purification). *Let S be a closed system satisfying the Physical Conservation of Information. Let A be an agent in S, and let $B = A'$ be its maximal adversary. Then, for every state $\rho \in \mathsf{St}(S_A)$, there exists a pure state $\psi \in \mathsf{St}(S)$, called the purification of ρ, such that $\rho = \mathrm{Tr}_B[\psi]$. Moreover, the purification* of ρ is essentially unique: if $\psi' \in \text{St}(S)$ is another pure state with $\text{Tr}_B[\psi'] = \rho$, then there exists a reversible transformation $\mathcal{U}_B \in \mathsf{G}_B$ such that $\psi' = \mathcal{U}_B \psi$.

Proof. By construction, the states of system S_A are orbits of states of system S under the adversarial group G_B. By Equation (71), every two states $\psi, \psi' \in \text{St}(S)$ in the same orbit are connected by an element of G_B. □

Note that the notion of purification used here is more general than the usual notion of purification in quantum information and quantum foundations. The most important difference is that system S_A need not be a factor in a tensor product. Consider the example of the coherent superpositions vs. classical mixtures (Section 4.3). There, systems S_A and S_B coincide, their states are classical probability distributions, and the purifications are coherent superpositions. Two purifications of the same classical state $\mathbf{p} = (p_1, p_2, \ldots, p_d)$ are two rank-one projectors $|\psi\rangle\langle\psi|$ and $|\psi'\rangle\langle\psi'|$ corresponding to unit vectors of the form

$$|\psi\rangle = \sum_n \sqrt{p_n}\, e^{i\theta_n} |n\rangle \quad \text{and} \quad |\psi'\rangle = \sum_n \sqrt{p_n}\, e^{i\theta'_n} |n\rangle. \tag{74}$$

One purification can be obtained from the other by applying a diagonal unitary matrix. Specifically, one has

$$|\psi'\rangle = U_B |\psi\rangle \quad \text{with} \quad U_B = \sum_n e^{i(\theta'_n - \theta_n)} |n\rangle\langle n|. \tag{75}$$

For finite dimensional quantum systems, the notion of purification proposed here encompasses both the notion of entanglement and the notion of coherent superposition. The case of infinite dimensional systems will be discussed in the next subsection.

9.2. Purification in Systems Satisfying the Logical Conservation of Information

For infinite dimensional quantum systems, every density matrix can be purified, but not all purifications are connected by reversible transformations. Consider for example the unit vectors

$$|\psi\rangle_{AB} = \sqrt{1-x^2} \sum_{n=0}^{\infty} x^n |n\rangle_A \otimes |n\rangle_B \quad \text{and} \quad |\psi'\rangle_{AB} = \sqrt{1-x^2} \sum_{n=0}^{\infty} x^n |n\rangle_A \otimes |n+1\rangle_B, \tag{76}$$

for some $x \in [0,1)$.

For every fixed $x \neq 0$, there is one and only one operator V_B satisfying the condition $|\psi'\rangle_{AB} = (I_A \otimes V_B)|\psi\rangle_{AB}$, namely the shift operator $V_B = \sum_{n=0}^{\infty} |n+1\rangle\langle n|$. However, V_B is only an isometry, but not a unitary. This means that, if we define the states of system S_A as equivalence classes of pure states under local unitary equivalence, the two states $|\psi\rangle\langle\psi|$ and $|\psi'\rangle\langle\psi'|$ would end up into two different equivalence classes.

One way to address the problem is to relax the requirement of reversibility and to consider the monoid of isometries, defining

$$\text{Transf}(S) := \{V \cdot V^\dagger : V \in \text{Lin}(S),\ V^\dagger V = I\}. \tag{77}$$

Given two purifications of the same state, say $|\psi\rangle$ and $|\psi'\rangle$, it is possible to show that at least one of the following possibilities holds:

1. $|\psi'\rangle = (I_A \otimes V_B) |\psi\rangle$ for some isometry V_B acting on system S_B,
2. $|\psi\rangle = (I_A \otimes V_B) |\psi'\rangle$ for some isometry V_B acting on system S_B.

Unfortunately, this uniqueness property is not automatically valid in every system satisfying the Logical Conservation of Information. Still, we will now show a regularity condition, under which the uniqueness property is satisfied:

Definition 16. Let S be a system satisfying the Logical Conservation of Information, let $M \subseteq \mathsf{Transf}(S)$ be a monoid, and let $\mathsf{Deg}_M(\psi)$ be the set defined by

$$\mathsf{Deg}_M(\psi) = \{\mathcal{V}\psi : \quad \mathcal{V} \in M\}. \tag{78}$$

We say that the monoid $M \subseteq \mathsf{Transf}(S)$ is *regular* iff

1. for every pair of states $\psi, \psi' \in \mathsf{St}(S)$, the condition $\mathsf{Deg}_M(\psi) \cap \mathsf{Deg}_M(\psi') \neq \emptyset$ implies that there exists a transformation $\mathcal{U} \in M$ such that $\psi' = \mathcal{U}\psi$ or $\psi = \mathcal{U}\psi'$,
2. for every pair of transformations $\mathcal{V}, \mathcal{V}' \in M$, there exists a transformation $\mathcal{W} \in M$ such that $\mathcal{V} = \mathcal{W} \circ \mathcal{V}'$ or $\mathcal{V}' = \mathcal{W} \circ \mathcal{V}$.

The regularity conditions are satisfied in quantum theory by the monoid of isometries.

Example 5 (Isometric channels in quantum theory). *Let S be a quantum system with separable Hilbert space $\mathcal{H}$, of dimension $d \leq \infty$. Let $\mathsf{St}(S)$ the set of all pure quantum states, and let $\mathsf{Transf}(S)$ be the monoid of all isometric channels.*

We now show that the monoid $M = \mathsf{Transf}(S)$ is regular. The first regularity condition is immediate because for every pair of unit vectors $|\psi\rangle$ and $|\psi'\rangle$ there exists an isometry (in fact, a unitary) V such that $|\psi'\rangle = U|\psi\rangle$. Trivially, this implies the relation $|\psi'\rangle\langle\psi'| = U|\psi\rangle\langle\psi|U^\dagger$ at the level of quantum states and isometric channels.

Let us see that the second regularity condition holds. Let $V, V' \in \mathsf{Lin}(\mathcal{H})$ be two isometries on $\mathcal{H}$, and let $\{|i\rangle\}_{i=1}^d$ be the standard basis for $\mathcal{H}$. Then, the isometries V and V' can be written as

$$V = \sum_{i=1}^d |\phi_i\rangle\langle i| \quad \text{and} \quad V' = \sum_i |\phi'_i\rangle\langle i|, \tag{79}$$

where $\{|\phi_i\rangle\}_{i=1}^d$ and $\{|\phi'_i\rangle\}_{i=1}^d$ are orthonormal vectors (not necessarily forming bases for the whole Hilbert space $\mathcal{H}$). Define the subspaces $S = \mathsf{Span}\{|\phi_i\rangle\}_{i=1}^d$ and $S' = \mathsf{Span}\{|\phi'_i\rangle\}_{i=1}^d$, and let $\{|\psi_j\rangle\}_{j=1}^r$ and $\{|\psi'_j\rangle\}_{j=1}^{r'}$ be orthonormal bases for the orthogonal complements $S^\perp$ and $S'^\perp$, respectively. If $r \leq r'$, we define the isometry

$$W = \left(\sum_{i=1}^d |\phi'_i\rangle\langle\phi_i|\right) + \left(\sum_{j=1}^r |\psi'_j\rangle\langle\psi_j|\right), \tag{80}$$

and we obtain the condition $V' = WV$. Alternatively, if $r' \leq r$, we can define the isometry

$$W = \left(\sum_{i=1}^d |\phi_i\rangle\langle\phi'_i|\right) + \left(\sum_{j=1}^r |\psi_j\rangle\langle\psi'_j|\right), \tag{81}$$

and we obtain the condition $V = WV'$. At the level of isometric channels, we obtained the condition $\mathcal{V}' = \mathcal{W} \circ \mathcal{V}$ or the condition $\mathcal{V} = \mathcal{W} \circ \mathcal{V}'$, with $\mathcal{V}(\cdot) = V \cdot V^\dagger$, $\mathcal{V}'(\cdot) = V' \cdot V'^\dagger$, and $\mathcal{W}(\cdot) = W \cdot W^\dagger$.

The fact that the monoid of all isometric channels is regular implies that other monoids of isometric channels are also regular. For example, if the Hilbert space $\mathcal{H}$ has the tensor product structure $\mathcal{H} = \mathcal{H}_A \otimes \mathcal{H}_B$, then the monoid of local isometric channels, defined by isometries of the form $I_A \otimes V_B$, is regular. More generally, if the Hilbert space is decomposed as

$$\mathcal{H} = \bigoplus_k (\mathcal{H}_{A,k} \otimes \mathcal{H}_{B,k}), \tag{82}$$

then the monoid of isometric channels generated by isometries of the form

$$V = \bigoplus_k (I_{A,k} \otimes V_{B,k}) \tag{83}$$

is regular.

We are now in position to derive the purification property for general closed systems:

Proposition 15. *Let S be a closed system. Let A be an agent and let $B = A'$ be its maximal adversary. If $\mathrm{Act}(B;S)$ is a regular monoid, the condition $\mathrm{Tr}_B[\psi] = \mathrm{Tr}_B[\psi']$ implies that there exists some invertible transformation $\mathcal{V}_B \in \mathrm{Transf}(B;S)$ such that the relation $\psi' = \mathcal{V}_B \psi$ or the relation $\psi = \mathcal{V}_B \psi'$ holds.*

The proof is provided in Appendix H. In conclusion, we obtained the following

Corollary 1 (Purification). *Let S be a closed system, let A be an agent in S, and let $B = A'$ be its maximal adversary. If the monoid $\mathrm{Act}(B;S)$ is regular, then every state $\rho \in \mathrm{St}(S_A)$ has a purification $\psi \in \mathrm{St}(S)$, i.e., a state such that $\rho = \mathrm{Tr}_B[\psi]$. Moreover, the purification is essentially unique: if $\psi' \in \mathrm{St}(S)$ is another state with $\mathrm{Tr}_B[\psi] = \rho$, then there exists a reversible transformation $\mathcal{V}_B \in \mathrm{Act}(B;S)$ such that the relation $\psi' = \mathcal{V}_B \psi$ or the relation $\psi = \mathcal{V}_B \psi'$ holds.*

10. Example: Group Representations on Quantum State Spaces

We conclude the paper with a macro-example, involving group representations in closed-system quantum theory. The point of this example is to illustrate the general notion of purification introduced in this paper and to characterize the sets of mixed states associated with different agents.

As system S, we consider a quantum system with Hilbert space $\mathcal{H}_S$, possibly of infinite dimension. We let $\mathrm{St}(S)$ be the set of pure quantum states, and let $\mathsf{G}(S)$ be the group of all unitary channels. With this choice, the total system is closed and satisfies the Physical Conservation of Information.

Suppose that agent A is able to perform a group of transformations, such as e.g., the group of phase shifts on a harmonic oscillator, or the group of rotations of a spin j particle. Mathematically, we focus our attention on unitary channels arising from some representation of a given compact group G. Denoting the representation as $U : \mathsf{G} \to \mathrm{Lin}(\mathcal{H}_S), g \mapsto U_g$, the group of Alice's actions is

$$\mathsf{G}_A = \left\{ \mathcal{U}_g(\cdot) = U_g \cdot U_g^\dagger : \quad g \in \mathsf{G} \right\}. \tag{84}$$

The maximal adversary of A is the agent $B = A'$ who is able to perform all unitary channels $\mathcal{V}$ that commute with those in G_A, namely, the unitary channels in the group

$$\mathsf{G}_B := \left\{ \mathcal{V} \in \mathsf{G}(S): \quad \mathcal{V} \circ \mathcal{U}_g = \mathcal{U}_g \circ \mathcal{V} \quad \forall g \in \mathsf{G} \right\}. \tag{85}$$

Specifically, the channels $\mathcal{V}$ correspond to unitary operators V satisfying the relation

$$V U_g = \omega(V, g) \, U_g V \quad \forall g \in \mathsf{G}, \tag{86}$$

where, for every fixed V, the function $\omega(V, \cdot) : \mathsf{G} \to \mathbb{C}$ is a multiplicative character, i.e., a one-dimensional representation of the group G.

Note that, if two unitaries V and W satisfy Equation (86) with multiplicative characters $\omega(V, \cdot)$ and $\omega(W, \cdot)$, respectively, then their product VW satisfies Equation (86) with multiplicative character $\omega(VW, \cdot) = \omega(V, \cdot)\omega(W, \cdot)$. This means that the function $\omega : \mathsf{G}_B \times \mathsf{G} \to \mathbb{C}$ is a multiplicative bicharacter: $\omega(V, \cdot)$ is a multiplicative character for G for every fixed $V \in \mathsf{G}_B$, and, at the same time, $\omega(\cdot, g)$ is a multiplicative character for G_B for every fixed $g \in \mathsf{G}$.

The adversarial group G_B contains the commutant of the representation $U : g \mapsto U_g$, consisting of all the unitaries V such that

$$VU_g = U_g V \qquad \forall g \in G. \tag{87}$$

The unitaries in the commutant satisfy Equation (86) with the trivial multiplicative character $\omega(V, g) = 1, \forall g \in G$. In general, the adversarial group may contain other unitary operators, corresponding to non-trivial multiplicative characters. The full characterization of the adversarial group is provided by the following theorem:

Theorem 3. *Let G be a compact group, let $U : G \to \mathrm{Lin}(\mathcal{H})$ be a projective representation of G, and let G_A be the group of channels $G_A := \{U_g \cdot U_g^\dagger \quad g \in G\}$. Then, the adversarial group G_B is isomorphic to the semidirect product $A \ltimes U'$, where U' is the commutant of the set $\{U_g : g \in G\}$, and A is an Abelian subgroup of the group of permutations of $\mathrm{Irr}(U)$, the set of irreducible representations contained in the decomposition of the representation U_g.*

The proof is provided in Appendix I, and a simple example is presented in Appendix J.
In the following, we will illustrate the construction of the state space S_A in a the prototypical example where the group G is a compact connected Lie group.

Compact Connected Lie Groups

When G is a compact connected Lie group, the characterization of the adversarial group is simplified by the following theorem:

Theorem 4. *If G is a compact connected Lie group, then the Abelian subgroup A of Theorem 3 is trivial, and all the solutions of Equation (86) have $\omega(V, g) = 1 \, \forall g \in G$.*

The proof is provided in Appendix K.
For compact connected Lie groups, the the adversarial group coincides exactly with the commutant of the representation $U : G \to \mathrm{Lin}(\mathcal{H}_S)$. An explicit expression can be obtained in terms of the isotypic decomposition [78]

$$U_g = \bigoplus_{j \in \mathrm{Irr}(U)} \left(U_g^{(j)} \otimes I_{\mathcal{M}_j} \right), \tag{88}$$

where $\mathrm{Irr}(U)$ is the set of irreducible representations (irreps) of G contained in the decomposition of U, $U^{(j)} : g \mapsto U_g^{(j)}$ is the irreducible representation of G acting on the representation space $\mathcal{R}_j$, and $I_{\mathcal{M}_j}$ is the identity acting on the multiplicity space $\mathcal{M}_j$. From this expression, it is clear that the adversarial group G_B consists of unitary gates V of the form

$$V = \bigoplus_{j \in \mathrm{Irr}(U)} \left(I_{\mathcal{R}_j} \otimes V_j \right), \tag{89}$$

where $I_{\mathcal{R}_j}$ is the identity operator on the representation space $\mathcal{R}_j$, and V_j is a generic unitary operator on the multiplicity space $\mathcal{M}_j$.

In general, the agents A and $B = A'$ do not form a dual pair. Indeed, it is not hard to see that the maximal adversary of B is the agent $C = A''$ that can perform every unitary channel $\mathcal{U}(\cdot) = U \cdot U^\dagger$, where U is a unitary operator of the form

$$U = \bigoplus_{j \in \mathrm{Irr}(U)} \left(U_j \otimes I_{\mathcal{M}_j} \right), \tag{90}$$

U_j being a generic unitary operator on the representation space $\mathcal{R}_j$. When A and B form a dual par, the groups G_A and G_B are sometimes called *gauge groups* [79].

It is now easy to characterize the subsystem S_A. Its states are equivalence classes of pure states under the relation $|\psi\rangle\langle\psi| \simeq_A |\psi'\rangle\langle\psi'|$ iff

$$\exists U_B \in G_B \quad \text{such that} \quad |\psi'\rangle = U_B|\psi\rangle. \tag{91}$$

It is easy to see that two states in the same equivalence class must satisfy the condition

$$\text{Tr}_B(|\psi'\rangle\langle\psi'|) = \text{Tr}_B(|\psi\rangle\langle\psi|), \tag{92}$$

where the "partial trace over agent B" is Tr_B is the map

$$\text{Tr}_B(\rho) := \bigoplus_{j \in \text{Irr}(U)} \text{Tr}_{\mathcal{M}_j}[\Pi_j \rho \Pi_j], \tag{93}$$

Π_j being the projector on the subspace $\mathcal{R}_j \otimes \mathcal{M}_j$.

Conversely, it is possible to show that the state $\text{Tr}_B(|\psi\rangle\langle\psi|)$ completely identifies the equivalence class $[|\psi\rangle\langle\psi|]_A$.

Proposition 16. *Let $|\psi\rangle, |\psi'\rangle \in \mathcal{H}_S$ be two unit vectors such that $\text{Tr}_B(|\psi\rangle\langle\psi|) = \text{Tr}_B(|\psi'\rangle\langle\psi'|)$. Then, there exists a unitary operator $U_B \in G_B$ such that $|\psi'\rangle = U_B|\psi\rangle$.*

The proof is provided in Appendix L.

We have seen that the states of system S_A are in one-to-one correspondence with the density matrices of the form $\text{Tr}_B(|\psi\rangle\langle\psi|)$, where $|\psi\rangle \in \mathcal{H}_S$ is a generic pure state. Note that the rank of the density matrices ρ_j in Equation (A109) cannot be larger than the dimensions of the spaces $\mathcal{R}_j$ and $\mathcal{M}_j$, denoted as $d_{\mathcal{R}_j}$ and $d_{\mathcal{M}_j}$, respectively. Taking this fact into account, we can represent the states of S_A as

$$\text{St}(S_A) \simeq \left\{ \rho = \bigoplus_{j \in \text{Irr}(U)} p_j \rho_j : \rho_j \in \text{QSt}(\mathcal{R}_j), \text{Rank}(\rho_j) \leq \min\{d_{\mathcal{R}_j}, d_{\mathcal{M}_j}\} \right\}, \tag{94}$$

where $\{p_j\}$ is a generic probability distribution. The state space of system S_A is *not convex*, unless the condition

$$d_{\mathcal{M}_j} \geq d_{\mathcal{R}_j} \quad \forall j \in \text{Irr}(U) \tag{95}$$

is satisfied. Basically, in order to obtain a convex set of density matrices, we need the total system S to be "sufficiently large" compared to its subsystem S_A. This observation is a clue suggesting that the standard convex framework could be considered as the effective description of subsystems of "large" closed systems.

Finally, note that, in agreement with the general construction, the pure states of system S are "purifications" of the states of the system S_A. Every state of system S_A can be obtained from a pure state of system S by "tracing out" system S_B. Moreover, every two purifications of the same state are connected by a unitary transformation in G_B.

11. Conclusions

In this paper, we adopted rather minimalistic framework, in which a single physical system was described solely in terms of states and transformations, without introducing measurements. Or at least, without introducing measurements *in an explicit way*: of course, one could always interpret certain transformations as "measurement processes", but this interpretation is not necessary for any of the conclusions drawn in this paper.

Our framework can be interpreted in two ways. One way is to think of it as a fragment of the larger framework of operational-probabilistic theories [8,11–13], in which systems can be freely composed and measurements are explicitly described. The other way is to regard our framework as a dynamicist framework, meant to describe physical systems *per se*, independently of any observer. Both approaches are potentially fruitful.

On the operational-probabilistic side, it is interesting to see how the definition of subsystem adopted in this paper interacts with probabilities. For example, we have seen in a few examples that the state space of a subsystem is not always convex: convex combination of allowed states are not necessarily allowed states. It is then natural to ask: under which condition is convexity retrieved? In a different context, the non-trivial relation between convexity and the dynamical notion of system has been emerged in a work of Galley and Masanes [80]. There, the authors studied alternatives to quantum theory where the closed systems have the same states and the same dynamics of closed quantum systems, while the measurements are different from the quantum measurements. Among these theories, they found that quantum theory is the only theory where subsystems have a convex state space. These and similar clues are an indication that the interplay between dynamical notions and probabilistic notions plays an important role in determining the structure of physical theories. Studying this interplay is a promising avenue of future research.

On the opposite end of the spectrum, it is interesting to explore how far the measurement-free approach can reach. An interesting research project is to analyze the notions of subsystem, pure state, and purification, in the context of algebraic quantum field theory [22] and quantum statistical mechanics [32]. This is important because the notion of pure state as an extreme point of the convex set breaks down for type III von Neumann algebras [81], whereas the notions used in this paper (commutativity of operations, cyclicity of states) would still hold. Another promising clue is the existence of dual pairs of non-overlapping agents, which amounts to the requirement that the set of operations of each agent has trivial center and coincides with its double commutant. A similar condition plays an important role in the algebraic framework, where the operator algebras with trivial center are known as factors, and are at the basis of the theory of von Neumann algebras [82,83].

Finally, another interesting direction is to enrich the structure of system with additional features, such as a metric, quantifying the proximity of states. In particular, one may consider a strengthened formulation of the Conservation of Information, in which the physical transformations are required not only to be invertible, but also to preserve the distances. It is then interesting to consider how the metric on the pure states of the whole system induces a metric on the subsystems, and to search for relations between global metric and local metric. Also in this case, there is a promising precedent, namely the work of Uhlmann [84], which led to the notion of fidelity [85]. All these potential avenues of future research suggest that the notions investigated in this work may find application in a variety of different contexts, and for a variety of interpretational standpoints.

Acknowledgments: It is a pleasure to thank Gilles Brassard and Paul Raymond-Robichaud for stimulating discussions on their recent work [66], Adán Cabello, Markus Müller, and Matthias Kleinmann for providing motivation to the problem of deriving subsystems, Mauro D'Ariano and Paolo Perinotti for the invitation to contribute to this Special Issue, and Christopher Timpson and Adam Coulton for an invitation to present at the Oxford Philosophy of Physics Seminar Series, whose engaging atmosphere stimulated me to think about extensions of the Purification Principle. I am also grateful to the three referees of this paper for useful suggestions, and to Robert Spekkens, Doreen Fraser, Lídia del Rio, Thomas Galley, John Selby, Ryszard Kostecki, and David Schmidt for interesting discussions during the revision of the original manuscript. This work is supported by the Foundational Questions Institute through grant FQXi-RFP3-1325, the National Natural Science Foundation of China through grant 11675136, the Croucher Foundation, the Canadian Institute for Advanced Research (CIFAR), and the Hong Research Grant Council through grant 17326616. This publication was made possible through the support of a grant from the John Templeton Foundation. The opinions expressed in this publication are those of the authors and do not necessarily reflect the views of the John Templeton Foundation. The authors also acknowledge the hospitality of Perimeter Institute for Theoretical Physics. Research at Perimeter Institute is supported by the Government of Canada through the Department of Innovation, Science and Economic Development Canada and by the Province of Ontario through the Ministry of Research, Innovation and Science.

Conflicts of Interest: The author declares no conflict of interest.

Appendix A. Proof That Definitions (20) and (21) Are Well-Posed

We give only the proof for definition (20), as the other proof follows the same argument.

Proposition A1. *If the transformations $S, \tilde{S}, T, \tilde{T} \in \text{Act}(A;S)''$ are such that $[S]_A = [\tilde{S}]_A$ and $[T]_A = [\tilde{T}]_A$, then $[S \circ T]_A = [\tilde{S} \circ \tilde{T}]_A$.*

Proof. Let $(S_1, S_2, \ldots, S_m) \subset \text{Act}(A;S)''$ and $(T_1, T_2, \ldots, T_n) \subset \text{Act}(A;S)''$ be two finite sequences such that

$$S_1 = S, \quad S_m = \tilde{S}, \quad \text{Deg}_{A'}(S_i) \cap \text{Deg}_{A'}(S_{i+1}) \neq \emptyset \quad \forall i \in \{1, \ldots, m-1\},$$
$$T_1 = T, \quad T_n = \tilde{T}, \quad \text{Deg}_{A'}(T_j) \cap \text{Deg}_{A'}(T_{j+1}) \neq \emptyset \quad \forall j \in \{1, \ldots, n-1\}. \tag{A1}$$

Without loss of generality, we assume that the two finite sequences have the same length $m = n$. When this is not the case, one can always add dummy entries and ensure that the two sequences have the same length: for example, if $m < n$, one can always define $S_i := S_m$ for all $i \in \{m+1, \ldots, n\}$.

Equation (A1) mean that for every i and j there exist transformations $B_i, \tilde{B}_i, C_j, \tilde{C}_j \in \text{Act}(A;S)'$ such that

$$B_i \circ S_i = \tilde{B}_i \circ S_{i+1},$$
$$C_j \circ T_j = \tilde{C}_j \circ T_{j+1}. \tag{A2}$$

Using the above equalities for $i = j$, and using the fact that transformations in $\text{Act}(A;S)'$ commute with transformations in $\text{Act}(A;S)''$, we obtain

$$(B_i \circ C_i) \circ (S_i \circ T_i) = (B_i \circ S_i) \circ (C_i \circ T_i)$$
$$= (\tilde{B}_i \circ S_{i+1}) \circ (\tilde{C}_i \circ T_{i+1})$$
$$= (\tilde{B}_i \circ \tilde{C}_i) \circ (S_{i+1} \circ T_{i+1}). \tag{A3}$$

In short, we proved that

$$\text{Deg}_{A'}(S_i \circ T_i) \cap \text{Deg}_{A'}(S_{i+i} \circ T_{i+1}) \neq \emptyset \quad \forall i \in \{1, \ldots, n-1\}. \tag{A4}$$

To conclude, observe that the sequence $(S_1 \circ T_1, S_2 \circ T_2, \ldots, S_n \circ T_n)$ satisfies $S_1 \circ T_1 = S \circ T$, $S_n \circ T_n = \tilde{S} \circ \tilde{T}$, and Equation (A4). By definition, this means that the transformations $S \circ T$ and $\tilde{S} \circ \tilde{T}$ are in the same equivalence class. □

Appendix B. The Commutant of the Local Channels

Here, we show that the commutant of the quantum channels of the form $\mathcal{A} \otimes \mathcal{I}_B$ consists of quantum channels of the form $\mathcal{I}_A \otimes \mathcal{B}$.

Let $\mathcal{C} \in \text{Chan}(S)$ be a quantum channel that commutes with all channels of the form $\mathcal{A} \otimes \mathcal{I}_B$, with $\mathcal{A} \in \text{Chan}(A)$. For a fixed unit vector $|\alpha\rangle \in \mathcal{H}_A$, consider the erasure channel $\mathcal{A}_\alpha \in \text{Chan}(A)$ defined by

$$\mathcal{A}_\alpha(\rho) = |\alpha\rangle\langle\alpha| \, \text{Tr}[\rho] \quad \forall \rho \in \text{Lin}(A). \tag{A5}$$

Then, the commutation condition $C \circ (\mathcal{A}_\alpha \otimes \mathcal{I}_B) = (\mathcal{A}_\alpha \otimes \mathcal{I}_B) \circ C$ implies

$$\begin{aligned}
C\Big(|\alpha\rangle\langle\alpha| \otimes |\beta\rangle\langle\beta|\Big) &= C\Big[\big(\mathcal{A}_\alpha \otimes \mathcal{I}_B\big)\big(|\alpha\rangle\langle\alpha| \otimes |\beta\rangle\langle\beta|\big)\Big] \\
&= \big(\mathcal{A}_\alpha \otimes \mathcal{I}_B\big)\Big[C\big(|\alpha\rangle\langle\alpha| \otimes |\beta\rangle\langle\beta|\big)\Big] \\
&= |\alpha\rangle\langle\alpha| \otimes \operatorname{Tr}_A\Big[C\big(|\alpha\rangle\langle\alpha| \otimes |\beta\rangle\langle\beta|\big)\Big] \qquad \forall |\beta\rangle \in \mathcal{H}_B.
\end{aligned} \qquad (A6)$$

Tracing over B on both sides of Equation (A6), we obtain

$$\operatorname{Tr}_B\Big[C\big(|\alpha\rangle\langle\alpha| \otimes |\beta\rangle\langle\beta|\big)\Big] = |\alpha\rangle\langle\alpha|. \qquad (A7)$$

The above relation implies that the state $C\big(|\alpha\rangle\langle\alpha| \otimes |\beta\rangle\langle\beta|\big)$ is of the form

$$C\big(|\alpha\rangle\langle\alpha| \otimes |\beta\rangle\langle\beta|\big) = |\alpha\rangle\langle\alpha| \otimes \mathcal{B}(|\beta\rangle\langle\beta|), \qquad (A8)$$

for some suitable channel $\mathcal{B} \in \operatorname{Chan}(B)$. Since $|\alpha\rangle$ and $|\beta\rangle$ are arbitrary, we obtained $C = \mathcal{I}_A \otimes \mathcal{B}$.

Appendix C. Subsystems Associated to Finite Dimensional Von Neumann Algebras

Here, we prove the statements made in the main text about quantum channels with Kraus operators in a given algebra.

Appendix C.1. The Commutant of $\operatorname{Chan}(A)$

The purpose of this subsection is to prove the following theorem:

Theorem A1. *Let* A *be a von Neumann subalgebra of* $M_d(\mathbb{C})$, $d < \infty$, *and let* $\operatorname{Chan}(A)$ *be the set of quantum channels with Kraus operators in* A. *Then, the commutant of* $\operatorname{Chan}(A)$ *is the set of channels with Kraus operators in the algebra* A'. *In formula,*

$$\operatorname{Chan}(A)' = \operatorname{Chan}(A'). \qquad (A9)$$

The proof consists of a few lemmas, provided in the following.

Lemma A1. *Every channel* $\mathcal{D} \in \operatorname{Chan}(A)'$ *must satisfy the condition*

$$\mathcal{P}_l \circ \mathcal{D} \circ \mathcal{P}_k = 0 \qquad \forall l \neq k, \qquad (A10)$$

where $\mathcal{P}_k$ *is the CP map* $\mathcal{P}_k(\cdot) := \Pi_k \cdot \Pi_k$, *and* Π_k *is the projector on the subspace* $\mathcal{H}_{A_k} \otimes \mathcal{H}_{B_k}$ *in Equation (31).*

Proof. Consider the quantum channel $C \in \operatorname{Chan}(A)$ defined as

$$C := \bigoplus_k \Big(|\alpha_k\rangle\langle\alpha_k| \operatorname{Tr}_{A_k} \otimes \mathcal{I}_{B_k}\Big) \circ \mathcal{P}_k, \qquad (A11)$$

where each $|\alpha_k\rangle$ is a generic (but otherwise fixed) unit vector in $\mathcal{H}_{A_k}$ and $\mathcal{I}_{B_k}$ is the identity map on $\mathrm{Lin}(\mathcal{H}_{B_k})$. By definition, every channel $\mathcal{D} \in \mathrm{Chan}(A)'$ must satisfy the condition $\mathcal{C} \circ \mathcal{D} = \mathcal{D} \circ \mathcal{C}$. In particular, we must have

$$\begin{aligned} \mathcal{D}(|\alpha_k\rangle\langle\alpha_k| \otimes |\beta_k\rangle\langle\beta_k|) &= (\mathcal{D} \circ \mathcal{C})(|\alpha_k\rangle\langle\alpha_k| \otimes |\beta_k\rangle\langle\beta_k|) \\ &= (\mathcal{C} \circ \mathcal{D})(|\alpha_k\rangle\langle\alpha_k| \otimes |\beta_k\rangle\langle\beta_k|) \\ &= \bigoplus_l \Big(|\alpha_l\rangle\langle\alpha_l| \otimes \mathrm{Tr}_{A_l}\big[(\mathcal{P}_l \circ \mathcal{D})(|\alpha_k\rangle\langle\alpha_k| \otimes |\beta_k\rangle\langle\beta_k|) \big] \Big). \end{aligned} \quad \text{(A12)}$$

Applying the CP map $\mathcal{P}_l$ on both sides of the above equality, we obtain the relation

$$(\mathcal{P}_l \circ \mathcal{D})(|\alpha_k\rangle\langle\alpha_k| \otimes |\beta_k\rangle\langle\beta_k|) = |\alpha_l\rangle\langle\alpha_l| \otimes \mathcal{M}_l(|\alpha_k\rangle\langle\alpha_k| \otimes |\beta_k\rangle\langle\beta_k|), \quad \text{(A13)}$$

where $\mathcal{M}_l$ is the map from $M_d(\mathbb{C})$ to $\mathrm{Lin}(\mathcal{H}_{A_l})$ defined as $\mathcal{M}_l := \mathrm{Tr}_{A_l} \circ \mathcal{P}_l \circ \mathcal{D}$.

Note that the right-hand side of Equation (A13) depends on the choice of vector $|\alpha_l\rangle$, which is arbitrary. On the other hand, the left-hand side does not depend on $|\alpha_l\rangle$. Hence, the only way that the two sides of Equation (A13) can be equal for $k \neq l$ is that they are both equal to 0. Moreover, since $|\alpha_k\rangle$ and $|\beta_k\rangle$ are arbitrary vectors in $\mathcal{H}_{A_k}$ and $\mathcal{H}_{B_k}$, respectively, Equation (A13) implies the relation

$$(\mathcal{P}_l \circ \mathcal{D})(\rho) = 0 \quad \forall \rho \in \mathrm{Lin}(\mathcal{H}_{A_k} \otimes \mathcal{H}_{B_k}), \quad \forall l \neq k. \quad \text{(A14)}$$

Since ρ is an arbitrary operator in $\mathrm{Lin}(\mathcal{H}_{A_k} \otimes \mathcal{H}_{B_k})$, we conclude that the relation $\mathcal{P}_l \circ \mathcal{D} \circ \mathcal{P}_k = 0$ holds for every $l \neq k$. □

Lemma A2. *Every channel $\mathcal{D} \in \mathrm{Chan}(A)'$ must satisfy the conditions*

$$\mathcal{D} \circ \mathcal{P}_k = \mathcal{P}_k \circ \mathcal{D} \circ \mathcal{P}_k \quad \forall k \quad \text{(A15)}$$

and

$$\mathcal{P}_k \circ \mathcal{D} = \mathcal{P}_k \circ \mathcal{D} \circ \mathcal{P}_k \quad \forall k. \quad \text{(A16)}$$

In short: $\mathcal{D} \circ \mathcal{P}_k = \mathcal{P}_k \circ \mathcal{D}$ for every k.

Proof. Define $\mathcal{D}_k := \mathcal{D} \circ \mathcal{P}_k$. Then, the Cauchy–Schwarz inequality yields

$$\begin{aligned} \left| \langle \phi | \Pi_i \mathcal{D}_k(\rho) \Pi_j | \phi \rangle \right| &\leq \sqrt{\langle \phi | \Pi_i \mathcal{D}_k(\rho) \Pi_i | \phi \rangle \langle \phi | \Pi_j \mathcal{D}_k(\rho) \Pi_j | \phi \rangle} \\ &\leq \sqrt{\langle \phi | (\mathcal{P}_i \circ \mathcal{D} \circ \mathcal{P}_k)(\rho) | \phi \rangle \langle \phi | (\mathcal{P}_j \circ \mathcal{D} \circ \mathcal{P}_k)(\rho) | \phi \rangle}. \end{aligned} \quad \text{(A17)}$$

Thanks to Lemma A1, we know the right-hand side is 0 unless $i = j = k$. Since the vector $|\phi\rangle$ is are arbitrary, the condition $|\langle \phi | \Pi_i \mathcal{D}_k(\rho) \Pi_j | \phi \rangle| = 0$ implies the relation $\Pi_i \mathcal{D}_k(\rho) \Pi_j = 0$. Using this fact, we obtain the relation

$$\begin{aligned} (\mathcal{D} \circ \mathcal{P}_k)(\rho) &= \mathcal{D}_k(\rho) \\ &= \sum_{i,j} \Pi_i \mathcal{D}_k(\rho) \Pi_j \\ &= \Pi_k \mathcal{D}_k(\rho) \Pi_k \\ &= (\mathcal{P}_k \circ \mathcal{D} \circ \mathcal{P}_k)(\rho), \end{aligned} \quad \text{(A18)}$$

valid for arbitrary density matrices ρ, and therefore for arbitrary matrices in $M_d(\mathbb{C})$. In conclusion, Equation (A16) holds.

The proof of Equation (A15) is analogous to that of Equation (A16), with the only difference that it uses the *adjoint map*, which for a generic linear map $\mathcal{L} : \text{Lin}(\mathcal{H}_S) \to \text{Lin}(\mathcal{H}_S)$ is defined by the relation

$$\text{Tr}[\mathcal{L}^\dagger(O)\,\rho] := \text{Tr}[O\,\mathcal{L}(\rho)] \qquad \forall O \in M_d(\mathbb{C})\,,\ \forall \rho \in M_d(\mathbb{C})\,. \tag{A19}$$

Specifically, we define the map $\widetilde{\mathcal{D}}_k := \mathcal{P}_k \circ \mathcal{D}$. Then, we obtain the relation

$$\begin{aligned}
\left| \langle \phi | \widetilde{\mathcal{D}}_k(\Pi_i \rho \Pi_j) | \phi \rangle \right| &= \left| \text{Tr}\left[\widetilde{\mathcal{D}}_k^\dagger(|\phi\rangle\langle\phi|) \Pi_i \rho \Pi_j \right] \right| \\
&= \left| \text{Tr}\left[\left(\sqrt{\widetilde{\mathcal{D}}_k^\dagger(|\phi\rangle\langle\phi|)} \Pi_i \sqrt{\rho} \right) \left(\sqrt{\rho} \Pi_j \sqrt{\widetilde{\mathcal{D}}_k^\dagger(|\phi\rangle\langle\phi|)} \right) \right] \right| \\
&\le \sqrt{ \text{Tr}\left[\widetilde{\mathcal{D}}_k^\dagger(|\phi\rangle\langle\phi|) \Pi_i \rho \Pi_i \right] \text{Tr}\left[\widetilde{\mathcal{D}}_k^\dagger(|\phi\rangle\langle\phi|) \Pi_j \rho \Pi_j \right] } \\
&= \sqrt{ \langle \phi | \widetilde{\mathcal{D}}_k(\Pi_i \rho \Pi_i) | \phi \rangle \, \langle \psi | \widetilde{\mathcal{D}}_k(\Pi_j \rho \Pi_j) | \psi \rangle } \\
&= \sqrt{ \langle \phi | (\mathcal{P}_k \circ \mathcal{D} \circ \mathcal{P}_i)(\rho) | \phi \rangle \, \langle \psi | (\mathcal{P}_k \circ \mathcal{D} \circ \mathcal{P}_j)(\rho) | \psi \rangle }\,,
\end{aligned} \tag{A20}$$

where the right-hand side is 0 unless $i = j = k$ (cf. Lemma A2). Since the condition $|\langle \phi | \widetilde{\mathcal{D}}_k(\Pi_i \rho \Pi_j) | \phi \rangle| = 0$, $\forall |\phi\rangle \in \mathcal{H}_S$ implies the condition $\widetilde{\mathcal{D}}_k(\Pi_i \rho \Pi_j) = 0$, we obtained the relation

$$\widetilde{\mathcal{D}}_k(\Pi_i \rho \Pi_j) = 0 \quad \text{unless} \quad i = j = k\,. \tag{A21}$$

Using this fact, we obtain the equality

$$\begin{aligned}
(\mathcal{P}_k \circ \mathcal{D})(\rho) &= \widetilde{\mathcal{D}}_k(\rho) \\
&= \sum_{i,j} \widetilde{\mathcal{D}}_k(\Pi_i \rho \Pi_j) \\
&= (\widetilde{\mathcal{D}}_k \circ \mathcal{P}_k)(\rho) \\
&= (\mathcal{P}_k \circ \mathcal{D} \circ \mathcal{P}_k)(\rho)\,.
\end{aligned} \tag{A22}$$

Since the equality holds for every ρ, this proves Equation (A16). □

Lemma A2 guarantees that the linear map $\mathcal{D} \circ \mathcal{P}_k$ sends $\text{Lin}(\mathcal{R}_k \otimes \mathcal{M}_k)$ into itself. It is also easy to see that the map $\mathcal{D} \circ \mathcal{P}_k$ has a simple form:

Lemma A3. *For every channel* $\mathcal{D} \in \text{Chan}(A)'$, *one has*

$$\mathcal{D} \circ \mathcal{P}_k = (\mathcal{I}_{A_k} \otimes \mathcal{B}_k) \circ \mathcal{P}_k \qquad \forall k, \tag{A23}$$

where $\mathcal{I}_{A_k}$ *is the identity map from* $\text{Lin}(\mathcal{H}_{A_k})$ *to itself, and* $\mathcal{B}_k$ *is a quantum channel from* $\text{Lin}(\mathcal{H}_{A_k})$ *to itself.*

Proof. Straightforward extension of the proof in Appendix B. □

Using the notion of adjoint, we can now prove the following

Lemma A4. *For every channel* $\mathcal{D} \in \text{Chan}(A)'$, *the adjoint* $\mathcal{D}^\dagger$ *preserves the elements of the algebra* A, *namely* $\mathcal{D}^\dagger(C) = C$ *for all* $C \in A$.

Proof. Let C be a generic element of A. By Equation (31), one has the equality

$$C = \bigoplus_k (C_k \otimes I_{B_k}) = \bigoplus_k \mathcal{P}_k(C)\,. \tag{A24}$$

Using Lemma A3 and the definition of adjoint, we obtain

$$\begin{aligned}
\text{Tr}[\mathcal{D}^\dagger(C)\rho] &= \text{Tr}[C\,\mathcal{D}(\rho)] \\
&= \sum_k \text{Tr}[\mathcal{P}_k(C)\,\mathcal{D}(\rho)] \\
&= \sum_k \text{Tr}[C\,(\mathcal{P}_k \circ \mathcal{D})(\rho)] \\
&= \sum_k \text{Tr}[C\,(\mathcal{P}_k \circ \mathcal{D} \circ \mathcal{P}_k)(\rho)] \\
&= \sum_k \text{Tr}\left[\mathcal{P}_k(C)\,(\mathcal{D} \circ \mathcal{P}_k)(\rho)\right] \\
&= \sum_k \text{Tr}\left\{(C_k \otimes I_{B_k})\,[(\mathcal{I}_{A_k} \otimes \mathcal{B}_k) \circ \mathcal{P}_k](\rho)\right\},
\end{aligned} \quad (A25)$$

having used Lemma A3 in the last equality. Then, we use the fact that the channel $\mathcal{B}_k$ is trace-preserving, and therefore its adjoint $\mathcal{B}_k^\dagger$ preserves the identity. Using this fact, we can continue the chain of equalities as

$$\begin{aligned}
\text{Tr}[\mathcal{D}^\dagger(C)] &= \sum_k \text{Tr}\left\{[C_k \otimes \mathcal{B}_k^\dagger(I_{B_k})]\,\mathcal{P}_k(\rho)\right\} \\
&= \sum_k \text{Tr}\left[(C_k \otimes I_{B_k})\,\mathcal{P}_k(\rho)\right] \\
&= \sum_k \text{Tr}\left[\mathcal{P}_k(C_k \otimes I_{B_k})\,\rho\right] \\
&= \text{Tr}\left[\left(\bigoplus_k C_k \otimes I_{B_k}\right)\rho\right] \\
&= \text{Tr}[C\rho],
\end{aligned} \quad (A26)$$

having used Equation (A24) in the last equality. Since the equality holds for every density matrix ρ, we proved the equality $\mathcal{D}^\dagger(C) = C$. □

We are now in position to prove Theorem A1.

Proof of Theorem A1. Let $\mathcal{D}$ be a quantum channel in Chan(A)'. Then, Lemma A4 guarantees that the adjoint $\mathcal{D}^\dagger$ preserves all operators in the algebra A. Then, a result due to Lindblad [86] guarantees that all the Kraus operators of $\mathcal{D}$ belong to the algebra A'. This proves the inclusion Chan(A)' ⊆ Chan(A').

The converse inclusion is immediate: if a channel $\mathcal{D}$ belongs to Chan(A'), it commutes with all channels in Chan(A) thanks to the block diagonal form of the Kraus operators (cf. Equations (32) and (33)). □

Appendix C.2. States of Subsystems Associated to Finite Dimensional Von Neumann algebras

Here, we provide the proof of Proposition 5, adopting the notation B := A'.
The proof uses the following lemma:

Lemma A5 (No signalling condition). *For every channel $\mathcal{D} \in \text{Chan}(B)$, one has $\text{Tr}_B \circ \mathcal{D} = \text{Tr}_B$.*

Proof. By definition, the partial trace channel Tr_B can be written as

$$\text{Tr}_B = \bigoplus_k (\mathcal{I}_{A_k} \otimes \text{Tr}_{B_k}) \circ \mathcal{P}_k. \quad (A27)$$

For every channel $\mathcal{D} \in \mathrm{Chan}(B)$, we have

$$\begin{aligned}
\mathrm{Tr}_B \circ \mathcal{D} &= \bigoplus_k (\mathcal{I}_{A_k} \otimes \mathrm{Tr}_{B_k}) \circ \mathcal{P}_k \circ \mathcal{D} \\
&= \bigoplus_k (\mathcal{I}_{A_k} \otimes \mathrm{Tr}_{B_k}) \circ (\mathcal{I}_{A_k} \otimes \mathcal{B}_k) \circ \mathcal{P}_k \\
&= \bigoplus_k \left[\mathcal{I}_{A_k} \otimes (\mathrm{Tr}_{B_k} \circ \mathcal{B}_k) \right] \circ \mathcal{P}_k \\
&= \bigoplus_k (\mathcal{I}_{A_k} \otimes \mathrm{Tr}_{B_k}) \circ \mathcal{P}_k \\
&= \mathrm{Tr}_B,
\end{aligned} \qquad (A28)$$

where the second equality follows from Lemma A3, and the third equality follows from the fact that $\mathcal{B}_k$ is trace-preserving. □

Proof of Proposition 5. Suppose that ρ and σ are equivalent for A. By definition, this means that there exists a finite sequence $(\rho_1, \rho_2, \ldots, \rho_n)$ such that

$$\rho_1 = \rho, \qquad \rho_n = \sigma, \qquad \text{and} \qquad \mathrm{Deg}_B(\rho_i) \cap \mathrm{Deg}_B(\rho_{i+1}) \neq \emptyset \quad \forall i \in \{1, 2, \ldots, n-1\}. \qquad (A29)$$

The condition of non-trivial intersection implies that, for every $i \in \{1, 2, \ldots, n-1\}$, one has

$$\mathcal{D}_i(\rho_i) = \widetilde{\mathcal{D}}_i(\rho_{i+1}), \qquad (A30)$$

where $\mathcal{D}_i$ and $\widetilde{\mathcal{D}}_i$ are two quantum channels in $\mathrm{Chan}(B)$. Tracing over B on both sides we obtain the relation

$$(\mathrm{Tr}_B \circ \mathcal{D}_i)(\rho_i) = (\mathrm{Tr}_B \circ \widetilde{\mathcal{D}}_i)(\rho_{i+1}), \qquad (A31)$$

and, thanks to Lemma A5, $\mathrm{Tr}_B[\rho_i] = \mathrm{Tr}_B[\rho_{i+1}]$. Since the equality holds for every $i \in \{1, \ldots, n-1\}$, we obtained the condition $\mathrm{Tr}_B[\rho] = \mathrm{Tr}_B[\sigma]$. In summary, if two states ρ and σ are equivalent for A, then $\mathrm{Tr}_B[\rho] = \mathrm{Tr}_B[\sigma]$.

To prove the converse, it is enough to define the channel $\mathcal{D}_0 \in \mathrm{Chan}(B)$ as

$$\mathcal{D}_0(\rho) := \bigoplus_k \mathrm{Tr}_{B_k}[\mathcal{P}_k(\rho)] \otimes \beta_k, \qquad (A32)$$

where each β_k is a fixed (but otherwise generic) density matrix in $\mathrm{Lin}(\mathcal{H}_{B_k})$. Now, if the equality $\mathrm{Tr}_B[\rho] = \mathrm{Tr}_B[\sigma]$ holds, then also the equality $\mathcal{D}_0(\rho) = \mathcal{D}_0(\sigma)$ holds. This proves that the intersection between $\mathrm{Deg}_B(\rho)$ and $\mathrm{Deg}_B(\sigma)$ is non-empty, and therefore ρ and σ are equivalent for A. □

Appendix C.3. Transformations of Subsystems Associated to Finite Dimensional von Neumann algebras

Here, we prove that all transformations of system S_A are of the form $\mathcal{A} = \bigoplus_k \mathcal{A}_k$, where each $\mathcal{A}_k$ is a quantum channel from $\mathrm{Lin}(\mathcal{H}_{A_k})$ to itself. The proof is based on the following lemmas:

Lemma A6. *For every channel $\mathcal{C} \in \mathrm{Chan}(A)$, one has the relation*

$$\mathcal{P}_k \circ \mathcal{C} = (\mathcal{A}_k \otimes \mathcal{I}_{B_k}) \circ \mathcal{P}_k, \qquad (A33)$$

where $\mathcal{A}_k$ is a quantum channel from $\mathrm{Lin}(\mathcal{H}_{A_k})$ to itself.

Proof. Let

$$C(\rho) = \sum_i C_i \rho C_i^\dagger, \qquad C_i = \bigoplus_k (C_{ik} \otimes I_{B_k}) \tag{A34}$$

be a Kraus representation of channel C. The preservation of the trace amounts to the condition

$$\begin{aligned} I &= \sum_i C_i^\dagger C_i \\ &= \bigoplus_k \left(\sum_i C_{ik}^\dagger C_{ik} \otimes I_{B_k} \right), \end{aligned} \tag{A35}$$

which implies

$$\sum_i C_{ik}^\dagger C_{ik} = I_{A_k} \qquad \forall k. \tag{A36}$$

Now, we have

$$\begin{aligned} (\mathcal{P}_k \circ \mathcal{C})(\rho) &= \sum_i (C_{ik} \otimes I_{B_k}) \mathcal{P}_k(\rho) (C_{ik} \otimes I_{B_k})^\dagger \\ &= (\mathcal{A}_k \otimes \mathcal{I}_{B_k}) [\mathcal{P}_k(\rho)], \end{aligned} \tag{A37}$$

where the channel $\mathcal{A}_k$ is defined as

$$\mathcal{A}_k(\sigma) := \sum_i C_{ik} \sigma C_{ik}^\dagger \qquad \forall \sigma \in \mathrm{Lin}(\mathcal{H}_{A_k}). \tag{A38}$$

Since the density matrix ρ in Equation (A37) is arbitrary, we proved the relation $\mathcal{P}_k \circ \mathcal{C} = (\mathcal{A}_k \otimes \mathcal{I}_{B_k}) \circ \mathcal{P}_k$. □

Lemma A7. *For two channels $\mathcal{C}, \mathcal{C}' \in \mathrm{Chan}(A)$, let $\mathcal{A}_k$ and $\mathcal{A}_k'$ be the quantum channels defined in Lemma A6. Then, the following are equivalent:*

1. $\mathrm{Tr}_B \circ \mathcal{C} = \mathrm{Tr}_B \circ \mathcal{C}'$,
2. $\mathcal{A}_k = \mathcal{A}_k'$ *for every k.*

Proof. $2 \Longrightarrow 1$. For channel $\mathcal{C}$, we have

$$\begin{aligned} \mathrm{Tr}_B \circ \mathcal{C} &= \bigoplus_k (\mathcal{I}_{A_k} \otimes \mathrm{Tr}_{B_k}) \circ \mathcal{P}_k \circ \mathcal{C} \\ &= \bigoplus_k (\mathcal{I}_{A_k} \otimes \mathrm{Tr}_{B_k}) \circ (\mathcal{A}_k \otimes \mathcal{I}_{B_k}) \circ \mathcal{P}_k \\ &= \bigoplus_k (\mathcal{A}_k \otimes \mathrm{Tr}_{B_k}) \circ \mathcal{P}_k. \end{aligned} \tag{A39}$$

Similarly, for channel $\mathcal{C}'$, we have

$$\mathrm{Tr}_B \circ \mathcal{C}' = \bigoplus_k (\mathcal{A}_k' \otimes \mathrm{Tr}_{B_k}) \circ \mathcal{P}_k. \tag{A40}$$

Clearly, if $\mathcal{A}_k$ and $\mathcal{A}_k'$ are equal for every k, then the partial traces $\mathrm{Tr}_B \circ \mathcal{C}$ and $\mathrm{Tr}_B \circ \mathcal{C}'$ are equal.

$1 \Longrightarrow 2$. Suppose that partial traces $\mathrm{Tr}_B \circ \mathcal{C}$ and $\mathrm{Tr}_B \circ \mathcal{C}'$ are equal. Then, Equations (A39) and (A40) imply the equality

$$(\mathcal{A}_k \otimes \mathrm{Tr}_{B_k}) \circ \mathcal{P}_k = (\mathcal{A}_k' \otimes \mathrm{Tr}_{B_k}) \circ \mathcal{P}_k \qquad \forall k. \tag{A41}$$

In turn, the above equality implies $\mathcal{A}_k = \mathcal{A}'_k$, $\forall k$, as one can easily verify by applying both sides of Equation (A41) to a generic product operator $X_k \otimes Y_k$, with $X_k \in \mathrm{Lin}(\mathcal{H}_{A_k})$ and $Y_k \in \mathrm{Lin}(\mathcal{H}_{B_k})$. □

Lemma A8. *Two channels $\mathcal{C}, \mathcal{C}' \in \mathrm{Chan}(A)$ are equivalent for A if and only if $\mathrm{Tr}_B \circ \mathcal{C} = \mathrm{Tr}_B \circ \mathcal{C}'$.*

Proof. Suppose that $\mathcal{C}$ and $\mathcal{C}'$ are equivalent for A. By definition, this means that there exists a finite sequence $(\mathcal{C}_1, \mathcal{C}_2, \ldots, \mathcal{C}_n) \subset \mathrm{Chan}(A)$ such that

$$\mathcal{C}_1 = \mathcal{C}, \quad \mathcal{C}_n = \mathcal{C}', \quad \mathrm{Deg}_B(\mathcal{C}_i) \cap \mathrm{Deg}_B(\mathcal{C}_{i+1}) \neq \emptyset \quad \forall i \in \{1, \ldots, n-1\}. \tag{A42}$$

This means that, for every i, there exist two channels $\mathcal{D}_i, \widetilde{\mathcal{D}}_i \in \mathrm{Chan}(B)$ such that

$$\mathcal{D}_i \circ \mathcal{C}_i = \widetilde{\mathcal{D}}_i \circ \mathcal{C}_{i+1}. \tag{A43}$$

Tracing over B on both sides, we obtain

$$\mathrm{Tr}_B \circ \mathcal{D}_i \circ \mathcal{C}_i = \mathrm{Tr}_B \circ \widetilde{\mathcal{D}}_i \circ \mathcal{C}_{i+1}, \tag{A44}$$

and, using the no signalling condition of Lemma A5,

$$\mathrm{Tr}_B \circ \mathcal{C}_i = \mathrm{Tr}_B \circ \mathcal{C}_{i+1}. \tag{A45}$$

Since the above relation holds for every i, we obtained the equality $\mathrm{Tr}_B \circ \mathcal{C} = \mathrm{Tr}_B \circ \mathcal{C}'$. Conversely, suppose that $\mathrm{Tr}_B \circ \mathcal{C} = \mathrm{Tr}_B \circ \mathcal{C}'$. Then, Lemma A7 implies the equality

$$\mathcal{A}_k = \mathcal{A}'_k \quad \forall k, \tag{A46}$$

where $\mathcal{A}_k$ and $\mathcal{A}'_k$ are the quantum channels defined in Lemma A6.

Now, let $\mathcal{D}_0$ be the channel in $\mathrm{Chan}(B)$ defined in Equation (A32). By definition, we have

$$\begin{aligned}
\mathcal{D}_0 \circ \mathcal{C} &= \bigoplus_k (\mathcal{I}_{A_k} \otimes \beta_k \; \mathrm{Tr}_{B_k}) \circ \mathcal{P}_k \circ \mathcal{C} \\
&= \bigoplus_k (\mathcal{I}_{A_k} \otimes \beta_k \; \mathrm{Tr}_{B_k}) \circ (\mathcal{A}_k \otimes \mathcal{I}_{B_k}) \circ \mathcal{P}_k \\
&= \bigoplus_k (\mathcal{A}_k \otimes \beta_k \; \mathrm{Tr}_{B_k}) \circ \mathcal{P}_k.
\end{aligned} \tag{A47}$$

Similarly, we have

$$\mathcal{D}_0 \circ \mathcal{C} = \bigoplus_k (\mathcal{A}'_k \otimes \beta_k \; \mathrm{Tr}_{B_k}) \circ \mathcal{P}_k. \tag{A48}$$

Since $\mathcal{A}_k$ and $\mathcal{A}'_k$ are equal for every k, we conclude that $\mathcal{D}_0 \circ \mathcal{C}$ is equal to $\mathcal{D}_0 \circ \mathcal{C}'$. This means that the intersection between $\mathrm{Deg}(\mathcal{C})$ and $\mathrm{Deg}(\mathcal{C}')$ is non-empty, and, therefore $\mathcal{C}$ is equivalent to $\mathcal{C}'$ modulo B. □

Combining Lemmas A7 and A8, we obtain the following corollary:

Corollary A1. *For two channels $\mathcal{C}, \mathcal{C}' \in \mathrm{Chan}(A)$, let $\mathcal{A}_k$ and $\mathcal{A}'_k$ be the quantum channels defined in Lemma A6. Then, the following are equivalent:*

1. *$\mathcal{C}$ and $\mathcal{C}'$ are equivalent for A,*
2. *$\bigoplus_k \mathcal{A}_k = \bigoplus_k \mathcal{A}'_k$.*

Proof. By Lemma A8, $\mathcal{C}$ and $\mathcal{C}'$ are equivalent for A if and only if the condition $\mathrm{Tr}_B \circ \mathcal{C} = \mathrm{Tr}_B \circ \mathcal{C}'$ holds. By Lemma A7, the condition $\mathrm{Tr}_B \circ \mathcal{C} = \mathrm{Tr}_B \circ \mathcal{C}'$ holds if and only if one has $\mathcal{A}_k = \mathcal{A}'_k$ for every k. In turn, the latter condition holds if and only if the equality $\bigoplus_k \mathcal{A}_k = \bigoplus_k \mathcal{A}'_k$ holds. □

In summary, the transformations of system S_A are characterized as

$$\mathrm{Transf}(S_A) = \bigoplus_k \mathrm{Chan}(A_k), \tag{A49}$$

where $\mathrm{Chan}(A_k)$ is the set of all quantum channels from $\mathrm{Lin}(\mathcal{H}_{A_k})$ to itself.

To conclude, we observe that the transformations of S_A act in the expected way. To this purpose, we consider the *restriction map*

$$\pi_A : \mathrm{Chan}(A) \to \bigoplus_k \mathrm{Chan}(A_k), \quad \mathcal{C} \mapsto \bigoplus_k \mathcal{A}_k, \tag{A50}$$

where $\mathcal{A}_k$ is defined as in Lemma A6.

Using the restriction map, we can prove the following propositions:

Proposition A2. *For every channel $\mathcal{C} \in \mathrm{Chan}(A)$, we have the relation*

$$\mathrm{Tr}_B \circ \mathcal{C} = \pi_A(\mathcal{C}) \circ \mathrm{Tr}_B . \tag{A51}$$

In words, evolving system S with $\mathcal{C}$ and then computing the local state of system S_A is the same as computing the local state of system S_A and then evolving it with $\pi_A(\mathcal{C})$.

Proof. Using Lemma A6, the proof is straightforward:

$$\begin{aligned}
\mathrm{Tr}_B \circ \mathcal{C} &= \bigoplus_k (\mathcal{I}_{A_k} \otimes \mathrm{Tr}_{B_k}) \circ \mathcal{P}_k \circ \mathcal{C} \\
&= \bigoplus_k (\mathcal{I}_{A_k} \otimes \mathrm{Tr}_{B_k}) \circ (\mathcal{A}_k \otimes \mathcal{I}_{B_k}) \circ \mathcal{P}_k \\
&= \bigoplus_k \mathcal{A}_k \circ (\mathcal{I}_{A_k} \otimes \mathrm{Tr}_{B_k}) \circ \mathcal{P}_k \\
&= \left(\bigoplus_k \mathcal{A}_k \right) \circ \left[\bigoplus_l (\mathcal{I}_{A_l} \otimes \mathrm{Tr}_{B_l}) \circ \mathcal{P}_l \right] \\
&= \pi_A(\mathcal{C}) \circ \mathrm{Tr}_B .
\end{aligned} \tag{A52}$$

□

Proposition A3. *For every pair of channels $\mathcal{C}_1, \mathcal{C}_2 \in \mathrm{Chan}(A)$, we have the homomorphism relation*

$$\pi_A(\mathcal{C}_1 \circ \mathcal{C}_2) = \pi_A(\mathcal{C}_1) \circ \pi_A(\mathcal{C}_2) . \tag{A53}$$

Proof. Let us write the channels $\pi_A(\mathcal{C}_1)$, $\pi_A(\mathcal{C}_2)$, and $\pi_A(\mathcal{C}_1 \circ \mathcal{C}_2)$ as

$$\pi_A(\mathcal{C}_1) = \bigoplus_k \mathcal{A}_{1k}, \quad \pi_A(\mathcal{C}_2) = \bigoplus_k \mathcal{A}_{2k}, \quad \text{and} \quad \pi_A(\mathcal{C}_1 \circ \mathcal{C}_2) = \bigoplus_k \mathcal{A}_{12k} . \tag{A54}$$

With this notation, we have

$$\begin{aligned}(\mathcal{A}_{12k} \otimes \mathcal{I}_{B_k}) \circ \mathcal{P}_k &= \mathcal{P}_k \circ \mathcal{C}_1 \circ \mathcal{C}_2 \\ &= (\mathcal{A}_{1k} \otimes \mathcal{I}_{B_k}) \circ \mathcal{P}_k \circ \mathcal{C}_2 \\ &= (\mathcal{A}_{1k} \otimes \mathcal{I}_{B_k}) \circ (\mathcal{A}_{2k} \otimes \mathcal{I}_{B_k}) \circ \mathcal{P}_k \\ &= \left[(\mathcal{A}_{1k} \circ \mathcal{A}_{2k}) \otimes \mathcal{I}_{B_k}\right] \circ \mathcal{P}_k \qquad \forall k. \end{aligned} \qquad (A55)$$

From the above equation, we obtain the equality $\mathcal{A}_{12k} = \mathcal{A}_{1k} \circ \mathcal{A}_{2k}$ for all k. In turn, this equality implies the desired result:

$$\begin{aligned}\pi_A(\mathcal{C}_1) \circ \pi_A(\mathcal{C}_2) &= \left(\bigoplus_k \mathcal{A}_{1k}\right) \circ \left(\bigoplus_l \mathcal{A}_{2l}\right) \\ &= \bigoplus_k \mathcal{A}_{1k} \circ \mathcal{A}_{2k} \\ &= \bigoplus_k \mathcal{A}_{12k} \\ &= \pi_A(\mathcal{C}_1 \circ \mathcal{C}_2). \end{aligned} \qquad (A56)$$

□

Appendix D. Basis-Preserving and Multiphase-Covariant Channels

Appendix D.1. Proof of Theorem 1

Here, we prove that the monoid of multiphase covariant channels on S (denoted as MultiPCov(S)) and the monoid of basis-preserving channels on S (denoted as BPres(S)) are one the commutant of the other.

The proof uses a few lemmas, the first of which is fairly straightforward:

Lemma A9. BPres$(S)' \subseteq$ MultiPCov(S).

Proof. Every unitary channel of the form $\mathcal{U}_\theta = U_\theta \cdot U_\theta^\dagger$ is basis-preserving, and therefore every channel $\mathcal{C}$ in the commutant of BPres(S) must commute with it. By definition, this means that $\mathcal{C}$ is multiphase covariant. □

To prove the converse inclusion, we use the following characterization of multiphase covariant channels:

Lemma A10 (Characterization of MultiPCov(S)). *A channel* $\mathcal{M} \in$ Chan(S) *is multiphase covariant if and only if it has a Kraus representation of the form*

$$\mathcal{M}(\rho) = \sum_{i=1}^r M_i \rho M_i^\dagger + \sum_{k=1}^d \sum_{j \neq k} p(j|k) |j\rangle\langle k| \rho |k\rangle\langle j|, \qquad (A57)$$

where each operator M_i is diagonal in the computational basis, and each $p(j|k)$ is non-negative.

Proof. Let $M \in \text{Lin}(\mathcal{H}_S \otimes \mathcal{H}_S)$ be the Choi operator of channel $\mathcal{M}$. For a multiphase covariant channel, the Choi operator must satisfy the commutation relation [87,88]

$$[M, U_\theta \otimes \overline{U}_\theta] = 0 \qquad \forall \theta \in [0, 2\pi)^{\otimes d}. \qquad (A58)$$

This condition implies that M must have the form

$$M = \sum_{s,t} M_{ss,tt} \, |s\rangle\langle t| \otimes |s\rangle\langle t| + \sum_{k} \sum_{j \neq k} M_{jk,jk} \, |j\rangle\langle j| \otimes |k\rangle\langle k|, \qquad (A59)$$

where the $d \times d$ matrix $[\Gamma_{s,t}] := [M_{ss,tt}]_{s,t \in \{1,\ldots,d\}}$ is positive semidefinite and each coefficient $M_{st,st}$ is non-negative. Then, Equation (A57) follows from diagonalizing the matrix Γ and using the relation $\mathcal{M}(\rho) = \mathrm{Tr}[M\,(I \otimes \rho^T)]$, where ρ^T is the transpose of ρ in the computational basis. □

From Equation (A57), one can show every multiphase covariant channel commutes with every basis-preserving channel:

Lemma A11. $\mathsf{MultiPCov}(S) \subseteq \mathsf{BPres}(S)'$.

Proof. Let $\mathcal{B} \in \mathsf{BPres}(S)$ be a generic basis-preserving channel, and let $\mathcal{M} \in \mathsf{MultiPCov}(S)$ be a generic multiphase covariant channel. Using the characterization of Equation (A57), we obtain

$$\begin{aligned}
\mathcal{M} \circ \mathcal{B}(\rho) &= \sum_i M_i \mathcal{B}(\rho) M_i^\dagger + \sum_k \sum_{j \neq k} p(j|k) |j\rangle\langle k| \mathcal{B}(\rho) |k\rangle\langle j| \\
&= \sum_i \mathcal{B}(M_i \rho M_i^\dagger) + \sum_k \sum_{j \neq k} p(j|k) |j\rangle\langle k| \mathcal{B}(\rho) |k\rangle\langle j| \\
&= \sum_i \mathcal{B}(M_i \rho M_i^\dagger) + \sum_k \sum_{j \neq k} p(j|k) |j\rangle \langle k|\rho|k\rangle \langle j| \\
&= \sum_i \mathcal{B}(M_i \rho M_i^\dagger) + \sum_k \sum_{j \neq k} p(j|k) \mathcal{B}(|j\rangle\langle j|) \langle k|\rho|k\rangle \\
&= \mathcal{B}\left(\sum_i M_i \rho M_i^\dagger + \sum_k \sum_{j \neq k} p(j|k) |j\rangle \langle k|\rho|k\rangle \langle j| \right) \\
&= \mathcal{B} \circ \mathcal{M}(\rho) \qquad \forall \rho \in \mathsf{Lin}(S). \qquad (A60)
\end{aligned}$$

The second equality used the fact that the Kraus operators of $\mathcal{B}$ are diagonal in the computational basis [71,72] and therefore commute with each operator M_i. The third equality uses the relation $\langle k|\mathcal{B}(\rho)|k\rangle = \langle k|\rho|k\rangle$, following from the fact that $\mathcal{B}$ preserves the computational basis [71,72]. □

Summarizing, we have shown that the multiphase covariant channels are the commutant of the basis-preserving channels:

Corollary A2. $\mathsf{MultiPCov}(S) = \mathsf{BPres}(S)'$.

Note that Corollary A2 implies the relation

$$\mathsf{MultiPCov}(S)' = \mathsf{BPres}(S)'' \supseteq \mathsf{BPres}(S). \qquad (A61)$$

To conclude the proof of Theorem 1, we prove the converse inclusion:

Lemma A12. $\mathsf{MultiPCov}(S)' \subseteq \mathsf{BPres}(S)$.

Proof. A special case of multiphase covariant channel is the erasure channel $\mathcal{M}_k$ defined by $\mathcal{M}_k(\rho) = |k\rangle\langle k|$ for every $\rho \in \mathsf{Lin}(S)$. For a generic channel $\mathcal{C} \in \mathsf{MultiPCov}(S)'$, one must have

$$\mathcal{C}(|k\rangle\langle k|) = \mathcal{C} \circ \mathcal{M}_k(|k\rangle\langle k|) = \mathcal{M}_k \circ \mathcal{C}(|k\rangle\langle k|) = |k\rangle\langle k|. \qquad (A62)$$

Since the above condition must hold for every k, the channel $\mathcal{C}$ must be basis-preserving. □

Combining Lemma A12 and Equation (A61), we obtain:

Corollary A3. $\mathsf{MultiPCov}(S)' = \mathsf{BPres}(S)$.

Putting Corollaries A2 and A3 together, we have an immediate proof of Theorem 1.

Appendix D.2. Proof of Equation (55)

Here, we show that the transformations on system S_A are classical channels. To construct the transformations of S_A, we have to partition the double commutant of $\mathsf{Act}(A;S) = \mathsf{MultiPCov}(S)$ into equivalence classes.

First, recall that $\mathsf{MultiPCov}(S)'' = \mathsf{MultiPCov}(S)$ (by Theorem 1). Then, note the following property:

Lemma A13. *If two channels* $\mathcal{M}, \widetilde{\mathcal{M}} \in \mathsf{MultiPCov}(S)$ *satisfy the condition*

$$\langle k | \mathcal{M}(|j\rangle\langle j|) | k \rangle = \langle k | \widetilde{\mathcal{M}}(|j\rangle\langle j|) | k \rangle, \tag{A63}$$

then $[\mathcal{M}]_{A'} = [\widetilde{\mathcal{M}}]_{A'}$.

Proof. Define the completely dephasing channel $\mathcal{D} = \sum_k |k\rangle\langle k| \cdot |k\rangle\langle k|$. Clearly, $\mathcal{D}$ is basis-preserving. Using the idempotence relation $\mathcal{D} \circ \mathcal{D} = \mathcal{D}$, we obtain

$$\begin{aligned}
\left(\mathcal{D} \circ \mathcal{M}\right)(\rho) &= \left(\mathcal{D} \circ \mathcal{D} \circ \mathcal{M}\right)(\rho) \\
&= \left(\mathcal{D} \circ \mathcal{M} \circ \mathcal{D}\right)(\rho) \\
&= \left(\mathcal{D} \circ \mathcal{M}\right)\left(\sum_j |j\rangle\langle j| \langle j|\rho|j\rangle\right) \\
&= \sum_j \langle j|\rho|j\rangle \, \mathcal{D}\left(\mathcal{M}(|j\rangle\langle j|)\right) \\
&= \sum_{j,k} \langle j|\rho|j\rangle \, \langle k|\mathcal{M}(|j\rangle\langle j|)|k\rangle \, |k\rangle\langle k|.
\end{aligned} \tag{A64}$$

Likewise, we have

$$\left(\mathcal{D} \circ \widetilde{\mathcal{M}}\right)(\rho) = \sum_{j,k} \langle j|\rho|j\rangle \, \langle k|\widetilde{\mathcal{M}}(|j\rangle\langle j|)|k\rangle \, |k\rangle\langle k|. \tag{A65}$$

If condition (A63) holds, then the equality $\mathcal{D} \circ \mathcal{M} = \mathcal{D} \circ \widetilde{\mathcal{M}}$ holds, meaning that $\mathrm{Deg}(\mathcal{M})$ and $\mathrm{Deg}(\widetilde{\mathcal{M}})$ have non-empty intersection. Hence, $\mathcal{M}$ and $\widetilde{\mathcal{M}}$ must be in the same equivalence class. □

The converse of Lemma A13 holds:

Lemma A14. *If two channels* $\mathcal{M}, \widetilde{\mathcal{M}} \in \mathsf{MultiPCov}(S)$ *are in the same equivalence class, then they must satisfy condition* (A63).

Proof. If $\mathcal{M}$ and $\widetilde{\mathcal{M}}$ are in the same equivalence class, then there exists a finite sequence $(\mathcal{M}_1, \mathcal{M}_2, \ldots, \mathcal{M}_n)$ such that

$$\mathcal{M}_1 = \mathcal{M}, \quad \mathcal{M}_n = \widetilde{\mathcal{M}}, \quad \forall i \in \{1, \ldots, n-1\} \; \exists \mathcal{B}_i, \widetilde{\mathcal{B}}_i \in \mathsf{BPres}(S): \quad \mathcal{B}_i \circ \mathcal{M}_i = \widetilde{\mathcal{B}}_i \circ \mathcal{M}_{i+1}.$$

The above condition implies

$$\begin{aligned}l\langle k|\,\mathcal{M}_i(\rho)\,|k\rangle = \mathrm{Tr}[\mathcal{M}_i(\rho)\,|k\rangle\langle k|] = \langle k|\,\mathcal{B}_i \circ \mathcal{M}_i(\rho)\,|k\rangle &= \langle k|\,\tilde{\mathcal{B}}_i \circ \mathcal{M}_{i+1}(\rho)\,|k\rangle \\ &= \langle k|\,\mathcal{M}_{i+1}(\rho)\,|k\rangle,\end{aligned} \quad (A66)$$

for all $i \in \{1, \ldots, n-1\}$ and for all $\rho \in \mathrm{Lin}(\rho)$. In particular, choosing $\rho = |j\rangle\langle j|$ we obtain

$$\langle k|\,\mathcal{M}_i(|j\rangle\langle j|)\,|k\rangle = \langle k|\,\mathcal{M}_{i+1}(|j\rangle\langle j|)\,|k\rangle \quad \forall i \in \{1,\ldots,n-1\}, \forall j,k \in \{1,\ldots,d\}. \quad (A67)$$

Hence, Equation (A63) follows. □

Appendix E. Classical Systems and the Resource Theory of Coherence

Here, we consider agents who have access to various types of free operations in the resource theory of coherence. We start from the types of operations that give rise to classical systems, and then show two examples that do not have this property.

Appendix E.1. Operations That Lead to Classical Subsystems

Consider the following monoids of operations

1. *Strictly incoherent operations [41]*, i.e., quantum channels $\mathcal{T}$ with the property that, for every Kraus operator T_i, the map $\mathcal{T}_i(\cdot) = T_i \cdot T_i$ satisfies the condition $\mathcal{D} \circ \mathcal{T}_i = \mathcal{T}_i \circ \mathcal{D}$, where $\mathcal{D}$ is the completely dephasing channel.
2. *Dephasing covariant operations [38–40]*, i.e., quantum channels $\mathcal{T}$ satisfying the condition $\mathcal{D} \circ \mathcal{T} = \mathcal{T} \circ \mathcal{D}$.
3. *Phase covariant channels [40]*, i.e., quantum channels $\mathcal{T}$ satisfying the condition $\mathcal{T} \circ \mathcal{U}_\varphi = \mathcal{U}_\varphi \circ \mathcal{T}$, $\forall \varphi \in [0, 2\pi)$, where $\mathcal{U}_\varphi$ is the unitary channel associated with the unitary matrix $U_\varphi = \sum_k e^{ik\varphi} |k\rangle\langle k|$.
4. *Physically incoherent operations [38,39]*, i.e., quantum channels that are convex combinations of channels $\mathcal{T}$ admitting a Kraus representation where each Kraus operator T_i is of the form

$$T_i = U_{\pi_i}\,U_{\theta_i}\,P_i, \quad (A68)$$

where U_{π_i} is a unitary that permutes the elements of the computational basis, U_{θ_i} is a diagonal unitary, and P_i is a projector on a subspace spanned by a subset of vectors in the computational basis.
5. *Classical channels* i.e., channels satisfying $\mathcal{T} = \mathcal{D} \circ \mathcal{T} \circ \mathcal{D}$.

We now show that all the above operations define classical subsystems according to our construction.

The first ingredient in the proof is the observation that each of the monoids 1–5 contains the monoid of classical channels. Then, we can apply the following lemma:

Lemma A15. *Let* $\mathsf{M} \subseteq \mathrm{Chan}(\mathsf{S})$ *be a monoid of quantum channels, and let* M' *be its commutant. If* M *contains the monoid of classical channels, then* M' *is contained in the set of basis-preserving channels.*

Proof. Consider the erasure channel $\mathcal{C}_k$ defined by $\mathcal{C}_k(\rho) := |k\rangle\langle k|\,\mathrm{Tr}[\rho]$, $\forall \rho \in \mathrm{Lin}(\mathcal{H}_S)$. Clearly, the erasure channel is a classical channel. Then, every channel $\mathcal{B} \in \mathsf{M}'$ must satisfy the condition

$$\mathcal{B}(|k\rangle\langle k|) = \mathcal{B} \circ \mathcal{C}_k(|k\rangle\langle k|) = \mathcal{C}_k \circ \mathcal{B}(|k\rangle\langle k|) = |k\rangle\langle k|. \quad (A69)$$

Since k is generic, this implies that $\mathcal{B}$ must be basis-preserving. □

Furthermore, we have the following

Lemma A16. *Let* $\text{Act}(A;S) \subseteq \text{Chan}(S)$ *be a set of quantum channels that contains the monoid of classical channels. If two quantum states* $\rho, \sigma \in \text{St}(S)$ *are equivalent for A, then they must have the same diagonal entries. Equivalently, they must satisfy* $\mathcal{D}(\rho) = \mathcal{D}(\sigma)$.

Proof. Same as the first part of the proof of Proposition 7. Suppose that Condition 1 holds, meaning that there exists a sequence $(\rho_1, \rho_2, \ldots, \rho_n)$ such that

$$\rho_1 = \rho, \qquad \rho_n = \sigma, \qquad \forall i \in \{1, \ldots, n-1\} \; \exists \mathcal{B}_i, \widetilde{\mathcal{B}}_i \in \text{Act}(B;S) : \mathcal{B}_i(\rho_i) = \widetilde{\mathcal{B}}_i(\rho_{i+1}), \qquad (A70)$$

where $\mathcal{B}_i$ and $\widetilde{\mathcal{B}}_i$ are channels in the commutant $\text{Act}(A;S)'$. The above equation implies

$$\langle k | \mathcal{B}_i(\rho_i) | k \rangle = \langle k | \widetilde{\mathcal{B}}_i(\rho_{i+1}) | k \rangle. \qquad (A71)$$

Now, we know that the commutant $\text{Act}(A;S)'$ consists of basis-preserving channels (Lemma A15). Since every basis-preserving channel satisfies the relation $\langle k | \mathcal{B}(\rho) | k \rangle = \langle k | \rho | k \rangle$ [71,72], we obtain that all the density matrices $(\rho_1, \rho_2, \ldots, \rho_n)$ must have the same diagonal entries, namely $\mathcal{D}(\rho_1) = \mathcal{D}(\rho_2) = \cdots = \mathcal{D}(\rho_n)$. □

Now, we observe that the completely dephasing channel $\mathcal{D}$ is contained in the commutant of all the monoids 1–5. This fact is evident for the monoids 1, 2 and 5, where the commutation with $\mathcal{D}$ holds by definition. For the monoid 3, the commutation with $\mathcal{D}$ has been proven in [38,39], and for the monoid 4 it has been proven in [40].

Since $\mathcal{D}$ is contained in the commutant of all the monoids 1–5, we can use the following obvious fact:

Lemma A17. *Let* $\text{Act}(A;S) \subseteq \text{Chan}(S)$ *be a monoid of quantum channels and suppose that its commutant* $\text{Act}(A;S)'$ *contains the dephasing channel* $\mathcal{D}$. *If two quantum states* $\rho, \sigma \in \text{St}(S)$ *satisfy* $\mathcal{D}(\rho) = \mathcal{D}(\sigma)$, *then they are equivalent for A.*

Proof. Trivial consequence of the definition. □

Combining Lemmas A16 and A17, we obtain the following

Proposition A4. *Let* $\text{Act}(A;S) \subseteq \text{Chan}(S)$ *be a monoid of quantum channels on system S. If* $\text{Act}(A;S)$ *contains the monoid of classical channels, and if the the commutant* $\text{Act}(A;S)'$ *contains the completely dephasing channel* $\mathcal{D}$, *then two states* $\rho, \sigma \in \text{St}(S)$ *are equivalent for A if and only if* $\mathcal{D}(\rho) = \mathcal{D}(\sigma)$.

Proof. Same as the proof of Proposition 7. □

Proposition A4 implies that the states of the subsystem S_A are in one-to-one correspondence with diagonal density matrices. Since the conditions of the proposition are satisfied by all the monoids 1–5, each of these monoids defines the same state space.

The same result holds for the transformations:

Proposition A5. *Let* $\text{Act}(A;S) \subseteq \text{Chan}(S)$ *be a monoid of quantum channels. If* $\text{Act}(A;S)$ *contains the monoid of classical channels, and if the the commutant* $\text{Act}(A;S)'$ *contains the completely dephasing channel* $\mathcal{D}$, *then two transformations* $\mathcal{S}, \mathcal{T} \in \text{Transf}(S)$ *are equivalent for A if and only if* $\mathcal{D} \circ \mathcal{T} \circ \mathcal{D} = \mathcal{D} \circ \mathcal{T} \circ \mathcal{D}$.

Proof. Same as the proofs of Lemmas A13 and A14. □

Proposition A5 implies that the transformations of subsystem S_A can be identified with classical channels. Hence, system S_A is exactly the d-dimensional classical subsystem of the quantum system S. In summary, each of the monoids 1–5 defines the same d-dimensional classical subsystem.

Appendix E.2. Operations That Do Not Lead to Classical Subsystems

Here, we show that our construction does not associate classical subsystems with the monoids of incoherent and maximally incoherent operations. To start with, we recall the definitions of these two subsets:

1. The *maximally incoherent operations* are the quantum channels $\mathcal{T}$ that map diagonal density matrices to diagonal density matrices, namely $\mathcal{T} \circ \mathcal{D} = \mathcal{D} \circ \mathcal{T} \circ \mathcal{D}$, where $\mathcal{D}$ is the completely dephasing channel.
2. The *Incoherent operations* are the quantum channels $\mathcal{T}$ with the property that, for every Kraus operator T_i, the map $\mathcal{T}_i(\cdot) = T_i \cdot T_i$ sends diagonal matrices to diagonal matrices, namely $\mathcal{T}_i \circ \mathcal{D} = \mathcal{D} \circ \mathcal{T}_i \circ \mathcal{D}$.

Note that each set of operations contains the set of classical channels. Hence, the commutant of each set of operation consists of (some subset of) basis-preserving channels (by Lemma A15).

Moreover, both sets of operations 1 and 2 contain the set of quantum channels $\mathcal{C}_\psi$ defined by the relation

$$\mathcal{C}_\psi(\rho) = |1\rangle\langle 1| \langle \psi|\rho|\psi\rangle + \frac{I - |1\rangle\langle 1|}{d-1} \operatorname{Tr}[(I - |\psi\rangle\langle \psi|)\rho] \qquad \forall \rho \in \operatorname{Lin}(\mathcal{H}_S), \tag{A72}$$

where $|\psi\rangle \in \mathcal{H}_S$ is a fixed (but otherwise arbitrary) unit vector. The fact that both monoids contain the channels $\mathcal{C}_\psi$ implies a strong constraint on their commutants:

Lemma A18. *The only basis-preserving quantum quantum channel $\mathcal{B} \in \operatorname{BPres}(S)$ satisfying the property $\mathcal{B} \circ \mathcal{C}_\psi = \mathcal{C}_\psi \circ \mathcal{B}$ for every $|\psi\rangle \in \mathcal{H}_S$ is the identity channel.*

Proof. The commutation property implies the relation

$$(\mathcal{C}_\psi \circ \mathcal{B})(|\psi\rangle\langle\psi|) = (\mathcal{B} \circ \mathcal{C}_\psi)(|\psi\rangle\langle\psi|)$$
$$= \mathcal{B}(|1\rangle\langle 1|)$$
$$= |1\rangle\langle 1|, \tag{A73}$$

where we used the fact that $\mathcal{B}$ is basis-preserving. Tracing both sides of the equality with the projector $|1\rangle\langle 1|$, we obtain the relation

$$1 = \langle 1|(\mathcal{C}_\psi \circ \mathcal{B})(|\psi\rangle\langle\psi|)|1\rangle$$
$$= \langle \psi| \mathcal{B}(|\psi\rangle\langle\psi|) |\psi\rangle, \tag{A74}$$

the second equality following from the definition of channel $\mathcal{C}_\psi$. In turn, Equation (A74) implies the relation $\mathcal{B}(|\psi\rangle\langle\psi|) = |\psi\rangle\langle\psi|$. Since $|\psi\rangle$ is arbitrary, this means that $\mathcal{B}$ must be the identity channel. □

In summary, the commutant of the set of incoherent channels consists only of the identity channel, and so is the the commutant of the set of maximally incoherent channels. Since the commutant is trivial, the equivalence classes are trivial, meaning that the subsystem S_A has exactly the same states and the same transformations of the original system S. In short, the subsystem associated with the incoherent (or maximally incoherent) channels is the full quantum system.

Appendix F. Enriching the Sets of Transformations

Here, we provide a mathematical construction that enlarges the sets of transformations in the "baby category" with objects S, S_A, and S_B. This construction provides a realization of a catagorical structure known as splitting of idempotents [73,74].

As we have seen in the main text, our basic construction does not provide transformations from the subsystem S_A to the global system S. One could introduce such transformations by hand, by defining an *embedding* [63]:

Definition A1. *An* embedding *of S_A into S is a map $\mathcal{E}_A : \text{St}(S_A) \to \text{St}(S)$ satisfying the property*

$$\text{Tr}_B \circ \mathcal{E}_A = \mathcal{I}_{S_A} . \tag{A75}$$

In other words, $\mathcal{E}_A$ associates a representative to every equivalence class $\rho \in \text{St}(S_A)$.

A priori, embeddings need not be physical processes. Consider the example of a classical system, viewed as a subsystem of a closed quantum system as in Section 4.3. An embedding would map each classical probability distribution $(p_1, p_2, \ldots, p_d)$ into a pure quantum state $|\psi\rangle = \sum_k c_k |k\rangle$ satisfying the condition $|c_k|^2 = p_k$ for all $k \in \{1, \ldots, d\}$. If the embedding were a physical transformation, there would be a way to physically transform every classical probability distributions into a corresponding pure quantum state, a fact that is impossible in standard quantum theory.

When building a new physical theory, one could *postulate* that there exists an embedding $\mathcal{E}_A$ that is physically realizable. In that case, the transformations from S_A to S would be those in the set

$$\text{Transf}(S_A \to S) = \left\{ \mathcal{T} \circ \mathcal{E}_A : \mathcal{T} \in \text{Transf}(S) \right\}, \tag{A76}$$

and similarly for the transformations from S_B to S. The transformations from S_A to S_B would be those in the set

$$\text{Transf}(S_A \to S_B) = \left\{ \text{Tr}_A \circ \mathcal{T} \circ \mathcal{E}_A : \mathcal{T} \in \text{Transf}(S) \right\}, \tag{A77}$$

and similarly for the transformations from S_B to S_A. In that new theory, the old set of transformations from S_A should be replaced by the new set:

$$\widetilde{\text{Transf}}(S_A) = \left\{ \text{Tr}_B \circ \mathcal{T} \circ \mathcal{E}_A : \mathcal{T} \in \text{Transf}(S) \right\}, \tag{A78}$$

so that the structure of category is preserved. Similarly, the old set of transformations from S_B to S_B should be replaced by the new set .

$$\widetilde{\text{Transf}}(S_B) = \left\{ \text{Tr}_A \circ \mathcal{T} \circ \mathcal{E}_B : \mathcal{T} \in \text{Transf}(S) \right\}. \tag{A79}$$

When this is done, the embeddings define two idempotent morphisms $\mathcal{P}_A := \mathcal{E}_A \circ \text{Tr}_B$ and $\mathcal{P}_B := \mathcal{E}_B \circ \text{Tr}_A$, i.e., two morphisms satisfying the conditions

$$\mathcal{P}_A \circ \mathcal{P}_A = \mathcal{P}_A \quad \text{and} \quad \mathcal{P}_B \circ \mathcal{P}_B = \mathcal{P}_B . \tag{A80}$$

The partial trace and the embedding define a *splitting of idempotents*, in the sense of Refs. [73,74]. The splitting of idempotents was considered in the categorical framework as a way to define general decoherence maps, and, more specifically, decoherence maps to classical subsystems [74,89].

Appendix G. The Total System as a Subsystem

For every system satisfying the Non-Overlapping Agents Requirement, the system S can be regarded as a subsystem:

Proposition A6. Let S be a system satisfying the Non-Overlapping Agents Requirement, let $A_{\max}$ be the maximal agent, and $S_{A_{\max}}$ be the associated subsystem. Then, one has $S_{A_{\max}} \simeq S$, meaning that there exist two isomorphisms $\gamma : \text{St}(S) \to \text{St}(S_{A_{\max}})$ and $\delta : \text{Transf}(S) \to \text{Transf}(S_{A_{\max}})$ satisfying the condition

$$\gamma(\mathcal{T}\psi) = \delta(\mathcal{T})\gamma(\psi), \qquad \forall \psi \in \text{St}(S), \forall \mathcal{T} \in \text{Transf}(S). \tag{A81}$$

Proof. The Non-Overlapping Agents Requirement guarantees that the commutant $\text{Act}(A_{\max}; S)'$ contains only the identity transformation. Hence, the equivalence class $[\psi]_{A_{\max}}$ contains only the state ψ. Hence, the partial trace $\text{Tr}_{A'_{\max}} : \psi \mapsto [\psi]_{A_{\max}}$ is a bijection from $\text{St}(S)$ to $\text{St}(S_{A_{\max}})$. Similarly, the equivalence class $[\mathcal{T}]_{A_{\max}}$ contains only the transformation $\mathcal{T}$. Hence, the restriction $\pi_{A_{\max}} : \mathcal{T} \mapsto [\mathcal{T}]_{A_{\max}}$ is a bijective function between $\text{Transf}(S)$ and $\text{Transf}(S_{A_{\max}})$. Such a function is an homomorphism of monoids, by Equation (20). Setting $\delta := \pi_{A_{\max}}$ and $\gamma := \text{Tr}_{A'_{\max}}$, the condition (A81) is guaranteed by Equation (21). □

Appendix H. Proof of Proposition 15

By definition, the condition $\text{Tr}_B[\psi] = \text{Tr}_B[\psi']$ holds if and only if there exists a finite sequence $(\psi_1, \psi_2, \ldots, \psi_n)$ such that

$$\psi_1 = \psi, \quad \psi_n = \psi', \quad \forall i \in \{1, \ldots, n-1\} \; \exists \mathcal{V}_i, \tilde{\mathcal{V}}_i \in \text{Act}(B; S) : \mathcal{V}_i \psi_i = \tilde{\mathcal{V}}_i \psi_{i+1}. \tag{A82}$$

Our goal is to prove that there exists an adversarial action $\mathcal{V}_B \in \text{Act}(B; S)$ such that the relation $\psi' = \mathcal{V}_B \psi$ or $\psi = \mathcal{V}_B \psi'$ holds.

We will proceed by induction on n, starting from the base case $n = 2$. In this case, we have $\text{Deg}_B(\psi) \cap \text{Deg}_B(\psi') \neq \emptyset$. Then, the first regularity condition implies that there exists a transformation $\mathcal{V}_B \in \text{Act}(B; S)$ such that at least one of the relations $\mathcal{V}_B \psi = \psi'$ and $\psi = \mathcal{V}_B \psi'$ holds. This proves the validity of the base case.

Now, suppose that the induction hypothesis holds for all sequences of length n, and suppose that ψ and ψ' are equivalent through a sequence of length $n+1$, say $(\psi_1, \psi_2, \ldots, \psi_n, \psi_{n+1})$. Applying the induction hypothesis to the sequence $(\psi_1, \psi_2, \ldots, \psi_n)$, we obtain that there exists a transformation $\mathcal{V} \in \text{Act}(B; S)$ such that at least one of the relations $\psi_n = \mathcal{V}\psi$ and $\psi = \mathcal{V}\psi_n$ holds. Moreover, applying the induction hypothesis to the pair (ψ_n, ψ_{n+1}) we obtain that there exists a transformation $\mathcal{V}' \in \text{Act}(B; S)$ such that $\psi_{n+1} = \mathcal{V}'\psi_n$, or $\psi_n = \mathcal{V}'\psi_{n+1}$. Hence, there are four possible cases:

1. $\psi_n = \mathcal{V}\psi$ and $\psi_{n+1} = \mathcal{V}'\psi_n$. In this case, we have $\psi_{n+1} = (\mathcal{V}' \circ \mathcal{V})\psi$, which proves the desired statement.
2. $\psi_n = \mathcal{V}\psi$ and $\psi_n = \mathcal{V}'\psi_{n+1}$. In this case, we have $\mathcal{V}\psi = \mathcal{V}'\psi_{n+1}$, or equivalently $\text{Deg}_B(\psi) \cap \text{Deg}_B(\psi_{n+1}) \neq \emptyset$. Applying the induction hypothesis to the sequence (ψ, ψ_{n+1}), we obtain the desired statement.
3. $\psi = \mathcal{V}\psi_n$ and $\psi_{n+1} = \mathcal{V}'\psi_n$. Using the second regularity condition, we obtain that there exists a transformation $\mathcal{W} \in \text{Act}(B; S)$ such that at least one of the relations $\mathcal{V} = \mathcal{W} \circ \mathcal{V}'$ and $\mathcal{V}' = \mathcal{W} \circ \mathcal{V}$ holds. Suppose that $\mathcal{V} = \mathcal{W} \circ \mathcal{V}'$. In this case, we have

$$\psi = \mathcal{V}\psi_n = (\mathcal{W} \circ \mathcal{V}')\psi_n = \mathcal{W}\psi_{n+1}. \tag{A83}$$

Alternatively, suppose that $\mathcal{V}' = \mathcal{W} \circ \mathcal{V}$. In this case, we have

$$\psi_{n+1} = \mathcal{V}'\psi_n = (\mathcal{W} \circ \mathcal{V})\psi_n = \mathcal{W}\psi. \tag{A84}$$

In both cases, we proved the desired statement.

4. $\psi = \mathcal{V}\psi_n$ and $\psi_n = \mathcal{V}'\psi_{n+1}$. In this case, we have $\psi = (\mathcal{V} \circ \mathcal{V}')\psi_{n+1}$, which proves the desired statement.

□

Appendix I. Characterization of the Adversarial Group

Here, we provide the proof of Theorem 3, proving a canonical decomposition of the elements of the adversarial group. The proof proceeds in a few steps:

Lemma A19 (Canonical form of the elements of the adversarial group). *Let $U : g \mapsto U_g$ be a projective representation of the group G, let $\mathrm{Irr}(U)$ be the set of irreducible representations contained in the isotypic decomposition of U, and let $\omega : G \to \mathbb{C}$ be a multiplicative character of G. Then, the commutation relation*

$$V U_g = \omega(g) \, U_g V \qquad \forall g \in G \tag{A85}$$

holds iff

1. *The map $U^{(j)} \mapsto \omega \, U^{(j)}$ is a permutation of the set $\mathrm{Irr}(U)$, denoted as $\pi : \mathrm{Irr}(U) \to \mathrm{Irr}(U)$. In other words, for every irrep $U^{(j)}$ with $j \in \mathrm{Irr}(U)$, the irrep $\omega \, U^{(j)}$ is equivalent to an irrep $k \in \mathrm{Irr}(U)$, and the correspondence between j and k is bijective.*
2. *The multiplicity spaces $\mathcal{M}_j$ and $\mathcal{M}_{\pi(j)}$ have the same dimension.*
3. *The unitary operator V has the canonical form $V = U_\pi V_0$, where V_0 is an unitary operator in the commutant U' and U_π is a permutation operator satisfying*

$$U_\pi \left(\mathcal{R}_j \otimes \mathcal{M}_j \right) = \left(\mathcal{R}_{\pi(j)} \otimes \mathcal{M}_{\pi(j)} \right) \qquad \forall j \in \mathrm{Irr}(U). \tag{A86}$$

Proof. Let us use the isotypic decomposition of U, as in Equation (88). We define

$$V_{j,k} := \Pi_j V \Pi_k, \tag{A87}$$

where Π_j (Π_k) is the projector onto $\mathcal{R}_j \otimes \mathcal{M}_j$ ($\mathcal{R}_k \otimes \mathcal{M}_k$). Then, Equation (A85) is equivalent to the condition

$$V_{j,k} \left(U_g^{(k)} \otimes I_{\mathcal{M}_k} \right) = \omega(g) \left(U_g^{(j)} \otimes I_{\mathcal{M}_j} \right) V_{jk}, \qquad \forall g \in G, \forall j, k, \tag{A88}$$

which in turn is equivalent to the condition

$$\langle \alpha | V_{j,k} | \beta \rangle \, U_g^{(k)} = \omega(g) \, U_g^{(j)} \, \langle \alpha | V_{j,k} | \beta \rangle, \qquad \forall g \in G, \forall j, k, \forall |\alpha\rangle \in \mathcal{M}_j, \forall |\beta\rangle \in \mathcal{M}_k, \tag{A89}$$

where $\langle \alpha | V_{j,k} | \beta \rangle$ is a shorthand for the partial matrix element $(I_{\mathcal{R}_j} \otimes \langle \alpha |) V_{j,k} (I_{\mathcal{R}_k} \otimes |\beta\rangle)$.

Equation (A89) means that each operator $\langle \alpha | V_{j,k} | \beta \rangle$ intertwines the two representations $U^{(k)}$ and $\omega \, U^{(j)}$. Recall that each representation is irreducible. Hence, the second Schur's lemma [78] implies that $\langle \alpha | V_{j,k} | \beta \rangle$ is zero if the two representations are not equivalent. Note that there can be *at most* one value of j such that $U^{(k)}$ is equivalent to $\omega \, U^{(j)}$. If such a value exists, we denote it as $j = \pi(k)$. By construction, the function $\pi : \mathrm{Irr}(U) \to \mathrm{Irr}(U)$ must be injective.

When $j = \pi(k)$, the first Schur's lemma [78] guarantees that the operator $\langle \alpha | V_{\pi(k),k} | \beta \rangle$ is proportional to the partial isometry $T_{\pi(k),k}$ that implements the equivalence of the two representations. Let us write

$$\langle \alpha | V_{\pi(k),k} | \beta \rangle = M_{\alpha,\beta} \, T_{\pi(k),k}, \tag{A90}$$

for some $M_{\alpha,\beta}^{(k)} \in \mathbb{C}$. Note also that, since the left-hand side is sesquilinear in $|\alpha\rangle$ and $|\beta\rangle$, the right-hand side should also be sesquilinear. Hence, we can find an operator $M_{\pi(k),k} : \mathcal{M}_k \to \mathcal{M}_{\pi(k)}$ such that $M_{\alpha,\beta}^{(k)} = \langle \alpha | M_{\pi(k),k} | \beta \rangle$. Putting everything together, the operator V can be written as

$$V = \bigoplus_{k \in \mathrm{Irr}(U)} \left(T_{\pi(k),k} \otimes M_{\pi(k),k} \right). \tag{A91}$$

Now, the operator V must be unitary, and, in particular, it should satisfy the condition $VV^\dagger = I$, which reads

$$\bigoplus_{k \in \mathrm{Irr}(U)} \left(I_{\mathcal{R}_{\pi(k)}} \otimes M_{\pi(k),k} M_{\pi(k),k}^\dagger \right) = I. \tag{A92}$$

The above condition implies that: (i) the function π must be surjective, and (ii) the operator $M_{\pi(k),k}$ must be a co-isometry. From the relation $V^\dagger V$, we also obtain that $M_{\pi(k),k}$ must be an isometry. Hence, $M_{\pi(k)}$ is unitary.

Summarizing, the condition (A85) can be satisfied only if there exists a permutation $\pi : \mathrm{Irr}(U) \to \mathrm{Irr}(U)$ such that, for every j,

1. the irreps $\omega\, U^{(k)}$ and $U^{\pi(k)}$ are equivalent,
2. the multiplicity spaces $\mathcal{M}_k$ and $\mathcal{M}_{\pi(k)}$ are unitarily isomorphic.

Fixing a unitary isomorphism $S_{\pi(k),k} : \mathcal{M}_k \to \mathcal{M}_{\pi(k)}$, we can write every element of the adversarial group in the canonical form $V = U_\pi V_0$, where U_π is the permutation operator

$$U_\pi = \bigoplus_{k \in \mathrm{Irr}(U)} \left(T_{\pi(k),k} \otimes S_{\pi(k),k} \right), \tag{A93}$$

and V_0 is an element of the commutant U', i.e., a generic unitary operator of the form

$$V_0 = \bigoplus_{k \in \mathrm{Irr}(U)} \left(I_j \otimes V_{0,k} \right). \tag{A94}$$

Conversely, if a permutation π exists with the properties that for every $k \in \mathrm{Irr}(U)$

1. $\omega\, U^{(k)}$ and $U^{(\pi(k))}$ are equivalent irreps,
2. $\mathcal{M}_k$ and $\mathcal{M}_{\pi(k)}$ are unitarily equivalent,

and if the operator V has the form $V = U_\pi V_0$, with U_π and V_0 as in Equations (A93) and (A94), then V satisfies the commutation relation (A85). □

We have seen that every element of the adversarial group can be decomposed into the product of a permutation operator, which permutes the irreps, and an operator in the commutant of the original group representation $U : G \to \mathrm{Lin}(\mathcal{H})$. We now observe that the allowed permutations have an additional structure: they must form an Abelian group, denoted as A.

Lemma A20. *The permutations π arising from Equation (A85) with a generic multiplicative character $\omega(V, \cdot)$ form an Abelian subgroup A of the group of all permutations of $\mathrm{Irr}(U)$.*

Proof. Let V and W be two elements of the adversarial group G_B, let $\omega(V, \cdot)$ and $\omega(W, \cdot)$ be the corresponding characters, and let π_V and π_W be the permutations associated with $\omega(V, \cdot)$ and $\omega(W, \cdot)$ as in Theorem A19, i.e., through the relation

$$\begin{aligned} j = \pi_V(k) &\iff U^{(j)} \text{ is equivalent to } \omega(V, \cdot)\, U^{(k)}, \\ j = \pi_W(k) &\iff U^{(j)} \text{ is equivalent to } \omega(W, \cdot)\, U^{(k)}. \end{aligned} \tag{A95}$$

Now, the element VW is associated with the permutation $\pi_V \circ \pi_W$, while the element WV is associated with the permutation $\pi_W \circ \pi_V$. On the other hand, the characters obey the equality

$$\omega(VW, g) = \omega(V, g)\omega(W, g) = \omega(WV, g) \qquad \forall g \in G. \tag{A96}$$

Hence, we conclude that $\pi_V \circ \pi_W$ and $\pi_W \circ \pi_V$ are, in fact, the same permutation. Hence, the elements of the adversarial group must correspond to an Abelian subgroup of the permutations of $\mathrm{Irr}(U)$. □

Combining Lemmas A19 and A20, we can now prove Theorem 3.

Proof of Theorem 3. For different permutations in A, we can choose the isomorphisms $S_{\pi(k),k} : \mathcal{M}_k \to \mathcal{M}_{\pi(k)}$ such that the following property holds:

$$S_{\pi_2 \circ \pi_2(k),k} = S_{\pi_2(\pi_1(k)), \pi_1(k)} \, S_{\pi_1(k),k}, \qquad \forall \pi_1, \pi_2 \in A. \tag{A97}$$

When this is done, the unitary operators U_π defined in Equation (A93) form a faithful representation of the Abelian group A. Using the canonical decomposition of Theorem A19, every element of $V \in G_B$ is decomposed *uniquely* as $V = U_\pi V_0$, where V_0 is an element of the commutant U'. Note also that the commutant U' is a normal subgroup of the adversarial group: indeed, for every element $V \in G_B$ we have $VU'V^\dagger = U'$. Since U' is a normal subgroup and the decomposition $V = U_\pi V_0$ is unique for every $V \in G_B$, it follows that the adversarial group G_B is the semidirect product $A \ltimes U'$. □

Appendix J. Example: The Phase Flip Group

Consider the Hilbert space $\mathcal{H}_S = \mathbb{C}^2$, and suppose that agent A can only perform the identity channel and the phase flip channel $\mathcal{Z}$, defined as

$$\mathcal{Z}(\cdot) = Z \cdot Z, \qquad Z = |0\rangle\langle 0| - |1\rangle\langle 1|. \tag{A98}$$

Then, the actions of agent A correspond to the unitary representation

$$U : \mathbb{Z}_2 \to \mathrm{Lin}(S), \qquad k \mapsto U_k = Z^k. \tag{A99}$$

The representation can be decomposed into two irreps, corresponding to the one-dimensional subspaces $\mathcal{H}_0 = \mathrm{Span}\{|0\rangle\}$ and $\mathcal{H}_1 = \mathrm{Span}\{|1\rangle\}$. The corresponding irreps, denoted by

$$\begin{aligned} \omega_0 &: \mathbb{Z}_2 \to \mathbb{C}, \quad \omega(k) = 1, \\ \omega_1 &: \mathbb{Z}_2 \to \mathbb{C}, \quad \omega(k) = (-1)^k, \end{aligned} \tag{A100}$$

are the only two irreps of the group and are multiplicative characters.

The condition $VU_k = U_k V$ yields the solutions

$$V = e^{i\theta_0}|0\rangle\langle 0| + e^{i\theta_1}|1\rangle\langle 1|, \qquad \theta_0, \theta_1 \in [0, 2\pi), \tag{A101}$$

corresponding to the commutant U'. The condition $VU_k = (-1)^k U_k V$ yields the solutions

$$V = e^{i\theta_0}|0\rangle\langle 1| + e^{i\theta_1}|1\rangle\langle 0|, \qquad \theta_0, \theta_1 \in [0, 2\pi). \tag{A102}$$

It is easy to see that the adversarial group G_B acts irreducibly on $\mathcal{H}_S$.

Let us consider now the subsystem S_A. The states of S_A are equivalence classes under the relation

$$|\psi\rangle \simeq_A |\psi'\rangle \qquad \exists V \in G_B : \quad |\psi'\rangle = V|\psi\rangle. \tag{A103}$$

It is not hard to see that the equivalence class of the state $|\psi\rangle$ is uniquely determined by the *unordered* pair $\{|\langle 0|\psi\rangle|, |\langle 1|\psi\rangle|\}$. In other words, the state space of system S_A is

$$\mathsf{St}(S_A) = \left\{ \{p, 1-p\}, : \quad p \in [0,1] \right\}. \tag{A104}$$

Note that, in this case, the state space is not a convex set of density matrices. Instead, it is the quotient of the set of diagonal density matrices, under the equivalence relation that two matrices with the same spectrum are equivalent.

Finally, note that the transformations of system S_A are trivial: since the adversarial group G_B contains the group G_A, the group $\mathsf{G}(S_A) = \pi_A(G_A)$ is trivial, namely

$$\mathsf{G}(S_A) = \{\mathcal{I}_{S_A}\}. \tag{A105}$$

Appendix K. Proof of Theorem 4

Let G be a connected Lie group, and let $\mathfrak{g}$ be the Lie algebra. Since G is connected, the exponential map reaches every element of the group, namely $\mathsf{G} = \exp[i\mathfrak{g}]$.

Let $h \in \mathsf{G}$ be a generic element of the group, written as $h = \exp[iX]$ for some $X \in \mathfrak{g}$, and consider the one-parameter subgroup $\mathsf{H} = \{\exp[i\lambda X], \lambda \in \mathbb{R}\}$. For a generic element $g \in \mathsf{H}$, the corresponding unitary operator can be expressed as $U_g = \exp[i\lambda K]$, where $K \in \mathsf{Lin}(S)$ is a suitable self-adjoint operator. Similarly, the multiplicative character has the form $\omega(g) = \exp[i\lambda \mu]$, for some real number $\mu \in \mathbb{R}$.

Now, every element V of the adversarial group must satisfy the relation

$$V \exp[i\lambda K] = \exp[i\lambda (K + \mu I_S)] V \qquad \forall \lambda \in \mathbb{R}, \tag{A106}$$

or equivalently,

$$\exp[i\lambda K] = V^\dagger \exp[i\lambda (K + \mu I_S)] V \qquad \forall \lambda \in \mathbb{R}. \tag{A107}$$

Since the operators $\exp[i\lambda K]$ and $\exp[i\lambda (K + \mu I_S)]$ are unitarily equivalent, they must have the same spectrum. This is only possible if the operators K and $K + \mu I_S$ have the same spectrum, which happens only if $\mu = 0$.

Now, recall that the one-parameter Abelian subgroup H is generic. Since every element of G is contained in some one-parameter Abelian subgroup H, we showed that $\omega(g) = 1$ for every $g \in \mathsf{G}$.

To conclude the proof, observe that the map $U^{(j)} \mapsto \omega U^{(j)}$ is the identity, and therefore induces the trivial permutation on the set of irreps $\mathsf{Irr}(U)$. Hence, the group of permutations A induced by multiplication by ω contains only the identity element. □

Appendix L. Proof of Proposition 16

Proof. It is enough to decompose the two states as

$$|\psi\rangle = \bigoplus_{j \in \mathsf{Irr}(U)} \sqrt{p_j} |\psi_j\rangle \quad \text{and} \quad |\psi'\rangle = \bigoplus_{j \in \mathsf{Irr}(U)} \sqrt{p'_j} |\psi'_j\rangle, \tag{A108}$$

where $|\psi_j\rangle$ and $|\psi_j'\rangle$ are unit vectors in $\mathcal{R}_j \otimes \mathcal{M}_j$. Using this decomposition, we obtain

$$\mathcal{T}_B(|\psi\rangle\langle\psi|) = \bigoplus_{j \in \mathsf{Irr}(U)} p_j \rho_j \quad \text{and} \quad \mathcal{T}_B(|\psi\rangle\langle\psi|) = \bigoplus_{j \in \mathsf{Irr}(U)} p_j' \rho_j', \tag{A109}$$

where ρ_j (ρ_j') is the marginal of $|\psi_j\rangle$ ($|\psi_j'\rangle$) on system $\mathcal{R}_j$. It is then clear that the equality $\mathcal{T}_B(|\psi\rangle\langle\psi|) = \mathcal{T}_B(|\psi'\rangle\langle\psi'|)$ implies $p_j = p_j'$ and $\rho_j = \rho_j'$ for every j. Since the states $|\psi_j\rangle$ and $|\psi_j'\rangle$ have the same marginal on system $\mathcal{R}_j$, there must exist a unitary operator $U_j : \mathcal{M}_j \to \mathcal{M}_j$ such that

$$|\psi_j'\rangle = (I_{\mathcal{R}_j} \otimes U_j)|\psi_j\rangle. \tag{A110}$$

We can then define the unitary gate

$$U_B = \bigoplus_{j \in \mathsf{Irr}(U)} \left(I_{\mathcal{R}_j} \otimes U_j \right), \tag{A111}$$

which satisfies the property $U_B|\psi\rangle = |\psi'\rangle$. By the characterization of Equation (89), U_B is an element of G_B. □

References

1. Nielsen, M.; Chuang, I. Quantum information and computation. *Nature* **2000**, *404*, 247.
2. Kitaev, A.Y.; Shen, A.; Vyalyi, M.N. *Classical and Quantum Computation*; Number 47; American Mathematical Society: Providence, RI, USA, 2002.
3. Einstein, A.; Podolsky, B.; Rosen, N. Can quantum-mechanical description of physical reality be considered complete? *Phys. Rev.* **1935**, *47*, 777. [CrossRef]
4. Schrödinger, E. Discussion of probability relations between separated systems. In *Mathematical Proceedings of the Cambridge Philosophical Society*; Cambridge University Press: Cambrdige, UK, 1935; Volume 31, pp. 555–563.
5. Hardy, L. Quantum theory from five reasonable axioms. *arXiv* **2001**, arXiv:quant-ph/0101012.
6. Barnum, H.; Barrett, J.; Leifer, M.; Wilce, A. Generalized no-broadcasting theorem. *Phys. Rev. Lett.* **2007**, *99*, 240501. [CrossRef] [PubMed]
7. Barrett, J. Information processing in generalized probabilistic theories. *Phys. Rev. A* **2007**, *75*, 032304. [CrossRef]
8. Chiribella, G.; D'Ariano, G.; Perinotti, P. Probabilistic theories with purification. *Phys. Rev. A* **2010**, *81*, 062348. [CrossRef]
9. Barnum, H.; Wilce, A. Information processing in convex operational theories. *Electron. Notes Theor. Comput. Sci.* **2011**, *270*, 3–15. [CrossRef]
10. Hardy, L. Foliable operational structures for general probabilistic theories. In *Deep Beauty: Understanding the Quantum World through Mathematical Innovation*; Halvorson, H., Ed.; Cambridge University Press: Cambrdige, UK, 2011; p. 409.
11. Hardy, L. A formalism-local framework for general probabilistic theories, including quantum theory. *Math. Struct. Comput. Sci.* **2013**, *23*, 399–440. [CrossRef]
12. Chiribella, G. Dilation of states and processes in operational-probabilistic theories. In Proceedings of the 11th workshop on Quantum Physics and Logic, Kyoto, Japan, 4–6 June 2014; Coecke, B., Hasuo, I., Panangaden, P., Eds.; Electronic Proceedings in Theoretical Computer Science; Volume 172, pp. 1–14.
13. Chiribella, G.; D'Ariano, G.M.; Perinotti, P. Quantum from principles. In *Quantum Theory: Informational Foundations and Foils*; Springer: Dordrecht, The Netherlands, 2016; pp. 171–221.
14. Hardy, L. Reconstructing quantum theory. In *Quantum Theory: Informational Foundations and Foils*; Springer: Dordrecht, The Netherlands, 2016; pp. 223–248.
15. Mauro D'Ariano, G.; Chiribella, G.; Perinotti, P. Quantum Theory from First Principles. In *Quantum Theory from First Principles*; D'Ariano, G.M., Chiribella, G., Perinotti, P., Eds.; Cambridge University Press: Cambridge, UK, 2017.

16. Abramsky, S.; Coecke, B. A categorical semantics of quantum protocols. In Proceedings of the 19th Annual IEEE Symposium on Logic in Computer Science, Turku, Finland, 17 July 2004; pp. 415–425.
17. Coecke, B. Kindergarten quantum mechanics: Lecture notes. In Proceedings of the AIP Conference Quantum Theory: Reconsideration of Foundations-3, Växjö, Sweden, 6–11 June 2005; American Institute of Physics: Melville, NY, USA, 2006; Volume 810, pp. 81–98.
18. Coecke, B. Quantum picturalism. *Contemp. Phys.* **2010**, *51*, 59–83. [CrossRef]
19. Abramsky, S.; Coecke, B. Categorical quantum mechanics. In *Handbook of Quantum Logic and Quantum Structures: Quantum Logic*; Elsevier Science: New York, NY, USA, 2008; pp. 261–324.
20. Coecke, B.; Kissinger, A. *Picturing Quantum Processes*; Cambridge University Press: Cambridge, UK, 2017.
21. Selinger, P. A survey of graphical languages for monoidal categories. In *New Structures for Physics*; Springer: Berlin/Heidelberg, Germany, 2010; pp. 289–355.
22. Haag, R. *Local Quantum Physics: Fields, Particles, Algebras*; Springer: Berlin/Heidelberg, Germany, 2012.
23. Viola, L.; Knill, E.; Laflamme, R. Constructing qubits in physical systems. *J. Phys. A Math. Gen.* **2001**, *34*, 7067. [CrossRef]
24. Zanardi, P.; Lidar, D.A.; Lloyd, S. Quantum tensor product structures are observable induced. *Phys. Rev. Lett.* **2004**, *92*, 060402. [CrossRef] [PubMed]
25. Palma, G.M.; Suominen, K.A.; Ekert, A.K. Quantum computers and dissipation. *Proc. R. Soc. Lond. A* **1996**, *452*, 567–584. [CrossRef]
26. Zanardi, P.; Rasetti, M. Noiseless quantum codes. *Phys. Rev. Lett.* **1997**, *79*, 3306. [CrossRef]
27. Lidar, D.A.; Chuang, I.L.; Whaley, K.B. Decoherence-free subspaces for quantum computation. *Phys. Rev. Lett.* **1998**, *81*, 2594. [CrossRef]
28. Knill, E.; Laflamme, R.; Viola, L. Theory of quantum error correction for general noise. *Phys. Rev. Lett.* **2000**, *84*, 2525. [CrossRef] [PubMed]
29. Zanardi, P. Stabilizing quantum information. *Phys. Rev. A* **2000**, *63*, 012301. [CrossRef]
30. Kempe, J.; Bacon, D.; Lidar, D.A.; Whaley, K.B. Theory of decoherence-free fault-tolerant universal quantum computation. *Phys. Rev. A* **2001**, *63*, 042307. [CrossRef]
31. Zanardi, P. Virtual quantum subsystems. *Phys. Rev. Lett.* **2001**, *87*, 077901. [CrossRef] [PubMed]
32. Bratteli, O.; Robinson, D.W. *Operator Algebras and Quantum Statistical Mechanics 1*; Springer: Berlin/Heidelberg, Germany, 1987.
33. Kraemer, L.; Del Rio, L. Operational locality in global theories. *arXiv* **2017**, arXiv:1701.03280.
34. Åberg, J. Quantifying superposition. *arXiv* **2006**, arXiv:quant-ph/0612146.
35. Baumgratz, T.; Cramer, M.; Plenio, M. Quantifying coherence. *Phys. Rev. Lett.* **2014**, *113*, 140401. [CrossRef] [PubMed]
36. Levi, F.; Mintert, F. A quantitative theory of coherent delocalization. *New J. Phys.* **2014**, *16*, 033007. [CrossRef]
37. Winter, A.; Yang, D. Operational resource theory of coherence. *Phys. Rev. Lett.* **2016**, *116*, 120404. [CrossRef] [PubMed]
38. Chitambar, E.; Gour, G. Critical examination of incoherent operations and a physically consistent resource theory of quantum coherence. *Phys. Rev. Lett.* **2016**, *117*, 030401. [CrossRef] [PubMed]
39. Chitambar, E.; Gour, G. Comparison of incoherent operations and measures of coherence. *Phys. Rev. A* **2016**, *94*, 052336. [CrossRef]
40. Marvian, I.; Spekkens, R.W. How to quantify coherence: Distinguishing speakable and unspeakable notions. *Phys. Rev. A* **2016**, *94*, 052324. [CrossRef]
41. Yadin, B.; Ma, J.; Girolami, D.; Gu, M.; Vedral, V. Quantum processes which do not use coherence. *Phys. Rev. X* **2016**, *6*, 041028. [CrossRef]
42. Chiribella, G.; D'Ariano, G.; Perinotti, P. Informational derivation of quantum theory. *Phys. Rev. A* **2011**, *84*, 012311. [CrossRef]
43. Hardy, L. Reformulating and reconstructing quantum theory. *arXiv* **2011**, arXiv:1104.2066.
44. Masanes, L.; Müller, M.P. A derivation of quantum theory from physical requirements. *New J. Phys.* **2011**, *13*, 063001. [CrossRef]
45. Dakic, B.; Brukner, C. Quantum Theory and Beyond: Is Entanglement Special? In *Deep Beauty: Understanding the Quantum World through Mathematical Innovation*; Halvorson, H., Ed.; Cambridge University Press: Cambridge, UK, 2011; pp. 365–392.

46. Masanes, L.; Müller, M.P.; Augusiak, R.; Perez-Garcia, D. Existence of an information unit as a postulate of quantum theory. *Proc. Natl. Acad. Sci. USA* **2013**, *110*, 16373–16377. [CrossRef] [PubMed]
47. Wilce, A. Conjugates, Filters and Quantum Mechanics. *arXiv* **2012**, arXiv:1206.2897.
48. Barnum, H.; Müller, M.P.; Ududec, C. Higher-order interference and single-system postulates characterizing quantum theory. *New J. Phys.* **2014**, *16*, 123029. [CrossRef]
49. Chiribella, G.; D'Ariano, G.; Perinotti, P. Quantum Theory, namely the pure and reversible theory of information. *Entropy* **2012**, *14*, 1877–1893. [CrossRef]
50. Chiribella, G.; Yuan, X. Quantum theory from quantum information: The purification route. *Can. J. Phys.* **2013**, *91*, 475–478. [CrossRef]
51. Chiribella, G.; Scandolo, C.M. Conservation of information and the foundations of quantum mechanics. In *EPJ Web of Conferences*; EDP Sciences: Les Ulis, France, 2015; Volume 95, p. 03003.
52. Chiribella, G.; Scandolo, C.M. Entanglement and thermodynamics in general probabilistic theories. *New J. Phys.* **2015**, *17*, 103027. [CrossRef]
53. Chiribella, G.; Scandolo, C.M. Microcanonical thermodynamics in general physical theories. *New J. Phys.* **2017**, *19*, 123043. [CrossRef]
54. Chiribella, G.; Scandolo, C.M. Entanglement as an axiomatic foundation for statistical mechanics. *arXiv* **2016**, arXiv:1608.04459.
55. Lee, C.M.; Selby, J.H. Generalised phase kick-back: The structure of computational algorithms from physical principles. *New J. Phys.* **2016**, *18*, 033023. [CrossRef]
56. Lee, C.M.; Selby, J.H. Deriving Grover's lower bound from simple physical principles. *New J. Phys.* **2016**, *18*, 093047. [CrossRef]
57. Lee, C.M.; Selby, J.H.; Barnum, H. Oracles and query lower bounds in generalised probabilistic theories. *arXiv* **2017**, arXiv:1704.05043.
58. Susskind, L. *The Black Hole War: My Battle with Stephen Hawking to Make the World Safe for Quantum Mechanics*; Hachette UK: London, UK, 2008.
59. Takesaki, M. *Theory of Operator Algebras I*; Springer: New York, NY, USA, 1979.
60. Barnum, H.; Knill, E.; Ortiz, G.; Somma, R.; Viola, L. A subsystem-independent generalization of entanglement. *Phys. Rev. Lett.* **2004**, *92*, 107902. [CrossRef] [PubMed]
61. Barnum, H.; Knill, E.; Ortiz, G.; Viola, L. Generalizations of entanglement based on coherent states and convex sets. *Phys. Rev. A* **2003**, *68*, 032308. [CrossRef]
62. Barnum, H.; Ortiz, G.; Somma, R.; Viola, L. A generalization of entanglement to convex operational theories: entanglement relative to a subspace of observables. *Int. J. Theor. Phys.* **2005**, *44*, 2127–2145. [CrossRef]
63. Del Rio, L.; Kraemer, L.; Renner, R. Resource theories of knowledge. *arXiv* **2015**, arXiv:1511.08818.
64. Del Rio, L. Resource Theories of Knowledge. Ph.D. Thesis, ETH Zürich, Zürich, Switzerland, 2015. [CrossRef]
65. Kraemer Gabriel, L. Restricted Agents in Thermodynamics and Quantum Information Theory. Ph.D. Thesis, ETH Zürich, Zürich, Switzerland, 2016. [CrossRef]
66. Brassard, G.; Raymond-Robichaud, P. The equivalence of local-realistic and no-signalling theories. *arXiv* **2017**, arXiv:1710.01380.
67. Holevo, A.S. *Statistical Structure of Quantum Theory*; Springer: Berlin/Heidelberg, Germany, 2003; Volume 67.
68. Kraus, K. *States, Effects and Operations: Fundamental Notions of Quantum Theory*; Springer: Berlin/Heidelberg, Germany, 1983.
69. Haag, R.; Schroer, B. Postulates of quantum field theory. *J. Math. Phys.* **1962**, *3*, 248–256. [CrossRef]
70. Haag, R.; Kastler, D. An algebraic approach to quantum field theory. *J. Math. Phys.* **1964**, *5*, 848–861. [CrossRef]
71. Buscemi, F.; Chiribella, G.; D'Ariano, G.M. Inverting quantum decoherence by classical feedback from the environment. *Phys. Rev. Lett.* **2005**, *95*, 090501. [CrossRef] [PubMed]
72. Buscemi, F.; Chiribella, G.; D'Ariano, G.M. Quantum erasure of decoherence. *Open Syst. Inf. Dyn.* **2007**, *14*, 53–61. [CrossRef]
73. Selinger, P. Idempotents in dagger categories. *Electron. Notes Theor. Comput. Sci.* **2008**, *210*, 107–122. [CrossRef]
74. Coecke, B.; Selby, J.; Tull, S. Two Roads to Classicality. *Electron. Proc. Theor. Comput. Sci.* **2018**, *266*, 104–118. [CrossRef]

75. Coecke, B.; Lal, R. Causal categories: relativistically interacting processes. *Found. Phys.* **2013**, *43*, 458–501. [CrossRef]
76. Coecke, B. Terminality implies no-signalling... and much more than that. *New Gener. Comput.* **2016**, *34*, 69–85. [CrossRef]
77. Chiribella, G. Distinguishability and copiability of programs in general process theories. *Int. J. Softw. Inform.* **2014**, *8*, 209–223.
78. Fulton, W.; Harris, J. *Representation Theory: A First Course*; Springer: Berlin/Heidelberg, Germany, 2013; Volume 129.
79. Marvian, I.; Spekkens, R.W. A generalization of Schur–Weyl duality with applications in quantum estimation. *Commun. Math. Phys.* **2014**, *331*, 431–475. [CrossRef]
80. Galley, T.D.; Masanes, L. Impossibility of mixed-state purification in any alternative to the Born Rule. *arXiv* **2018**, arXiv:1801.06414.
81. Yngvason, J. Localization and entanglement in relativistic quantum physics. In *The Message of Quantum Science*; Springer: Berlin/Heidelberg, Germany, 2015; pp. 325–348.
82. Murray, F.J.; Neumann, J.V. On rings of operators. *Ann. Math.* **1936**, *37*, 116–229. [CrossRef]
83. Murray, F.J.; von Neumann, J. On rings of operators. II. *Trans. Am. Math. Soc.* **1937**, *41*, 208–248. [CrossRef]
84. Uhlmann, A. The transition probability in the state space of a *-algebra. *Rep. Math. Phys.* **1976**, *9*, 273–279. [CrossRef]
85. Jozsa, R. Fidelity for mixed quantum states. *J. Mod. Opt.* **1994**, *41*, 2315–2323. [CrossRef]
86. Lindblad, G. A general no-cloning theorem. *Lett. Math. Phys.* **1999**, *47*, 189–196. [CrossRef]
87. D'Ariano, G.M.; Presti, P.L. Optimal nonuniversally covariant cloning. *Phys. Rev. A* **2001**, *64*, 042308. [CrossRef]
88. Chiribella, G.; D'Ariano, G.; Perinotti, P.; Cerf, N. Extremal quantum cloning machines. *Phys. Rev. A* **2005**, *72*, 042336. [CrossRef]
89. Coecke, B.; Selby, J.; Tull, S. Categorical Probabilistic Theories. *Electron. Proc. Theor. Comput. Sci.* **2018**, *266*, 367–385.

© 2018 by the author. Licensee MDPI, Basel, Switzerland. This article is an open access article distributed under the terms and conditions of the Creative Commons Attribution (CC BY) license (http://creativecommons.org/licenses/by/4.0/).

Article

Ruling out Higher-Order Interference from Purity Principles

Howard Barnum [1,2,*], Ciarán M. Lee [3,*], Carlo Maria Scandolo [4,*] and John H. Selby [4,5,*]

1. Centre for the Mathematics of Quantum Theory (QMATH), Department of Mathematical Sciences, University of Copenhagen, DK-2100 Copenhagen, Denmark
2. Department of Physics and Astronomy, University of New Mexico, Albuquerque, NM 87131, USA
3. Department of Physics, University College London, London WC1E 6BT, UK
4. Department of Computer Science, University of Oxford, Oxford OX1 3QD, UK
5. Department of Physics, Imperial College London, London SW7 2AZ, UK
* Correspondence: hnbarnum@aol.com (H.B.); ciaran.lee@ucl.ac.uk (C.M.L.); carlomaria.scandolo@cs.ox.ac.uk (C.M.S.); john.selby08@imperial.ac.uk (J.H.S.)

Academic Editors: Giacomo Mauro D'Ariano and Paolo Perinotti
Received: 21 April 2017; Accepted: 22 May 2017; Published: 1 June 2017

Abstract: As first noted by Rafael Sorkin, there is a limit to quantum interference. The interference pattern formed in a multi-slit experiment is a function of the interference patterns formed between pairs of slits; there are no genuinely new features resulting from considering three slits instead of two. Sorkin has introduced a hierarchy of mathematically conceivable *higher-order* interference behaviours, where classical theory lies at the first level of this hierarchy and quantum theory theory at the second. Informally, the order in this hierarchy corresponds to the number of slits on which the interference pattern has an irreducible dependence. Many authors have wondered why quantum interference is limited to the second level of this hierarchy. Does the existence of higher-order interference violate some natural physical principle that we believe should be fundamental? In the current work we show that such principles can be found which limit interference behaviour to second-order, or "quantum-like", interference, but that do not restrict us to the entire quantum formalism. We work within the operational framework of generalised probabilistic theories, and prove that any theory satisfying Causality, Purity Preservation, Pure Sharpness, and Purification—four principles that formalise the fundamental character of purity in nature—exhibits at most second-order interference. Hence these theories are, at least conceptually, very "close" to quantum theory. Along the way we show that systems in such theories correspond to Euclidean Jordan algebras. Hence, they are self-dual and, moreover, multi-slit experiments in such theories are described by pure projectors.

Keywords: higher-order interference; generalised probabilistic theories; Euclidean Jordan algebras

1. Introduction

Described by Feynman as "impossible, *absolutely* impossible, to explain in any classical way" [1] (volume 1, chapter 37), quantum interference is a distinctive signature of non-classicality. However, as first noted by Rafael Sorkin [2,3], there is a limit to this interference; in contrast to the case of two slits, the interference pattern formed in a three slit experiment *can* be written as a linear combination of two and one slit patterns. Sorkin has introduced a hierarchy of mathematically conceivable *higher-order* interference behaviours, where classical theory lies at the first level of this hierarchy and quantum theory theory at the second. Informally, the order in this hierarchy corresponds to the number of slits on which the interference pattern has an irreducible dependence.

Many authors have wondered why quantum interference is limited to the second level of this hierarchy [2,4–13]. Does the existence of higher-order interference violate some natural physical principle that we believe should be fundamental [14]? In the current work we show that such natural principles can be found which limit interference behaviour to second-order, or "quantum-like", interference, but that do not restrict us to the entire quantum formalism.

We work in the framework of general probabilistic theories [15–28]. This framework is general enough to accommodate essentially arbitrary operational theories, where an operational theory specifies a set of laboratory devices which can be connected together in different ways, and assigns probabilities to different experimental outcomes. Investigating how the structural and information-theoretic features of a given theory in this framework depend on different physical principles deepens our physical and intuitive understanding of such features. Indeed, many authors [20,22,23,28,29] have derived the entire structure of finite-dimensional quantum theory from simple information-theoretic axioms—reminiscent of Einstein's derivation of special relativity from two simple physical principles. So far, ruling out higher-order interference has required thermodynamic arguments. Indeed, by combining the results and axioms of Refs. [30,31], higher-order interference could be ruled out in theories satisfying the combined axioms. In this paper we show that we can prove this in a more direct way from first principles, using only the axioms of Ref. [30].

Many experimental investigations have searched for divergences from quantum theory by looking for higher-order interference [32–36]. These experiments involved passing a particle through a physical barrier with multiple slits and comparing the interference patterns formed on a screen behind the barrier when different subsets of slits are closed. Given this set-up, one would expect that the physical theory being tested should possess transformations that correspond to the action of blocking certain subsets of slits. Moreover, blocking all but two subsets of slits should not affect states which can pass through either slit. This intuition suggests that these transformations should correspond to projectors.

Many operational probabilistic theories do not possess such a natural mathematical interpretation of multi-slit experiments; indeed many theories do not admit well-defined projectors [9]. Here, we show that there exist natural information-theoretic principles that both imply the existence of the projector structure, and rule out third-, and higher-, order interference. The principles that ensure this structure are Causality, Purity Preservation, Pure Sharpness, and Purification. These formalise intuitive ideas about the fundamental role of purity in nature. More formally, we show that such theories possess a self-dualising inner product, and that there exist pure projectors which represent the opening and closing of slits in a multi-slit experiment. Barnum, Müller and Ududec have shown that in any self-dual theory in which such projectors exist for every face, if projectors map pure states to pure states, then there can be at most second-order interference [4] (Proposition 29). The conjunction of our new results and the principle of Purity Preservation implies the conditions of Barnum et al.'s proposition. Hence sharp theories with purification do not exhibit higher-order interference. In fact we prove a stronger result, that the systems in such theories are Euclidean Jordan algebras which have been studied in quantum foundations [4,13,37].

This paper is organised as follows. In Section 2 we review the basics of the operational probabilistic theory framework. In Section 3 we formally define higher-order interference. In Section 4 we define sharp theories with purification and review relevant known results. In Section 5 we present and prove our new results. Finally, in Section 6, we offer some suggestions on how new experiments might be devised to observe higher-order interference.

2. Framework

We will describe theories in the framework of operational-probabilistic theories (OPTs) [19,20,24,29,38–40], arising from the marriage of category theory [41–46] with probabilities. The foundation of this framework is the idea that any successful physical theory must provide an account of experimental data. Hence, such theories should have an operational description in terms of such experiments.

The OPT framework is based on the graphical language of circuits, describing experiments that can be performed in a laboratory with physical systems connecting together physical processes, which are denoted as wires and boxes respectively. The systems/wires are labelled with a *type* denoted A,

B, C, ... For example, the type given to a quantum system is the dimension of the Hilbert space describing the system. The processes/boxes are then viewed as transformations with some input and output systems/wires. For instance, in quantum theory these correspond to quantum instruments. We now give a brief introduction to the important concepts in this formalism.

2.1. States, Transformations, and Effects

A fundamental tenant of the OPT framework is composition of systems and physical processes. Given two systems A and B, they can be combined into a composite system, denoted by A ⊗ B. Physical processes can be composed to build circuits, such as

$$\rho \quad \begin{array}{c} A \\ B \end{array} \boxed{\mathcal{A}} \begin{array}{c} A' \\ B' \end{array} \boxed{\mathcal{B}} \begin{array}{c} A' \\ B' \end{array} \boxed{\mathcal{A}'} \begin{array}{c} A'' \\ \end{array} \boxed{a} \\ b \end{array}. \tag{1}$$

Processes with no inputs (such as ρ in the above diagram) are called *states*, those with no outputs (such as a and b) are called *effects* and, those with both inputs and outputs (such as $\mathcal{A}, \mathcal{A}', \mathcal{B}$) are called transformations. We define:

1. St (A) as the set of states of system A,
2. Eff (A) as the set of effects on A,
3. Transf (A, B) as the set of transformations from A to B, and Transf (A) as the set of transformations from A to A,
4. $\mathcal{B} \circ \mathcal{A}$ (or $\mathcal{B}\mathcal{A}$, for short) as the sequential composition of two transformations $\mathcal{A}$ and $\mathcal{B}$, with the input of $\mathcal{B}$ matching the output of $\mathcal{A}$,
5. $\mathcal{A} \otimes \mathcal{B}$ as the parallel composition (or tensor product) of the transformations $\mathcal{A}$ and $\mathcal{B}$.

OPTs include a particular system, the trivial system I, representing the lack of input or output for a particular device.

Hence, states (resp. effects) are transformations with the trivial system as input (resp. output). Circuits with no external wires, like the circuit in Equation (1), are called scalars and are associated with probabilities. We will often use the notation $(a|\rho)$ to denote the circuit

$$(a|\rho) := \quad \boxed{\rho} \; \underline{\; A \;} \; \boxed{a},$$

and of the notation $(a|\mathcal{C}|\rho)$ to denote the circuit

$$(a|\mathcal{C}|\rho) := \quad \boxed{\rho} \; \underline{\; A \;} \; \boxed{\mathcal{C}} \; \underline{\; B \;} \; \boxed{a}.$$

The fact that scalars are probabilities and so are real numbers induces a notion of a sum of transformations, so that the sets St (A), Transf (A, B), and Eff (A) become spanning sets of real vector spaces, denoted by $\text{St}_\mathbb{R}$ (A), $\text{Transf}_\mathbb{R}$ (A, B), and $\text{Eff}_\mathbb{R}$ (A). In this work we will restrict our attention to *finite* systems, i.e., systems for which the vector space spanned by states is finite-dimensional for all systems. Operationally this assumption means that one need not perform an infinite number of distinct experiments to fully characterise a state. Restricting ourselves to non-negative real numbers, we have the convex cone of states and of effects, denoted by St_+ (A) and Eff_+ (A) respectively. We moreover make the assumption that the set of states is close. Operationally this is justified by the fact that up to any experimental error a state space is indistinguishable from its closure.

The composition of states and effects leads naturally to a norm. This is defined, for states ρ as $\|\rho\| := \sup_{a \in \text{Eff}(A)} (a|\rho)$, and similarly for effects a as $\|a\| := \sup_{\rho \in \text{St}(A)} (a|\rho)$. The set of normalised states (resp. effects) of system A is denoted by St_1 (A) (resp. Eff_1 (A)).

Transformations are characterised by their action on states of composite systems: if $\mathcal{A}, \mathcal{A}' \in$ Transf (A, B), we have that $\mathcal{A} = \mathcal{A}'$ if and only if

$$\left(\!\rho\!\begin{array}{c}A\\ \hline S\end{array}\!\!\boxed{\mathcal{A}}\!\!\begin{array}{c}B\\ \end{array}\right) = \left(\!\rho\!\begin{array}{c}A\\ \hline S\end{array}\!\!\boxed{\mathcal{A}'}\!\!\begin{array}{c}B\\ \end{array}\right), \qquad (2)$$

for every system S and every state $\rho \in \text{St}(A \otimes S)$. However it follows that [19] effects (resp. states) are completely defined by their action on states (resp. effects) of a single system.

Equality on states of the single system A is, in general, not enough to discriminate between $\mathcal{A}$ and $\mathcal{A}'$, as is the case for quantum theory over real Hilbert spaces [47]. However, for the scope of the present article, which focuses on single-system properties, we often concern ourselves with equality on single system.

Definition 1. *Two transformations* $\mathcal{A}, \mathcal{A}' \in \text{Transf}(A, B)$ *are equal on single system, denoted by* $\mathcal{A} \doteq \mathcal{A}'$, *if* $\mathcal{A}\rho = \mathcal{A}'\rho$ *for all states* $\rho \in \text{St}(A)$.

2.2. Tests and Channels

In general, the boxes corresponding to physical processes come equipped with classical pointers. When used in an experiment, the final position of the a given pointer indicates the particular process which occurred for that box in that run. In general, this procedure can be non-deterministic. These non-deterministic processes are described by *tests* [19,39]: a test from A to B is a collection of transformations $\{\mathcal{C}_i\}_{i \in X}$ from A to B, where X is the set of outcomes. If A (resp. B) is the trivial system, the test is called a *preparation-test* (resp. *observation-test*). If the set of outcomes X has a single element, we say that the test is *deterministic*, because only one transformation can occur. Deterministic transformations will be called *channels*.

A channel $\mathcal{U}$ from A to B is *reversible* if there exists another channel $\mathcal{U}^{-1}$ from B to A such that $\mathcal{U}^{-1}\mathcal{U} = \mathcal{I}_A$ and $\mathcal{U}\mathcal{U}^{-1} = \mathcal{I}_B$, where $\mathcal{I}_S$ is the identity transformation on system S. If there exists a reversible channel transforming A into B, we say that A and B are *operationally equivalent*, denoted as $A \simeq B$. The composition of systems is required to be *symmetric*, meaning that $A \otimes B \simeq B \otimes A$. Physically, this means that for every pair of systems there exists a reversible channel swapping them. A state χ is called *invariant* if $\mathcal{U}\chi = \chi$ for all reversible channels $\mathcal{U}$.

A particularly useful class of observation-tests allows for the following.

Definition 2. *The states* $\{\rho_i\}_{i \in X}$ *are called* perfectly distinguishable *if there exists an observation-test* $\{a_i\}_{i \in X}$ *such that* $(a_i|\rho_j) = \delta_{ij}$ *for all* $i, j \in X$.

Moreover, if there is no other state ρ_0 *such that the states* $\{\rho_i\}_{i \in X} \cup \{\rho_0\}$ *are perfectly distinguishable, the set* $\{\rho_i\}_{i \in X}$ *is said* maximal.

2.3. Pure Transformations

There are various different ways to define pure transformations, for example in terms of resources [30,48–51] or "side information" [39,52]. Informally pure transformations correspond to an experimenter having maximal control of or information about a process. Here, we formalise this notion by defining the notion of a *coarse-graining* [19]. Coarse-graining is the operation of joining two or more outcomes of a test into a single outcome. More precisely, a test $\{\mathcal{C}_i\}_{i \in X}$ is a coarse-graining of the test $\{\mathcal{D}_j\}_{j \in Y}$ if there is a partition $\{Y_i\}_{i \in X}$ of Y such that, for all $i \in X$

$$\mathcal{C}_i = \sum_{j \in Y_i} \mathcal{D}_j$$

In this case, we say that the test $\{\mathcal{D}_j\}_{j \in Y}$ is a *refinement* of the test $\{\mathcal{C}_i\}_{i \in X}$, and that the transformations $\{\mathcal{D}_j\}_{j \in Y_i}$ are a refinement of the transformation $\mathcal{C}_i$. A transformation $\mathcal{C} \in \text{Transf}(A, B)$ is *pure* if it has only trivial refinements, namely refinements $\{\mathcal{D}_j\}$ of the form $\mathcal{D}_j = p_j \mathcal{C}$, where $\{p_j\}$ is a probability distribution. We denote the sets of pure transformations, pure states, and pure effects as

PurTransf (A, B), PurSt (A), and PurEff (A) respectively. Similarly, PurSt$_1$ (A), and PurEff$_1$ (A) denote *normalised* pure states and effects respectively. Non-pure states are called *mixed*.

Definition 3. *Let $\rho \in \text{St}_1$ (A). A normalised state σ is contained in ρ if we can write $\rho = p\sigma + (1-p)\tau$, where $p \in (0,1]$ and τ is another normalised state.*

Clearly, no states are contained in a pure state. On the other edge of the spectrum we have complete states.

Definition 4. *A state $\omega \in \text{St}_1$ (A) is complete if every state is contained in it.*

Definition 5. *We say that two transformations $\mathcal{A}, \mathcal{A}' \in$ Transf (A, B) are equal upon input of the state $\rho \in \text{St}_1$ (A) if $\mathcal{A}\sigma = \mathcal{A}'\sigma$ for every state σ contained in ρ. In this case we will write $\mathcal{A} =_\rho \mathcal{A}'$.*

2.4. Causality

A natural requirement of a physical theory is that it is *causal*, that is, no signals can be sent from the future to the past. In the OPT framework this is formalised as follows:

Axiom 1 (Causality [19,39])**.** *The probability that a transformation occurs is independent of the choice of tests performed on its output.*

Causality is equivalent to the requirement that, for every system A, there exists a unique deterministic effect u_A on A (or simply u, when no ambiguity can arise) [19]. Owing to the uniqueness of the deterministic effect, the marginals of a bipartite state can be uniquely defined as:

$$\left(\rho_A \!-\!\!\!\!\begin{array}{c} A \end{array}\!\!\!\!-\right) := \left(\rho_{AB}\!-\!\!\!\!\begin{array}{c} A \\ B \end{array}\!\!\!\!-\!\!\!u\right),$$

Moreover, this uniqueness forbids the ability to signal [19,53]. We will denote by $\text{Tr}_B \rho_{AB}$ the marginal on system A, in analogy with the notation used in the quantum case. We will stick to the notation Tr in formulas where the deterministic effect is applied directly to a state, e.g., $\text{Tr}\,\rho := (u|\rho)$.

In a causal theory it is easy to see that the norm of a state takes the form $\|\rho\| = \text{Tr}\,\rho$, and that a state can be prepared deterministically if and only if it is normalised.

3. Higher-Order Interference

The definition of higher-order interference we shall present in this section takes its motivation from the set-up of multi-slit interference experiments. In such experiments a particle passes through slits in a physical barrier and is detected at a screen. By repeating the experiment many times, one builds up a pattern on the screen. To determine if this experiment exhibits interference one compares this pattern to those produced when certain subsets of the slits are blocked. In quantum theory, for example, the two-slit experiment exhibits interference as the pattern formed with both slits open is not equal to the sum of the one-slit patterns.

Consider the state of the particle just before it passes through the slits. For every slit, there should exist states such that the particle is definitely found at that slit, if measured. Mathematically, this means that there is a face [4] of the state space, such that all states in this face give unit probability for the "yes" outcome of the two-outcome measurement "is the particle at this slit?". Recall that a face is a convex set with the property that if $px + (1-p)y$, for $0 \leq p \leq 1$, is an element then x and y are also elements. These faces will be labelled F_i, one for each of the n slits $i \in \{1, \ldots, n\}$. As the slits should be perfectly distinguishable, the faces associated with each slit should be perfectly distinguishable, or orthogonal. One can additionally ask coarse-grained questions of the form "Is the particle found among a certain subset of slits, rather than somewhere else?". The set of states that give outcome "yes" with probability one must contain all the faces associated with each slit in the subset. Hence the face associated with the subset of slits $I \subseteq \{1,\ldots,n\}$ is the smallest face containing each face in this subset $F_I := \bigvee_{i \in I} F_i$, where the operation $\bigvee$ is the least upper bound of the lattice of faces where the ordering is provided by subset inclusion of one face within another. The face F_I contains all those states which can be found among the slits contained in I. The experiment is "complete" if all states in the state space (of a given system A) can be found among some subset of slits. That is, if $F_{12\cdots n} = \text{St}(A)$.

An n-slit experiment requires a system that has n orthogonal faces F_i, with $i \in \{1,\ldots,n\}$. Consider an effect E associated with finding a particle at a particular point on the screen. We now formally define an n-slit experiment.

Definition 6. *An n-slit experiment is a collection of effects e_I, where $I \subseteq \{1,\ldots,n\}$, such that*

$$(e_I|\rho) = (E|\rho), \quad \forall \rho \in F_I, \text{ and}$$
$$(e_I|\rho) = 0, \quad \forall \rho \text{ where } \rho \perp F_I.$$

The effects introduced in the above definition arise from the conjunction of blocking off the slits $\{1,\ldots,n\} \setminus I$ and applying the effect E. If the particle was prepared in a state such that it would be unaffected by the blocking of the slits (i.e., $\rho \in F_I$) then we should have $(e_I|\rho) = (E|\rho)$. If instead the particle is prepared in a state which is guaranteed to be blocked (i.e., $\rho' \perp F_I$) then the particle should have no probability of being detected at the screen, i.e., $(e_I|\rho') = 0$.

The relevant quantities for the existence of various orders of interference are [2,9,13,15]:

$$I_1 := (E|\rho), \tag{3}$$
$$I_2 := (E|\rho) - (e_1|\rho) - (e_2|\rho), \tag{4}$$
$$I_3 := (E|\rho) - (e_{12}|\rho) - (e_{23}|\rho) - (e_{31}|\rho) + (e_1|\rho) + (e_2|\rho) + (e_3|\rho), \tag{5}$$
$$I_n := \sum_{\emptyset \neq I \subseteq \{1,\ldots,n\}} (-1)^{n-|I|} (e_I|\rho), \tag{6}$$

for some state ρ, and defining $e_{\{1,\ldots,n\}} := E$.

Definition 7. *A theory has n-th order interference if there exists a state ρ and an effect E such that $I_n \neq 0$.*

In a slightly different formal setting, it was shown in [2] that $I_n = 0 \implies I_{n+1} = 0$, so if there is no nth order interference, there will be no $(n+1)$th order interference; the argument of [2] applies here.

It should be noted that there appears to be a lot of freedom in choosing a set of effects $\{e_I\}$ to test for the existence of higher-order interference. Indeed, in arbitrary generalised theories this appears to be the case [9]. However, it is natural to ask whether there exists physical transformations T_I in the theory which correspond to leaving the subset of slits I open and blocking the rest. Hence a unique e_I is assigned to each fixed E defined as $e_I = ET_I$. Ruling out the existence of higher-order interference then reduces to proving certain properties of the T_I. This will turn out to be the case in sharp theories with purification.

4. Sharp Theories with Purification

In this section we present the definition and important properties of sharp theories with purification. They were originally introduced in [30,49,54] for the analysis of the foundations of thermodynamics and statistical mechanics.

Sharp theories with purification are causal theories defined by three axioms. The first axiom—Purity Preservation—states that no information can leak when two pure transformations are composed:

Axiom 2 (Purity Preservation [55]). *Sequential and parallel compositions of pure transformations yield pure transformations.*

The second axiom—Pure Sharpness—guarantees that every system possesses at least one elementary property.

Axiom 3 (Pure Sharpness [54]). *For every system there exists at least one pure effect occurring with unit probability on some state.*

These axioms are satisfied by both classical and quantum theory. Our third axiom—Purification—signals the departure from classicality, and characterises when a physical theory admits a level of description where all deterministic processes are pure and reversible.

Given a normalised state $\rho_A \in \mathsf{St}_1(A)$, a normalised pure state $\Psi \in \mathsf{PurSt}_1(A \otimes B)$ is a *purification* of ρ_A if

$$\left(\Psi \begin{array}{c} A \\ B\ u \end{array} \right. = \rho_A \quad A\ ;$$

in this case B is called the *purifying system*. We say that a pure state $\Psi \in \mathsf{PurSt}(A \otimes B)$ is an *essentially unique purification* of its marginal ρ_A [39] if every other pure state $\Psi' \in \mathsf{PurSt}(A \otimes B)$ satisfying the purification condition must be of the form

$$\left(\Psi' \begin{array}{c} A \\ B \end{array} \right. = \left(\Psi \begin{array}{c} A \\ B\ \mathcal{U}\ B \end{array} \right. ,$$

for some reversible channel $\mathcal{U}$.

Axiom 4 (Purification [19,39]). *Every state has a purification. Purifications are essentially unique.*

Quantum theory, both on complex and real Hilbert spaces, satisfies Purification, and also Spekkens' toy model [56]. Examples of sharp theories with purification besides quantum theory include fermionic quantum theory [57,58], a superselected version of quantum theory known as doubled quantum theory [49], and a recent extension of classical theory with the theory of codits [30].

Properties of Sharp Theories With Purifications

Sharp theories with purifications enjoy some nice properties, which were mainly derived in Refs. [30,54]. The first property is that every non-trivial system admits perfectly distinguishable states [54], and that all maximal sets of pure states have the same cardinality [30].

Proposition 1. *For every system A there is a positive integer d_A, called the* dimension *of A, such that all maximal sets of pure states have d_A elements.*

Note that we will omit the subscript A when the context is clear.

In sharp theories with purification every state can be diagonalised, i.e., written as a convex combination of perfectly distinguishable pure states (cf. Refs. [30,54]).

Theorem 5. *Every normalised state $\rho \in \mathsf{St}_1(A)$ of a non-trivial system can be decomposed as*

$$\rho = \sum_{i=1}^{d} p_i \alpha_i,$$

where $\{p_i\}_{i=1}^{d}$ is a probability distribution, and $\{\alpha_i\}_{i=1}^{d}$ is a pure maximal set. Moreover, given ρ, $\{p_i\}_{i=1}^{d}$ is unique up to rearrangements.

Such a decomposition is called a *diagonalisation* of ρ, the p_i's are the *eigenvalues* of ρ, and the α_i's are the *eigenstates*. Theorem 5 implies that the eigenvalues of a state are unique, and independent of its diagonalisation. Sharp theories with purification have a unique invariant state χ [19], which can be diagonalised as $\chi = \frac{1}{d} \sum_{i=1}^{d} \alpha_i$, where $\{\alpha_i\}_{i=1}^{d}$ is *any* pure maximal set [30]. Furthermore, the diagonalisation result of Theorem 5 can be extended to every vector in $\text{St}_\mathbb{R}(A)$, but here the eigenvalues will be generally real numbers [30].

One of the most important consequences for this paper of the axioms defining sharp theories with purification is a duality between normalised pure states and normalised pure effects.

Theorem 6 (States-effects duality [30,54]). *For every system A, there is a bijective correspondence †: $\text{PurSt}_1(A) \to \text{PurEff}_1(A)$ such that if $\alpha \in \text{PurSt}_1(A)$, $\alpha^\dagger$ is the unique normalised pure effect such that $(\alpha^\dagger|\alpha) = 1$. Furthermore this bijection can be extended by linearity to an isomorphism between the vector spaces $\text{St}_\mathbb{R}(A)$ and $\text{Eff}_\mathbb{R}(A)$.*

With a little abuse of notation we will use † also to denote the inverse map $\text{PurEff}_1(A) \to \text{PurSt}_1(A)$, by which, if $a \in \text{PurEff}_1(A)$, $a^\dagger$ is the unique pure state such that $(a|a^\dagger) = 1$. Pure maximal sets $\{\alpha_i\}_{i=1}^d$ have the property that $\sum_{i=1}^d \alpha_i^\dagger = u$ [30].

A diagonalisation result holds for vectors of $\text{Eff}_\mathbb{R}(A)$ as well [30]: they can be written as $X = \sum_{i=1}^d \lambda_i \alpha_i^\dagger$, where $\{\alpha_i\}_{i=1}^d$ is a pure maximal set. Again, the λ_i's are uniquely defined given X.

Another result that will be made use of in the following sections is the following. It was shown to hold in Ref. [30], and expresses the possibility of constructing non-disturbing measurements [20,59,60].

Proposition 2. *Given a system A, let $a \in \text{Eff}(A)$ be an effect such that $(a|\rho) = 1$, for some $\rho \in \text{St}_1(A)$. Then there exists a pure transformation $\mathcal{T} \in \text{PurTransf}(A)$ such that $\mathcal{T} =_\rho \mathcal{I}$, with $(u|\mathcal{T}|\sigma) \leq (a|\sigma)$, for every state $\sigma \in \text{St}_1(A)$.*

Note that the pure transformation $\mathcal{T}$ is non-disturbing on ρ because it acts as the identity on ρ and on all states contained in it. In other words, whenever we have an effect occurring with unit probability on some state ρ, we can always find a transformation that does not disturb ρ (i.e., a non-disturbing, non-demolition measurement) [30].

Finally, a property that we will use often is a sort of no-restriction hypothesis for tests, derived in [20] (Corollary 4).

Proposition 3. *A collection of transformations $\{\mathcal{A}_i\}_{i \in X}$ is a valid test if and only if $\sum_{i \in X} u\mathcal{A}_i = u$. A collection of effects $\{a_i\}_{i \in X}$ is a valid observation-test if and only if $\sum_{i \in X} a_i = u$.*

5. Sharp Theories with Purification Have No Higher-Order Interference

Here we will show that sharp theories with purification do not exhibit higher-order interference. Our proof strategy will be to show that results of [4], which rule out the existence of higher-order interference from certain assumptions, hold in sharp theories with purification. To this end, we will first prove that these theories are self-dual, and that they admit *pure* orthogonal projectors which satisfy certain properties, compatible with the setting presented in Section 3.

5.1. Self-Duality

Now we will prove that sharp theories with purification are self-dual. Recall that a theory is *self-dual* if for every system A there is an inner product $\langle \bullet, \bullet \rangle$ on $\text{St}_\mathbb{R}(A)$ such that $\xi \in \text{St}_+(A)$ if and only if $\langle \xi, \eta \rangle \geq 0$ for every $\eta \in \text{St}_+(A)$. To show that, we need to find a self-dualising inner product on $\text{St}_\mathbb{R}(A)$ for every system A. The dagger will provide us with a good candidate. First we need the following lemma.

Lemma 1. *Let $a \in \text{Eff}_1(A)$ be a normalised effect. Then a is of the form $a = \sum_{i=1}^r \alpha_i^\dagger$, with $r \leq d$, and the pure states $\{\alpha_i\}_{i=1}^r$ are perfectly distinguishable.*

Proof. We know that every effect a can be written as $a = \sum_{i=1}^{r} \lambda_i \alpha_i^\dagger$, where $r \leq d$, the pure states $\{\alpha_i\}_{i=1}^{r}$ are perfectly distinguishable, and for every $i \in \{1,\ldots,r\}$, $\lambda_i \in (0,1]$. Since the state space is closed, and a is normalised, then there exists a (normalised) state ρ such that $(a|\rho) = 1$. One has

$$1 = (a|\rho) = \sum_{i=1}^{r} \lambda_i \left(\alpha_i^\dagger \middle| \rho\right).$$

Now, $\left(\alpha_i^\dagger \middle| \rho\right) \geq 0$, and $\sum_{i=1}^{r} \left(\alpha_i^\dagger \middle| \rho\right) \leq 1$ because

$$\sum_{i=1}^{r} \left(\alpha_i^\dagger \middle| \rho\right) \leq \sum_{i=1}^{d} \left(\alpha_i^\dagger \middle| \rho\right) = \operatorname{Tr}\rho = 1,$$

where we have used the fact that $\sum_{i=1}^{d} \alpha_i^\dagger = u$. Then $\sum_{i=1}^{r} \lambda_i \left(\alpha_i^\dagger \middle| \rho\right) \leq \lambda_{\max}$, where $\lambda_{\max}$ is the maximum of the λ_i's. Therefore, $\lambda_{\max} \geq 1$, which implies $\lambda_{\max} = 1$. Now, the condition

$$\sum_{i=1}^{r} \lambda_i \left(\alpha_i^\dagger \middle| \rho\right) = \lambda_{\max}$$

means that $\lambda_i = \lambda_{\max} = 1$ for all $i \in \{1,\ldots,r\}$. □

In the above, we call r the *rank* of the normalised effect. We can use this result to prove the following.

Lemma 2. *For every system A, the map*

$$\langle \xi, \eta \rangle := \left(\xi^\dagger \middle| \eta\right),$$

for every $\xi, \eta \in \mathsf{St}_\mathbb{R}(A)$ is an inner product on $\mathsf{St}_\mathbb{R}(A)$.

Proof. The map $\langle \bullet, \bullet \rangle$ is clearly bilinear by construction, because the dagger is also linear. Let us show that it is positive-definite. Take a non-null vector $\xi \in \mathsf{St}_\mathbb{R}(A)$, and diagonalise it as $\xi = \sum_{i=1}^{d} x_i \alpha_i$. Then

$$\langle \xi, \xi \rangle = \left(\xi^\dagger \middle| \xi\right) = \sum_{i,j=1}^{d} x_i x_j \left(\alpha_i^\dagger \middle| \alpha_j\right) = \sum_{i=1}^{d} x_i^2 > 0,$$

where we have used the fact that for perfectly distinguishable pure states $\left(\alpha_i^\dagger \middle| \alpha_j\right) = \delta_{ij}$ [30].

The hard part is to prove that this bilinear map is symmetric, namely $\langle \xi, \eta \rangle = \langle \eta, \xi \rangle$, for every $\xi, \eta \in \mathsf{St}_\mathbb{R}(A)$. Let us define a new (double) dagger ‡. The double dagger of a normalised state ρ is an effect $\rho^\ddagger$ whose action on normalised states σ is defined as

$$\left(\rho^\ddagger \middle| \sigma\right) := \left(\sigma^\dagger \middle| \rho\right), \tag{7}$$

where † is the dagger of Theorem 6. Note that Equation (7) is enough to characterise $\rho^\ddagger$ completely, and it guarantees that $\rho^\ddagger$ is a mathematically well-defined effect, because it is linear and $\left(\sigma^\dagger \middle| \rho\right) \in [0,1]$. Consider now ρ and σ to be a normalised *pure* state ψ. Then $\left(\psi^\ddagger \middle| \psi\right) = \left(\psi^\dagger \middle| \psi\right) = 1$, this means that $\alpha^\ddagger$ is normalised. If we manage to show that $\psi^\ddagger$ is pure, then by Theorem 6 we can conclude that $\psi^\ddagger = \psi^\dagger$. By Lemma 1, $\psi^\ddagger$ is of the form $\psi^\ddagger = \sum_{i=1}^{r} \alpha_i^\dagger$, with $r \leq d$, and the pure states $\{\alpha_i\}_{i=1}^{r}$ are perfectly distinguishable. Clearly $\psi^\ddagger$ is pure if and only if $r = 1$. To prove it, first let us evaluate $\psi^\ddagger$ on χ:

$$\left(\psi^\ddagger \middle| \chi\right) = \left(\chi^\dagger \middle| \psi\right) = \frac{1}{d}\operatorname{Tr}\psi = \frac{1}{d}, \tag{8}$$

as prescribed by Equation (7). Now, since $\psi^\ddagger = \sum_{i=1}^r \alpha_i^\dagger$, we have

$$\left(\psi^\ddagger \middle| \chi\right) = \sum_{i=1}^r \left(\alpha_i^\dagger \middle| \chi\right) = \frac{r}{d}, \qquad (9)$$

because $\left(\alpha_i^\dagger \middle| \chi\right) = \frac{1}{d}$ for every i [30]. A comparison between Equations (8) and (9), shows that $r = 1$. This means that $\psi^\ddagger$ is pure, whence $\psi^\ddagger = \psi^\dagger$. Now we can show that the double dagger $\ddagger$ actually coincides with the dagger of Theorem 6. Indeed, given a state ρ, diagonalise it as $\rho = \sum_{i=1}^d p_i \alpha_i$. One can easily show that the double dagger of Equation (7) is linear, so we have $\rho^\ddagger = \sum_{i=1}^d p_i \alpha_i^\ddagger$, but we have just proved that $\alpha_i^\ddagger = \alpha_i^\dagger$ for pure states, so $\rho^\ddagger = \sum_{i=1}^d p_i \alpha_i^\dagger = \rho^\dagger$. This means that $\ddagger = \dagger$, and that Equation (7) is nothing but a redefinition of the usual dagger. This means for every normalised states we have

$$\left(\rho^\dagger \middle| \sigma\right) = \left(\sigma^\dagger \middle| \rho\right), \qquad (10)$$

and this extends linearly to all vectors $\xi, \eta \in \mathsf{St}_\mathbb{R}(A)$. We have proved that $\langle \bullet, \bullet \rangle$ is symmetric, and this concludes the proof. □

Note that the above result immediately yields the "symmetry of transition probabilities" as defined in Ref. [61,62].

Now we prove that this inner product is invariant under reversible transformations.

Proposition 4. *For every $\xi, \eta \in \mathsf{St}_\mathbb{R}(A)$ and every reversible channel $\mathcal{U}$ one has*

$$\langle \mathcal{U}\xi, \mathcal{U}\eta \rangle = \langle \xi, \eta \rangle.$$

Proof. To prove the statement, let us first prove that for a normalised pure state α one has $(\mathcal{U}\alpha)^\dagger = \alpha^\dagger \mathcal{U}^{-1}$, for every reversible channel $\mathcal{U}$. $\alpha^\dagger \mathcal{U}^{-1}$ is a pure effect and one has $(\alpha^\dagger \mathcal{U}^{-1} | \mathcal{U}\alpha) = (\alpha^\dagger | \alpha) = 1$. By the uniqueness of the dagger for normalised pure states, $\alpha^\dagger \mathcal{U}^{-1} = (\mathcal{U}\alpha)^\dagger$. This can be extended by linearity to all vectors ξ in $\mathsf{St}_\mathbb{R}(A)$, so $(\mathcal{U}\xi)^\dagger = \xi^\dagger \mathcal{U}^{-1}$. Therefore, when we compute $\langle \mathcal{U}\xi, \mathcal{U}\eta \rangle$, we have

$$\langle \mathcal{U}\xi, \mathcal{U}\eta \rangle = \left(\xi^\dagger \middle| \mathcal{U}^{-1}\mathcal{U} \middle| \eta\right) = \left(\xi^\dagger \middle| \eta\right) = \langle \xi, \eta \rangle.$$

□

The fact that $\langle \bullet, \bullet \rangle$ is an inner product allows us to define an additional norm in sharp theories with purification: if $\xi \in \mathsf{St}_\mathbb{R}(A)$, define the *dagger norm* as

$$\|\xi\|_\dagger := \sqrt{\langle \xi, \xi \rangle}.$$

See Appendix A.1 for an extended discussion on the properties of this norm.

Now we are ready to state the core of this subsection.

Proposition 5. *Sharp theories with purification are self-dual.*

Proof. Given a system A, we need to prove that $\xi \in \mathsf{St}_\mathbb{R}(A)$ is in $\mathsf{St}_+(A)$ if and only if $\langle \xi, \eta \rangle \geq 0$ for all $\eta \in \mathsf{St}_+(A)$. Note that $\xi \in \mathsf{St}_+(A)$ if and only if it can be diagonalised as $\xi = \sum_{i=1}^d x_i \alpha_i$, where the x_i's are all non-negative.

Necessity. Suppose $\xi \in \mathsf{St}_+(A)$, and take any $\eta \in \mathsf{St}_+(A)$, diagonalised as $\eta = \sum_{i=1}^d y_i \beta_i$. Then we have

$$\langle \xi, \eta \rangle = \sum_{i,j=1}^d x_i y_j \left(\alpha_i^\dagger \middle| \beta_j\right) \geq 0$$

because all the terms x_i, y_j, and $\left(\alpha_i^\dagger \middle| \beta_j\right)$ are non-negative.

Sufficiency. Take $\xi \in \mathsf{St}_{\mathbb{R}}(A)$, and assume that $\langle \xi, \eta \rangle \geq 0$ for all $\eta \in \mathsf{St}_+(A)$. Assume ξ is diagonalised as $\xi = \sum_{i=1}^{d} x_i \alpha_i$, where the x_i's are generic real numbers. We wish to prove that all the x_i's are non-negative. Then

$$\langle \xi, \eta \rangle = \sum_{i,j=1}^{d} x_i \left(\alpha_i^\dagger \middle| \eta \right) \geq 0.$$

Recalling that for perfectly distinguishable pure states one has $\left(\alpha_i^\dagger \middle| \alpha_j \right) = \delta_{ij}$ [30], it is enough to take η to be one of the states $\{\alpha_i\}_{i=1}^{d}$ to conclude that $x_i \geq 0$ for every $i \in \{1,\ldots,d\}$, meaning that $\xi \in \mathsf{St}_+(A)$. □

The self-dualising inner product, besides being a nice mathematical tool, has some operational meaning, because it provides a measure of the distinguishability of states, as explained in Appendix A.2. Moreover, it is the starting point for extending the dagger to all transformations. This is done in Appendix B.

5.2. Existence of Pure Orthogonal Projectors

Now we show that we have orthogonal projectors on every face of the state space. A consequence of diagonalisation is that all faces are generated by perfectly distinguishable *pure* states. Indeed, every face F is generated by a state ω in its relative interior. ω can be diagonalised as $\omega = \sum_{i=1}^{r} p_i \alpha_i$, where $r \leq d$, and $p_i > 0$ for $i \in \{1,\ldots,r\}$. By definition of face, this means that the states $\{\alpha_i\}_{i=1}^{r}$ are in F, and therefore generate F. Consequently, there is an effect a that picks out the whole face as the set of states ρ such that $(a|\rho) = 1$. In the specific case considered above, it is $a = \sum_{i=1}^{r} \alpha_i^\dagger$. Such faces are called *exposed*.

Therefore the study of faces of sharp theories with purification reduces to the study of normalised effects. Thanks to Lemma 1, it is enough to consider subsets of pure maximal sets. Pick a pure maximal set $\{\alpha_i\}_{i=1}^{d}$, and consider a subset I of $\{1,\ldots,d\}$. The subset I flags the slits that are open in the experiment. Setting $a_I := \sum_{i\in I} \alpha_i^\dagger$, we can define the two faces

1. $F_I := \{\rho \in \mathsf{St}_1(A) : (a_I|\rho) = 1\}$;
2. $F_I^\perp := \{\rho \in \mathsf{St}_1(A) : (a_I|\rho) = 0\}$,

in analogy with those of Definition 6. Clearly the effect $a_I^\perp := \sum_{i\notin I} \alpha_i^\dagger$ defines the orthogonal face $F_I^\perp$, as it occurs with probability one on the states of $F_I^\perp$. Note that each of the effects $\{\alpha_i^\dagger\}_{i\notin I}$ occurs with zero probability on the states of F_I.

Definition 8. *An orthogonal projector (in the sense of [20]) on the face F_I is a transformation $P_I \in \mathsf{Transf}(A)$ such that*

- *if $\rho \in F_I$, then $P_I \rho = \rho$;*
- *if $\rho \in F_I^\perp$, then $P_I \rho = 0$.*

We can prove the existence of a projector at least in one case, when $I = \{1,\ldots,d\}$. In this case $a_I = u$, so $F_I = \mathsf{St}_1(A)$, and $F_I^\perp = \emptyset$. Then it is enough to take $P_I \doteq \mathcal{I}$. However, sharp theories with purification admit projectors on *every* face.

Proposition 6. *Sharp theories with purification have pure projectors on every face F_I. Furthermore one has $u P_I = a_I$.*

Proof. Suppose ρ is any state in F_I, then $(a_I|\rho) = 1$. By Proposition 2 we know that there is a *pure* transformation P_I such that $P_I \rho = \rho$ for every $\rho \in F_I$. We also have $(u|P_I|\sigma) \leq (a_I|\sigma)$, so if $\sigma \in F_I^\perp$, we have $(u|P_I|\sigma) = 0$, whence $P_I \sigma = 0$.

To prove that $uP_\mathsf{I} = a_\mathsf{I}$, first note that $\psi^\dagger P_\mathsf{I} = \psi^\dagger$ for every pure state $\psi \in F_\mathsf{I}$. Indeed $\psi^\dagger P_\mathsf{I}$ is pure by Purity Preservation, and we have $(\psi^\dagger|P_\mathsf{I}|\psi) = (\psi^\dagger|\psi) = 1$ because $P_\mathsf{I}\psi = \psi$ by definition. By Theorem 6, we have $\psi^\dagger P_\mathsf{I} = \psi^\dagger$. Furthermore, $\varphi^\dagger P_\mathsf{I} = 0$ for a pure state $\varphi \in F_\mathsf{I}^\perp$. Indeed, consider

$$\left(\varphi^\dagger \middle| P_\mathsf{I} \middle| \chi\right) = \frac{1}{d}\sum_{i \in \mathsf{I}}\left(\varphi^\dagger \middle| P_\mathsf{I} \middle| \alpha_i\right) + \frac{1}{d}\sum_{i \notin \mathsf{I}}\left(\varphi^\dagger \middle| P_\mathsf{I} \middle| \alpha_i\right).$$

The second term vanishes because $\alpha_i \in F_\mathsf{I}^\perp$ for $i \notin \mathsf{I}$. The first term vanishes because $P_\mathsf{I}\alpha_i = \alpha_i$ for $i \in \mathsf{I}$, and φ is perfectly distinguishable from any of the α_i's for $i \in \mathsf{I}$ by means of the observation-test $\{u - a_\mathsf{I}, a_\mathsf{I}\}$, implying $(\varphi^\dagger|\alpha_i) = 0$ [30]. This means that $\varphi^\dagger P_\mathsf{I}$ occurs with zero probability on all states contained in χ, and since χ is complete [19], $\varphi^\dagger P_\mathsf{I} = 0$. Now, when we calculate uP_I, we separate the contribution arising from states in orthogonal faces:

$$uP_\mathsf{I} = \sum_{i \in \mathsf{I}} \alpha_i^\dagger P_\mathsf{I} + \sum_{i \notin \mathsf{I}} \alpha_i^\dagger P_\mathsf{I} = \sum_{i \in \mathsf{I}} \alpha_i^\dagger = a_\mathsf{I}$$

This concludes the proof. □

In other words, P_I occurs with the same probability as a_I, thus satisfying one of the desiderata of Section 3. Moreover, extending some of the results in the Proof 6 by linearity, we obtain the dual statements of Definition 8, namely

- $\rho^\dagger P_\mathsf{I} = \rho^\dagger$ if $\rho \in F_\mathsf{I}$
- $\rho^\dagger P_\mathsf{I} = 0$ if $\rho \in F_\mathsf{I}^\perp$

Another consequence of Proposition 6 is that projectors actually project on their associated face, viz. for every normalised state ρ, $P_\mathsf{I}\rho = \lambda\sigma$, where σ is in F_I, and $\lambda = (a_\mathsf{I}|\rho)$. Indeed, $\lambda = (u|P_\mathsf{I}|\rho) = (a_\mathsf{I}|\rho)$. If $\lambda \neq 0$, which means $\rho \notin F_\mathsf{I}^\perp$, then and $(a_\mathsf{I}|\sigma) = \frac{1}{\lambda}(a_\mathsf{I}|P_\mathsf{I}|\rho)$. However, we know that $a_\mathsf{I}P_\mathsf{I} = a_\mathsf{I}$, so $(a_\mathsf{I}|\sigma) = 1$, showing that $\sigma \in F_\mathsf{I}$.

Furthermore, we can show that every projector P_I has a *complement* $P_\mathsf{I}^\perp$, which is the projector associated with the effect $a_\mathsf{I}^\perp = \sum_{i \notin \mathsf{I}} \alpha_i^\dagger$, which defines the orthogonal face $F_\mathsf{I}^\perp$. Clearly $P_\mathsf{I}^\perp \rho = (a_\mathsf{I}^\perp|\rho)\sigma$, with $\sigma \in F_\mathsf{I}^\perp$. In particular, $P_\mathsf{I}^\perp\rho$ vanishes if and only if $\rho \in F_\mathsf{I}$.

These properties are the starting point for proving the idempotence of projectors.

Proposition 7. *Given a fixed pure maximal set $\{\alpha_i\}_{i=1}^d$ and $\mathsf{I} \subseteq \{1,\ldots,d\}$, one has $P_\mathsf{I}^2 \doteq P_\mathsf{I}$. Moreover, if J is another subset of $\{1,\ldots,d\}$ disjoint from I, then $P_\mathsf{I}P_\mathsf{J} \doteq 0$.*

Proof. Recall that for every state ρ, $P_\mathsf{I}\rho = \lambda\sigma$, where σ is in F_I. Now, P_I leaves σ invariant by definition, so

$$P_\mathsf{I}^2\rho = \lambda P_\mathsf{I}\sigma = \lambda\sigma,$$

so $P_\mathsf{I}^2 \doteq P_\mathsf{I}$. To prove the other property, note that if I and J are disjoint, they define orthogonal faces. Indeed, suppose $\rho \in F_\mathsf{I}$, then

$$1 = \mathrm{Tr}\,\rho = (a_\mathsf{I}|\rho) + (a_\mathsf{J}|\rho) + \sum_{i \notin \mathsf{I} \cup \mathsf{J}}\left(\alpha_i^\dagger \middle| \rho\right),$$

which implies $(a_\mathsf{J}|\rho) = 0$ because $(a_\mathsf{I}|\rho) = 1$. Hence $\rho \in F_\mathsf{J}^\perp$. Now, given *any* normalised state ρ, $P_\mathsf{I}P_\mathsf{J}\rho = 0$ because $P_\mathsf{J}\rho$ is proportional to a state in $F_\mathsf{I}^\perp$. This proves that $P_\mathsf{I}P_\mathsf{J} \doteq 0$. □

This result shows that, once a pure maximal set $\{\alpha_i\}_{i=1}^d$ is fixed, whenever we have a partition $\{\mathsf{I}_j\}$ of $\{1,\ldots,d\}$, the test $\left\{P_{\mathsf{I}_j}\right\}$ is a von Neumann measurement. The only thing left to check is that

$\sum_j u P_{1_j} = u$, which is a sufficient condition for a set of transformations to be a test in sharp theories with purification. This is satisfied because, recalling Proposition 6,

$$\sum_j u P_{1_j} = \sum_j a_{1_j} = \sum_{i=1}^d a_i^\dagger = u.$$

Because of the properties proved above, von Neumann measurements are repeatable and minimally disturbing measurements in the sense of Refs. [59,63]. Indeed, $a_{1_j} P_{1_j} = a_{1_j}$, and

$$a_{1_j} \sum_k P_{1_k} = a_{1_j} P_{1_j} + \sum_{k \neq j} a_{1_j} P_{1_k} = a_{1_j},$$

because for $k \neq j$ the P_{1_k}'s project on faces orthogonal to F_{1_j}.

The next proposition concerns the interplay between orthogonal projectors and the dagger.

Proposition 8. *For every normalised state ρ, and for every projector P_1 on a face F_1, one has $(P_1\rho)^\dagger = \rho^\dagger P_1$.*

Proof. First of all, note that $0 \leq \|P_1\rho\| \leq 1$, and it vanishes if and only if $\rho \in F_1^\perp$. If $\rho \in F_1^\perp$, then $\rho^\dagger P_1 = 0$, so the statement is trivially true. Now suppose $\|P_1\rho\| > 0$. We will first prove the statement for normalised pure states ψ, then it is sufficient to extend it by linearity to all states. We will make use of the uniqueness of the dagger for normalised pure states. Then the statement is equivalent to proving

$$\left(\frac{P_1 \psi}{\|P_1 \psi\|} \right)^\dagger = \frac{\psi^\dagger P_1}{\|P_1 \psi\|},$$

Noting that the term in brackets is a *normalised* pure state (by Purity Preservation), and that the RHS is a pure effect (again by Purity Preservation), by the uniqueness of the dagger for normalised pure states (cf. Theorem 6), it is enough to prove that

$$\frac{(\psi^\dagger P_1 | P_1 \psi)}{\|P_1 \psi\|^2} = 1;$$

in other words that $(\psi^\dagger P_1 | P_1 \psi) = \|P_1 \psi\|^2$. Recall that $P_1^2 \doteq P_1$ (Proposition 7), so $(\psi^\dagger P_1 | P_1 \psi) = (\psi^\dagger | P_1 | \psi)$. Now, $P_1 \psi = \|P_1 \psi\| \psi'$, where ψ' is a pure state in F_1. We have $(\psi^\dagger P_1 | P_1 \psi) = \|P_1 \psi\| (\psi^\dagger | \psi')$. We only need to prove that $(\psi^\dagger | \psi') = \|P_1 \psi\|$. Recall that $(\psi^\dagger | \psi') = (\psi'^\dagger | \psi)$ by Lemma 2, and that $\psi'^\dagger P_1 = \psi'^\dagger$ as $\psi' \in F_1$, thus

$$(\psi^\dagger | \psi') = (\psi'^\dagger | P_1 | \psi) = \|P_1 \psi\| (\psi'^\dagger | \psi') = \|P_1 \psi\|.$$

By the uniqueness of the dagger for normalised pure states we conclude that $\left(\frac{P_1 \psi}{\|P_1 \psi\|} \right)^\dagger = \frac{\psi^\dagger P_1}{\|P_1 \psi\|}$, namely $(P_1 \psi)^\dagger = \psi^\dagger P_1$. □

A consequence of this proposition is that orthogonal projectors play nicely with the inner product of Lemma 2, namely for every $\xi, \eta \in \text{St}_\mathbb{R}(A)$ one has

$$\langle P_1 \xi, \eta \rangle = \langle \xi, P_1 \eta \rangle. \tag{11}$$

In other words, projections are symmetric with respect to the inner product.

The last property we need is a generalisation of the results of Proposition 7.

Proposition 9. *Fixing a pure maximal set $\{\alpha_i\}_{i=1}^d$, and considering $\mathsf{I}, \mathsf{J} \subseteq \{1, \ldots, d\}$, we have $P_\mathsf{I} P_\mathsf{J} \doteq P_{\mathsf{I} \cap \mathsf{J}}$.*

Proof. First let us prove that

$$P_I P_J \rho = \|P_I P_J \rho\| \, \rho' \tag{12}$$

for every normalised state ρ, where $\rho' \in F_{I \cap J}$. Let us show that $\|P_I P_J \rho\| = (a_{I \cap J}|\rho)$. By Proposition 6, $(u|P_I P_J|\rho) = (a_I|P_J|\rho)$. Now, recalling that $a_I = \sum_{i \in I} \alpha_i^\dagger$,

$$(a_I|P_J|\rho) = \sum_{i \in I \cap J} \left(\alpha_i^\dagger \big| P_J \big| \rho\right) + \sum_{i \in I \setminus J} \left(\alpha_i^\dagger \big| P_J \big| \rho\right) = \sum_{i \in I \cap J} \left(\alpha_i^\dagger \big| \rho\right) = (a_{I \cap J}|\rho),$$

where we have used the fact that $\alpha_i^\dagger P_J = \alpha_i^\dagger$ if $i \in J$, and $\alpha_i^\dagger P_J = 0$ if $i \notin J$. If $\rho \in F_{I \cap J}^\perp$, both the LHS and the RHS of Equation (12) vanish, and the statement is trivially satisfied. Now, let us assume $\rho \notin F_{I \cap J}^\perp$, in this case $(a_{I \cap J}|\rho) > 0$. We wish to prove that $(a_{I \cap J}|P_I P_J|\rho) = (a_{I \cap J}|\rho)$. Recalling the expression of $a_{I \cap J}$, we have

$$\sum_{i \in I \cap J} \left(\alpha_i^\dagger \big| P_I P_J \big| \rho\right) = \sum_{i \in I \cap J} \left(\alpha_i^\dagger \big| P_J \big| \rho\right) = \sum_{i \in I \cap J} \left(\alpha_i^\dagger \big| \rho\right) = (a_{I \cap J}|\rho),$$

again by the properties of P_I and P_J. This means that $P_I P_J$ maps every normalised state to a state of $F_{I \cap J}$, up to normalisation.

Now let us prove that $(P_I P_J)^2 \doteq P_I P_J$. First note that $F_{I \cap J} \subseteq F_I$. Indeed, suppose $\rho \in F_{I \cap J}$, then

$$(a_I|\rho) = \sum_{i \in I \cap J} \left(\alpha_i^\dagger \big| \rho\right) + \sum_{i \in I \setminus J} \left(\alpha_i^\dagger \big| \rho\right) = (a_{I \cap J}|\rho) = 1,$$

where we have used the fact that $(\alpha_i^\dagger|\rho) = 0$ if $i \notin I \cap J$. By a similar argument, $F_{I \cap J} \subseteq F_J$. Now, $P_I P_J \rho = \|P_I P_J \rho\| \rho'$, with $\rho' \in F_{I \cap J}$. Then $(P_I P_J)^2 \rho = \|P_I P_J \rho\| P_I P_J \rho'$. However, $\rho' \in F_J$, so $P_J \rho' = \rho'$, and, similarly, $\rho' \in F_I$, so $P_I \rho' = \rho'$. Consequently,

$$(P_I P_J)^2 \rho = \|P_I P_J \rho\| \rho' = P_I P_J \rho,$$

proving that $(P_I P_J)^2 \doteq P_I P_J$.

Now let us prove that for every $\xi \in \mathrm{St}_{\mathbb{R}}(A)$, we have $(P_I P_J \xi)^\dagger = \xi^\dagger P_I P_J$. Following the lines of proof of Proposition 8, let us show that this is true when ξ is a normalised pure state ψ. This boils down to showing that

$$\left(\psi^\dagger P_I P_J \big| P_I P_J \psi\right) = \|P_I P_J \psi\|^2 .$$

The proof goes on as for Proposition 8, noting that if $\psi' \in F_{I \cap J}$, then $\psi'^\dagger P_I P_J = \psi'^\dagger$ because $\psi'^\dagger P_I = \psi'^\dagger$ as $\psi' \in F_I$, and, similarly, $\psi'^\dagger P_J = \psi'^\dagger$ as $\psi' \in F_J$. Eventually we find that for pure states $(P_I P_J \psi)^\dagger = \psi^\dagger P_I P_J$, and by linearity this means that $(P_I P_J \xi)^\dagger = \xi^\dagger P_I P_J$.

A consequence of this property is that $\langle P_I P_J \xi, \eta \rangle = \langle \xi, P_I P_J \eta \rangle$, for all $\xi, \eta \in \mathrm{St}_{\mathbb{R}}(A)$. These linear maps on $\mathrm{St}_{\mathbb{R}}(A)$ are such that $\mathrm{St}_{\mathbb{R}}(A) = \mathrm{im}\, P_I P_J \oplus \ker P_I P_J$, and $\ker P_I P_J$ is the orthogonal subspace to $\mathrm{im}\, P_I P_J$, hence it is uniquely defined once $\mathrm{im}\, P_I P_J$ is fixed. Note that for any projector P_I we have $\mathrm{im}\, P_I = \mathrm{span}\, F_I$, and we have just proved that $\mathrm{im}\, P_I P_J = \mathrm{span}\, F_{I \cap J} = \mathrm{im}\, P_{I \cap J}$. Having the same image, and consequently the same kernel, $P_I P_J$ and $P_{I \cap J}$ agree on a basis of $\mathrm{St}_{\mathbb{R}}(A)$, therefore they agree also on all states of A, meaning that $P_I P_J \doteq P_{I \cap J}$. □

5.3. Main Result

Proposition 29 of [4] asserts that theories satisfying two postulates, Strong Symmetry and Projectivity, have higher-order interference if and only if their projectors (in our terminology here) preserve purity. A close examination of its proof, and those of all lemmas and propositions used in its proof—notably Lemma 22 and Propositions 18, 25, 26, and 28 of [4]—reveals that only premises weaker than the conjunction of Strong Symmetry and Projectivity are used: self-duality, the "spectral-like decomposition" of effects as in Lemma 1 above, the fact that faces are determined by subsets of maximal distinguishable sets of states as in Section 5.2 above, the existence of projectors onto each face in the sense of Definition 8 above, and the fact that these are symmetric with respect to the self-dualising inner product (i.e., orthogonal projectors), and satisfy Proposition 9 above. We have established these weaker premises for sharp theories with purification, and moreover, we have established in Proposition 6 that their projectors preserve purity, so we have proved:

Theorem 7. *In any sharp theory with purification there can be no nth order interference for $n \geq 3$.*

5.4. Jordan-Algebraic Structure

Our results also imply that systems, and therefore also the "subsystems" associated with their faces, are operationally equivalent to finite-dimensional Jordan-algebraic systems. These are systems A for which $\mathsf{St}_+(A)$ is the cone of squares in a finite-dimensional Euclidean Jordan algebra (EJA) and $\mathsf{Eff}_+(A)$ is identified with the same cone, with evaluation of effects on states given by the inner product and the Jordan unit as the deterministic effect. (See [37] for more on Jordan algebraic operational systems, and [61] for a mathematical treatment.)

Theorem 8. *In a sharp theory with purification, every system A has both $\mathsf{St}_+(A)$ and $\mathsf{Eff}_+(A)$ isomorphic to the cone of squares in a Euclidean Jordan algebra (EJA) via isomorphisms S and T such that $(a|\rho) = \langle Ta, S\rho \rangle$, where $\langle \bullet, \bullet \rangle$ is the canonical inner product on the EJA, and T takes the deterministic effect to the Jordan unit.*

Proof. The proof uses results of Alfsen and Shultz [64], for which we refer to [61]. Theorem 9.33 in [61] implies that finite-dimensional systems with symmetry of transition probabilities (STP), a type of projection operator they call "compression" associated with every face, and whose compressions preserve purity, have state spaces affinely isomorphic to the state spaces of Euclidean Jordan algebras. Sharp theories with purification satisfy STP, as noted following Lemma 2 above. Our projectors are easily shown to be examples of compressions by the same argument as in Theorem 17 of [4]; this argument uses only properties satisfied by our projectors (the same ones needed in the proof of Theorem 7, except for Purity Preservation) and does not need Strong Symmetry. As shown above, our projectors also preserve purity. □

Since faces of Jordan-algebraic systems are also Jordan-algebraic (to see this, combine a result of Iochum [65] (Theorem 5.32 in [61]), whose finite dimensional case is that all faces of EJAs are the positive part of the images of compressions, with the facts (cf. pp. 22–26 of [61]) that every face of the cone of squares is the image of such a compression P ([61], Lemma 1.39), and also a Jordan subalgebra whose unit is the image of the order unit under P ([61], Proposition 1.43).), so are the faces of state spaces in sharp theories with purification. However, it is not the case that in sharp theories with purification, each face of a system is necessarily isomorphic to a stand-alone system of the theory (an *object* of the category, in the categorical formulation), but, it is always possible to extend the theory such that they are. Every category has a *Cauchy completion*: this is a minimal extension of the category such that every idempotent morphism $\pi : A \to A$ can be written as a retraction-section pair, i.e., as the composition $\pi = \sigma \circ \rho$, with $\rho : A \to B$ and $\sigma : B \to A$, such that the reverse composition $\rho \circ \sigma$ is the identity morphism on B. When the idempotents are projectors P like the ones we consider here, B will be a system isomorphic to the face $\mathrm{im}_+(P)$. Of course, since there may be idempotents beyond the projectors onto faces (for example, decoherence of a set of orthogonal subspaces, or damping to a fixed state, in quantum theory), Cauchy completion of an operational theory T may add many objects in addition to ones isomorphic to faces of systems of T; indeed, for many operational theories (e.g., ones possessing idempotent decoherence maps) this will add some classical systems. This is indeed the case for quantum theory where the Cauchy completion leads to the category of finite-dimensional C*-algebras and completely positive maps [66]. The Cauchy completion can be thought of as adding in all operationally accessible systems that can be simulated on the physical system via a consistent restriction on the allowed states, effects and transformations. The Cauchy completion of a sharp theory with purification will likely satisfy the Ideal Compression postulate by virtue of containing the faces that are images of orthogonal projectors; but there are also non-Cauchy complete theories that satisfy it, e.g., the category CPM of finite-dimensional quantum systems and CP maps, in which all systems, and also all images of orthogonal projectors as defined above, are fully coherent quantum systems, but there are no classical systems.

In [37], some categories, including dagger-compact-closed categories, of Jordan algebraic systems were constructed; these categories are equivalent to operational theories as we use the term here. Although sharp theories with purification also have Jordan algebraic state and effect spaces, it is interesting to note that some of the explicit examples in [30,49] involve composites different from those that would be obtained in the categories considered in [37] for systems with the same state spaces. On the other hand, the category combining real and quaternionic systems in [37] does *not* satisfy Purity Preservation by parallel composition and hence falls outside the class of sharp theories with purification, although its filters do preserve purity. Of course, the failure of Purity Preservation by parallel composition seems likely to allow phenomena like the nonextensiveness of entropy when products of states are taken, which could warrant focusing on sharp theories with purification in thermodynamically motivated work such as [30].

That Jordan-algebraic systems lack higher-order interference was shown by Barnum and Ududec ([12]; announced in [67]) and by Niestegge [68]; combining this with Theorem 8 gives another way to see that our results on sharp theories with purification imply the absence of higher-order interference. Moreover, as not all EJAs satisfy our postulates, it is clear that our postulates are sufficient but not necessary conditions for ruling out higher-order interfence.

6. Discussion and Conclusions

We proved that in sharp theories with purification multi-slit experiments must have a pure projector structure and, moreover, such theories exhibit at most second-order interference. Hence these theories are, at least conceptually, very "close" to quantum theory. Moreover, recent work has shown that sharp theories with purification are close to quantum theory in terms of other physical and information processing features. Indeed, such theories possess quantum-like contextuality behaviour [59,63], quantum-like computation [7,8], and quantum-like thermodynamic Properties [30,49,54]. Recall from Section 4 that quantum theory is not the only example of a generalised probabilistic theory satisfying these principles. Hence Causality, Purity Preservation, Pure Sharpness, and Purification do not recover the entire quantum formalism.

However, if one were to introduce the Ideal Compression and Local Discriminability principles of the reconstruction of quantum theory due to Chiribella, D'Ariano, and Perinotti [20], one would indeed regain the entire quantum formalism. Indeed, both additional principles are necessary: Local Discriminability to preclude real quantum theory and Ideal Compression to preclude the contrived—yet admissible—example of the theory in which all systems are composites of qubits. Sharp theories with purification thus serve as a fertile test-bed for physics that is conceptually quite close to that predicted by the quantum world, but which may diverge from it in certain small, yet interesting, ways.

Finding Higher Order Interference

To date there has been no experiment that has found higher-order interference, at least, none that cannot be explained by taking into account the fact that the "sets of histories are not mutually exclusive" [2,35]. However, this might be due to the specific experimental set-up employed, rather than a fundamental preclusion of higher-order interference in nature. We show here that many of the properties needed to rule out observing higher-order interference are in fact quite natural assumptions which appear to be suggested by the experimental set-up employed. This suggests that the experimental set-up itself may implicitly rule out observing higher-order interference from the outset.

The main result of the current work is that sharp theories with purification can never exhibit higher-order interference in any experiment. However, in a wider class of theories, we still will not observe higher-order interference in a particular experiment if the following three conditions are met;

hence, to have any chance of observing higher-order interference, experiments must be designed in order to try to violate these conditions.

1. The transformations corresponding to blocking slits satisfy: $T_I T_J = T_{I \cap J}$. By this we mean that they share several properties with the projectors P_I of Section 5: if we define the effects $a_I = uT_I$ and the faces F_I and $F_I^\perp$ as in Section 5.2, i.e., as the 1-set and 0-set of a_I, then the T_I are assumed to be *orthogonal projectors* in the sense of Definition 8, and to be both idempotent and "orthogonal" ($T_I T_J = 0$) if I and J are disjoint (as in Proposition 7).
2. The T_I's map pure states to pure states
3. The T_I's are self-adjoint.

The first of these is generally expected as only those slits belonging to both I and J will not be blocked by either T_I or T_J, and so should hold in this experimental set-up for any theory that can describe it.

The second assumption, which is also natural given the multi-slit set-up, is that, in an idealised scenario, the slits should not introduce fundamental noise. That is, if an input state ρ is pure, i.e., has no classical noise associated with it, then $T_I \rho$ should also be pure. Hence it appears natural to assume that T_I maps pure states to pure states. Violating this principle by just adding noise to the experiment does not seem likely to demonstrate higher-order interference. A more plausible way to violate this however would be if the particle passing through the slits were to become entangled with some degree of freedom associated with them, if we do not have access to this degree of freedom then this would send a pure input to a mixed state.

The final assumption is far less general than the others, as it places a constraint on the theory. That is, to even discuss whether a transformation is self-adjoint (cf. also Appendix B), one requires that the theory itself be self-dual. To fully understand what this assumption entails, one needs an operational or physical interpretation of the self-dualising inner product (see [69] for an example of such an interpretation). However, intuitively this notion reflects the inherent symmetry of the experimental set-up. Here one could consider propagation from the source to the effect or from the effect to the source as being "dual" to one another and, moreover, that the physical blocking of slits has an equivalent effect in either situation. That is, the assumption of self-adjointness corresponds to the statement that the projector has an equivalent action on the effects associated with a particular slit as it does on the states which can pass through them.

If an experiment satisfies these assumptions then for any self-dual theory it was shown in [4] (Proposition 29) that we will not see higher-order interference in this experiment. Hence any set of physical principles which ensure these assumptions hold will rule out higher-order interference. Because the mathematical assumptions involved in formalising a multi-slit experiment are so natural when interpreted operationally, perhaps one should search for higher-order interference in set-ups that don't seem to preclude it from the outset. This could involve "asymmetric" multi-slit set-ups that are not obviously time-symmetric in an arbitrary generalised probabilistic theory. One could also consider experiments that search for higher-order phases [8], a reformulation of higher-order interference that makes no reference to projectors and hence does not preclude certain generalised theories from the outset. The assumption that nature is self-dual could also be rejected; this poses the question as to whether it is possible to find a direct experimental test of this principle.

Acknowledgments: The authors thank J. Barrett for useful discussions and J. J. Barry for encouragement while writing the current paper. This work was supported by EPSRC grants through the Controlled Quantum Dynamics Centre for Doctoral Training, the UCL Doctoral Prize Fellowship (project number 534936), and an Oxford doctoral training scholarship, and also by Oxford-Google DeepMind Graduate Scholarship. We also acknowledge financial support from the European Research Council (ERC Grant Agreement No. 337603), the Danish Council for Independent Research (Sapere Aude) and VILLUM FONDEN via the QMATH Centre of Excellence (Grant No. 10059). This work began while the authors were attending the "Formulating and Finding Higher-order Interference" workshop at the Perimeter Institute. Research at Perimeter Institute is supported by the Government of Canada through the Department of Innovation, Science and Economic Development Canada and by the Province of Ontario through the Ministry of Research, Innovation and Science.

Author Contributions: All authors contributed equally to the present work.

Conflicts of Interest: The authors declare no conflict of interest.

Appendix A. Norms and Fidelity

Appendix A.1. Operational Norm and Dagger Norm

In Ref. [19] the operational norm for every vector $\xi \in \text{St}_\mathbb{R}(A)$ was introduced:

$$\|\xi\| := \sup_{a \in \text{Eff}(A)} (a|\xi) - \inf_{a \in \text{Eff}(A)} (a|\xi)$$

As pointed out in [19], in quantum theory the operational norm coincides with the trace norm. The analogy is apparent also in sharp theories with purification.

Proposition A1. *Let $\xi \in \text{St}_\mathbb{R}(A)$ be diagonalised as $\xi = \sum_{i=1}^{d} x_i \alpha_i$. Then $\|\xi\| = \sum_{i=1}^{d} |x_i|$.*

Proof. Let us separate the terms with non-negative eigenvalues from the terms with negative eigenvalues, so that we can write $\xi = \xi_+ - \xi_-$, where $\xi_+ := \sum_{x_i \geq 0} x_i \alpha_i$, and $\xi_- = \sum_{x_i < 0} (-x_i) \alpha_i$. Clearly, $\xi_+, \xi_- \in \text{St}_+(A)$. In order to achieve the supremum of $(a|\xi)$ we must have $(a|\xi_-) = 0$. Moreover,

$$(a|\xi_+) = \sum_{x_i \geq 0} x_i (a|\alpha_i) \leq \sum_{x_i \geq 0} x_i$$

since $(a|\alpha_i) \leq 1$ for every i. The supremum of $(a|\xi_+)$ is achieved by $a = \sum_{x_i \geq 0} \alpha_i^\dagger$. Hence $\sup_a (a|\xi) = \sum_{x_i \geq 0} x_i$. By a similar argument, one shows that $\inf_a (a|\xi) = \sum_{x_i < 0} x_i$. Therefore

$$\|\xi\| = \sum_{x_i \geq 0} x_i + \sum_{x_i < 0} (-x_i) = \sum_{i=1}^{d} |x_i|.$$

□

For $p \geq 1$, the p-norm of a vector $\mathbf{x} \in \mathbb{R}^d$ is defined as $\|\mathbf{x}\|_p := \left(\sum_{i=1}^{d} |x_i|^p \right)^{\frac{1}{p}}$, thus we have $\|\xi\| = \|\mathbf{x}\|_1$, where $\mathbf{x}$ is the spectrum of ξ.

In sharp theories with purification we have an additional norm, the dagger norm, defined in Section 5.1. The dagger norm of a vector $\xi \in \text{St}_\mathbb{R}(A)$ is $\|\xi\|_\dagger = \sqrt{\sum_{i=1}^{d} x_i^2}$, where the x_i's are the eigenvalues of ξ. It is obvious from the very definition that $\|\xi\|_\dagger = \|\mathbf{x}\|_2$. Thanks to these results following from diagonalisation, we can derive the standard bounds between the two norms, by making use of the well-known bounds $\|\mathbf{x}\|_2 \leq \|\mathbf{x}\|_1 \leq \sqrt{d} \|\mathbf{x}\|_2$, which imply

$$\|\xi\|_\dagger \leq \|\xi\| \leq \sqrt{d} \|\xi\|_\dagger. \tag{A1}$$

Note that, unlike Ref. [70], here the bounds are derived without assuming Bit Symmetry [4,71].

If we take ξ to be a normalised state ρ, its eigenvalues form a probability distribution, and we have $\|\rho\|_\dagger \leq 1$, with equality if and only if ρ is pure. Note that $\|\rho\|_\dagger$ is a Schur-convex function [72] of the eigenvalues of ρ, so it is a purity monotone [30]. As such, it attains its minimum on the invariant state, which is $\|\chi\|_\dagger = \frac{1}{\sqrt{d}}$, so for every normalised state one has

$$\frac{1}{\sqrt{d}} \leq \|\rho\|_\dagger \leq 1,$$

consistently with the bounds (A1). The square of the dagger norm, still a Schur-convex function, was called *purity* in Refs. [70,73]. Consequently $1 - \|\rho\|_\dagger^2$ is a measure of mixedness, sometimes called the *impurity* $I(\rho)$ of ρ. The impurity can be extended to subnormalised states by defining it as $I(\rho) := (\text{Tr}\,\rho)^2 - \|\rho\|_\dagger^2$ [4].

The two norms behave differently under channels applied to states. In Ref. [19] it was shown that in causal theories the operational norm of a state ρ is preserved by channels: $\|\mathcal{C}\rho\| = \|\rho\|$ for every channel $\mathcal{C}$, because channels are such that $u\mathcal{C} = u$.

Instead the dagger norm shows a different behaviour. To describe it, it is useful to divide channels into two classes: unital and non-unital channels [49].

Definition A1. *A channel $\mathcal{D} \in \text{Transf}(A, B)$ is unital if $\mathcal{D}\chi_A = \chi_B$.*

Unital channels do not increase the dagger norm of states.

Proposition A2. *If $\mathcal{D}$ is a unital channel, then $\|\mathcal{D}\rho\|_\dagger \leq \|\rho\|_\dagger$, for every normalised state ρ.*

Proof. Unital channels can be chosen as free operations for the resource theory of purity [49]. In Ref. [49] it was shown that the spectrum of $\mathcal{D}\rho$ is majorised by the spectrum of ρ (see Ref. [72] for a definition of majorisation and Schur-convex functions). Since the dagger norm is a Schur-convex function, we have $\|\mathcal{D}\rho\|_\dagger \leq \|\rho\|_\dagger$. □

Clearly if $\mathcal{D}$ is reversible, the dagger norm is preserved, by Proposition 4.

For non-unital channels there is at least one state—the invariant state χ—for which the dagger norm increases. Indeed, if $\mathcal{C}$ is non-unital, χ is majorised by $\mathcal{C}\chi$, whence $\|\chi\|_\dagger \leq \|\mathcal{C}\chi\|_\dagger$. Is it true, then, that non-unital channels increase the dagger norm of all states? The answer is clearly negative. Consider the non-unital channel mapping all states to a fixed *mixed* state $\rho_0 \neq \chi$. For some states, e.g., the invariant state, the dagger norm will increase, for others, e.g., pure states, the dagger norm will decrease because it is a purity monotone. In short, for non-unital channels there is no uniform behaviour of the dagger norm.

Appendix A.2. Dagger Fidelity

The inner product defined in Section 5.1 allows us to define a fidelity-like quantity, called the *dagger fidelity*.

Definition A2. *Given two normalised states ρ and σ, the dagger fidelity is defined as*

$$F_\dagger(\rho, \sigma) = \frac{\langle \rho, \sigma \rangle}{\|\rho\|_\dagger \|\sigma\|_\dagger}.$$

The dagger fidelity measures the overlap between two states. It shares some properties with the fidelity in quantum theory (cf. for instance Ref. [74]), despite *not* coinciding with it. The first, obvious one, is that $F_\dagger(\rho, \sigma) = F_\dagger(\sigma, \rho)$.

To prove the other properties we need the following lemma, generalising one of the results of Ref. [30].

Lemma A1. *Let $\{\rho_i\}_{i=1}^n$ be perfectly distinguishable states. Then $(\rho_i^\dagger | \rho_j) = \|\rho_i\|_\dagger^2 \delta_{ij}$.*

Proof. Clearly what we need to prove is that $(\rho_i^\dagger | \rho_j) = 0$ if $i \neq j$. Let $\{a_i\}_{i=1}^n$ be the perfectly distinguishing test, and let ρ_i be diagonalised as $\rho_i = \sum_{k=1}^{r_i} p_{k,i} \alpha_{k,i}$, where $p_{k,i} > 0$ for all $k = 1, \ldots, r$. We have $(a_i | \rho_i) = 1$, hence by Proposition 2 there exists a non-disturbing pure transformation $\mathcal{T}_i$ such that $\mathcal{T}_i =_{\rho_i} \mathcal{I}$. Specifically, we have that $\mathcal{T}_i \alpha_{k,i} = \alpha_{k,i}$. Moreover if $i \neq j$, we have $(u | \mathcal{T}_i | \rho_j) \leq (a_i | \rho_j) = 0$, whence $(u | \mathcal{T}_i | \rho_j) = 0$. This means that $\mathcal{T}_i \rho_j = 0$ for all $j \neq i$.

Now, consider

$$\left(\alpha_{k,i}^\dagger \middle| \mathcal{T}_i \middle| \alpha_{k,i}\right) = \left(\alpha_{k,i}^\dagger \middle| \alpha_{k,i}\right) = 1,$$

where we have used the fact that $\mathcal{T}_i \alpha_{k,i} = \alpha_{k,i}$. Since $\alpha_{k,i}^\dagger \mathcal{T}_i$ is a pure effect, it must be $\alpha_{k,i}^\dagger \mathcal{T}_i = \alpha_{k,i}^\dagger$ by Theorem 6. By linearity we have $\rho_i^\dagger \mathcal{T}_i = \rho_i^\dagger$. Now, using this fact, for all $j \neq i$

$$\left(\rho_i^\dagger \middle| \rho_j\right) = \left(\rho_i^\dagger \middle| \mathcal{T}_i \middle| \rho_j\right) = 0,$$

because $\mathcal{T}_i \rho_j = 0$. □

Recalling that $\left(\rho^\dagger \middle| \sigma\right) = \langle \rho, \sigma \rangle$, this lemma means that perfectly distinguishable states form an orthogonal set. Specifically, if the states are pure, the set is orthonormal.

The following proposition extends and generalises the properties of the self-dualising inner product of Ref. [71].

Proposition A3. *The dagger fidelity has the following properties, for all normalised states ρ and σ.*

1. $0 \leq F_\dagger(\rho, \sigma) \leq 1$;
2. $F_\dagger(\rho, \sigma) = 0$ if and only if ρ and σ are perfectly distinguishable;
3. $F_\dagger(\rho, \sigma) = 1$ if and only if $\rho = \sigma$;
4. $F_\dagger(\mathcal{U}\rho, \mathcal{U}\sigma) = F_\dagger(\rho, \sigma)$, for every reversible channel $\mathcal{U}$.

Proof. Let us prove the various properties.

1. Recall that $\langle \rho, \sigma \rangle = \left(\rho^\dagger \middle| \sigma\right) \geq 0$, whence $F_\dagger(\rho, \sigma) \geq 0$. Moreover, by Schwarz inequality, $\langle \rho, \sigma \rangle \leq \|\rho\|_\dagger \|\sigma\|_\dagger$, so $F_\dagger(\rho, \sigma) \leq 1$.
2. Suppose ρ and σ are perfectly distinguishable, then by Lemma A1 $\langle \rho, \sigma \rangle = 0$, implying $F_\dagger(\rho, \sigma) = 0$. Now suppose $F_\dagger(\rho, \sigma) = 0$; then $\langle \rho, \sigma \rangle = 0$. Let $\rho = \sum_{i=1}^r p_i \alpha_i$ be a diagonalisation of ρ, with $p_i > 0$, for all $i = 1, \ldots, r$, and $r \leq d$. We have $\sum_{i=1}^r p_i \left(\alpha_i^\dagger \middle| \sigma\right) = 0$, which means that $\left(\alpha_i^\dagger \middle| \sigma\right) = 0$ for $i = 1, \ldots, r$. This means that we can build an observation-test that distinguishes ρ and σ perfectly by taking $\{a, u - a\}$, where $a = \sum_{i=1}^r \alpha_i^\dagger$.
3. Clearly, if $\rho = \sigma$, $\langle \rho, \sigma \rangle = \|\rho\|_\dagger^2$, whence $F_\dagger(\rho, \sigma) = 1$. Conversely, suppose $F_\dagger(\rho, \sigma) = 1$. This means that $\langle \rho, \sigma \rangle = \|\rho\|_\dagger \|\sigma\|_\dagger$. By Schwarz inequality, this is true if and only if $\rho = \lambda \sigma$, for some $\lambda \in \mathbb{R}$. Since both states are normalised, $\lambda = 1$, yielding $\rho = \sigma$.
4. This property follows by Proposition 4, because the inner product and the dagger norm are invariant under reversible channels.

□

Note that Property 3 captures the sharpness of the dagger for all normalised states [69].
A property involving tensor product of states is the following.

Proposition A4. *For all normalised states $\rho_1, \rho_2, \sigma_1, \sigma_2$ one has*

$$F_\dagger(\rho_1 \otimes \rho_2, \sigma_1 \otimes \sigma_2) = F_\dagger(\rho_1, \sigma_1) F_\dagger(\rho_2, \sigma_2)$$

The proof needs the following easy lemma.

Lemma A2. *Let $\rho, \sigma \in \mathsf{St}_1(A)$, then $(\rho \otimes \sigma)^\dagger = \rho^\dagger \otimes \sigma^\dagger$.*

Proof. Let us prove the result for ρ and σ pure, the general result will follow by linearity. By Purity Preservation, $\rho \otimes \sigma$ and $\rho^\dagger \otimes \sigma^\dagger$ are pure, and one has $\left(\rho^\dagger \otimes \sigma^\dagger \middle| \rho \otimes \sigma\right) = 1$. By Theorem 6, $(\rho \otimes \sigma)^\dagger = \rho^\dagger \otimes \sigma^\dagger$. □

Now comes the actual proof.

Proof of Proposition A4. We have

$$F_+ (\rho_1 \otimes \rho_2, \sigma_1 \otimes \sigma_2) = \frac{\langle \rho_1 \otimes \rho_2, \sigma_1 \otimes \sigma_2 \rangle}{\|\rho_1 \otimes \rho_2\|_+ \|\sigma_1 \otimes \sigma_2\|_+}.$$

Now, by Lemma A2,

$$\langle \rho_1 \otimes \rho_2, \sigma_1 \otimes \sigma_2 \rangle = \left(\rho_1^\dagger \otimes \rho_2^\dagger \middle| \sigma_1 \otimes \sigma_2 \right) = \left(\rho_1^\dagger \middle| \sigma_1 \right) \left(\rho_2^\dagger \middle| \sigma_2 \right) = \langle \rho_1, \sigma_1 \rangle \langle \rho_2, \sigma_2 \rangle.$$

Furthermore,

$$\|\rho_1 \otimes \rho_2\|_+ = \sqrt{\langle \rho_1 \otimes \rho_2, \rho_1 \otimes \rho_2 \rangle} = \sqrt{\langle \rho_1, \rho_1 \rangle \langle \rho_2, \rho_2 \rangle} = \|\rho_1\|_+ \|\rho_2\|_+.$$

Putting everything together,

$$F_+ (\rho_1 \otimes \rho_2, \sigma_1 \otimes \sigma_2) = \frac{\langle \rho_1, \sigma_1 \rangle}{\|\rho_1\|_+ \|\sigma_1\|_+} \cdot \frac{\langle \rho_2, \sigma_2 \rangle}{\|\rho_2\|_+ \|\sigma_2\|_+} = F_+ (\rho_1, \sigma_1) F_+ (\rho_2, \sigma_2).$$

□

Appendix B. Dagger of All Transformations

Inspired by the results of Lemma 2, in sharp theories with purification, we can extend the dagger to all transformations, a feature often present in process theories [44,45,69,75].

Definition A3. *Given the transformation* $\mathcal{A} \in$ Transf (A, B), *its dagger (or adjoint) is a linear transformation* $\mathcal{A}^\dagger$ *from B to A defined as*

$$\left(\rho \;\middle|\; \mathcal{A}^\dagger \right)_{B\;A\;S} = \left(\left(\;_{S}^{A}\; \mathcal{A} \;_{}^{B}\; \rho^\dagger \right) \right)^\dagger, \tag{A2}$$

for every system S, and every state $\rho \in \mathrm{St}_1 (B \otimes S)$.

This definition specifies the dagger of a transformation completely, thanks to Equation (2). Note that Lemma 2 allows us to formulate Equation (10) in term of effects and their dagger:

$$\left(a \middle| b^\dagger \right) = \left(b \middle| a^\dagger \right)$$

for all effects a, and b. In this way, Definition A3 can be recast in equivalent terms by taking b as the term in round brackets in the RHS of Equation (A2). This yields

$$\left(\rho \;\middle|\; \mathcal{A}^\dagger \;\middle|\; E \right) = \left(E^\dagger \;\middle|\; \mathcal{A} \;\middle|\; \rho^\dagger \right), \tag{A3}$$

for every system S, every state $\rho \in \mathrm{St}_1 (B \otimes S)$, and every effect $E \in \mathrm{Eff} (A \otimes S)$.

The dagger of a transformation may not be a physical transformation, i.e., it may send physical states to non-physical ones. Indeed, the action of $\mathcal{A}^\dagger \otimes \mathcal{I}$ on a generic state (the LHS of Equation (A2)) is defined as the dagger of an effect. However, not all daggers of effects are physical states. For instance, take the deterministic effect $u = \sum_{i=1}^d \alpha_i^\dagger$, where $\{\alpha_i\}_{i=1}^d$ is a pure maximal set. Its dagger is $u^\dagger = \sum_{i=1}^d \alpha_i = d\chi$, which is a supernormalised (and hence non-physical) state.

For channels, we can give a necessary condition for the existence of a physical dagger of the channel.

Proposition A5. Let $\mathcal{C} \in \text{Transf}(A, B)$ be a channel. If $\mathcal{C}^\dagger$ is a physical transformation, then $\mathcal{C}$ is unital, and $\mathcal{C}^\dagger$ itself is a unital channel.

Proof. If $\mathcal{C}^\dagger$ is a physical transformation, then, for every normalised state $\rho \in \text{St}_1(B)$, we have $\|\mathcal{C}^\dagger \rho\| \leq 1$, or in other words, $(u|\mathcal{C}^\dagger|\rho) \leq 1$. By Equation (A3), $(u|\mathcal{C}^\dagger|\rho) = (\rho^\dagger|\mathcal{C}|u^\dagger)$, so the condition $\|\mathcal{C}^\dagger \rho\| \leq 1$ is equivalent to

$$\left(\rho^\dagger|\mathcal{C}|\chi\right) \leq \frac{1}{d}, \tag{A4}$$

with equality if and only if $\mathcal{C}^\dagger$ is a channel. Suppose by contradiction that $\mathcal{C}$ is not unital, then $\mathcal{C}\chi = \rho_0 \neq \chi$. Diagonalise ρ_0 as $\rho_0 = \sum_{i=1}^d p_i \alpha_i$, where $p_1 \geq p_2 \geq \ldots \geq p_d \geq 0$, and $p_1 > \frac{1}{d}$. Then taking ρ to be α_1 in $(\rho^\dagger|\mathcal{C}|\chi)$ yields p_1, but $p_1 > \frac{1}{d}$, contradicting Equation (A4).

Being $\mathcal{C}$ unital, we have that

$$\left(\rho^\dagger|\mathcal{C}|\chi\right) = \left(\rho^\dagger|\chi\right) = \frac{1}{d}\text{Tr}\,\rho = \frac{1}{d},$$

showing that $\mathcal{C}^\dagger$ is itself a channel. Let us prove it is unital. The action of $\mathcal{C}^\dagger$ on χ is defined in Equation (A2), so

$$\mathcal{C}^\dagger \chi = \left(\chi^\dagger \mathcal{C}\right)^\dagger = \frac{1}{d}(u\mathcal{C})^\dagger = \frac{1}{d}u^\dagger = \chi,$$

where we have used the fact that $\mathcal{C}$ is a channel, so $u\mathcal{C} = u$. This proves that $\mathcal{C}^\dagger$ is unital. □

We can prove that the dagger of a transformation has some nice properties.

Proposition A6. For every transformation $\mathcal{A} \in \text{Transf}(A, B)$, one has $\left(\mathcal{A}^\dagger\right)^\dagger = \mathcal{A}$.

Proof. By Equation (A3) given any system S, any state $\rho \in \text{St}_1(A \otimes S)$, and any effect $E \in \text{Eff}(B \otimes S)$, we have

$$\begin{array}{c}\rho \!-\!\!\boxed{\begin{array}{c}A\\ (\mathcal{A}^\dagger)^\dagger \\ S\end{array}}\!\!-\!\!\boxed{\begin{array}{c}B\\ E\end{array}} = E^\dagger \!-\!\!\boxed{\begin{array}{c}B\\ \mathcal{A}^\dagger \\ S\end{array}}\!\!-\!\!\boxed{\begin{array}{c}A\\ \rho^\dagger\end{array}}\end{array}. \tag{A5}$$

A linear extension of Equation (A3) to cover the case when $E^\dagger$ is not a physical state, applied to the RHS of Equation (A5) yields

$$E^\dagger \!-\!\!\boxed{\begin{array}{c}B\\ \mathcal{A}^\dagger \\ S\end{array}}\!\!-\!\! \rho^\dagger = \rho \!-\!\!\boxed{\begin{array}{c}A\\ \mathcal{A} \\ S\end{array}}\!\!-\!\! E.$$

Comparing this with Equation (A5), we get the thesis. □

We can give a characterisation of the dagger of reversible channels, which are unital channels.

Proposition A7. If $\mathcal{U} \in \text{Transf}(A, B)$ is a reversible channel, $\mathcal{U}^\dagger = \mathcal{U}^{-1}$.

Proof. We have

$$\rho \!-\!\!\boxed{\begin{array}{c}B\\ \mathcal{U}^\dagger \\ S\end{array}}\!\!-\!\! E = E^\dagger \!-\!\!\boxed{\begin{array}{c}A\\ \mathcal{U} \\ S\end{array}}\!\!-\!\! \rho^\dagger,$$

for any S, ρ, E. Recalling Lemma 2, the RHS is $\langle \rho, (\mathcal{U} \otimes \mathcal{I}) E^\dagger \rangle$. By Proposition 4 $\langle \rho, (\mathcal{U} \otimes \mathcal{I}) E^\dagger \rangle = \langle (\mathcal{U}^{-1} \otimes \mathcal{I}) \rho, E^\dagger \rangle$, and by symmetry of the inner product we have that

$$\left\langle \left(\mathcal{U}^{-1}\otimes\mathcal{I}\right)\rho,E^{\dagger}\right\rangle = \left\langle E^{\dagger},\left(\mathcal{U}^{-1}\otimes\mathcal{I}\right)\rho\right\rangle = \begin{array}{c}\rho\end{array}\!\!\begin{array}{|c|}\hline B\ \mathcal{U}^{-1}\ A\\ \hline S\end{array}\!\!\begin{array}{|c|}\hline \\ E\\ \hline\end{array},$$

whence the thesis follows. □

In particular we have that the dagger of the SWAP channel between two systems is the SWAP with the input and output systems reversed.

The orthogonal projectors of Section 5.2, on the other hand, are self-adjoint on single system.

Proposition A8. *Given the orthogonal projector P_1 on a face F_1, we have $P_1^{\dagger} \doteq P_1$.*

Proof. For every ρ and E, we have $(E|P_1^{\dagger}|\rho) = (\rho^{\dagger}|P_1|E^{\dagger})$. The RHS is $\langle \rho, P_1 E^{\dagger}\rangle$. By the properties of projectors,

$$\langle \rho, P_1 E^{\dagger}\rangle = \langle P_1\rho, E^{\dagger}\rangle = \langle E^{\dagger}, P_1\rho\rangle = (E|P_1|\rho).$$

This shows that $P_1^{\dagger} \doteq P_1$. □

Finally we prove some properties of the dagger with respect to compositions. We need an easy lemma first.

Lemma A3. *For every $\mathcal{A} \in$ Transf (A, B), every system S, and every vector $\zeta \in$ St$_{\mathbb{R}}$ $(A \otimes S)$ we have*

$$\left(\begin{array}{c}\zeta\end{array}\!\!\begin{array}{|c|}\hline A\ \mathcal{A}\ B\\ \hline S\end{array}\right)^{\dagger} = \begin{array}{|c|}\hline B\ \mathcal{A}^{\dagger}\ A\\ \hline S\end{array}\!\!\begin{array}{c}\zeta^{\dagger}\end{array}.$$

Proof. Recall that $\mathcal{A} = \left(\mathcal{A}^{\dagger}\right)^{\dagger}$; by Definition A3 we have $\left(\mathcal{A}^{\dagger}\right)^{\dagger}\zeta = \left(\zeta^{\dagger}\mathcal{A}^{\dagger}\right)^{\dagger}$

$$\begin{array}{c}\zeta\end{array}\!\!\begin{array}{|c|}\hline A\ \mathcal{A}\ B\\ \hline S\end{array} = \begin{array}{c}\zeta\end{array}\!\!\begin{array}{|c|}\hline A\ \left(\mathcal{A}^{\dagger}\right)^{\dagger}\ B\\ \hline S\end{array} = \left(\begin{array}{|c|}\hline B\ \mathcal{A}^{\dagger}\ A\\ \hline S\end{array}\!\!\begin{array}{c}\zeta^{\dagger}\end{array}\right)^{\dagger}.$$

Taking the dagger of this equation yields the desired result. □

Now we can state the main results. The first concerns sequential composition.

Proposition A9. *For all transformations $\mathcal{A} \in$ Transf (A, B), $\mathcal{B} \in$ Transf (B, C), one has $(\mathcal{BA})^{\dagger} = \mathcal{A}^{\dagger}\mathcal{B}^{\dagger}$.*

Proof. Take any system S, any state $\rho \in$ St$_1$ $(C \otimes S)$, and any effect $E \in$ Eff $(A \otimes S)$. By Equation (A3) we have

$$\begin{array}{c}\rho\end{array}\!\!\begin{array}{|c|}\hline C\ (\mathcal{BA})^{\dagger}\ A\\ \hline S\end{array}\!\!\begin{array}{c}E\end{array} = \begin{array}{c}E^{\dagger}\end{array}\!\!\begin{array}{|c|}\hline A\ \mathcal{BA}\ C\\ \hline S\end{array}\!\!\begin{array}{c}\rho^{\dagger}\end{array} = \begin{array}{c}E^{\dagger}\end{array}\!\!\begin{array}{|c|}\hline A\ \mathcal{A}\ B\ \mathcal{B}\ C\\ \hline S\end{array}\!\!\begin{array}{c}\rho^{\dagger}\end{array}.$$

Define ζ as $\zeta := (\mathcal{A}\otimes\mathcal{I})E^{\dagger}$, so

$$\begin{array}{c}\rho\end{array}\!\!\begin{array}{|c|}\hline C\ (\mathcal{BA})^{\dagger}\ A\\ \hline S\end{array}\!\!\begin{array}{c}E\end{array} = \begin{array}{c}\zeta\end{array}\!\!\begin{array}{|c|}\hline B\ \mathcal{B}\ C\\ \hline S\end{array}\!\!\begin{array}{c}\rho^{\dagger}\end{array} = \begin{array}{c}\rho\end{array}\!\!\begin{array}{|c|}\hline C\ \mathcal{B}^{\dagger}\ B\\ \hline S\end{array}\!\!\begin{array}{c}\zeta^{\dagger}\end{array}.$$

By Lemma A3 $\zeta^\dagger = [(\mathcal{A} \otimes \mathcal{I}) E^\dagger]^\dagger = E(\mathcal{A}^\dagger \otimes \mathcal{I})$, then

$$\begin{array}{c}\rho \underline{}_S^C \boxed{(\mathcal{BA})^\dagger}^A E = \rho \underline{}_S^C \boxed{\mathcal{B}^\dagger}^B \boxed{\mathcal{A}^\dagger}^A E,\end{array}$$

therefore $(\mathcal{BA})^\dagger = \mathcal{A}^\dagger \mathcal{B}^\dagger$. □

Finally the dagger respects parallel composition. Again we need a lemma.

Lemma A4. *For every $\mathcal{A} \in \mathrm{Transf}\,(\mathrm{A},\mathrm{B})$, every systems S and S', we have $(\mathcal{I}_S \otimes \mathcal{A} \otimes \mathcal{I}_{S'})^\dagger = \mathcal{I}_S \otimes \mathcal{A}^\dagger \otimes \mathcal{I}_{S'}$.*

Proof. As a first step, let us prove that, for every system S, we have $(\mathcal{A} \otimes \mathcal{I}_S)^\dagger = \mathcal{A}^\dagger \otimes \mathcal{I}_S$. Take any system S', any state $\rho \in \mathrm{St}_1\,(\mathrm{B} \otimes \mathrm{S} \otimes \mathrm{S}')$, and any effect $E \in \mathrm{Eff}\,(\mathrm{A} \otimes \mathrm{S} \otimes \mathrm{S}')$, Equation (A3) yields

$$\rho \boxed{(\mathcal{A} \otimes \mathcal{I})^\dagger} E = E^\dagger \boxed{\mathcal{A} \otimes \mathcal{I}} \rho^\dagger = E^\dagger \boxed{\mathcal{A}} \rho^\dagger.$$

Specialising Equation (A3) to the case of a composite system, we have

$$E^\dagger \boxed{\mathcal{A}} \rho^\dagger = \rho \boxed{\mathcal{A}^\dagger} E,$$

whence we conclude that $(\mathcal{A} \otimes \mathcal{I}_S)^\dagger = \mathcal{A}^\dagger \otimes \mathcal{I}_S$.

Now let us prove that, for every system S, $(\mathcal{I}_S \otimes \mathcal{A})^\dagger = \mathcal{I}_S \otimes \mathcal{A}^\dagger$. Note that

$$\boxed{\mathcal{A}} = \mathrm{SWAP}\ \boxed{\mathcal{A}}\ \mathrm{SWAP}.$$

By Proposition A9, and recalling what we have just proved, we have

$$\left(\boxed{\mathcal{A}}\right)^\dagger = \mathrm{SWAP}\ \boxed{\mathcal{A}^\dagger}\ \mathrm{SWAP} = \boxed{\mathcal{A}^\dagger}.$$

To get the thesis, note that $(\mathcal{I}_S \otimes \mathcal{A} \otimes \mathcal{I}_{S'})^\dagger = [(\mathcal{I}_S \otimes \mathcal{A}) \otimes \mathcal{I}_{S'}]^\dagger$. We have just proved that

$$[(\mathcal{I}_S \otimes \mathcal{A}) \otimes \mathcal{I}_{S'}]^\dagger = (\mathcal{I}_S \otimes \mathcal{A})^\dagger \otimes \mathcal{I}_{S'},$$

and that $(\mathcal{I}_S \otimes \mathcal{A})^\dagger = \mathcal{I}_S \otimes \mathcal{A}^\dagger$, therefore we conclude that $(\mathcal{I}_S \otimes \mathcal{A} \otimes \mathcal{I}_{S'})^\dagger = \mathcal{I}_S \otimes \mathcal{A}^\dagger \otimes \mathcal{I}_{S'}$. □

Proposition A10. *Let $\mathcal{A} \in \mathrm{Transf}\,(\mathrm{A},\mathrm{B})$, and $\mathcal{B} \in \mathrm{Transf}\,(\mathrm{C},\mathrm{D})$. We have $(\mathcal{A} \otimes \mathcal{B})^\dagger = \mathcal{A}^\dagger \otimes \mathcal{B}^\dagger$.*

Proof. Take any system S, any state $\rho \in \mathrm{St}_1\,(\mathrm{B} \otimes \mathrm{D} \otimes \mathrm{S})$, and any effect $E \in \mathrm{Eff}\,(\mathrm{A} \otimes \mathrm{C} \otimes \mathrm{S})$, we have

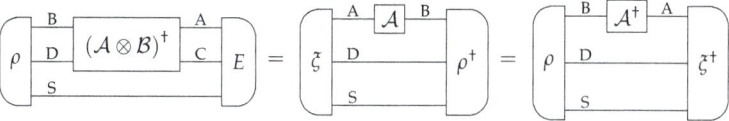

Now define $\xi := (\mathcal{I}_A \otimes \mathcal{B} \otimes \mathcal{I}_S) E^\dagger$, hence

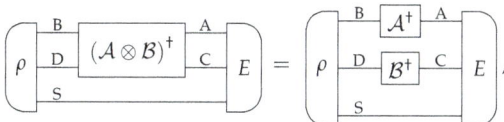

By Lemmas A3 and A4, we have that $\xi^\dagger = E \left(\mathcal{I}_A \otimes \mathcal{B}^\dagger \otimes \mathcal{I}_S \right)$, so

whence the thesis. □

This means that the dagger respects the composition of diagrams, and corresponds to the action of flipping a diagram with respect to a vertical axis.

References

1. Feynman, R.P.; Leighton, R.; Sands, M. *The Feynman Lectures on Physics. The Definitive and Extended Edition*; Addison Wesley: Boston, MA, USA, 2005.
2. Sorkin, R.D. Quantum mechanics as quantum measure theory. *Mod. Phys. Lett. A* **1994**, *9*, 3119–3127.
3. Sorkin, R.D. *Quantum Classical Correspondence: The 4th Drexel Symposium on Quantum Nonintegrability*; Chapter Quantum Measure Theory and Its Interpretation; International Press: Boston, MA, USA, 1997; pp. 229–251.
4. Barnum, H.; Müller, M.P.; Ududec, C. Higher-order interference and single-system postulates characterizing quantum theory. *New J. Phys.* **2014**, *16*, 123029.
5. Bolotin, A. On the ongoing experiments looking for higher-order interference: What are they really testing? *arXiv* **2016**, arXiv:1611.06461.
6. Dakić, B.; Paterek, T.; Brukner, Č. Density cubes and higher-order interference theories. *New J. Phys.* **2014**, *16*, 023028.
7. Lee, C.M.; Selby, J.H. Deriving grover's lower bound from simple physical principles. *New J. Phys.* **2016**, *18*, 093047.
8. Lee, C.M.; Selby, J.H. Generalised phase kick-back: The structure of computational algorithms from physical principles. *New J. Phys.* **2016**, *18*, 033023.
9. Lee, C.M.; Selby, J.H. Higher-order interference in extensions of quantum theory. *Found. Phys.* **2017**, *47*, 89–112.
10. Niestegge, G. Three-slit experiments and quantum nonlocality. *Found. Phys.* **2013**, *43*, 805–812.
11. Ududec, C. Perspectives on the Formalism of Quantum Theory. Ph.D. Thesis, University of Waterloo, Waterloo, ON, Canada, 2012.
12. Ududec, C.; Barnum, H.; Emerson, J. Probabilistic Interference in Operational Models. 2009, in preparation.
13. Ududec, C.; Barnum, H.; Emerson, J. Three slit experiments and the structure of quantum theory. *Found. Phys.* **2011**, *41*, 396–405.
14. Lee, C.M.; Selby, J.H. A no-go theorem for theories that decohere to quantum mechanics. *arXiv* **2017**, arXiv:1701.07449.

15. Barnum, H.; Barrett, J.; Leifer, M.; Wilce, A. Generalized no-broadcasting theorem. *Phys. Rev. Lett.* **2007**, *99*, 240501.
16. Barnum, H.; Wilce, A. Information processing in convex operational theories. *Electron. Notes Theor. Comput. Sci.* **2011**, *270*, 3–15.
17. Barrett, J. Information processing in generalized probabilistic theories. *Phys. Rev. A* **2007**, *75*, 032304.
18. Barrett, J.; de Beaudrap, N.; Hoban, M.J.; Lee, C.M. The computational landscape of general physical theories. *arXiv* **2017**, arXiv:1702.08483.
19. Chiribella, G.; D'Ariano, G.M.; Perinotti, P. Probabilistic theories with purification. *Phys. Rev. A* **2010**, *81*, 062348.
20. Chiribella, G.; D'Ariano, G.M.; Perinotti, P. Informational derivation of quantum theory. *Phys. Rev. A* **2011**, *84*, 012311.
21. Chiribella, G.; Spekkens, R.W. (Eds.) *Quantum Theory: Informational Foundations and Foils*; Fundamental Theories of Physics; Springer: Dordrecht, The Netherlands, 2016; Volume 181.
22. Dakić, B.; Brukner, Č. *Quantum Theory and Beyond: Is Entanglement Special*; Cambridge University Press: Cambridge, UK, 2011; pp. 365–392.
23. Hardy, L. Quantum Theory From Five Reasonable Axioms. *arXiv* **2001**, arXiv:quant-ph/0101012.
24. Hardy, L. *Foliable Operational Structures for General Probabilistic Theories*; Cambridge University Press: Cambridge, UK, 2011; pp. 409–442.
25. Lee, C.M.; Barrett, J. Computation in generalised probabilistic theories. *New J. Phys.* **2015**, *17*, 083001.
26. Lee, C.M.; Hoban, M.J. Bounds on the power of proofs and advice in general physical theories. *Proc. R. Soc. A* **2016**, *472*, 20160076.
27. Lee, C.M.; Hoban, M.J. The information content of systems in general physical theories. In Proceedings of the 7th International Workshop on Physics and Computation, Manchester, UK, 14 July 2016; Volume 214, pp. 22–28.
28. Masanes, L.; Müller, M.P. A derivation of quantum theory from physical requirements. *New J. Phys.* **2011**, *13*, 063001.
29. Hardy, L. Reformulating and reconstructing quantum theory. *arXiv* **2011**, arXiv:1104.2066.
30. Chiribella, G.; Scandolo, C.M. Entanglement as an axiomatic foundation for statistical mechanics. *arXiv* **2016**, arXiv:1608.04459.
31. Krumm, M.; Barnum, H.; Barrett, J.; Müller, M.P. Thermodynamics and the structure of quantum theory. *New J. Phys.* **2017**, *19*, 043025.
32. Jin, F.; Liu, Y.; Geng, J.; Huang, P.; Ma, W.; Shi, M.; Duan, C.; Shi, F.; Rong, X.; Du, J. Experimental test of born's rule by inspecting third-order quantum interference on a single spin in solids. *Phys. Rev. A* **2017**, *95*, 012107.
33. Kauten, T.; Keil, R.; Kaufmann, T.; Pressl, B.; Brukner, Č.; Weihs, G. Obtaining tight bounds on higher-order interferences with a 5-path interferometer. *New J. Phys.* **2017**, *19*, 033017.
34. Park, D.K.; Moussa, O.; Laflamme, R. Three path interference using nuclear magnetic resonance: A test of the consistency of born's rule. *New J. Phys.* **2012**, *14*, 113025.
35. Sinha, A.; Vijay, A.H.; Sinha, U. On the superposition principle in interference experiments. *Sci. Rep.* **2015**, *5*, 10304.
36. Sinha, U.; Couteau, C.; Jennewein, T.; Laflamme, R.; Weihs, G. Ruling out multi-order interference in quantum mechanics. *Science* **2010**, *329*, 418–421.
37. Barnum, H.; Graydon, M.; Wilce, A. Composites and categories of Euclidean Jordan algebras. *arXiv* **2016**, arXiv:1606.09331.
38. Chiribella, G. Dilation of states and processes in operational-probabilistic theories. In Proceedings of the 11th workshop on Quantum Physics and Logic, Kyoto, Japan, 4–6 June 2014; Volume 172, pp. 1–14.
39. Chiribella, G.; D'Ariano, G.M.; Perinotti, P. *Quantum Theory: Informational Foundations and Foils*; Chapter Quantum from Principles; Springer: Dordrecht, The Netherlands, 2016; pp. 171–221.
40. Hardy, L. *Quantum Theory: Informational Foundations and Foils*; Chapter Reconstructing Quantum Theory; Springer: Dordrecht, The Netherlands, 2016; pp. 223–248.
41. Abramsky, S.; Coecke, B. A categorical semantics of quantum protocols. In Proceedings of the 19th Annual IEEE Symposium on Logic in Computer Science, Turku, Finland, 13–17 July 2004; pp. 415–425.
42. Coecke, B. Kindergarten quantum mechanics: Lecture notes. *AIP Conf. Proc.* **2006**, *810*, 81–98.

43. Coecke, B. Quantum picturalism. *Contemp. Phys.* **2010**, *51*, 59.
44. Coecke, B.; Duncan, R.; Kissinger, A.; Wang, Q. *Quantum Theory: Informational Foundations and Foils*; Chapter Generalised Compositional Theories and Diagrammatic Reasoning; Springer: Dordrecht, The Netherlands, 2016; pp. 309–366.
45. Coecke, B.; Kissinger, A. *Picturing Quantum Processes: A First Course in Quantum Theory and Diagrammatic Reasoning*; Cambridge University Press: Cambridge, UK, 2017.
46. Selinger, P. A survey of graphical languages for monoidal categories. In *New Structures for Physics*; Coecke, B., Ed.; Springer: Berlin, Germany, 2011; pp. 289–356.
47. Wootters, W.K. Local accessibility of quantum states. In *Complexity, Entropy and the Physics of Information*; Zurek, W.H., Ed.; Westview Press: Boulder, CO, USA, 1990; pp. 39–46.
48. Chiribella, G.; Scandolo, C.M. Entanglement and thermodynamics in general probabilistic theories. *New J. Phys.* **2015**, *17*, 103027.
49. Chiribella, G.; Scandolo, C.M. Purity in microcanonical thermodynamics: A tale of three resource theories. *arXiv* **2016**, arXiv:1608.0446.
50. Gour, G.; Müller, M.P.; Narasimhachar, V.; Spekkens, R.W.; Yunger Halpern, N. The resource theory of informational nonequilibrium in thermodynamics. *Phys. Rep.* **2015**, *583*, 1–58.
51. Horodecki, M.; Horodecki, P.; Oppenheim, J. Reversible transformations from pure to mixed states and the unique measure of information. *Phys. Rev. A* **2003**, *67*, 062104.
52. Selby, J.H.; Coecke, B. Leaks: Quantum, classical, intermediate, and more. *Entropy* **2017**, *19*, 174.
53. Coecke, B. Terminality implies non-signalling. In Proceedings of the 11th workshop on Quantum Physics and Logic, Kyoto, Japan, 4–6 June 2014; Volume 172, pp. 27–35.
54. Chiribella, G.; Scandolo, C.M. Operational axioms for diagonalizing states. In Proceedings of the 12th International Workshop on Quantum Physics and Logic, Oxford, UK, 15–17 July 2015; Volume 195, pp. 96–115.
55. Chiribella, G.; Scandolo, C.M. Conservation of information and the foundations of quantum mechanics. *EPJ Web Conf.* **2015**, *95*, 03003.
56. Disilvestro, L.; Markham, D. Quantum protocols within Spekkens' toy model. *Phys. Rev. A* **2017**, *95*, 052324.
57. D'Ariano, G.M.; Manessi, F.; Perinotti, P.; Tosini, A. Fermionic computation is non-local tomographic and violates monogamy of entanglement. *Europhys. Lett.* **2014**, *107*, 20009.
58. D'Ariano, G.M.; Manessi, F.; Perinotti, P.; Tosini, A. The Feynman problem and fermionic entanglement: Fermionic theory versus qubit theory. *Int. J. Mod. Phys. A* **2014**, *29*, 1430025.
59. Chiribella, G.; Yuan, X. Bridging the gap between general probabilistic theories and the device-independent framework for nonlocality and contextuality. *Inf. Comput.* **2016**, *250*, 15–49.
60. Pfister, C.; Wehner, S. An information-theoretic principle implies that any discrete physical theory is classical. *Nat. Commun.* **2013**, *4*, 1851.
61. Alfsen, E.M.; Shultz, F.W. *Geometry of State Spaces of Operator Algebras*; Mathematics Theory & Applications; Birkhäuser: Basel, Switzerland, 2003.
62. Barnum, H.; Barrett, J.; Krumm, M.; Müller, M.P. Entropy, majorization and thermodynamics in general probabilistic theories. In Proceedings of the 12th International Workshop on Quantum Physics and Logic, Oxford, UK, 15–17 July 2015; Volume 195, pp. 43–58.
63. Chiribella, G.; Yuan, X. Measurement sharpness cuts nonlocality and contextuality in every physical theory. *arXiv* **2014**, arXiv:1404.3348.
64. Alfsen, E.M.; Shultz, F.W. State spaces of Jordan algebras. *Acta Math.* **1978**, *140.1*, 155–190.
65. Iochum, B. *Cônes Autopolaires et Algèbres de Jordan*; Lecture Notes in Mathematics; Springer: Berlin/Heidelberg, Germany, 1984; Volume 1049, doi: 10.1007/BFb0071358. (In French)
66. Coecke, B.; Selby, J.; Tull, S. Two roads to classicality. *arXiv* **2017**, arXiv:1701.07400.
67. Barnum, H. Spectrality as a Tool for Quantum Reconstruction: Higher-Order Interference, Jordan State Space Characterizations, Aug. 2009. Talk Given at the Conference "Reconstructing Quantum Theory", August 9–11, Perimeter Institute for Theoretical Physics. Available online: http://pirsa.org/09080016/ (accessed on 26 May 2017).
68. Niestegge, G. Conditional probability, three-slit experiments, and the jordan algebra structure of quantum mechanics. *Adv. Math. Phys.* **2012**, *2012*, 156573.
69. Selby, J.H.; Coecke, B. Process-theoretic characterisation of the hermitian adjoint. *arXiv* **2016**, arXiv:1606.05086.

70. Müller, M.P.; Oppenheim, J.; Dahlsten, O.C.O. The black hole information problem beyond quantum theory. *J. High Energy Phys.* **2012**, *2012*, 9.
71. Müller, M.P.; Ududec, C. Structure of reversible computation determines the self-duality of quantum theory. *Phys. Rev. Lett.* **2012**, *108*, 130401.
72. Marshall, A.W.; Olkin, I.; Arnold, B.C. *Inequalities: Theory of Majorization and Its Applications*; Springer Series in Statistics; Springer: New York, NY, USA, 2011.
73. Müller, M.P.; Dahlsten, O.C.O.; Vedral, V. Unifying typical entanglement and coin tossing: On randomization in probabilistic theories. *Commun. Math. Phys.* **2012**, *316*, 441–487.
74. Wilde, M.M. *Quantum Information Theory*, 2nd ed.; Cambridge University Press: Cambridge, UK, 2017.
75. Selinger, P. Dagger compact closed categories and completely positive maps. *Electron. Notes Theor. Comput. Sci.* **2007**, *170*, 139–163.

© 2017 by the authors. Licensee MDPI, Basel, Switzerland. This article is an open access article distributed under the terms and conditions of the Creative Commons Attribution (CC BY) license (http://creativecommons.org/licenses/by/4.0/).

Article
Leaks: Quantum, Classical, Intermediate and More

John Selby [1] and Bob Coecke [2],*

[1] Department of Physics, Imperial College London, Kensington, London SW7 2AZ, UK; john.selby08@imperial.ac.uk
[2] Department of Computer Science, University of Oxford, Oxford OX1 3PA, UK
* Correspondence: Bob.Coecke@cs.ox.ac.uk; Tel.: +44-7881-333990

Academic Editors: Giacomo Mauro D'Ariano and Paolo Perinotti
Received: 26 January 2017; Accepted: 12 April 2017; Published: 19 April 2017

Abstract: We introduce the notion of a leak for general process theories and identify quantum theory as a theory with minimal leakage, while classical theory has maximal leakage. We provide a construction that adjoins leaks to theories, an instance of which describes the emergence of classical theory by adjoining decoherence leaks to quantum theory. Finally, we show that defining a notion of purity for processes in general process theories has to make reference to the leaks of that theory, a feature missing in standard definitions; hence, we propose a refined definition and study the resulting notion of purity for quantum, classical and intermediate theories.

Keywords: process theory; classical limit; purity

1. Introduction

Can we explain why the world is quantum by finding some sense in which quantum theory is an optimal theory? Broadcasting distinguishes quantum theory from classical theory in that quantum states cannot be broadcast [1], but neither can the states of many other theories [2,3]. Non-locality is a measure of non-classicality, and quantum theory is non-local, but not maximally so [4]. Therefore, is there some manner in which we can uniquely single out quantum theory? In this paper, we show that quantum theory is a leak-free theory, whilst classical theory is maximally leaking. We formalise the notion of a leak, which can roughly be thought of as a 'one-sided broadcasting map', within the process-theoretic framework [3,5,6] as a particular type of process, which, as the name suggests, accounts for leaking state-data into the environment.

Moreover, there is a natural way to introduce leaks to any theory, and by doing so, we obtain new theories. We call this the leak construction. In particular, classical theory can be obtained from quantum theory in this manner, where, in this example, the leaking is then nothing but decoherence [7,8]. Hence, the concept of a leak allows us to generalise decoherence to arbitrary process theories. Besides classical theory, any theory characterised by some finite-dimensional C*-algebra can be obtained in this manner from quantum theory. In fact, as we show in a follow-up paper [9], only C*-algebras can be obtained in this manner. Leaks therefore capture the operational content of finite-dimensional C*-algebras on-the-nose, in a manner that does not involve any additive structure, nor a *-operation.

Finally, we observe that defining purity of processes in process theories with leaks is problematic; in particular, this is the case for classical theory. Making explicit use of the concept of a leak, we therefore propose a new definition that makes sense for arbitrary processes in arbitrary process theories.

Related Work

As explained in detail in the follow-up paper [9], the leak construction is related to the "constructions of classical system types" in [10–12]. More specifically, in the case of quantum theory, we exactly obtain the same result, but in a much simpler way, with much less use of structure and guided by a clear operational meaning. The notion of a leak is closely related to the decomposability of a state-space [13] in the generalised probabilistic theory framework as, at least under some standard assumptions, such as the "no-restriction hypothesis", each is equivalent to the existence of a non-disturbing measurement as discussed in [14].

2. Process Theories with Discarding...

A process theory [3,6] is a collection of systems that are represented by wires and processes that are represented by boxes with wires as inputs (at the bottom) and outputs (at the top). Moreover, when we plug these boxes together:

the resulting diagram should also be a process. To be mathematically more precise, the data that make up a diagram are:

- the boxes that appear in the diagram and
- how these boxes are wired together, including the overall ordering of inputs/outputs.

Hence, two diagrams are equal when these data match up.

By a circuit [3,6], we mean a diagram that can be constructed by means of the obvious operations of parallel composition $\otimes$ and sequential composition $\circ$ of boxes. For example, the following diagrams is a circuit:

Composite systems, denoted $A \otimes B$, then simply arise by pairing wires:

$$\left| A \otimes B \right. := \left| A \right. \left| B \right.$$

Remark 1. *A process theory with circuits as diagrams can also be defined as a strict symmetric monoidal category. Strictness means that associativities and unit laws hold on-the-nose, unlike the symmetric monoidal categories of concrete mathematical models where non-trivial associativity and unit natural isomorphisms are required. Fortunately, by Mac Lane's strictification theorem [15], every such category is categorically equivalent (although not isomorphic) to a strict one, which means that for all practical purposes, it can be thought of as a strict one.*

A state is a process without inputs; an effect is a process without outputs; and a number is a process with neither inputs nor outputs. One special number is the empty diagram:

which in most theories coincides with the number one.

Throughout this paper, for each system in a process theory, we postulate the existence of a discarding effect, which is interpreted just as its name indicates and which is denoted as:

We also make the natural assumption that discarding effects compose:

$$\overline{\overline{\top}}_{A \otimes B} := \overline{\overline{\top}}_A \; \overline{\overline{\top}}_B \qquad (1)$$

A process f is causal if we have:

$$\overline{\overline{\top}} \circ f = \overline{\overline{\top}} \qquad (2)$$

and a theory is causal if all of the processes of the theory are causal. Therefore, except for the fact that it composes, discarding is not subject to any defining constraints. In a sense, its behaviour is entirely implicit within its role within the defining equation of causality. In particular, by Equation (2) where f is taken to be an effect, it immediately follows that the only effects in a causal theory are the discarding effects. In this form, the axiom of causality traces back to [16]. When restricting to causal processes, a process theory is non-signalling [17]; hence, the causality of a theory is vital to guarantee compatibility with relativity.

Example 1 (Classical probability theory). *When viewing probability theory as a process theory, systems are n-state classical systems and boxes are n × m stochastic matrices, and so, in particular, states are probability distributions. Discarding is given by marginalisation, and so, causality boils down to the fact that the entries of a probability distribution add up to one and that the entries in each column of a stochastic matrix add up to one.*

Example 2 (Quantum theory). *Quantum theory as a process theory has finite dimensional Hilbert spaces $\mathcal{H}$ as its systems and completely positive trace preserving (CPTP) maps:*

$$\xi : \mathcal{B}(\mathcal{H}) \to \mathcal{B}(\mathcal{H}')$$

as its processes. Causality for density operators means having trace one and for completely positive maps means being trace-preserving. One can also include classical data as additional systems, and then measurements and controlled operations are also processes. If this is the case, we will often denote the classical systems as dotted wires to distinguish them from quantum wires. Specifically, measurements are processes from quantum to classical systems where the probabilities of obtaining the different outcomes are encoded in the classical system. Causality then implies that, for projective measurements, the projectors form a resolution of the identity and, for general measurements that the POVM elements sum to discarding. A full description and a pedagogical introduction to this theory is in [3,6,18].

Typically, as will be the case in the examples below, we will want to describe both causal and non-causal processes. We therefore will still, for each system, have a discarding map, which specifies the causal processes, but there will also be other processes that will not satisfy Equation (2). There are two main reasons for this. The first is to allow us to discuss events, i.e., processes that we cannot make happen deterministically, but that can occur as a particular outcome in some experiment; therefore, allowing us to obtain the probability of obtaining a specific outcome, which, in particular, allows us, via suitable renormalisation, to describe post-selection. The second reason is mathematical simplicity: it is often much easier to define the process theory, or various structures within it, in the non-causal setting and then to restrict to the causal sub-theory when necessary.

Example 3 (Non-causal extension of quantum theory). *To describe non-causal processes in quantum theory, rather than taking processes as completely positive trace-preserving maps, we instead just require that they are completely positive. It is very standard within quantum theory to consider such processes, for example Dirac bras are non-causal or, more generally, individual POVM elements are non-causal.*

An important tool is the Choi–Jamiolkowski isomorphism between transformations and bipartite states. One direction of this isomorphism can be realised causally, using the Bell state, which we represent, up to a normalisation factor, with a cup-shaped wire:

$$\frac{1}{D} \cup \qquad \text{where} \qquad D := \overline{\overline{}}\,\overline{\overline{}}$$

which allows us to "bend wires up":

$$\boxed{f} \;\mapsto\; \frac{1}{D}\,\boxed{f}$$

This associates with each (causal) process a (causal) bipartite state. The other direction is however not realisable causally, as it relies on the Bell effect, which we represent with a cap-shaped wire:

$$\rho \;\mapsto\; D\,\rho$$

The fact that this is an isomorphism provides us with the following intuitive diagrammatic rule (justifying the representation of these as a cup and cap):

$$\cap \cup \;=\; | \tag{3}$$

It is then clear that the cap cannot be causal (even up to a rescaling) as, if it were, then the identity transformation would be separable, i.e.:

$$| \;=\; \cap \cup \;=\; \overline{\overline{}}\,\overline{\overline{}}\,\cup \;=\; \rho \;\overline{\overline{}}$$

where in the second step we relied on the fact that by causality, all effects must be discarding, so in particular, the cap, as well as Equation (1).

Example 4 (Non-causal extension of classical theory). *We can similarly extend classical theory, taking processes as $n \times m$ matrices with positive real elements as opposed to stochastic matrices. This again allows us to discuss particular outcomes of measurements, which may not happen with certainty, and moreover, gives us a classical equivalent of the Choi–Jamiolkowski isomorphism where rather than using the Bell state and effect, we use the perfectly correlated state and effect, again denoted by a cup and a cap. These can be defined in terms of the orthonormal basis states and effects as:*

$$\frac{1}{n}\,\triangle_i\,\triangle_j \;=\; \frac{\delta_{ij}}{n} \qquad \text{and} \qquad \triangledown^i\,\triangledown^j \;=\; \delta_{ij}$$

respectively. It is simple to check that these also satisfy Equation (3) as we would expect from the choice of the diagrammatic representation.

3. ... and Leaks

Definition 1. *A leak is a process:*

$$\tag{4}$$

which has discarding as a right counit, that is:

$$\tag{5}$$

Proposition 1. *All leaks are causal.*

Proof. Causality of a leak means:

and this equation is obtained by discarding the outputs in (5). □

When we have multiple leaks around, we may often represent them with different colours to distinguish them.

Proposition 2. *Leaks compose to give leaks.*

Proof. Sequential composition of leaks is again a leak:

since we have:

and the same goes for parallel composition:

since we have:

$$\begin{array}{c}\text{[diagram: } L_1 \otimes L_2 \text{ over } A \otimes B\text{]} = \text{[diagram: } L_1, L_2 \text{ over } A, B\text{]} = \text{[diagram: identities on } A, B\text{]} = \text{[identity on } A \otimes B\text{]}\end{array}$$

□

For classical probability theory, copying of support elements provides a leak:

$$\vee : X \to X \times X :: x \mapsto (x,x)$$

since if we discard a copy, we are back with what we started off with. In fact, strictly speaking, what we are dealing with here is not a copying operation since while it copies pure classical states, it does not do that for impure ones. What it is instead is broadcasting, that is besides Equation (5), discarding is also a left counit for the leaking process:

$$\text{[diagram]} = \text{[identity]} \qquad (6)$$

Note that this requires $L := A$ in Equation (4). This is the maximal possible leak for any system, as all of the information about the ingoing state is leaked out.

On the other hand, quantum theory does not allow for broadcasting [1]. In fact, the only kind of leak quantum theory admits is constant leaking. This immediately follows from the following fact about quantum processes, which states that any dilation of a pure process, i.e., representation as a process with an extra output that is discarded, must separate:

Proposition 3. *For pure quantum processes f, we have:*

$$\boxed{f} = \boxed{g}\!\!\!\top \implies \boxed{g}\!\!\!\top = \boxed{f}\,\,\boxed{\rho}\!\!\!\vee \qquad (7)$$

with ρ causal. That is, if a reduced process f is pure, then the process g we started from must separate.

Proof. See, e.g., [3,6]. □

Hence, since the identity is pure, by Proposition 3, it follows from the defining equation of a leak (5) that any leak for quantum theory must be constant, that is of the form:

$$\text{[identity]}\,\,\boxed{\rho}\!\!\!\vee \qquad (8)$$

where we need to take the state to be causal:

$$\overline{\boxed{\rho}\!\!\!\vee} = \qquad (9)$$

Remark 2. *In quantum theory, Proposition 3 can actually be taken as a definition of the purity of processes, that is a quantum process f is pure if and only if all dilations of f separate. However, in theories with non-constant leaks, this definition must be revised as we discuss in detail in Section 7.*

Of course, (8) is also a leak for classical probability theory, and another example arises by combining broadcasting and a constant:

   (10)

At least qualitatively, quantum theory can therefore be described as a minimally-leaking theory, as all leaks are constant leaks, whilst classical theory is maximally leaking, as for each system, there is a maximal leak. We will now provide qualitative substance to this claim.

4. Quality of a Leak

For the sake of simplicity of the argument, we will restrict ourselves to a special kind of process theories that admit the notion of a feedback wire. Explicitly spelling out the process-theoretic characterisation of a feedback wire as in [19] goes beyond the scope of this paper. It suffices to know that they exist in both quantum and classical theory, where they can be constructed in the obvious way using the cups and caps of Examples 3 and 4. The behaviour of such a feedback wire is that of a wire of the shape:

for which we have the obvious equations, such as:

In particular, by means of such a wire, we can feed an output of a process back into it as an input:

i.e., we create a feedback-loop.

Feedback-loops allow us to ask questions, such as how closely will some outgoing data match an ingoing data. In particular, for the case of leaks where $L = A$, we can measure how closely the leaked data matches the original (while ignoring the output) via the following diagram:

However, what tends to be more useful, particularly in the case where $L \neq A$, is not asking precisely how well does the outgoing data match the ingoing, but how well does the outgoing data encode the ingoing data. For example, all of the information could be there just scrambled up or encoded in some other system type. We therefore want to consider maximising over potential restoration maps $r : L \to A$, where r is taken to be causal. We call this notion the quality of a leak:

$$\mathcal{Q}[\leak] := \mathrm{Max}_r \left[\diagram \right]$$

If the structure of the numbers in a process theory is sufficiently rich, e.g., they are the real numbers or probabilities, one can moreover renormalise this quantity as follows:

$$\frac{\mathcal{Q}[\leak] - }{\bigcirc - } \qquad (11)$$

where the circle indicates the feedback-loop applied to the identity. As a leak, the quality of broadcasting is one, since we have:

$$\diagram \stackrel{(6)}{=} \bigcirc$$

while for constant leaks, it is zero, since we have:

$$\diagram = \diagram \stackrel{(9)}{=}$$

We therefore see that quantum theory is a minimally leaking theory as the renormalised quality for any leak is zero, whilst classical theory is maximal as every system has a leak with renormalised quality of one. In the next section, we consider how to increase the amount of leaking for a theory, providing a process-theoretic perspective on the quantum to classical transition.

Example 5. *If a process theory admits sums (cf. [6] or Appendix A), then set:*

$$\leak := c\, \leak + q\, \diagram$$

with $c + q = 1$. Now, quality in the form Equation (11) is c.

5. A Representation for All Classical-Quantum Leaks

We already characterised all quantum leaks as being constant leaks; we next characterise all classical leaks.

Proposition 4. *All classical leaks are of the form:*

$$\diagram_{A}^{A\ L} = \diagram_{A}^{A\ L} \qquad (12)$$

where l is any causal classical process.

Proof. First, let us define, using the non-causal "cap" of Example 4:

$$\tikz \;:=\; \boxed{l}$$

Despite the fact that this is defined using a non-causal process, the composite process l is actually causal:

$$\boxed{l} \;=\; \cdots \;=\; \cdots \;=\; \top$$

We can then use the matrix representation of the leak (see Appendix A):

$$\tikz \;:=\; \sum_{ijk} \Delta_k^{ij} \;\tikz$$

where $\Delta_k^{ij} \in \mathbb{R}^+$. The leak condition then implies that:

$$\sum_j \Delta_k^{ij} = \delta_k^i$$

and so:

$$\Delta_k^{ij} = \Delta_k^{ij} \delta_k^i$$

Then, we can check that Equation (12) is indeed satisfied:

$$\tikz \;=\; \tikz \;=\; \sum_{ijk} \Delta_k^{ij} \delta_k^i \;\tikz \;=\; \sum_{ijk} \Delta_k^{ij} \;\tikz \;=\; \tikz$$

□

We can now also characterise all leaks for composite classical-quantum systems:

Proposition 5. *Denoting the classical system by a dotted line and the quantum system by a solid line; all composite classical quantum systems have leaks of the form:*

$$\tikz \;=\; \boxed{L} \tag{13}$$

where L is any causal process from classical to quantum systems.

Proof. Note that any composite leak defines a quantum leak as:

$$\frac{1}{D}$$

and therefore, as we know all quantum leaks separate:

where ρ defines a classical leak as:

and so putting this together, we have:

□

The bottom line is that all of these leaks involve the copying leak as the fundamental ingredient. This is not all too surprising, since, as we showed in the previous section, it stands for maximal leakage. The processes l and L then play the role of reducing the leakage, with as the extremal cases l and L being constant, producing a constant leak.

6. The Leak-Construction

We now show how one can construct new process theories from old ones by introducing leaks. This is done by inserting particular processes of the old theory of the form (15) on all of the wires. The processes (14), to which we refer in the old theory as pre-leaks, then become leaks in the new theory. Hence, the leak construction turns pre-leaks into leaks.

Theorem 1. *Given any process theory and for each system a causal process:*

(14)

which is such that the following process is idempotent:

(15)

and which are chosen coherently for composite systems:

$$
\begin{array}{c}
\text{[diagram]}
\end{array}
\qquad (16)
$$

we can construct a new process theory in which each process (14) is a leak for the system A. This construction goes as follows:

- systems stay the same;
- one restricts processes to those of the form:

$$
\text{[diagram]} \qquad (17)
$$

Proof. By causality of (14):

$$
\text{[diagram]} \qquad (18)
$$

discarding is preserved by the leak-construction. Given the form Equation (17) of the processes in the theory and due to the idempotence of Equation (15), plain wires have taken the form Equation (15), so the defining equation of a leak Equation (5) is satisfied. To consider the pre-leak in the new theory, we must apply the leak construction Equation (17), and using the condition for composites Equation (16), we get the following process in the new theory:

$$
\text{[diagram]}
$$

which is indeed a leak in the new theory:

$$
\text{[diagram]} \stackrel{(18)}{=} \text{[diagram]} \stackrel{(15)}{=} \text{[diagram]}
$$

which is the form of a plain wire in the new theory, and so, this construction does turn pre-leaks into leaks. It is moreover straightforward to see that we again obtain a process theory. □

Sometimes the leak-construction does nothing, in particular, when the pre-leaks are already leaks:

Example 6 (Trivial). *A simple example of the leak construction is the one where the pre-leaks are taken to already be leaks, since then (17) will reduce to the processes f themselves.*

The main motivating example for this construction is of course the following:

Example 7 (Decoherence). *The leak construction for the pre-leak:*

$$\phi : \mathcal{B}(\mathcal{H}) \to \mathcal{B}(\mathcal{H} \otimes \mathcal{H}) :: |i\rangle\langle i| \mapsto |i\rangle\langle i| \otimes |i\rangle\langle i|$$

applied to the process theory of quantum processes (i.e., Example 2), we obtain classical probability theory (i.e., Example 1).

In the above construction, it is really the idempotents rather than the specific pre-leaks that determine the theory that is obtained. We can therefore have several different perspectives on the "cause" of this idempotent, by considering different pre-leaks from which it could be obtained. Firstly, we can always take the trivial case, where the pre-leak is just the idempotent itself, i.e., taking the leaked system as the empty system. There are however three alternate forms that always exist in quantum theory and that are more insightful.

Example 8. *Firstly we can consider the purification f of the idempotent, in the sense of [16]:*

$$\phi = \boxed{f}$$

This corresponds to the idea that information can never be fundamentally destroyed, only discarded, and so, we can see this leaking of information into some causally-separated system leading to decoherence. Another standard way to represent a general process is, via Stinespring dilation [20], as a reversible interaction with an environment:

$$\phi = \boxed{U}_{s}$$

and so, we can equivalently view decoherence as arising due to a reversible interaction with some uncontrolled environment [8]. A final example, suggested to us by Rob Spekkens, is that the idempotent can be viewed as describing a system that lacks a reference frame [21]; the leaked system would then correspond to the reference system itself. This is the subject of ongoing work and is discussed in the Conclusion.

Example 7 leaves open the question whether there are any theories that can be obtained from this leak construction in between classical and quantum theory. This question is solved in a forthcoming paper where the key result is the following theorem:

Theorem 2. *The leak construction applied to quantum processes (i.e., Example 2) gives all C*-algebras and C*-algebras only.*

Therefore, despite the weak structure of a leak, for the specific case of quantum theory, we obtain precisely the C*-algebras via the leak construction. This leads one to contemplate the view that the operational essence of (finite dimensional) C*-algebras is entirely captured by leaks and that the additional structure of C*-algebras is merely an artefact of the Hilbert space representation.

Remark 3. *The leak-construction does not apply to Example 5, since only for $c = 0, 1$, we have idempotence of (15).*

Remark 4. *For a process theory in which all systems are compositions $A^{\otimes n}$ of one atomic system A, it suffices to pick a single process (14) for the system A (where L_A will be of the form $A^{\otimes n}$, since all other such processes arise then by coherence (16)).*

Remark 5. *If a pre-leak with $L := A$ is co-associative:*

$$\text{(diagram)}$$

then the idempotence of (15) follows from causality of the pre-leak.

Remark 6. *The construction in Theorem 1, when modified by not fixing a pre-leak for each type, but rather considering all pairs of a system and a corresponding pre-leak, is known as the Karoubi envelope, or Cauchy completion, or splitting of idempotents. More details on this can be found in [9].*

7. Process-Purity from Leaks

In this section, we consider how leaks relate to purity in process theories. The purity (or lack of purity) of a state is a fundamental concept in quantum theory and is equally important in most approaches to generalised physical theories. However, there is no reason to consider this as solely a property for states, but should be considered for all processes in a theory. Indeed, lack of knowledge about a process, the noisiness of a channel and detection errors on a POVM-element all correspond to process-impurities. We will show that defining such a property for general theories, and classical theory in particular, requires leaks.

In Reference [22], Chiribella et al. introduce the notion of side-information; this can be thought of as information that is lost during a process that, in principle, could be possessed by some other agent. The use of this in cryptographic scenarios is clear, where the side-information can be thought of as being possessed by an eavesdropper attempting to influence or gain information about some cryptographic protocol. Diagrammatically, this side information is depicted as:

$$f = g \quad \text{Side information about process } f$$

Lack of side-information for a process would imply that g must separate such that the side-information is independent of the process f. Indeed, this must be the case for any such g, i.e.:

$$f = g \quad \Longrightarrow \quad g = f \, \rho \tag{19}$$

or in other words, all dilations of f must separate. As mentioned in Remark 2, the separability of dilations (cf. Proposition 3) has been proposed as a definition of process-purity. Indeed for the case of quantum theory, this corresponds to the expected notion of purity, that is that the CPTP map must be Kraus rank 1. Remarkably, however, in the form of (19), this definition does not extend to general processes of classical probability theory. In fact, nor does it do so for any theory that has broadcasting:

Proposition 6. *If a non-trivial theory has broadcasting and one defines purity by means of (19), then plain wires (i.e., identity processes) are not pure.*

Proof. Assuming identities are pure and applying (19) to the defining equation of a leak (5), we obtain:

$$\quad (20)$$

that is, it is a constant leak. However, then, from the second defining equation of broadcasting, we obtain:

that is, each plain wire is a constant process, and hence, the theory is trivial, since as a consequence, all processes must then be constant since for (causal) processes, we have:

Hence, in a non-trivial theory with broadcasting, identities cannot be pure in the sense of (19). □

From the first part of this proof, namely that this definition of purity implies that leaks must be constant, it follows that this issue arises in any theories with non-constant leaks. We can think of this as the fact that, if a system has a leak, then there is irreducible side-information contained within the system itself:

Fortunately, leaks also allow us to fix this problem. Firstly, let us suppose that a theory has leaks and also has a pure process f. Then, clearly, the following is a dilation of f:

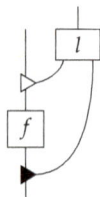

where l is causal. One may therefore consider explicitly bringing leaks into play in the definition of purity. A first step in this direction is to weaken (19) as follows:

$$\boxed{f} = \boxed{g} \quad \Longrightarrow \quad \exists \blacktriangleright, \blacktriangleright \& \boxed{l} : \boxed{g} = \boxed{f} \qquad (21)$$

However, now, we have the opposite problem: all classical processes, including all states, are pure! (See Appendix B). It is clear that we are missing a constraint. The original idea was that for a process to be pure, it should have no side-information that some eavesdropper could take advantage of. However, we have shown that for some systems, there is irreducible side-information represented by leakage. Therefore, to ensure that the eavesdropper cannot gain information or influence the process, we must demand that the process does not interact with this irreducible side-information, such that leaking before or after is equivalent:

$$\forall \blacktriangleright \exists \blacktriangleright \text{ and } \forall \blacktriangleright \exists \blacktriangleright \text{ such that } \boxed{f} = \boxed{f} \qquad (22)$$

Hence, we propose the following definition of process-purity, which packages these two conditions, (21) and (22), into a neat form:

Definition 2. *f is pure if and only if:*

$$\boxed{f} = \boxed{g} \quad \Longrightarrow \quad \exists \blacktriangleright \& \blacktriangleright : \boxed{g} = \boxed{f} = \boxed{f} \qquad (23)$$

This ensures that the only side-information is this irreducible kind, i.e., system leakage, and moreover, that pure processes do not interact with this irreducible side information. To further motivate this definition, we will show that it provides a sensible definition for quantum, classical and composite systems. However, first, note that for states, this definition reduces to:

Example 9. *A state ψ is pure if we have:*

$$\boxed{\psi} = \boxed{\sigma} \quad \Longrightarrow \quad \boxed{\sigma} = \boxed{\psi}\boxed{\rho}$$

This is the same as the original definition, and so, we see that it is only for general processes that this new definition is necessary. Similarly, in the case of quantum theory, it is only the first condition that provides a non-trivial constraint:

Example 10 (quantum purity). *As for quantum theory, the only leaks are constant leaks, Condition (21) in Definition 2 reduces to (19), while Condition (22) becomes trivial.*

Whilst, in the classical case, as we have mentioned above, (21) is satisfied by all classical processes, and so, it is only (22) that needs to be considered:

Example 11 (classical purity). *All pure classical processes, between an n and m state system, are of the form:*

$$\tag{24}$$

where we can define the 'upside-down broadcasting map' by:

and the black/white dot is any process that satisfies:

$$\quad\quad \text{and} \quad\quad$$

Proof. We prove here that pure classical processes must be of this form and leave the proof that any process of this form is pure to Appendix C.

First consider the condition:

$$\forall \quad \exists \quad \text{such that} \quad =$$

for the special case where:

$$=$$

and using the standard form for classical leaks to write:

$$:=$$

Then, we can show that:

$$= \quad = \quad =$$

This implies that, for all i and j:

$$\begin{array}{c}\boxed{f}\end{array} = \begin{array}{c}\boxed{f}\ \boxed{l}\end{array}$$

so, for each i and j:

$$f_i^j := \boxed{f} = 0 \quad \text{or} \quad l_i^j := \boxed{l} = 1$$

Causality of l then implies that, for each j, there can only be a single i where $l_i^j = 1$, and so, for all other i, we must have $f_i^j = 0$. This means that in each row of f_i^j, there is at most a single non-zero element.

We can run through this argument in the opposite direction using the condition:

$$\forall \quad \exists \quad \text{such that} \quad \boxed{f} = \boxed{f}$$

which shows that f_i^j can have at most a single non-zero element in each column. This is precisely what is enforced by the black/white dot in the above form; the value of the non-zero elements is then determined by the state r. Hence, we can write f in the desired form. □

Example 12. *If we consider purity for causal classical processes, then we find that the pure processes are those that are reversible (i.e., are isometries).*

Proof. The definition of purity, and the standard form for classical leaks, requires that:

$$\boxed{f} = \boxed{f}^{\,l}$$

and so, we have:

$$\big| = \big| = \boxed{f} = \boxed{f}^{\,l} = \boxed{f}^{\,l}$$

Therefore, f is reversible in the sense that it has a left-inverse, i.e., l. □

Finally, we consider the composite case, where the conjunction of (21) and (22) is necessary:

Example 13 (Composite classical-quantum purity). *Pure processes are:*

(25)

where we denote the classical system with a dotted line, and:

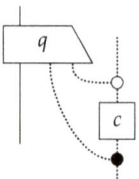 (26)

is pure for all i.

Proof. Again, we prove the interesting direction here that pure processes on composite systems must be of this form and, again, leave the other direction to Appendix C.

Note that a generic process can be written as:

An (almost) identical argument to the classical case shows that if this is pure, it can be written as:

We therefore move on to considering the other part of the definition of purity, that is that any dilation can be written as a leak; that means that any dilation of this process can be written as:

Now, note that any collection of dilations of the processes:

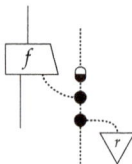

defines a dilation of the whole process, which must be able to be written as a leak:

Therefore, each g_i must separate, and hence, the f_i are each pure quantum processes. □

An immediate consequence of this is the following.

Proposition 7. *The pure quantum to classical or classical to quantum maps are separable.*

Proof. First note that,

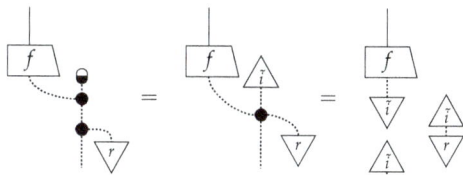

Then, using the above result regarding pure maps for composite quantum classical systems, we have,

Similarly, we obtain separability for pure quantum to classical maps. □

This means that there is no pure way to transform between classical and quantum information.

8. Conclusions

In this paper, we introduce the concept of leaks to generalised process theories. The definition of which can be thought of as a "one-way" broadcasting map. These prove to be very useful for understanding various aspects of quantum theory from a physically well-motivated perspective. In particular, we show:

- that quantum theory is a leak-free theory, whilst classical theory is maximally leaking, giving a clear separation between the theories for which quantum theory is optimal.
- how to construct sub-theories via a "leak construction", which can be thought of as the sub-theories that can be obtained from a dynamical decoherence mechanism. For quantum theory, we can obtain classical theory, composite quantum classical theory and, generally, finite dimensional C*-algebras from this construction [9].
- a characterisation of the leaks and pure processes for quantum, classical and composite systems; in particular, we demonstrate that there is no pure way to transform quantum systems into classical systems or vice versa.
- that leaks are essential to defining purity of processes; we therefore introduce a novel definition of purity of processes, which makes sense both for quantum theory and for classical theory.

Future Work

In this paper, we have shown how classical theory emerges from quantum theory due to the leak construction, providing a process-theoretic perspective on why the world on a large scale appears to us to be classical. It is natural to ask: Is there some deeper theory of nature than quantum theory from which quantum theory emerges in an analogous way? This is the subject of a forthcoming paper [23]. A second, related question, would be to ask: What does it imply about a theory if it can obtain classical theory via a leak construction; is the ability for this to happen in quantum theory special or is this a generic feature of general theories?

We have also shown that quantum theory is minimally leaking and classical theory maximally; moreover, if we start from a process theory describing finite dimensional C*-algebras, then quantum theory is singled out as the unique minimally-leaking theory. Can this idea lead to a complete reconstruction of quantum theory [24]?

As mentioned in Example 8, one interpretation of the leak construction is as a way to represent systems for which there is a missing reference frame, that is we can write the pre-leak as ([21], Section IVB):

$$\bigtriangledown = \int_G dg \; U_g \; \bigtriangledown_g$$

where G is a group associated with a reference frame for a particular degree of freedom, U_g is the representation of G on the system of interest and g the state of the reference system. Note, however, that making sense of this integral for general symmetry groups requires the reference be an infinite dimensional quantum system and so is beyond the scope of this paper. One could replace, at least for compact groups, the integral by a finite convex mixture (using the results of [16], Corollary 33 from Caratheodory's theorem), for which the resulting idempotent would be the same. This can be thought of as there only being a finite set of possible orientations for the reference frame. However, a comprehensive understanding of the connections here would require consideration of the infinite dimensional case. Moreover, we know that in the finite dimensional case, the leak construction leads to C*-algebraic systems only; however, it remains an interesting open question as to what the leak construction leads to for infinite dimensional systems.

Acknowledgments: We thank Aleks Kissinger, Dan Marsden, Rob Spekkens and Sean Tull for useful feedback. John Selby was supported by the EPSRC (Engineering and Physical Sciences Research Council) through the Controlled Quantum Dynamics Centre for Doctoral Training, and Bob Coecke is supported by the U.S. Air Force Office of Scientific Research.

Author Contributions: Both authors contributed equally to all aspects of this work. Both authors have read and approved the final manuscript.

Conflicts of Interest: The authors declare no conflict of interest.

Appendix A. Mathematical Tools for Proofs

Definition A1. *Sums can be defined by the fact that they distribute over diagrams, that is:*

$$\sum_i \left(\begin{array}{c} c \\ b_i \\ a \end{array} \right) = \sum_i \left(\begin{array}{c} c \\ b_i \\ a \end{array} \right)$$

In particular, in classical probability theory, we can take sums of diagrams where the sum is the standard sum of matrices. In fact, this provides us with a matrix calculus for our diagrams. In particular, we have a basis and co-basis for each system, denoted:

$$\left\{ \bigtriangledown_i \right\}_{i=1}^n \quad \text{and} \quad \left\{ \bigtriangleup^j \right\}_{j=1}^n$$

respectively, such that they are orthonormal:

$$\bigtriangleup_i^j = \delta_{ij}$$

Then, this provides a decomposition of the identity

$$\left| \;\; = \;\; \sum_i \;\; \vcenter{\hbox{$\triangledown_i \triangle^i$}} \right.$$

which allows us to write any process as:

$$\boxed{f} \;\; = \;\; \sum_i \sum_j \vcenter{\hbox{$\triangledown_i \;\boxed{f}\; \triangle^j$}} \;\; := \;\; \sum_{ij} f^i_j \vcenter{\hbox{$\triangledown_i \triangle^j$}}$$

where it is simple to check that sequential composition then coincides with a matrix multiplication, parallel composition with the matrix tensor product and diagrammatic sum with the sum of matrices.

Definition A2. *For each classical system type, we have a family of spiders diagrammatically defined by, firstly:*

and secondly, that the symmetries of the representation as spiders are respected. Alternately, spiders can be defined via the matrix representation as:

$$\vcenter{\hbox{spider}} \;\; := \;\; \sum_i \vcenter{\hbox{$\triangledown^i \triangle_i \; \triangledown^i \triangle_i$}}$$

This family of maps is particularly important as, for classical theory at least, they allow us to define various concepts that we have used throughout the paper in a unified way. Firstly, the broadcasting map can now be seen as just an example spider with one input and two outputs, but moreover, we have:

$$\vcenter{\hbox{spider-cup}} = \vcenter{\hbox{cup}} \qquad \vcenter{\hbox{spider-cap}} = \vcenter{\hbox{cap}} \qquad \vcenter{\hbox{spider-dot}} = \vcenter{\hbox{ground}}$$

The feedback-loop we introduced can also now be interpreted as the composite of two spiders:

We moreover want to consider a way to join spiders of different dimensionality (denoted by using a different colour), which is exactly what the black/white dots achieve.

Definition A3. *Diagrammatically, the black/white dots are any process satisfying:*

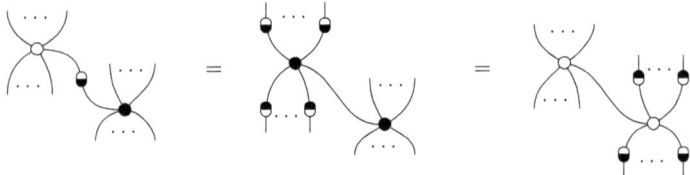

which is equivalent to how they were introduced in Example 11. Alternatively, their matrix representation is:

$$
\begin{array}{c} m \\ \bigg\downarrow \\ n \end{array} := \sum_{i=1}^{l} \begin{array}{c} \pi_1 \\ \nabla_i \\ \triangle_i \\ \pi_2 \end{array} \tag{A1}
$$

requiring that $l \leq \mathrm{Min}(n,m)$ *and* π_i *are arbitrary permutations of the basis elements. These are then just matrices with elements* $\{0,1\}$ *with at most a single one in each row and column.*

Appendix B. Dilations of Classical Processes

Any dilation of a classical process f can be written as:

$$\boxed{f} = \boxed{F}\!\!\!\!\overline{\overline{}} \;\; \Longrightarrow \;\; \boxed{F} = \boxed{f}\,l$$

to check this define l by its matrix elements as:

$$
l^{j}_{ik} := \begin{cases} 1 & \text{if } f^{k}_{i} = 0 \\ \dfrac{F^{kj}_{i}}{f^{k}_{i}} & \text{otherwise} \end{cases}
$$

and then, it is simple to check this satisfies the above equation and, moreover, is causal.

Appendix C. Pure Quantum-Classical Composite Processes

We need to prove that our definition of purity, i.e., Conditions (22) and (21), is satisfied by any process of the form:

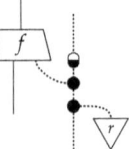

where:

$$\boxed{f_i} := \boxed{f}_{\nabla_i}$$

is pure for all i.

That (22) is satisfied is a straightforward proof once the following observation, easily verified by a straightforward calculation, is made:

$$\forall\, \boxed{l}\ \exists\, \boxed{\tilde{l}}$$

such that l and $\tilde{l}$ are both causal and:

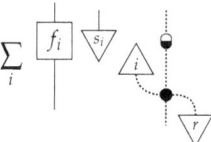

To check this, simply define $\tilde{l}$ as:

$$\boxed{\tilde{l}} := \boxed{l} + \sum_{j \in J} \nabla^s_j$$

where $J = \mathrm{Ker}\,[\phi]$.

That (21) is satisfied is also simple if, using the purity of the f_i, we can write the dilation as:

$$\sum_i \boxed{f_i}\,\nabla^{s_i}_i\,\cdots\nabla_r$$

and note that this can be written as a leak by defining:

$$\mathcal{V} := \boxed{l} \qquad \text{where} \qquad \boxed{l} := \sum_i \nabla^{s_i}_i$$

References

1. Barnum, H.; Caves, C.M.; Fuchs, C.A.; Jozsa, R.; Schumacher, B. Noncommuting mixed states cannot be broadcast. *Phys. Rev. Lett.* **1996**, *76*, 2818.
2. Barnum, H.; Barrett, J.; Leifer, M.; Wilce, A. A generalized no-broadcasting theorem. *Phys. Rev. Lett.* **2007**, *99*, 240501.
3. Coecke, B.; Kissinger, A. Categorical quantum mechanics I: Causal quantum processes. In *Categories for the Working Philosopher*; Landry, E., Ed.; Oxford University Press: Oxford, UK, 2016.
4. Popescu, S.; Rohrlich, D. Quantum nonlocality as an axiom. *Found. Phys.* **1994**, *24*, 379–385.
5. Abramsky, S.; Coecke, B. A categorical semantics of quantum protocols. In *Proceedings of the 19th Annual IEEE Symposium on Logic in Computer Science (LICS)*, Washington, DC, USA, 13–17 July 2004; pp. 415–425.

6. Coecke, B.; Kissinger, A. *Picturing Quantum Processes. A First Course in Quantum Theory and Diagrammatic Reasoning*; Cambridge University Press: Cambridge, UK, 2016.
7. Kuperberg, G. The capacity of hybrid quantum memory. *IEEE Trans. Inf. Theory* **2003**, *49*, 1465–1473.
8. Zurek, W.H. Quantum darwinism. *Nat. Phys.* **2009**, *5*, 181–188.
9. Coecke, B.; Selby, J.; Tull, S. Two roads to classicality. *arXiv* **2017**, arXiv:1701.07400.
10. Selinger, P. Idempotents in Dagger Categories (Extended Abstract). *Electron. Notes Theor. Comput. Sci.* **2008**, *210*, 107–122.
11. Heunen, C.; Kissinger, A.; Selinger, P. Completely positive projections and biproducts. In Proceedings of the 10th International Workshop on Quantum Physics and Logic, Barcelona, Spain, 17–19 July 2013.
12. Cunningham, O.; Heunen, C. Axiomatizing complete positivity. *arXiv* **2015**, arXiv:1506.02931.
13. Barrett, J. Information processing in generalized probabilistic theories. *Phys. Rev. A* **2007**, *75*, 032304.
14. Richens, J.; Selby, J.; Al-Safi, S. Entanglement is an inevitable feature of any non-classical theory. *arXiv* **2016**, arXiv:1610.00682.
15. Mac Lane, S. *Categories for the Working Mathematician*; Springer: Berlin/Heidelberg, Germany, 1998.
16. Chiribella, G.; D'Ariano, G.M.; Perinotti, P. Probabilistic theories with purification. *Phys. Rev. A* **2010**, *81*, 062348.
17. Coecke, B. Terminality implies non-signalling. *arXiv* **2014**, arXiv:1405.3681.
18. Coecke, B.; Kissinger, A. Categorical quantum mechanics II: Classical-quantum interaction. *arXiv* **2016**, arXiv:1605.08617.
19. Joyal, A.; Street, R.; Verity, D. Traced monoidal categories. In *Mathematical Proceedings of the Cambridge Philosophical Society*; Cambridge University Pressess: Cambridge, UK, 1996; Volume 119, pp. 447–468.
20. Stinespring, W.F. Positive functions on C*-algebras. *Proc. Am. Math. Soc.* **1955**, *6*, 211–216.
21. Bartlett, S.D.; Rudolph, T.; Spekkens, R.W. Reference frames, superselection rules, and quantum information. *Rev. Mod. Phys.* **2007**, *79*, 555.
22. Chiribella, G.; D'Ariano, G.M.; Perinotti, P. Quantum from principles. In *Quantum Theory: Informational Foundations and Foils*; Springer: Berlin/Heidelberg, Germany, 2016; pp. 171–221.
23. Lee, C.M.; Selby, J.H. A no-go theorem for post-quantum theories that decohere to quantum theory. *arXiv* **2017**, arXiv:1701.07449.
24. Selby, J.; Scandolo, C.M.; Coecke, B. Quantum theory from diagrammatic postulates. Forthcoming submitted.

© 2017 by the authors. Licensee MDPI, Basel, Switzerland. This article is an open access article distributed under the terms and conditions of the Creative Commons Attribution (CC BY) license (http://creativecommons.org/licenses/by/4.0/).

Article

Measurement Uncertainty Relations for Position and Momentum: Relative Entropy Formulation

Alberto Barchielli [1,2,3], Matteo Gregoratti [1,2,*] and Alessandro Toigo [1,3]

1   Dipartimento di Matematica, Politecnico di Milano, Piazza Leonardo da Vinci 32, I-20133 Milano, Italy; alberto.barchielli@polimi.it (A.B.); alessandro.toigo@polimi.it (A.T.)
2   Istituto Nazionale di Alta Matematica (INDAM-GNAMPA), 00185 Roma, Italy
3   Istituto Nazionale di Fisica Nucleare (INFN), Sezione di Milano, 20133 Milano, Italy
*   Correspondence: matteo.gregoratti@polimi.it; Tel.: +39-0223994569

Received: 26 May 2017; Accepted: 21 June 2017; Published: 24 June 2017

Abstract: Heisenberg's uncertainty principle has recently led to general measurement uncertainty relations for quantum systems: incompatible observables can be measured jointly or in sequence only with some unavoidable approximation, which can be quantified in various ways. The relative entropy is the natural theoretical quantifier of the information loss when a 'true' probability distribution is replaced by an approximating one. In this paper, we provide a lower bound for the amount of information that is lost by replacing the distributions of the sharp position and momentum observables, as they could be obtained with two separate experiments, by the marginals of any smeared joint measurement. The bound is obtained by introducing an entropic error function, and optimizing it over a suitable class of covariant approximate joint measurements. We fully exploit two cases of target observables: (1) n-dimensional position and momentum vectors; (2) two components of position and momentum along different directions. In (1), we connect the quantum bound to the dimension n; in (2), going from parallel to orthogonal directions, we show the transition from highly incompatible observables to compatible ones. For simplicity, we develop the theory only for Gaussian states and measurements.

Keywords: measurement uncertainty relations; relative entropy; position; momentum

PACS: 03.65.Ta, 03.65.Ca, 03.67.-a, 03.65.Db

MSC: 81P15, 81P16, 94A17, 81P45

1. Introduction

Uncertainty relations for position and momentum [1] have always been deeply related to the foundations of Quantum Mechanics. For several decades, their axiomatization has been of 'preparation' type: an inviolable lower bound for the widths of the position and momentum distributions, holding in any quantum state. Such kinds of uncertainty relations, which are now known as *preparation uncertainty relations* (PURs) have been later extended to arbitrary sets of $n \geq 2$ observables [2–5]. All PURs trace back to the celebrated Robertson's formulation [6] of Heisenberg's uncertainty principle: for any two observables, represented by self-adjoint operators A and B, the product of the variances of A and B is bounded from below by the expectation value of their commutator; in formulae, $\mathrm{Var}_\rho(A)\mathrm{Var}_\rho(B) \geq \frac{1}{4}|\mathrm{Tr}\{\rho[A,B]\}|^2$, where Var_ρ is the variance of an observable measured in any system state ρ. In the case of position Q and momentum P, this inequality gives Heisenberg's relation $\mathrm{Var}_\rho(Q)\mathrm{Var}_\rho(P) \geq \frac{\hbar^2}{4}$. About 30 years after Heisenberg and Robertson's formulation, Hirschman attempted a first statement of position and momentum uncertainties in terms of informational quantities. This led him to a formulation of PURs based on Shannon entropy [7]; his bound was later refined [8,9], and extended to discrete observables [10]. Also other entropic quantities have been used [11]. We refer to [12,13] for an extensive review on entropic PURs.

However, Heisenberg's original intent [1] was more focused on the unavoidable disturbance that a measurement of position produces on a subsequent measurement of momentum [14–21]. Trying to give a better understanding of his idea, more recently new formulations were introduced, based on a 'measurement' interpretation of uncertainty, rather than giving bounds on the probability distributions of the target observables. Indeed, with the modern development of the quantum theory of measurement and the introduction of *positive operator valued measures* and *instruments* [3,22–26], it became possible to deal with approximate measurements of incompatible observables and to formulate *measurement uncertainty relations* (MURs) for position and momentum, as well as for more general observables. The MURs quantify the degree of approximation (or inaccuracy and disturbance) made by replacing the original incompatible observables with a *joint approximate measurement* of them. A very rich literature on this topic flourished in the last 20 years, and various kinds of MURs have been proposed, based on distances between probability distributions, noise quantifications, conditional entropy, etc. [12,14–21,27–32].

In this paper, we develop a new information-theoretical formulation of MURs for position and momentum, using the notion of the *relative entropy* (or *Kullback-Leibler divergence*) of two probabilities. The relative entropy $S(p\|q)$ is an informational quantity which is precisely tailored to quantify the amount of information that is lost by using an approximating probability q in place of the target one p. Although classical and quantum relative entropies have already been used in the evaluation of the performances of quantum measurements [24,27,30,33–40], their first application to MURs is very recent [41].

In [41], only MURs for discrete observables were considered. The present work is a first attempt to extend that information-theoretical approach to the continuous setting. This extension is not trivial and reveals peculiar problems, that are not present in the discrete case. However, the nice properties of the relative entropy, such as its scale invariance, allow for a satisfactory formulation of the entropic MURs also for position and momentum.

We deal with position and momentum in two possible scenarios. Firstly, we consider the case of n-dimensional position and momentum, since it allows to treat either scalar particles, or vector ones, or even the case of multi-particle systems. This is the natural level of generality, and our treatment extends without difficulty to it. Then, we consider a couple made up of one position and one momentum component along two different directions of the n-space. In this case, we can see how our theory behaves when one moves with continuity from a highly incompatible case (parallel components) to a compatible case (orthogonal ones).

The continuous case needs much care when dealing with arbitrary quantum states and approximating observables. Indeed, it is difficult to evaluate or even bound the relative entropy if some assumption is not made on probability distributions. In order to overcome these technicalities and focus on the quantum content of MURs, in this paper we consider only the case of Gaussian preparation states and Gaussian measurement apparatuses [2,4,5,42–45]. Moreover, we identify the class of the approximate joint measurements with the class of the joint POVMs satisfying the same symmetry properties of their target position and momentum observables [3,23]. We are supported in this assumption by the fact that, in the discrete case [41], symmetry covariant measurements turn out to be the best approximations without any hypothesis (see also [17,19,20,29,32] for a similar appearance of covariance within MURs for different uncertainty measures).

We now sketch the main results of the paper. In the vector case, we consider approximate joint measurements M of the position $Q \equiv (Q_1, \ldots, Q_n)$ and the momentum $P \equiv (P_1, \ldots, P_n)$. We find the following entropic MUR (Theorem 5, Remark 14): for every choice of two positive thresholds

ϵ_1, ϵ_2, with $\epsilon_1 \epsilon_2 \geq \hbar^2/4$, there exists a Gaussian state ρ with position variance matrix $A^\rho \geq \epsilon_1 \mathbb{1}$ and momentum variance matrix $B^\rho \geq \epsilon_2 \mathbb{1}$ such that

$$S(Q_\rho \| M_{1,\rho}) + S(P_\rho \| M_{2,\rho}) \geq n \,(\log e) \left\{ \ln\left(1 + \frac{\hbar}{2\sqrt{\epsilon_1 \epsilon_2}}\right) - \frac{\hbar}{\hbar + 2\sqrt{\epsilon_1 \epsilon_2}} \right\} \quad (1)$$

for all Gaussian approximate joint measurements M of Q and P. Here Q_ρ and P_ρ are the distributions of position and momentum in the state ρ, and M_ρ is the distribution of M in the state ρ, with marginals $M_{1,\rho}$ and $M_{2,\rho}$; the two marginals turn out to be noisy versions of Q_ρ and P_ρ. The lower bound is strictly positive and it linearly increases with the dimension n. The thresholds ϵ_1 and ϵ_2 are peculiar of the continuous case and they have a classical explanation: the relative entropy $S(p\|q) \to +\infty$ if the variance of p vanishes faster than the variance of q, so that, given M, it is trivial to find a state ρ enjoying (1) if arbtrarily small variances are allowed. What is relevant in our result is that the total loss of information $S(Q_\rho \| M_{1,\rho}) + S(P_\rho \| M_{2,\rho})$ exceeds the lower bound even if we forbid target distributions with small variances.

The MUR (1) shows that there is no Gaussian joint measurement which can approximate arbitrarily well both Q and P. The lower bound (1) is a consequence of the incompatibility between Q and P and, indeed, it vanishes in the classical limit $\hbar \to 0$. Both the relative entropies and the lower bound in (1) are scale invariant. Moreover, for fixed ϵ_1 and ϵ_2, we prove the existence and uniqueness of an optimal approximate joint measurement, and we fully characterize it.

In the scalar case, we consider approximate joint measurements M of the position $Q_u = u \cdot Q$ along the direction u and the momentum $P_v = v \cdot P$ along the direction v, where $u \cdot v = \cos \alpha$. We find two different entropic MURs. The first entropic MUR in the scalar case is similar to the vector case (Theorem 3, Remark 11). The second one is (Theorem 1):

$$S(Q_{u,\rho} \| M_{1,\rho}) + S(P_{v,\rho} \| M_{2,\rho}) \geq c_\rho(\alpha), \quad (2)$$

$$c_\rho(\alpha) = (\log e) \left\{ \ln\left(1 + \frac{\hbar |\cos \alpha|}{2\sqrt{\mathrm{Var}\,(Q_{u,\rho})\,\mathrm{Var}\,(P_{v,\rho})}}\right) - \frac{\hbar |\cos \alpha|}{\hbar |\cos \alpha| + 2\sqrt{\mathrm{Var}\,(Q_{u,\rho})\,\mathrm{Var}\,(P_{v,\rho})}} \right\},$$

for all Gaussian states ρ and all Gaussian joint approximate measurements M of Q_u and P_v. This lower bound holds for every Gaussian state ρ without constraints on the position and momentum variances $\mathrm{Var}\,(Q_{u,\rho})$ and $\mathrm{Var}\,(P_{v,\rho})$, it is strictly positive unless u and v are orthogonal, but it is state dependent. Again, the relative entropies and the lower bound are scale invariant.

The paper is organized as follows. In Section 2, we introduce our target position and momentum observables, we discuss their general properties and define some related quantities (spectral measures, mean vectors and variance matrices, PURs for second order quantum moments, Weyl operators, Gaussian states). Section 3 is devoted to the definitions and main properties of the relative and differential (Shannon) entropies. Section 4 is a review on the entropic PURs in the continuous case [7–9,46], with a particular focus on their lack of scale invariance. This is a flaw due to the very definition of differential entropy, and one of the reasons that lead us to introduce relative entropy based MURs. In Section 5 we construct the covariant observables which will be used as approximate joint measurements of the position and momentum target observables. Finally, in Section 6 the main results on MURs that we sketched above are presented in detail. Some conclusions are discussed in Section 7.

2. Target Observables and States

Let us start with the usual position and momentum operators, which satisfy the canonical commutation rules:

$$Q \equiv (Q_1, \ldots, Q_n), \qquad P \equiv (P_1, \ldots, P_n), \qquad [Q_i, P_j] = i\hbar \delta_{ij}. \quad (3)$$

Each of the vector operators has n components; it could be the case of a single particle in one or more dimensions ($n = 1, 2, 3$), or several scalar or vector particles, or the quadratures of n modes of the electromagnetic field. We assume the Hilbert space $\mathcal{H}$ to be irreducible for the algebra generated by the canonical operators $\boldsymbol{Q}$ and $\boldsymbol{P}$. An *observable* of the quantum system $\mathcal{H}$ is identified with a *positive operator valued measure* (POVM); in the paper, we shall consider observables with outcomes in $\mathbb{R}^k$ endowed with its Borel σ-algebra $\mathcal{B}(\mathbb{R}^k)$. The use of POVMs to represent observables in quantum theory is standard and the definition can be found in many textbooks [22,23,26,47]; the alternative name "non-orthogonal resolutions of the identity" is also used [3–5]. Following [5,23,26,31], a *sharp observable* is an observable represented by a *projection valued measure* (pvm); it is standard to identify a sharp observable on the outcome space $\mathbb{R}^k$ with the k self-adjoint operators corresponding to it by spectral theorem. Two observables are *jointly measurable* or *compatible* if there exists a POVM having them as marginals. Because of the non-vanishing commutators, each couple Q_i, P_i, as well as the vectors $\boldsymbol{Q}$, $\boldsymbol{P}$, are not jointly measurable.

We denote by $\mathcal{T}(\mathcal{H})$ the trace class operators on $\mathcal{H}$, by $\mathcal{S} \subset \mathcal{T}(\mathcal{H})$ the subset of the statistical operators (or states, preparations), and by $\mathcal{L}(\mathcal{H})$ the space of the linear bounded operators.

2.1. Position and Momentum

Our target observables will be either n-dimensional position and momentum (*vector case*) or position and momentum along two different directions of $\mathbb{R}^n$ (*scalar case*). The second case allows to give an example ranging with continuity from maximally incompatible observables to compatible ones.

2.1.1. Vector Observables

As target observables we take $\boldsymbol{Q}$ and $\boldsymbol{P}$ as in (3) and we denote by $\mathsf{Q}(A), \mathsf{P}(B)$, $A, B \in \mathcal{B}(\mathbb{R}^n)$, their pvm's, that is

$$Q_i = \int_{\mathbb{R}^n} x_i \mathsf{Q}(\mathrm{d}\boldsymbol{x}), \qquad P_i = \int_{\mathbb{R}^n} p_i \mathsf{P}(\mathrm{d}\boldsymbol{p}). \qquad (4)$$

Then, the distributions in the state $\rho \in \mathcal{S}$ of a sharp position and a sharp momentum measurements (denoted by Q_ρ and P_ρ) are absolutely continuous with respect to the Lebesgue measure; we denote by $f(\bullet|\rho)$ and $g(\bullet|\rho)$ their probability densities: $\forall A, B \in \mathcal{B}(\mathbb{R}^n)$,

$$\mathsf{Q}_\rho(A) = \mathrm{Tr}\{\rho \mathsf{Q}(A)\} = \int_A f(\boldsymbol{x}|\rho) \mathrm{d}\boldsymbol{x}, \qquad \mathsf{P}_\rho(B) = \mathrm{Tr}\{\rho \mathsf{P}(B)\} = \int_B g(\boldsymbol{p}|\rho) \mathrm{d}\boldsymbol{p}. \qquad (5)$$

In the Dirac notation, if $|\boldsymbol{x}\rangle$ and $|\boldsymbol{p}\rangle$ are the improper position and momentum eigenvectors, these densities take the expressions $f(\boldsymbol{x}|\rho) = \langle \boldsymbol{x}|\rho|\boldsymbol{x}\rangle$ and $g(\boldsymbol{p}|\rho) = \langle \boldsymbol{p}|\rho|\boldsymbol{p}\rangle$, respectively. The mean vectors and the variance matrices of these distributions will be given in (7) and (8).

2.1.2. Scalar Observables

As target observables we take the position along a given direction $\boldsymbol{u}$ and the momentum along another given direction $\boldsymbol{v}$:

$$Q_u = \boldsymbol{u} \cdot \boldsymbol{Q}, \qquad P_v = \boldsymbol{v} \cdot \boldsymbol{P}, \qquad \text{with} \quad \boldsymbol{u}, \boldsymbol{v} \in \mathbb{R}^n, \quad |\boldsymbol{u}| = |\boldsymbol{v}| = 1, \quad \boldsymbol{u} \cdot \boldsymbol{v} = \cos \alpha. \qquad (6)$$

In this case we have $[Q_u, P_v] = i\hbar \cos \alpha$, so that Q_u and P_v are not jointly measurable, unless the directions $\boldsymbol{u}$ and $\boldsymbol{v}$ are orthogonal.

Their pvm's are denoted by Q_u and P_v, their distributions in a state ρ by $\mathsf{Q}_{u,\rho}$ and $\mathsf{P}_{v,\rho}$, and their corresponding probability densities by $f_u(\bullet|\rho)$ and $g_v(\bullet|\rho)$: $\forall A, B \in \mathcal{B}(\mathbb{R})$,

$$\mathsf{Q}_{u,\rho}(A) = \mathrm{Tr}\{\mathsf{Q}_u(A)\rho\} = \int_A f_u(x|\rho) \, \mathrm{d}x, \qquad \mathsf{P}_{v,\rho}(B) = \mathrm{Tr}\{\mathsf{P}_v(A)\rho\} = \int_B g_v(p|\rho) \, \mathrm{d}p.$$

Of course, the densities in the scalar case are marginals of the densities in the vector case. Means and variances will be given in (11).

2.2. Quantum Moments

Let $\mathcal{S}_2$ be the set of states for which the second moments of position and momentum are finite:

$$\mathcal{S}_2 := \left\{ \rho \in \mathcal{S} : \int_{\mathbb{R}^n} |x|^2 f(x|\rho) dx < +\infty, \int_{\mathbb{R}^n} |p|^2 g(p|\rho) dp < +\infty \right\}.$$

Then, the mean vector and the variance matrix of the position Q in the state $\rho \in \mathcal{S}_2$ are

$$a_i^\rho := \int_{\mathbb{R}^n} x_i f(x|\rho) dx \equiv \text{Tr}\{\rho Q_i\},$$

$$A_{ij}^\rho := \int_{\mathbb{R}^n} \left(x_i - a_i^\rho\right)\left(x_j - a_j^\rho\right) f(x|\rho) dx \equiv \text{Tr}\left\{\rho \left(Q_i - a_i^\rho\right)\left(Q_j - a_j^\rho\right)\right\}, \quad (7)$$

while for the momentum P we have

$$b_i^\rho := \int_{\mathbb{R}^n} p_i g(p|\rho) dp \equiv \text{Tr}\{\rho P_i\},$$

$$B_{ij}^\rho := \int_{\mathbb{R}^n} \left(p_i - b_i^\rho\right)\left(p_j - b_j^\rho\right) g(p|\rho) dp \equiv \text{Tr}\left\{\rho \left(P_i - b_i^\rho\right)\left(P_j - b_j^\rho\right)\right\}. \quad (8)$$

For $\rho \in \mathcal{S}_2$ it is possible to introduce also the mixed 'quantum covariances'

$$C_{ij}^\rho := \text{Tr}\left\{\rho \, \frac{(Q_i - a_i^\rho)(P_j - b_j^\rho) + (P_j - b_j^\rho)(Q_i - a_i^\rho)}{2}\right\}. \quad (9)$$

Since there is no joint measurement for the position Q and momentum P, the quantum covariances C_{ij}^ρ are not covariances of a joint distribution, and thus they do not have a classical probabilistic interpretation.

By means of the moments above, we construct the three real $n \times n$ matrices A^ρ, B^ρ, C^ρ, the $2n$-dimensional vector μ^ρ and the symmetric $2n \times 2n$ matrix V^ρ, with

$$\mu^\rho := \begin{pmatrix} a^\rho \\ b^\rho \end{pmatrix}, \quad V^\rho := \begin{pmatrix} A^\rho & C^\rho \\ (C^\rho)^T & B^\rho \end{pmatrix}. \quad (10)$$

We say V^ρ is the *quantum variance matrix* of position and momentum in the state ρ. In [2] dimensionless canonical operators are considered, but apart from this, our matrix V^ρ corresponds to their "noise matrix in real form"; the name "variance matrix" is also used [44,48].

In a similar way, we can introduce all the moments related to the position Q_u and momentum P_v introduced in (6). For $\rho \in \mathcal{S}_2$, the means and variances are respectively

$$u \cdot a^\rho, \quad \text{Var}(Q_{u,\rho}) = u \cdot A^\rho u, \quad v \cdot b^\rho, \quad \text{Var}(P_{v,\rho}) = v \cdot B^\rho v. \quad (11)$$

Similarly to (9), we have also the 'quantum covariance' $u \cdot C^\rho v \equiv v \cdot (C^\rho)^T u$. Then, we collect the two means in a single vector and we introduce the variance matrix:

$$\mu_{u,v}^\rho := \begin{pmatrix} u \cdot a^\rho \\ v \cdot b^\rho \end{pmatrix}, \quad V_{u,v}^\rho := \begin{pmatrix} u \cdot A^\rho u & u \cdot C^\rho v \\ u \cdot C^\rho v & v \cdot B^\rho v \end{pmatrix}. \quad (12)$$

Proposition 1. Let $V = \begin{pmatrix} A & C \\ C^T & B \end{pmatrix}$ be a real symmetric $2n \times 2n$ block matrix with the same dimensions of a quantum variance matrix. Define

$$V_\pm := \begin{pmatrix} A & C \pm i\frac{\hbar}{2}\mathbf{1} \\ C^T \mp i\frac{\hbar}{2}\mathbf{1} & B \end{pmatrix} \equiv V \pm \frac{i}{2}\Omega, \quad \text{with} \quad \Omega := \begin{pmatrix} 0 & \hbar\mathbf{1} \\ -\hbar\mathbf{1} & 0 \end{pmatrix}. \tag{13}$$

Then

$$V = V^\rho \text{ for some state } \rho \in \mathcal{S}_2 \iff V_+ \geq 0 \iff V_- \geq 0. \tag{14}$$

In this case we have: $V \geq 0$, $A > 0$, $B > 0$, and

$$(u' \cdot Au')(v' \cdot Bv') \geq (v' \cdot Cu')^2 + \frac{\hbar^2}{4}(v' \cdot u')^2, \quad \forall u' \in \mathbb{R}^n, \quad \forall v' \in \mathbb{R}^n. \tag{15}$$

The inequalities (14) for $V_\pm$ tell us exactly when a (positive semi-definite) real matrix V is the quantum variance matrix of position and momentum in a state ρ. Moreover, they are the multidimensional version of the usual uncertainty principle expressed through the variances [2,3,5], hence they represent a form of PURs. The block matrix Ω in the definition of $V_\pm$ is useful to compress formulae involving position and momentum; moreover, it makes simpler to compare our equations with their frequent dimensionless versions (with $\hbar = 1$) in the literature [43,44].

Proof. Equivalences (14) are well known, see e.g., [3] (Section 1.1.5), [5] (Equation (2.20)), and [2] (Theorem 2). Then $V = \frac{1}{2}V_+ + \frac{1}{2}V_- \geq 0$.

By using the real block vector $\begin{pmatrix} \alpha u' \\ \beta v' \end{pmatrix}$, with arbitrary $\alpha, \beta \in \mathbb{R}$ and given $u', v' \in \mathbb{R}^n$, the semi-positivity (14) implies

$$\begin{pmatrix} u' \cdot Au' & u' \cdot Cv' \pm i\frac{\hbar}{2}u' \cdot v' \\ v' \cdot C^T u' \mp i\frac{\hbar}{2}v' \cdot u' & v' \cdot Bv' \end{pmatrix} \geq 0, \quad \forall u' \in \mathbb{R}^n, \quad \forall v' \in \mathbb{R}^n,$$

which in turn implies $A \geq 0$, $B \geq 0$ and (15). Then, by choosing $u' = v' = u_i$, where $u_1, \ldots, u_n$ are the eigenvectors of A (since A is a real symmetric matrix, $u_i \in \mathbb{R}^n$ for all i), one gets the strict positivity of all the eigenvalues of A; analogously, one gets $B > 0$. □

Inequality (15) for $u' = u$ and $v' = v$ becomes the uncertainty rule à la Robertson [6] for the observables in (6) (a position component and a momentum component spanning an arbitrary angle α):

$$\text{Var}(Q_{u,\rho})\text{Var}(P_{v,\rho}) \geq (v \cdot C^\rho u)^2 + \frac{\hbar^2}{4}(\cos\alpha)^2. \tag{16}$$

Inequality (16) is equivalent to

$$V_{u,v}^\rho \pm \frac{i\hbar}{2}\cos\alpha \begin{pmatrix} 0 & 1 \\ -1 & 0 \end{pmatrix} \geq 0. \tag{17}$$

Since $V_\pm$ are block matrices, their positive semi-definiteness can be studied by means of the Schur complements [49–51]. However, as $V_\pm$ are complex block matrices with a very peculiar structure, special results hold for them. Before summarizing the properties of $V_\pm$ in the next proposition, we need a simple auxiliary algebraic lemma.

Lemma 1. Let A and B be complex self-adjoint matrices such that $A \geq B \geq 0$. Then $\det A \geq \det B \geq 0$, and the equality $\det A = \det B$ holds iff $A = B$.

Proof. Let $\lambda_i^\downarrow(A)$ and $\lambda_i^\downarrow(B)$ be the ordered decreasing sequences of the eigenvalues of A and B, respectively. Then, by Weyl's inequality, $A \geq B \geq 0$ implies $\lambda_i^\downarrow(A) \geq \lambda_i^\downarrow(B) \geq 0$ for every i [52] (Section III.2). This gives the first statement. Moreover, if $A \geq B \geq 0$ and $\det A = \det B$, we get $\lambda_i^\downarrow(A) = \lambda_i^\downarrow(B)$ for every i. Then $A = B$ because $A - B \geq 0$ and $\text{Tr}\{A - B\} = 0$. □

Proposition 2. Let $V = \begin{pmatrix} A & C \\ C^T & B \end{pmatrix}$ be a real symmetric $2n \times 2n$ matrix with the same dimensions of a quantum variance matrix. Then $V_+ \geq 0$ (or, equivalently, $V_- \geq 0$) if and only if $A > 0$ and

$$B \geq \left(C^T \mp \frac{i\hbar}{2}\mathbb{1}\right) A^{-1} \left(C \pm \frac{i\hbar}{2}\mathbb{1}\right) \equiv C^T A^{-1} C + \frac{\hbar^2}{4} A^{-1} \mp \frac{i\hbar}{2}\left(A^{-1}C - C^T A^{-1}\right). \tag{18}$$

In this case we have

$$B \geq C^T A^{-1} C + \frac{\hbar^2}{4} A^{-1} \geq \frac{\hbar^2}{4} A^{-1} > 0. \tag{19}$$

Moreover, we have also the following properties for the various determinants:

$$(\det A)(\det B) \geq \det V = (\det A)\det\left(B - C^T A^{-1} C\right) \geq \left(\frac{\hbar}{2}\right)^{2n}, \tag{20}$$

$$\det V = \left(\frac{\hbar}{2}\right)^{2n} \Leftrightarrow B = C^T A^{-1} C + \frac{\hbar^2}{4} A^{-1} \Rightarrow CA = AC^T, \tag{21}$$

$$(\det A)(\det B) = \left(\frac{\hbar}{2}\right)^{2n} \Leftrightarrow B = \frac{\hbar^2}{4} A^{-1}, \ C = 0. \tag{22}$$

By interchanging A with B and C with C^T in (18)–(22) equivalent results are obtained.

Proof. Since we already know that $V_+ \geq 0$ implies the invertibility of A, the equivalence between (14) and (18) with $A > 0$ follows from [49] (Theorem 1.12 p. 34) (see also [50] (Theorem 11.6) or [51] (Lemma 3.2)).

In (19), the first inequality follows by summing up the two inequalities in (18). The last two ones are immediate by the positivity of A^{-1}.

The equality in (20) is Schur's formula for the determinant of block matrices ([49], Theorem 1.1 p. 19). Then, the first inequality is immediate by the lemma above and the trivial relation $B \geq B - C^T A^{-1} C$; the second one follows from (19):

$$B - C^T A^{-1} C \geq \frac{\hbar^2}{4} A^{-1} \Rightarrow \det\left(B - C^T A^{-1} C\right) \geq \det\left(\frac{\hbar^2}{4} A^{-1}\right) = \frac{(\hbar/2)^{2n}}{\det A}.$$

The equality $\det V = \left(\frac{\hbar}{2}\right)^{2n}$ is equivalent to $\det\left(B - C^T A^{-1} C\right) = \det\left(\frac{\hbar^2}{4} A^{-1}\right)$; since the latter two determinants are evaluated on ordered positive matrices by (19), they coincide if and only if the respective arguments are equal (Lemma 1); this shows the equivalence in (21). Then, by (18), the self-adjoint matrix $\frac{i\hbar}{2}\left(A^{-1}C - C^T A^{-1}\right)$ is both positive semi-definite and negative semi-definite; hence it is null, that is, $CA = AC^T$.

Finally, $B = \frac{\hbar^2}{4} A^{-1}$ gives $(\det A)(\det B) = \left(\frac{\hbar}{2}\right)^{2n}$ trivially. Conversely, $(\det A)(\det B) = \left(\frac{\hbar}{2}\right)^{2n}$ implies $\det B = \det\left(B - C^T A^{-1} C\right)$ by (20); since $B \geq B - C^T A^{-1} C \geq 0$ by (19), Lemma 1 then implies $C^T A^{-1} C = 0$ and so $C = 0$. □

By (18) and (19), every time three matrices A, B, C define the quantum variance matrix of a state ρ, the same holds for A, B, $\tilde{C} = 0$. This fact can be used to characterize when two positive matrices

A and B are the diagonal blocks of some quantum variance matrix, or two positive numbers c_Q and c_P are the position and momentum variances of a quantum state along the two directions u and v.

Proposition 3. *Two real matrices $A > 0$ and $B > 0$, having the dimension of the square of a length and momentum, respectively, are the diagonal blocks of a quantum variance matrix V^ρ if and only if*

$$B \geq \frac{\hbar^2}{4} A^{-1}.$$

Two real numbers $c_Q > 0$ and $c_P > 0$, having the dimension of the square of a length and momentum, respectively, are such that $c_Q = \mathrm{Var}(Q_{u,\rho})$ and $c_P = \mathrm{Var}(P_{v,\rho})$ for some state ρ if and only if

$$c_Q c_P \geq \left(\frac{\hbar}{2} \cos\alpha\right)^2.$$

Proof. For A and B, the necessity follows from (19). The sufficiency comes from (18) by choosing $V^\rho = \begin{pmatrix} A & 0 \\ 0 & B \end{pmatrix}$.

For c_Q and c_P, the necessity follows from (15). The sufficiency comes from (18) with $V^\rho = \begin{pmatrix} A & 0 \\ 0 & B \end{pmatrix}$ and for example the following choices of A and B:

- if $\cos\alpha = \pm 1$, we take $A = c_Q \mathbb{1}$ and $B = c_P \mathbb{1}$;
- if $\cos\alpha = 0$, we let

$$A = c_Q uu^T + \frac{\hbar^2}{4c_P} vv^T + A' \qquad B = \frac{\hbar^2}{4c_Q} uu^T + c_P vv^T + B',$$

where A' and B' are any two scalar multiples of the orthogonal projection onto $\{u,v\}^\perp$ satisfying $B' \geq \frac{\hbar^2}{4} A'^{-1}$ when restricted to $\{u,v\}^\perp$;

- if $\cos\alpha \notin \{0,\pm 1\}$, we choose

$$A = c_Q \left[uu^T - \frac{1}{\cos\alpha}(uv^T + vu^T) + \frac{2}{(\cos\alpha)^2} vv^T \right] + A'$$

$$B = \frac{c_P}{(\sin\alpha)^4} \left[\frac{(\sin\alpha)^2 + (\cos\alpha)^4}{(\cos\alpha)^2} uu^T - \frac{1}{\cos\alpha}(uv^T + vu^T) + vv^T \right] + B',$$

where A' and B' are as in the previous item.

In the last two cases, we chose A and B in such a way that $B = \frac{c_Q c_P}{(\cos\alpha)^2} A^{-1}$ when restricted to the linear span of $\{u,v\}$. □

2.3. Weyl Operators and Gaussian States

In the following, we shall introduce Gaussian states, Gaussian observables and covariant observables on the phase-space. In all these instances, the Weyl operators are involved; here we recall their definition and some properties (see e.g., [4] (Section 5.2) or [5] (Section 12.2), where, however, the definition differs from ours in that the Weyl operators are composed with the map Ω^{-1} of (13)).

Definition 1. *The Weyl operators are the unitary operators defined by*

$$W(x,p) := \exp\left\{\frac{i}{\hbar}(p \cdot Q - x \cdot P)\right\} = \prod_{j=1}^n e^{\frac{i}{\hbar}(p_j Q_j - x_j P_j)} = \prod_{j=1}^n \left(e^{\frac{i}{\hbar} p_j Q_j} e^{-\frac{i}{\hbar} x_j P_j} e^{-\frac{i x_j p_j}{2\hbar}}\right). \quad (23)$$

The Weyl operators (23) satisfy the composition rule

$$W(x_1, p_1)W(x_2, p_2) = \exp\left\{-\frac{i}{2\hbar}(x_1 \cdot p_2 - x_2 \cdot p_1)\right\} W(x_1 + x_2, p_1 + p_2);$$

in particular, this implies the commutation relation

$$W(x_1, p_1)W(x_2, p_2) = \exp\left\{-i\begin{pmatrix}x_1^T & p_1^T\end{pmatrix}\Omega^{-1}\begin{pmatrix}x_2\\p_2\end{pmatrix}\right\} W(x_2, p_2)W(x_1, p_1). \tag{24}$$

These commutation relations imply the translation property

$$W(x,p)^* Q_i W(x,p) = Q_i + x_i, \qquad W(x,p)^* P_i W(x,p) = P_i + p_i, \qquad i = 1, \ldots, n; \tag{25}$$

due to this property, the Weyl operators are also known as *displacement operators*.
With a slight abuse of notation, we shall sometimes use the identification

$$W(x,p) \equiv W\left(\begin{pmatrix}x\\p\end{pmatrix}\right), \tag{26}$$

where $\begin{pmatrix}x\\p\end{pmatrix}$ is a block column vector belonging to the phase-space $\mathbb{R}^n \times \mathbb{R}^n \equiv \mathbb{R}^{2n}$; here, the first block x is a position and the second block p is a momentum.

By means of the Weyl operators, it is possible to define the characteristic function of any trace-class operator.

Definition 2. *For any operator $\rho \in \mathcal{T}(\mathcal{H})$, its characteristic function is the complex valued function $\hat{\rho}: \mathbb{R}^{2n} \to \mathbb{C}$ defined by*

$$\hat{\rho}(w) := \mathrm{Tr}\left\{\rho W(-\Omega w)\right\}, \qquad w \equiv \begin{pmatrix}k\\l\end{pmatrix}. \tag{27}$$

Note that k is the inverse of a length and l is the inverse of a momentum, so that w is a block vector living in the space $\mathbb{R}^{2n} \equiv \mathbb{R}^n \times \mathbb{R}^n$ regarded as the dual of the phase-space.

Instead of the characteristic function, sometimes the so called Weyl transform $\mathrm{Tr}\left\{W(x,p)\rho\right\}$ is introduced [4,44].

By [4] (Proposition 5.3.2, Theorem 5.3.3), we have $\hat{\rho}(w) \in L^2(\mathbb{R}^{2n})$ and the following trace formula holds: $\forall \rho, \sigma \in \mathcal{T}(\mathcal{H})$,

$$\mathrm{Tr}\{\sigma^* \rho\} = \left(\frac{\hbar}{2\pi}\right)^n \int_{\mathbb{R}^{2n}} \overline{\hat{\sigma}(w)}\, \hat{\rho}(w)\, dw. \tag{28}$$

As a corollary [4] (Corollary 5.3.4), we have that a state $\rho \in \mathcal{S}$ is pure if and only if

$$\left(\frac{\hbar}{2\pi}\right)^n \int_{\mathbb{R}^{2n}} |\hat{\rho}(w)|^2\, dw = 1.$$

By [53] (Lemma 3.1) or [26] (Proposition 8.5.(e)), the trace formula also implies

$$\frac{1}{(2\pi\hbar)^n}\int_{\mathbb{R}^{2n}} W(x,p)\rho W(x,p)^*\, dx dp = \mathrm{Tr}\{\rho\}\mathbb{1}, \qquad \forall \rho \in \mathcal{T}(\mathcal{H}). \tag{29}$$

Moreover, the following inversion formula ensures that the characteristic function $\hat{\rho}$ completely characterizes the state ρ [4] (Corollary 5.3.5):

$$\rho = \left(\frac{\hbar}{2\pi}\right)^n \int_{\mathbb{R}^{2n}} W(\Omega w)\, \hat{\rho}(w)\, dw, \qquad \forall \rho \in \mathcal{T}(\mathcal{H}).$$

The last two integrals are defined in the weak operator topology.

Finally, for $\rho \in \mathcal{S}_2$, the moments (7)–(10) can be expressed as in [4] (Section 5.4):

$$-i\, \frac{\partial \hat{\rho}(w)}{\partial w_i}\bigg|_0 = \mu_i^\rho, \qquad -\frac{\partial^2 \hat{\rho}(w)}{\partial w_i \partial w_j}\bigg|_0 = V_{ij}^\rho + \mu_i^\rho \mu_j^\rho. \tag{30}$$

Definition 3 ([2–5,44,48]). *A state ρ is Gaussian if*

$$\begin{aligned}
\hat{\rho}(w) &= \exp\left\{ iw^T \mu^\rho - \frac{1}{2} w^T V^\rho w \right\} \\
&= \exp\left\{ i(k \cdot a^\rho + l \cdot b^\rho) - \frac{1}{2}(k \cdot A^\rho k + l \cdot B^\rho l) - k \cdot C^\rho l \right\},
\end{aligned} \tag{31}$$

for a vector $\mu^\rho \in \mathbb{R}^{2n}$ and a real $2n \times 2n$ matrix V^ρ such that $V_+^\rho \geq 0$.

The condition $V_+^\rho \geq 0$ is necessary and sufficient in order that the function (31) defines the characteristic function of a quantum state [4] (Theorem 5.5.1), [5] (Theorem 12.17). Therefore, Gaussian states are exactly the states whose characteristic function is the exponential of a second order polynomial [4] (Equation (5.5.49)), [5] (Equation (12.80)).

We shall denote by $\mathcal{G}$ the set of the Gaussian states; we have $\mathcal{G} \subset \mathcal{S}_2 \subset \mathcal{S}$. By (30), the vectors a^ρ, b^ρ and the matrices A^ρ, B^ρ, C^ρ characterizing a Gaussian state ρ are just its first and second order quantum moments introduced in (7)–(9). By (31), the corresponding distributions of position and momentum are Gaussian, namely

$$Q_\rho = \mathcal{N}(a^\rho; A^\rho), \quad Q_{u,\rho} = \mathcal{N}(u \cdot a^\rho; u \cdot A^\rho u), \quad P_\rho = \mathcal{N}(b^\rho; B^\rho), \quad P_{v,\rho} = \mathcal{N}(v \cdot b^\rho; v \cdot B^\rho v). \tag{32}$$

Proposition 4 (Pure Gaussian states). *For $\rho \in \mathcal{G}$, we have $\det V^\rho = \left(\frac{\hbar}{2}\right)^{2n}$ if and only if ρ is pure.*

Proof. The trace formula (28) and (31) give $\mathrm{Tr}\{\rho^2\} = \frac{(\hbar/2)^n}{\sqrt{\det V^\rho}}$, and this implies the statement. □

Proposition 5 (Minimum uncertainty states). *For $\rho \in \mathcal{S}_2$, we have $(\det A^\rho)(\det B^\rho) = \left(\frac{\hbar}{2}\right)^{2n}$ if and only if ρ is a pure Gaussian state and it factorizes into the product of minimum uncertainty states up to a rotation of $\mathbb{R}^n$.*

Proof. If $(\det A^\rho)(\det B^\rho) = \left(\frac{\hbar}{2}\right)^{2n}$, then the equivalence (22) gives $B^\rho = \frac{\hbar^2}{4}(A^\rho)^{-1}$, so that the variance matrices A^ρ and B^ρ have a common eigenbasis $u_1, \ldots, u_n$. Thus, all the corresponding couples of position Q_{u_i} and momentum P_{u_i} have minimum uncertainties: $\mathrm{Var}(Q_{u_i})\, \mathrm{Var}(P_{u_i}) = \frac{\hbar^2}{4}$. Therefore, if we consider the factorization of the Hilbert space $\mathcal{H} = \mathcal{H}_1 \otimes \cdots \otimes \mathcal{H}_n$ corresponding to the basis $u_1, \ldots, u_n$, all the partial traces of the state ρ on each factor $\mathcal{H}_i$ are minimum uncertainty states. Since for $n = 1$ the minimum uncertainty states are pure and Gaussian, the state ρ is a pure product Gaussian state.

The converse is immediate. □

3. Relative and Differential Entropies

In this paper, we will be concerned with entropic quantities of classical type [54–56]. We express them in 'bits', that is we use the base-2 logarithms: $\log a \equiv \log_2 a$.

We deal only with probabilities on the measurable space $(\mathbb{R}^n, \mathcal{B}(\mathbb{R}^n))$ which admit densities with respect to the Lebesgue measure. So, we define the relative entropy and differential entropy only for such probabilities; moreover, we list only the general properties used in the following.

3.1. Relative Entropy or Kullback-Leibler Divergence

The fundamental quantity is the *relative entropy*, also called *information divergence, discrimination information, Kullback-Leibler divergence* or *information* or *distance* or *discrepancy*. The relative entropy of a probability p with respect to a probability q is defined for any couple of probabilities p, q on the same probability space.

Given two probabilities p and q on $(\mathbb{R}^n, \mathcal{B}(\mathbb{R}^n))$ with densities f and g, respectively, the *relative entropy* of p with respect to q is

$$S(p\|q) = \int_{\mathbb{R}^n} f(x) \log \frac{f(x)}{g(x)} \, dx. \tag{33}$$

The value $+\infty$ is allowed for $S(p\|q)$; the usual convention $0 \log(0/0) = 0$ is understood. The relative entropy (33) is the amount of information that is lost when q is used to approximate p [54] (p. 51). Of course, if x is dimensioned, then the densities f and g have the same dimension (that is, the inverse of x), and the argument of the logarithm is dimensionless, as it must be.

Proposition 6 ([56], Theorem 8.6.1). *The following properties hold.*

(i) $S(p\|q) \geq 0$.
(ii) $S(p\|q) = 0 \iff p = q \iff f = g$ a.e..
(iii) $S(p\|q)$ is invariant under a change of the unit of measurement.
(iv) If $p = \mathcal{N}(a; A)$ and $q = \mathcal{N}(b; B)$ with invertible variance matrices A and B, then

$$2 S(p\|q) = (\log e) \left\{ (a - b) \cdot B^{-1} (a - b) + \text{Tr} \left\{ B^{-1} A - \mathbb{1} \right\} \right\} + \log \frac{\det B}{\det A}. \tag{34}$$

As $S(p\|q)$ is scale invariant, it quantifies a relative error for the use of q as an approximation of p, not an absolute one.

Let us employ the relative entropy to evaluate the effect of an additive Gaussian noise $\nu \sim \mathcal{N}(b; \beta^2)$ on an independent Gaussian random variable X. If $X \sim \mathcal{N}(a; \alpha^2)$, then $X + \nu \sim \mathcal{N}(a + b; \alpha^2 + \beta^2)$, and the relative entropy of the true distribution of X with respect to its disturbed version $X + \nu$ is

$$S(X\|X+\nu) = \frac{\log e}{2} \frac{b^2 - \beta^2}{\alpha^2 + \beta^2} + \frac{1}{2} \log \frac{\alpha^2 + \beta^2}{\alpha^2}.$$

This expression vanishes if the noise becomes negligible with respect to the true distribution, that is if $\beta^2/\alpha^2 \to 0$ and $b^2/\alpha^2 \to 0$. On the other hand, $S(X\|X + \nu)$ diverges if the noise becomes too strong with respect to the true distribution, or, in other words, if the true distribution becomes too peaked with respect to the noise, that is, $\beta^2/\alpha^2 \to +\infty$ or $b^2/\alpha^2 \to +\infty$.

3.2. Differential Entropy

The *differential entropy* of an absolutely continuous random vector X with a probability density f is

$$H(X) := - \int_{\mathbb{R}^n} f(x) \log f(x) dx.$$

This quantity is commonly used in the literature, even if it lacks many of the nice properties of the Shannon entropy for discrete random variables. For example, $H(X)$ is not scale invariant, and it can be negative [56] (p. 244).

Since the density f enters in the logarithm argument, the definition of $H(X)$ is meaningful only when f is dimensionless, which is the same as X being dimensionless. Note that, if X is dimensioned and $c > 0$ is a real parameter making $\widetilde{X} = cX$ a dimensionless random variable, then

$$H(\widetilde{X}) = - \int_{\mathbb{R}^n} \frac{f(u/c)}{c^n} \log \frac{f(u/c)}{c^n} du = - \int_{\mathbb{R}^n} f(x) \log \frac{f(x)}{c^n} dx.$$

In the following, we shall consider the differential entropy only for dimensionless random vectors X.

Proposition 7 ([56], Section 8.6). *The following properties hold.*

(i) *If X is an absolutely continuous random vector with variance matrix A, then*

$$H(X) \leq \frac{1}{2} \log\left((2\pi e)^n \det A\right) = \frac{n}{2} \log(2\pi e) + \frac{1}{2} \operatorname{Tr} \log A.$$

The equality holds iff X is Gaussian with variance matrix A and arbitrary mean vector a.

(ii) *If $X = (X_1, \ldots, X_n)$ is an absolutely continuous random vector, then*

$$H(X) \leq \sum_{i=1}^{n} H(X_i).$$

The equality holds iff the components $X_1, \ldots, X_n$ are independent.

Remark 1. *In property (i) we have used the following well-known matrix identity, which follows by diagonalization:*

$$\log \det A = \operatorname{Tr} \log A, \quad \forall A > 0.$$

Remark 2. *Property (i) yields that the differential entropy of a Gaussian random variable $X \sim \mathcal{N}(a; \alpha^2)$ is*

$$H(X) = \frac{1}{2} \log\left(2\pi e \alpha^2\right),$$

which is an increasing function of the variance α^2, and thus it is a measure of the uncertainty of X. Note that $H(X) \geq 0$ iff $\alpha^2 \geq 1/(2\pi e)$.

4. Entropic PURs for Position and Momentum

The idea of having an entropic formulation of the PURs for position and momentum goes back to [7–9]. However, we have just seen that, due to the presence of the logarithm, the Shannon differential entropy needs dimensionless probability densities. So, this leads us to introduce dimensionless versions of position and momentum.

Let $\lambda > 0$ be a dimensionless parameter and $\varkappa$ a second parameter with the dimension of a mass times a frequency. Then, we introduce the dimensionless versions of position and momentum:

$$\widetilde{Q} := \sqrt{\frac{\varkappa}{\hbar}} Q, \quad \widetilde{P} = \frac{\lambda}{\sqrt{\hbar \varkappa}} P \quad \Rightarrow \quad \left[\widetilde{Q}_i, \widetilde{P}_j\right] = i\lambda \delta_{ij}. \tag{35}$$

We use a unique dimensional constant $\varkappa$, in order to respect rotation symmetry and do not distinguish different particles. Anyway, there is no natural link between the parameter multiplying Q and the parameter multiplying P; this is the reason for introducing λ. As we see from the commutation rules, the constant λ plays the role of a dimensionless version of $\hbar$; in the literature on PURs, often $\lambda = 1$ is used [8,9,12,46].

4.1. Vector Observables

Let $\tilde{\mathsf{Q}}$ and $\tilde{\mathsf{P}}$ be the pvm's of $\tilde{Q}$ and $\tilde{P}$; then, $\tilde{\mathsf{Q}}_\rho$ and $\tilde{\mathsf{P}}_\rho$ are their probability distributions in the state ρ. The total preparation uncertainty is quantified by the sum of the two differential entropies $H(\tilde{\mathsf{Q}}_\rho) + H(\tilde{\mathsf{P}}_\rho)$. For $\rho \in \mathcal{G}$, by Proposition 7 we get

$$H(\tilde{\mathsf{Q}}_\rho) + H(\tilde{\mathsf{P}}_\rho) = n \log(\pi e \lambda) + \frac{1}{2} \log\left[\left(\frac{4}{\hbar^2}\right)^n (\det A^\rho)(\det B^\rho)\right]. \tag{36}$$

In the case of product states of minimum uncertainty, we have $(\det A^\rho)(\det B^\rho) = \left(\hbar^2/4\right)^n$; then, by taking (20) into account, we get

$$\inf_{\rho \in \mathcal{G}} \left\{H(\tilde{\mathsf{Q}}_\rho) + H(\tilde{\mathsf{P}}_\rho)\right\} = n \log(\pi e \lambda). \tag{37}$$

Thus, the bound (37) arises from quantum relations between Q and P; indeed, there would be no lower bound for (36) if we could take both $\det A^\rho$ and $\det B^\rho$ arbitrarily small.

By item (ii) of Proposition 7, the differential entropy for the distribution of a random vector is smaller than the sum of the entropies of its marginals; however, the final bound (37) is a tight bound for both $H(\tilde{\mathsf{Q}}_\rho) + H(\tilde{\mathsf{P}}_\rho)$ and $\sum_{i=1}^n H(\tilde{\mathsf{Q}}_{i,\rho}) + \sum_{i=1}^n H(\tilde{\mathsf{P}}_{i,\rho})$.

By the results of [8,9], the same bound (37) is obtained even if the minimization is done over all the states, not only the Gaussian ones.

The uncertainty result (37) depends on λ, this being a consequence of the lack of scale invariance of the differential entropy; note that the bound is positive if and only if $\lambda > 1/(\pi e)$. Sometimes in the literature the parameter $\hbar$ appears in the argument of the logarithm [27,30]; this fact has to be interpreted as the appearance of a parameter with the numerical value of $\hbar$, but without dimensions. In this sense the formulation (37) is consistent with both the cases with $\lambda = 1$ or $\lambda = \hbar$. Sometimes the smaller bound $\ln 2\pi$ appears in place of $\log \pi e$ [10]; this is connected to a state dependent formulation of the entropic PUR [12] (Section V.B).

4.2. Scalar Observables

The dimensionless versions of the scalar observables introduced in (6) are

$$\tilde{Q}_u = \sqrt{\frac{\varkappa}{\hbar}} Q_u, \quad \tilde{P}_v = \frac{\lambda}{\sqrt{\hbar \varkappa}} P_v \quad \Rightarrow \quad \left[\tilde{Q}_u, \tilde{P}_v\right] = i\lambda \cos\alpha. \tag{38}$$

We denote by $\tilde{\mathsf{Q}}_{u,\rho}$ and $\tilde{\mathsf{P}}_{v,\rho}$ the associated distributions in the state ρ. For $\rho \in \mathcal{S}_2$, the respective means and variances are

$$\sqrt{\frac{\varkappa}{\hbar}} u \cdot a^\rho, \quad \frac{\lambda}{\sqrt{\hbar \varkappa}} v \cdot b^\rho, \quad \mathrm{Var}(\tilde{\mathsf{Q}}_{u,\rho}) = \frac{\varkappa}{\hbar} u \cdot A^\rho u, \quad \mathrm{Var}(\tilde{\mathsf{P}}_{v,\rho}) = \frac{\lambda^2}{\hbar \varkappa} v \cdot B^\rho v,$$

with $\sqrt{\mathrm{Var}(\tilde{\mathsf{Q}}_{u,\rho}) \mathrm{Var}(\tilde{\mathsf{P}}_{v,\rho})} \geq \lambda |\cos\alpha|/2$.

As in the vector case, the total preparation uncertainty is quantified by the sum of the two differential entropies $H(\tilde{\mathsf{Q}}_{u,\rho}) + H(\tilde{\mathsf{P}}_{v,\rho})$. For $\rho \in \mathcal{G}$, Proposition 7 gives

$$H(\tilde{\mathsf{Q}}_{u,\rho}) + H(\tilde{\mathsf{P}}_{v,\rho}) = \log\left(2\pi e \sqrt{\mathrm{Var}(\tilde{\mathsf{Q}}_{u,\rho}) \mathrm{Var}(\tilde{\mathsf{P}}_{v,\rho})}\right). \tag{39}$$

Then, we have the lower bound

$$\inf_{\rho \in \mathcal{G}} \left\{H(\tilde{\mathsf{Q}}_{u,\rho}) + H(\tilde{\mathsf{P}}_{v,\rho})\right\} = \log(\pi e \lambda |\cos\alpha|) = \frac{1 + \ln(\pi|\lambda \cos\alpha|)}{\ln 2}, \tag{40}$$

which depends on λ, but not on $\varkappa$. Of course, because of (39), for Gaussian states a lower bound for the sum $H(\widetilde{Q}_{u,\rho}) + H(\widetilde{P}_{v,\rho})$ is equivalent to a lower bound for the product $\mathrm{Var}(\widetilde{Q}_{u,\rho})\, \mathrm{Var}(\widetilde{P}_{v,\rho})$. By the generalization of the results of [8,9] given in [46], the bound (40) is obtained also when the minimization is done over all the states.

Let us note that the bound in (40) is positive for $|\lambda \cos \alpha| > 1/(\pi e)$, and it goes to $-\infty$ for $\alpha \to \pi/2$, which is the case of compatible $Q_{u,\rho}$ and $P_{v,\rho}$. In the case $\alpha = 0$, the bound (40) is the same as (37) for $n = 1$.

5. Approximate Joint Measurements of Position and Momentum

In order to deal with MURs for position and momentum observables, we have to introduce the class of approximate joint measurements of position and momentum, whose marginals we will compare with the respective sharp observables. As done in [3,4,18,57], it is natural to characterize such a class by requiring suitable properties of covariance under the group of space translations and velocity boosts: namely, by *approximate joint measurement of position and momentum* we will mean any POVM on the product space of the position and momentum outcomes sharing the same covariance properties of the two target sharp observables. As we have already discussed, two approximation problems will be of our concern: the approximation of the position and momentum vectors (vector case, with outcomes in the phase-space $\mathbb{R}^n \times \mathbb{R}^n$), and the approximation of one position and one momentum component along two arbitrary directions (scalar case, with oucomes in $\mathbb{R} \times \mathbb{R}$). In order to treat the two cases altogether, we consider POVMs with outcomes in $\mathbb{R}^m \times \mathbb{R}^m \equiv \mathbb{R}^{2m}$, which we call *bi-observables*; they correspond to a measurement of m position components and m momentum components. The specific covariance requirements will be given in the Definitions 5–7.

In studying the properties of probability measures on $\mathbb{R}^k$, a very useful notion is that of the characteristic function, that is, the Fourier cotransform of the measure at hand; the analogous quantity for POVMs turns out to have the same relevance. Different names have been used in the literature to refer to the characteristic function of POVMs, or, more generally, quantum instruments, such as characteristic operator or operator characteristic function [3,24,34,44,58–62]. As a variant, also the symplectic Fourier transform quite often appears [5] (Section 12.4.3). The characteristic function has been used, for instance, to study the quantum analogues of the infinite-divisible distributions [3,34,58–60,62] and measurements of Gaussian type [5,44,61]. Here, we are interested only in the latter application, as our approximating bi-observables will typically be Gaussian. Since we deal with bi-observables, we limit our definition of the characteristic function only to POVMs on $\mathbb{R}^m \times \mathbb{R}^m$, which have the same number of variables of position and momentum type.

Being measures, POVMs can be used to construct integrals, whose theory is presented e.g., in [26] (Section 4.8) and [4] (Section 2.9, Proposition 2.9.1).

Definition 4. *Given a bi-observable* $\mathsf{M} : \mathcal{B}(\mathbb{R}^{2m}) \to \mathcal{L}(\mathcal{H})$, *the characteristic function of* M *is the operator valued function* $\widehat{\mathsf{M}} : \mathbb{R}^{2m} \to \mathcal{L}(\mathcal{H})$, *with*

$$\widehat{\mathsf{M}}(k,l) = \int_{\mathbb{R}^{2m}} e^{i(k \cdot x + l \cdot p)} \mathsf{M}(dx dp). \tag{41}$$

In this definition the dimensions of the vector variables k and l are the inverses of a length and momentum, respectively, as in the definition of the characteristic function of a state (27). This definition is given so that $\mathrm{Tr}\left\{\widehat{\mathsf{M}}(k,l)\rho\right\}$ is the usual characteristic function of the probability distribution M_ρ on $\mathbb{R}^{2m}$.

5.1. Covariant Vector Observables

In terms of the pvm's (4), the translation property (25) is equivalent to the symmetry properties

$$W(x,p) Q(A) W(x,p)^* = Q(A+x), \qquad W(x,p) P(B) W(x,p)^* = P(B+p), \qquad \forall A, B \in \mathcal{B}(\mathbb{R}^n),$$

and they are taken as the transformation property defining the following class of POVMs on $\mathbb{R}^{2n}$ [23,26,44,53,57].

Definition 5. *A covariant phase-space observable is a bi-observable* $\mathsf{M} : \mathcal{B}(\mathbb{R}^{2n}) \to \mathcal{L}(\mathcal{H})$ *satisfying the covariance relation*

$$W(x,p)\mathsf{M}(Z)W(x,p)^* = \mathsf{M}\left(Z + \begin{pmatrix} x \\ p \end{pmatrix}\right), \qquad \forall Z \in \mathcal{B}(\mathbb{R}^{2n}), \quad \forall x, p \in \mathbb{R}^n.$$

We denote by $\mathcal{C}$ *the set of all the covariant phase-space observables.*

The interpretation of covariant phase-space observables as approximate joint measurements of position and momentum is based on the fact that their marginal POVMs

$$\mathsf{M}_1(A) = \mathsf{M}(A \times \mathbb{R}^n), \qquad \mathsf{M}_2(B) = \mathsf{M}(\mathbb{R}^n \times B), \qquad A, B \in \mathcal{B}(\mathbb{R}^n),$$

have the same symmetry properties of Q and P, respectively. Although Q and P are not jointly measurable, the following well-known result says that there are plenty of covariant phase-space observables [4] (Theorem 4.8.3), [63,64]. In (43) below, we use the parity operator Π on $\mathcal{H}$, which is such that

$$\Pi W(x,p)\Pi = W(-x,-p) = W(x,p)^*. \tag{42}$$

Proposition 8. *The covariant phase-space observables are in one-to-one correspondence with the states on* $\mathcal{H}$, *so that we have the identification* $\mathcal{S} \sim \mathcal{C}$; *such a correspondence* $\sigma \leftrightarrow \mathsf{M}^\sigma$ *is given by*

$$\mathsf{M}^\sigma(B) = \int_B \mathsf{M}^\sigma(x,p)\,dxdp, \qquad \forall B \in \mathcal{B}(\mathbb{R}^{2n}),$$
$$\mathsf{M}^\sigma(x,p) = \frac{1}{(2\pi\hbar)^n} W(x,p)\Pi\sigma\Pi W(x,p)^*. \tag{43}$$

The characteristic function (41) of a measurement $\mathsf{M}^\sigma \in \mathcal{C}$ has a very simple structure in terms of the characteristic function (27) of the corresponding state $\sigma \in \mathcal{S}$.

Proposition 9. *The characteristic function of* $\mathsf{M}^\sigma \in \mathcal{C}$ *is given by*

$$\widehat{\mathsf{M}}^\sigma(k,l) = W(-\Omega w)\,\hat{\sigma}(w), \qquad w \equiv \begin{pmatrix} k \\ l \end{pmatrix} \in \mathbb{R}^{2n}, \tag{44}$$

and the characteristic function of the probability M^σ_ρ *is*

$$\operatorname{Tr}\left\{\widehat{\mathsf{M}}^\sigma(k,l)\rho\right\} = \hat{\rho}(w)\hat{\sigma}(w). \tag{45}$$

In (44) we have used the identification (26). The characteristic function of a state is introduced in (27).

Proof. By the commutation relations (24) we have

$$W(-\hbar l, \hbar k)W(x,p)W(-\hbar l, \hbar k)^* = e^{i(k \cdot x + l \cdot p)}W(x,p).$$

Then, we get

$$\widehat{M^\sigma}(k,l) = \frac{1}{(2\pi\hbar)^n} \int_{\mathbb{R}^{2n}} e^{i(k\cdot x + l\cdot p)} W(x,p)\Pi\sigma\Pi W(x,p)^* dx dp$$

$$= \frac{1}{(2\pi\hbar)^n} \int_{\mathbb{R}^{2n}} W(-\hbar l, \hbar k) W(x,p) W(-\hbar l, \hbar k)^* \Pi\sigma\Pi W(x,p)^* dx dp$$

$$= W(-\hbar l, \hbar k) \operatorname{Tr}\{W(-\hbar l, \hbar k)^* \Pi\sigma\Pi\},$$

where we used the formula (29). By (42) and the definition (27), we get (44). Again by (27), we get (45). □

In terms of probability densities, measuring M^σ on the state ρ yields the density function $h^\sigma(x,p|\rho) = \operatorname{Tr}\{M^\sigma(x,p)\rho\}$. Then, by (45), the densities of the marginals $M_{1,\rho}^\sigma$ and $M_{2,\rho}^\sigma$ are the convolutions

$$h_1^\sigma(\bullet|\rho) = f(\bullet|\rho) * f(\bullet|\sigma), \qquad h_2^\sigma(\bullet|\rho) = g(\bullet|\rho) * g(\bullet|\sigma), \qquad (46)$$

where f and g are the sharp densities introduced in (5). By the arbitrariness of the state ρ, the marginal POVMs of M^σ turn out to be the convolutions (or 'smearings')

$$M_1^\sigma(A) \int_A dx \int_{\mathbb{R}^n} f(x-x'|\sigma)Q(dx'), \qquad M_2^\sigma(B) \int_B dp \int_{\mathbb{R}^n} g(p-p'|\sigma)P(dp'),$$

(see e.g., [23] (Section III, Equations (2.48) and (2.49))).

Let us remark that the distribution of the approximate position observable M_1^σ in a state ρ is the distribution of the sum of two independent random vectors: the first one is distributed as the sharp position Q in the state ρ, the second one is distributed as the sharp position Q in the state σ. In this sense, the approximate position M_1^σ looks like a sharp position plus an independent noise given by σ. Of course, a similar fact holds for the momentum. However, this statement about the distributions can not be extended to a statement involving the observables. Indeed, since Q and P are incompatible, nobody can jointly observe M^σ, Q and P, so that the convolutions (46) do not correspond to sums of random vectors that actually exist when measuring M^σ.

5.2. Covariant Scalar Observables

Now we focus on the class of approximate joint measurements of the observables Q_u and P_v representing position and momentum along two possibly different directions u and v (see Section 2.1.2). As in the case of covariant phase-space observables, this class is defined in terms of the symmetries of its elements: we require them to transform as if they were joint measurements of Q_u and P_v. Recall that Q_u and P_v denote the spectral measures of Q_u, P_v.

Due to the commutation relation (24), the following covariance relations hold

$$W(x,p)Q_u(A)W(x,p)^* = Q_u(A + u\cdot x), \qquad W(x,p)P_v(B)W(x,p)^* = P_v(B + v\cdot p),$$

for all $A, B \in \mathcal{B}(\mathbb{R})$ and $x, p \in \mathbb{R}^n$. We employ covariance to define our class of approximate joint measurements of Q_u and P_v.

Definition 6. *A (u,v)-covariant bi-observable is a POVM $M : \mathcal{B}(\mathbb{R}^2) \to \mathcal{L}(\mathcal{H})$ such that*

$$W(x,p)M(Z)W(x,p)^* = M\left(Z + \begin{pmatrix} u\cdot x \\ v\cdot p \end{pmatrix}\right), \qquad \forall Z \in \mathcal{B}(\mathbb{R}^2), \quad \forall x, p \in \mathbb{R}^n.$$

We denote by $\mathcal{C}_{u,v}$ the class of such bi-observables.

So, our approximate joint measurements of Q_u and P_v will be all the bi-observables in the class $\mathcal{C}_{u,v}$.

Example 1. *The marginal of a covariant phase-space observable M^σ along the directions $\boldsymbol{u}$ and $\boldsymbol{v}$ is a $(\boldsymbol{u}, \boldsymbol{v})$-covariant bi-observable. Actually, it can be proved that, if $\cos \alpha \neq 0$, all $(\boldsymbol{u}, \boldsymbol{v})$-covariant bi-observables can be obtained in this way.*

It is useful to work with a little more generality, and merge Definitions 5 and 6 into a single notion of covariance.

Definition 7. *Suppose J is a $k \times 2n$ real matrix. A POVM $\mathsf{M} : \mathcal{B}(\mathbb{R}^k) \to \mathcal{L}(\mathcal{H})$ is a J-covariant observable on $\mathbb{R}^k$ if*

$$W(x,p) \mathsf{M}(Z) W(x,p)^* = \mathsf{M}\left(Z + J \begin{pmatrix} x \\ p \end{pmatrix}\right), \quad \forall Z \in \mathcal{B}(\mathbb{R}^k), \quad \forall x, p \in \mathbb{R}^n.$$

Thus, approximate joint observables of Q_u and P_v are just J-covariant observables on $\mathbb{R}^2$ for the choice of the $2 \times 2n$ matrix

$$J = \begin{pmatrix} u^T & 0^T \\ 0^T & v^T \end{pmatrix}. \tag{47}$$

On the other hand, covariant phase-space observables constitute the class of $\mathbb{1}_{2n}$-covariant observables on $\mathbb{R}^{2n}$, where $\mathbb{1}_{2n}$ is the identity map of $\mathbb{R}^{2n}$.

5.3. Gaussian Measurements

When dealing with Gaussian states, the following class of bi-observables quite naturally arises.

Definition 8. *A POVM $\mathsf{M} : \mathcal{B}(\mathbb{R}^{2m}) \to \mathcal{L}(\mathcal{H})$ is a Gaussian bi-observable if*

$$\widehat{\mathsf{M}}(k,l) = W\left(-\Omega (J^\mathsf{M})^T \begin{pmatrix} k \\ l \end{pmatrix}\right) \exp\left\{i \begin{pmatrix} k^T & l^T \end{pmatrix} \begin{pmatrix} a^\mathsf{M} \\ b^\mathsf{M} \end{pmatrix} - \frac{1}{2} \begin{pmatrix} k^T & l^T \end{pmatrix} V^\mathsf{M} \begin{pmatrix} k \\ l \end{pmatrix}\right\} \tag{48}$$

for two vectors $a^\mathsf{M}, b^\mathsf{M} \in \mathbb{R}^m$, a real $2m \times 2n$ matrix J^M and a real symmetric $2m \times 2m$ matrix V^M satisfying the condition

$$V^\mathsf{M} \pm \frac{i}{2} J^\mathsf{M} \Omega (J^\mathsf{M})^T \geq 0. \tag{49}$$

*We set $\mu^\mathsf{M} = \begin{pmatrix} a^\mathsf{M} \\ b^\mathsf{M} \end{pmatrix}$. The triple $(\mu^\mathsf{M}, V^\mathsf{M}, J^\mathsf{M})$ is the set of the **parameters** of the Gaussian observable M.*

In this definition, the vector a^M has the dimension of a length, and b^M of a momentum; similarly, the matrices $J^\mathsf{M}, V^\mathsf{M}$ decompose into blocks of different dimensions. The condition (49) is necessary and sufficient in order that the function (48) defines the characteristic function of a POVM.

For unbiased Gaussian measurements, i.e., Gaussian bi-observables with $a^\mathsf{M} = b^\mathsf{M} = 0$, the previous definition coincides with the one of [5] (Section 12.4.3). It is also a particular case of the more general definition of Gaussian observables on arbitrary (not necessarily symplectic) linear spaces that is given in [43,44]. We refer to [5,44] for the proof that Equation (48) is actually the characteristic function of a POVM.

Measuring the Gaussian observable M on the Gaussian state ρ yields the probability distribution M_ρ whose characteristic function is

$$\mathrm{Tr}\{\widehat{M}(k,l)\rho\} = \widehat{\rho}\left((J^M)^T\begin{pmatrix}k\\l\end{pmatrix}\right)\exp\left\{i\begin{pmatrix}k^T & l^T\end{pmatrix}\begin{pmatrix}a^M\\b^M\end{pmatrix} - \frac{1}{2}\begin{pmatrix}k^T & l^T\end{pmatrix}V^M\begin{pmatrix}k\\l\end{pmatrix}\right\}$$

$$= \exp\left\{i\begin{pmatrix}k^T & l^T\end{pmatrix}\left[\begin{pmatrix}a^M\\b^M\end{pmatrix} + J^M\begin{pmatrix}a^\rho\\b^\rho\end{pmatrix}\right] - \frac{1}{2}\begin{pmatrix}k^T & l^T\end{pmatrix}\left[V^M + J^M V^\rho (J^M)^T\right]\begin{pmatrix}k\\l\end{pmatrix}\right\};$$

hence the output distribution is Gaussian,

$$M_\rho = \mathcal{N}\left(J^M \mu^\rho + \mu^M;\, J^M V^\rho (J^M)^T + V^M\right). \tag{50}$$

5.3.1. Covariant Gaussian Observables

For Gaussian bi-observables, J-covariance has a very easy characterization.

Proposition 10. *Suppose M is a Gaussian bi-observable on $\mathbb{R}^{2m}$ with parameters (μ^M, V^M, J^M). Let J be any $2m \times 2n$ real matrix. Then, the POVM M is a J-covariant observable if and only if $J^M = J$.*

Proof. For $x, p \in \mathbb{R}^n$, we let M' and M'' be the two POVMs on $\mathbb{R}^{2m}$ given by

$$M'(Z) = W(x,p) M(Z) W(x,p)^*, \qquad M''(Z) = M\left(Z + J\begin{pmatrix}x\\p\end{pmatrix}\right), \qquad \forall Z \in \mathcal{B}(\mathbb{R}^{2m}).$$

By the commutation relations (24) for the Weyl operators, we immediately get

$$\widehat{M'}(k,l) = W(x,p)\widehat{M}(k,l)W(x,p)^* = \exp\left\{-i\begin{pmatrix}x^T & p^T\end{pmatrix}\Omega^{-1}\left[-\Omega(J^M)^T\begin{pmatrix}k\\l\end{pmatrix}\right]\right\}\widehat{M}(k,l)$$

$$= \exp\left\{-i\begin{pmatrix}k^T & l^T\end{pmatrix}J^M\begin{pmatrix}x\\p\end{pmatrix}\right\}\widehat{M}(k,l);$$

we have also

$$\widehat{M''}(k,l) = \int_{\mathbb{R}^{2m}}\exp\left\{i\begin{pmatrix}k^T & l^T\end{pmatrix}\left[\begin{pmatrix}x'\\p'\end{pmatrix} - J\begin{pmatrix}x\\p\end{pmatrix}\right]\right\}M(dx'dp')$$

$$= \exp\left\{-i\begin{pmatrix}k^T & l^T\end{pmatrix}J\begin{pmatrix}x\\p\end{pmatrix}\right\}\widehat{M}(k,l).$$

Since $\widehat{M}(k,l) \neq 0$ for all k, l, by comparing the last two expressions we see that $M' = M''$ if and only if

$$\exp\left\{-i\begin{pmatrix}k^T & l^T\end{pmatrix}J^M\begin{pmatrix}x\\p\end{pmatrix}\right\} = \exp\left\{-i\begin{pmatrix}k^T & l^T\end{pmatrix}J\begin{pmatrix}x\\p\end{pmatrix}\right\}, \qquad \forall x, p \in \mathbb{R}^n, \quad \forall k, l \in \mathbb{R}^m,$$

which in turn is equivalent to $J^M = J$. □

Vector Observables

Let us point out the structure of the Gaussian approximate joint measurements of Q and P.

Proposition 11. A bi-observable $M^\sigma \in \mathcal{C}$ is Gaussian if and only if the state σ is Gaussian. In this case, the covariant bi-observable M^σ is Gaussian with parameters

$$\mu^{M^\sigma} = \mu^\sigma, \qquad V^{M^\sigma} = V^\sigma, \qquad J^{M^\sigma} = \mathbb{1}_{2n}.$$

Proof. By comparing (31), (44) and (48), and using the fact that $W(x_1, p_1) \propto W(x_2, p_2)$ if and only if $x_1 = x_2$ and $p_1 = p_2$, we have the first statement. Then, for $\sigma \in \mathcal{G}$, we see immediately that M^σ is a Gaussian observable with the above parameters. □

We call $\mathcal{C}^G$ the class of the Gaussian covariant phase-space observables. By (50), observing M^σ on a Gaussian state $\rho \in \mathcal{G}$ yields the normal probability distribution $M^\sigma_\rho = \mathcal{N}(\mu^\rho + \mu^\sigma; V^\rho + V^\sigma)$, with marginals

$$M^\sigma_{1,\rho} = \mathcal{N}(a^\rho + a^\sigma; A^\rho + A^\sigma), \qquad M^\sigma_{2,\rho} = \mathcal{N}(b^\rho + b^\sigma; B^\rho + B^\sigma). \tag{51}$$

When $a^\sigma = 0$ and $b^\sigma = 0$, we have an *unbiased measurement*.

Scalar Observables

We now study the Gaussian approximate joint measurements of the target observables Q_u and P_u defined in (6).

Proposition 12. A Gaussian bi-observable M with parameters (μ^M, V^M, J^M) is in $\mathcal{C}_{u,v}$ if and only if $J^M = J$, where J is given by (47). In this case, the condition (49) is equivalent to

$$V^M_{11} \geq 0, \qquad V^M_{22} \geq 0, \qquad V^M_{11} V^M_{22} \geq \frac{\hbar^2}{4}(\cos \alpha)^2 + (V^M_{12})^2. \tag{52}$$

Proof. The first statement follows from Proposition 10. Then, the matrix inequality (49) reads

$$V^M \pm \frac{i\hbar}{2} \begin{pmatrix} 0 & \cos \alpha \\ -\cos \alpha & 0 \end{pmatrix} \geq 0,$$

which is equivalent to (52). □

We write $\mathcal{C}^G_{u,v}$ for the class of the Gaussian (u,v)-covariant phase-space observables. An observable $M \in \mathcal{C}^G_{u,v}$ is thus characterized by the couple (μ^M, V^M). From (50) with $J^M = J$ given by (47), we get that measuring $M \in \mathcal{C}^G_{u,v}$ on a Gaussian state ρ yields the probability distribution $M_\rho = \mathcal{N}\left(\mu^\rho_{u,v} + \mu^M; V^\rho_{u,v} + V^M\right)$ with $\mu^\rho_{u,v}$ and $V^\rho_{u,v}$ given by (12). Its marginals with respect to the first and second entry are, respectively,

$$M_{1,\rho} = \mathcal{N}\left(u \cdot a^\rho + a^M; \operatorname{Var}(Q_{u,\rho}) + V^M_{11}\right), \qquad M_{2,\rho} = \mathcal{N}\left(v \cdot b^\rho + b^M; \operatorname{Var}(P_{v,\rho}) + V^M_{22}\right). \tag{53}$$

Example 2. Let us construct an example of an approximate joint measurement of Q_u and P_v, by using a noisy measurement of position along u followed by a sharp measurement of momentum along v. Let Δ be a positive real number yielding the precision of the position measurement, and consider the POVM M on $\mathbb{R}^2$ given by

$$M(A \times B) = \frac{1}{\sqrt{2\pi\Delta}} \int_A \exp\left\{-\frac{(x - Q_u)^2}{4\Delta}\right\} P_v(B) \exp\left\{-\frac{(x - Q_u)^2}{4\Delta}\right\} dx, \qquad \forall A, B \in \mathcal{B}(\mathbb{R}).$$

The characteristic function of M *is*

$$\widehat{\mathsf{M}}(k,l) = \frac{1}{\sqrt{2\pi\Delta}} \int_{\mathbb{R}} e^{ikx} \exp\left\{-\frac{(x-Q_u)^2}{4\Delta}\right\} \left[\int_{\mathbb{R}} e^{ilp} P_v(\mathrm{d}p)\right] \exp\left\{-\frac{(x-Q_u)^2}{4\Delta}\right\} \mathrm{d}x$$

$$= \frac{1}{\sqrt{2\pi\Delta}} \int_{\mathbb{R}} \exp\left\{ikx - \frac{(x-Q_u)^2}{4\Delta}\right\} e^{ilP_v} \exp\left\{-\frac{(x-Q_u)^2}{4\Delta}\right\} \mathrm{d}x$$

$$= \frac{e^{ilP_v}}{\sqrt{2\pi\Delta}} \int_{\mathbb{R}} \exp\left\{ikx - \frac{(x-Q_u+\hbar l u \cdot v)^2}{4\Delta}\right\} \exp\left\{-\frac{(x-Q_u)^2}{4\Delta}\right\} \mathrm{d}x$$

$$= \frac{1}{\sqrt{2\pi\Delta}} \exp\left\{ilP_v - \frac{(\hbar l \cos\alpha)^2}{8\Delta}\right\} \int_{\mathbb{R}} \exp\left\{ikx - \frac{(x-Q_u+\hbar l \cos\alpha/2)^2}{2\Delta}\right\} \mathrm{d}x$$

$$= \exp\left\{ilP_v + ik\left(Q_u + \frac{\hbar l \cos\alpha}{2}\right) - \frac{\Delta}{2}k^2 - \frac{(\hbar \cos\alpha)^2}{8\Delta}l^2\right\}$$

$$= W(-\hbar l v, \hbar k u) \exp\left\{-\frac{\Delta}{2}k^2 - \frac{(\hbar \cos\alpha)^2}{8\Delta}l^2\right\}.$$

Therefore, M *is a Gaussian bi-observable with parameters* $a^M = 0$, $b^M = 0$ *and* $J^M = J$, *where* J *is given by* (47) *and* $V^M_{11} = \Delta$, $V^M_{22} = \frac{(\hbar \cos\alpha)^2}{4\Delta}$ *and* $V^M_{12} = 0$. *This implies* $\mathsf{M} \in \mathcal{C}^G_{u,v}$; *in particular, the set* $\mathcal{C}^G_{u,v}$ *is non-empty. Moreover, the lower bound* $V^M_{11} V^M_{22} = \frac{\hbar^2}{4}(\cos\alpha)^2$ *is attained, cf.* (52).

Example 3. *Let us consider the case* $\alpha = \pm\pi/2$; *now the target observables* Q_u *and* P_v *are compatible and we can define a pvm* M *on* $\mathbb{R}^2$ *by setting* $\mathsf{M}(A \times B) = Q_u(A)P_v(B)$ *for all* $A, B \in \mathcal{B}(\mathbb{R})$. *Its characteristic function is*

$$\widehat{\mathsf{M}}(k,l) = \int_{\mathbb{R}} e^{ikx} Q_u(\mathrm{d}x) \int_{\mathbb{R}} e^{ilp} P_v(\mathrm{d}p) = e^{i(kQ_u + lP_v)} = W(-\hbar l v, \hbar k u).$$

Then, $\mathsf{M} \in \mathcal{C}^G_{u,v}$ *with parameters* $a^M = 0$, $b^M = 0$, $V^M = 0$ *and* $J^M = J$ *given by* (47). *Note that* M *can be regarded as the limit case of the observables of the previous example when* $\cos\alpha = 0$ *and* $\Delta \downarrow 0$.

6. Entropic MURs for Position and Momentum

In the case of two discrete target observables, in [41] we found an entropic bound for the precision of their approximate joint measurements, which we named *entropic incompatibility degree*. Its definition followed a three steps procedure. Firstly, we introduced an *error function*: when the system is in a given state ρ, such a function quantifies the total amount of information that is lost by approximating the target observables by means of the marginals of a bi-observable; the error function is nothing else than the sum of the two relative entropies of the respective distributions. Then, we considered the worst possible case by maximizing the error function over ρ, thus obtaining an *entropic divergence* quantifying the approximation error in a state independent way. Finally, we got our index of the incompatibility of the two target observables by minimizing the entropic divergence over all bi-observables. In particular, when symmetries are present, we showed that the minimum is attained at some covariant bi-observables. So, the covariance followed as a byproduct of the optimization procedure, and was not a priori imposed upon the class of approximating bi-observables.

As we shall see, the extension of the previous procedure to position and momentum target observables is not straightforward, and peculiar problems of the continuous case arise. In order to overcome them, in this paper we shall fully analyse only a case in which explicit computations can be done: Gaussian preparations, and Gaussian bi-observables, which we a priori assume to be covariant. We conjecture that the final result should be independent of these simplifications, as we shall discuss in Section 7.

As we said in Section 5, by "approximate joint measurement" we mean "a bi-observable with the 'right' covariance properties".

6.1. Scalar Observables

Given the directions u and v, the target observables are Q_u and P_v in (6) with pvm's Q_u and P_v. For $\rho \in \mathcal{G}$ with parameters (μ^ρ, V^ρ) given in (10), the target distributions $Q_{u,\rho}$ and $P_{v,\rho}$ are normal with means and variances (11).

An approximate joint measurements of Q_u and P_v is given by a covariant bi-observable $M \in \mathcal{C}_{u,v}$; then, we denote its marginals with respect to the first and second entry by M_1 and M_2, respectively. For a Gaussian covariant bi-observable $M \in \mathcal{C}^G_{u,v}$ with parameters (μ^M, V^M), the distribution of M in a Gaussian state ρ is normal,

$$M_\rho = \mathcal{N}\left(\mu^\rho_{u,v} + \mu^M; V^\rho_{u,v} + V^M\right),$$

so that its marginal distributions $M_{1,\rho}$ and $M_{2,\rho}$ are normal with means $u \cdot a^\rho + a^M$ and $v \cdot b^\rho + b^M$ and variances

$$\mathrm{Var}\,(M_{1,\rho}) = \mathrm{Var}\,(Q_{u,\rho}) + V^M_{11}, \qquad \mathrm{Var}\,(M_{2,\rho}) = \mathrm{Var}\,(P_{v,\rho}) + V^M_{22}. \tag{54}$$

Let us recall that $|u|=1$, $|v|=1$, $u \cdot v = \cos \alpha$, and that by (16) and (52), we have

$$\mathrm{Var}\,(Q_{u,\rho})\,\mathrm{Var}\,(P_{v,\rho}) \geq \frac{\hbar^2}{4}(\cos \alpha)^2, \qquad V^M_{11} V^M_{22} \geq \frac{\hbar^2}{4}(\cos \alpha)^2. \tag{55}$$

6.1.1. Error Function

The relative entropy is the amount of information that is lost when an approximating distribution is used in place of a target one. For this reason, we use it to give an informational quantification of the error made in approximating the distributions of sharp position and momentum by means of the marginals of a joint covariant observable.

Definition 9. *Given the preparation $\rho \in \mathcal{S}$ and the covariant bi-observable $M \in \mathcal{C}_{u,v}$, the error function for the scalar case is the sum of the two relative entropies:*

$$S(\rho, M) := S(Q_{u,\rho} \| M_{1,\rho}) + S(P_{v,\rho} \| M_{2,\rho}). \tag{56}$$

The relative entropy is invariant under a change of the unit of measurement, so that the error function is scale invariant, too; indeed, it quantifies a relative error, not an absolute one. In the Gaussian case the error function can be explicitly computed.

Proposition 13 (Error function for the scalar Gaussian case). *For $\rho \in \mathcal{G}$ and $M \in \mathcal{C}^G_{u,v}$, the error function is*

$$S(\rho, M) = \frac{\log e}{2}\left[s(x) + s(y) + \Delta(\rho, M)\right], \tag{57}$$

where

$$x := \frac{V^M_{11}}{\mathrm{Var}\,(Q_{u,\rho})}, \qquad y := \frac{V^M_{22}}{\mathrm{Var}\,(P_{v,\rho})}, \qquad \Delta(\rho, M) := \frac{(a^M)^2}{\mathrm{Var}\,(M_{1,\rho})} + \frac{(b^M)^2}{\mathrm{Var}\,(M_{2,\rho})},$$

and $s : [0, +\infty) \to [0, +\infty)$ is the following C^∞ strictly increasing function with $s(0) = 0$:

$$s(x) := \ln(1+x) - \frac{x}{1+x}. \tag{58}$$

Proof. The statement follows by a straightforward combination of (32), (34), (53) and (56). □

Note that the error function does not depend on the mixed covariances $u \cdot C^\rho v$ and V^M_{12}. Note also that, if we select a possible approximation M, then the error function $S(\rho, M)$ decreases for states ρ

with increasing sharp variances Var $(Q_{u,\rho})$ and Var $(P_{v,\rho})$: the loss of information decreases when the sharp distributions make the approximation error negligible. Finally, note that

$$s(x) + s(y) = \ln[(1+x)(1+y)] + (1+x)^{-1} + (1+y)^{-1} - 2,$$

$$1 + x = \frac{\text{Var}(M_{1,\rho})}{\text{Var}(Q_{u,\rho})}, \qquad 1 + y = \frac{\text{Var}(M_{2,\rho})}{\text{Var}(P_{v,\rho})}.$$

This means that, apart from the term $\Delta(\rho, M)$ due to the bias, our error function $S(\rho, M)$ only depends on the two ratios "variance of the approximating distribution over variance of the target distribution". Thus, in order to optimize the error function, one has to optimize these two ratios.

We use formula (57) to firstly give a state dependent MUR, and then, following the scheme of [41], a state independent MUR.

A lower bound for the error function can be found by minimizing it over all possible approximate joint measurements of Q_u and P_v. First of all, let us remark that this minimization makes sense because we consider only (u,v)-covariant bi-observables: if we minimized over all possible bi-observables, then the minimum would be trivially zero for every given preparation ρ. Indeed, the trivial bi-observable $M(A \times B) = Q_{u,\rho}(A)P_{v,\rho}(B) \mathbb{1}$ yields $S(\rho, M) = 0$.

When minimizing the error function over all (u,v)-covariant bi-observables, both the minimum and the best measurement attaining it are state dependent. When $\alpha = \pm \pi/2$, the two target observables are compatible, so that their joint measurement trivially exists (see Example 3) and we get $\inf_{M \in \mathcal{C}_{u,v}} S(\rho, M) = 0$. In order to have explicit results for any angle α, we consider only the Gaussian case.

Theorem 1 (State dependent MUR, scalar observables). *For every $\rho \in \mathcal{G}$ and $M \in \mathcal{C}_{u,v}^G$,*

$$S(Q_{u,\rho} \| M_{1,\rho}) + S(P_{v,\rho} \| M_{2,\rho}) \geq c_\rho(\alpha), \tag{59}$$

where the lower bound is

$$c_\rho(\alpha) = s(z_\rho) \log e$$

$$= (\log e) \left\{ \ln\left(1 + \frac{\hbar |\cos \alpha|}{2\sqrt{\text{Var}(Q_{u,\rho}) \text{Var}(P_{v,\rho})}}\right) - \frac{\hbar |\cos \alpha|}{\hbar |\cos \alpha| + 2\sqrt{\text{Var}(Q_{u,\rho}) \text{Var}(P_{v,\rho})}} \right\}, \tag{60}$$

with

$$z_\rho := \frac{\hbar |\cos \alpha|}{2\sqrt{\text{Var}(Q_{u,\rho}) \text{Var}(P_{v,\rho})}} \in [0,1]. \tag{61}$$

The lower bound is tight and the optimal measurement is unique: $c_\rho(\alpha) = S(\rho, M_*)$, for a unique $M_* \in \mathcal{C}_{u,v}^G$; such a Gaussian (u,v)-covariant bi-observable is characterized by

$$\mu^{M_*} = 0, \quad V_{12}^{M_*} = 0, \quad V_{11}^{M_*} = \frac{\hbar}{2}\sqrt{\frac{\text{Var}(Q_{u,\rho})}{\text{Var}(P_{v,\rho})}} |\cos \alpha|, \quad V_{22}^{M_*} = \frac{\hbar}{2}\sqrt{\frac{\text{Var}(P_{v,\rho})}{\text{Var}(Q_{u,\rho})}} |\cos \alpha|. \tag{62}$$

Proof. As already discussed, the case $\cos \alpha = 0$ is trivial. If $\cos \alpha \neq 0$, we have to minimize the error function (57) over M. First of all we can eliminate the positive term $\Delta(\rho, M)$ by taking an unbiased measurement. Then, since s is an increasing function, by the second condition in (55) we can also take $V_{11}^{M_*} V_{22}^{M_*} = \frac{\hbar^2}{4} (\cos \alpha)^2$. This implies $V_{12}^{M_*} = 0$ by (52). In this case the error function (57) reduces to

$$S(\rho, M_*) = \frac{\log e}{2} \left(s(x) + s(z_\rho^2/x) \right), \qquad x = \frac{V_{11}^{M_*}}{\text{Var}(Q_{u,\rho})},$$

with z_ρ given by (61); by the first of (55), we have $z_\rho \in (0,1]$.

Now, we can minimize the error function with respect to x by studying its first derivative:

$$\frac{d}{dx}\left(s(x) + s(z_\rho^2/x)\right) = \frac{x}{(1+x)^2} - \frac{z_\rho^4}{x(z_\rho^2+x)^2} = \frac{\left(x^2 - z_\rho^2\right)\left(x^2 + 2z_\rho^2 x + z_\rho^2\right)}{x\left(z_\rho^2 + x\right)^2 (1+x)^2}.$$

Having $x > 0$, we immediately get that $x = z_\rho$ gives the unique minimum. Thus

$$S(\rho, M) \geq S(\rho, M_*) = s(z_\rho)\log e = \log(1 + z_\rho) - \frac{z_\rho}{1 + z_\rho}\log e,$$

and

$$V_{11}^{M_*} = z_\rho \operatorname{Var}(Q_{u,\rho}) \equiv \frac{\hbar}{2}\sqrt{\frac{\operatorname{Var}(Q_{u,\rho})}{\operatorname{Var}(P_{v,\rho})}}\,|\cos\alpha|, \quad V_{22}^{M_*} = z_\rho \operatorname{Var}(P_{v,\rho}) \equiv \frac{\hbar}{2}\sqrt{\frac{\operatorname{Var}(P_{v,\rho})}{\operatorname{Var}(Q_{u,\rho})}}\,|\cos\alpha|,$$

which conclude the proof. □

Remark 3. *The minimum information loss $c_\rho(\alpha)$ depends on both the preparation ρ and the angle α. When $\alpha \neq \pm\pi/2$, that is when the target observables are not compatible, $c_\rho(\alpha)$ is strictly grater than zero. This is a peculiar quantum effect: given ρ, u and v, there is no Gaussian approximate joint measurement of Q_u and P_v that can approximate them arbitrarily well. On the other side, in the limit $\alpha \to \pm\pi/2$, the lower bound $c_\rho(\alpha)$ goes to zero; so, the case of commuting target observables is approached with continuity.*

Remark 4. *The lower bound $c_\rho(\alpha)$ goes to zero also in the classical limit $\hbar \to 0$. This holds for every angle α and every Gaussian state ρ.*

Remark 5. *Another case in which $c_\rho(\alpha) \to 0$ is the limit of large uncertainty states, that is, if we let the product $\operatorname{Var}(Q_{u,\rho})\operatorname{Var}(P_{v,\rho}) \to +\infty$: our entropic MUR disappears because, roughly speaking, the variance of (at least) one of the two target observables goes to infinity, its relative entropy vanishes by itself, and an optimal covariant bi-observable M_* has to take care of (at most) only the other target observable.*

Remark 6. *Actually, something similar to the previous remark happens also at the macroscopic limit, and does not require the measuring instrument to be an optimal one; indeed, unbiasedness is enough in this case. This happens because the error function $S(\rho, M)$ quantifies a relative error; even if the measurement approximation M is fixed, such an error can be reduced by suitably changing the preparation ρ. Indeed, if we consider the position and momentum of a macroscopic particle, for instance the center of mass of many particles, it is natural that its state has much larger position and momentum uncertainties than the intrinsic uncertainties of the measuring instrument; that is, $\frac{V_{11}^M}{\operatorname{Var}(Q_{u,\rho})} \ll 1$ and $\frac{V_{22}^M}{\operatorname{Var}(P_{v,\rho})} \ll 1$, implying that the error function (57) is negligible. In practice, this is a classical case: the preparation has large position and momentum uncertainties and the measuring instrument is relatively good. In this situation we do not see the difference between the joint measurement of position and momentum and their separate sharp observations.*

Remark 7. *The optimal approximating joint measurement $M_* \in \mathcal{C}_{u,v}^G$ is unique; by (62) it depends on the preparation ρ one is considering, as well as on the directions u and v. A realization of M_* is the measuring procedure of Example 2.*

Remark 8. *The MUR (59) is scale invariant, as both the error function $S(\rho, M)$ and the lower bound $c_\rho(\alpha)$ are such.*

Remark 9. For $\cos \alpha \neq 0$, we get $\inf_{M \in \mathcal{C}_{u,v}^G} S(\rho, M) = s(z_\rho) \log e$, where z_ρ is defined by (61). As z_ρ ranges in the interval $(0,1]$, the quantity $\inf_{M \in \mathcal{C}_{u,v}^G} S(\rho, M)$ takes all the values in the interval $\left(0, 1 - \frac{\log e}{2}\right]$, so that

$$\sup_{\rho \in \mathcal{G}} \inf_{M \in \mathcal{C}_{u,v}^G} S(\rho, M) = 1 - \frac{\log e}{2}. \tag{63}$$

In order to get this result, we needed $\cos \alpha \neq 0$; however, the final result does not depend on α. Therefore, in the $\sup_\rho \inf_M$-approach of (63), the continuity from quantum to classical is lost.

6.1.2. Entropic Divergence of Q_u, P_v from M

Now we want to find an entropic quantification of the error made in observing $M \in \mathcal{C}_{u,v}$ as an approximation of Q_u and P_v in an arbitrary state ρ. The procedure of [41], already suggested in [19] (Section VI.C) for a different error function, is to consider the worst case by maximizing the error function over all the states. However, in the continuous framework this is not possible for the error function (56); indeed, from (57) we get $\sup_{\rho \in \mathcal{G}} S(\rho, M) = +\infty$ even if we restrict to unbiased covariant bi-observables.

Anyway, the reason for $S(\rho, M)$ to diverge is classical: it depends only on the continuous nature of Q_u and P_v, without any relation to their (quantum) incompatibility. Indeed, as we noted in Section 3.1, if an instrument measuring a random variable $X \sim \mathcal{N}(a; \alpha^2)$ adds an independent noise $\nu \sim \mathcal{N}(b; \beta^2)$, thus producing an output $X + \nu \sim \mathcal{N}(a + b; \alpha^2 + \beta^2)$, then the relative entropy $S(X \| X + \nu)$ diverges for $\alpha^2 \to 0$; this is what happens if we fix the noise and we allow for arbitrarily peaked preparations. Thus, the sum $S(Q_{u,\rho} \| M_{1,\rho}) + S(P_{v,\rho} \| M_{2,\rho})$ diverges if, fixed M, we let $\text{Var}(Q_{u,\rho})$ or $\text{Var}(P_{v,\rho})$ go to 0.

The difference between the classical and quantum frameworks emerges if we bound from below the variances of the sharp position and momentum observables. Indeed, in the classical framework we have $\inf_{b,\beta^2} \sup_{\alpha^2 \geq \epsilon} S(X \| X + \nu) = 0$ for every $\epsilon > 0$; the same holds for the sum of two relative entropies if no relation exists between the two noises. On the contrary, in the quantum framework the entropic MURs appear due to the relation between the position and momentum errors occurring in any approximate joint measurement.

In order to avoid that $S(\rho, M) \to +\infty$ due to merely classical effects, we thus introduce the following subset of the Gaussian states:

$$\mathcal{G}_\epsilon^{u,v} := \{\rho \in \mathcal{G} : \text{Var}(Q_{u,\rho}) \geq \epsilon_1, \text{Var}(P_{v,\rho}) \geq \epsilon_2\}, \qquad \epsilon_i > 0, \tag{64}$$

and we evaluate the error made in approximating Q_u and P_v with the marginals of a (u, v)-covariant bi-observable by maximizing the error function over all these states.

Definition 10. *The* Gaussian ϵ-entropic divergence *of Q_u, P_v from $M \in \mathcal{C}_{u,v}$ is*

$$D_\epsilon^G(Q_u, P_v \| M) := \sup_{\rho \in \mathcal{G}_\epsilon^{u,v}} S(\rho, M). \tag{65}$$

For Gaussian M, depending on the choice of the thresholds ϵ_1 and ϵ_2, the divergence $D_\epsilon^G(Q_u, P_v \| M)$ can be easily computed or at least bounded.

Theorem 2. *Let the bi-observable $M \in \mathcal{C}_{u,v}^G$ be fixed.*

(i) *For $\epsilon_1 \epsilon_2 \geq \frac{\hbar^2}{4}(\cos \alpha)^2$, the divergence $D_\epsilon^G(Q_u, P_v \| M)$ is given by*

$$D_\epsilon^G(Q_u, P_v \| M) = S(\rho_\epsilon(u, v), M) = \frac{\log e}{2}[s(x_\epsilon) + s(y_\epsilon) + \Delta(\epsilon; M)], \tag{66}$$

where $\rho_\epsilon(u,v)$ is any Gaussian state with $\text{Var}\left(Q_{u,\rho_\epsilon(u,v)}\right) = \epsilon_1$ and $\text{Var}\left(P_{v,\rho_\epsilon(u,v)}\right) = \epsilon_2$, and

$$x_\epsilon := \frac{V_{11}^M}{\epsilon_1}, \quad y_\epsilon := \frac{V_{22}^M}{\epsilon_2}, \quad \Delta(\epsilon;\sigma) := \frac{(a^M)^2}{V_{11}^M + \epsilon_1} + \frac{(b^M)^2}{V_{22}^M + \epsilon_2}.$$

(ii) For $\epsilon_1 \epsilon_2 < \frac{\hbar^2}{4}(\cos\alpha)^2$, the divergence $D_\epsilon^G(Q_u, P_v \| M)$ is bounded from below by

$$D_\epsilon^G(Q_u, P_v \| M) \geq S(\rho_\epsilon(u,v), M) = \frac{\log e}{2}\left[s(x_\epsilon) + s(y_\epsilon) + \Delta(\epsilon; M)\right], \quad (67)$$

where $\rho_\epsilon(u,v)$ is any Gaussian state with $\text{Var}\left(Q_{u,\rho_\epsilon(u,v)}\right) = \epsilon_1$ and $\text{Var}\left(P_{v,\rho_\epsilon(u,v)}\right) = \frac{\hbar^2}{4\epsilon_1}(\cos\alpha)^2$, and

$$x_\epsilon := \frac{V_{11}^M}{\epsilon_1}, \quad y_\epsilon := \frac{4\epsilon_1 V_{22}^M}{\hbar^2(\cos\alpha)^2}, \quad \Delta(\epsilon;\sigma) := \frac{(a^M)^2}{V_{11}^M + \epsilon_1} + \frac{(b^M)^2}{V_{22}^M + \frac{\hbar^2}{4\epsilon_1}(\cos\alpha)^2}.$$

The existence of the above states $\rho_\epsilon(u,v)$ is guaranteed by Proposition 3.

Proof. By Proposition 3, maximizing the error function over the states in $\mathcal{G}_\epsilon^{u,v}$ is the same as maximizing (57) over the parameters $\text{Var}(Q_{u,\rho})$ and $\text{Var}(P_{v,\rho})$ satisfying (55) and (64) (note that in the bias $\Delta(\rho, M)$, the variances $\text{Var}\,M_{1,\rho}$ and $\text{Var}\,M_{2,\rho}$ depend on $\text{Var}(Q_{u,\rho})$ and $\text{Var}(P_{v,\rho})$ by (54)).

(i) In the case $\epsilon_1\epsilon_2 \geq \frac{\hbar^2}{4}(\cos\alpha)^2$, the thresholds themselves satisfy Heisenberg uncertainty relation, and so equality (66) follows from the expression (57) and the fact the functions $s(x)$, $s(y)$, $\Delta(\rho, M)$ are decreasing in $\text{Var}(Q_{u,\rho})$ and $\text{Var}(P_{v,\rho})$.

(ii) In the case $\epsilon_1\epsilon_2 < \frac{\hbar^2}{4}(\cos\alpha)^2$, we have to take into account the relation (55) for $\text{Var}(Q_{u,\rho})$ and $\text{Var}(P_{v,\rho})$: the supremum of $S(\rho, M)$ is achieved when $\text{Var}(Q_{u,\rho})\text{Var}(P_{v,\rho}) = \frac{\hbar^2}{4}(\cos\alpha)^2$, with $\text{Var}(Q_{u,\rho}) \geq \epsilon_1$ and $\text{Var}(P_{v,\rho}) \geq \epsilon_2$. Then inequality (67) follows by choosing $\text{Var}(Q_{u,\rho}) = \epsilon_1$ and $\text{Var}(P_{v,\rho}) = \frac{\hbar^2}{4\epsilon_1}(\cos\alpha)^2$. □

Remark 10. *The conditions on the states $\rho_\epsilon(u,v)$ do not depend on M, but only on the parameters defining $\mathcal{G}_\epsilon^{u,v}$. Thus, in the case $\epsilon_1\epsilon_2 \geq \frac{\hbar^2}{4}(\cos\alpha)^2$, any choice of $\rho_\epsilon(u,v)$ yields a state which is the worst one for every Gaussian approximate joint measurement M.*

6.1.3. Entropic Incompatibility Degree of Q_u and P_v

The last step is to optimize the state independent ϵ-entropic divergence (65) over all the approximate joint measurements of Q_u and P_v. This is done in the next definition.

Definition 11. *The Gaussian ϵ-entropic incompatibility degree of Q_u, P_v is*

$$c_{\text{inc}}^G(Q_u, P_v; \epsilon) := \inf_{M \in \mathcal{C}_{u,v}^G} D_\epsilon^G(Q_u, P_v \| M) \equiv \inf_{M \in \mathcal{C}_{u,v}^G} \sup_{\rho \in \mathcal{G}_\epsilon^{u,v}} S(\rho, M). \quad (68)$$

Again, depending on the choice of the thresholds ϵ_1 and ϵ_2, the entropic incompatibility degree $c_{\text{inc}}^G(Q_u, P_v; \epsilon)$ can be easily computed or at least bounded.

Theorem 3. (i) For $\epsilon_1\epsilon_2 \geq \frac{\hbar^2}{4}(\cos\alpha)^2$, the incompatibility degree $c_{inc}^G(Q_u,P_v;\epsilon)$ is given by

$$c_{inc}^G(Q_u,P_v;\epsilon) = (\log e)\left\{\ln\left(1+\frac{\hbar|\cos\alpha|}{2\sqrt{\epsilon_1\epsilon_2}}\right) - \frac{\hbar|\cos\alpha|}{2\sqrt{\epsilon_1\epsilon_2}+\hbar|\cos\alpha|}\right\}. \tag{69}$$

The infimum in (68) is attained and the optimal measurement is unique, in the sense that

$$c_{inc}^G(Q_u,P_v;\epsilon) = D_\epsilon^G(Q_u,P_v\|M_\epsilon) \tag{70}$$

for a unique $M_\epsilon \in \mathcal{C}_{u,v}^G$; such a bi-observable is characterized by

$$a^{M_\epsilon}=0,\quad b^{M_\epsilon}=0,\quad V_{11}^{M_\epsilon}=\frac{\hbar}{2}\sqrt{\frac{\epsilon_1}{\epsilon_2}}|\cos\alpha|,\quad V_{22}^{M_\epsilon}=\frac{\hbar}{2}\sqrt{\frac{\epsilon_2}{\epsilon_1}}|\cos\alpha|,\quad V_{12}^{M_\epsilon}=0. \tag{71}$$

(ii) For $\epsilon_1\epsilon_2 < \frac{\hbar^2}{4}(\cos\alpha)^2$, the incompatibility degree $c_{inc}^G(Q_u,P_v;\epsilon)$ is bounded from below by

$$c_{inc}^G(Q_u,P_v;\epsilon) \geq (\log e)\left\{\ln(2)-\frac{1}{2}\right\}. \tag{72}$$

The latter bound is

$$(\log e)\left\{\ln(2)-\frac{1}{2}\right\} = S(\rho_\epsilon(u,v),M_\epsilon) = \inf_{M\in\mathcal{C}_{u,v}^G} S(\rho_\epsilon(u,v),M), \tag{73}$$

where the state $\rho_\epsilon(u,v)$ is defined in item (ii) of Theorem 2 and M_ϵ is the bi-observable in $\mathcal{C}_{u,v}^G$ such that

$$a^{M_\epsilon}=0,\quad b^{M_\epsilon}=0,\quad V_{11}^{M_\epsilon}=\epsilon_1,\quad V_{22}^{M_\epsilon}=\frac{\hbar^2}{4\epsilon_1}(\cos\alpha)^2,\quad V_{12}^{M_\epsilon}=0. \tag{74}$$

Proof. (i) In the case $\epsilon_1\epsilon_2 \geq \frac{\hbar^2}{4}(\cos\alpha)^2$, due to (66), the proof is the same as that of Theorem 1 with the replacements $\text{Var}(Q_{u,\rho}) \mapsto \epsilon_1$ and $\text{Var}(P_{v,\rho}) \mapsto \epsilon_2$.

(ii) In the case $\epsilon_1\epsilon_2 < \frac{\hbar^2}{4}(\cos\alpha)^2$, starting from (67), the proof is the same as that of Theorem 1 with the replacements $\text{Var}(Q_{u,\rho}) \mapsto \epsilon_1$ and $\text{Var}(P_{v,\rho}) \mapsto \frac{\hbar^2}{4\epsilon_1}(\cos\alpha)^2$. □

Remark 11 (State independent MUR, scalar observables). By means of the above results, we can formulate a state independent entropic MUR for the position Q_u and the momentum P_v in the following way. Chosen two positive thresholds ϵ_1 and ϵ_2, there exists a preparation $\rho_\epsilon(u,v) \in \mathcal{G}_\epsilon^{u,v}$ (introduced in Theorem 2) such that, for all Gaussian approximate joint measurements M of Q_u and P_v, we have

$$S(Q_{u,\rho_\epsilon(u,v)}\|M_{1,\rho_\epsilon(u,v)}) + S(P_{v,\rho_\epsilon(u,v)}\|M_{2,\rho_\epsilon(u,v)})$$

$$\geq \begin{cases}(\log e)\left\{\ln\left(1+\frac{\hbar|\cos\alpha|}{2\sqrt{\epsilon_1\epsilon_2}}\right)-\frac{\hbar|\cos\alpha|}{2\sqrt{\epsilon_1\epsilon_2}+\hbar|\cos\alpha|}\right\}, & \text{if } \epsilon_1\epsilon_2 \geq \frac{\hbar^2}{4}(\cos\alpha)^2, \\ (\log e)\left\{\ln(2)-\frac{1}{2}\right\}, & \text{if } \epsilon_1\epsilon_2 < \frac{\hbar^2}{4}(\cos\alpha)^2.\end{cases} \tag{75}$$

The inequality follows by (66) and (69) in the case $\epsilon_1\epsilon_2 \geq \frac{\hbar^2}{4}(\cos\alpha)^2$, and (73) in the case $\epsilon_1\epsilon_2 < \frac{\hbar^2}{4}(\cos\alpha)^2$.

What is relevant is that, for every approximate joint measurement M, the total information loss $S(\rho,M)$ does exceed the lower bound (75) even if the set of states $\mathcal{G}_\epsilon^{u,v}$ forbids preparations ρ with too peaked target distributions. Indeed, without the thresholds ϵ_1, ϵ_2, it would be trivial to exceed the lower bound (75), as we noted in Section 6.1.2.

We also remark that, chosen ϵ_1 and ϵ_2, we found a single state $\rho_\epsilon(u,v)$ in $\mathcal{G}_\epsilon^{u,v}$ that satisfies (75) for every M, so that $\rho_\epsilon(u,v)$ is a 'bad' state for all Gaussian approximate joint measurements of position and momentum.

When $\epsilon_1 \epsilon_2 \geq \frac{\hbar^2}{4}(\cos\alpha)^2$, the optimal approximate joint measurement M_ϵ is unique in the class of Gaussian (u,v)-covariant bi-observables; it depends only on the class of preparations $\mathcal{G}_\epsilon^{u,v}$: it is the best measurement for the worst choice of the preparation in the class $\mathcal{G}_\epsilon^{u,v}$.

Remark 12. *The entropic incompatibility degree $c_{\text{inc}}^G(Q_u, P_v; \epsilon)$ is strictly positive for $\cos\alpha \neq 0$ (incompatible target observables) and it goes to zero in the limits $\alpha \to \pm\pi/2$ (compatible observables), $\hbar \to 0$ (classical limit), and $\epsilon_1 \epsilon_2 \to \infty$ (large uncertainty states).*

Remark 13. *The scale invariance of the relative entropy extends to the error function $S(\rho, M)$, hence to the divergence $D_\epsilon^G(Q_u, P_v \| M)$ and the entropic incompatibility degree $c_{\text{inc}}^G(Q_u, P_v; \epsilon)$, as well as the entropic MUR (75).*

6.2. Vector Observables

Now the target observables are Q and P given in (3), with pvm's Q and P; the approximating bi-observables are the covariant phase-space observables $\mathcal{C}$ of Definition 5. Each bi-observable $M \in \mathcal{C}$ is of the form $M = M^\sigma$ for some $\sigma \in \mathcal{S}$, where M^σ is given by (43). $\mathcal{C}^G$ is the subset of the Gaussian bi-observables in $\mathcal{C}$, and $M^\sigma \in \mathcal{C}^G$ if and only if σ is a Gaussian state.

We proceed to define the analogues of the scalar quantities introduced in Sections 6.1.1–6.1.3. In order to do it, in the next proposition we recall some known results on matrices.

Proposition 14 ([50–52,65]). *Let M_1 and M_2 be $n \times n$ complex matrices such that $M_1 > M_2 > 0$. Then, we have $0 < M_1^{-1} < M_2^{-1}$. Moreover, if $s : \mathbb{R}_+ \to \mathbb{R}$ is a strictly increasing continuous function, we have $\text{Tr}\{s(M_1)\} > \text{Tr}\{s(M_2)\}$.*

6.2.1. Error Function

Definition 12. *Given the preparation $\rho \in \mathcal{S}$ and the covariant phase-space observable M^σ, with $\sigma \in \mathcal{S}$, the error function for the vector case is the sum of the two relative entropies:*

$$S(\rho, M^\sigma) := S(Q_\rho \| M_{1,\rho}^\sigma) + S(P_\rho \| M_{2,\rho}^\sigma). \tag{76}$$

As in the scalar case, the error function is scale invariant, it quantifies a relative error, and we always have $S(\rho, M^\sigma) > 0$ because position and momentum are incompatible. Indeed, since the marginals of a bi-observable $M^\sigma \in \mathcal{C}$ turn out to be convolutions of the respective sharp observables Q and P with some probability densities on $\mathbb{R}^n$, $Q_\rho \neq M_{1,\rho}^\sigma$ and $P_\rho \neq M_{2,\rho}^\sigma$ for all states ρ; this is an easy consequence, for instance, of Problem 26.1, p. 362, in [66].

In the Gaussian case the error function can be explicitly computed.

Proposition 15 (Error function for the vector Gaussian case). *For $\rho, \sigma \in \mathcal{G}$, the error function has the two equivalent expressions:*

$$S(\rho, M^\sigma) = \frac{\log e}{2}\left[\text{Tr}\left\{s(E_{\rho,\sigma}) + s(F_{\rho,\sigma})\right\} + a^\sigma \cdot (A^\rho + A^\sigma)^{-1} a^\sigma + b^\sigma \cdot (B^\rho + B^\sigma)^{-1} b^\sigma\right] \tag{77a}$$

$$= \frac{\log e}{2}\left[\text{Tr}\left\{s(N_{\rho,\sigma}^{-1}) + s(R_{\rho,\sigma}^{-1})\right\} + a^\sigma \cdot (A^\rho + A^\sigma)^{-1} a^\sigma + b^\sigma \cdot (B^\rho + B^\sigma)^{-1} b^\sigma\right], \tag{77b}$$

where the function s is defined in (58), and

$$E_{\rho,\sigma} := (A^\rho)^{-1/2} A^\sigma (A^\rho)^{-1/2}, \qquad F_{\rho,\sigma} := (B^\rho)^{-1/2} B^\sigma (B^\rho)^{-1/2}, \tag{78a}$$

$$N_{\rho,\sigma} := (A^\sigma)^{-1/2} A^\rho (A^\sigma)^{-1/2}, \qquad R_{\rho,\sigma} := (B^\sigma)^{-1/2} B^\rho (B^\sigma)^{-1/2}. \tag{78b}$$

Proof. First of all, recall that

$$Q_\rho = \mathcal{N}(a^\rho; A^\rho), \qquad \mathsf{M}_{1,\rho}^\sigma = \mathcal{N}(a^\rho + a^\sigma; A^\rho + A^\sigma)$$
$$P_\rho = \mathcal{N}(b^\rho; B^\rho), \qquad \mathsf{M}_{2,\rho}^\sigma = \mathcal{N}(b^\rho + b^\sigma; B^\rho + B^\sigma).$$

A direct application of (34) yields

$$S(Q_\rho \| \mathsf{M}_{1,\rho}^\sigma) = \frac{1}{2} \log \frac{\det(A^\rho + A^\sigma)}{\det A^\rho} + \frac{\log e}{2} \left[\mathrm{Tr}\left\{ (A^\rho + A^\sigma)^{-1} A^\rho - \mathbb{1} \right\} + a^\sigma \cdot (A^\rho + A^\sigma)^{-1} a^\sigma \right].$$

We can transform this equation by using

$$\frac{\det(A^\sigma + A^\rho)}{\det A^\rho} = \det\left[(A^\rho)^{-1/2} (A^\sigma + A^\rho)(A^\rho)^{-1/2} \right] = \det(\mathbb{1} + E_{\rho,\sigma}),$$

$$\ln \det(\mathbb{1} + E_{\rho,\sigma}) = \mathrm{Tr}\left\{ \ln(\mathbb{1} + E_{\rho,\sigma}) \right\},$$

$$\mathrm{Tr}\left\{ (A^\rho + A^\sigma)^{-1} A^\rho - \mathbb{1} \right\} = \mathrm{Tr}\left\{ (A^\rho)^{1/2} (A^\rho + A^\sigma)^{-1} (A^\rho)^{1/2} - \mathbb{1} \right\} = -\mathrm{Tr}\left\{ (\mathbb{1} + E_{\rho,\sigma})^{-1} E_{\rho,\sigma} \right\}.$$

This gives

$$S(Q_\rho \| \mathsf{M}_{1,\rho}^\sigma) = \frac{\log e}{2} \left[\mathrm{Tr}\{ s(E_{\rho,\sigma})\} + a^\sigma \cdot (A^\rho + A^\sigma)^{-1} a^\sigma \right].$$

In the same way a similar expression is obtained for $S(P_\rho \| \mathsf{M}_{2,\rho}^\sigma)$ and (77a) is proved. On the other hand, by using

$$\ln \frac{\det(A^\sigma + A^\rho)}{\det A^\rho} = \ln \frac{\det(\mathbb{1} + N_{\rho,\sigma})}{\det N_{\rho,\sigma}} = \ln \det\left(\mathbb{1} + N_{\rho,\sigma}^{-1}\right) = \mathrm{Tr}\left\{ \ln\left(\mathbb{1} + N_{\rho,\sigma}^{-1}\right) \right\},$$

$$\mathrm{Tr}\left\{ (A^\rho + A^\sigma)^{-1} A^\rho - \mathbb{1} \right\} = -\mathrm{Tr}\left\{ (A^\rho + A^\sigma)^{-1} A^\sigma \right\} = -\mathrm{Tr}\left\{ \left(\mathbb{1} + N_{\rho,\sigma}^{-1}\right)^{-1} N_{\rho,\sigma}^{-1} \right\},$$

and the analogous expressions involving B^ρ and $R_{\rho,\sigma}$, one gets (77b). □

State Dependent Lower Bound

In principle, a state dependent lower bound for the error function could be found by analogy with Theorem 1, by taking again the infimum over all joint covariant measurements, that is $\inf_\sigma S(\rho, \mathsf{M}^\sigma)$. By considering only Gaussian states ρ and measurements M^σ, from (18), (77a) and (78a), the infimum over $\sigma \in \mathcal{G}$ can be reduced to an infimum over the matrices A^σ:

$$\inf_{\sigma \in \mathcal{G}} S(\rho, \mathsf{M}^\sigma) = \frac{\log e}{2} \inf_{A^\sigma} \mathrm{Tr}\left\{ s\left((A^\rho)^{-1/2} A^\sigma (A^\rho)^{-1/2} \right) + s\left(\frac{\hbar^2}{4} (B^\rho)^{-1/2} (A^\sigma)^{-1} (B^\rho)^{-1/2} \right) \right\}.$$

The above equality follows since the monotonicity of s (Proposition 14) implies that the trace term in (77a) attains its minimum when $B^\sigma = \frac{\hbar^2}{4}(A^\rho)^{-1}$. However, it remains an open problem to explicitly compute the infimum over the matrices A^σ when the preparation ρ is arbitrary.

Nevertheless, the computations can be done at least for a preparation ρ_* of minimum uncertainty (Proposition 5). Indeed, by (22) we get

$$\inf_{\sigma \in \mathcal{G}} S(\rho_*, \mathsf{M}^\sigma) = \frac{\log e}{2} \inf_{A^\sigma} \mathrm{Tr}\left\{ s(E_{\rho,\sigma}) + s\left(E_{\rho,\sigma}^{-1}\right) \right\}.$$

Now we can diagonalize $E_{\rho,\sigma}$ and minimize over its eigenvalues; since $s(x) + s(x^{-1})$ attains its minimum value at $x = 1$, this procedure gives $E_{\rho,\sigma} = \mathbb{1}$. So, by denoting by σ_* the state giving the minimum, we have

$$A^{\sigma_*} = A^{\rho_*}, \qquad B^{\sigma_*} = B^{\rho_*} = \frac{\hbar^2}{4}(A^{\rho_*})^{-1}, \tag{79}$$

$$\inf_{\sigma \in \mathcal{G}} S(\rho_*, M^\sigma) = S(\rho_*, M^{\sigma_*}) = ns(1)\log e. \tag{80}$$

For an arbitrary $\rho \in \mathcal{G}$, we can use the last formula to deduce an upper bound for $\inf_{\sigma \in \mathcal{G}} S(\rho, M^\sigma)$. Indeed, if ρ_* is a minimum uncertainty state with $A^{\rho_*} = A^\rho$, then $B^\rho \geq \frac{\hbar^2}{4}(A^\rho)^{-1} = B^{\rho_*}$ by (19), and, using again the state σ_* of (79), we find

$$\inf_{\sigma \in \mathcal{G}} S(\rho, M^\sigma) \leq S(\rho, M^{\sigma_*}) \leq S(\rho_*, M^{\sigma_*}) = ns(1)\log e.$$

The second inequality in the last formula follows from (77b), (78b) and the monotonicity of s (Proposition 14).

6.2.2. Entropic Divergence of Q, P from M^σ

In order to define a state independent measure of the error made in regarding the marginals of M^σ as approximations of Q and P, we can proceed along the lines of the scalar case in Section 6.1.2. To this end, we introduce the following vector analogue of the Gaussian states defined in (64):

$$\mathcal{G}_\epsilon := \{\rho \in \mathcal{G} : A^\rho \geq \epsilon_1 \mathbb{1}, \ B^\rho \geq \epsilon_2 \mathbb{1}\}, \qquad \epsilon \equiv (\epsilon_1, \epsilon_2), \quad \epsilon_i > 0. \tag{81}$$

In the vector case, Definition 10 then reads as follows.

Definition 13. *The* Gaussian ϵ-entropic divergence *of* Q, P *from* $M^\sigma \in \mathcal{C}$ *is*

$$D_\epsilon^G(Q, P \| M^\sigma) := \sup_{\rho \in \mathcal{G}_\epsilon} S(\rho, M^\sigma). \tag{82}$$

As in the scalar case, when M^σ is Gaussian, depending on the choice of the product $\epsilon_1 \epsilon_2$, we can compute the divergence $D_\epsilon^G(Q, P \| M^\sigma)$ or at least bound it from below.

Theorem 4. *Let the bi-observable* $M^\sigma \in \mathcal{C}^G$ *be fixed.*

(i) *For* $\epsilon_1 \epsilon_2 \geq \frac{\hbar^2}{4}$, *the divergence* $D_\epsilon^G(Q, P \| M^\sigma)$ *is given by*

$$D_\epsilon^G(Q, P \| M^\sigma) = S(\rho_\epsilon, M^\sigma) = \frac{\log e}{2}\Big[\mathrm{Tr}\left\{s\left(A^\sigma/\epsilon_1\right) + s\left(B^\sigma/\epsilon_2\right)\right\}$$
$$+ a^\sigma \cdot (A^\sigma + \epsilon_1 \mathbb{1})^{-1} a^\sigma + b^\sigma \cdot (B^\sigma + \epsilon_2 \mathbb{1})^{-1} b^\sigma\Big], \tag{83}$$

where ρ_ϵ is any Gaussian state with $A^{\rho_\epsilon} = \epsilon_1 \mathbb{1}$ and $B^{\rho_\epsilon} = \epsilon_2 \mathbb{1}$.

(ii) *For* $\epsilon_1 \epsilon_2 < \frac{\hbar^2}{4}$, *the divergence* $D_\epsilon^G(Q, P \| M^\sigma)$ *is bounded from below by*

$$D_\epsilon^G(Q, P \| M^\sigma) \geq S(\rho_\epsilon, M^\sigma) = \frac{\log e}{2}\Big[\mathrm{Tr}\left\{s\left(A^\sigma/\epsilon_1\right) + s\left(4\epsilon_1 B^\sigma/\hbar^2\right)\right\}$$
$$+ a^\sigma \cdot (A^\sigma + \epsilon_1 \mathbb{1})^{-1} a^\sigma + b^\sigma \cdot \left(B^\sigma + \frac{\hbar^2}{4\epsilon_1}\mathbb{1}\right)^{-1} b^\sigma\Big], \tag{84}$$

where ρ_ϵ is any Gaussian state with $A^{\rho_\epsilon} = \epsilon_1 \mathbb{1}$ and $B^{\rho_\epsilon} = \dfrac{\hbar^2}{4\epsilon_1}\mathbb{1}$.

Proof. (i) In the case $\epsilon_1\epsilon_2 \geq \dfrac{\hbar^2}{4}$, for $\rho \in \mathcal{G}_\epsilon$ we have $N_{\rho,\sigma} \geq \epsilon_1(A^\sigma)^{-1}$ and $R_{\rho,\sigma} \geq \epsilon_2(B^\sigma)^{-1}$; by Proposition 14 we get

$$\mathrm{Tr}\{s(N_{\rho,\sigma}^{-1})\} \leq \mathrm{Tr}\{s(A^\sigma/\epsilon_1)\}, \qquad \mathrm{Tr}\{s(R_{\rho,\sigma}^{-1})\} \leq \mathrm{Tr}\{s(B^\sigma/\epsilon_2)\},$$

$$(A^\rho + A^\sigma)^{-1} \leq (\epsilon_1 \mathbb{1} + A^\sigma)^{-1}, \qquad (B^\rho + B^\sigma)^{-1} \leq (\epsilon_2 \mathbb{1} + B^\sigma)^{-1}.$$

By using these inequalities in the expression (77b), we get (83).

(ii) In the case $\epsilon_1\epsilon_2 < \dfrac{\hbar^2}{4}$, the lower bound (84) follows by evaluating $S(\rho, M^\sigma)$ at the state $\rho = \rho_\epsilon \in \mathcal{G}_\epsilon$ with $A^{\rho_\epsilon} = \epsilon_1 \mathbb{1}$ and $B^{\rho_\epsilon} = \dfrac{\hbar^2}{4\epsilon_1}\mathbb{1}$. □

Note that ρ_ϵ does not depend on σ, but only on the parameters defining $\mathcal{G}_\epsilon$: again, in the case $\epsilon_1\epsilon_2 \geq \dfrac{\hbar^2}{4}$, the error attains its maximum at a state which is independent of the approximate measurement.

6.2.3. Entropic Incompatibility Degree of Q and P

By analogy with Section 6.1.3, we can optimize the ϵ-entropic divergence over all the approximate joint measurements of Q and P.

Definition 14. *The Gaussian ϵ-entropic incompatibility degree of Q and P is*

$$c_{\mathrm{inc}}^G(Q,P;\epsilon) := \inf_{\sigma \in \mathcal{G}} D_\epsilon^G(Q,P\|M^\sigma) \equiv \inf_{\sigma \in \mathcal{G}} \sup_{\rho \in \mathcal{G}_\epsilon} S(\rho, M^\sigma). \qquad (85)$$

Again, depending on the product $\epsilon_1\epsilon_2$, we can compute or at least bound $c_{\mathrm{inc}}^G(Q,P;\epsilon)$ from below.

Theorem 5. (i) *For $\epsilon_1\epsilon_2 \geq \dfrac{\hbar^2}{4}$, the incompatibility degree $c_{\mathrm{inc}}^G(Q,P;\epsilon)$ is given by*

$$c_{\mathrm{inc}}^G(Q,P;\epsilon) = n\,(\log e)\left\{\ln\left(1 + \dfrac{\hbar}{2\sqrt{\epsilon_1\epsilon_2}}\right) - \dfrac{\hbar}{2\sqrt{\epsilon_1\epsilon_2} + \hbar}\right\}. \qquad (86)$$

The infimum in (85) is attained and the optimal measurement is unique, in the sense that

$$c_{\mathrm{inc}}^G(Q,P;\epsilon) = D_\epsilon^G(Q,P\|M^{\sigma_\epsilon}) \qquad (87)$$

for a unique $\sigma_\epsilon \in \mathcal{G}$; such a state is the minimal uncertainty state characterized by

$$a^{\sigma_\epsilon} = 0, \qquad b^{\sigma_\epsilon} = 0, \qquad A^{\sigma_\epsilon} = \dfrac{\hbar}{2}\sqrt{\dfrac{\epsilon_1}{\epsilon_2}}\,\mathbb{1}, \qquad B^{\sigma_\epsilon} = \dfrac{\hbar}{2}\sqrt{\dfrac{\epsilon_2}{\epsilon_1}}\,\mathbb{1}, \qquad C^{\sigma_\epsilon} = 0. \qquad (88)$$

(ii) *For $\epsilon_1\epsilon_2 < \dfrac{\hbar^2}{4}(\cos\alpha)^2$, the incompatibility degree $c_{\mathrm{inc}}^G(Q,P;\epsilon)$ is bounded from below by*

$$c_{\mathrm{inc}}^G(Q,P;\epsilon) \geq n(\log e)\left\{\ln(2) - \dfrac{1}{2}\right\}. \qquad (89)$$

The latter bound is

$$n(\log e)\left\{\ln(2) - \frac{1}{2}\right\} = S(\rho_\epsilon, M^{\sigma_\epsilon}) = \inf_{\sigma \in \mathcal{G}} S(\rho_\epsilon, M^\sigma), \qquad (90)$$

where the preparation ρ_ϵ is defined in item (ii) of Theorem 4 and σ_ϵ is the state in $\mathcal{G}$ such that

$$a^{\sigma_\epsilon} = 0, \quad b^{\sigma_\epsilon} = 0, \quad A^{\sigma_\epsilon} = \epsilon_1 \mathbb{1}, \quad B^{\sigma_\epsilon} = \frac{\hbar^2}{4\epsilon_1} \mathbb{1}, \quad C^{\sigma_\epsilon} = 0. \qquad (91)$$

Proof. (i) In the case $\epsilon_1 \epsilon_2 \geq \frac{\hbar^2}{4}$, from the expression (83) we get immediately $a^{\sigma_\epsilon} = 0$, $b^{\sigma_\epsilon} = 0$ and by (19) we have $B^\sigma \geq \frac{\hbar^2}{4}(A^\sigma)^{-1}$. So, by (83) and Propositions 3 and 14, we get $B^\sigma = \frac{\hbar^2}{4}(A^\sigma)^{-1}$, and

$$\inf_{\sigma \in \mathcal{G}} \sup_{\rho \in \mathcal{G}_\epsilon} S(\rho, M^\sigma) = \frac{\log e}{2} \inf_{A^\sigma} \operatorname{Tr}\left\{s(A^\sigma/\epsilon_1) + s\left(\frac{\hbar^2}{4\epsilon_2}(A^\sigma)^{-1}\right)\right\}.$$

By minimizing over all the eigenvalues of A^σ, we get the minimum (86), which is attained if and only if A^σ is as in (88). Hence, A^{σ_ϵ} and B^{σ_ϵ} are as in (88). This implies that any optimal state σ_ϵ is a minimum uncertainty state; so, $C^{\sigma_\epsilon} = 0$ and the state σ_ϵ is unique.

(ii) In the case $\epsilon_1 \epsilon_2 < \frac{\hbar^2}{4}$, by (19) and Proposition 14, inequality (84) implies

$$\inf_{\sigma \in \mathcal{G}} \sup_{\rho \in \mathcal{G}_\epsilon} S(\rho, M^\sigma) \geq \frac{\log e}{2} \inf_{A^\sigma} \operatorname{Tr}\left\{s(A^\sigma/\epsilon_1) + s\left(\epsilon_1(A^\sigma)^{-1}\right)\right\}.$$

By minimizing over all the eigenvalues of A^σ, we get (89). Then (89) holds for ρ_ϵ as in item (ii) of Theorem 4 and σ_ϵ in (91). □

Remark 14 (State independent MUR, vector observables). *By means of the above results, we can formulate the following state independent entropic MUR for the position $\mathbf{Q}$ and momentum $\mathbf{P}$. Chosen two positive thresholds ϵ_1 and ϵ_2, there exists a preparation $\rho_\epsilon \in \mathcal{G}_\epsilon$ (introduced in Theorem 4) such that, for all Gaussian approximate joint measurements M^σ of $\mathbf{Q}$ and $\mathbf{P}$, we have*

$$S(Q_{\rho_\epsilon} \| M^\sigma_{1,\rho_\epsilon}) + S(P_{\rho_\epsilon} \| M^\sigma_{2,\rho_\epsilon})$$

$$\geq \begin{cases} n(\log e)\left\{\ln\left(1 + \frac{\hbar}{2\sqrt{\epsilon_1 \epsilon_2}}\right) - \frac{\hbar}{2\sqrt{\epsilon_1 \epsilon_2} + \hbar}\right\}, & \text{if } \epsilon_1 \epsilon_2 \geq \frac{\hbar^2}{4}, \\ n(\log e)\left\{\ln(2) - \frac{1}{2}\right\}, & \text{if } \epsilon_1 \epsilon_2 < \frac{\hbar^2}{4}. \end{cases} \qquad (92)$$

The inequality follows by (83) and (86) for $\epsilon_1 \epsilon_2 \geq \frac{\hbar^2}{4}$, and (90) for $\epsilon_1 \epsilon_2 < \frac{\hbar^2}{4}$.

Thus, also in the vector case, for every approximate joint measurement M^σ, the total information loss $S(\rho, M^\sigma)$ does exceed the lower bound (92) even if $\mathcal{G}_\epsilon$ forbids preparations ρ with too peaked target distributions. Moreover, chosen ϵ_1 and ϵ_2, one can fix again a single 'bad' state ρ_ϵ in $\mathcal{G}_\epsilon$ that satisfies (92) for all Gaussian approximate joint measurements M^σ of $\mathbf{Q}$ and $\mathbf{P}$.

Whenever $\epsilon_1 \epsilon_2 \geq \frac{\hbar^2}{4}$, the optimal approximating joint measurement M^{σ_ϵ} is unique in the class of Gaussian covariant bi-observables; it corresponds to a minimum uncertainty state σ_ϵ which depends only on the chosen class of preparations $\mathcal{G}_\epsilon$, that is, on the thresholds ϵ_1 and ϵ_2: M^{σ_ϵ} is the best measurement for the worst choice of the preparation in that class.

Remark 15. *For $n = 1$, the vector lower bound in (92) reduces to the scalar lower bound found in (75) for two parallel directions $\mathbf{u}$ and $\mathbf{v}$; for $n \geq 1$, the bound linearly increases with n.*

Remark 16. The entropic incompatibility degree $c_{\text{inc}}^G(Q_u, P_v; \epsilon)$ is strictly positive for $\cos \alpha \neq 0$ (incompatible target observables) and it goes to zero in the limit $\alpha \to \pm \pi/2$ (compatible observables), $\hbar \to 0$ (classical limit), and $\epsilon_1 \epsilon_2 \to \infty$ (large uncertainty states).

Remark 17. Similarly to Remark 6 for scalar target observables, also the MUR (92) is actually ineffective for macroscopic systems. Indeed, suppose we are concerned with position and momentum of a macroscopic particle, say the center of mass of a multi-particle system (in this case $n = 3$). The states ρ which can be prepared in practice have macroscopic widths, say $\rho \in \mathcal{G}_\epsilon$ with 'large' thresholds ϵ and $\epsilon_1 \epsilon_2 \gg \hbar^2/4$. Then, we consider a measuring instrument M^{σ_*} having a high precision with respect to this class of states, but not necessarily attaining a precision near the quantum limits. For instance, let us take $\mathsf{M}^{\sigma_*} \in \mathcal{C}^G$ with $A^{\sigma_*} = \delta_1 \mathbb{1}$, $B^{\sigma_*} = \delta_2 \mathbb{1}$, and $0 < \delta_1 \ll \epsilon_1$, $0 < \delta_2 \ll \epsilon_2$; we assume M^{σ_*} is also unbiased: $a^{\sigma_*} = 0$, $b^{\sigma_*} = 0$. Obviously, $\delta_1 \delta_2 \geq \hbar^2/4$ must hold. Then, $\forall \rho \in \mathcal{G}_\epsilon$ by (77a) and (78a) we have

$$E_{\rho, \sigma_*} = \frac{\delta_1}{A^\rho} \leq \frac{\delta_1}{\epsilon_1} \mathbb{1}, \qquad F_{\rho, \sigma_*} = \frac{\delta_2}{B^\rho} \leq \frac{\delta_2}{\epsilon_2} \mathbb{1},$$

$$0 < S(\rho, \mathsf{M}^{\sigma_*}) = \frac{\log e}{2} \operatorname{Tr} \{s(E_{\rho, \sigma_*}) + s(F_{\rho, \sigma_*})\} \leq \frac{n \log e}{2} \left[s(\delta_1/\epsilon_1) + s(\delta_2/\epsilon_2)\right].$$

By (58) the function s is increasing and it behaves as $s(x) \simeq x^2/2$ in a neighborhood of zero; in the present case $\delta_1/\epsilon_1 \ll 1$ and $\delta_2/\epsilon_2 \ll 1$, thus implying that the error function is negligible. This is practically a 'classical' case: the preparation has 'large' position and momentum uncertainties and the measuring instrument is 'relatively good'. In this situation we do not see the difference between the joint measurement of position and momentum and their separate sharp distributions. Of course the bound (92) continues to hold, but it is also negligible since $\epsilon_1 \epsilon_2 \gg \hbar^2/4$.

Remark 18. Also in the vector case, the scale invariance of the relative entropy extends to the error function $S(\rho, \mathsf{M}^\sigma)$, the divergence $D_\epsilon^G(Q, P \| \mathsf{M}^\sigma)$ and the entropic incompatibility degree $c_{\text{inc}}^G(Q, P; \epsilon)$, as well as the entropic MUR (92). Indeed, let us consider the dimensionless versions of position and momentum (35) and their associated projection valued measures $\widetilde{Q}$, $\widetilde{P}$ introduced in Section 4. Accordingly, we rescale the joint measurement M^σ of (43) in the same way, obtaining the POVM

$$\widetilde{\mathsf{M}}^\sigma(B) = \int_B \widetilde{M}^\sigma(\widetilde{x}, \widetilde{p}) \mathrm{d}\widetilde{x} \mathrm{d}\widetilde{p},$$

$$\widetilde{M}^\sigma(\widetilde{x}, \widetilde{p}) = \frac{1}{(2\pi\lambda)^n} \exp\left\{\frac{\mathrm{i}}{\lambda} \left(\widetilde{p} \cdot \widetilde{Q} - \widetilde{x} \cdot \widetilde{P}\right)\right\} \Pi \sigma \Pi \exp\left\{-\frac{\mathrm{i}}{\lambda} \left(\widetilde{p} \cdot \widetilde{Q} - \widetilde{x} \cdot \widetilde{P}\right)\right\}.$$

Here, both the vector variables $\widetilde{x}$ and $\widetilde{p}$, as well as the components of the Borel set B, are dimensionless. By the scale invariance of the relative entropy, the error function takes the same value as in the dimensioned case:

$$S(\widetilde{Q}_\rho \| \widetilde{\mathsf{M}}_{1,\rho}^\sigma) + S(\widetilde{P}_\rho \| \widetilde{\mathsf{M}}_{2,\rho}^\sigma) = S(Q_\rho \| \mathsf{M}_{1,\rho}^\sigma) + S(P_\rho \| \mathsf{M}_{2,\rho}^\sigma). \tag{93}$$

Then, the scale invariance holds for the entropic divergence and incompatibility degree, too:

$$D_{\widetilde{\epsilon}}^G(\widetilde{Q}, \widetilde{P} \| \widetilde{\mathsf{M}}^\sigma) = D_\epsilon^G(Q, P \| \mathsf{M}^\sigma), \qquad c_{\text{inc}}^G(\widetilde{Q}, \widetilde{P}; \widetilde{\epsilon}) = c_{\text{inc}}^G(Q, P; \epsilon),$$

where $\widetilde{\epsilon}_1 := \frac{\varkappa \epsilon_1}{\hbar}$ and $\widetilde{\epsilon}_2 := \frac{\lambda^2 \epsilon_2}{\varkappa \hbar}$. In particular $\widetilde{\epsilon}_1 \widetilde{\epsilon}_2 \geq \frac{\lambda^2}{4} \iff \epsilon_1 \epsilon_2 \geq \frac{\hbar^2}{4}$ and, in this case, we have

$$n (\log e) s\left(\frac{\lambda}{2\sqrt{\widetilde{\epsilon}_1 \widetilde{\epsilon}_2}}\right) = c_{\text{inc}}^G(\widetilde{Q}, \widetilde{P}; \widetilde{\epsilon}) = c_{\text{inc}}^G(Q, P; \epsilon) = n (\log e) s\left(\frac{\hbar}{2\sqrt{\epsilon_1 \epsilon_2}}\right).$$

7. Conclusions

We have extended the relative entropy formulation of MURs given in [41] from the case of discrete incompatible observables to a particular instance of continuous target observables, namely the position and momentum vectors, or two components of them along two possibly non parallel directions. The entropic MURs we found share the nice property of being scale invariant and well-behaved in the classical and macroscopic limits. Moreover, in the scalar case, when the angle spanned by the position and momentum components goes to $\pm\pi/2$, the entropic bound correctly reflects their increasing compatibility by approaching zero with continuity.

Although our results are limited to the case of Gaussian preparation states and covariant Gaussian approximate joint measurements, we conjecture that the bounds we found still hold for arbitrary states and general (not necessarily covariant or Gaussian) bi-observables. Let us see with some more detail how this should work in the case when the target observables are the vectors $\mathbf{Q}$ and $\mathbf{P}$.

The most general procedure should be to consider the error function $S(\mathbf{Q}_\rho \| \mathsf{M}_{1,\rho}) + S(\mathbf{P}_\rho \| \mathsf{M}_{2,\rho})$ for an arbitrary POVM M on $\mathbb{R}^n \times \mathbb{R}^n$ and any state $\rho \in \mathcal{S}$. First of all, we need states for which neither the position nor the momentum dispersion are too small; the obvious generalization of the test states (81) is

$$\mathcal{S}_\epsilon := \{\rho \in \mathcal{S}_2 : A^\rho \geq \epsilon_1 \mathbb{1}, \ B^\rho \geq \epsilon_2 \mathbb{1}\}, \qquad \epsilon_i > 0.$$

Then, the most general definitions of the entropic divergence and incompatibility degree are:

$$D_\epsilon(\mathbf{Q}, \mathbf{P} \| \mathsf{M}) := \sup_{\rho \in \mathcal{S}_\epsilon} \left[S(\mathbf{Q}_\rho \| \mathsf{M}_{1,\rho}) + S(\mathbf{P}_\rho \| \mathsf{M}_{2,\rho}) \right], \tag{94}$$

$$c_{\text{inc}}(\mathbf{Q}, \mathbf{P}; \epsilon) := \inf_{\mathsf{M}} D_\epsilon(\mathbf{Q}, \mathbf{P} \| \mathsf{M}). \tag{95}$$

It may happen that $\mathbf{Q}_\rho$ is not absolutely continuous with respect to $\mathsf{M}_{1,\rho}$, or $\mathbf{P}_\rho$ with respect to $\mathsf{M}_{2,\rho}$; in this case, the error function and the entropic divergence take the value $+\infty$ by definition. So, we can restrict to bi-observables that are (weakly) absolutely continuous with respect to the Lebesgue measure. However, the true difficulty is that, even with this assumption, here we are not able to estimate (94), hence (95). It could be that the symmetrization techniques used in [17,19] can be extended to the present setting, and one can reduce the evaluation of the entropic incompatibility index to optimizing over all covariant bi-observables. Indeed, in the present paper we a priori selected only covariant approximating measurements; we would like to understand if, among all approximating measurements, the relative entropy approach selects covariant bi-observables by itself. However, even if M is covariant, there remains the problem that we do not know how to evaluate (94) if ρ and M are not Gaussian. It is reasonable to expect that some continuity and convexity arguments should apply, and the bounds in Theorem 5 might be extended to the general case by taking dense convex combinations. Also the techniques used for the PURs in [8,9] could be of help in order to extend what we did with Gaussian states to arbitrary states. This leads us to conjecture:

$$c_{\text{inc}}(\mathbf{Q}, \mathbf{P}; \epsilon) = c_{\text{inc}}^G(\mathbf{Q}, \mathbf{P}; \epsilon). \tag{96}$$

Conjecture (96) is also supported since the uniqueness of the optimal approximating bi-observable in Theorem 5(i) is reminiscent of what happens in the discrete case of two Fourier conjugated mutually unbiased bases (MUBs); indeed, in the latter case, the optimal bi-observable is actually unique among all the bi-observables, not only the covariant ones (see [41] (Theorem 5)).

Similar considerations obviously apply also to the case of scalar target observables. We leave a more deep investigation of equality (96) to future work.

As a final consideration, one could be interested in finding error/disturbance bounds involving sequential measurements of position and momentum, rather than considering all their possible approximate joint measurements. As sequential measurements are a proper subset of the set of all the bi-observables, optimizing only over them should lead to bounds that are greater than c_{inc}.

This is the reason for which in [41] an error/disturbance entropic bound, denoted by c_{ed} and dinstinct from c_{inc}, was introduced. However, it was also proved that the equality $c_{inc} = c_{ed}$ holds when one of the target observables is discrete and sharp. Now, in the present paper, only sharp target observables are involved; although the argument of [41] can not be extended to the continuous setting, the optimal approximating joint observables we found in Theorems 3(i) and 5(i) *actually are* sequential measurements. Indeed, the optimal bi-observable in Theorem 3(i) is one of the POVMs described in Examples 2 and 3 (see (74)); all these bi-observables have a (trivial) sequential implementation in terms of an unsharp measurement of Q_u followed by sharp P_v. On the other hand, in the vector case, it was shown in ([67], Corollary 1) that all covariant phase-space observables can be obtained as a sequential measurement of an unsharp version of the position Q followed by the sharp measurement of the momentum P. Therefore, $c_{inc} = c_{ed}$ also for target position and momentum observables, in both the scalar and vector case.

Author Contributions: The three authors equally contributed to the paper.

Conflicts of Interest: The authors declare no conflict of interest.

References

1. Heisenberg, W. Über den anschaulichen Inhalt der quantentheoretischen Kinematik und Mechanik. *Zeitschr. Phys.* **1927**, *43*, 172–198.
2. Simon, R.; Mukunda, N.; Dutta, B. Quantum-noise matrix for multimode systems: $U(n)$ invariance, squeezing, and normal forms. *Phys. Rev. A* **1994**, *49*, 1567–1583.
3. Holevo, A.S. *Statistical Structure of Quantum Theory*; Lecture Notes in Physics Monographs 67; Springer: Berlin, Germany, 2001.
4. Holevo, A.S. *Probabilistic and Statistical Aspects of Quantum Theory*; Quaderni della Normale; Edizioni della Normale: Pisa, Italy, 2011.
5. Holevo, A.S. *Quantum Systems, Channels, Information*; De Gruiter: Berlin, Germany, 2012.
6. Robertson, H. The uncertainty principle. *Phys. Rev.* **1929**, *34*, 163–164.
7. Hirschman, I.I. A note on entropy. *Am. J. Math.* **1957**, *79*, 152–156.
8. Beckner, W. Inequalities in Fourier analysis. *Ann. Math.* **1975**, *102*, 159–182.
9. Białynicki-Birula, I.; Mycielski, J. Uncertainty relations for information entropy in wave machanics. *Commun. Math. Phys.* **1975**, *44*, 129–132.
10. Maassen, H.; Uffink, J.B.M. Generalized entropic uncertainty relations. *Phys. Rev. Lett.* **1988**, *60*, 1103–1106.
11. Gibilisco, P.; Isola, T. On a refinement of Heisenberg uncertainty relation by means of quantum Fisher information. *J. Math. Anal. Appl.* **2011**, *375*, 270–275.
12. Coles, P.J.; Berta, M.; Tomamichel, M.; Whener, S. Entropic uncertainty relations and their applications. *Rev. Mod. Phys.* **2017**, *89*, 015002.
13. Wehner, S.; Winter, A. Entropic uncertainty relations—A survey. *New J. Phys.* **2010**, *12*, 025009.
14. Ozawa, M. Position measuring interactions and the Heisenberg uncertainty principle. *Phys. Lett. A* **2002**, *299*, 1–7.
15. Ozawa, M. Physical content of Heisenberg's uncertainty relation: Limitation and reformulation. *Phys. Lett. A* **2003**, *318*, 21–29.
16. Ozawa, M. Universally valid reformulation of the Heisenberg uncertainty principle on noise and disturbance in measurement. *Phys. Rev. A* **2003**, *67*, 042105.
17. Werner, R.F. The uncertainty relation for joint measurement of position and momentum. *Quantum Inf. Comput.* **2004**, *4*, 546–562.
18. Busch, P.; Heinonen, T.; Lahti, P. Heisenberg's Uncertainty Principle. *Phys. Rep.* **2007**, *452*, 155–176.
19. Busch, P.; Lahti, P.; Werner, R. Measurement uncertainty relations. *J. Math. Phys.* **2014**, *55*, 042111.
20. Busch, P.; Lahti, P.; Werner, R. Quantum root-mean-square error and measurement uncertainty relations. *Rev. Mod. Phys.* **2014**, *86*, 1261–1281.
21. Ozawa, M. Heisenberg's original derivation of the uncertainty principle and its universally valid reformulations. *Curr. Sci.* **2015**, *109*, 2006–2016.
22. Davies, E.B. *Quantum Theory of Open Systems*; Academic: London, UK, 1976.

23. Busch, P.; Grabowski, M.; Lahti, P. *Operational Quantum Physics*; Springer: Berlin, Germany, 1997.
24. Barchielli, A.; Gregoratti, M. *Quantum Trajectories and Measurements in Continuous Time: The Diffusive Case*; Lecture Notes in Physics; Springer: Berlin/Heidelberg, Germany, 2009; Volume 782.
25. Heinosaari, T.; Ziman, M. *The Mathematical Language of Quantum Theory: From Uncertainty to Entanglement*; Cambridge University Press: Cambridge, UK, 2012.
26. Busch, P.; Lahti, P.; Pellonpää, J.-P.; Ylinen, K. *Quantum Measurement*; Springer: Berlin, Germany, 2016.
27. Buscemi, F.; Hall, M.J.W.; Ozawa, M.; Wilde, M.M. Noise and disturbance in quantum measurements: An information-theoretic approach. *Phys. Rev. Lett.* **2014**, *112*, 050401.
28. Busch, P.; Heinosaari, T.; Schultz, J.; Stevens, N. Comparing the degrees of incompatibility inherent in probabilistic physical theories. *Europhys. Lett.* **2013**, *103*, 10002.
29. Busch, P.; Lahti, P.; Werner, R. Proof of Heisenberg's error-disturbance relation. *Phys. Rev. Lett.* **2013**, *111*, 160405.
30. Coles, P.J.; Furrer, F. State-dependent approach to entropic measurement-disturbance relations. *Phys. Lett. A* **2015**, *379*, 105–112.
31. Heinosaari, T.; Schultz, J.; Toigo, A.; Ziman, M. Maximally incompatible quantum observables. *Phys. Lett. A* **2014**, *378*, 1695–1699.
32. Werner, R.F. Uncertainty relations for general phase spaces. *Front. Phys.* **2016**, *11*, 110305.
33. Buscemi, F.; Das, S.; Wilde, M.M. Approximate reversibility in the context of entropy gain, information gain, and complete positivity. *Phys. Rev. A* **2016**, *93*, 062314.
34. Barchielli, A.; Lupieri, G. Instrumental processes, entropies, information in quantum continual measurements. *Quantum Inf. Comput.* **2004**, *4*, 437–449.
35. Barchielli, A.; Lupieri, G. Instruments and channels in quantum information theory. *Opt. Spectrosc.* **2005**, *99*, 425–432.
36. Barchielli, A.; Lupieri, G. Quantum measurements and entropic bounds on information transmission. *Quantum Inf. Comput.* **2006**, *6*, 16–45.
37. Barchielli, A.; Lupieri, G. Instruments and mutual entropies in quantum information. *Banach Center Publ.* **2006**, *73*, 65–80.
38. Barchielli, A.; Lupieri, G. Entropic bounds and continual measurements. In *Quantum Probability and Infinite Dimensional Analysis*; QP-PQ: Quantum Probability and White Noise Analysis; Accardi, L., Freudenberg, W., Schürmann, M., Eds.; World Scientific: Singapore, 2007; Volume 20, pp. 79–89.
39. Barchielli, A.; Lupieri, G. Information gain in quantum continual measurements. In *Quantum Stochastic and Information*; Belavkin, V.P., Guţă, M., Eds.; World Scientific: Singapore, 2008; pp. 325–345.
40. Maccone, L. Entropic information-disturbance tradeoff. *EPL* **2007**, *77*, 40002.
41. Barchielli, A.; Gregoratti, M.; Toigo, A. Measurement uncertainty relations for discrete observables: Relative entropy formulation. *arXiv* **2016**, arXiv:1608.01986.
42. Braunstein, S.L.; van Loock, P. Quantum information with continuous variables. *Rev. Mod. Phys.* **2005**, *77*, 513–577.
43. Heinosaari, T.; Kiukas, J.; Schultz, J. Breaking Gaussian incompatibility on continuous variable quantum systems. *J. Math. Phys.* **2015**, *56*, 082202.
44. Kiukas, J.; Schultz, J. Informationally complete sets of Gaussian measurements. *J. Phys. A Math. Theor.* **2013**, *46*, 485303.
45. Weedbrook, C.; Pirandola, S.; García-Patrón, R.; Cerf, N.J.; Ralph, T.C.; Shapiro, J.H.; Lloyd, S. Gaussian quantum information. *Rev. Mod. Phys.* **2012**, *84*, 621–669.
46. Huang, Y. Entropic uncertainty relations in multidimensional position and momentum spaces. *Phys. Rev. A* **2011**, *83*, 052124.
47. Heinosaari, T.; Miyadera, T.; Ziman, M. An invitation to quantum incompatibility. *J. Phys. A Math. Theor.* **2016**, *49*, 123001.
48. Simon, R.; Sudarshan, E.C.G.; Mukunda, N. Gaussian-Wigner distributions in quantum mechanics and optics. *Phys. Rev. A* **1987**, *36*, 3868–3880.
49. Horn, R.A.; Zhang, F. Basic Properties of the Schur Complement. In *The Schur Complement and Its Applications*; Zhang, F., Ed.; Numerical Methods and Algorithms; Springer: Berlin, Germany, 2005; pp. 17–46.
50. Petz, D. *Quantum Information Theory and Quantum Statistics*; Springer: Berlin, Germany, 2008.

51. Carlen, E. Trace Inequalities and Quantum Entropy: An Introductory Course. In *Entropy and the Quantum*; Contemporary Mathematics; American Mathematical Society: Providence, RI, USA, 2010; Volume 529, pp. 73–140.
52. Bhatia, R. *Matrix Analysis*; Springer: New York, NY, USA, 1997.
53. Werner, R.F. Quantum harmonic analysis on phase spaces. *J. Math. Phys.* **1983**, *25*, 1404–1411.
54. Burnham, K.P.; Anderson, D.R. *Model Selection and Multimodel Inference—A Practical Information—Theoretic Approach*; Springer: New York, NY, USA, 2002.
55. Topsøe, F. Basic concepts, identities and inequalities—The toolkit of Information Theory. *Entropy* **2011**, *3*, 162–190.
56. Cover, T.M.; Thomas, J.A. *Elements of Information Theory*, 2nd ed.; Wiley: Hoboken, NJ, USA, 2006.
57. Carmeli, C.; Heinonen, T.; Toigo, A. Position and momentum observables on R and on R^3. *J. Math. Phys.* **2004**, *45*, 2526–2539.
58. Barchielli, A.; Lupieri, G. Quantum stochastic calculus, operation valued stochastic processes and continual measurements in quantum mechanics. *J. Math. Phys.* **1985**, *26*, 2222–2230.
59. Barchielli, A.; Lupieri, G. A quantum analogue of Hunt's representation theorem for the generator of convolution semigroups on Lie groups. *Probab. Theory Rel. Fields* **1991**, *88*, 167–194.
60. Barchielli, A.; Holevo, A.S.; Lupieri, G. An analogue of Hunt's representation theorem in quantum probability. *J. Theor. Probab.* **1993**, *6*, 231–265.
61. Holevo, A.S. Investigations in the General Theory of Statistical Decisions. *Proc. Steklov Inst. Math.* **1978**, *124*, 1–140.
62. Holevo, A.S. Infinitely divisible measurements in quantum probability theory. *Theory Probab. Appl.* **1986**, *31*, 493–497.
63. Cassinelli, G.; De Vito, E.; Toigo, A. Positive operator valued measures covariant with respect to an irreducible representation. *J. Math. Phys.* **2003**, *44*, 4768–4775.
64. Kiukas, J.; Lahti, P.; Ylinen, K. Normal covariant quantization maps. *J. Math. Anal. Appl.* **2006**, *319*, 783–801.
65. Ohya, M.; Petz, D. *Quantum Entropy and Its Use*; Springer: Berlin, Germany, 1993.
66. Billingsley, P. *Probability and Measure*, 2nd ed.; Wiley: New York, NY, USA, 1986.
67. Carmeli, C.; Heinonen, T.; Toigo, A. Sequential measurements of conjugate observables. *J. Phys. A Math. Theor.* **2011**, *44*, 285304.

© 2017 by the authors. Licensee MDPI, Basel, Switzerland. This article is an open access article distributed under the terms and conditions of the Creative Commons Attribution (CC BY) license (http://creativecommons.org/licenses/by/4.0/).

Article

Planck-Scale Soccer-Ball Problem: A Case of Mistaken Identity

Giovanni Amelino-Camelia [1,2]

[1] Dipartimento di Fisica, Università di Roma "La Sapienza", P.le A. Moro 2, 00185 Roma, Italy; amelino@roma1.infn.it
[2] Istituto Nazionale di Fisica Nucleare (INFN), Sez. Roma1, P.le A. Moro 2, 00185 Roma, Italy

Received: 23 April 2017; Accepted: 18 July 2017; Published: 2 August 2017

Abstract: Over the last decade, it has been found that nonlinear laws of composition of momenta are predicted by some alternative approaches to "real" 4D quantum gravity, and by all formulations of dimensionally-reduced (3D) quantum gravity coupled to matter. The possible relevance for rather different quantum-gravity models has motivated several studies, but this interest is being tempered by concerns that a nonlinear law of addition of momenta might inevitably produce a pathological description of the total momentum of a macroscopic body. I here show that such concerns are unjustified, finding that they are rooted in failure to appreciate the differences between two roles for laws composition of momentum in physics. Previous results relied exclusively on the role of a law of momentum composition in the description of spacetime locality. However, the notion of total momentum of a multi-particle system is not a manifestation of locality, but rather reflects translational invariance. By working within an illustrative example of quantum spacetime, I show explicitly that spacetime locality is indeed reflected in a nonlinear law of composition of momenta, but translational invariance still results in an undeformed linear law of addition of momenta building up the total momentum of a multi-particle system.

Keywords: quantum foundations; relativity; quantum gravity

1. Introduction

An emerging characteristic of quantum-gravity research over the last decade has been a gradual shift of focus toward manifestations of the Planck scale on momentum space, particularly pronounced in some approaches to quantum gravity. For some research lines based on spacetime noncommutativity, several momentum-space structures have been in focus, including the possibility of deformed laws of composition of momenta, which shall be here of interest. While deformed laws of composition of momenta are found to be inevitable in some approaches based on spacetime noncommutativity (e.g., [1–6]), the situation is less certain in the loop-quantum-gravity approach. For "real" 4D loop quantum gravity, the relevant issues are partly obscured by our present limited understanding of the semiclassical limit of that theory [7], but some indirect arguments suggest that a nonlinear law of composition of momenta might arise [8,9]. These arguments find further strength in results on 3D loop quantum gravity, where the simplifications afforded by that dimensionally-reduced model allow one to rigorously show that indeed the nonlinearities on momentum space are present (e.g., [10]). Actually, evidence is growing that in all alternative formulations of 3D quantum gravity coupled to matter there are nonlinearities in momentum space, including nonlinear laws of composition of momenta (e.g., [11]). The role played by nonlinearities on momentum space is also noteworthy in two recently-proposed approaches to the quantum-gravity problem: the one based on group field theory [12] and the one based on the relative-locality framework [13].

Due to the lack of experimental guidance, a variety of approaches to quantum gravity are being developed, and in most cases the different approaches have very little in common. This of course endows with additional reasons of interest any result which is found to apply to more than one approach. Indeed, there has been growing interest in the conceptual implications and possible phenomenological implications [14] of nonlinear laws on momentum space and particularly nonlinear laws of composition of momenta. However, this interest is being tempered by concerns that a nonlinear law of addition of momenta might inevitably produce a pathological description of the total momentum of a macroscopic body [15–23] (also see References [24–26] for a related discussion focused within the novel relative-locality framework). This issue has often been labelled as the "soccer-ball problem" [17]: the quantum-gravity pictures lead one to expect nonlinearities of the law of composition of momentum which are suppressed by the Planck scale ($\sim 10^{28}$ eV) and would be unobservably small for particles at energies we presently can access, but in the analysis of a macroscopic body (e.g., a soccer ball), one might have to add up very many of such minute nonlinearities, ultimately obtaining results in conflict with observations [15–23].

If this so-called "soccer-ball problem" really was a scientific problem (a case of actual conflict with experimental data), we could draw rather sharp conclusions about several areas of quantum-gravity research. Perhaps most notably we should consider as ruled out large branches of research on quantum-gravity based on spacetime noncommutativity and we should consider the whole effort of research on dimensionally-reduced 3D quantum gravity as completely unreliable in forming an intuition for "real" 4D quantum gravity. However, I here show that previous discussions of this soccer-ball problem [15–26] failed to appreciate the differences between two roles for laws of composition of momentum in physics. Previous results supporting a nonlinear law of addition of momenta relied exclusively on the role of a law of momentum composition in the description of spacetime locality. The notion of total momentum of a multi-particle system is not a manifestation of locality, but rather reflects translational invariance in interacting theories. After being myself confused about these issues for quite some time [17] I feel I am now in a position to articulate the needed discussion at a completely general level. However, considering the tone and content of the bulk of literature that precedes this contribution of mine I find it is best to opt here instead for a very explicit discussion based on illustrative examples of calculations performed within a specific simple model affected by nonlinearities for a law of composition of momenta. The model I focus on has 2 + 1-dimensional pure-spatial κ-Minkowski noncommutativity [1–6], with the time coordinate left unaffected by the deformation and the two spatial coordinates, x_1 and x_2, governed by

$$[x_1, x_2] = i\ell x_1 \tag{1}$$

(with the deformation scale ℓ expected to be of the order of the inverse of the Planck scale).

In the next section I briefly review within this example of quantum spacetime previous arguments showing that spacetime locality is reflected in a nonlinear law of composition of momenta. Then, Section 3 takes off from known results on translational invariance for κ-Minkowski noncommutative spacetimes and builds on those to achieve the first ever example of translationally-invariant interacting two-particle system in κ-Minkowski. This allows me to explicitly verify that the conserved charge associated with that translational invariance (the total momentum of the two-particle system) adds linearly the momenta of the two particles involved. Section 4 offers some closing remarks.

2. Soccer-Ball Problem and Sum of Momenta from Locality

The ingredients needed for seeing a nonlinear law of composition of momenta emerging from noncommutativity of type (1) are very simple. Essentially, one needs only to rely on results establishing that functions of coordinates governed by (1) still admit a rather standard Fourier expansion (e.g., [1,2])

$$\Phi(x) = \int d^4k \, \tilde{\Phi}(k) \, e^{ik_\mu x^\mu}$$

and that the notion of integration on such a noncommutative space preserves many of the standard properties including [1,3]

$$\int d^4x \, e^{ik_\mu x^\mu} = (2\pi)^4 \delta^{(4)}(k). \tag{2}$$

It is a rather standard exercise for practitioners of spacetime noncommutativity to use these tools in order to enforce locality within actions describing classical fields. For example, one might want to introduce in the action the product of three (possibly identical, but in general different) fields, Φ, Ψ, Y, insisting on locality in the sense that the three fields be evaluated "at the same quantum point x"; i.e., $\Phi(x) \Psi(x) Y(x)$. There is still no consensus on how one should formulate the more interesting quantum-field version of such theories, and it remains unclear to which extent and in which way our ordinary notion of locality is generalized by the requirement of evaluating "at the same quantum point x" fields intervening in a product such as $\Phi(x) \Psi(x) Y(x)$. Nonetheless, for the classical-field case there is a sizable body of literature consistently adopting this prescription for locality. Important for my purposes here is the fact that with such a prescription, locality inevitably leads to a nonlinear law of composition of momenta, as I show explicitly in the following example:

$$\int d^4x \, \Phi(x) \, \Psi(x) \, Y(x) = \tag{3}$$
$$= \int d^4x \int d^4k \int d^4p \int d^4q \, \tilde\Phi(k) \, \tilde\Psi(p) \, \tilde Y(q) \, e^{ik_\mu x^\mu} e^{ip_\nu x^\nu} e^{iq_\rho x^\rho}$$
$$= \int d^4x \int d^4k \int d^4p \int d^4q \, \tilde\Phi(k) \, \tilde\Psi(p) \, \tilde Y(q) e^{i(k \oplus p \oplus q)_\mu x^\mu}$$
$$= (2\pi)^4 \int d^4k \int d^4p \int d^4q \, \tilde\Phi(k) \, \tilde\Psi(p) \, \tilde Y(q) \, \delta^{(4)}(k \oplus p \oplus q)$$

where $\oplus$ is such that

$$(k \oplus p)_0 = k_0 + p_0 \tag{4}$$

$$(k \oplus p)_2 = k_2 + p_2 \tag{5}$$

$$(k \oplus p)_1 = \frac{k_2 + p_2}{1 - e^{\ell(k_2 + p_2)}} \left[\frac{1 - e^{\ell k_2}}{k_2 e^{\ell p_2}} k_1 + \frac{1 - e^{\ell p_2}}{p_2} p_1 \right] \tag{6}$$

This result is rooted in one of the most studied aspects of such noncommutative spacetimes, which is their "generalized star product" [1–3]. This is essentially a characterization of the properties of products of exponentials induced by rules of noncommutativity of type (1). Specifically, one easily arrives at (3) (with $\oplus$ such that, in particular, (6) holds) by just observing that from the defining commutator (1) it follows that (Equation (7) is a particular example of application of the Baker-Campbell-Hausdorff formula for products of exponentials of noncommuting variables. In general, the Baker–Campbell–Hausdorff formula involves an infinite series of nested commutators, but the case of noncommutativity (1) is one of the cases for which the series of nested commutators can be resummed explicitly [2,3]) [2,3]:

$$\log \left[\exp(ik_2 x_2 + ik_1 x_1) \exp(ip_2 x_2 + ip_1 x_1) \right] = \tag{7}$$
$$= ix_2(p_2 + k_2) + ix_1 \frac{k_2 + p_2}{1 - e^{\ell(k_2 + p_2)}} \left(\frac{1 - e^{\ell k_2}}{k_2 e^{\ell p_2}} k_1 + \frac{1 - e^{\ell p_2}}{p_2} p_1 \right)$$

The so-called soccer-ball problem concerns the acceptability of laws of composition of type (6). Since one assumes that the deformation scale ℓ is on the order of the inverse of the Planck scale, applying (6) to microscopic/fundamental particles has no sizable consequences: of course (6) gives us back to good approximation $(k \oplus p)_1 \simeq k_1 + p_1$ whenever $|\ell k_2| \ll 1$ and $|\ell p_2| \ll 1$. However, if a law of composition such as (6) should be used also when we add very many microparticle momenta in obtaining the total momentum of a multiparticle system (such as a soccer ball), then the final result could be pathological [15–26] even when each microparticle in the system has momentum much smaller than $1/\ell$.

3. Sum of Momenta from Translational Invariance

As clarified in the brief review of known results given in the previous section, a nonlinear law of composition of momenta arises in characterizations of locality, as a direct consequence of the form of some star products. My main point here is that a different law of composition of momenta is produced by the analysis of translational invariance, and it is this other law of composition of momenta which is relevant for the characterization of the total momentum of a multi-particle system. Here too I shall use only known facts about the peculiarities of translation transformations in certain noncommutative spacetimes, but exploit them to obtain results that had not been derived before—indeed, results relevant for the description of the total momentum of a multi-particle system.

A first hint that translation transformations should be modified [4–6] in certain noncommutative spacetimes comes from noticing that (1) is incompatible with the standard Heisenberg relations $[p_j, x_k] = i\delta_{jk}$. Indeed, if one adopts (1) and $[p_j, x_k] = i\delta_{jk}$, one then easily finds that some Jacobi identities are not satisfied. The relevant Jacobi identities are satisfied if one allows for a modification of the Heisenberg relations which balances for the noncommutativity of the coordinates:

$$[p_1, x_1] = i \,, \quad [p_2, x_1] = 0 \,, \quad [p_2, x_2] = i \,, \tag{8}$$

$$[p_1, x_2] = -i\ell p_1 \,, \tag{9}$$

One easily finds that by combining (1), (8), and (9), all Jacobi identities are satisfied [4–6].

Additional intuition for these nonstandard properties of the momenta p_j comes from actually looking at which formulation of translation transformations preserves the form of the noncommutativity of coordinates (1). Evidently, the standard description

$$x_2 \to x_2' = x_2 + a_2 \,, \quad x_1 \to x_1' = x_1 + a_1$$

is not a symmetry of (1):

$$[x_1', x_2'] = [x_1 + a_1, x_2 + a_2] = i\ell x_1 = i\ell(x_1' - a_1) \tag{10}$$

Unsurprisingly, what does work is the description of translation transformations using as generators the p_j of (8) and (9), which as stressed above satisfy the Jacobi-identity criterion. These deformed translation transformations take the form

$$x_1' = x_1 - ia_1[p_1, x_1] - ia_2[p_2, x_1] = x_1 + a_1 \,,$$
$$x_2' = x_2 - ia_1[p_1, x_2] - ia_2[p_2, x_2] = x_2 + a_2 - \ell a_1 p_1 \tag{11}$$

and indeed are symmetries of the commutation rules (1):

$$[x_1', x_2'] = [x_1 + a_1, x_2 + a_2 - \ell a_1 p_1] =$$
$$= i\ell x_1 - \ell a_1 [x_1, p_1] = i\ell(x_1 + a_1) = i\ell x_1' \tag{12}$$

All this about translation transformations in certain noncommutative spacetimes is well known (e.g., [4–6]). The part which I am here going to contribute is to show how this is relevant for the mentioned much-debated issue about the total momentum of a multi-particle system. My starting point is that in order for us to be able to even contemplate the total momentum of a multiparticle system, we must be dealing with a case where translational invariance is ensured: total momentum is the conserved charge for a translationally invariant multi-particle system. Surely the introduction of translationally invariant multi-particle systems must involve some subtleties due to the noncommutativity of coordinates, and these subtleties are directly connected to the new properties of translation transformations (9), but they are not directly connected to the properties of the star product (7) and the associated law of composition of momenta (6). For my purposes, also considering the heated debate that precedes this contribution of mine, it is best to show the implications of this point very simply and explicitly, focusing on a system of two particles interacting via a harmonic potential.

I start by noticing that evidently one does not achieve translational invariance through a description of the form

$$\mathcal{H}_{non-transl} = \frac{(p_1^A)^2}{2m} + \frac{(p_2^A)^2}{2m} + \frac{(p_1^B)^2}{2m} + \frac{(p_2^B)^2}{2m} + $$
$$+ \frac{1}{2}\rho[(x_1^A - x_1^B)^2 + (x_2^A - x_2^B)^2] \tag{13}$$

where indices A and B label the two particles involved in the interaction via the harmonic potential. As stressed above, translation transformations consistent with the coordinate noncommutativity (1) must be such that (see (11)) $x_1 \to x_1 + a_1$ and $x_2 \to x_2 + a_2 - \ell a_1 p_1$, and as a result by writing the harmonic potential with $(x_1^A - x_1^B)^2 + (x_2^A - x_2^B)^2$, one does not achieve translational invariance.

One does get translational invariance by adopting instead

$$\mathcal{H} = \frac{(p_1^A)^2}{2m} + \frac{(p_2^A)^2}{2m} + \frac{(p_1^B)^2}{2m} + \frac{(p_2^B)^2}{2m} + $$
$$+ \frac{1}{2}\rho[(x_1^A - x_1^B)^2 + (x_2^A + \ell x_1^A p_1^A - x_2^B - \ell x_1^B p_1^B)^2] \tag{14}$$

This is trivially invariant under translations generated by p_2, which simply produce $x_1 \to x_1$ and $x_2 \to x_2 + a_2$. It is also invariant under translations generated by p_1, since they produce $x_1 \to x_1 + a_1$ and $x_2 \to x_2 - \ell a_1 p_1$, so that $x_2 + \ell x_1 p_1$ is left unchanged:

$$x_2 + \ell x_1 p_1 \to x_2 - \ell a_1 p_1 + \ell(x_1 + a_1)p_1 = x_2 + \ell x_1 p_1$$

It is interesting for my purposes to see which conserved charge is associated with this invariance under translations of the hamiltonian $\mathcal{H}$. This conserved charge will describe the total momentum of the two-particle system governed by $\mathcal{H}$ (i.e., the center-of-mass momentum). It is easy to see that this conserved charge is just the standard $\vec{p}^A + \vec{p}^B$. For the second component, one trivially finds that indeed

$$[p_2^A + p_2^B, \mathcal{H}] = 0$$

and the same result also applies to the first component:

$$[p_1^A + p_1^B, \mathcal{H}] \propto [p_1^A + p_1^B, (x_1^A - x_1^B)^2] + $$
$$+ [p_1^A + p_1^B, (x_2^A + \ell x_1^A p_1^A - x_2^B - \ell x_1^B p_1^B)^2] = $$
$$= [p_1^A + p_1^B, (x_2^A + \ell x_1^A p_1^A - x_2^B - \ell x_1^B p_1^B)^2] \propto$$
$$\propto [p_1^A + p_1^B, x_2^A + \ell x_1^A p_1^A - x_2^B - \ell x_1^B p_1^B]$$
$$= -i\ell p_1^A + i\ell p_1^A + i\ell p_1^B - i\ell p_1^B = 0 \tag{15}$$

where the only non-trivial observation I have used is that (1) leads to $[p_1, x_2 + \ell x_1 p_1] = -i\ell p_1 + i\ell p_1 = 0$.

The result (15) shows that indeed $\vec{p}^A + \vec{p}^B$ is the momentum of the center of mass of my translationally-invariant two-particle system; i.e., it is the total momentum of the system.

The concerns about total momentum that had been voiced in discussions of the Planck-scale soccer-ball problem were rooted in the different sum of momenta relevant for locality, the $\oplus$ sum discussed in the previous section. It was feared that one should obtain the total momentum by combining single-particle momenta with the nonlinear $\oplus$ sum. The result (15) shows that this expectation was incorrect. One can also directly verify that indeed $\vec{p}^A \oplus \vec{p}^B$ is not a conserved charge for my translationally-invariant two-particle system, and specifically, taking into account (6), one finds that

$$[(\vec{p}^A \oplus \vec{p}^B)_1, \mathcal{H}] \neq 0$$

This completes my thesis, but in closing this section I should warn readers of the fact that while the picture emerging from my analysis is rather compelling, one should not forget that the interpretation of the notion of total momentum in a noncommutative spacetime remains affected by some open issues (see Reference [14] and references therein). Even the physical meaning of having noncommutative spacetime coordinates is still being debated. In the shadow of these interpretational issues, we cannot even be sure that the Hamiltonian of Equation (14) has physical (observable) consequences different from an ordinary harmonic-oscillator theory. Nonetheless, my analysis contributes to this ongoing debate by exposing two notions of momentum conservation: one connected to locality, and one connected with translational invariance. Evidently, if interpreted in standard way, these two notions could be mutually incompatible: in the analysis of a chain of events one might naturally want to insist on overall total-momentum conservation, but in some parts of the chain of events the conserved quantity might be the one coming from locality, while in other parts of the chain of events the conserved quantity might the one coming from translational invariance. Addressing this apparent puzzle might require a totally new interpretation of the notion of momentum of a particle in a quantum spacetime, while failing to address it might be a mortal blow to the whole research area. While in part my results are sub judice because of these interpretational issues, my analysis nonetheless firmly establishes the main conceptual point I am making, which concerns the differences between "composition of momentum appearing in locality analyses" and "composition of momentum appearing in translational-invariance analyses"—two notions which are usually confused with each other due to the fact that in a classical spacetime they coincide.

4. Implications and Outlook

The results here reported suggest that—at least within the framework of κ-Minkowski spacetime noncommutativity, there might be no "soccer-ball problem". I am confident that analogous results will emerge in other similar formalisms, but of course dedicated analyses are needed. A case of particular interest might be that of the Snyder model of spacetime noncommutativity [27], which is already known to have a complicated interplay with translational invariance: the original model of Reference [27] is not invariant under translations, but a variant with an extra dimension recovers translational invariance [28].

The Hamiltonian of Equation (14) is the only one I managed to find which is invariant under the translation transformations (11), but I do not have any proof of uniqueness. It would be interesting to consider other Hamiltonians that are invariant under (11) and give the ordinary harmonic-oscillator Hamiltonian in the $\ell \to 0$ limit.

As usual in physics, attempts to generalize a theory also help us understand the theory itself: the analysis I here reported makes us appreciate how our current theories are built on a non-trivial correspondence between the momentum-space manifestations of locality and translational invariance. This can be viewed from a different perspective by reconsidering the fact that in Galilean relativity all laws of composition of momenta and velocities are linear, and there is a linear relationship between velocity and momentum. Within Galilean-relativistic theories, one could choose to never speak of momentum and work exclusively in terms of velocities, with apparently a single linear law of composition of velocities. In our current post-Galilean theories, the relationship between momentum and velocity is non-linear, and we then manage to appreciate differences between composition laws (in our current theories all laws of composition of momenta remain linear, but velocities are composed non-linearly).

I must also comment on the fact that aspects of my analysis pertaining to translational invariance were confined to a first-quantized system. This came out of necessity since several grey areas remain for the formulation of second quantization with κ-Minkowski noncommutativity. As a matter of fact, I here provided the first ever translationally-invariant formulation of an interacting theory in κ-Minkowski. All previous attempts had been made within quantum field theory, and led to unsatisfactory results, particularly concerning global translational invariance. Perhaps the results I here reported could provide guidance for improving upon previous attempts at formulating interacting quantum field theories in κ-Minkowski. In particular, it might be appropriate to make room for some novel notion of "coincidence of points"—a possibility which had not been considered in previous attempts. I see a hint pointing in this direction in the structure of my translationally-invariant harmonic potential: unlike standard Harmonic potentials, the potential in my Equation (14) does not vanish when the coordinates of the particles coincide: the potential in Equation (14) vanishes for $x_1^A = x_1^B$ and $x_2^A = x_2^B$ only if the momenta also coincide ($p_1^A = p_1^B$). This is reminiscent of some results obtained within the recently-proposed relative-locality framework [13], where the only meaningful notion of "coincidence" is a phase-space notion (not a notion that could be formulated exclusively in spacetime). This suggests that one could perhaps improve upon previous attempts to formulate interacting quantum field theories in κ-Minkowski by exploiting quantum-field-theory results being developed [29] for the relative-locality framework.

Another direction for future studies which might bring some enlightenment concerns building interacting theories with full relativistic covariance. Herein I focused on translation transformations because it was sufficient for the purposes of my study, but it would be interesting to ask what additional constraints would arise if one insists on full relativistic covariance (including boosts and spatial rotations) rather than just translational invariance. For the law of composition of momenta based on locality, a fully consistent relativistic picture is already known [13,14,29], and its consistency with κ-Minkowski noncommutativity is well established. Important insight might be gained by establishing whether or not analogous results are available for the law of composition of momenta based on translational invariance of my interacting Hamiltonian.

Conflicts of Interest: The author declares no conflict of interest.

References

1. Majid, S. Meaning of Noncommutative Geometry and the Planck-Scale Quantum Group. *Lect. Notes Phys.* **2000**, *541*, 227.
2. Kosinski, P.; Lukierski, J.; Maslanka, P. Local Field Theory ON κ-Minkowski Space, Star Products and Noncommutative Translations. *Czechoslov. J. Phys.* **2000**, *50*, 1283–1290.
3. Agostini, A.; Lizzi, F.; Zampini, A. Generalized Weyl systems and kappa-Minkowski space. *Mod. Phys. Lett. A* **2002**, *17*, 2105–2126.
4. Lukierski, J.; Ruegg, H.; Zakrzewski, W.J. Classical and Quantum Mechanics of Free κ Relativistic Systems. *Ann. Phys.* **1995**, *243*, 90–116.
5. Amelino-Camelia, G.; Lukierski, J.; Nowicki, A. Distance Measurement and κ-Deformed Propagation of Light and Heavy Probes. *Int. J. Mod. Phys. A* **1999**, *14*, 4575–4588.
6. Kowalski-Glikman, J.; Nowak, S. Doubly Special Relativity theories as different bases of κ-Poincaré algebra. *Phys. Lett. B* **2002**, *539*, 126–132.
7. Rovelli, C. Loop Quantum Gravity. *Living Rev. Relativ.* **2008**, *11*, 5.
8. Smolin, L. Quantum gravity with a positive cosmological constant. *arXiv* **2002**, arXiv:hep-th/0209079.
9. Amelino-Camelia, G.; Smolin, L.; Starodubtsev, A. Quantum symmetry, the cosmological constant and Planck scale phenomenology. *Class. Quant. Grav.* **2004**, *21*, 3095–3110.
10. Noui, K. Three Dimensional Loop Quantum Gravity: Particles and the Quantum Double. *J. Math. Phys.* **2006**, *47*, 102501.
11. Freidel, L.; Livine, E.R. 3-D quantum gravity and non-commutative quantum field theory. *Phys. Rev. Lett.* **2006**, *96*, 221301.
12. Oriti, D.; Ryan, J. Group field theory formulation of 3D quantum gravity coupled to matter fields. *Class. Quant. Grav.* **2006**, *23*, 6543–6576.

13. Amelino-Camelia, G.; Freidel, L.; Kowalski-Glikman, J.; Smolin, L. The principle of relative locality. *Phys. Rev. D* **2011**, *84*, 084010.
14. Amelino-Camelia, G. Quantum Spacetime Phenomenology. *Living Rev. Relativ.* **2013**, *16*, 5.
15. Lukierski, J. From noncommutative space-time to quantum relativistic symmetries with fundamental mass parameter. In Proceedings of the Second International Symposium on Quantum Theory (QTS2), Krakow, Poland, 18–21 July 2001.
16. Maggiore, M. The Atick-Witten free energy, closed tachyon condensation and deformed Poincare' symmetry. *Nucl. Phys. B* **2002**, *69*, 647.
17. Amelino-Camelia, G. Doubly-Special Relativity: First Results and Key Open Problems. *Int. J. Mod. Phys. D* **2002**, *11*, 1643.
18. Kowalski-Glikman, J. Introduction to Doubly Special Relativity. *Lect. Notes Phys.* **2005**, *669*, 131–159.
19. Girelli, F.; Livine, E.R. Physics of Deformed Special Relativity. *Braz. J. Phys.* **2005**, *35*, 432–438.
20. Jacobson, T.; Liberati, S.; Mattingly, D. Lorentz violation at high energy: Concepts, phenomena and astrophysical constraints. *Ann. Phys.* **2006**, *321*, 150–196.
21. Hossenfelder, S. Multi-Particle States in Deformed Special Relativity. *Phys. Rev. D* **2007**, *75*, 105005.
22. Mignemi, S. Doubly special relativity and translation invariance. *Phys. Lett. B* **2009**, *672*, 186–189.
23. Magpantay, J.A. Dual doubly special relativity. *Phys. Rev. D* **2011**, *84*, 024016.
24. Amelino-Camelia, G.; Freidel, L.; Kowalski-Glikman, J.; Smolin, L. Relative locality and the soccer ball problem. *Phys. Rev. D* **2011**, *84*, 087702.
25. Hossenfelder, S. Comment on "Relative locality and the soccer ball problem". *Phys. Rev. D* **2013**, *88*, 028701.
26. Amelino-Camelia, G.; Freidel, L.; Kowalski-Glikman, J.; Smolin, L. Noisy soccer balls. *Phys. Rev. D* **2013**, *88*, 028702.
27. Snyder, H.S. Quantized Space-Time. *Phys. Rev.* **1947**, *71*, 38.
28. Yang, C.N. On Quantized Space-Time. *Phys. Rev.* **1947**, *72*, 874.
29. Freidel, L.; Rempel, T. Scalar Field Theory in Curved Momentum Space. *arXiv* **2013**, arXiv:1312.3674.

 © 2017 by the authors. Licensee MDPI, Basel, Switzerland. This article is an open access article distributed under the terms and conditions of the Creative Commons Attribution (CC BY) license (http://creativecommons.org/licenses/by/4.0/).

Article

Structure of Multipartite Entanglement in Random Cluster-Like Photonic Systems

Mario Arnolfo Ciampini [1,*], Paolo Mataloni [1] and Mauro Paternostro [2]

[1] Dipartimento di Fisica, Sapienza Università di Roma, Piazzale Aldo Moro 5, Rome 00185, Italy; paolo.mataloni@uniroma1.it
[2] Centre for Theoretical Atomic, Molecular and Optical Physics, School of Mathematics and Physics, Queen's University Belfast, Belfast BT7 1NN, UK; m.paternostro@qub.ac.uk
* Correspondence: marioarnolfo.ciampini@uniroma1.it; Tel.: +39-06-4991-3526

Received: 24 July 2017; Accepted: 2 September 2017; Published: 5 September 2017

Abstract: Quantum networks are natural scenarios for the communication of information among distributed parties, and the arena of promising schemes for distributed quantum computation. Measurement-based quantum computing is a prominent example of how quantum networking, embodied by the generation of a special class of multipartite states called cluster states, can be used to achieve a powerful paradigm for quantum information processing. Here we analyze randomly generated cluster states in order to address the emergence of correlations as a function of the density of edges in a given underlying graph. We find that the most widespread multipartite entanglement does not correspond to the highest amount of edges in the cluster. We extend the analysis to higher dimensions, finding similar results, which suggest the establishment of small world structures in the entanglement sharing of randomised cluster states, which can be exploited in engineering more efficient quantum information carriers.

Keywords: cluster states; multipartite entanglement; percolation

1. Introduction

In 1929, the Hungarian author Karinthy famously set out the concept of *six degrees of separation* [1], the conjecture according to which any two living entities on Earth are distant by no more than five intermediate steps. This concept was reprised and developed later on more rigorous sociological and statistical grounds. Remarkably, for instance, a variation of the *six degrees* was unveiled by the group of Barabasi in 1999 [2], who predicted that any page in the World Wide Web can be reached from any other one with only nineteen intermediate steps (or clicks) on average.

As counterintuitive as this result might look, they are actually based on a very solid concept in graph theory, namely the emergence of *small worlds* from connected networks. A small-world network is a type of mathematical graph in which most nodes are not neighbours of one another, but can be reached from every other one by a small number of steps that actually grows logarithmically with the number of nodes themselves. The six and nineteen degrees of separation highlighted above are different yet similar manifestations of the emergence of small worlds in a network.

Can these concepts be exported to the quantum domain? While the theory of quantum networks has found fertile applications in quantum communication [3] and ground-breaking results in the proposal of quantum repeaters for the faithful long-haul transport of quantum information [4,5], the implications of the emergence of small worlds have been far less explored, and mostly confined to studies of excitation-transport and the analysis of the transition from localised to delocalised regimes in spatially extended interacting-particle models [6,7].

Here, inspired by the analogy between classical network bonds and the correlations set between two elements of a given network of quantum particles, we aim at exploring different aspects.

In particular, motivated by the current experimental state-of-the-art in linear optics, which makes available controllable networks of interconnected information carriers, we address the emergence of typical lengths in the entanglement established by a random set of unitary gates applied to the elements of a given graph. In particular, we focus on a particular class of operations and networks, i.e., those typically put in place in the procedure for the creation of so-called cluster states, which are resources for measurement-based quantum computing [8].

Such computational paradigm, which has been demonstrated equivalent to any circuital quantum computing protocol, is of fundamental importance in quantum information processing. Linear-optics measurement-based quantum information processing has emerged as a promising avenue for the exploration of controllable quantum protocols. Encoding and entangling qubits in more than one degree of freedom of photons is a promising avenue for the generation of medium-to-large scale photonic cluster states: hyperentanglement-based protocols have so far allowed for the creation of cluster states of up to 6 qubits [9], which have been used to validate fundamental one-way quantum algorithms [10,11].

In this paper, by randomising the application of the elementary gates needed to engineer a cluster state of a given size, we induce the establishment of small worlds in the underlying network of a given physical system, and address how the spreading of entanglement across the network itself is affected by the degree of stochasticity of such gates. We unveil an interesting hierarchy with which entanglement appears in subnetworks of growing size: only a sufficient degree of determinism allows for the settling of multipartite entanglement within a given cluster lattice, the threshold for k-element entanglement depending neatly on the number of elements k itself. Moreover, we illustrate a fundamental difference between the phenomenology illustrated in this paper and recently introduced concepts of classical entanglement percolation [12].

The significance of this study goes beyond the context set by cluster states and measurement-based quantum information processing and addresses the fundamental concept of entanglement [13]. In fact, the emergence of different lengths at which bipartite and multipartite entanglement emerge from a set of entangling transformations applied to the elements of a given network, provides insightful information on the entanglement sharing structure. In turn, such information could be used to design better resources for quantum information protocols, obtained by applying only a small subset of entangling operations than the whole one determined by the size of the network itself and nevertheless bearing entanglement-sharing properties very close to those of the fully connected network.

The remainder of this paper is organised as follows. In Section 2.1 we present randomly generated cluster states as the platform for our investigation; in Section 2.2 we focus our attention to four-qubit cluster states, presenting a rich analysis on the interplay between stochasticity of the gates used to set the network and the settling of bipartite and multipartite entanglement. In Section 2.3 we extend our analysis to larger networks.

2. Results

2.1. Theoretical Framework

The approach that we use in order to investigate the core question of our work can be schematised as follows:

1. We set the value of the threshold q and generate a suitable number of random variables $p_{ij} \in [0,1]$, which embody the probabilities to apply the gate $\text{CPHASE}_{i,j}(\pi)$ to the pair of qubits (e_i, e_j).
2. We compare p_{ij} to q. Should it be $p_{ij} < q$ ($p_{ij} > q$), $\text{CPHASE}_{i,j}(\pi)$ is (not) applied. We exhaust the number of all inequivalent pairs of qubits in the network. This produces the network state $|\psi\rangle_\Sigma$, where $\Sigma = \{e_1, \ldots, e_N\}$ is the set of qubits of the register.
3. We compute the reduced density matrices $\rho_\sigma = \text{Tr}_{\Sigma \setminus \sigma}[|\psi\rangle\langle\psi|_\Sigma]$ that are obtained upon tracing the overall state over all qubits but those in the subset $\sigma \in \Sigma$.
4. We calculate the percent fraction of such reductions that are entangled at the set value of q.

5. In order to eliminate any dependence on the specific random pattern of applications of the joint gate, we repeat the procedure above for a number $Q \gg 1$ of instances.
6. When Q is reached, we change q and repeat the protocol from point 1 to 5.

Needless to say, the number of applications of $\text{CPHASE}_{i,j}(\pi)$ at a set value of the threshold depends strongly on the actual value of q itself: the larger the chosen value of q, the higher the number of gate applications. This is illustrated in Figure 1, where we show the different configurations achieved for a network of $N=8$ elements for $q=0.2, 0.5$ and 1, which is associated with a fully connected graph. It is important to remark that, in our notation as well as in Figure 1, a bond connecting elements e_i and e_j only means that gate $\text{CPHASE}_{i,j}(\pi)$ was applied, and does not imply the existence of entanglement between such elements.

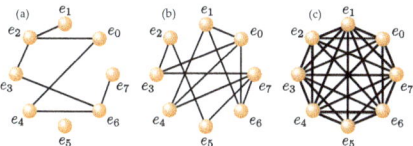

Figure 1. Example of instances of an $N=8$ qubits random cluster states. For (**a**–**c**) we have taken $q=0.2$, $q=0.5$, and $q=1$ respectively.

Scope of our investigation is ascertaining the phenomenology of distribution of (in general) multipartite entanglement across a given network. In particular, we will focus on the possible emergence of special values of q that are associated with the onset of multipartite entanglement, and the characterisation of such quantum correlations. The inherently random nature of the resource states that we consider makes any analytical prediction difficult to be drawn and provides the necessary motivations for the statistical approach that, instead, will be used in the analysis that follows. Notwithstanding its limited analytical power, we find such investigation both powerful and insightful.

As a side remark we mention that, as we have in mind a linear-optics implementation, which to date is one of the most promising and successful platforms for the engineering of cluster-state resources, in our analysis we will not account for any effect of dissipation on the random states that are generated using the protocol illustrated above, as photon losses are negligible in such a setting.

2.2. Analysis of the Entanglement Structure in a Random Four-Qubit State

We start our analysis by focusing on an intuitive figure of merit that is nevertheless able to provide crucial information on the distribution of entanglement across one of the random graph states discussed above, namely state purity. We thus proceed to compute the purity

$$\mathcal{P}_\sigma = \text{Tr}_\sigma[\rho_\sigma^2] \in [0,1] \qquad (1)$$

of the reduced density matrix ρ_σ, and use the fact that, given the overall pure nature of $|\psi\rangle_\Sigma$, a value of $\mathcal{P}_\sigma < 1$ necessarily implies entanglement in the bipartition $(\Sigma\backslash\sigma)|\sigma$. We have thus implemented the protocol illustrated in Section 2.1 by calculating, in step 4, the percentage of reductions with $\mathcal{P}_\sigma < 1$.

In order to illustrate the salient features of our analysis, we now address explicitly the case of $N = 4$, for which $\Sigma = \{e_1, \ldots, e_4\}$. The state that would be produced by applying CPHASE$_{i,j}(\pi)$ gates to every pair of qubits in the network, which would correspond to chosing $q = 1$, reads

$$\begin{aligned}
|\psi\rangle_\Sigma &= \frac{1}{\sqrt{2}} \hat{H}_{e_4}(|\phi_+\rangle_{e_1 e_4}|\phi_-\rangle_{e_2 e_3} + |\psi_+\rangle_{e_1 e_4}|\psi_-\rangle_{e_2 e_3}) \\
&= \frac{1}{\sqrt{2}} \hat{H}_{e_3}(|\phi_-\rangle_{e_1 e_2}|\phi_+\rangle_{e_3 e_4} - |\psi_+\rangle_{e_1 e_2}|\psi_-\rangle_{e_3 e_4}) \\
&= \frac{1}{\sqrt{2}} \hat{H}_{e_2}(|\phi_-\rangle_{e_1 e_3}|\phi_+\rangle_{e_2 e_4} - |\psi_+\rangle_{e_1 e_2}|\psi_-\rangle_{e_2 e_4}) \\
&= \frac{1}{\sqrt{2}} \hat{H}_{e_1}(|\phi_-\rangle_{e_1 e_2}|\phi_+\rangle_{e_3 e_4} - |\psi_+\rangle_{e_1 e_2}|\psi_-\rangle_{e_3 e_4})
\end{aligned} \quad (2)$$

where $\hat{H}_{e_j}$ is the Hadamard gate on qubit e_j and we have introduced the Bell states $|\phi_\pm\rangle_{e_i e_j} = (|00\rangle \pm |11\rangle)_{e_i e_j}/\sqrt{2}$, $|\psi_\pm\rangle_{e_i e_j} = (|01\rangle \pm |10\rangle)_{e_i e_j}/\sqrt{2}$. The orthogonality of Bell states ensures that entanglement exists in the three inequivalent bipartition $(e_i, e_j)|(e_k, e_l)$. Moreover, it is equally straightforward to check that any single-qubit reduction is maximally mixed. Therefore, also the bipartitions $e_i|(e_j, e_k, e_l)$ are entangled. This implies that for $q = 1$ we expect all six bipartitions that can be identified to be inseparable and the state to be genuinely multipartite entangled. The purity of the associated reduced states is thus necessarily smaller than one. However, for $q < 1$ the number of mixed-state reduction is not necessarily as large as six, and our calculations aim at quantifying the percentage of such reduced states as q is varied.

The results of such calculations are presented in Figure 2 (blue and red dots), where each data point is the result of an average over $Q = 5000$ random instances, a sample-size that was large enough to ensure convergence of the numerics. The error bars attached to each point show the uncertainty associated to the averages, calculated as the standard deviation of each Q-sized sample and divided by $\sqrt{Q}$. Clearly, for $q = 0$ the state of the network is deterministically found to be the factorised initial state $\otimes_{j=1}^4 |+\rangle_{e_j}$, while for $q = 1$ we retrieve the result anticipated above (Equation (2)). In between such extreme situations, the number of inseparable two-vs.-two and one-vs.-three qubits bipartitions (equivalently, mixed two-qubit and one-qubit states) grows monotonically with q, albeit at slightly different rates. In particular, we find that the percentage fraction of inseparable two-vs.-two (three-vs.-one) qubits bipartitions exceeds 99.9% at $q = 0.82 \pm 0.01$ ($q = 0.89 \pm 0.01$), as shown by the vertical dashed line marked as T_2 (T_3) in Figure 2. The nominal positions (uncertainties) of $T_{2,3}$ have been obtained as the average (standard deviations) over 100 analytical non-linear interpolations of the results of our simulations, each producing the functions $f_{2,3}(q)$ (whose averages are shown by the blue and red lines in Figure 2) that have been used to solve numerically the equations $f_{2,3}(q) = 99.9$. Quite clearly, $T_2 \neq T_3$ beyond statistical errors, which implies that the random network at hand requires a higher threshold in q to produce a complete set of inseparable one-vs.-three qubits bipartitions.

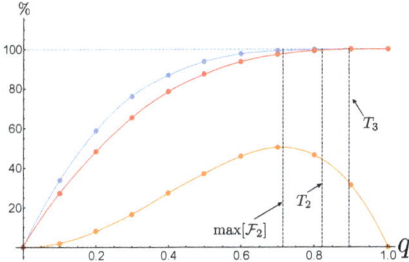

Figure 2. We study the percentage fraction of mixed-state reductions that can be identified in a network of $N = 4$ elements, against the threshold parameter q. The blue (red) dots show the results of the numerical experiment aimed at quantifying the fraction of mixed two-qubit (one-qubit) reductions. The orange points identify the values of the percentage fraction $\mathcal{F}_2$ of two-qubit reductions whose purity is exactly $1/4$. The solid lines are non-linear interpolations of the data points. Each point is the result of an average over a sample of $Q = 5000$ elements. Error bars show the standard deviations associated with such averages. Dashed lines $T_{2,3}$ identify the value of q at which the number of mixed two- and one-qubit reductions is at least 99.9% of the possible ones. The line labelled $\max[\mathcal{F}_2]$ identifies the value of q at which the maximum of $\mathcal{F}_2$ occurs.

Needless to say, the empirical rule of "no free lunch" applies here as well: the establishment of multipartite entanglement in the network under scrutiny has to come at the expenses of something else, in light of the monogamy of entanglement. The specific algorithm at hand allows us to explore who pays the toll represented by the establishment of genuine multipartite entanglement in the random network.

In particular, we expect bipartite entanglement to be affected by the emergence of multipartite one. Such expectation is corroborated by the analysis summarized by the orange dots and curve in Figure 2, which show the percentage fraction $\mathcal{F}_2$ of two-vs.-two qubits reductions of random states at a given value of q that have purity exactly equal to $1/4$, which is the lowest a two-qubit state can achieve and witnesses maximum entanglement across the $(e_i, e_j)|(e_k, e_l)$ bipartition. Quite intuitively, $\mathcal{F}_2$ grows at small values of q: a low threshold implies very small probability to apply multiple CPHASE gates, which inevitably favours the construction of maximally entangled two-qubit states. For $q \simeq 1$, we have a large probability that one qubit is affected by multiple CPHASE gates. Intuitively, this should be able to set strong multipartite entanglement and deplete the degree of bipartite one, and we expect $\mathcal{F}_2$ to decrease accordingly. Indeed, we know that at $q = 1$ we have a genuinely multipartite entangled. The orange dots in Figure 2 confirm such expectation, and show the occurrence of a maximum of $\mathcal{F}_2$ that is close, yet not identical, to the chosen thresholds $T_{2,3}$ discussed above (we have that $\max[\mathcal{F}_2]$ occurs at $q = 0.72 \pm 0.01$).

Of course, counting for the number of reductions that are in mixed states does not provide full information about multipartite nature of the entanglement that is established among the elements of the network. We remind that a pure N-partite state is called genuinely multipartite entangled if it is not separable with respect to any of the possible bipartitions of its N elements. One can thus check the multipartite nature of the entanglement of a given pure state by *counting* the number of separable bipartitions that can be drawn. As each instance of our random sample is a pure state, we have decided to approach this task by using the N-partite generalisation of negativity defined as

$$\mathcal{E}_N = \sqrt[N]{\Pi_{\{\sigma\}} \mathcal{E}_{\sigma|\Sigma\setminus\sigma}}, \tag{3}$$

where $\mathcal{E}_{\sigma|\Sigma\backslash\sigma}$ is the negativity of the partially transposed density matrix of the bipartition $\sigma|\Sigma\backslash\sigma$ and the product extends to all the bipartitions. We recall the definition of negativity as

$$\mathcal{E}_{\sigma|\Sigma\backslash\sigma} = \max[0, -2\sum_j \lambda_j^-] \qquad (4)$$

with $\{\lambda_j^-\}$ the set of negative eigenvalues of the partially transposed (with respect to any of the subparties) density matrix of the bipartition $\sigma|\Sigma\backslash\sigma$. The geometric average upon which Equation (3) is built is null whenever at least one of the bipartitions of the network is positive under partial transposition. Therefore, for pure states, only if all bipartitions are certified inseparable according to the partial transposition criterion is the state of the network genuinely multipartite entangled. The situation is much more difficult when mixed states are considered, for which the non-nullity of the quantity in Equation (3) is no guarantee of the existence of genuine multipartite entanglement in a given state [14].

Figure 3 shows the behavior of $\mathcal{E}_4$ against q. While for $q > 0$ we always have four-partite entanglement (in line with the finding in Figure 2), it is remarkable that $q = 1$ is not associated with the largest degree of four-partite negativity, which actually occurs at $q = 0.72 \pm 0.01$.

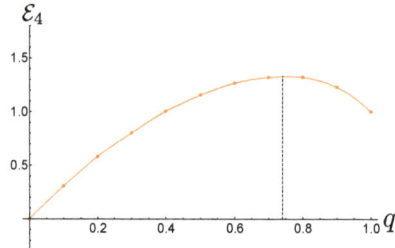

Figure 3. Average four-partite negativity $\mathcal{E}_4$ plotted against q obtained for a sample of $Q = 5000$ random network states. The error bars are the standard deviations associated with the averages. The orange solid line is a non-linear interpolating function whose maximum is achieved at $q = 0.72 \pm 0.01$ (vertical dashed line).

We continue the assessment of the four-partite case by pointing out the differences between the average behavior of the figures of merit addressed herein and the values taken by such indicators over the *average state* of the network. The latter is defined as the state obtained upon mediating over Q random instances of network states. Formally, by assuming all instances to be equally likely to occur (which is entailed by choosing the probabilities to apply gates $\text{CPHASE}_{i,j}(\pi)$ uniformly), the physical state of the system is described by the density matrix

$$\rho_\Sigma = \frac{1}{Q}\sum_{j=1}^Q |\psi\rangle\langle\psi|_{\Sigma,j}, \qquad (5)$$

where $|\psi\rangle\langle\psi|_{\Sigma,j}$ is the jth random state of the Q-sized sample.

With the exception of the cases associated with $q = 0, 1$ (when we sum identically prepared states), by averaging we lose the purity of the network state: $\mathcal{P}_\Sigma$ reaches values as low as $\simeq 0.14$ for $q = 0.5$ (cf. Inset (a) of Figure 4), which is however larger than the minimum purity $1/16$ achievable by a four-qubit state. Despite being mixed, the average state of the network preserves significant quantum coherences as quantified by the measure proposed in [15] and formalised as

$$\mathcal{C} = \sum_{i\neq j}|(\rho_\Sigma)_{ij}| \qquad (6)$$

with $|(\rho_\Sigma)_{ij}|$ the off-diagonal elements of the density matrix ρ_Σ. The behavior of $\mathcal{C}$ against q is shown in Inset (b) in Figure 4: a minimum of the measure of coherence is achieved in correspondence of the minimum purity. However, such a minimum is strictly non-null, thus leaving open the possibility of dealing with a (mixed) state of the network exhibiting a non-trivial entanglement structure. Such a possibility is confirmed by the analysis of $\mathcal{E}_4$ (cf. main panel of Figure 4), which is a growing function of q (similar trends are exhibited by both the two-vs.-two qubits entanglement $\mathcal{E}_{(e_i,e_j)|(e_k,e_l)}$, and the one-vs.-three qubits one $\mathcal{E}_{(e_i)|(e_j,e_k,e_l)}$). Nothing remarkable in the behavior of $\mathcal{E}_4$ appears to be related to the value of $q = 0.5$, although the function changes concavity in correspondence to such a value of the probability threshold. It should be noticed that, as anticipated, in such an average-state case $\mathcal{E}_N$ cannot be interpreted as a quantifier of genuine multipartite entanglement. Indeed, the revelation of multipartite entanglement in general multiparty mixed states requires a more refined approach (see [16] for a recent assessment of this point and the provision of useful criteria). Nevertheless, this figure of merit is still very useful for our analysis, as it provides valuable information on the average amount of bipartite entanglement within the statistically average stage of the network, and we will thus make further use of $\mathcal{E}_N$ in the remainder of this work. Finally, the non-nullity of either $\mathcal{E}_{(e_i)|(e_j,e_k,e_l)}$'s or $\mathcal{E}_{(e_i,e_j)|(e_k,e_l)}$'s does not exclude the possibility of facing bound entanglement (i.e., non-distillable entanglement) of the negative-partial-transposition nature [17] in those bipartitions, an issue that goes beyond the scopes of this work.

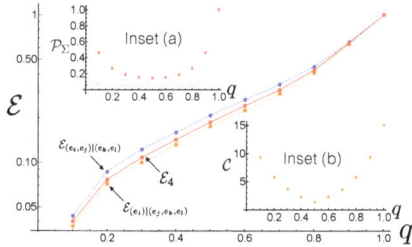

Figure 4. Main panel: Logarithmic plot of the entanglement within the average estate ρ_Σ of an $N = 4$ random network against the threshold probability q. The red dots show the value taken by the four-partite negativity $\mathcal{E}_4$, while the blue and orange ones are for the entanglement within the bipartitions $(e_i, e_j)|(e_k, e_l)$ and $(e_i)|(e_j, e_k, e_l)$. The lines connecting the dots are simply guides to the eye. Inset (a): Purity $\mathcal{P}_\Sigma$ of the average state against q. The dashed horizontal line shows the minimum purity of a four-qubit state. Inset (b): Values taken by the measure of coherence $\mathcal{C}$ against the threshold probability.

To finish the study of this paradigmatic case, we report in the main panel of Figure 5 the behavior of $\mathcal{E}_3$ in the four three-qubit reduced states that can be singled out from our network. We have used the tripartite version of Equation (3) to quantify the entanglement and changed our notation so as to make explicit the triplets of elements of the network that we ave considered. Moreover, by tracing out two elements, we have evaluated the residual two-qubit entanglement, whose average across the six two-qubit reductions is displayed in the inset of Figure 5. The general trend of such figures of merit follows the expectation that, in the large-q region, the entanglement in the reduction is depleted to favour the emergence of multipartite one. Moreover, their quantitative value is, in general, very small. A point of notice is that the peak of three- and two-qubit negativity does not occur at the same value of q, thus suggesting an interesting hierarchy of values of q at which the various structures of entanglement across the system are triggered or destroyed.

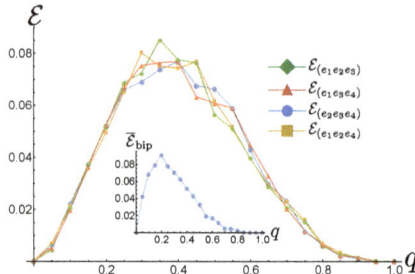

Figure 5. Main panel: $\mathcal{E}_3$ in the three-qubit reductions (extracted from an $N = 4$ network) identified in the legend, plotted against q. Each plot is an average over $Q = 5000$ realisation of the random network state (we omit the error bars for clarity of presentation). Inset: Mean bipartite negativity $\overline{\mathcal{E}}_{bip}$ averaged over the six two-qubit reduced states that can be singled out from our network. Same conditions as in the main panel.

2.3. Enlarging the Size of the Network

We now assess the features of larger networks of qubits, addressing questions that are akin to those assessed in Section 2.2. Features similar to those showcased in the four-qubit network are present in all the higher-dimensional systems that we have studied through our simulations. For instance, Figures 6 and 7 display the same behaviors highlighted in Figures 2 and 4, respectively. Rather than reporting qualitatively similar plots for larger networks, in Table 1 we present the threshold values of q at which progressively larger reductions of the state of the network are mixed.

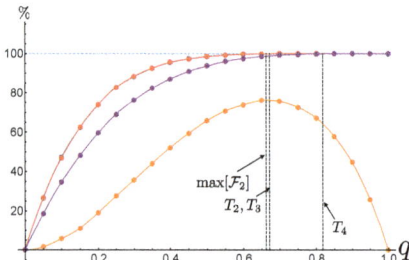

Figure 6. We study the percentage fraction of mixed-state reductions that can be identified in a network of $N = 5$ elements, against the threshold parameter q. The red dots show the results of the numerical experiment aimed at quantifying the fraction of mixed two- and three-qubit reductions, which actually coincide. The purple dots show the results for the one-qubit reductions. The orange points identify the values of the percentage fraction $\mathcal{F}_2$ of two-qubit reductions whose purity is exactly $1/4$. The solid lines are non-linear interpolations of the data points. Each point is the result of an average over a sample of $Q = 10^4$ elements. Error bars show the standard deviations associated with such averages. Dashed lines $T_{2,3}$ (T_4) identify the value of q at which the number of mixed two- and three-qubit (one-qubit) reductions is at least 99.9% of the possible ones. The line labelled $\max[\mathcal{F}_2]$ identifies the value of q at which the maximum of $\mathcal{F}_2$ occurs.

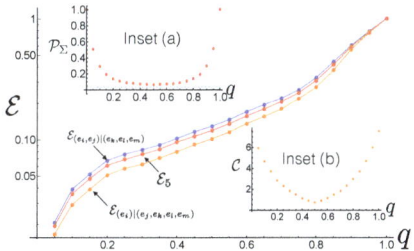

Figure 7. Main panel: Logarithmic plot of the entanglement within the average estate ρ_Σ of an $N=5$ random network against the threshold probability q. The red dots show the value taken by $\mathcal{E}_5$, while the blue and orange ones are for the entanglement within the bipartitions $(e_i,e_j)|(e_k,e_l,e_m)$ and $(e_i)|(e_j,e_k,e_l,e_m)$. The lines connecting the dots are simply guides to the eye. Inset (a): Purity $\mathcal{P}_\Sigma$ of the average state against q. The dashed horizontal line shows the minimum purity of a four-qubit state. Inset (b): Values taken by the measure of coherence $\mathcal{C}$ against the threshold probability.

The trend is clear: as we look into larger networks, the value of T_k ($k=2,3,\dots$) decreases.

Table 1. The table shows the threshold value of q at which the fraction of progressively larger reductions in an N-element random network is at least 99.9%. Black squares stands for unavailable data at that size of the network. As before, $\max[\mathcal{F}_2]$ is the value of q at which the maximum of $\mathcal{F}_2$ occurs.

N	4	5	6	...	9
$\max \mathcal{F}_2$	0.72	0.66	0.64		0.40
T_2	0.82	0.67	0.57		0.39
T_3	0.89	0.67	0.54		0.31
T_4	■	0.818	0.57		0.27
T_5	■	■	0.75		0.27
T_6	■	■	■		0.31
T_7	■	■	■		0.39
T_8	■	■	■		0.40

2.4. Entanglement Percolation

It is interesting to compare our analysis to entanglement percolation, a concept akin to classical bond percolation introduced in [12]. Consider a graph of particles akin to one of those addressed in this paper. This time, though, a link between two elements implies the presence of entanglement between them. Ref. [12] shows the existence of a minimum amount of entanglement between any two elements of the network needed to establish a perfect quantum channel between distant (not directly connected) elements, with significant (non-exponentially decaying) probability.

This is fundamentally different from our situation, where instead we point out the existence of a minimum probability to randomly apply a two-qubit gate in a network associated with the establishment of a genuinely multipartite entangled state of the network. Our threshold does not guarantee the existence of a long-distance entangled channel between arbitrarily chosen elements of the network. In fact, non-nearest-neighbour elements of a cluster state are not necessarily entangled, their entanglement being in general dependent on the geometry of the underlying network.

In order to ascertain if a value of q exists above which long-haul entanglement is set in the network, we computed the negativity of the reduced state of the qubits that have the largest number of intermediate sites between them, at a given value of N. This is analogous to the study presented in the inset of Figure 5, although instead of an average over all the possible two-qubit reductions, here we consider now only a specific reduction. Figure 8 shows the results valid for the case of $N=6$, for which we address the entanglement between elements e_1 and e_4. We have considered the percentage of reductions of such elements with a non-zero value of negativity against the value of q. Quite clearly, such a percentage remains always very small, regardless of q, showing that no classical entanglement percolation effect occurs, as there is no value of q at which long-distance entanglement within the network is set deterministically. The results should be considered as canonical, qualitatively valid regardless of the actual choice of N, and indicative of the profound differences between the situation addressed here and the study in [12].

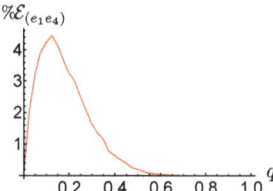

Figure 8. Percentage of reduced states of elements e_1 and e_4 of an $N = 6$ random network exhibiting a non-zero negativity, plotted against q, for a sample of 5×10^4 states.

3. Discussion

We have studied the entanglement sharing structure among the elements of a qubit network subjected to probabilistic CPHASE gates. We have highlighted the existence of statistically inequivalent thresholds in the probability of application of the gates for the settling of entanglement in various subsets of network elements, thus unveiling an interesting hierarchy in the entanglement distribution pattern of a given network. The phenomenology that we have highlighted cannot be understood in terms of the statistical properties of an intuitive, yet too naive, reference state such as the one obtained by averaging overall the elements of the random set of states generated in our numerical experiments: the above-mentioned hierarchy is a statistical feature of random networks rather than a property of the statistically average state of the network. Remarkably, *small worlds* structures in the entanglement sharing of the random set of network states appear to emerge. This is an interesting feature that deserves more attention and upon which we plan to focus our forthcoming (theoretical and experimental) efforts.

Acknowledgments: Mario Arnolfo Ciampini acknowledges support from QUCHIP-Quantum Simulation on a Photonic Chip, FETPROACT-3-2014, Grant agreement no: 641039, Mauro Paternostro acknowledges support from the SFI-DfE Investigator Programme (grant 15/IA/2864), and the Royal Society.

Author Contributions: Mario Arnolfo Ciampini conceived the idea, Mario Arnolfo Ciampini and Mauro Paternostro performed the simulation and analysed the data, Mauro Paternostro and Paolo Mataloni interpreted the results, all authors contributed to drafting and revisioning the manuscript.

Conflicts of Interest: The authors declare no conflict of interest.

References

1. Karinthy, F. Láncszemek. In *Minden Másképpen van*, 1929. Available online: http://mek.oszk.hu/15500/15588/15588.pdf (accessed on 5 September 2017). (In Hungarian)
2. Albert, R.; Jeong, H.; Barabasi, A.-L. Internet: Diameter of the World-Wide Web. *Nature* **1999**, *401*, 130–131.
3. Kimble, H.J. The quantum internet. *Nature* **2008**, *453*, 1023–1030.
4. Munro, W.J.; Harrison, K.A.; Stephens, A.M.; Devitt, S.J.; Nemoto, K. From quantum multiplexing to high-performance quantum networking. *Nat. Photonics* **2010**, *4*, 792–796.
5. Epping, M.; Kampermann, H.; Bruß, D. Robust entanglement distribution via quantum network coding. *New J. Phys.* **2016**, *18*, 103052.
6. Zhu, C. P.; Xiong, S.-J. Localization-delocalization transition of electron states in a disordered quantum small-world network. *Phys. Rev. B* **2000**, *62*, 14780.
7. Giraud, O.; Georgeot, B.; Shepelyansky, D.L. Tuning clustering in random networks with arbitrary degree distributions. *Phys. Rev. E* **2005**, *72*, 036203.

8. Briegel, H.J.; Browne, D.E.; Dür, W.; Raussendorf, R.; Van den Nest, M. Measurement-based quantum computation. *Nat. Phys.* **2009**, *5*, 19.
9. Vallone, G.; Donati, G.; Ceccarelli, R.; Mataloni, P. Six-qubit two-photon hyperentangled cluster states: Characterization and application to quantum computation. *Phys. Rev. A* **2010**, *81*, 052301
10. Vallone, G.; Pomarico, E.; De Martini, F.; Mataloni, P. One-way quantum computation with two-photon multiqubit cluster states. *Phys. Rev. A* **2008**, *78*, 042335.
11. Ciampini, M.A.; Orieux, A.; Paesani, S.; Sciarrino, F.; Corrielli, G.; Crespi, A.; Ramponi, R.; Osellame, R.; Mataloni, P. Path-polarization hyperentangled and cluster states of photons on a chip. *Light Sci. Appl.* **2016**, *5*, e16064.
12. Acín, A.; Cirac, J.I.; Lewenstein, M. Entanglement Percolation in Quantum Networks. *Nat. Phys.* **2007**, *3*, 256.
13. Horodecki, R.; Horodecki, P.; Horodecki, M.; Horodecki, K. Quantum entanglement. *Rev. Mod. Phys.* **2009**, *81*, 865.
14. Huber, M.; Mintert, F.; Gabriel, A.; Hiesmayr, B.C. Detection of high-dimensional genuine multipartite entanglement of mixed states. *Phys. Rev. Lett.* **2010**, *104*, 210501.
15. Baumgratz, T.; Cramer, M.; Plenio, M.B. Quantifying Coherence. *Phys. Rev. Lett.* **2014**, *113*, 140401.
16. Lancien, C.; Gühne, O.; Sengupta, R.; Huber, M. Relaxations of separability in multipartite systems: Semidefinite programs, witnesses and volumes. *J. Phys. A Math. Theor.* **2015**, *48*, 505302.
17. Horodecki, P.; Horodecki, R. Distillation and bound entanglement. *Quant. Inf. Comp.* **2001**, *1*, 45.

© 2017 by the authors. Licensee MDPI, Basel, Switzerland. This article is an open access article distributed under the terms and conditions of the Creative Commons Attribution (CC BY) license (http://creativecommons.org/licenses/by/4.0/).

Article

Non-Causal Computation

Ämin Baumeler [1] and Stefan Wolf [2,*]

[1] Faculty of Informatics, Università della Svizzera italiana, 6900 Lugano, Switzerland; baumea@usi.ch
[2] Facoltà indipendente di Gandria, 6978 Gandria, Switzerland
* Correspondence: wolfs@usi.ch; Tel.: +41-58-666-4000

Received: 11 May 2017; Accepted: 30 June 2017; Published: 2 July 2017

Abstract: Computation models such as circuits describe sequences of computation steps that are carried out *one after the other*. In other words, algorithm design is traditionally subject to the restriction imposed by a fixed causal order. We address a novel computing paradigm beyond quantum computing, replacing this assumption by mere logical consistency: We study *non-causal circuits*, where a fixed time structure *within* a gate is locally assumed whilst the global causal structure *between* the gates is dropped. We present examples of logically consistent non-causal circuits outperforming all causal ones; they imply that suppressing loops entirely is more restrictive than just avoiding the contradictions they can give rise to. That fact is already known for correlations as well as for communication, and we here extend it to *computation*.

Keywords: physical computing models; complexity classes; causality

1. Introduction

Computations, understood as realized through Turing machines, billiard or ballistic computers [1], circuits, lists of computer instructions, or otherwise, are often designed to have a linear (i.e., causal) time flow: After a fundamental operation is carried out, the program counter moves to the next operation, and so forth. Surely, this is in agreement with our everyday experience; after you finish to read this sentence, you continue to the next (hopefully), or do something else (in that case: goodbye!). *What sorts of computation become admissible if one drops the assumption of a linear time flow and reduces it to mere logical consistency?* One could imagine that a linear time flow restricts computation strictly *beyond* what would be allowed for the purely logical point of view. Indeed, we show this to be true. If the assumption of a linear time flow is dropped, a variable of the computational device could depend on "past" as well as "future" computation steps. Such a dependence can be interpreted as *loops* in the time flow, e.g., generated by a closed timelike curve [2]. There are two fundamental issues that might make loops logically inconsistent. One is the liability to the *grandfather antinomy*. In a loop-like information flow, multiple contradicting values could potentially be assigned to a variable—the variable is *overdetermined*. The other issue is *underdetermination*: a variable could take multiple consistent values, yet the model of computation *cannot predict* which actual value it takes. This underdetermination is also known as the *information antinomy*. To overcome both issues, we restrict ourselves to models of computation where the assumption of a linear time flow is dropped and replaced by the assumption of *logical consistency*: All variables are neither overdetermined nor underdetermined. We call such models of computation *non-causal*. Our main result is that non-causal models of computation are *strictly* more powerful than the traditional causal ones. Therefore, causality is a stronger assumption than logical consistency in the context of computation. Similar results are also known with respect to *quantum computation* [3–7], correlations [5,8–11] as well as communication [12]. As we will show later, such circuits are "programmed" by introducing a *contradiction* if an *undesired* result is found. This is like guessing the solution to a problem and killing the own grandfather in the event that the guess was wrong (similar to "quantum suicide" [13] or "anthropic computing" [14]).

The article is structured as follows. First, we discuss the assumption of logical consistency in more depth, then we describe a non-causal circuit model of computation and give a few examples of problems that can be solved more efficiently. We continue by describing other non-causal models of computations: the non-causal Turing machine and non-causal billiard computer. We conclude by showing how to efficiently find a satisfying assignment to a SAT formula if the number of satisfying assignments is *previously known*.

2. Logical Consistency

Let ρ_t be the ensemble of all variables (also called *state*) of a computational model at a time t. In general, ρ_t depends on $\rho_{t-1}, \rho_{t-2}, \ldots$. Without loss of generality, assume that ρ_t depends on ρ_{t-1} only (i.e., the computation is described by a Markov chain). These dependencies are depicted in Figure 1a. In a non-causal model, however, the values that are assigned to the variables at time t could in principle depend on "future" time-steps; e.g., the assignment ρ_0 could depend on ρ_m, which results in a Markovian "bracelet" or circle (see Figure 1b).

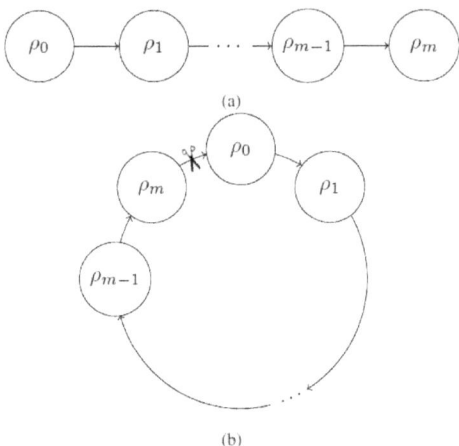

Figure 1. Causal and non-causal computation. The arrows point in the direction of computation. (a) The values that are assigned to the variables of a computational model at time t depend on ρ_{t-1}. (b) Cyclic dependencies of the values that are assigned to the variables at different steps during the computation.

A computational model is *not overdetermined* if and only if the values that are assigned to the variables do not contradict each other. This is equivalent to the existence of a fixed point [15] of the Markov chain that results from cutting the "bracelet" at an arbitrary position (see Figure 1b). Let f be a function that describes the behaviour of this Markov chain. Then, the computational model is *not overdetermined* if and only if $\exists x : f(x) = x$.

A computational model is *not underdetermined* if and only if there exists at most one fixed point [15]:

$$|\{x \mid x = f(x)\}| \leq 1.$$

Logical consistency is identified [15] with no overdetermination and no underdetermination; i.e., the existence of a *unique* fixed point:

$$\exists! x : f(x) = x.$$

3. Non-Causal Circuit Model

A circuit consists of gates that are interconnected with wires. In the traditional circuit model, back-connections (i.e., a cyclic path through a graph where gates are identified with nodes and wires are identified with edges) are either forbidden or interpreted as *feedback* channels. An example of a feedback channel is an autopilot system in an aircraft that, depending on the measured altitude, adjusts the rudder and the power setting to maintain the desired altitude, at the same time avoiding a stall. Here, we interpret back-connections or loops differently. Whilst in the above scenario the feedback gets introduced at a *later* point in the computation, the back-action in a non-causal circuit effects the system at an *earlier* point. Such a back-action can be interpreted as acting into the past. Another interpretation is that every gate has its own time (clock), but no global time is assumed—this interpretation stems from the studies of correlations without causal order [5,8]. Such an interpretation might be more pleasing: Here, "earlier" is understood *logically*, and the assumption of a global causal order is simply replaced by logical consistency.

A *non-causal* circuit consists of gates that can be interconnected arbitrarily by wires, as long as the circuit as a whole remains logically consistent. An example of a circuit that is overdetermined and an example of a circuit that leads to the information antinomy (under-determined) are given in Figure 2.

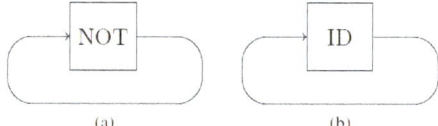

(a) (b)

Figure 2. (a) Overdetermined circuit: The bit 0 is mapped to 1 and vice versa; i.e., there is no consistent assignment of a value that travels on the wire. (b) Information antinomy: Both 0 and 1 could potentially travel on the wire, yet the circuit does *not specify* which.

We model a gate G by a Markov matrix $\hat{G}$ with 0–1 entries. Without loss of generality, assume that the input and output dimensions of a gate are equal. The Markov matrix of the ID gate on a single bit (see Figure 2b) is

$$\mathbb{1} = \begin{pmatrix} 1 & 0 \\ 0 & 1 \end{pmatrix},$$

and the Markov matrix of the NOT gate on a single bit (see Figure 2a) is

$$\hat{N} = \begin{pmatrix} 0 & 1 \\ 1 & 0 \end{pmatrix}.$$

Values are modeled by vectors; e.g., in a binary setting, the value 0 is represented by the vector $(1,0)^T$ and the value 1 is represented by the vector $(0,1)^T$. In general, an n-dimensional variable with value i is modeled by the n-dimensional vector i with a 1 at position i, and where all other entries are 0. A gate is applied to a value via the matrix-vector multiplication; i.e., the output of G on input a is $x = \hat{G}a$. Let F and G be two gates. The Markov matrix of the parallel composition of both gates is $\hat{F} \otimes \hat{G}$. They are composed sequentially with a wire that takes the d-dimensional output of F and forwards it as input to G. By this, we obtain a new gate $H = G \circ F$ which represents the sequential composition. The sequentially composed gate is

$$\hat{H} = \sum_{v=0}^{d-1} \hat{G} v v^T \hat{F} = \hat{G}\hat{F}.$$

By using these rules of composition, a *causal* circuit can always be modeled by a single gate. A *closed* circuit is a circuit where all wires are connected to gates on both sides. Let H be the gate that describes the composition of all gates for a given causal circuit. We can transform any such circuit into a closed non-causal circuit by connecting all outputs from H with all inputs to H. A *logically consistent* closed circuit is thus a circuit where a *unique* assignment of a value c to the looping wire exists:

$$c = \hat{H}c \iff c^T \hat{H} c = 1. \tag{1}$$

In other words, the described closed circuit is logically consistent if and only if the diagonal of $\hat{H}$ consists of 0's with a single 1. The position of the 1-entry represents the fixed point and the value c on the looping wire. Note that for a given closed circuit, the gate H is not unique, but might depend on where the "cut" is introduced. An *open* circuit is a circuit where some wires are not connected to a gate on one side. Thus, such a circuit has either an input a, an output x, or both. A logically consistent open circuit, therefore, is a circuit where for *any* choice of input a, a *unique* assignment of a value c to the looping wire and to the output x exists, such that

$$(x \otimes c)^T \hat{H}(a \otimes c) = 1,$$

where the second output from H is looped to the second input to H.

Let c_a be the value on the looping wire of a logically consistent open circuit $\mathcal{C}$ with input a. We can transform $\mathcal{C}$ into a family $\{\mathcal{C}_i\}_{0 \leq i < d}$ of logically consistent *closed* circuits such that the value on the same looping wire of $\mathcal{C}_i$ is c_i. The circuit $\mathcal{C}_i$ is constructed by attaching the gate

$$\hat{D}_i = \sum_{v=0}^{d-1} i^T v$$

to the input and output wires of $\mathcal{C}$ (see Figure 3a,b). The gate D_i unconditionally outputs the value i.

There is an ambiguity on which wires are regarded as "looping". We show that two different representations H and H' of the same closed non-causal circuit $\mathcal{C}$ yield the same computation (the difference between H and H' is the identification of the looping wires). Different H and H' that represent the *same* non-causal circuit $\mathcal{C}$ can be written as $H = Q \circ R$ and $H' = R \circ Q$. For H, the looping wires are those that exit Q and enter R, and for H', vice versa. From Equation (1), we have

$$\exists! c : c^T \hat{H} c = c^T \hat{Q} \hat{R} c = c^T \hat{Q} \left(\sum_e e e^T \right) \hat{R} c = 1.$$

Since R is deterministic, the value of e is uniquely determined. Thus, we obtain

$$\exists! c : c^T \hat{Q} e_* e_*^T \hat{R} c = 1,$$

where e_* is the specific value on the wire exiting R and entering Q. Conversely,

$$\exists! e' : e'^T \hat{H}' e' = e'^T \hat{R} \hat{Q} e'$$
$$= e'^T \hat{R} \left(\sum_{c'} c' c'^T \right) \hat{Q} e'$$
$$= e'^T \hat{R} c'_* c'^T_* \hat{Q} e' = 1,$$

holds. The only way H and H' each have a *unique* fixed point is with the identification $e_* = e'$. Therefore, both representations H and H' assign the same values to the wires. By the above translation from *open* to *closed* circuits, we see that the same reasoning can be applied to open circuits.

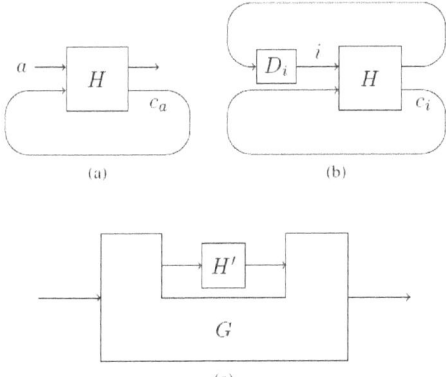

Figure 3. (a) Open circuit C with input a. (b) Closed circuit C_i with $a = i \to c_a = c_i$. (c) The big box represents a non-causal comb (note that combs obey causality; the higher-order transformations described here are equivalent to combs, yet where the causality assumption is dropped) that transforms a gate (H') to a new gate, the composition.

Above, we considered *deterministic* Markov processes. It is natural to extend this model to probabilistic processes (i.e., stochastic matrices). The logical consistency condition in that case—as studied in Ref. [15]—is

$$\operatorname{Tr} \hat{H} = 1, \qquad (2)$$
$$\forall i, j : \hat{H}_{i,j} \geq 0,$$

that is, the diagonal of $\hat{H}$ consists of non-negative numbers (probabilities) that add up to 1. Equation (2) can be interpreted as "the average number of fixed points is 1". To see this, we decompose H as a convex combination of *deterministic* matrices

$$\hat{H} = \sum_i p_i \hat{H}_i,$$

where for all i, $\hat{H}_i$ is deterministic. Then, Equation (2) states

$$\operatorname{Tr} \hat{H} = \sum_i p_i \operatorname{Tr} \hat{H}_i = 1.$$

For an arbitrary deterministic matrix $\hat{D}$, the expression $\operatorname{Tr} \hat{D}$ represents the *number* of fixed points, with which we arrive at the stated interpretation.

An open non-causal circuit can be represented by a non-causal comb [5] G which is a higher-order transformation—G transforms the gate H' to a new gate (see Figure 3c). The non-causal comb G, for instance, could connect the output from H' with the input of H', as long as the composition remains logically consistent.

4. Computational Advantage

The logical consistency requirement forces the value on a looping wire to be the unique fixed point of the transformation. This can be exploited for *finding fixed points* of a black box, which yields an advantage in higher-order computation. Suppose we are given a black box B that takes (produces) a d-dimensional input (output) and has a *unique* fixed point x previously unknown to us. As a Markov matrix, B is

$$\hat{B} = \sum_{i=0}^{d-1} e_i i^T, \quad \text{with } |\{i \,|\, e_i = i\}| = 1.$$

Our task is to find the fixed point x in as few queries as possible. If we solve this task with a causal circuit, then, in the worst case, $d-1$ queries are needed. In contrast, with a non-causal circuit, a *single* query suffices. The reason for this is that the black box is queried with the fixed point only. Any other query would lead to a logical contradiction, and therefore does not occur. For that purpose, we just connect the output of B with the input of B and use a second wire to read out the value (see Figure 4a). This circuit is logically consistent because

$$\forall a, \exists ! c, x : (x \otimes c)^T \hat{C} (\mathbb{1} \otimes \hat{B})(a \otimes c)$$
$$= (x \otimes c)^T \hat{C} (a \otimes \hat{B} c) = 1,$$

where $\hat{C}$ is the CNOT gate and $\mathbb{1}$ is the identity. However, this construction only works if B has a *unique* fixed point. Suppose B_2 has *two* fixed points. In that case, the circuit from Figure 4b can be used to find both fixed points with two queries. In addition to short-cutting the black boxes, we need to introduce a gate G that ensures a *unique* fixed point of the whole circuit. The gate G works in the following way:

$$\hat{G} = \sum_{e, c-a < c'-b} (a \otimes b \otimes c \otimes c' \otimes 0)(a \otimes b \otimes c \otimes c' \otimes e)^T +$$
$$\sum_{e, c-a \geq c'-b} (a \otimes b \otimes c \otimes c' \otimes \bar{e})(a \otimes b \otimes c \otimes c' \otimes e)^T,$$

where e is binary, $\bar{e} = e \oplus 1$, the addition is carried out modulo 2, and $\mathbf{0}$ is a 2-dimensional vector representing the value 0. In words, if the value c on the upper wire is less than the value on the lower wire c', and e is 0, then we get a fixed point on the third wire of G (variable e in Figure 4b). Otherwise, the bit on the third wire gets flipped—no fixed point. This guarantees that all loops together have a *unique* fixed point. Ironically, the gate G suppresses certain fixed points on the previous loops by introducing a logical *inconsistency* at a later point in the circuit. This resembles "anthropic computing" [14], where one guesses the solution to a problem and commits suicide if the guess was wrong—a recipe to solve NP-complete problems in the relative-state interpretation of quantum mechanics [16] and where consciousness follows only those branches where the programmer remains alive. Such a construction can be used to find the fixed points of a black box with a *few* fixed points and where the number of fixed points is *known*. For a large number n of fixed points (e.g., $n = d/2$), we can use the probabilistic approach to non-causal circuits. Let B_n be a black box with n fixed points and input and output spaces of dimension d. The Markov matrix of B_n is

$$\hat{B}_n = \sum_{i=0}^{d-1} e_i i^T, \quad \text{with } |\{i \,|\, e_i = i\}| = n.$$

We construct a randomized gate where the average number of fixed points is one:

$$\hat{B}' = \frac{1}{n} \hat{B}_n + \frac{n-1}{n} \hat{N},$$

with

$$\hat{N} = \sum_{i=0}^{n-1} \bar{i} i^T, \quad \bar{i} = i \oplus 1.$$

The gate $\hat{N}$ can be understood as a d-dimensional generalization of the NOT gate for bits: The input is increased by one modulo d. Such an $\hat{N}$ has *no* fixed points. The mixture $\hat{B}'$ is logically consistent, because

$$\mathrm{Tr}\left(\frac{1}{n}\hat{B}_n + \frac{n-1}{n}\hat{N}\right) = \frac{1}{n}\mathrm{Tr}\,\hat{B}_n + \frac{n-1}{n}\mathrm{Tr}\,\hat{N} = 1.$$

This means that we can use the circuit from Figure 4a to find a random fixed point of B_n.

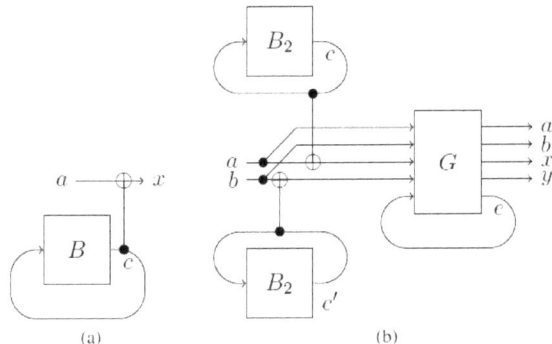

Figure 4. Fixed point search for a black box with one and a black box with two fixed points. (**a**) The output x is the fixed point c added to the input a. (**b**) Circuit for finding a fixed point for a black box with *two* fixed points.

We apply these tools to find solutions to instances of search problems with a *known* number of solutions, and where a guess for a solution can be verified efficiently by a verifier V. In other words, we can find solutions to NP search problems, yet where the number of solutions to an instance must be known to us in advance. Note that the following construction does not solve a decision problem, but rather *finds* the solution. Suppose an instance I to a problem Π has a *unique* solution. We replace the gate B of Figure 4a with a new gate V' that acts in the following way: it takes a guess c for a solution to $\Pi(I)$ as input, and runs V to verify c. If V accepts c, then V' outputs c, and otherwise, V' outputs $c \oplus 1$, where the addition is carried out modulo d. Such a circuit has a unique fixed point c which equals the solution of $\Pi(I)$. This, for instance, could be applied to a SAT formula, where a *unique* assignment of values to variables exist which make the formula true. Note that this approach does not prove an advantage in finding satisfying assignments for SAT formulas, even if the number of these satisfying assignments is previously known; currently, we do not know how difficult or easy it is to solve such instances *causally*.

5. Other Non-Causal Computational Models

We briefly discuss non-causal Turing machines and non-causal billiard computers. A Turing machine T has a tape, a read/write head, and an internal state machine. After every read instruction, the state machine moves to the next internal state, and thereby decides what to write and where to move the head to. A non-causal Turing machine is a machine where parts of the tape are not "within time": "Future" (from the head's point of view) *write* instructions influence "past" *read* instructions. A symbol that is written at time t to position j could be read at time $t' < t$ form position j; i.e., symbols can be read "before" they are written. As with other self-referential systems, this leads to problems that can be solved if we enforce the condition of logical consistency, as discussed above. Another issue is that multiple *write* instructions could *overwrite* the value on position j. This leaves open the question of what value is read. We can overcome this issue by running the Turing machine in a reversible fashion and by generating a history tape [17], where *no* memory position gets overwritten. An example of a non-causal Turing machine is where the *history* tape is non-causal in the sense that symbols can be read "before" they are written.

The billiard computer is a model of computation on a billiard table [1]. Before the computation starts, obstacles are placed on the table in such a way that the induced reflections of the balls and the collisions among the balls result in the desired computation. A non-causal version of a billiard computer is a billiard table where the holes are connected with closed timelike curves (CTCs) [2] that are logically consistent. Now, a billiard ball could also collide with its younger self; this introduces a non-causal effect. Echeverria, Klinkhammer, and Thorne [2] showed that solutions to CTC-dynamics that are not overdetermined exist. However, all solutions that they found are underdetermined. The non-causal circuits presented in this work indicate that logically consistent non-causal billiard computers are also admissible.

6. Conclusions and Open Questions

We show that models of computation where parts of the output of a computation are (re)used as input to the *same* computation are logically possible. Furthermore, such a model of computation helps to solve certain tasks more efficiently. The question is how much more powerful this new model of computation is, and whether uncomputable tasks become computable when compared to the standard circuit model. A strong restriction of the model is that before one can find a fixed point, one needs to know the number of fixed points. For instance, if we want to find a satisfying assignment for a SAT formula F with variables $x_0, x_1, \ldots$, we first need to know the number of satisfying assignments—otherwise, we do not know how to construct the circuit. Ironically, this means that to solve a SAT problem without any promise, we first need to solve a problem that is believed to be much harder: a #SAT problem. One might want to apply the Valiant–Vazirani [18] method to $F' = F \vee (x_0 \wedge x_1 \wedge \ldots)$ to reduce the number of satisfying assignments to 1 (the reason why we modify F to F' is to guarantee satisfiability). The problem that we are left with is that we do *not* know whether the output F'' of the Valiant–Vazirani method has a unique satisfying assignment or not—the reduction is probabilistic. Therefore, we cannot plug F'' into a circuit like the one shown in Figure 4a to find the fixed point.

A model of computation similar to but more general than ours is based on Deutsch's [19] CTCs. Aaronson and Watrous [20] showed that the classical special case of Deutsch's model can solve problems in PSPACE efficiently. However, in Deutsch's model, in contrast to ours, the information antinomy arises. Deutsch mitigates this issue by defining that the value on the looping wire is the uniform mixture of all solutions. This introduces a non-linearity into Deutsch's model: the output of a circuit depends non-linearly on the input. A consequence of this is that—in the quantum version—quantum states can be cloned [21]. As it is linear, the model studied here is not exposed to such consequences.

Acknowledgments: We thank Mateus Araújo, Veronika Baumann, Cyril Branciard, Časlav Brukner, Fabio Costa, Paul Erker, Adrien Feix, Arne Hansen, Alberto Montina, Christopher Portmann, and Benno Salwey for helpful discussions. This work was supported by the Swiss National Science Foundation (SNF), the National Centre of Competence in Research "Quantum Science and Technology" (QSIT) and the COST action on Fundamental Problems in Quantum Physics.

Author Contributions: Ämin Baumeler carried out this research and wrote this article, which is also part of his Ph.D., based on discussions with Stefan Wolf. Both authors have read and approved the final manuscript.

Conflicts of Interest: The authors declare no conflict of interest.

References

1. Fredkin, E.; Toffoli, T. Conservative logic. *Int. J. Theor. Phys.* **1982**, *21*, 219–253.
2. Echeverria, F.; Klinkhammer, G.; Thorne, K.S. Billiard balls in wormhole spacetimes with closed timelike curves: Classical theory. *Phys. Rev. D* **1991**, *44*, 1077–1099.

3. Chiribella, G. Perfect discrimination of no-signalling channels via quantum superposition of causal structures. *Phys. Rev. A* **2012**, *86*, 040301.
4. Colnaghi, T.; D'Ariano, G.M.; Facchini, S.; Perinotti, P. Quantum computation with programmable connections between gates. *Phys. Lett. A* **2012**, *376*, 2940–2943.
5. Chiribella, G.; D'Ariano, G.M.; Perinotti, P.; Valiron, B. Quantum computations without definite causal structure. *Phys. Rev. A* **2013**, *88*, 022318.
6. Araújo, M.; Costa, F.; Brukner, Č. Computational Advantage from Quantum-Controlled Ordering of Gates. *Phys. Rev. Lett.* **2014**, *113*, 250402.
7. Procopio, L.M.; Moqanaki, A.; Araújo, M.; Costa, F.; Alonso Calafell, I.; Dowd, E.G.; Hamel, D.R.; Rozema, L.A.; Brukner, Č.; Walther, P. Experimental superposition of orders of quantum gates. *Nat. Commun.* **2015**, *6*, 7913.
8. Oreshkov, O.; Costa, F.; Brukner, Č. Quantum correlations with no causal order. *Nat. Commun.* **2012**, *3*, 1092.
9. Baumeler, Ä.; Feix, A.; Wolf, S. Maximal incompatibility of locally classical behavior and global causal order in multiparty scenarios. *Phys. Rev. A* **2014**, *90*, 042106.
10. Baumeler, Ä.; Wolf, S. The space of logically consistent classical processes without causal order. *New J. Phys.* **2016**, *18*, 013036.
11. Branciard, C.; Araújo, M.; Feix, A.; Costa, F.; Brukner, Č. The simplest causal inequalities and their violation. *New J. Phys.* **2016**, *18*, 013008.
12. Feix, A.; Araújo, M.; Brukner, Č. Quantum superposition of the order of parties as a communication resource. *Phys. Rev. A* **2015**, *92*, 052326.
13. Tegmark, M. The Interpretation of Quantum Mechanics: Many Worlds or Many Words? *Fortschr. Phys.* **1998**, *46*, 855–862.
14. Aaronson, S. Guest Column: NP-complete problems and physical reality. *ACM SIGACT News* **2005**, *36*, 30–52.
15. Baumeler, Ä.; Wolf, S. Device-independent test of causal order and relations to fixed-points. *New J. Phys.* **2016**, *18*, 035014.
16. Everett, H. "Relative State" Formulation of Quantum Mechanics. *Rev. Mod. Phys.* **1957**, *29*, 454–462.
17. Bennett, C.H. Logical Reversibility of Computation. *IBM J. Res. Dev.* **1973**, *17*, 525–532.
18. Valiant, L.G.; Vazirani, V.V. NP is as easy as detecting unique solutions. *Theor. Comput. Sci.* **1986**, *47*, 85–93.
19. Deutsch, D. Quantum mechanics near closed timelike lines. *Phys. Rev. D* **1991**, *44*, 3197–3217.
20. Aaronson, S.; Watrous, J. Closed timelike curves make quantum and classical computing equivalent. *Proc. R. Soc. A Math. Phys. Eng. Sci.* **2009**, *465*, 631–647.
21. Brun, T.A.; Wilde, M.M.; Winter, A. Quantum State Cloning Using Deutschian Closed Timelike Curves. *Phys. Rev. Lett.* **2013**, *111*, 190401.

© 2017 by the authors. Licensee MDPI, Basel, Switzerland. This article is an open access article distributed under the terms and conditions of the Creative Commons Attribution (CC BY) license (http://creativecommons.org/licenses/by/4.0/).

Article
The Many Classical Faces of Quantum Structures

Chris Heunen

School of Informatics, University of Edinburgh, 10 Crichton Street, Edinburgh EH8 9AB, UK; chris.heunen@ed.ac.uk; Tel.: +44-131-650-5132

Academic Editors: Giacomo Mauro D'Ariano, Paolo Perinotti, Jay Lawrence and Giorgio Kaniadakis
Received: 9 January 2017; Accepted: 23 March 2017; Published: 29 March 2017

Abstract: Interpretational problems with quantum mechanics can be phrased precisely by only talking about empirically accessible information. This prompts a mathematical reformulation of quantum mechanics in terms of classical mechanics. We survey this programme in terms of algebraic quantum theory.

Keywords: algebraic quantum theory; C*-algebra; gelfand duality; classical context; bohrification

1. Introduction

The mathematical formalism of quantum mechanics is open to interpretation. For example, the possibility of deterministic hidden variables, the uncertainty principle, the measurement problem, and the reality of the wave function, are all up for debate. (The first and the last of course have rigorous restrictions: hidden variables by the Bell inequalities [1] and the Kochen–Specker theorem [2], discussed below, and reality of the wave function by the Pusey–Barrett–Rudolph theorem [3].) Classical mechanics shares none of those interpretational questions. This article surveys a mathematical reformulation of quantum mechanics in terms of classical mechanics, intended to bring the interpretational issues with the former to a head. This programme proposes to replace the usual notion of state space of a quantum-mechanical system by a new one, in a way that avoids the interpretational questions above and leaves classical systems unaffected:

- known obstructions to hidden variable interpretations merely say that states cannot be located with exact precision in the state space, and are circumvented via open regions of states;
- the uncertainty principle cannot be expressed and therefore poses no interpretational problem;
- the measurement problem is obviated because the new notion of state space incorporates all classical data resulting from possible measurements.

If we also take dynamics into account, the new notion of configuration space, called an active lattice:

- yields the same predictions as traditional quantum mechanics.

This programme branches into a number of related themes, spread over the literature; see the extensive bibliography. The aim of this article is to bring all these active developments together to give an overview. There are hardly any new results. Instead, the novelty lies in rephrasing foundations to give an accessible, coherent, and complete overview of the current state-of-the-art. To do so, we will have to be rather brief and refer to references for many technical details. Nevertheless, there is a novel contribution regarding topological structure of the new notion of configuration space. We will use an n-level physical system as a running example to illustrate new notions (though many results have exceptions for $n \leq 2$, and most interesting features occur in infinite dimension). The rest of this introduction summarizes the framework and discusses four salient features, before giving an overview of the rest of this article.

1.1. Algebraic Quantum Theory

The traditional formalism of quantum theory holds that the *(pure) state space* is a Hilbert space H, that *(sharp) observables* correspond to self-adjoint operators on that Hilbert space, and that *(undisturbed) evolution* corresponds to unitary operators. Algebraic quantum theory instead takes the observables as primitive, and the state space is a derived notion. Self-adjoint operators combine with unitaries to give all bounded operators, and these form a so-called C*-algebra $B(H)$. However, superselection rules mandate that not all self-adjoint operators correspond to valid observables. Thus, one considers arbitrary C*-algebras, rather than only those of the form $B(H)$. Nevertheless, it turns out that any C*-algebra A embeds into $B(H)$ for some Hilbert space H, and in that sense C*-algebra theory faithfully captures quantum theory. Finally, one could impose extra conditions on a C*-algebra, leading to so-called AW*-algebras, and W*-algebras, also known as von Neumann algebras. A good example to keep in mind is the algebra $\mathbb{M}_n(\mathbb{C})$ of n-by-n complex matrices, that models (the observables of) an n-level system, or direct sums $\mathbb{M}_{n_1}(\mathbb{C}) \oplus \cdots \oplus \mathbb{M}_{n_k}(\mathbb{C})$.

To pass from pure to mixed states (density matrices), from sharp to unsharp observables (positive operator valued measurements), and from undisturbed evolution to including measurement (quantum channels), the traditional formalism prescribes completely positive maps. These find their natural home in the algebraic formulation. States of a C*-algebra A can then be recovered as unital (completely) positive maps $A \to \mathbb{C}$. Observables with n outcomes are unital (completely) positive maps $\mathbb{C}^n \to A$; sharp observables correspond to homomorphisms. Evolution is described by a completely positive map $A \to A$; undisturbed evolution corresponds to a homomorphism. Indeed, if $A = \mathbb{M}_n(\mathbb{C})$, then states $A \to \mathbb{C}$ are precisely density matrices; observables $\mathbb{C}^n \to A$ are precisely positive operator valued measurements with n outcomes; completely positive maps $A \to A$ are precisely those that map density matrices to density matrices; and homomorphisms $A \to A$ are precisely the linear functions that map pure states to pure states.

For more information on algebraic quantum theory, see [4–13].

1.2. Gelfand Duality

The advantage of algebraic quantum theory is that it places quantum mechanics on the same footing as classical mechanics. The *(pure) state space* in classical mechanics can be any locally compact Hausdorff topological space X, *(sharp) observables* are continuous functions $X \to \mathbb{R}$, and *evolution* is given by homeomorphisms $X \to X$. This leads to the C*-algebra $C_0(X)$ of continuous complex-valued functions on X vanishing at infinity; for compact X, we write $C(X)$. A simple example is the algebra $\mathbb{C}^n$, where X is a discrete space with n points. Indeed, in that case there are n (pure) states; (sharp) observables are precisely vectors in $\mathbb{R}^n$; and (deterministic) evolutions are just functions $n \to n$.

Again, we can pass from classical mechanics to the probabilistic setting of statistical mechanics by considering completely positive maps. States of $C(X)$ can be recovered as unital (completely) positive maps $C(X) \to \mathbb{C}$ as before; pure states $x \in X$ correspond to homomorphisms. Observables with m outcomes are (completely) positive maps $\mathbb{C}^m \to C(X)$, and sharp observables correspond to homomorphisms. Stochastic evolution is described by a (completely) positive map $C(X) \to C(X)$; deterministic evolution corresponds to a homomorphism. Indeed, for X, the discrete space with n points, states $C(X) \to \mathbb{C}$ are precisely probability distributions on n points; observables $\mathbb{C}^m \to C(X)$ with m outcomes are precisely m-tuples of probability distributions on n points summing to one; sharp observables $\mathbb{C}^m \to C(X)$ are just functions $m \to n$; and evolutions $C(X) \to C(X)$ are simply stochastic m-by-n matrices.

Note that multiplication in $C(X)$ is *commutative*, whereas $B(H)$ was *noncommutative*. Gelfand duality says that any commutative C*-algebra C is of the form $C(X)$ for some compact Hausdorff space X, called its *spectrum* and written as $\mathrm{Spec}(C)$. That is, $C \cong C(\mathrm{Spec}(C))$ and $X \cong \mathrm{Spec}(C(X))$. Moreover, this gives a dual equivalence of categories: if $f \colon X \to Y$ is a continuous function then $C(f) \colon C(Y) \to C(X)$ is a homomorphism, and conversely, if $f \colon C \to D$ is a homomorphism, then $\mathrm{Spec}(f) \colon \mathrm{Spec}(D) \to \mathrm{Spec}(C)$ is a continuous function. Thus, C*-algebra theory is often regarded as *noncommutative topology*. In the case of a discrete space X with n points, this simply says that up to isomorphism $\mathbb{C}^n$ is the only commutative C*-algebra of dimension n, and that functions $n \to n$ are the only way to describe deterministic evolutions.

For more information, we refer to [14–17] in addition to references above.

1.3. Bohr's Doctrine of Classical Concepts

To summarize, both classical systems and quantum systems are first-class citizens that can interact in the algebraic framework. Classical systems are commutative algebras C, and quantum systems are noncommutative ones A. An example interaction is measurement, given by maps $C \to A$. For n-level systems, a measurement with m outcomes is a map $\mathbb{C}^m \to \mathbb{M}_n(\mathbb{C})$. Having no superfluous outcomes in $\mathrm{Spec}(C)$ of the measurement corresponds to the injectivity of these maps. So the information that all possible measurements can give us about a possibly noncommutative algebra A is its collection $\mathcal{C}(A)$ of commutative subalgebras C. In other words, all empirically accessible information in a quantum system is encoded in its family of classical subsystems. This observation is known as the *doctrine of classical concepts* and dates back to Bohr [18,19]. For an n-level system $A = \mathbb{M}_n(\mathbb{C})$, elements of $\mathcal{C}(A)$ indeed correspond to all possible measurement setups: the ways of choosing an orthonormal basis of $\mathbb{C}^n$ and a partition of an n-element set with m equivalence classes for outcomes.

The main aim of this paper is to survey what can be said about the quantum structure A based on its many classical faces $\mathcal{C}(A)$, explaining the title.

1.4. The Kadison–Singer Problem

A case in point is the long-standing but recently solved Kadison–Singer problem [20,21]. In a noncommutative C*-algebra, not all observables are compatible, in the sense that they can be measured simultaneously (without uncertainty). What can at most be measured in an experiment are those observables in a single commutative subalgebra. The best an experimenter can do is repeat the experiment to determine the values of those observables, giving a pure state of that commutative subalgebra. Ideally, this tomography procedure should determine the state of the entire system. Indeed, there are various protocols for performing such tomography on n-level systems that have been experimentally verified [22].

The Kadison–Singer result says that this procedure indeed works in the discrete case. Let H be a Hilbert space of countable dimension. Then $B(H)$ has a discrete maximal commutative subalgebra $\ell^\infty(\mathbb{N})$ consisting of operators that are diagonal in a fixed basis. The precise result is that a pure state of $\ell^\infty(\mathbb{N})$ extends *uniquely* to a pure state of $B(H)$. Thus, (the state of) a quantum system is characterized by what we can learn about it from experiments, giving a positive outlook on Bohr's doctrine of classical concepts.

1.5. The Kochen–Specker Theorem

Nevertheless, Bohr's doctrine of classical concepts should be interpreted carefully. It does not say that collections of states of each classical subsystem assemble to a state of the quantum system. That is ruled out by the Kochen–Specker theorem. In physical terms, local deterministic hidden variables are impossible; one cannot assign definite values to all observables of a quantum system in a noncontextual way, i.e., giving coherent states on classical subsystems. In mathematical terms, Gelfand duality does not extend to noncommutative algebras via $\mathcal{C}(A)$; this will be discussed in more detail in Section 2. More precisely, the zero map is the only function $\mathbb{M}_n(\mathbb{C}) \to C(X)$ that restricts to homomorphisms $C \to C(X)$ for each $C \in \mathcal{C}(\mathbb{M}_n(\mathbb{C}))$ when $n \geq 3$. That is, there is no way to assign measurement outcomes in $\mathbb{R}^m$ to all possible positive operator valued measures on an n-level system with m outcomes in a consistent way. This extends to more general noncommutative A that do not contain a subalgebra $\mathbb{M}_2(\mathbb{C})$. See [2,11,23].

1.6. Overview of This Article

Section 2 continues in more depth the discussion of the structure of quantum systems from the perspective just sketched. In particular, it covers exactly how much of A can be reconstructed from $\mathcal{C}(A)$, and makes precise the link between the Kochen–Specker theorem and noncommutative Gelfand duality. Section 3 shows how to interpret a quantum system A as a classical system via $\mathcal{C}(A)$ by changing the rules of the ambient set theory, and discusses the surrounding interesting interpretational issues. Section 4 considers fine-graining. Increasing chains of classical subsystems give more and more information about the quantum system. We discuss $\mathcal{C}(A)$ from this information-theoretic point of view, called *domain theory*. Section 5 explains how to incorporate dynamics into $\mathcal{C}(A)$, turning it into a so-called *active lattice*. It turns out that this extra information does make $\mathcal{C}(A)$ into a full invariant, from which one can reconstruct A. This raises interesting interpretational questions: its active lattice can be regarded as a *configuration space* that completely determines a quantum system. By encoding more than static hidden variables, it circumvents the obstructions of Section 2. To obtain an equivalence for quantum systems like Gelfand duality did for classical ones, it thus suffices to characterize the active lattices arising this way. This is examined in Section 6. Finally, Section 7 considers to what extent the successes of the doctrine of classical concepts in the previous sections are due to the use of algebraic quantum theory, and to what extent they generalize to other formulations.

2. Invariants

Bohr's doctrine of classical concepts teaches that a quantum system can only be empirically understood through its classical subsystems. These classical subsystems should therefore contain all the physically relevant information about the quantum system.

Definition 1. *For a unital C*-algebra A, write $\mathcal{C}(A)$ for its family of commutative unital C*-subalgebras C (with the same unit as A). We may think of it either as partially ordered set by inclusion, or as a diagram that remembers that the points of the partially ordered set are C*-algebras C.*

For example, the partially ordered set $\mathcal{C}(A)$ of a 2-level system $A = \mathbb{M}_2(\mathbb{C})$ has Hasse diagram

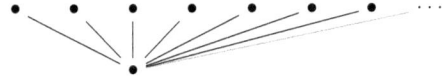

with a point on the upper level for each unitary in $U(2)$.

The question is then: how does the mathematical formalism of the quantum theory of A translate into terms of $\mathcal{C}(A)$? For example, it turns out that the entropy of a state of A can be reconstructed from the entropies of its restriction to $\mathcal{C}(A)$ [24], see also [25]. Ideally, we would like to completely reconstruct A from $\mathcal{C}(A)$. A priori, $\mathcal{C}(A)$ is merely an invariant of A. This section investigates how strong an invariant it is. The first step is to realize that, from $\mathcal{C}(A)$, we can reconstruct A as a set, as well as operations between commuting elements. This can be made precise by the notion of a *piecewise C*-algebra*, which is basically a C*-algebra that forgot how to add or multiply noncommuting operators.

Definition 2. *A piecewise C*-algebra consists of a set A with*

- *a reflexive and symmetric binary (commeasurability) relation $\odot \subseteq A \times A$;*
- *elements $0, 1 \in A$;*
- *a (total) involution $*: A \to A$;*
- *a (total) function $\cdot : \mathbb{C} \times A \to A$;*
- *a (total) function $\|-\| : A \to \mathbb{R}$;*
- *(partial) binary operations $+, \cdot : \odot \to A$;*

such that every set $S \subseteq A$ of pairwise commeasurable elements is contained in a set $T \subseteq A$ of pairwise commeasurable elements that forms a commutative C*-algebra under the above operations.

Of course, any commutative C*-algebra is a piecewise C*-algebra. More generally, the normal elements (those commuting with their own adjoint) of any C*-algebra A form a piecewise C*-algebra. For an n-level system $A = \mathbb{M}_n(\mathbb{C})$, the piecewise C*-algebra consists of all normal n-by-n matrices, together with their norms and adjoints, as well as the knowledge of how commuting elements add and multiply. Notice that $\mathcal{C}(A)$ makes perfect sense for any piecewise C*-algebra A. To make precise how we can reconstruct the piecewise structure of A from $\mathcal{C}(A)$, we will use the language of *category theory* [26]. C*-algebras, with *-homomorphisms between them, form a category. We can also make piecewise C*-algebras into a category with the following arrows: (total) functions $f \colon A \to B$ that preserve commeasurability and the algebraic operations, whenever defined.

The precise notion we need is that of a *colimit*. Suffice to say here, a colimit, when it exists, is a universal solution that compatibly pastes together a given diagram into a single object. Thinking of A as the whole and $\mathcal{C}(A)$ as its parts, we would like to know whether the whole is determined by the parts. The following theorem says that $\mathcal{C}(A)$ indeed contains enough information to reconstruct A as a piecewise C*-algebra.

Theorem 1 ([27]). *Every piecewise C*-algebra is the colimit of its commutative C*-subalgebras in the category of piecewise C*-algebras.*

This means that the diagram $\mathcal{C}(A)$ determines the piecewise C*-algebra A: if $\mathcal{C}(A)$ and $\mathcal{C}(B)$ are isomorphic diagrams, then A and B are isomorphic piecewise C*-algebras. Moreover, the previous theorem gives a concrete way to reconstruct A from $\mathcal{C}(A)$. For the n-level system $A = \mathbb{M}_n(\mathbb{C})$, this means we can reconstruct from $\mathcal{C}(A)$ the normal n-by-n matrices, as well as sums and products of commuting ones. An important point to note here is that the reconstruction is happening in the setting of piecewise C*-algebras. We could not have taken the colimit in the category of commutative C*-algebras instead. Indeed, one way to reformulate the Kochen–Specker theorem in terms of colimits is the following. The following reformulation might not look much like the original, but it is nevertheless equivalent, and more suited to our purposes; see also ([2], p. 66).

Theorem 2 ([2,28]). *If $n \geq 3$, then the colimit of $\mathcal{C}(\mathbb{M}_n(\mathbb{C}))$ in the category of commutative C*-algebras is the degenerate, 0-dimensional, C*-algebra.*

In fact, the colimit of $\mathcal{C}(A)$ degenerates for many more C*-algebras A than just $\mathbb{M}_n(\mathbb{C})$, such as any C*-algebra of the form $\mathbb{M}_n(B)$ for some C*-algebra B, or any W*-algebra that has no direct summand $\mathbb{C}$ or $\mathbb{M}_2(\mathbb{C})$ [29,30].

As mentioned in the introduction, Gelfand duality is a *functor* from the category of commutative C*-algebras to the category of compact Hausdorff topological spaces. That is, a systematic way to assign a space to a C*-algebra, that respects functions. Interpreted physically: any classical system is determined by a configuration space in a way that respects operations on the system. The previous theorem can be used to show that there is no such configuration space determining quantum systems—at least, if the notion of configuration space is to be a *conservative extension* of the classical notion. The latter can be made precise as a continuous functor from the category of compact Hausdorff spaces to some category with a degenerate space like the empty set, more precisely, a strict initial object 0.

Theorem 3 ([29]). *Suppose there exist a category conservatively extending that of compact Hausdorff spaces and a functor F completing the following square.*

$$\begin{array}{ccc}
\text{commutative C*-algebras} & \xrightarrow{\text{Spec}} & \text{compact Hausdorff spaces} \\
\cap & & \cap \\
\text{C*-algebras} & \xrightarrow{F} & ?
\end{array}$$

Then $F(\mathbb{M}_n(\mathbb{C})) = 0$ for $n \geq 3$. In particular, F cannot be a dual equivalence.

Asking the functor on the right to be continuous is appropriate to model the classical limit of quantum systems converging to a classical one, because then the state space of the product of two limiting classical systems should be computed as the classical limit of the joint quantum systems. In fact, the proof in [29] holds if the category on the bottom right has limits, and the functor on the right reflects them. However, one might still wonder if it is reasonable to ask the diagram to commute on the nose. Instead, we could ask it to commute up to a natural isomorphism. This is precisely the way out we will explore in Sections 3 and 5.

This rules out many possible quantum configuration spaces that have been proposed for the bottom right role in the square; in particular many generalized notions of topological spaces, such as sets, topological spaces themselves, pointfree topological spaces, ringed spaces, quantales, toposes, categories of sheaves, and many more [28,29,31]. In particular, the *state space* of a C*-algebra, as discussed in the introduction, will not do for us, even though it is one of the most important tools associated with a C*-algebra [32]. That explains why we deliberately talk about "configuration spaces". In the classical case, the two notions coincide. The previous theorem shows that serious notions of quantum configuration space must be less conservative. This points the way towards good candidates: Sections 3 and 5 will cover two that do fit the bill.

The question of noncommutative extensions of Gelfand duality is also very interesting from a purely mathematical perspective. As mentioned in the introduction, C*-algebra theory can be regarded as noncommutative topology. Adding more structure than mere topology leads to *noncommutative geometry*, which is a rich field of study [33]. However, it takes place entirely on the algebraic side. Finding the right notion of quantum configuration space could reintroduce geometric intuition, which is usually very powerful [34,35]. For example, in certain cases, extensions of $\mathcal{C}(A)$ can be used to compute the *K-theory* of A, which is a way to study homotopies of the configuration space underlying A, that includes many local-to-global principles [36]. Similarly, closed *ideals* of a W*-algebra A, that are important because they correspond to open subsets in the classical case, are in bijection with certain piecewise ideals of $\mathcal{C}(A)$ [37].

So far, we have considered $\mathcal{C}(A)$ as a *diagram* of parts of the whole. We finish this section by considering it as a mere partially ordered set, where we forget that elements have the structure of commutative C*-algebras. That is, we only consider the shape of how the parts fit together. This information is already enough to determine the piecewise structure of A, but as a *Jordan algebra*. (In fact, considering $\mathcal{C}(A)$ as a mere partially ordered set gives precisely the same information as considering it as a diagram [38]. This justifies Definition 1.) The self-adjoint elements of a C*-algebra form a Jordan algebra under the product $a \circ b = \frac{1}{2}(ab + ba)$; this even gives a so-called JB-algebra. In fact, any JB-algebra is a subalgebra of the direct sum of one of this form and an exceptional one, such as quaternionic matrices $\mathbb{M}_3(\mathbb{H})$ [39]. For example, the n-level system gives the JB-algebra of hermitian n-by-n matrices multiplied via anticommutators. Piecewise Jordan algebras and their homomorphisms are defined analogously to Definition 2. The structure of quantum observables leads naturally to the axioms of Jordan algebras [8] (Modern mathematical physics tends to prefer C*-algebras, as their theory is slightly less complicated, and the connections to Jordan algebras are so tight anyway [39].) The following theorem justifies that point of view.

Theorem 4 ([40]). *Let A and B be C*-algebras. If $\mathcal{C}(A)$ and $\mathcal{C}(B)$ are isomorphic partially ordered sets, then A and B are isomorphic as piecewise Jordan algebras.*

A little more can be said. Any isomorphism $f: \mathcal{C}(A) \to \mathcal{C}(B)$ is implemented by an isomorphism $g: A \to B$ of piecewise Jordan algebras, in the sense that $f(C) = \{g(c) \mid c \in C\}$. In fact, this g is unique, unless A is either $\mathbb{C}^2$ or $\mathbb{M}_2(\mathbb{C})$. For *AW*-algebras* more is true because of Gleason's theorem, that we will meet in Section 5, we can actually reconstruct the full linear structure rather than just the piecewise linear structure. (An *AW*-algebra* is a C*-algebra A that has enough projections, in the sense that every $C \in \mathcal{C}(A)$ is the closed linear span of its projections, and those projections work together well, in the sense that orthogonal families in the partially ordered set of projections have least upper bounds [7,41]. See also Section 5. They are more general than W*-algebras, and much of the theory of W*-algebra generalizes to AW*-algebras, such as the type decomposition. An n-level system $A = \mathbb{M}_n(\mathbb{C})$ forms a W*-algebra, and hence also an AW*-algebra.) Type I_2 AW*-algebras are those of the form $\mathbb{M}_2(C)$ for a commutative AW*-algebra C. AW*-algebras with a type I_2 direct summand correspond to the exceptional case $n = 2$ in the Kochen–Specker Theorem 2. We will call them *atypical*, and algebras without a type I_2 direct summand *typical*, as we will meet this exception often. An n-level system is typical when $n \geq 3$.

Corollary 1 ([42,43]). *Let A and B be typical AW*-algebras. If $\mathcal{C}(A)$ and $\mathcal{C}(B)$ are isomorphic partially ordered sets, then A and B are isomorphic as Jordan algebras.*

Whereas the C*-algebra product is associative but need not be commutative, the Jordan product is commutative but need not be associative; commutative C*-subalgebras correspond to associative Jordan subalgebras. Indeed, the previous theorem generalizes to Jordan algebras in those terms [44].

3. Toposes

In this section, we consider $\mathcal{C}(A)$ as a diagram. That is, we regard it as an operation that assigns to each classical subsystem $C \in \mathcal{C}(A)$ of the quantum system A a classical system C. What kind of operation is this diagram $C \mapsto C$? We can think of it as a set $S(C)$ that varies with the context $C \in \mathcal{C}(A)$. Moreover, this *contextual set* respects coarse-graining: if $C \subset D$, then $S(C) \subseteq S(D)$. That is, when the measurement context C grows to include more observables, the information contained in the set $S(C)$ assigned to it grows along accordingly. For example, for an 2-level system $A = \mathbb{M}_2(\mathbb{C})$, this comes down to a choice of set $S(u)$ for each unitary $u \in U(2)$, that all include a fixed set $S(0)$. Hence, these contextual sets are functors S from $\mathcal{C}(A)$, now regarded as a partially ordered set, to the category of sets and functions. The totality of all such functors forms a category. In fact, contextual sets form a particularly nice category, namely a *topos*.

A topos is a category that shares a lot of the properties of the category of sets and functions. In particular, one can *do mathematics inside* a topos: we may think about objects of a topos as sets, that we may specify and manipulate using logical formulae. Of course, this internal perspective comes with some caveats. Most notably, if a proof is to hold in the internal language of any topos, it has to be *constructive*: we are not allowed to use the axiom of choice or proofs by contradiction, and have to be careful about real numbers. We cannot go into more detail here, but for more information on topos theory, see [45].

One particular object of interest in the topos of contextual sets over $\mathcal{C}(A)$ is our canonical contextual set $C \mapsto C$. It turns out that, according to the logic of the topos of contextual sets, this object is a *commutative* C*-algebra.

Theorem 5 ([19]). *Let A be a C*-algebra. In the topos of contextual sets over $\mathcal{C}(A)$, the canonical contextual set $C \mapsto C$ is a commutative C*-algebra.*

This procedure is called *Bohrification*:

1. Start with a quantum system A.
2. Change the logical rules of set theory by moving to the topos of contextual sets over $\mathcal{C}(A)$.

3. The quantum system A turns into a classical one given by the canonical contextual set $C \mapsto C$.

See also [46].

Thus, we may study the quantum system A as if it were a classical system. Of course, we lose the same information as in the previous section. For example, we can only hope to reconstruct the Jordan structure of A from the contextual set $C \mapsto C$. Nevertheless, placing it in a topos of its peers opens up many possibilities. In particular, we may try to find a configuration space inside the topos. It turns out that Gelfand duality can be formulated so that its proof is constructive, and hence applies inside the topos. This involves talking about *locales* rather than topological spaces. We may think of a locale as a topological space that forgot it had points.

More precisely, a locale may be thought of as the partially ordered family of open sets of a topological space, but without a carrier set of points. Most of topology can be formulated to work for locales as well. Again, we cannot go into more detail here, but for more information on locales see [47].

Corollary 2 ([48]). *Let A be a C*-algebra. In the topos of contextual sets over $\mathcal{C}(A)$, there is a compact Hausdorff locale X such that the canonical contextual set is of the form $C(X)$.*

For example, if A is the 2-level system $\mathbb{M}_2(\mathbb{C})$, then X is the contextual set S that assigns to $u \in U(2)$ the orthonormal basis of $\mathbb{C}^2$ corresponding to u, and that assigns to 0 the zero vector in $\mathbb{C}^2$, where $S(u)$ locally carries the structure of a 2-element discrete space, and $S(0)$ carries the structure of a 1-element discrete space. We will call this locale X the *spectral contextual set*. In general, it is not just the contextual set $C \mapsto \text{Spec}(C)$. However, it does resemble that if we think about *bundles* instead of contextual sets [49,50]: a bundle is a map of locales into the locale of ideals of $\mathcal{C}(A)$, and by restricting the intuitionistic logic of a topos further to so-called geometric logic, the bundle corresponding to the spectral contextual set does have fibre $\text{Spec}(C)$ over C. Also, if we reverse the partial order on $\mathcal{C}(A)$, the assignment $C \mapsto \text{Spec}(C)$ plays the role of the canonical contextual set. So there are two approaches:

- Either one uses $\mathcal{C}(A)$; the canonical contextual set $C \mapsto C$ is a commutative C*-algebra, and the spectral contextual set X does not take a canonical form [19,51–55].
- Or one uses the opposite order; the spectral contextual set X is a locale of the canonical form $C \mapsto \text{Spec}(C)$, and the commutative C*-algebra $C(X)$ does not take a canonical form [56–59].

For a comparison, see [60]. For this overview article, the choice of direction does not matter so much. In any case, X is an object inside the topos of contextual sets, and as such we may reason about it as a locale. In particular, we may wonder whether it is a topological space, that is, whether it does in fact have enough points. It turns out that the Kochen–Specker Theorem 2 can be reformulated as saying that not only does X not have enough points, in fact it has no points at all. In terms of bundles: the canonical bundle has no global sections. This illustrates the need for locales rather than topological spaces.

Proposition 1 ([23]). *Let A be a C*-algebra satisfying the Kochen–Specker Theorem 2. In the topos of contextual sets over $\mathcal{C}(A)$, the spectral contextual set has no points.*

Thus, Bohrification turns a quantum system A into a locale X inside the topos of contextual sets over $\mathcal{C}(A)$. There is an equivalence between locales X inside such a topos over $\mathcal{C}(A)$, and certain continuous functions from a locale $\text{Spec}(A)$ to $\mathcal{C}(A)$ *outside* the topos [61]. This gives a way to cut out the whole topos detour, and assign to the quantum system A a configuration space that we will temporarily call $\text{Spec}(A)$ for the rest of this section.

Proposition 2 ([62]). *For any C*-algebra A, the internal locale X is determined by a continuous function from some locale $\text{Spec}(A)$ to $\mathcal{C}(A)$.*

In many cases, Spec(A) will in fact have enough points, i.e., will be a topological space [60,62]—despite Proposition 1. The construction $A \mapsto$ Spec(A) circumvents the obstruction of Theorem 3 for several reasons. First, when the C*-algebra A is commutative, Spec(A) turns out to be a locale based on $\mathcal{C}(A)$, rather than on A itself; therefore what we are currently denoting by Spec(A) does not match the Gelfand spectrum of A. Second, the construction $A \mapsto$ Spec(A) is only partially functorial: if we regard $\mathcal{C}(A)$ as a locale, the construction only respects functions that reflect commutativity [27], and to get functorality we have to regard $\mathcal{C}(A)$ as a *localed topos*, that is, a topos with a locale in it [63].

We can only touch on it briefly here, but one of the main features of building the topos of contextual sets over $\mathcal{C}(A)$ and distilling the configuration space Spec(A) is that they encode a contextual *logic*. This logic is intuitionistic, and therefore very different from traditional *quantum logic* [52]. The latter concerns the set Proj(A) of yes–no questions on the quantum system A; more precisely, the set of sharp observables with two outcomes. These correspond to *projections*: $p \in A$ satisfying $p^2 = p = p^*$. They are partially ordered by $p \leq q$ when $pq = p$, which should be read as saying that p implies q. Similarly, least upper bounds in Proj(A) are logical disjunctions [11]. In an n-level system $A = \mathbb{M}_n(\mathbb{C})$, projections correspond to subspaces of $\mathbb{C}^n$, regarded logically as the set of (pure) states where the proposition is true; the order becomes inclusion of subspaces; and the disjunction of subspaces is their linear span. AW*-algebras A are determined to a great extent by their projections, and indeed the quantum logic Proj(A) carries precisely the same amount of information as $\mathcal{C}(A)$ [64]. For more information about this topos-theoretic approach to quantum logic, we refer to [19,49,51–54,56–58].

To connect contextual sets to probabilities and the Born rule, we have to translate states of A into some notion based on the spectral contextual set X, and observables of A into some notion based on the canonical contextual set $C \mapsto C$. For the latter, one has to resort to approximations, as not every $a \in A$ will be present in each $C \in \mathcal{C}(A)$; this process is sometimes called *daseinisation* [57]. The former has a satisfying solution in terms of *piecewise states*: piecewise linear (completely) positive maps $A \to \mathbb{C}$.

Theorem 6 ([23,51,65]). *There is a bijective correspondence between piecewise states on an AW* algebra A, and states of the canonical contextual set $C \mapsto C$ inside the topos of contextual sets over $\mathcal{C}(A)$.*

(The cited references consider W*-algebras, but the proof holds for AW*-algebras because Corollary 5 does so, see Section 5. The same goes for the references in Corollary 3.) By Gleason's theorem (see Section 5), we can say more for AW*-algebras. See also [25].

Corollary 3 ([66,67]). *There is a bijective correspondence between states of a typical AW*-algebra A, and states of the canonical contextual set $C \mapsto C$ inside the topos of contextual sets over $\mathcal{C}(A)$.*

In the n-level system $A = \mathbb{M}_n(\mathbb{C})$ for $n \geq 3$, this means that n-by-n density matrices correspond precisely to a choice of probability distribution over m points that is consistent over all unitaries $u \in U(n)$ and partitions of n points into m equivalence classes.

Combining daseinisation with the above results gives rise to a contextual Born rule, justifying the Bohrification procedure of Theorem 5 [50]. Summarizing, we can formulate the physics of the quantum system A completely in terms of $\mathcal{C}(A)$ and its topos of contextual sets, and work within there as if dealing with a classical system.

To end this section, let us mention some other related work. The "amount of nonclassicality" of the contextual logic discussed of A measures the computational power of the quantum system A [68]. For philosophical aspects of Bohrification and related constructions, see [69,70]. Similar contextual ideas have been used to model quantum numbers [71]. Transferring C*-algebras between different toposes has been used successfully before in so-called Boolean-valued analysis [72–74]. Finally, contextuality and the Kochen–Specker theorem can be formulated more generally than in algebraic quantum theory [75].

4. Domains

The partially ordered set $\mathcal{C}(A)$ of empirically accessible classical contexts C of a quantum system A embodies *coarse-graining*. As in the introduction, we think of each $C \in \mathcal{C}(A)$ as consisting of compatible observables that we can measure together in a single experiment. Larger experiments, involving more observables, should give us more information, and this is reflected in the partial order: if $C \subseteq D$, then D contains more observables, and hence provides more information. If A itself is noncommutative, the best we can do is approximate it with larger and larger commutative subalgebras C. This sort of informational approximation is studied in computer science under the name *domain theory* [76,77]. This section discusses the domain-theoretic properties of $\mathcal{C}(A)$. Domain theory is mostly concerned with partial orders where every element can be approximated by finite ones, as those are the ones we can measure in practice, leading to the following definitions.

Definition 3. *A partially ordered set $(\mathcal{C}, \leq)$ is directed complete when every ascending chain $\{D_i\}$ has a least upper bound $\bigvee_i D_i$. An element C approximates D, written $C \ll D$, when $D \leq \bigvee_i D_i$ implies $C \leq D_i$ for any chain $\{D_i\}$ and some i. An element C is finite when $C \ll C$. A continuous domain is a directed complete partially ordered set, every element of which satisfies $D = \bigvee \{C \mid C \ll D\}$. An algebraic domain is a directed complete partially ordered set, every element of which is approximated by finite ones: $D = \bigvee \{C \mid C \ll C \leq D\}$.*

Lemma 1 ([65,78])**.** *If A is a C*-algebra, then $\mathcal{C}(A)$ is a directed complete partially ordered set, in which $\bigvee_i C_i$ is the norm-closure of $\bigcup_i C_i$.*

We saw in Section 2 that $\mathcal{C}(A)$ captures precisely the structure of A as a (piecewise) Jordan algebra. Order-theoretic techniques give an alternative proof of Corollary 1. First, we can recognize the dimension of A from $\mathcal{C}(A)$. Recall that a partially ordered set is *Artinian* when: every nonempty subset has a minimal element; every nonempty filtered subset has a least element; every descending sequence $C_1 \geq C_2 \geq \cdots$ eventually becomes constant. The dual notion, satisfying an ascending chain condition, is called *Noetherian*.

Proposition 3 ([79])**.** *A C*-algebra A is finite-dimensional if and only if $\mathcal{C}(A)$ is Artinian, if and only if $\mathcal{C}(A)$ is Noetherian.*

Indeed, in an n-level system $A = \mathbb{M}_n(\mathbb{C})$, elements $C \in \mathcal{C}(A)$ correspond to a choice of unitary $u \in U(n)$ and a partition of n points into m equivalence classes. Because $C \subseteq D$ when the partition for D is finer than that for C, the partially ordered set $\mathcal{C}(A)$ can only have strictly increasing chains of length at most n.

By the Artin–Wedderburn theorem, we know that any finite-dimensional C*-algebra A is a finite direct sum of matrix algebras $\mathbb{M}_{n_i}(\mathbb{C})$. It is therefore specified up to isomorphism by the numbers $\{n_i\}$, which we can extract from the partially ordered set $\mathcal{C}(A)$. A partially ordered set $\mathcal{C}$ is called *directly indecomposable* when $\mathcal{C} = \mathcal{C}_1 \times \mathcal{C}_2$ implies that either $\mathcal{C}_1$ or $\mathcal{C}_2$ is a singleton set.

Proposition 4 ([79,80])**.** *If $A = \bigoplus_{i=1}^{n} \mathbb{M}_{n_i}(\mathbb{C})$, then the C*-subalgebras $\mathbb{M}_{n_i}(\mathbb{C})$ correspond to directly indecomposable partially ordered subsets $\mathcal{C}_i$ of $\mathcal{C}(A)$, and furthermore n_i is the length of a maximal chain in $\mathcal{C}_i$.*

The previous proposition does not generalize to arbitrary C*-algebras, which need not have a decomposition as a direct sum of factors. One might expect that $\mathcal{C}(A)$ is a domain when A is *approximately finite-dimensional*, as this would match with the intuition of approximation using practically obtainable information. However, there also needs to be a large enough supply of projections for this to work; see also Section 3. It turns out that the correct notion is that of *scattered* C*-algebras [81], that is, C*-algebras A for which every positive map $A \to \mathbb{C}$ is a sum of pure ones. The n-level system $A = \mathbb{M}_n(\mathbb{C})$ is scattered.

Theorem 7 ([38]). *A C*-algebra A is scattered if and only if $\mathcal{C}(A)$ is a continuous domain if and only if $\mathcal{C}(A)$ is an algebraic domain.*

Compare this to the situation using commutative W*-subalgebras $\mathcal{V}(A)$ of a W*-algebra A: $\mathcal{V}(A)$ is a continuous or algebraic domain only when A is finite-dimensional [78]. Connecting back to Theorem 6 and Corollary 3, let us notice that $\mathbb{C}$ can also be regarded as a domain using the interval topology: smaller intervals approximate an ideal complex number better than larger ones. Moreover, (piecewise) states $A \to \mathbb{C}$ respect such approximations: the induced functions from $\mathcal{C}(A)$ to the interval domain on $\mathbb{C}$ are *Scott continuous* [65,78].

There are several topologies with which one could adorn $\mathcal{C}(A)$. As any partially ordered set, it carries the order topology. We have just mentioned the Scott topology on directed complete partially ordered sets. For the purposes of information approximation that we are interested in, there is the *Lawson topology*, which refines both the Scott topology and the order topology. If the domain is continuous, the topological space will be Hausdorff. The topological space will be compact for so-called FS-domains, which $\mathcal{C}(A)$ happens to be.

Corollary 4 ([77]). *For a scattered C*-algebra A, the Lawson topology makes $X = \mathcal{C}(A)$ compact Hausdorff. Hence to each scattered C*-algebra A we may assign a commutative C*-algebra $C(X)$.*

The assignment $A \mapsto C(\mathcal{C}(A))$ is not functorial, does not leave commutative C*-algebras invariant, and of course only works for scattered C*-algebras A in the first place [38]. Hence there is no contradiction with Theorem 3.

One can also furnish $\mathcal{C}(A)$ with a topology inspired by the topology of A itself. We will use the topology induced by the following variation on the *Hausdorff metric*; similar variations are named after Banach–Mazur, Kadets [82], Gromov–Hausdorff, Effros–Maréchal [83], and Kadison–Kastler [84]. See also [85]. Define the distance between $C, D \in \mathcal{C}(A)$ to be

$$d(C,D) = \max\left(\sup_{\substack{c \in C \\ \|c\| \leq 1}} \inf_{\substack{d \in D \\ \|d\| \leq 1}} \|c - d\|, \sup_{\substack{d \in D \\ \|d\| \leq 1}} \inf_{\substack{c \in C \\ \|c\| \leq 1}} \|c - d\|\right).$$

Now if C and D are generated by projections p and q, and A is represented on a Hilbert space H, then

$$\|p - q\| = \sup_{\substack{x \in H \\ \|x\| \leq 1}} \|p(x) - q(x)\| = \sup_{\substack{x \in p(H) \\ \|x\| \leq 1}} \|x - q(x)\| = \sup_{\substack{x \in p(H) \\ \|x\| \leq 1}} \inf_{\substack{y \in q(H) \\ \|y\| \leq 1}} \|x - y\|$$

is the Hausdorff distance between $p(H)$ and $q(H)$. It follows that the distance between C and D is $\max(\|p - q\|, \|(1 - p) - q\|, \|p - (1 - q)\|, \|(1 - p) - (1 - q)\|) = \max(\|p - q\|, \|(1 - p) - q\|)$. This topology on $\mathcal{C}(A)$ matches the case of the 2-level system $A = \mathbb{M}_2(\mathbb{C})$, where $\mathcal{C}(A)$ is in bijection with the one-point compactification of the real projective plane $\mathbb{RP}^2$ [50].

5. Dynamics

So far, we have only considered kinematics of the quantum system A, by looking for configuration spaces based on $\mathcal{C}(A)$. It is clear, however, that $\mathcal{C}(A)$ in itself is not enough to reconstruct all of A. For a counterexample, observe that any C*-algebra A has an opposite C*-algebra A^{op} in which the multiplication is reversed. Clearly, $\mathcal{C}(A)$ and $\mathcal{C}(A^{\text{op}})$ are isomorphic as partially ordered sets, but there exist C*-algebras A that are not isomorphic to A^{op} as C*-algebras [86]. So we need to add more information to $\mathcal{C}(A)$ to be able to reconstruct A as a C*-algebra, which is the topic of this section. To do so, we bring dynamics into the picture. For motivation of why dynamics and configuration spaces should go together, see also [87].

We begin by viewing dynamics as a time-dependent group of evolutions. The traditional view is that the 1-parameter group consists of unitary evolutions of the Hilbert space. For an n-level system, these 1-parameter groups are continuous homomorphisms $\mathbb{R} \to U(n)$. In algebraic quantum theory, it becomes a 1-parameter group of isomorphisms $A \to A$ of the C*-algebra.

The group $\mathrm{Aut}(A)$ inherits the pointwise norm topology from A, that has subbasis

$$\{g \in \mathrm{Aut}(A) \mid \forall a \in S \colon \|f(a) - g(a)\| < \varepsilon > \|f(a) - g(1-a)\|\}$$

for $f \in \mathrm{Aut}(A)$, $\varepsilon > 0$, and $S \subseteq A$ finite, and makes conjugation $U(A) \to \mathrm{Aut}(A)$ continuous [88]. We can similarly consider 1-parameter groups of isomorphisms $\mathcal{C}(A) \to \mathcal{C}(A)$ of partially ordered sets.

Similarly, $\mathrm{Aut}(\mathcal{C}(A))$ becomes a topological group with subbasis

$$\{g \in \mathrm{Aut}(\mathcal{C}(A)) \mid \forall C \in S \colon d(f(C), g(D)) < \varepsilon\}$$

for $f \in \mathrm{Aut}(\mathcal{C}(A))$, $\varepsilon > 0$, and finite sets S of atoms of $\mathcal{C}(A)$.

Definition 4. *Let A be a C*-algebra. A 1-parameter group on A is a continuous injection $\varphi \colon \mathbb{R} \to \mathrm{Aut}(A)$, that assigns to each $t \in \mathbb{R}$ an isomorphism $\varphi_t \colon A \to A$ of C*-algebras, satisfying $\varphi_0 = 1$ and $\varphi_{t+s} = \varphi_t \circ \varphi_s$. A 1-parameter group on $\mathcal{C}(A)$ is a continuous injection $\alpha \colon \mathbb{R} \to \mathrm{Aut}(\mathcal{C}(A))$, that assigns to each $t \in \mathbb{R}$ an isomorphism $\alpha_t \colon \mathcal{C}(A) \to \mathcal{C}(A)$ of partially ordered sets, satisfying $\alpha_{t+s} = \alpha_t \circ \alpha_s$.*

The following theorem shows that both notions in fact coincide. A *factor* is an algebra with trivial center, that is, a single superselection sector: the n-level system $\mathbb{M}_n(\mathbb{C})$ is a factor, but $\mathbb{M}_m(\mathbb{C}) \oplus \mathbb{M}_n(\mathbb{C})$ is not, because its center is two-dimensional. More precisely, the following theorem shows that the only freedom between the two notions in the previous definition lies in permutations of the center, because $\mathrm{Aut}(A) \simeq \mathrm{Aut}(\mathcal{C}(A))$ for typical AW*-factors.

Theorem 8 ([89,90]). *Let A be a typical AW*-factor. Any 1-parameter group on $\mathcal{C}(A)$ is induced by a 1-parameter group on A, and vice versa.*

So C*-dynamics of A can be completely justified in terms of $\mathcal{C}(A)$. This also justifies our choice of the topology on $\mathcal{C}(A)$ induced by the Hausdorff metric. See also [91]. Equilibrium states are described in algebraic quantum theory by *Kubo–Martin–Schwinger* states, and these can be described in terms of $\mathcal{C}(A)$ as well, see [92].

We now switch gear. By Stone's theorem, 1-parameter groups of unitaries e^{ith} in certain W*-algebras correspond to self-adjoint (possibly unbounded) observables h. Thus, we may forget about the explicit dependence on a time parameter and consider single self-adjoint elements of C*-algebras. In fact, we will mostly be interested in *symmetries*: self-adjoint unitary elements $s = s^* = s^{-1}$.

Symmetries are tightly linked to projections. Every projection p gives rise to a symmetry $1 - 2p$, and every symmetry s comes from a projection $(1-s)/2$. As they are unitary, the symmetries of a C*-algebra A generate a subgroup $\mathrm{Sym}(A)$ of the unitary group. For a commutative C*-algebra $A = C(X)$, symmetries compose, so that $\mathrm{Sym}(A)$ consists of symmetries only. For an n-level system $A = \mathbb{M}_n(\mathbb{C})$, it turns out that $\mathrm{Sym}(A)$ consists of those unitaries $u \in U(n)$ whose determinant is 1 or -1. This 'orientation' is what we will add to $\mathcal{C}(A)$ to make it into a full invariant of A. See also [93].

Having enough symmetries means having enough projections. Therefore, we now consider AW*-algebras rather than general C*-algebras. For commutative AW*-algebras $C(X)$, the Gelfand spectrum X is not just compact Hausdorff, but *Stonean*, or *extremally disconnected*, in the sense that the closure of an open set is still open. (For comparison, the Lawson topology in Corollary 4 is totally disconnected, in the sense that connected components are singleton sets, which is weaker than Stonean).

Gelfand duality restricts to commutative AW*-algebras and Stonean spaces. Another way to put this is to say that the projections $\mathrm{Proj}(A)$ of a commutative AW*-algebra A form a *complete*

Boolean algebra, and vice versa, every complete Boolean algebra gives a commutative AW*-algebra. The appropriate homomorphisms between AW*-algebras are *normal*, meaning that they preserve least upper bounds of projections [94]. There are versions of Definition 2 for piecewise AW*-algebras, and piecewise complete Boolean algebras, too [94]. One could also define a piecewise Stonean space, but the following lemma suffices here.

Lemma 2 ([94]). *The category of piecewise complete Boolean algebras and the category of piecewise AW*-algebras are equivalent.*

The *orthocomplement* $p \mapsto 1 - p$ makes sense for the projections $\text{Proj}(A)$ of any C*-algebra A. We can now make precise what equivariance under symmetries achieves: it makes the difference between being able to recover Jordan structure and C*-algebra structure.

Proposition 5 ([43,94]). *Let A and B be typical AW*-algebras, and suppose that $f \colon \text{Proj}(A) \to \text{Proj}(B)$ preserve least upper bounds and orthocomplements. Then f extends to a Jordan homomorphism $A \to B$. It extends to a homomorphism if additionally $f\bigl((1-2p)(1-2q)\bigr) = \bigl(1-2f(p)\bigr)\bigl(1-2f(q)\bigr)$.*

To arrive at a good configuration space for A, we package all this information up. We saw that $\text{Proj}(A)$ embedded in $\text{Sym}(A)$. Conversely, $\text{Sym}(A)$ acts on $\text{Proj}(A)$: a symmetry s and a projection p give rise to a new projection sps. In this way, $\text{Proj}(A)$ *acts on itself*, and we may forget about $\text{Sym}(A)$. Including this action leads to the notion of an *active lattice* $\text{AProj}(A)$. More precisely, an active lattice consists of a complete orthomodular lattice P, a group G generated by $1 - 2p$ for $p \in P$ within the unitary group of the piecewise AW*-algebra $A(P)$ with projections P, and an action of G on P that becomes conjugation on $A(P)$. The active lattice of an n-level system $A = \mathbb{M}_n(\mathbb{C})$ has, for P, the lattice of subspaces of $\mathbb{C}^n$; for G, the group $\{u \in U(n) \mid \det(u) = \pm 1\}$; the injection $P \to G$ sends $V \subseteq \mathbb{C}^n$ to the reflection in V; and $u \in G$ acts on $V \in P$ as $uVu^* = \{uvu^* \mid v \in V\} \subseteq \mathbb{C}^n$. For morphisms of active lattices, we refer to [94], but let us point out that thanks to Lemma 2 they can be phrased in terms of projections alone, just like the above definition of the active lattice itself. See also [95]. We can now make precise that we can reconstruct an AW*-algebra A from its active lattice $\text{AProj}(A)$. Up to now, we have mostly considered reconstructions of the form "if some structures based on A and B are isomorphic, then so are A and B". The following theorem gives a much stronger form of reconstruction. Recall that a functor F is *fully faithful* when it gives a bijection between morphisms $A \to B$ and $F(A) \to F(B)$.

Theorem 9 ([94]). *The functor that assigns to an AW*-algebra A its active lattice $\text{AProj}(A)$ is fully faithful.*

It follows immediately that if A and B are AW*-algebras with isomorphic active lattices $\text{AProj}(A) \cong \text{AProj}(B)$, then $A \cong B$ are isomorphic AW*-algebras. That is, its active lattice completely determines an AW*-algebra. We can therefore think of them as configuration spaces. As mentioned before, $\text{Proj}(A)$ contains precisely the same information as $\mathcal{C}(A)$, so we could phrase active lattices in terms of $\mathcal{C}(A)$ as well. This configuration space circumvents the obstruction of Theorem 3, because active lattices are not a conservative extension of the "passive lattices" coming from compact Hausdorff spaces. Another thing to note about the previous theorem is that it has no need to except atypical cases such as $\mathbb{M}_2(\mathbb{C})$. Finally, let us point out that functoriality of $A \mapsto \text{AProj}(A)$ is nontrivial [96].

To get a good notion of configuration space for general quantum systems, we would eventually like to pass from AW*-algebras to C*-algebras. One way to think about this step is as refining an underlying carrying set to a topological space, that is, moving from algebras $\ell^\infty(X)$ of all (bounded) functions on the set X to algebras $C(X)$ of continuous functions on the topological space X. One might hope that AW*-algebras or W*-algebras play the former role in a noncommutative generalization, and to some extent this works [97,98]. Unfortunately, the Kadison–Singer problem raises rigorous obstructions to the most obvious noncommutative generalization of such a "discretization" of C*-algebras to AW*-algebras [99].

Nevertheless, AW*-algebras are pleasant to work with. Their theory is entirely algebraic, whereas the theory of (commutative) W*-algebras involves a good deal of measure theory. For example, Gelfand spectra of commutative AW*-algebras are Stonean spaces, whereas Gelfand spectra of commutative W*-algebras are so-called hyperstonean spaces; they additionally have to satisfy a measure-theoretic condition that seems divorced from topology. A similar downside occurs with projections: the projection lattice of a commutative W*-algebra is not just a complete Boolean algebra, it additionally has to satisfy a measure-theoretic condition. In particular, projections of an enveloping AW*-algebra should correspond to certain ideals in a C*-algebra, without needing measure-theoretic intricacies.

Much of the theory of W*-algebra finds its natural home in AW*-algebras at any rate. As a case in point, consider *Gleason's theorem*. It states that any probability measure on $\text{Proj}(\mathbb{M}_n(\mathbb{C}))$ extends to a positive linear function $\mathbb{M}_n(\mathbb{C}) \to \mathbb{C}$ when $n > 2$. Roughly speaking, any quantum probability measure μ is of the form $\mu(p) = \text{Tr}(\rho p)$ for some density matrix ρ. In the algebraic formulation, any probability measure $\text{Proj}(A) \to \mathbb{C}$ extends to a state $A \to \mathbb{C}$, for an n-level system $A = \mathbb{M}_n(\mathbb{C})$ [100]. One can replace A by an arbitrary W*-algebra, and one can even replace $\mathbb{C}$ by an arbitrary operator algebra B [101,102]. Thanks to Proposition 5, Gleason's theorem generalizes to many typical AW*-algebras A, such as those of so-called homogeneous type I, and those generated by two projections, which leads to the following corollary, that supports many results in Sections 2 and 3.

Corollary 5 ([43]). *Any normal piecewise Jordan homomorphism between typical AW*-algebras is a Jordan homomorphism.*

6. Characterization

Now that we have seen that most of the algebraic quantum theory of A can be phrased in terms of $\mathcal{C}(A)$ only, let us try to axiomatize $\mathcal{C}(A)$ itself. Given any partially ordered set, when is it of the form $\mathcal{C}(A)$ for some quantum system A? An answer to this question would, for example, make Theorem 9 into an equivalence of categories, bringing configuration spaces for quantum systems on a par with Gelfand duality for classical systems. An axiomatization would also open up the possibility of generalizations, that might go beyond algebraic quantum theory.

We start with the classical case, of commutative C*-algebras $C(X)$. By Gelfand duality, any $C \in \mathcal{C}(C(X))$ corresponds to a quotient $X/\sim$. In turn, the equivalence relation corresponds to a *partition* of X into equivalence classes. Partitions are partially ordered by refinement: if $C \subseteq D$, then any equivalence class in the partition corresponding to D is contained in an equivalence class of the partition corresponding to C. Hence axiomatizing $\mathcal{C}(C(X))$ comes down to axiomatizing *partition lattices*, and this has been well-studied, both in the finite-dimensional case [103,104], and in the general case [105]. The list of axioms is too long to reproduce here, but let us remark that it is based on a definition of *points* of the partition lattice. In the case of a finite partition lattice, the points are simply the *atoms*, that is, the minimal nonzero elements. So for a classical system $\mathbb{C}^n$ with n states, the elements of the partition lattice $\mathcal{C}(\mathbb{C}^n)^{\text{op}}$ are the ways to partition a set of n points into m equivalence classes; the atoms put two of the n points in an equivalence class and all the others in their own equivalence class of one point each. The other axioms are geometric in nature.

Lemma 3 ([64]). *A partially ordered set is isomorphic to $\mathcal{C}(C(X))$ for a compact Hausdorff space X if and only if it is opposite to a partition lattice whose points are in bijection with X.*

Thanks to (a variation of) Lemma 2, the same strategy applies to piecewise Boolean algebras B. Write $\mathcal{C}(B)$ for the partially ordered set of Boolean subalgebras of B. The *downset* of an element D of a partially ordered set consists of all elements $C \leq D$. In fact, the idea that any quantum logic (piecewise

Boolean algebra) should be seen as many classical sublogics (Boolean algebras) pasted together, is not new, and drives much of the research in that area [27,106–109].

Theorem 10 ([110])**.** *A partially ordered set is isomorphic to $\mathcal{C}(B)$ for a piecewise Boolean algebra B if and only if:*

- *it is an algebraic domain;*
- *any nonempty subset has a greatest lower bound;*
- *a set of atoms has an upper bound whenever each pair of its elements does;*
- *the downset of each compact element is isomorphic to the opposite of a finite partition lattice.*

In the case of a classical system with n states, B is the powerset of n points, and the above conditions merely say that $\mathcal{C}(B)^{\text{op}}$ is a partition lattice.

Just like in Section 3, if we consider $\mathcal{C}(B)$ as a diagram rather than a mere partially ordered set, we can reconstruct B. Starting from just the partially ordered set $\mathcal{C}(B)$, the same issues surface as in Sections 2 and 5, about Jordan structure verses full algebra structure. In the current piecewise Boolean setting, it can be solved neatly by adding an *orientation* to $\mathcal{C}(B)$ [110]. This comes down to making a consistent choice of atom in the Boolean subalgebras with two atoms, corresponding to the atypical cases for AW*-algebras before.

Returning to C*-algebras, Lemma 3 reduces the question of characterizing $\mathcal{C}(A)$ for a C*-algebra A to finding relationships between $\mathcal{C}(A)$ and $\mathcal{C}(C)$ for $C \in \mathcal{C}(A)$. One prototypical case where we know such a relationship is for the n-level system $A = \mathbb{M}_n(\mathbb{C})$. Namely, inspired by the previous section, there is an action of the unitary group $U(n)$ on $\mathcal{C}(A)$: if $u \in U(n)$ is some rotation, and $C \in \mathcal{C}(A)$ is diagonal in some basis, then also the rotation uCu^* is diagonal in the rotated basis and therefore is in $\mathcal{C}(A)$ again. In fact, any $C \in \mathcal{C}(A)$ will be a rotation of an element of $\mathcal{C}(A)$ that is diagonal in the standard basis. Therefore, we can recognize $\mathcal{C}(\mathbb{M}_n(\mathbb{C}))$ as a *semidirect product* of $\mathcal{C}(\mathbb{C}^n)$ and $U(n)$. Such semidirect products can be axiomatized; for details, we refer to [64]. This can be generalized to C*-algebras A that have a *weakly terminal* commutative C*-subalgebra D, in the sense that any $C \in \mathcal{C}(A)$ allows an injection $C \to D$. This includes all finite-dimensional C*-algebras, as well as algebras of all bounded operators on a Hilbert space. For example, for the n-level system $A = \mathbb{M}_n(\mathbb{C})$, the matrices that are diagonal in the standard basis form a terminal subalgebra $\mathbb{C}^n$.

However, the mere partially ordered set $\mathcal{C}(A)$ cannot detect this unitary action. For this we need injections rather than inclusions. Therefore, we now switch to a category $\mathcal{C}_\hookrightarrow(A)$ of commutative C*-subalgebras, with *injective ∗-homomorphisms* between them. For $A = \mathbb{M}_n(\mathbb{C})$, these morphisms consist of a rotation in $U(n)$ followed by an inclusion $\mathbb{C}^k \to \mathbb{C}^l$ with $k \leq l$. The following theorem characterizes this category $\mathcal{C}_\hookrightarrow(A)$ up to equivalence. This is the same as characterizing $\mathcal{C}(A)$ up to *Morita equivalence*, meaning that it determines the topos of contextual sets on $\mathcal{C}(A)$ discussed in Section 3 up to categorical equivalence, rather than determining $\mathcal{C}(A)$ itself up to equivalence. To phrase the following theorem, we introduce the monoid $S(X)$ of continuous surjections $X \to X$ on a compact Hausdorff space X. In the finite-dimensional case, this is just the symmetric group $S(n)$. Because of our switch from $\mathcal{C}(A)$ to $\mathcal{C}_\hookrightarrow(A)$, it plays the role of the unitary group we need.

Theorem 11 ([64])**.** *Suppose that a C*-algebra A has a weakly terminal commutative C*-subalgebra $C(X)$. A category is equivalent to $\mathcal{C}_\hookrightarrow(A)$ if and only if it is equivalent to a semidirect product of $\mathcal{C}(C(X))$ and $S(X)$.*

See also [111].

The unitary action can also be used to determine $\mathcal{C}(A)$ for small A such as $\mathbb{M}_n(\mathbb{C})$. Combining Lemma 3 with Theorem 11, we see that k-dimensional C in $\mathcal{C}(\mathbb{M}_n(\mathbb{C}))$ are parametrized by a partition of n into k nonempty parts together with an element of $U(n)$. Two such parameters induce the same subalgebra when the unitary permutes equal-sized parts of the partition. This can be handled neatly in terms of *Young tableaux* and *Grassmannians*, see [50,51].

Using this concrete parametrization of $\mathcal{C}(A)$ for $A = \mathbb{M}_n(\mathbb{C})$, to characterize $\mathcal{C}(A)$ it would suffice to characterize the unitary group $U(A)$. Surprisingly, this question is open, even in the finite-dimensional case. All that seems to be known is that, up to isomorphism, $U(1)$ is the unique nondiscrete locally compact Hausdorff group all of whose proper closed subgroups are finite [112]. This characterization does not generalize to finite dimensions higher than one, although closed subgroups have received study in the infinite-dimensional case [113]. The unitary group $U(n)$ is also, up to isomorphism, the unique irreducible subgroup of $\mathrm{GL}(n)$ the trace of whose elements is bounded [114]. It is known that unitary groups of C*-algebras cannot be countably classified [115]. Finally, the characterization of $\mathcal{C}(B(H))$ for Hilbert spaces H could give rise to a description of the category of Hilbert spaces in terms of generators and relations [116].

7. Generalizations

As mentioned in the introduction, the idea to describe quantum structures in terms of their classical substructures applies very generally. This final section discusses to what extent algebraic quantum theory is special, by considering a generalization as an example of another framework.

Namely, we consider *categorical quantum mechanics* [117]. This approach formulates quantum theory in terms of the category of Hilbert spaces, and then abstracts away to more general categories with the same structures. Specifically, what is retained is the notion of a *tensor product* to be able to build compound systems, the notion of *entanglement* in the form of objects that form a duality under the tensor product, and the notion of *reversibility* in the sense that every map between Hilbert spaces has an adjoint in the reverse direction. It turns out that these primitives suffice to derive a lot of quantum-mechanical features, such as scalars, the Born rule, no-cloning, quantum teleportation, and complementarity. As a case in point, one can define so-called *Frobenius algebras* in any category with this structure, which is important because of the following proposition.

Theorem 12 ([118,119]). *Finite-dimensional C*-algebras correspond to Frobenius algebras in the category of Hilbert spaces.*

The point is that these notions make sense in *any* category with a tensor product, entanglement, and reversibility. A different example of such a category is that of sets with relations between them. That is, objects are sets X, and arrows $X \to Y$ are relations $R \subseteq X \times Y$. For the tensor product, we take the Cartesian product of sets, which makes every object dual to itself and thereby fulfilling the structure of entanglement, and time reversibility is given by taking the opposite relation $R^\dagger \subseteq Y \times X$. Two relations $R \subseteq X \times Y$ and $S \subseteq Y \times Z$ compose to $S \circ R = \{(x,z) \mid \exists y\colon (x,y) \in R, (y,z) \in S\}$. We may regard this as a toy example of *possibilistic quantum theory*: rather than complex matrices, we now care about entries ranging over $\{0,1\}$. A *groupoid* is a small category, every arrow of which is an isomorphism; they may be considered as a multi-object generalization of groups.

Theorem 13 ([120]). *Frobenius algebras in the category of sets and relations correspond to groupoids.*

Algebraic quantum theory, as set out in the introduction, makes perfect sense in categories such as sets and relations as well [121]. However, in this generality, it is not true that all classical subsystems determine a quantum system at all. The previous theorem provides a counterexample. In commutative groupoids, there can only be arrows $X \to X$, for arrows $g\colon X \to Y$ between different objects cannot commute with their inverse, as $g \circ g^{-1} = 1_Y$ and $g^{-1} \circ g = 1_X$. Therefore, any arrow between different objects in a groupoid can never be recovered from any commutative subgroupoid.

Similarly, quantum logic, as discussed in Section 3, makes perfect sense in this general categorical setting [122]. Moreover, it matches neatly with algebraic quantum theory via taking projections [123]. However, it is no longer true that commutative subalgebras correspond to Boolean sublattices. Again, a counterexample can be found using Theorem 13 [124].

One could object that commutativity might be too narrow a notion of classicality. However, consider broadcastability instead: classical information can be broadcast, but quantum information cannot. More precisely, a Frobenius algebra A is *broadcastable* when there exists a completely positive map $A \to A \otimes A$ such that both partial traces are the identity $A \to A$. Again, this makes perfect sense in general categories. It turns out that the broadcastable objects in the category of sets and relations are the groupoids that are totally disconnected, in the sense that there are no arrows $g \colon X \to Y$ between different objects [117]. So even with this more liberal operational notion of classicality, classical subsystems do not determine a quantum system.

This breaks a well-known information-theoretic characterization of quantum theory, that is phrased in terms of C*-algebras [125,126]. Hence there is something about (algebraic) quantum theory beyond the categorical properties of having tensor products, entanglement, and reversibility, that underwrites Bohr's doctrine of classical concepts. It relates to characterizing unitary groups, as discussed in Section 6. We close this overview by raising the interesting interpretational question of just what this defining property is.

Acknowledgments: Supported by EPSRC Fellowship EP/L002388/1.

Conflicts of Interest: The author declares no conflict of interest.

References

1. Bell, J.S. On the Einstein Podolsky Rosen paradox. *Physics* **1964**, *1*, 195–200.
2. Kochen, S.; Specker, E. The problem of hidden variables in quantum mechanics. *J. Math. Mech.* **1967**, *17*, 59–87.
3. Pusey, M.; Barrett, J.; Rudolph, T. On the reality of the quantum state. *Nat. Phys.* **2012**, *8*, 475–478.
4. Busch, P.; Grabowski, M.; Lahti, P.J. *Operational Quantum Physics*; Springer: Berlin/Heidelberg, Germany, 1995.
5. Keyl, M. Fundamentals of quantum information theory. *Phys. Rep.* **2002**, *369*, 431–548.
6. Kadison, R.V.; Ringrose, J.R. *Fundamentals of the Theory of Operator Algebras*; Number 15–16 in Graduate Studies in Mathematics; Academic Press: Cambridge, MA, USA, 1983.
7. Berberian, S.K. *Baer *-Rings*; Springer: Berlin/Heidelberg, Germany, 1972.
8. Emch, G.G. *Mathematical and Conceptual Foundations of 20th-Century Physics*, 1st ed.; North-Holland: Amsterdam, The Netherlands, 1984.
9. Davies, E.B. *Quantum Theory of Open Systems*; Academic Press: Cambridge, MA, USA, 1976.
10. Earman, J. Superselection rules for philosophers. *Erkenn* **2008**, *69*, 377–414.
11. Rédei, M. *Quantum Logic in Algebraic Approach*; Springer: Cham, The Netherlands, 1998.
12. Haag, R. *Local Quantum Physics*; Texts and Monographs in Physics; Springer: Berlin/Heidelberg, Germany, 1996.
13. Strocchi, F. *An Introduction to the Mathematical Structure of Quantum Mechanics*; World Scientific: Singapore, 2008.
14. Emch, G.G. *Algebraic Methods in Statistical Mechanics and Quantum Field Theory*; Wiley: Hoboken, NJ, USA, 1972.
15. Alberti, P.M.; Uhlmann, A. Existence and density theorems for stochastic maps on commutative C*-algebras. *Math. Nachr.* **1980**, *97*, 279–295.
16. Landsman, N.P. *Mathematical Topics between Classical and Quantum Mechanics*; Springer: Berlin/Heidelberg, Germany, 1998.
17. Weaver, N. *Mathematical Quantization*; Chapman & Hall: London, UK, 2001.
18. Bohr, N. Chapter Discussion with Einstein on epistemological problems in atomic physics. In *Albert Einstein: Philosopher-Scientist*; Cambridge University Press: Cambridge, UK, 1949.
19. Heunen, C.; Landsman, N.P.; Spitters, B. A topos for algebraic quantum theory. *Commun. Math. Phys.* **2009**, *291*, 63–110.
20. Kadison, R.V.; Singer, I.M. Extensions of pure states. *Am. J. Math.* **1959**, *81*, 383–400.
21. Marcus, A.; Spielman, D.A.; Srivastava, N. Interlacing families II: Mixed characteristic polynomials and the Kadison–Singer problem. *Ann. Math.* **2015**, *182*, 327–350.
22. Altepeter, J.B.; James, D.F.V.; Kwiat, P.G. Qubit quantum state tomography. In *Quantum State Estimation*; Springer: Berlin/Heidelberg, Germany, 2004.
23. Butterfield, J.; Isham, C.J. A topos perspective on the Kochen–Specker theorem: I. Quantum States as Generalized Valuations. *Int. J. Theor. Phys.* **1998**, *37*, 2669–2733.

24. Constantin, C.M.; Döring, A. Contextual entropy and reconstruction of quantum states. *arXiv* **2012**, arXiv:1208.2046.
25. Hamhalter, J.; Turilova, E. Orthogonal measures on state spaces and context structure of quantum theory. *Int. J. Theor. Phys.* **2016**, *55*, 3353–3365.
26. Mac Lane, S. *Categories for the Working Mathematician*, 2nd ed.; Springer: Berlin/Heidelberg, Germany, 1971.
27. Berg, B.; Heunen, C. Noncommutativity as a colimit. *Appl. Categorical Struct.* **2012**, *20*, 393–414.
28. Reyes, M.L. Obstructing extensions of the functor Spec to noncommutative rings. *Isr. J. Math.* **2012**, *192*, 667–698.
29. Berg, B.; Heunen, C. Extending obstructions to noncommutative functorial spectra. *Theory Appl. Categories* **2014**, *29*, 457–474.
30. Döring, A. Kochen–Specker theorem for von Neumann algebras. *Int. J. Theor. Phys.* **2005**, *44*, 139–160.
31. Reyes, M.L. Sheaves that fail to represent matrix rings. In *Ring theory and Its Applications*; American Mathematical Society: Providence, RI, USA, 2014; Volume 609, pp. 285–297.
32. Alfsen, E.M.; Shultz, F.W. *State Spaces of Operator Algebras: Basic Theory, Orientations, and C*-Products*; Birkhäuser: Basel, Switzerland, 2001.
33. Connes, A. *Noncommutative Geometry*; Academic Press: Cambridge, MA, USA, 1994.
34. Akemann, C.A. The general Stone–Weierstrass problem. *J. Funct. Anal.* **1969**, *4*, 277–294.
35. Giles, R.; Kummer, H. A non-commutative generalization of topology. *Indiana Univ. Math. J.* **1971**, *21*, 91–102.
36. De Silva, N. From topology to noncommutative geometry: K-theory. *arXiv* **2014**, arXiv:1408.1170.
37. De Silva, N.; Soares Barbosa, R. Partial and total ideals of von Neumann algebras. *arXiv* **2014**, arXiv:1408.1172.
38. Heunen, C.; Lindenhovius, A.J. Domains of commutative C*-subalgebras. In Proceedings of the 2015 30th Annual ACM/IEEE Symposium on Logic in Computer Science (LICS), Kyoto, Japan, 6–10 July 2015; pp. 450–461.
39. Hanche-Olsen, H.; Størmer, E. *Jordan Operator Algebras*; Pitman Advanced Publishing Program: Boston, MA, USA, 1984.
40. Hamhalter, J. Isomorphisms of ordered structures of abelian C*-subalgebras of C*-algebras. *J. Math. Anal. Appl.* **2011**, *383*, 391–399.
41. Kaplansky, I. Projections in Banach algebras. *Ann. Math.* **1951**, *53*, 235–249.
42. Döring, A.; Harding, J. Abelian subalgebras and the Jordan structure of von Neumann algebras. *arXiv* **2015**, arXiv:1009.4945.
43. Hamhalter, J. Dye's theorem and Gleason's theorem for AW*-algebras. *J. Math. Anal. Appl.* **2015**, *422*, 1103–1115.
44. Hamhalter, J.; Turilova, E. Structure of associative subalgebras of Jordan operator algebras. *Q. J. Math.* **2013**, *64*, 397–408.
45. Johnstone, P.T. *Sketches of an Elephant: A Topos Theory Compendium*; Clarendon Press: Oxford, UK, 2002.
46. Landsman, N.P. *Bohrification: From Classical Concepts to Commutative Operator Algebras*; Springer: Berlin/Heidelberg, Germany, 2017.
47. Johnstone, P.T. *Stone Spaces*; Number 3 in Cambridge Studies in Advanced Mathematics; Cambridge University Press: Cambridge, UK, 1982.
48. Banaschewski, B.; Mulvey, C.J. A globalisation of the Gelfand duality theorem. *Ann. Pure Appl. Log.* **2006**, *137*, 62–103.
49. Spitters, B.; Vickers, S.; Wolters, S. Gelfand spectra in Grothendieck toposes using geometric mathematics. *Electron. Proc. Theor. Comput. Sci.* **2014**, *158*, 77–107.
50. Fauser, B.; Raynaud, G.; Vickers, S. The Born rule as structure of spectral bundles. *Electron. Proc. Theor. Comput. Sci.* **2012**, *95*, 81–90.
51. Heunen, C.; Landsman, N.P.; Spitters, B. Bohrification. In *Deep Beauty: Understanding the Quantum World through Mathematical Innovation*, Halvorson, H., Ed.; Cambridge University Press: Cambridge, UK, 2011; pp. 217–313.
52. Caspers, M.; Heunen, C.; Landsman, N.P.; Spitters, B. Intuitionistic quantum logic of an n-level system. *Found. Phys.* **2009**, *39*, 731–759.
53. Heunen, C.; Landsman, N.P.; Spitters, B. Bohrification of operator algebras and quantum logic. *Synthese* **2012**, *186*, 719–752.
54. Wolters, S. Topos models for physics and topos theory. *J. Math. Phys.* **2013**, *55*, 082110.
55. Nuiten, J. Bohrification of local nets. *Electron. Proc. Theor. Comput. Sci.* **2011**, *95*, 211–218.

56. Döring, A.; Isham, C.J. Topos Methods in the Foundations of Physics. In *Deep Beauty: Understanding the Quantum World through Mathematical Innovation*, Halvorson, H., Ed.; Cambridge University Press: Cambridge, UK, 2011.
57. Döring, A.; Isham, C.J. New Structure for Physics; Chapter What is a thing? Topos theory in the founcations of physics. In *Lecture Notes in Physics*; Springer: Berlin/Heidelberg, Germany, 2011; Volume 813; pp. 753–940.
58. Döring, A.; Isham, C.J. A topos founcation for theories of physics. *J. Math. Phys.* **2008**, *49*, 053515.
59. Flori, C. *A First Course in Topos Quantum Theory*; Lecture Notes in Physics; Springer: Berlin/Heidelberg, Germany, 2013; Volume 868.
60. Wolters, S. A comparison of two topos-theoretic approaches to quantum theory. *Commun. Math. Phys.* **2013**, *317*, 3–53.
61. Joyal, A.; Tierney, M. *An Extension of the Galois Theory of Grothendieck (Memoirs of the American Mathematical Society)*; Proquest Info & Learning: Ann Arbor, MI, USA, 1984; Volume 51.
62. Heunen, C.; Landsman, N.P.; Spitters, B.; Wolters, S. The Gelfand spectrum of a noncommutative C*-algebra: A topos-theoretic approach. *J. Aust. Math. Soc.* **2011**, *90*, 39–52.
63. Berg, B.; Heunen, C. Erratum to: Noncommutativity as a colimit. *Appl. Categorical Struct.* **2013**, *21*, 103–104.
64. Heunen, C. Characterizations of categories of commutative C*-subalgebras. *Commun. Math. Phys.* **2014**, *331*, 215–238.
65. Spitters, B. The space of measurement outcomes as a spectral invariant for non-commutative algebras. *Found. Phys.* **2012**, *42*, 896–908.
66. De Groote, H.F. Observables IV: The presheaf perspective. *arXiv* **2007**, arXiv:0708.0677.
67. Döring, A. Quantum states and measures on the spectral presheaf. *Adv. Sci. Lett.* **2009**, *2*, 291–301.
68. Loveridge, L.; Dridi, R.; Raussendorf, R. Topos logic in measurement-based quantum computation. *Proc. R. Soc. A* **2015**, *471*, 20140716.
69. Heunen, C.; Landsman, N.P.; Spitters, B. The principle of general tovariance. *Int. Fall Workshop Geom. Phys.* **2008**, *1023*, 93–102.
70. Epperson, M.; Zafiris, E. *Foundations of Relational Realism: A Topological Approach to Quantum Mechanics and the Philosophy of Nature*; Lexington: Lanham, MD, USA, 2013.
71. Adelman, M.; Corbett, J.V. A sheaf model for intuitionistic quantum mechanics. *Appl. Categorical Struct.* **1995**, *3*, 79–104.
72. Takeuti, G. C*-algebras and Boolean-valued analysis. *Jpn. J. Math.* **1983**, *9*, 207–245.
73. Ozawa, M. A transfer principle from von Neumann algebras to AW*-algebras. *J. Lond. Math. Soc.* **1985**, *32*, 141–148.
74. Ozawa, M. A classification of type I AW*-algebras and Boolean-valued analysis. *J. Math. Soc. Jpn.* **1984**, *36*, 589–608.
75. Abramsky, S.; Brandenburger, A. The sheaf-theoretic structure of non-locality and contextuality. *New J. Phys.* **2011**, *13*, 113036.
76. Abramsky, S.; Jung, A. Domain Theory. In *Handbook of Logic in Computer Science*; Oxford University Press: Oxford, UK, 1994; Volume 3.
77. Gierz, G.; Hofmann, K.H.; Keimel, K.; Lawson, J.D.; Mislove, M.W.; Scott, D.S. *Continuous Lattices and Domains*; Number 93 in Encyclopedia of Mathematics and its Applications; Cambridge University Press: Cambridge, UK, 2003.
78. Döring, A.; Barbosa, R.S. Unsharp values, domains and topoi. In *Quantum Field Theory and Gravity: Conceptual and Mathematical Advances in the Search for a Unified Framework*; Springer: Berlin/Heidelberg, Germany, 2011; pp. 65–96.
79. Lindenhovius, A.J. Classifying finite-dimensional C*-algebras by posets of their commutative C*-subalgebras. *Int. J. Theor. Phys.* **2015**, *54*, 4615–4635.
80. Lindenhovius, A.J. $\mathcal{C}(A)$. Ph.D. Thesis, Radboud University, Nijmegen, The Netherlands, 5 July 2016.
81. Jensen, H.E. Scattered C*-algebras. *Math. Scand.* **1977**, *41*, 308–314.
82. Kalton, N.J.; Ostrovskii, M.I. Distances between Banach spaces. *Forum Math.* **1999**, *11*, 17–48.
83. Haagerup, U.; Winsløw, C. The Effros–Maréchal topology in the space of von Neumann algebras. *Am. J. Math.* **1998**, *120*, 567–617.
84. Kadison, R.V.; Kastler, D. Perturbations of von Neumann algebras I: Stability of type. *Am. J. Math.* **1972**, *94*, 38–54.

85. Chetcuti, E.; Hamhalter, J.; Weber, H. The order topology for a von Neumann algebra. *Stud. Math.* **2015**, *230*, 95–120.
86. Connes, A. A factor not anti-isomorphic to itself. *Ann. Math.* **1975**, *101*, 536–554.
87. Spekkens, R.W. The paradigm of kinematics and dynamics must yield to causal structure. Foundational Questions Institute essay contest winner. *arXiv* **2013**, arXiv:1209.0023.
88. Moffat, J. Groups of Automorphisms of Operator Algebras. Ph.D. Thesis, University of Newcastle upon Tyne, Newcastle, UK, 1974.
89. Hamhalter, J.; Turilova, E. Automorphisms of ordered structures of abelian parts of operator algebras and their role in quantum theory. *Int. J. Theor. Phys.* **2014**, *53*, 3333–3345.
90. Döring, A. Flows on generalised Gelfand spectra of nonabelian unital C*-algebras and time evolution of quantum systems. *arXiv* **2012**, arXiv:1212.4882
91. Heunen, C.; Lindenhovius, A.J. Domains of commutative C*-subalgebras. *arXiv* **2015**, arXiv:1504.02730.
92. Geloun, J.B.; Flori, C. Topos analogues of the KMS state. *arXiv* **2012**, arXiv:1207.0227.
93. Alfsen, E.M.; Shultz, F.W. Orientation in operator algebras. *Proc. Natl. Acad. Sci. USA* **1998**, *95*, 6596–6601.
94. Heunen, C.; Reyes, M.L. Active lattices determine AW*-algebras. *J. Math. Anal. Appl.* **2014**, *416*, 289–313.
95. Chevalier, G. Automorphisms of an orthomodular poset of projections. *Int. J. Theor. Phys.* **2005**, *44*, 985–998.
96. Heunen, C.; Reyes, M.L. Diagonalizing matrices over AW*-algebras. *J. Funct. Anal.* **2013**, *264*, 1873–1898.
97. Kornell, A. Quantum Collections. *arXiv* **2012**, arXiv:1202.2994.
98. Kornell, A. V*-algebras. *arXiv* **2015**, arXiv:1502.01516.
99. Heunen, C.; Reyes, M.L. On discretization of C*-algebras. *J. Oper. Theory* **2017**, *77*, 19–37.
100. Mackey, G.W. *The Mathematical Foundations of Quantum Mechanics*; W. A. Benjamin: New York, NY, USA, 1963.
101. Bunce, L.J.; Wright, J.D.M. The Mackey–Gleason problem. *Bull. Am. Math. Soc.* **1992**, *26*, 288–293.
102. Hamhalter, J. *Quantum Measure Theory*; Springer: Berlin/Heidelberg, Germany, 2004.
103. Birkhoff, G. *Lattice Theory*; American Mathematical Society: Providence, RI, USA, 1948.
104. Stonesifer, J.R.; Bogart, K.P. Characterizations of partition lattices. *Algebra Univers.* **1984**, *19*, 92–98.
105. Firby, P.A. Lattices and compactifications I. *Proc. Lond. Math. Soc.* **1973**, *27*, 22–50.
106. Gudder, S.P. Partial algebraic structures associated with orthomodular posets. *Pac. J. Math.* **1972**, *41*, 717–730.
107. Finch, P.D. On the structure of quantum logic. *J. Symb. Log.* **1969**, *34*, 415–425.
108. Hughes, R.I.G. Omnibus review. *J. Symb. Log.* **1985**, *50*, 558–566.
109. Scheibe, E. *The Logical Analysis of Quantum Mechanics*; Pergamon Press: Oxford, UK, 1973.
110. Heunen, C. Piecewise Boolean algebras and their domains. *Lect. Notes Comput. Sci.* **2014**, *8573*, 208–219.
111. Flori, C.; Fritz, T. Compositories and gleaves. *Theory Appl. Categories* **2016**, *31*, 928–988.
112. Morris, S.A. A characterization of the topological group of real numbers. *Bull. Aust. Math. Soc.* **1986**, *34*, 473–475.
113. Kadison, R.V. Infinite unitary groups. *Trans. Am. Math. Soc.* **1952**, *72*, 386–399.
114. Marcus, M.; Newman, M. Some results on unitary matrix groups. *Linear Algebra Its Appl.* **1970**, *3*, 173–178.
115. Kerr, D.; Lupini, M.; Phillips, N.C. Borel complexity and automorphisms of C*-algebras. *J. Funct. Anal.* **2015**, *268*, 3767–3789.
116. Heunen, C. On the functor ℓ^2. In *Computation, Logic, Games, and Quantum Foundations*; Springer: Berlin/Heidelberg, Germany, 2013; pp. 107–121.
117. Heunen, C.; Vicary, J. *Categories for Quantum Theory: An Introduction*; Oxford University Press: Oxford, UK, 2017.
118. Vicary, J. Categorical formulation of finite-dimensional quantum algebras. *Commun. Math. Phys.* **2011**, *304*, 765–796.
119. Abramsky, S.; Heunen, C. H*-algebras and nonunital Frobenius algebras: First steps in infinite-dimensional categorical quantum mechanics. *Clifford Lect. AMS Proc. Symp. Appl. Math.* **2012**, *71*, 1–24.
120. Heunen, C.; Contreras, I.; Cattaneo, A.S. Relative Frobenius algebras are groupoids. *J. Pure Appl. Algebra* **2013**, *217*, 114–124.
121. Coecke, B.; Heunen, C.; Kissinger, A. Categories of quantum and classical channels. *Quantum Inf. Process.* **2016**, *15*, 5179–5209.
122. Heunen, C.; Jacobs, B. Quantum logic in dagger kernel categories. *Order* **2010**, *27*, 177–212.
123. Heunen, C. Complementarity in categorical quantum mechanics. *Found. Phys.* **2012**, *42*, 856–873.
124. Coecke, B.; Heunen, C.; Kissinger, A. Chapter Compositional Quantum Logic. In *Computation, Logic, Games, and Quantum Foundations*; Springer: Berlin/Heidelberg, Germany, 2013; pp. 21–36.

125. Clifton, R.; Bub, J.; Halvorson, H. Characterizing quantum theory in terms of information-theoretic constraints. *Found. Phys.* **2003**, *33*, 1561–1591.
126. Heunen, C.; Kissinger, A. Can quantum theory be characterized by information-theoretic constraints? *arXiv* **2016**, arXiv:1604.05948.

© 2017 by the author. Licensee MDPI, Basel, Switzerland. This article is an open access article distributed under the terms and conditions of the Creative Commons Attribution (CC BY) license (http://creativecommons.org/licenses/by/4.0/).

Review

Quantum Theory from Rules on Information Acquisition

Philipp Andres Höhn [1,2]

[1] Vienna Center for Quantum Science and Technology, University of Vienna, Boltzmanngasse 5, 1090 Vienna, Austria; p.hoehn@univie.ac.at
[2] Institute for Quantum Optics and Quantum Information, Austrian Academy of Sciences, Boltzmanngasse 3, 1090 Vienna, Austria

Academic Editors: Giacomo Mauro D'Ariano and Paolo Perinotti
Received: 23 January 2017; Accepted: 17 February 2017; Published: 3 March 2017

Abstract: We summarize a recent reconstruction of the quantum theory of qubits from rules constraining an observer's acquisition of information about physical systems. This review is accessible and fairly self-contained, focusing on the main ideas and results and not the technical details. The reconstruction offers an informational explanation for the architecture of the theory and specifically for its correlation structure. In particular, it explains entanglement, monogamy and non-locality compellingly from limited accessible information and complementarity. As a by-product, it also unravels new 'conserved informational charges' from complementarity relations that characterize the unitary group and the set of pure states.

Keywords: reconstruction of quantum theory; entanglement; monogamy; quantum non-locality; conserved informational charges; limited information; complementarity; characterization of unitary group and state spaces

1. Introduction

Why is the physical world described by quantum theory? If we wish to sensibly address this question, we have to step beyond quantum theory and to consider it within a landscape of alternative theories. This, after all, permits us to ponder about how the world could have been different, possibly described by modifications of quantum theory. Such an endeavor forces us to leave the usual textbook formulation of quantum theory, and everything we take for granted about it, behind and to develop a more general language that also applies to alternative theories. Ideally, this language should be operational, encompassing the interactions of some observer with physical systems in a plethora of conceivable, physically-distinct worlds.

If we wish to also provide a possible answer to the above question, we then have to find physical properties of quantum theory that single it out, at least within the given landscape of alternatives. In particular, the goal should be to find an operational justification for the textbook axioms, i.e., ultimately for complex Hilbert spaces, unitary dynamics, tensor product structure for composite systems, Born rule, and so on. The result would be a reconstruction of quantum theory from operational axioms [1–10] and should ideally yield a better understanding of what quantum theory tells us about Nature; and why it is the way it is.

In this manuscript, we shall review and summarize how the quantum formalism for arbitrarily many qubits can be reconstructed from operational rules restricting an observer's acquisition of information about a set of observed systems [1,2]. The goal of this summary is to provide a didactical and easily-accessible overview of this reconstruction. Its underlying framework is especially engineered for unraveling the architecture of quantum theory, and so many reconstruction steps are instructive for understanding the origin of quantum properties. As we shall see, this reconstruction provides a transparent, informational explanation for the structure of qubit quantum theory and especially also for its paradigmatic features, such as entanglement, monogamy and non-locality. The approach also produces novel 'conserved informational charges', indeed appearing in quantum theory, that turn out to characterize the unitary group and the set of pure states and which might find practical applications in quantum information.

The premise of the summarized approach is to only speak about information that the observer has access to. It is thus purely operational and survives without any ontological commitments. This approach is inspired, in part, by Rovelli's relational quantum mechanics [11] and the Brukner–Zeilinger informational interpretation of quantum theory [12,13]; this successful reconstruction can be viewed as a completion of these ideas for qubit systems.

The rest of the manuscript is organized as follows. In Section 2, we review the landscape of alternative theories; in Section 3, we formulate the operational quantum axioms; in Section 4, we summarize the key steps of the reconstruction itself and, finally, conclude in Section 5.

2. Overview of a Landscape of Theories

We shall begin with an overview of a landscape of alternative theories, which has been developed in [1,2] to which we also refer for further details.

2.1. From Questions and Answers to Probabilities and States

Our first aim is to define a notion of a state both for a single system and an ensemble of systems.

Consider an observer O who interrogates an ensemble of (identically prepared [1]) systems $\{S_a\}_{a=1}^n$, coming out of a preparation device, with binary questions Q_i from some set $\mathcal{Q}$. For example, in the case of quantum theory, such a question could read "is the spin of the electron up in x-direction?" This set $\mathcal{Q}$ shall only contain repeatable questions in the sense that O will receive $m \in \mathbb{N}$ times the same answer whenever asking any $Q_i \in \mathcal{Q}$ m times in immediate succession to a single system S_a. We shall assume any S_a to always give a definite answer if asked some $Q_i \in \mathcal{Q}$, which moreover is not independent of S_a's preparation. Accordingly, $\mathcal{Q}$ can only contain physically-implementable questions, which are 'answerable' by the $\{S_a\}$ and not arbitrary logically conceivable binary questions. Furthermore, since we assume definite answers, we do not address the measurement problem. The answers to the $Q_i \in \mathcal{Q}$ given by the $\{S_a\}$ shall follow a specific statistics for each way of preparing the $\{S_a\}$ (for n sufficiently large). The set of all the possible answer statistics for all $Q_i \in \mathcal{Q}$ for all preparations is denoted by Σ.

O, being a good experimenter, has developed, through his experiments, a theoretical model for $\mathcal{Q}$ and Σ which he employs to interpret the outcomes of his interrogations (and to decide whether a question is in $\mathcal{Q}$ or not). This permits O to assign, for the next S_a to be interrogated, a prior probability y_i that S_a's answer to $Q_i \in \mathcal{Q}$ will be 'yes'. Namely, O determines y_i through a belief updating—in a broadly Bayesian spirit—according to his model of Σ, any prior information on the way of preparation and possibly to the frequencies of 'yes' answers to questions from $\mathcal{Q}$, which he may have recorded in previous interrogation runs on systems identically prepared to S_a. (We add "broadly" here as we also consider the typical laboratory situation of an ensemble of systems.) In particular, O may also not have carried out previous interrogations on systems identically prepared to S_a (e.g., if the ensemble contains only the single S_a) in which case, he will estimate the prior y_i for the single S_a solely according to his model of Σ and any prior information about the preparation (more on this and update rules will be discussed in Sections 2.3 and 2.4).

While $\mathcal{Q}$ need not necessarily contain all binary measurements that O could, in principle, perform on the $\{S_a\}$, we shall assume that $\mathcal{Q}$ is 'tomographically complete' in the sense that the $\{y_i\}_{\forall Q_i \in \mathcal{Q}}$ are sufficient to compute the probabilities for all other physically realizable measurements possibly not contained in the $\mathcal{Q}$, as well. Hence, the y_i encode everything O could possibly say about the future outcomes to arbitrary experiments on the $\{S_a\}$ in his laboratory. It will therefore be sufficient to henceforth restrict O to acquire information about the S_a solely through the $Q_i \in \mathcal{Q}$. It is also natural to identify O's 'catalog of knowledge' about the given S_a, i.e., the collection of $\{y_i\}_{\forall Q_i \in \mathcal{Q}}$, with the state of S_a relative to O. This is a state of information and an element of Σ. Conversely, any element in Σ assigns a probability y_i to all $Q_i \in \mathcal{Q}$. Thus, we identify Σ with the state space of S_a.

The state $\{y_i\}_{\forall Q_i \in \mathcal{Q}}$ is the prior state for the single S_a to be interrogated next, but also coincides with the state O assigns to the ensemble $\{S_a\}$ (which may only contain a single member) given that its members are identically prepared [1].

2.2. Time Evolution of O's "Catalog of Knowledge"

We permit O to subject the $\{S_a\}$ to interactions, which cause a state $\{y_i(t_0)\}_{\forall Q_i \in \mathcal{Q}}$ at time t_0 to evolve in time to another legitimate state. Any permitted time evolution shall be temporally translation invariant, thus defining a one-parameter map $T_{\Delta t}(\{y_i(t_0)\}_{\forall Q_i \in \mathcal{Q}}) = \{y_i(t_0 + \Delta t)\}_{\forall Q_i \in \mathcal{Q}}$ from Σ to itself, which only depends on the time interval Δt, but not on t_0. We denote by $\mathcal{T}$ the set of all time evolutions to which we allow O to expose the $\{S_a\}$.

Clearly, $\mathcal{T}$ is a further crucial ingredient of O's world model; his model for describing his interrogations with the $\{S_a\}$ is thus encoded in the triple $(\mathcal{Q}, \Sigma, \mathcal{T})$.

2.3. Convexity and State of No Information

It will be our challenge to unravel what O's world model is. This requires us to subject the triple $(\mathcal{Q}, \Sigma, \mathcal{T})$ to a number of further operational conditions that are 'natural' in the context of information acquisition with a broadly Bayesian spirit. Upon imposing the quantum postulates, this will turn out to restrict $\mathcal{Q}$ and $\mathcal{T}$ to incorporate only a 'natural' subset of all possible quantum measurements and time evolutions, namely projective binary measurements and unitaries, respectively (rather than arbitrary positive operator-valued measures (POVMs) and completely positive maps). However, this suffices for our purposes to reconstruct the textbook quantum formalism.

To account for the possibility of randomness in the method of preparation, we assume Σ to be convex. Consider a collection of identical systems (i.e., with identical $(\mathcal{Q}, \Sigma, \mathcal{T})$) that are not necessarily in identical states and for which O uses a cascade of biased coin tosses to decide which system to interrogate. Then O is enabled to assign a single prior state to this collection, which is a convex combination of their individual states.

Next, we assume the existence of a special method of preparation, which generates even completely random answer statistics over all $Q_i \in \mathcal{Q}$. This preparation is described by a special state in Σ, namely $y_i = \frac{1}{2}, \forall Q_i \in \mathcal{Q}$, and shall be called the state of no information. This distinguished state is a constraint on the pair $(\mathcal{Q}, \Sigma)$. (E.g., in quantum theory, the pair ({binary POVMs}, {density matrices}) does not satisfy this condition because there exist inherently biased POVMs, while ({projective binary measurements}, {density matrices}) does.) It plays two crucial roles: it defines (1) the prior state of S_a that O will start with in a Bayesian updating when he has no 'prior information' about the $\{S_a\}$ (except what his model $(\mathcal{Q}, \Sigma, \mathcal{T})$ is); and (2) an unambiguous notion of the (in-)dependence of questions (cf. Section 2.4), which otherwise would be state dependent. (E.g., in quantum theory, the questions $Q_{x_1} = $ "Is the spin of Qubit 1 up in x-direction?" and $Q_{x_2} = $ "Is the spin of Qubit 2 up in x-direction?" are independent relative to the completely mixed state, however not relative to a state with entanglement in x-direction.)

2.4. State Updating and (In)Dependence and Compatibility of Questions

There are two kinds of state update rules, one for the state of the ensemble $\{S_a\}$ (which coincides with the prior state assigned to the next S_a to be interrogated) and one for the posterior state of a given ensemble member S_a. In a single shot interrogation, O receives a single S_a, assigns a prior state to it according to his prior information (cf. Section 2.1), interrogates it with some questions from $\mathcal{Q}$ (without intermediate re-preparation) and, depending on the answers, updates the prior to a posterior state valid for this specific S_a only. This requires a consistent posterior state update rule, which permits O to update the probabilities y_i for all $Q_i \in \mathcal{Q}$ in a manner that respects the structure of Σ and the repeatability of questions (i.e., an answer $Q_i =$ 'yes' or 'no' must have a posterior $y_i = 1$ or 0 as a consequence, respectively). This is also a belief updating, but about the single S_a, and is not the same as in Sections 2.1 and 2.3. Specifically, the posterior state of S_a may differ significantly from its prior state if O has experienced an information gain on at least some $Q_i \in \mathcal{Q}$ (this will necessarily happen when complementary questions are involved; see below). This is the 'collapse' of the state: it is merely O's update of information about the specific S_a [1].

By contrast, in a multiple shot interrogation, O carries out a single shot interrogation on each member of an entire (identically prepared [1]) ensemble $\{S_a\}$ to do ensemble state tomography and estimate the state of the ensemble from his/her prior information about the preparation and the collection of posterior states from the single shot interrogations. With every further interrogated S_a, O updates the ensemble state, which coincides with the prior state of the next system from the ensemble to be interrogated. Accordingly, this requires a prior state update rule. This is the belief updating alluded to in Sections 2.1 and 2.3 about the ensemble $\{S_a\}$.

It will not be necessary to specify these two update rules in detail; we just assume O uses consistent ones. Specifically, given a posterior state update rule, we shall call $Q_i, Q_j \in \mathcal{Q}$

(maximally) independent if, after having asked Q_i to S in the state of no information, the posterior probability $y_j = \frac{1}{2}$. That is, if the answer to Q_i relative to the state of no information tells O 'nothing' about the answer to Q_j.

dependent if, after having asked Q_i to S in the state of no information, the posterior probability $y_j \neq \frac{1}{2}$ (if $y_j = 0$ or 1, they are maximally dependent). That is, if the answer to Q_i relative to the state of no information gives O at least partial information about the answer to Q_j.

(maximally) compatible if O may know the answers to both Q_i, Q_j simultaneously, i.e., if there exists a state in Σ such that y_i, y_j can be simultaneously zero or one.

(maximally) complementary if every state in Σ, which features $y_i = 0, 1$, necessarily implies $y_j = \frac{1}{2}$. Notice that complementarity implies independence (but not vice versa).

(One can also define partial compatibility similarly [1].) These relations shall be symmetric; e.g., Q_i is independent of Q_j if and only if Q_j is independent of Q_i, etc.

We impose a final condition on the posterior state update rule: if Q_i, Q_j are maximally compatible and independent, then asking Q_i shall not change y_j, i.e., O's information about Q_j.

2.5. Informational Completeness

The fundamental building blocks of the theories in the landscape that we are constructing are to be sets of pairwise independent questions. This will help to render the convoluted parametrization of a state by $\{y_i\}_{\forall Q_i \in \mathcal{Q}}$ more economical. Consider a set of pairwise independent questions $\mathcal{Q}_M := \{Q_1, \ldots, Q_D\}$; it is called maximal if no question from $\mathcal{Q} \setminus \mathcal{Q}_M$ can be added to $\mathcal{Q}_M$ without destroying the pairwise independence of its elements. We shall assume that any maximal $\mathcal{Q}_M$ is informationally complete in the sense that all $\{y_i\}_{\forall Q_i \in \mathcal{Q}}$ can be computed from the corresponding probabilities $\{y_i\}_{i=1}^D$ for all states in Σ. Any such $\mathcal{Q}_M$ features D elements [1] such that Σ becomes a D-dimensional convex set and states become vectors:

$$\vec{y} = \begin{pmatrix} y_1 \\ y_2 \\ \vdots \\ y_D \end{pmatrix}.$$

2.6. Information Measure

Our focus is O's acquisition of information, so we need to quantify O's information about the systems. Since $Q_i \in \mathcal{Q}$ is binary, we quantify O's information about S_a's answer to it by a function $\alpha(y_i)$

with $0 \leq \alpha(y_i) \leq 1$ bit and $\alpha(y) = 0$ bit $\Leftrightarrow y = \frac{1}{2}$ and $\alpha(1) = \alpha(0) = 1$ bit. O's total information about a S_a must be a function of the state; we make an additive ansatz:

$$I(\vec{y}) := \sum_{i=1}^{D} \alpha(y_i). \tag{1}$$

The quantum postulates will single out the specific function α.

Consider a set $\{Q_1, \ldots, Q_n\}$ of mutually (maximally) complementary questions. It is clear that whenever O has maximal information $\alpha(y_i) = 1$ bit about Q_i from this set, he must have zero bits of information about all other questions in the set. We require more generally that such a set cannot support more than one bit of information, regardless of the state:

$$\alpha(y_1) + \cdots + \alpha(y_n) \leq 1 \text{ bit} \tag{2}$$

for otherwise O could, for some states, reduce his total information about such a set by asking another question from it. These complementarity inequalities represent informational uncertainty relations that describe how the information gain about one question enforces an information loss about questions complementary to it (see also the state 'collapse' in Section 2.4).

2.7. Composite Systems and (Classical) Rules of Inference

O must be able to tell a composite system apart into its constituents purely by means of the information accessible to him through interrogation and thus ultimately by means of the question sets. Let systems S_A, S_B have question sets $\mathcal{Q}_A, \mathcal{Q}_B$. It is then natural to say that they define a composite system S_{AB} if any $Q_a \in \mathcal{Q}_A$ is maximally compatible with any $Q_b \in \mathcal{Q}_B$ and if:

$$\mathcal{Q}_{AB} = \mathcal{Q}_A \cup \mathcal{Q}_B \cup \tilde{\mathcal{Q}}_{AB}, \tag{3}$$

where $\tilde{\mathcal{Q}}_{AB}$ only contains composite questions, which are iterative compositions, $Q_a *_1 Q_b, Q_a *_2 (Q_{a'} *_3 Q_b), (Q_a *_4 Q_b) *_5 Q_{b'}, (Q_a *_6 Q_b) *_7 (Q_{a'} *_8 Q_{b'}), \ldots$, via some logical connectives $*_1, *_2, *_3, \cdots$, of individual questions $Q_a, Q_{a'}, \ldots \in \mathcal{Q}_A$ about S_A and $Q_b, Q_{b'}, \ldots \in \mathcal{Q}_B$ about S_B. This definition is extended recursively to composite systems with more than two subsystems.

Since O can never test the truthfulness of statements about the logical connectives of complementary questions through interrogations and since all propositions must have operational meaning, we shall permit O to logically connect two (possibly composite) questions directly with some $*$ only if they are compatible. For the same reason, O is allowed to apply classical rules of inference (in terms of Boolean logic) exclusively to sets of mutually-compatible questions.

We stress that this definition of composite systems is distinct from the usual state tensor product rule in generalized probabilistic theories coming from local tomography [3–5]. In particular, this composition rule admits non-locally tomographic composites (see Section 4.3).

2.8. Computing Probabilities and Questions as Vectors

Thanks to informational completeness, the probability function $Y(Q|\vec{y}) \in [0, 1]$ that $Q = $ 'yes', given the state $\vec{y}$, exists for all $Q \in \mathcal{Q}$ and $\vec{y} \in \Sigma$. As shown in [2], the exhibited structure yields:

$$Y(Q|\vec{y}) = Y(\vec{q}|\vec{y}) = \frac{1}{2}\left(\vec{q} \cdot (2\vec{y} - \vec{1}) + 1\right), \tag{4}$$

where $\vec{q} \in \mathbb{R}^D$ is a question vector encoding $Q \in \mathcal{Q}$ and $\vec{1}$ is a vector with each coefficient equal to one in the basis corresponding to $\mathcal{Q}_M$. This equation gives rise to (part of) the Born rule.

Suppose $Q, Q' \in \mathcal{Q}$ were both encoded by the same $\vec{q}$. Then, by (4), they would be probabilistically indistinguishable, and O must view them as logically equivalent. O is free to remove any such redundancy from his description of $\mathcal{Q}$ upon which every permissible question vector $\vec{q}$ will encode a unique $Q \in \mathcal{Q}$. Finally, for every $Q \in \mathcal{Q}$, there exists a state $\vec{y}_Q$, which is the updated posterior state of S_a after O received a 'yes' answer to the single question Q from S_a in the (prior) state of no information. O had zero bits of information before, and $\vec{y}_Q$ encodes a single independent question answer, so we naturally require that it encodes one independent bit. Hence, for every $Q \in \mathcal{Q}$, there exists $\vec{y}_Q \in \Sigma$ with $I(\vec{y}_Q) = 1$ bit, such that $Y(Q|\vec{y}_Q) = 1$. (In quantum theory, the $\vec{y}_Q$ will only turn out to be pure states for a single qubit; e.g., for two qubits and $Q = $ 'Is the spin of Qubit 1 up in z-direction?', represented by the rank-two projector $P_{z_1} = \frac{1}{2}(\mathbb{1} + \sigma_z \otimes \mathbb{1}_{2 \times 2})$, $\vec{y}_Q$ corresponds to the mixed state $\rho_{z_1} = \frac{1}{4}(\mathbb{1} + \sigma_z \otimes \mathbb{1}_{2 \times 2})$. Clearly, $\mathrm{tr}(P_{z_1} \rho_{z_1}) = 1$.)

3. The Quantum Principles as Rules Constraining O's Information Acquisition

In the sequel, we consider the most elementary of information carriers. Within the introduced landscape of theories, we now establish rules on O's acquisition of information that single out the quantum theory of a composite system S_N of $N \in \mathbb{N}$ qubits, modeled in our language by a triple $(\mathcal{Q}_N, \Sigma_N, \mathcal{T}_N)$. Effectively, these rules constitute a set of 'coordinates' for quantum theory on this landscape. The rules are spelled out first colloquially, then mathematically and are motivated in more detail in [1,2].

Empirically, the information accessible to an experimenter about (characteristic properties of) elementary systems is limited. For example, an experimenter may know one binary proposition about an electron (e.g., its spin in x-direction), but nothing fully independent of it (and similarly for a classical bit). We shall characterize a composition of N elementary systems according to how much information is, in principle, simultaneously available to O.

Rule 1. *(Limited information)* "The observer O can acquire maximally $N \in \mathbb{N}$ independent bits of information about the system S_N at any moment of time."
There exists a maximal set Q_i, $i = 1, \ldots, N$, of N mutually maximally independent and compatible questions in $\mathcal{Q}_N$.

O can thereby distinguish maximally 2^N states of S_N in a single shot interrogation.

However, empirically, elementary systems admit more independent propositions than what, due to the information limit, they are able to answer at a time. This is Bohr's complementarity. The unanswered properties must be random (and so 'in superposition') because the information limit makes it impossible to ascribe definite outcomes to them. For example, an experimenter may also inquire about the spin of the electron in y-direction. Yet doing so is at the total expense of his information about its spin in the x- and z-directions, and subsequent such measurements have random outcomes. For the N elementary systems, we assert the existence of complementarity.

Rule 2. *(Complementarity)* "The observer O can always get up to N new independent bits of information about the system S_N. However, whenever O asks S_N a new question, he experiences no net loss in his total amount of information about S_N."
There exists another maximal set Q'_i, $i = 1, \ldots, N$, of N mutually maximally independent and compatible questions in $\mathcal{Q}_N$, such that Q'_i, Q_i are maximally complementary and $Q'_i, Q_{j \neq i}$ are maximally compatible.

The peculiar mathematical form of Rule 2 becomes intuitive upon recalling that S_N is a composite system, such that complementarity should exist *per* elementary system [1].

Rules 1 and 2 are conceptually inspired by (non-technical) proposals made by Rovelli [11] and Zeilinger and Brukner [12,13]. These rules say nothing about what happens in-between interrogations. Naturally, we demand O not to gain or lose information without asking questions.

Rule 3. *(Information preservation)* "The total amount of information O has about (an otherwise non-interacting) S_N is preserved in-between interrogations."
$I(\vec{y})$ *is constant in time in-between interrogations for (an otherwise non-interacting) S_N.*

Hence, O's total information $I(\vec{y})$ is a 'conserved charge' of any time evolution $T_{\Delta t} \in \mathcal{T}_N$.

The more interactions to which O may subject S_N are available, the more ways in which any state may, in principle, change in time and, thus, the more 'interesting' O's world. We therefore demand that any time evolution is physically realizable as long as it is consistent with the other rules (since $\Sigma_N, \mathcal{T}_N$ are interdependent, this is distinct from 'maximizing the number' of states).

Rule 4. *(Time evolution)* "O's 'catalog of knowledge' about S_N evolves continuously in time in-between interrogations, and every consistent such evolution is physically realizable."

$\mathcal{T}_N$ *is the maximal set of transformations* $T_{\Delta t}$ *on states such that, for any fixed state* $\vec{y}$, $T_{\Delta t}(\vec{y})$ *is continuous in* Δt *and compatible with Principles 1–3 (and the structure of the theory landscape).*

(If we did not require this 'maximality' of $\mathcal{T}_N$, we would still ultimately obtain a linear, unitary evolution, but not necessarily the full unitary group. This is the sole reason for demanding 'maximality'. Note that Principles 3 and 4 are *not* equivalent to the axiom of 'continuous reversibility' of generalized probabilistic theories [3–5].)

We shall also allow O to ask any question to S_N which 'makes (probabilistic) sense'.

Rule 5. *(Question unrestrictedness)* "Every question that yields legitimate probabilities for every way of preparing S_N is physically realizable by O."

Every question vector $\vec{q} \in \mathbb{R}^{D_N}$ *that satisfies* $Y(\vec{q}|\vec{y}) \in [0,1]\ \forall\, \vec{y} \in \Sigma_N$ *and for which there exists* $\vec{y}_Q \in \Sigma_N$ *with* $I(\vec{y}_Q) = 1\ bit$, *such that* $Y(\vec{q}|\vec{y}_Q) = 1$ *corresponds to a* $Q \in \mathcal{Q}_N$.

(Without Principle 5, we would still obtain the structure of an informationally complete set $\mathcal{Q}_{M_N}$, finding that it encodes a basis of projective Pauli operator measurements [2]; Principle 5 legalizes *all* such measurements.)

These five rules turn out to leave two solutions for the triple $(\mathcal{Q}_N, \Sigma_N, \mathcal{T}_N)$. Remarkably, they cannot distinguish between complex and real numbers. Namely, the two solutions are qubit and rebit quantum theory, i.e., two-level systems over real Hilbert spaces [1,2]. Since the latter is both mathematically and physically a subcase of the former, these five rules can be regarded as sufficient. However, if one also wishes to discriminate rebits operationally, then an extra rule, adapted from [3–5] and imposed solely for this purpose (it is partially redundant), succeeds.

Rule 6. *(Tomographic locality)* "O can determine the state of the composite system S_N by interrogating only its subsystems."

As shown in [1,2], Rules 1–6 are equivalent to the textbook axioms. More precisely:

Claim. *The only solution to Rules 1–6 is qubit quantum theory where:*

- $\Sigma_N \simeq$ *convex hull of* $\mathbb{CP}^{2^N-1}$ *is the space of* $2^N \times 2^N$ *density matrices over* $\mathbb{C}^{2^N}$,
- *states evolve unitarily according to* $\mathcal{T}_N \simeq \mathrm{PSU}(2^N)$ *and the equation describing the state dynamics is (equivalent to) the von Neumann evolution equation,*
- $\mathcal{Q}_N \simeq \mathbb{CP}^{2^N-1}$ *is (isomorphic to) the set of projective measurements onto the* $+1$ *eigenspaces of N-qubit Pauli operators (a Hermitian operator on* $\mathbb{C}^{2^N}$ *is a Pauli operator iff it has two eigenvalues* ± 1 *of equal multiplicity), and the probability for* $Q \in \mathcal{Q}_N$ *to be answered with 'yes' in some state is given by the Born rule for projective measurements.*

4. Synopsis of the Reconstruction Steps and Key Results

Since this gives rise to a constructive derivation of the explicit architecture of qubit quantum theory, it involves a large number of individual steps compared to the rather abstract reconstructions [3–10]. However, this is also rewarding as it offers novel informational explanations for typical features of quantum theory, and so many reconstruction steps are actually quite instructive. We now provide a summary of key results and reconstruction steps from [1,2] (to which we refer for technical details) needed for proving the claim of the previous section.

4.1. Logical Connectives for Building Informationally Complete Sets

The first task is to build informationally complete sets $\mathcal{Q}_{M_N}$ [1]. The conjunction of Rules 1 and 2 implies that $\mathcal{Q}_{M_1} = \{Q_1, Q_2, \ldots, Q_{D_1}\}$ for a single elementary system must be a maximal mutually complementary set with $D_1 \geq 2$. We changed notation slightly compared to rules 1 and 2, labeling complementary questions by numbers, not primes. Of course, in quantum theory, $D_1 = 3$; the more involved $N = 2$ case will entail this. The structure (3) of a composite system implies that $\mathcal{Q}_{M_2}$ should contain individual questions about its subsystems. Continuing with a slight change of notation, we denote $\mathcal{Q}_{M_1}$ for System 1 by $\{Q_1, Q_2, \ldots, Q_{D_1}\}$ and for System 2 with a prime by $\{Q'_1, Q'_2, \ldots, Q'_{D_1}\}$. Apart from these individual questions, $\mathcal{Q}_{M_2}$ should contain composite questions $Q_i * Q'_j$ for some connective $*$. Pairwise independence of $\mathcal{Q}_{M_2}$ enforces that $*$ must satisfy the following truth table, where 'yes' = 1 and 'no' = 0 (Q_i, Q'_j are compatible) [1]:

Q_i	Q'_j	$Q_i * Q'_j$
0	1	a
1	0	a
1	1	b
0	0	b

$$a \neq b \qquad a, b \in \{0, 1\}. \qquad (5)$$

Hence, $*$ is either the XNOR $\leftrightarrow$ (for $a = 0, b = 1$) or its negation, the XOR $\oplus$ (for $a = 1, b = 0$). Up to an overall negation $\neg$, the two connectives are logically equivalent, and so, we henceforth make the convention to only build up composite questions (for informationally complete sets) using the XNOR. The composite question $Q_{ij} := Q_i \leftrightarrow Q'_j$ is a 'correlation question', representing "are the answers to Q_i, Q'_j the same?." Ultimately, in quantum theory, $\leftrightarrow$ will turn out to correspond to the tensor product $\otimes$ in $\sigma_i \otimes \sigma_j$ where σ_i is a Pauli matrix; Q_{ij} will then correspond to "are the spins of Qubit 1 in the i- and of Qubit 2 in the j-direction correlated?."

4.2. Question Graphs, Independence and Compatibility for $N = 2$ and Entanglement

It is convenient to represent questions graphically: individual questions are represented as vertices and bipartite correlation questions as edges between them. For instance, we may have:

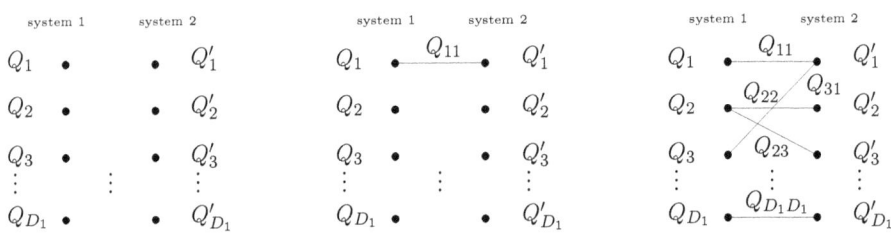

Since O is only allowed to connect compatible questions logically, there can be no edge between individual questions of the same system.

Using only Rules 1 and 2 and logical arguments, the following result is proven in [1]:

Lemma 1. Q_i, Q'_j, Q_{ij} are pairwise independent for all $i, j = 1, \ldots, D_1$ and will thus be part of an informationally complete set $\mathcal{Q}_{M_2}$. Furthermore:

(i) Q_i is compatible with Q_{ij}, $\forall j = 1, \ldots, D_1$ and complementary to Q_{kj}, $\forall k \neq i$ and $\forall j = 1, \ldots, D_1$. That is, graphically, an individual question Q_i is compatible with a correlation question Q_{ij} if and only if its corresponding vertex is a vertex of the edge corresponding to Q_{ij}. By symmetry, the analogous result holds for Q'_j.

(ii) Q_{ij} and Q_{kl} are compatible if and only if $i \neq k$ and $j \neq l$. That is, graphically, Q_{ij} and Q_{kl} are compatible if their corresponding edges do not intersect in a vertex and complementary if they intersect in one vertex.

For example, Q_1 in the third question graph above is compatible with Q_{11} and complementary to Q_{22}, while Q_{11} and Q_{22} are compatible and Q_{11} and Q_{31} are complementary.

This lemma has a striking consequence: *it implies entanglement*. Indeed, since, e.g., Q_{11} and Q_{22} are independent and compatible, O may spend his maximally accessible amount of $N = 2$ *independent* bits of information (Rule 1) over correlation questions only. Since non-intersecting edges do not share a common vertex, the lemma implies that no individual question is simultaneously compatible with two correlation questions that are compatible. Hence, when knowing the answers to Q_{11}, Q_{22}, O will be entirely ignorant about the individual questions; O has then maximal information about S_2, but purely composite information. This is entanglement in the very sense of Schrödinger ("...*the best possible knowledge of a whole does not necessarily include the best possible knowledge of all its parts...*" [14]). For example, in quantum theory, a state with $Q_{11} = Q_{22} =$ 'yes' will coincide with a Bell state having the spins of Qubits 1 and 2 correlated in x- and y-direction (and anti-correlated in z-direction). Of course, there is nothing special about Q_{11}, Q_{22}, and the argument works similarly for other composite question pairs and can be extended also to states with non-maximal entanglement (see [1] for details).

For systems with limited information content, *entanglement is therefore a direct consequence of complementarity*; without it there would be no independent and compatible composite questions sufficient to saturate the information limit [1]. For instance, two classical bits satisfy Rule 1, as well, but admit no complementarity so that $\mathcal{Q}^{cbit}_{M_2} = \{Q_1, Q'_1, Q_{11}\}$ and the maximum amount of $N = 2$ *independent* bits cannot be spent on composite questions only.

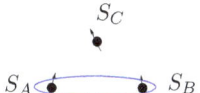

We also note that Rules 1 and 2 offer a simple, intuitive explanation for *monogamy of entanglement*. Consider, for a moment, $N = 3$ elementary systems S_A, S_B, S_C, and suppose S_A and S_B are maximally entangled (say, because O received the answer $Q_{11} = Q_{22} =$ 'yes' from S_{AB}). Noting that S_{AB} is a composite bipartite system inside the tripartite S_{ABC}, O has then already spent his maximal amount of information of $N = 2$ independent bits, which he may know about S_{AB} and can therefore not know anything else that is independent, including non-trivial correlations with S_C, about the pair. To saturate the $N = 3$ *independent* bit limit for the tripartite system S_{ABC}, he may then only inquire about individual information about S_C. This is monogamy in its extreme form: the maximally entangled pair S_{AB} cannot be entangled with any other system S_C. This heuristic argument can be made rigorous in terms of the compatibility and independence structure of questions for $N \geq 3$ and can be extended to the non-extremal case using informational *monogamy inequalities* [1].

4.3. A Logical Explanation for the Three-Dimensionality of the Bloch Ball

A key result of the reconstruction, proven in [1] is the following. Since its proof is instructive and representative for this approach, we shall rephrase it here.

Theorem 1. *$D_1 = 2$ or 3.*

Proof. Consider the $N = 2$ case. Lemma 1 implies that any maximal set of pairwise compatible correlation questions has D_1 elements. Indeed, there are maximally D_1 non-intersecting edges between the D_1 vertices of System 1 and the D_1 vertices of System 2; e.g., the D_1 'diagonal' Q_{ii}:

$$
\begin{array}{c}
\bullet \!\!-\!\! Q_{11} \!\!-\!\! \bullet \\
\bullet \!\!-\!\! Q_{22} \!\!-\!\! \bullet \\
\bullet \!\!-\!\! Q_{33} \!\!-\!\! \bullet \\
\vdots \\
\bullet \!\!-\!\! Q_{D_1 D_1} \!\!-\!\! \bullet
\end{array}
$$

are pairwise independent and compatible. The constraints on the posterior state update rule in Section 2.4 entail that they are also mutually compatible (Specker's principle) [1] such that O may simultaneously know the answers to all D_1 Q_{ii}. Since O may not know more than $N = 2$ independent bits (Rule 1), the D_1 Q_{ii} cannot be mutually independent if $D_1 > 2$. Thus, assuming the Q_{ii} are of equivalent status, the answers to any pair of them, say Q_{11}, Q_{22}, must imply the answers to all others, say Q_{ii}, $i = 3, \ldots, D_1$. Hence, $Q_{jj} = Q_{11} * Q_{22}$, $j \neq 1, 2$, for a connective $*$ that preserves pairwise independence of Q_{11}, Q_{22}, Q_{jj}. Reasoning as in (5) implies that either:

$$Q_{jj} = Q_{11} \leftrightarrow Q_{22}, \quad \text{or} \quad Q_{jj} = \neg(Q_{11} \leftrightarrow Q_{22}), \quad j = 3, \ldots, D_1 \qquad (6)$$

so that for $D_1 > 3$ Q_{jj}, $j = 3, \ldots, D_1$ could *not* be pairwise independent. Arguing identically for all other sets of D_1 pairwise independent and compatible Q_{ij}, we conclude that $D_1 \leq 3$. □

This theorem has several crucial repercussions. We may already suggestively call $D_1 = 2$ and $D_1 = 3$ the 'rebit' (two-level systems over real Hilbert spaces) and 'qubit' case, respectively. Reasoning as in (6) shows that the Q_{ij} are logically closed under $\leftrightarrow$; as demonstrated in [1]:

Theorem 2. *If $D_1 = 3$, then $\mathcal{Q}_{M_2} := \{Q_i, Q'_j, Q_{ij}\}_{i,j=1,2,3}$ is logically closed under $\leftrightarrow$ and, thus, constitutes an informationally complete set for $N = 2$ with $D_2 = 15$.*

If $D_1 = 2$, then $\mathcal{Q}_{M_2} = \{Q_i, Q'_j, Q_{ij}, Q_{11} \leftrightarrow Q_{22}\}_{i,j=1,2}$ is logically closed under $\leftrightarrow$ and, thus, constitutes an informationally complete set for $N = 2$ with $D_2 = 9$. Furthermore, $Q_{11} \leftrightarrow Q_{22}$ is complementary to the individual questions Q_i, Q'_j, $i, j = 1, 2$.

Indeed, $D_2 = 9, 15$ are the correct numbers of degrees of freedom for $N = 2$ rebits and qubits, respectively. However, since the composite question $Q_{11} \leftrightarrow Q_{22}$ is complementary to all individual questions in the rebit case (this is not true in the qubit case!), it is impossible for O to do ensemble state tomography by asking only individual questions Q_i, Q'_j, thereby violating Rule 6. We are left with the qubit case and shall henceforth ignore rebits (for rebits see [1]).

4.4. Ruling out Local Hidden Variables and the Correlation Structure for $N = 2$

Using (6) and repeating the argument leading to it for 'non-diagonal' Q_{ij} show that either:

$$Q_{11} \leftrightarrow Q_{22} = Q_{12} \leftrightarrow Q_{21}, \quad \text{or} \quad Q_{11} \leftrightarrow Q_{22} = \neg(Q_{12} \leftrightarrow Q_{21}). \qquad (7)$$

The first case (without relative negation) is the case of classical logic and compatible with local hidden variables for the individual questions Q_i, Q'_j. Namely, note that $Q_{11} \leftrightarrow Q_{22} = Q_{12} \leftrightarrow Q_{21}$ can be rewritten in terms of the individuals as:

$$(Q_1 \leftrightarrow Q'_1) \leftrightarrow (Q_2 \leftrightarrow Q'_2) = (Q_1 \leftrightarrow Q'_2) \leftrightarrow (Q_2 \leftrightarrow Q'_1). \qquad (8)$$

Suppose for a moment that Q_1, Q_1', Q_2, Q_2' had simultaneous definite values (although not accessible to O). It is easy to convince oneself that any distribution of simultaneous truth values over the Q_i, Q_j' satisfies (8) [1]. In fact, (8) is a classical logical identity and can be argued to follow from classical rules of inference [1]. However, it involves complementary individual questions, thereby violating our premise from Section 2.7 that O may apply classical rules of inference exclusively to mutually compatible questions. This classical case is thus ruled out.

One can check that the second case, $Q_{11} \leftrightarrow Q_{22} = \neg(Q_{12} \leftrightarrow Q_{21})$, does not admit a local hidden variable interpretation, but is consistent with the structure of the theory landscape and rules [1]. Since one of the two cases (7) must be true, we conclude that this second case holds. In fact, for any complementary pairs Q, Q' and Q'', Q''' such that both Q and Q' are compatible with both Q'', Q''', one finds similarly [1]:

$$(Q \leftrightarrow Q'') \leftrightarrow (Q' \leftrightarrow Q''') = \neg\left((Q \leftrightarrow Q''') \leftrightarrow (Q' \leftrightarrow Q'')\right). \tag{9}$$

This precludes to reason classically about the distribution of truth values over O's questions.

Equation (9) permits us to unravel the complete correlation structure for $\mathcal{Q}_{M_2}$. In fact, it turns out that there are two distinct representations of this correlation structure: one corresponding to quantum theory in its standard representation, the other to its 'mirror' representation, related by a passive (not a physical) transformation, reassigning $Q_1 \mapsto \neg Q_1$ (in quantum theory tantamount to a partial transpose on qubit 1) [1]. The two distinct representations turn out to be physically equivalent, and so, a convention has to be made. Choosing the 'standard' case and using (9), one finds that the compatibility and correlation structure of $\mathcal{Q}_{M_2}$ can be represented graphically as in Figure 1. For Q, Q', Q'' compatible, we shall henceforth distinguish between:

even correlation: if $Q = Q' \leftrightarrow Q''$ and
odd correlation: if $Q = \neg(Q' \leftrightarrow Q'')$.

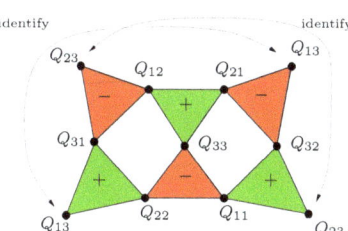 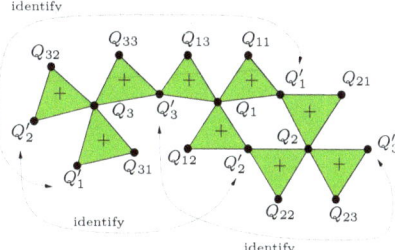

Figure 1. The compatibility and correlation structure of the informationally complete set $\mathcal{Q}_{M_2}$ for the $N = 2$ qubit case. Two questions are compatible if connected by a triangle edge and complementary otherwise. Red and green triangles denote odd and even correlation, respectively; e.g., $Q_{33} = \neg(Q_{11} \leftrightarrow Q_{22}) = Q_{12} \leftrightarrow Q_{21}$. (Taken from [1].)

One can easily check that quantum theory satisfies this correlation structure for projective spin measurements if one replaces $i = 1, 2, 3$ by x, y, z. For instance, $Q_{11} = Q_{22} =$ 'yes' implies, by Figure 1, the dependent $Q_{33} =$ 'no'. In quantum theory, this corresponds to the (unnormalized) Bell state with spin correlation in the x- and y-direction and anti-correlated spins in the z-direction:

$$|x_+ x_+\rangle - |x_- x_-\rangle = -i|y_+ y_+\rangle + i|y_- y_-\rangle = |z_+ z_-\rangle + |z_- z_+\rangle.$$

4.5. Compatibility, Independence and Informational Completeness for Arbitrary N

Consider N elementary systems in the 'qubit' ($D_1 = 3$) case and the XNOR conjunction:

$$Q_{\mu_1\mu_2\cdots\mu_N} := Q_{\mu_1} \leftrightarrow Q_{\mu_2} \leftrightarrow \cdots \leftrightarrow Q_{\mu_N} \tag{10}$$

of individual questions, where $\mu_a = 0, 1, 2, 3$ and $Q_0 :=$ 'yes'. The conjunction yields 'yes' and 'no' if an even and odd number of $Q_{\mu_a} =$ 'no', respectively, and thus, does not represent "are the answers to all Q_{μ_a} the same?." As shown in [1], these conjunctions are informationally complete:

Theorem 3. *(Qubits) The $4^N - 1$ questions $Q_{\mu_1\cdots\mu_N}$, $\mu = 0, 1, 2, 3$ (we deduct the trivial question $Q_{000\cdots000}$), are pairwise independent and logically closed under $\leftrightarrow$ and, thus, form an informationally complete set $\mathcal{Q}_{M_N}$ with $D_N = 4^N - 1$. Moreover, $Q_{\mu_1\cdots\mu_N}$ and $Q_{\nu_1\cdots\nu_N}$ are compatible if they differ by an even number (including zero) of non-zero indices and complementary otherwise.*

We note that an N-qubit density matrix has precisely $4^N - 1$ degrees of freedom.

4.6. Linear, Reversible Time Evolution and a Quadratic Information Measure

Thus far, the summarized results invoked only Rules 1 and 2 (and in one instance, Rule 6). Rules 3 and 4, on the other hand, can be demonstrated to entail a linear and reversible evolution of the generalized Bloch vector $\mathbb{R}^{4^N-1} \ni \vec{r} = 2\vec{y} - \vec{1}$ that already appeared in (4),

$$\vec{r}(\Delta t + t_0) = T(\Delta t)\,\vec{r}(t_0), \tag{11}$$

where $T(\Delta t) \subset \mathcal{T}_N$ defines a one-parameter matrix group [1]. Suppose $T(\Delta t), T'(\Delta t') \in \mathcal{T}_N$ correspond to two distinct interactions to which O may subject S_N. By Rule 4, $T(\Delta t) \cdot T'(\Delta t')$ must likewise be contained in $\mathcal{T}_N$, and since both T, T' are invertible, also the entire set $\mathcal{T}_N$ must be a group. We shall henceforth often represent states with Bloch vectors $\vec{r}$.

Rules 3 and 4, together with elementary operational conditions on the information measure, enforce it to be quadratic $\alpha(y_i) = (2y_i - 1)^2$ so that O's total information (1):

$$I_N(\vec{y}) = \sum_{i=1}^{4^N-1} (2y_i - 1)^2 = |\vec{r}|^2 \tag{12}$$

is simply the square norm of the Bloch vector [1]. Interestingly, this derivation would not work without the *continuity* of time evolution (Rule 4). Crucially, (12) is not the Shannon entropy (see [1] for a discussion about why the Shannon entropy is also conceptually not suitable for quantifying O's information). This reconstruction thereby corroborates an earlier proposal for a quadratic information measure for quantum theory by Brukner and Zeilinger [13,15,16].

This quadratic information measure becomes key for the remaining steps of the reconstruction. Given that (12) is a 'conserved charge' of time evolution (rule 3), we can already infer that $\mathcal{T}_N \subset \text{SO}$ $(4^N - 1)$ because time evolution must be connected to the identity.

4.7. Pure and Mixed States

Suppose O knows S_N's answers to N mutually compatible questions from $\mathcal{Q}_{M_N}$, thereby saturating the information limit of N independent bits (Rule 1). He will then also know the answers to each of their bipartite, tripartite, ..., and N-partite XNOR conjunctions which, by Theorem 3, are also in $\mathcal{Q}_{M_N}$ (and compatible). In total, he then knows the answers to:

$$\binom{N}{1} + \binom{N}{2} + \cdots + \binom{N}{N} = \sum_{i=1}^{N}\binom{N}{i} = 2^N - 1$$

questions from $\mathcal{Q}_{M_N}$. Thus, O's total information (12) is $2^N - 1$ bits in this case. It contains *dependent* bits of information because the questions in $\mathcal{Q}_{M_N}$ are pairwise, but not all mutually independent. Thanks to Rule 3, this is invariant under time evolution.

This allows us to distinguish two kinds of states [1]; $\vec{y}$ is called a:

pure state: if it is a state of maximal information and, hence, of maximal length:

$$I_N(\vec{y}) = \sum_{i=1}^{4^N-1} (2y_i - 1)^2 = (2^N - 1) \, \text{bits}, \qquad (13)$$

mixed state: if it is a state of non-maximal information,

$$0 \, \text{bit} \leq I_N(\vec{y}) = \sum_{i=1}^{4^N-1} (2y_i - 1)^2 < (2^N - 1) \, \text{bits}. \qquad (14)$$

The square length of the Bloch vector thus corresponds to the number of answered questions. The state of no information $\vec{y} = \frac{1}{2}\vec{1}$ has length zero bits.

As can be easily checked, quantum theory satisfies this characterization. In particular, an N-qubit density matrix, corresponding to a pure state, has a Bloch vector with square norm equal to $2^N - 1$. This peculiar mathematical fact now has a clear informational interpretation.

4.8. The Bloch Ball and Unitary Group for a Single Qubit from a Conserved Informational Charge

Since $D_1 = 3$ (cf. Section 4.3), we have that $\mathcal{Q}_{M_1} = \{Q_1, Q_2, Q_3\}$ is a maximal set of mutually complementary questions, i.e., no further $Q \in \mathcal{Q}_1$ can be added to $\mathcal{Q}_{M_1}$ without destroying mutual complementarity in the set (cf. Section 4.1). According to (13), a pure state satisfies:

$$I_{N=1}(\vec{y}) = r_1^2 + r_2^2 + r_3^2 = (2y_1 - 1)^2 + (2y_2 - 1)^2 + (2y_3 - 1)^2 = 1 \, \text{bit}. \qquad (15)$$

For later, we thus observe: *for pure states, the maximal mutually complementary set carries exactly 1 bit of information, and this is a conserved charge of time evolution (Rule 3).*

Rule 1 implies that, e.g., the pure state $\vec{y}_* = (1,0,0)$ exists in Σ_1, and we know $\mathcal{T}_1 \subset SO(3)$. However, it is clear that applying any $T \in SO(3)$ to $\vec{y}_*$, according to (11), yields only states that are also compatible with all Rules 1–3 (and the landscape). Hence, by Rule 4, we must actually have $\mathcal{T}_1 = SO(3) \simeq PSU(2)$. Clearly, $\mathcal{T}_1$ then generates all quantum pure states from $\vec{y}_*$, i.e., it yields the entire Bloch sphere (the image of any legal state under a legal time evolution is also a legal state). Recalling that Σ_1 is convex, we obtain that $\Sigma_1 = B^3 \simeq$ convex hull of $\mathbb{CP}^1$ is the entire unit Bloch ball with mixed states (14) lying inside; the completely mixed state equals the state of no information at the center. $\Sigma_1, \mathcal{T}_1$ coincide exactly with the set of density matrices $\rho = \frac{1}{2}(\mathbb{1} + \vec{r} \cdot \vec{\sigma})$ and the set of unitary transformations $\rho \mapsto U \rho U^\dagger$, $U \in SU(2)$, respectively, for a single qubit in its adjoint (i.e., Bloch vector) representation, where $\vec{\sigma} = (\sigma_1, \sigma_2, \sigma_3)$ is the vector of Pauli matrices. Finally, from the assumptions in Section 2.8 and Rule 5, it is also clear that $\mathcal{Q}_1 = \{\vec{q} \in \mathbb{R}^3 \,|\, |\vec{q}|^2 = 1 \, \text{bit}\} \simeq \mathbb{CP}^1$. This coincides with the set of projectors $P_{\vec{q}} = \frac{1}{2}(\mathbb{1} + \vec{q} \cdot \vec{\sigma})$ onto the $+1$ eigenspaces of the Pauli operators $\vec{q} \cdot \vec{\sigma}$. Noting that:

$$\text{Tr}(\rho \, P_{\vec{q}}) = \frac{1}{2}(1 + \vec{r} \cdot \vec{q}) \equiv Y(Q|\vec{y}) \qquad (16)$$

we also recover that (4) yields the Born rule for projective measurements. We thus have the claim of Section 3 for $N = 1$ (for details see [1,2]).

4.9. Unitary Group and Density Matrices for Two Qubits from Conserved Informational Charges

Also for $N = 2$, it is rewarding to consider maximal mutually complementary sets within $\mathcal{Q}_{M_2}$. Using Lemma 1, one can check that there are exactly six maximal complementarity sets containing five questions and twenty containing three [2]; e.g., two graphical representatives are:

$$\text{Pent}_1 = \{Q_{11}, Q_{12}, Q_{13}, Q_2, Q_3\}, \qquad \text{Tri}_1 = \{Q_{11}, Q_{12}, Q_3'\}.$$

The six maximal complementarity sets of five elements can be represented as a lattice of pentagons; see Figure 2 (which also contains four green triangles, each representing one of the twenty maximal complementarity sets of three questions) [2].

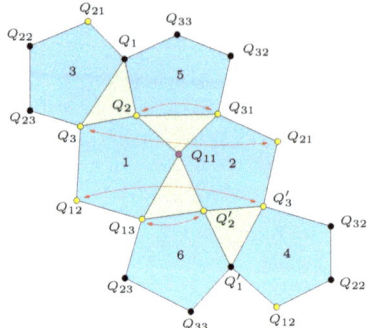

Figure 2. The six maximal complementarity sets represented as pentagons. Two questions are complementary if they share a pentagon or are connected by an edge and compatible otherwise. Every pentagon is connected to all of the other five because any $Q \in \mathcal{Q}_{M_2}$ is contained in precisely two pentagons. The red arrows represent the information swap (21) between Pentagons 1 and 2 that preserves all pentagon equalities (18) and defines the time evolution generator (22). (Figure adapted from [2]. Reprinted with permission from [P. Höhn and C. Wever, *Phys. Rev.* **A95**, 012102 2017.] Copyright (2017) by the American Physical Society.)

Each of these sets has to satisfy the complementarity inequalities (2); specifically $0\,\text{bits} \leq I(\text{Pent}_a) := \sum_{i \in \text{Pent}_a} r_i^2 \leq 1\,\text{bit}$ for the information carried by the five questions in pentagon a. Since any $Q \in \mathcal{Q}_{M_2}$ is contained in precisely two pentagons (cf. Figure 2), we find:

$$\sum_{a=1}^{6} I(\text{Pent}_a) = 2 \left(\sum_{i=1,2,3} (r_{i_1}^2 + r_{i_2}^2) + \sum_{i,j=1,2,3} r_{ij}^2 \right) = 2\, I_{N=2}(\vec{r}). \tag{17}$$

Noting that for pure states $I_{N=2}(\vec{r}_{\text{pure}}) = 3\,\text{bits}$ thus produces the pentagon equalities [2]:

$$\textbf{pure states:} \qquad I(\text{Pent}_a) \equiv 1\,\text{bit}, \qquad a = 1, \ldots, 6. \tag{18}$$

Any pure state must satisfy (18), and $\mathcal{T}_2$ evolves pure states to pure states (Rule 3). Hence, in analogy to $N = 1$: for pure states, these six maximal mutually complementary sets carry exactly one **bit** of information, and these are six conserved charges of time evolution. There are further interesting constraints on the distribution of O's information over $\mathcal{Q}_{M_2}$ [2].

It can be straightforwardly checked that quantum theory actually satisfies (18). Indeed, in the case of quantum theory, the identity for Pent_1 reads in more familiar language (pure states):

$$I(\text{Pent}_1) = \langle \sigma_2 \otimes \mathbb{1} \rangle^2 + \langle \sigma_3 \otimes \mathbb{1} \rangle^2 + \langle \sigma_1 \otimes \sigma_1 \rangle^2 + \langle \sigma_1 \otimes \sigma_2 \rangle^2 + \langle \sigma_1 \otimes \sigma_3 \rangle^2 = 1,$$

etc. Remarkably, these identities of quantum theory seem not to have been reported before in the literature. These novel conserved informational charges are a prediction of our reconstruction, underscoring the benefits of taking this informational approach. Additionally, these informational charges are indispensable for deriving the unitary group and the state space, as we shall now see.

Using that $I(\text{Pent}_a(\vec{r}))$ is conserved under $\mathcal{T}_2 \subset SO(15)$ entails (with new index $i = 1, \ldots, 15$):

$$\sum_{i \in \text{Pent}_a, 1 \leq j \leq 15} r_i G_{ij} r_j = 0, \qquad a = 1, \ldots 6, \tag{19}$$

where $T(\Delta t) = \exp(\Delta t G)$ for $G \in \mathfrak{so}(15)$ [2]. The correlation structure of Figure 1 enforces [2]:

$$G_{ij} = 0, \qquad \text{whenever } Q_i, Q_j \text{ are compatible.} \tag{20}$$

Each of the 15 $Q_i \in \mathcal{Q}_{M_2}$ is complementary to eight others, and since $G_{ij} = -G_{ji}$, there could be maximally 60 linearly independent G_{ij} of $\mathcal{T}_2$.

These are constructed as follows. For every pair of pentagons, there is a *unique* information swap transformation that preserves (18). For instance, the red arrows in Figure 2 represent the complete information swap between pentagons Pent_1 and Pent_2 ($\longleftrightarrow$ is not the XNOR):

$$r_2^2 \longleftrightarrow r_{31}^2 \;(\text{Pent}_5), \quad r_3^2 \longleftrightarrow r_{21}^2 \;(\text{Pent}_3), \quad r_{12}^2 \longleftrightarrow r_3'^2 \;(\text{Pent}_4), \quad r_{13}^2 \longleftrightarrow r_2'^2 \;(\text{Pent}_6) \tag{21}$$

that keeps all other components fixed. (18) are preserved because every swap in (21) occurs *within* a pentagon. The correlation structure of Figure 1 fixes the corresponding generator to [2]:

$$G_{ij}^{\text{Pent}_1, \text{Pent}_2} = \delta_{i2}\delta_{j(31)} - \delta_{i3}\delta_{j(21)} + \delta_{i(12)}\delta_{j3'} - \delta_{i(13)}\delta_{j2'} - (i \longleftrightarrow j). \tag{22}$$

One can repeat the argument for all 15 pentagon pairs, producing 15 linearly independent generators [2]. Remarkably, they turn out to coincide exactly with the adjoint representation of the 15 fundamental generators of SU(4) [2]. In particular, (22) is the generator of entangling unitaries leaving r_{11} invariant. The other 45 independent generators satisfying (20) are ruled out by the correlation structure so that $\mathcal{T}_2$ cannot be generated by anything else than these 15 pentagon swaps [2]. One can show that the exponentiation of (linear combinations of) these 15 pentagon swaps generates PSU(4) and that this group abides by all rules and forms a maximal subgroup of SO(15) [2]. Rule 4 then implies $\mathcal{T}_2 \simeq \text{PSU}(4)$, which is the correct set of unitary transformations $\rho \mapsto U \rho U^\dagger$, $U \in SU(4)$, for two qubits.

It turns out that the set of Bloch vectors satisfying all six pentagon equalities (18) and the conservation equations (19) for the 15 pentagon swaps splits into two sets on each of which $\mathcal{T}_2 = \text{PSU}(4)$ acts transitively [2]. These two sets correspond precisely to the two possible conventions of building up composite questions either using the XNOR or XOR (cf. Section 4.1) and are therefore physically equivalent. Adhering to the XNOR convention, we conclude that the surviving set of Bloch vectors solving (18) and (19) is the set of $N = 2$ states admitted by the rules. Indeed, it coincides exactly with the set of quantum pure states, which forms a $\mathbb{CP}^3$ of which PSU(4) is the isometry group [2]. Employing convexity of Σ_2, one finally finds:

$$\Sigma_2 = \text{closed convex hull of } \mathbb{CP}^3,$$

which is exactly the set of normalized 4×4 density matrices over $\mathbb{C}^2 \otimes \mathbb{C}^2$.

Concluding, the new conserved informational charges (18), in analogy to (15) for $N=1$, define both the unitary group and the set of states for two qubits (for neglected details, see [2]).

4.10. Unitaries and States for $N>2$ Elementary Systems

According to Theorem 3, Σ_N is (4^N-1)-dimensional and $\mathcal{T}_N \subset$ SO (4^N-1) (cf. Section 4.6). The reconstruction of the unitary group uses a universality result from quantum computation: two-qubit unitaries PSU(4) (between any pair) and single-qubit unitaries PSU(2) $\simeq$ SO(3) generate the full projective unitary group PSU (2^N) for N qubits [17,18]. Given that S_N is a composite system, all of these bipartite and local unitaries must be in $\mathcal{T}_N$. One can check that PSU (2^N) again abides by all rules and constitutes a maximal subgroup of SO (4^N-1) [2]. Thanks to Rule 4, this yields $\mathcal{T}_N \simeq$ PSU (2^N), which coincides with the set of unitary transformations on N-qubit density matrices. In analogy to the previous case, one obtains as the state space:

$$\Sigma_N = \text{closed convex hull of } \mathbb{CP}^{2^N-1},$$

which agrees with the set of normalized N-qubit density matrices (for details, see [2]).

4.11. Questions as Projective Measurements and the Born Rule

The assumptions in Section 2.8 and Rule 5 yield the following question set characterization [2]:

$$\mathcal{Q}_N \simeq \{\vec{q} \in \mathbb{R}^{4^N-1} \mid Y(\vec{q}|\vec{r}) \in [0,1] \,\forall \vec{r} \in \Sigma_N \text{ and } \vec{q} \text{ is a 1 bit quantum state}\}. \tag{23}$$

As shown in [2], this set is isomorphic to the set of projectors $P_{\vec{q}} = \frac{1}{2}(\mathbb{1} + \vec{q}\cdot\vec{\sigma})$ onto the +1 eigenspaces of the Pauli operators $\vec{q}\cdot\vec{\sigma} = \sum_{\mu_1\cdots\mu_N} q_{\mu_1\cdots\mu_N}\sigma_{\mu_1\cdots\mu_N}$, where $\sigma_{\mu_1\cdots\mu_N} = \sigma_{\mu_1} \otimes \cdots \otimes \sigma_{\mu_N}$ and $\sigma_0 = \mathbb{1}$. Noting that $q_{\mu_1\cdots\mu_N}$ corresponds to (10) reveals that the XNOR at the question level corresponds to the tensor product $\otimes$ at the operator level. One also finds that (16) again holds, such that (4) yields the Born rule for projective measurements for arbitrary N (for the neglected details and many further interesting properties of $\mathcal{Q}_N$, we refer to [2]).

4.12. The von Neumann Evolution Equation

We thus obtain qubit quantum theory in its adjoint (i.e., Bloch vector) representation. Lastly, we note that $\vec{r}(t) = T(t)\vec{r}(0)$ with $T(t) = e^{tG} \in$ PSU (2^N) is equivalent to the adjoint action:

$$\rho(t) = U(t)\rho(0)U^\dagger(t), \tag{24}$$

of $U(t) = e^{-iHt} \in$ SU(2^N) for some Hermitian operator H on $\mathbb{C}^{2^N}$, where $\rho(t) = \frac{1}{2^N}(\mathbb{1} + \vec{r}(t)\cdot\vec{\sigma})$ [2]. (24), in turn, is equivalent to $\rho(t)$ solving the von Neumann evolution equation:

$$i\frac{\partial \rho}{\partial t} = [H, \rho]. \tag{25}$$

We have therefore also recovered the correct time evolution equation for quantum states.

5. Conclusions

We have reviewed and summarized the key steps from [1,2] necessary to prove the claim of Section 3. This yields a reconstruction of the explicit formalism of qubit quantum theory from rules constraining an observer's acquisition of information about a system [1,2]. The derivation corroborates the consistency of interpreting the state as the observer's 'catalog of knowledge' and shows that it is sufficient to speak only about the information accessible to him for reproducing quantum theory. In fact, for qubits, this derivation accomplishes an informational reconstruction of the type proposed in

Rovelli's relational quantum mechanics [11] and in the Brukner-Zeilinger informational interpretation of quantum theory [12,13].

As a key benefit, this reconstruction also provides a novel informational explanation for the architecture of qubit quantum theory. In particular, it explains the logical structure of a basis of spin measurements, the dimensionality and structure of quantum state spaces, the correlation structure and the unitarity of time evolution from the perspective of information acquisition. This unravels previously unknown structural properties: conserved 'informational charges' from complementarity relations define and explain the unitary group and the set of pure states.

Acknowledgments: The author thanks Christopher S. P. Wever for an enjoyable collaboration on [2]. The project leading to this publication has received funding from the European Union's Horizon 2020 research and innovation program under the Marie Sklodowska-Curie Grant Agreement No. 657661.

Conflicts of Interest: The author declares no conflict of interest.

References

1. Höhn, P.A. Toolbox for reconstructing quantum theory from rules on information acquisition. *arXiv* **2014**, arXiv:1412.8323.
2. Höhn, P.A.; Wever, C.S.P. Quantum theory from questions. *Phys. Rev. A* **2017**, *95*, 012102.
3. Hardy, L. Quantum Theory From Five Reasonable Axioms. *arXiv* **2001**, arXiv:quant-ph/0101012.
4. Dakic, B.; Brukner, C. Quantum Theory and Beyond: Is Entanglement Special? In *Deep Beauty*; Halvorson, H., Ed.; Cambridge University Press: Cambridge, UK, 2011; p. 365.
5. Masanes, L.; Müller, M.P. A derivation of quantum theory from physical requirements. *New J. Phys.* **2011**, *13*, 063001.
6. Chiribella, G.; D'Ariano, G.M.; Perinotti, P. Informational derivation of quantum theory. *Phys. Rev. A* **2011**, *84*, 012311.
7. Barnum, H.; Müller, M.P.; Ududec, C. Higher-order interference and single-system postulates characterizing quantum theory. *New J. Phys.* **2014**, *16*, 123029.
8. De la Torre, G.; Masanes, L.; Short, A.J.; Müller, M.P. Deriving Quantum Theory from Its Local Structure and Reversibility. *Phys. Rev. Lett.* **2012**, *109*, 090403.
9. Goyal, P. From information geometry to quantum theory. *New J. Phys.* **2010**, *12*, 023012.
10. Appleby, M.; Fuchs, C.A.; Stacey, B.C.; Zhu, H. Introducing the Qplex: A Novel Arena for Quantum Theory. *arXiv* **2016**, arXiv:1612.03234.
11. Rovelli, C. Relational quantum mechanics. *Int. J. Theor. Phys.* **1996**, *35*, 1637–1678.
12. Zeilinger, A. A Foundational Principle for Quantum Mechanics. *Found. Phys.* **1999**, *29*, 631–643.
13. Brukner, C.; Zeilinger, A. Information and fundamental elements of the structure of quantum theory. In *Time, Quantum and Information*; Castell, L., Ischebeck, O., Eds.; Springer: Berlin/Heidelberg, Germany, 2003.
14. Schrödinger, E. Discussion of Probability Relations between Separated Systems. *Math. Proc. Camb. Philos. Soc.* **1935**, *31*, 555–563.
15. Brukner, C.; Zeilinger, A. Operationally Invariant Information in Quantum Measurements. *Phys. Rev. Lett.* **1999**, *83*, 3354.
16. Brukner, C.; Zeilinger, A. Conceptual inadequacy of the Shannon information in quantum measurements. *Phys. Rev. A* **2001**, *63*, 022113.
17. Bremner, M.J.; Dawson, C.M.; Dodd, J.L.; Gilchrist, A.; Harrow, A.W.; Mortimer, D.; Nielsen, M.A.; Osborne,T.J. Practical Scheme for Quantum Computation with Any Two-Qubit Entangling Gate. *Phys. Rev. Lett.* **2002**, *89*, 247902.
18. Harrow, A.W. Exact universality from any entangling gate without inverses. *Quant. Inf. Comput.* **2009**, *9*, 773–777.

© 2017 by the author. Licensee MDPI, Basel, Switzerland. This article is an open access article distributed under the terms and conditions of the Creative Commons Attribution (CC BY) license (http://creativecommons.org/licenses/by/4.0/).

Brief Report

Test of the Pauli Exclusion Principle in the VIP-2 Underground Experiment

Catalina Curceanu [1,2,3,*,‡,§], Hexi Shi [1,4,*,§], Sergio Bartalucci [1], Sergio Bertolucci [5], Massimiliano Bazzi [1], Carolina Berucci [1,4], Mario Bragadireanu [1,3], Michael Cargnelli [1,4], Alberto Clozza [1], Luca De Paolis [1], Sergio Di Matteo [6], Jean-Pierre Egger [7], Carlo Guaraldo [1], Mihail Iliescu [1], Johann Marton [1,4], Matthias Laubenstein [8], Edoardo Milotti [9], Marco Miliucci [1], Andreas Pichler [1,4], Dorel Pietreanu [1,3], Kristian Piscicchia [2,1], Alessandro Scordo [1], Diana Laura Sirghi [1,3], Florin Sirghi [1,3], Laura Sperandio [1], Oton Vazquez Doce [1,10], Eberhard Widmann [4] and Johann Zmeskal [1,4]

1. Laboratori Nazionali di Frascati, INFN, I-00044 Frascati, Italy; Sergio.bartalucci@lnf.infn.it (S.B.); massimiliano.bazzi@lnf.infn.it (Mas.B.); Carolina.Berucci@oeaw.ac.at (C.B.); Bragadireanu.Mario@lnf.infn.it (Mar.B.); michael.cargnelli@oeaw.ac.at (M.C.); alberto.clozza@lnf.infn.it (A.C.); luca.depaolis@lnf.infn.it (L.D.P.); carlo.guaraldo@lnf.infn.it (C.G.); Mihai.Iliescu@lnf.infn.it (M.I.); johann.marton@oeaw.ac.at (J.M.); marco.miliucci@lnf.infn.it (M.M.); Andreas.Pichler@oeaw.ac.at (A.P.); dorel.pietreanu@lnf.infn.it (D.P.); kristian.piscicchia@gmail.com (K.P.); alessandro.scordo@lnf.infn.it (A.S.); Diana.Laura.Sirghi@lnf.infn.it (D.L.S.); Sirghi.FlorinCatalin@lnf.infn.it (F.S.); laura.sperandio@lnf.infn.it (L.S.); oton.vazquez@universe-cluster.de (O.V.D.); johann.zmeskal@oeaw.ac.at (J.Z.)
2. CENTRO FERMI - Museo Storico della Fisica e Centro Studi e Ricerche 'Enrico Fermi', I-00184 Rome, Italy
3. Institutul National pentru Fizica si Inginerie Nucleara Horia Hulubbei, IFIN-HH, R-077125 Magurele, Romania
4. Stefan-Meyer-Institute for Subatomic Physics, Austrian Academy of Science, A-1090 Vienna, Austria; eberhard.widmann@oeaw.ac.at
5. Dipartimento di Fisica e Astronomia, Universitá di Bologna, I-40127 Bologna, Italy; Sergio.Bertolucci@Lnf.infn.it
6. Institut de Physique UMR CNRS-UR1 6251, Université de Rennes1, F-35042 Rennes, France; sergio.dimatteo@univ-rennes1.fr
7. Institut de Physique, Université de Neuchâtel, CH-2000 Neuenburg, Switzerland; jean-pierre.egger@net2000.ch
8. Laboratori Nazionali del Gran Sasso, INFN, I-67100 Assergi L'Aquila, Italy; Matthias.laubenstein@lngs.infn.it
9. Dipartimento di Fisica, Universitá di Trieste and INFN-Sezione di Trieste, I-34127 Trieste, Italy; edoardo.milotti@ts.infn.it
10. Excellence Cluster Universe, Technische Universität München, D-85748 Garching, Germany
* Correspondence: Catalina.Curceanu@lnf.infn.it (C.C.); hexi.shi@lnf.infn.it (H.S.); Tel.: +39-06-9403-2321 (C.C.)
† This paper is an extended version of our paper published in the XIV International Conference on Topics in Astroparticle and Underground Physics (TAUP2015), 7–11 September 2015, Torino, Italy.
‡ Current address: Laboratori Nazionali di Frascati, INFN, Via E. Fermi 40, I-00044, Frascati, Italy.
§ These authors contributed equally to this work.

Received: 29 April 2017; Accepted: 22 June 2017; Published: 24 June 2017

Abstract: The validity of the Pauli exclusion principle—a building block of Quantum Mechanics—is tested for electrons. The VIP (violation of Pauli exclusion principle) and its follow-up VIP-2 experiments at the Laboratori Nazionali del Gran Sasso search for X-rays from copper atomic transitions that are prohibited by the Pauli exclusion principle. The candidate events—if they exist—originate from the transition of a $2p$ orbit electron to the ground state which is already occupied by two electrons. The present limit on the probability for Pauli exclusion principle violation for electrons set by the VIP experiment is 4.7×10^{-29}. We report a first result from the VIP-2 experiment improving on the VIP limit, which solidifies the final goal of achieving a two orders of magnitude gain in the long run.

Keywords: Pauli exclusion principle; quantum foundations; X-ray spectroscopy; underground experiment; silicon drift detector

1. Introduction

The Pauli exclusion principle (PEP) states that in a system there cannot be two (or more) fermions with all quantum numbers identical, and is a fundamental principle in physics. The validity of the PEP is the basis of the periodic table of elements, electric conductivity in metals, the degeneracy pressure which makes white dwarfs and neutron stars stable, as well as many other phenomena in physics, chemistry, and biology. In quantum mechanics (QM), the states of particles are described in terms of wave functions. For identical particles, with respect to their permutation, the states are necessarily either symmetric for bosons, or antisymmetric for fermions. This "symmetrization postulate" [1] excludes the mixing of different symmetrization groups, and it is at the basis of the PEP. Messiah and Greenberg noted in [2] that this superselection rule "does not appear as a necessary feature of the QM description of nature". In this context, the violation of PEP is equivalent to the violation of spin-statistics [3], and experimentally to the existence of states of particles that follow statistics other than the fermionic or the bosonic ones.

Exhaustive reviews of the experimental and theoretical searches for a small violation of the PEP or the violation of spin-statistics can be found, for example, in [3,4]. We first point out that there is no established model in quantum field theory that can explicitly include small violations of the PEP. Secondly, although many experimental searches present limits for the violation, the parameters that quantify the limits are model/system-dependent and are not generally comparable. Moreover, in order to search for states that are in a mixed symmetry, it is crucial to introduce new states into the system, among which the PEP-violating states may be found. Ramberg and Snow [5] took this argument into account by running a high electric DC current through a copper conductor, and they searched for X-rays from transitions that are PEP-forbidden after electrons are captured by copper atoms. In particular, they searched for PEP-violating transitions from the $2p$ level to the $1s$ level, which is already occupied by two electrons. Due to the shielding effect of the additional electron in the ground level, the energy of such abnormal transitions will deviate from the copper $K\alpha$ X-ray at 8 keV by about 300 eV [6], which are distinguishable in precision spectroscopic measurements. Since the *new* electrons from the current are supposed to have no a-priori established symmetry with the electrons inside the copper atoms, the detection of the energy-shifted X-rays is an explicit indication of the violation of spin-statistics, and thus the violation of the PEP for electrons.

We want to mention that one known system in which the dichotomy of fermions and bosons does not work is in the two-dimensional condensed matter physics through the (fractional) quantum Hall effect [7]. Particles that are neither fermions nor bosons, and that may exist in electronic systems confined to two spatial dimensions have been constructed theoretically and investigated in the laboratory with great consistency with the theories as reviewed in [8]. The physics of this special system is exciting in itself and may provide hints to the searches for the violation of the PEP in other systems.

In Section 2, we will introduce the VIP (violation of Pauli exclusion principle) and VIP-2 experiments at Laboratori Nazionali del Gran Sasso (LNGS), and in Section 3 the first results from the physics run of VIP-2 in 2016, which already improved the best result previously achieved by the VIP experiment with 3 years of data collection. The paper ends with conclusions and future perspectives.

2. VIP-2 Experiment

The first experiment performed in the LNGS-INFN underground laboratory—the VIP experiment—used a similar method as that of Ramberg and Snow, and the same definition of the parameter to represent the probability that the PEP is violated, for a direct comparison of the experimental results. An improvement in sensitivity was achieved firstly by performing the experiment in the low radioactivity laboratory at LNGS, which has the advantage of the excellent shielding against cosmic rays. Secondly, the application of charge-coupled device (CCD) as the X-ray detector with a typical energy resolution of 320 eV at 8 keV increased the precision in the definition of the region of interest to search for anomalous X-rays. The VIP experiment set the limit for the probability of the PEP violation for electrons to be 4.7×10^{-29} [9–11].

By using new X-ray detectors and an active shielding of scintillators, the VIP-2 experiment plans to further improve the sensitivity by two orders of magnitude. The major improvements come from the change of the layout of the copper strip target and of the X-ray detectors, which allow a larger acceptance for the X-ray detection. Secondly, a DC current with 100 amperes is applied instead of 40 amperes, which introduces two times the new electrons into the copper strip. Finally, in addition to the improved passive shielding surrounding the setup to reduce the background generated by the environmental radiations, the use of silicon drift detectors (SDDs) as the X-ray detectors allows the implementation of an active shielding using scintillators, as illustrated in Figure 1a, which removes the background induced by the high-energy charged particles that are not shielded. More details of the detectors and the VIP-2 setup are given in [12–15].

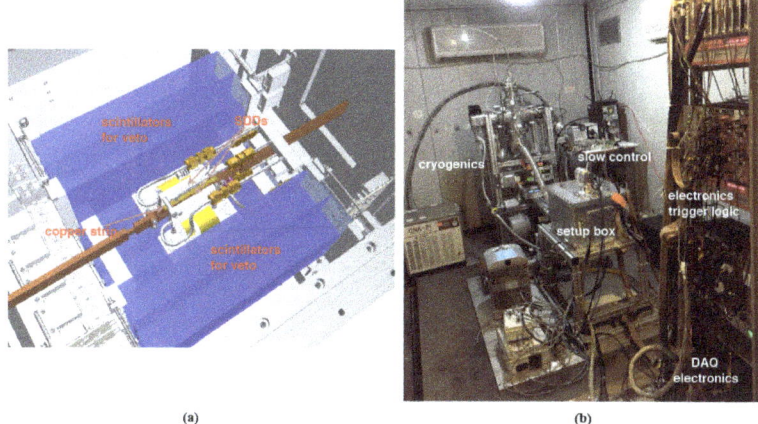

(a) (b)

Figure 1. (a) The design of the core components of the VIolation of Pauli exclusion principle 2 (VIP-2) setup, including the silicon drift detectors (SDDs) as the X-ray detector, the scintillators as active shielding with silicon photomultiplier readout; (b) a picture of the VIP-2 setup in operation at the underground laboratory of Gran Sasso.

The VIP-2 trigger logic was implemented using the Nuclear Instrumentation Module (NIM) standard modules, and it is defined by either an event at any SDD or a coincidence between two layers of the veto detector. A Versa Module Europa (VME) based data acquisition system for the detectors was constructed. It records the energy deposit of the six SDDs from the output of a CAEN 568 spectroscopy amplifier which processes the analog signals of the SDD preamplifier output. The charge to digital signals (QDC) of the 32 scintillator channels, and the timing information of the SDDs with respect to the main trigger are recorded in the data as well. The data acquisition computer transfers data from the VME whenever there is one event ready in the memories of the modules, and clears the registers of the VME when the data transfer is done. During the whole communication process between the computer and the VME controller, the trigger logic is prohibited from receiving further events. The user interface of the Labview-based data-taking program can be remotely accessed and controlled from the computer terminals outside the Gran Sasso laboratory.

The temperatures of the SDDs, the copper conductor, the cooling system, as well as the ambient temperature and vacuum pressure of the setup are monitored by a slow control system. The slow control which can be accessed from remote terminals also controls the DC power supply to switch on and off the current applied to the copper strip. A closed circuit chiller coupled to a cooling pad attached to the copper strips keeps a constant temperature below 25 Celsius of the strips when the DC current up to 100 A is applied. The temperature of the SDDs' holder frame had a change of less than 2 K when the 100 A current was applied to the copper strip. At this level of temperature variation, the effect of change in the energy resolution of the SDDs is negligible.

In November 2015, after having performed exhaustive tests in the laboratory, the VIP-2 setup was transported and mounted in the Gran Sasso underground laboratory, as shown in Figure 1b. After tuning and optimization, from October 2016 we started the first campaign of data taking with the complete detector system. The energy calibration of the SDDs was performed in in-situ, by placing a weak Iron-55 source covered by a 25 μm-thick titanium foil near the detectors. The manganese K-series X-rays from the source partly go through the foil and partly irradiate the foil, generating titanium K-series X-rays. These fluorescence X-rays are detected by the SDDs at an overall rate of about 2 Hz, and provide reference energy peaks to calibrate the digitized SDD signals to energy scale.

3. First VIP-2 Results

During the data collection from October to December 2016, the DC current was typically switched on for one week and off for the next. The energy calibrations for the SDDs were done for each data set corresponding to a period of about one week, and then summed separately over the whole data collection period of over two months, for 100 A current-on data and current-off data sets. The spectra that correspond to 34 days of effective data acquisition with 100 A current on and 28 days with current off are shown in Figure 2, in which the fluorescence lines of titanium and manganese are marked.

The environmental gamma radiations and high-energy charged particles can irradiate the copper conductor or the strip inside the setup, and the normal K-series X-rays from the de-excitation of the copper form the main background near the energy region of interest (ROI in Figure 2) from 7629 eV to 7829 eV, which is defined by the SDD energy resolution (200 eV full width at half maximum, FWHM) at the K_α copper transition (8.04 keV) near the expected value of the PEP violating transition. In order to obtain the number of events violating PEP in the ROI, the current-on spectrum was normalized to 28 days of data collection time, and then a subtraction with the current-off spectrum was performed. The numbers of X-rays in the region of interest were :

- with I = 100 A; N_X = 2222 ± 47 (for 34 days of data collection);
- with I = 0 A; N_X = 2181 ± 47 (28 days of data collection normalized to 34 days);
- numerical subtraction : ΔN_X = 41 ± 66 (normalized to 34 days of data collection time).

Following the similar notations used by Ramberg–Snow and the VIP experiment papers, the number of possible PEP violating events, ΔN_X, is related to the $\beta^2/2$ parameter giving the probability of PEP violation [16] :

$$\Delta N_X \geq \tfrac{1}{2}\beta^2 N_{new} \tfrac{1}{10} N_{int} \times (\text{detection efficiency factor}) \qquad (1)$$
$$= \frac{\beta^2 (\Sigma I \Delta t) D}{e\mu} \tfrac{1}{20} \times (\text{detection efficiency factor}).$$

Furthermore, the number of new electrons that pass through the conductor,

$$N_{new} = (1/e)\Sigma I \Delta t, \qquad (2)$$

is given by the electric charge e of the electron, the intensity I of the applied DC current, and the duration time Δt of the measurement. The minimum number of internal scattering processes between a new electron and the atoms of the copper lattice, N_{int}, is of order D/μ, where D is the length of the copper strip (10 cm), and μ is the mean free path of electrons in copper. We follow the same assumption used in the VIP paper [17], that the capture probability of a new electron by an atom of the copper lattice is greater than 1/10 of the scattering probability.

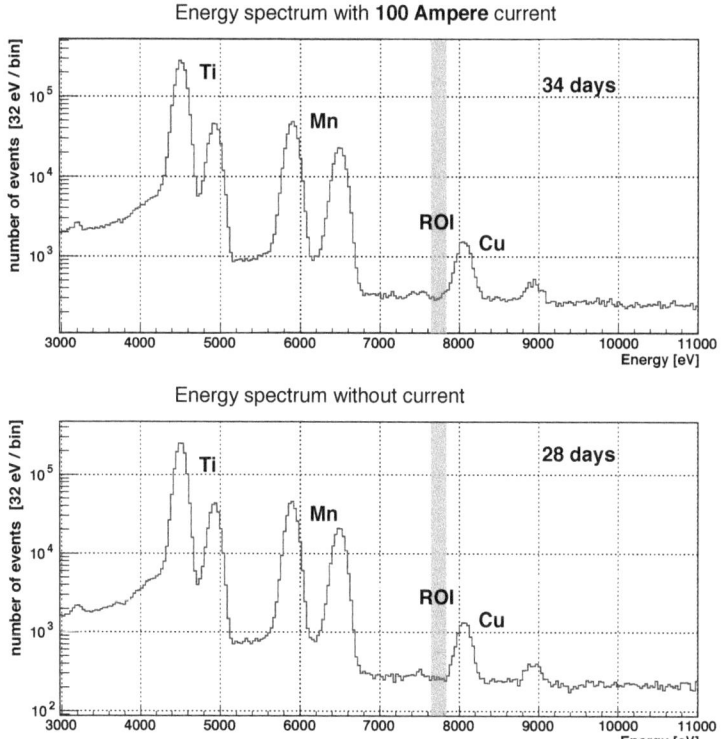

Figure 2. The energy spectra from all the SDDs, for data with and without applied DC current to the copper strip, taken during the physics run in late 2016 at the Laboratori Nazionali del Gran Sasso (LNGS).

The detection efficiency factor is evaluated with a Monte Carlo simulation based on Geant4.10 with realistic detector configuration, taking into account: the transmission rate of a copper K_α X-ray that originates at a random position inside the copper strip and reaches the surface; the geometrical acceptance of the photons coming from the surface of the copper stip arriving at the six SDD detectors; the detection efficiency of a copper K_α X-ray by the 450 µm-thick SDD unit, and the value is determined to be about 1%.

With D = 10 cm, $\mu = 3.9 \times 10^{-6}$ cm, $e = 1.602 \times 10^{-19}$ C, I = 100 A, and normalizing the measurement time with current to 34 days, using the three sigma upper bound of $\Delta N_X = 41 \pm 66$ to give a 99.7% C.L., we get an upper limit for the $\beta^2/2$ parameter:

$$\frac{\beta^2}{2} \leq \frac{3 \times 66}{4.7 \times 10^{30}} = 4.2 \times 10^{-29}. \tag{3}$$

4. Conclusions and Future Perspectives

The first VIP-2 physics run from two months of data collection already gave a better limit than the VIP result obtained from three years of running.

In Figure 3, we show all the past experimental results of the PEP violation tests for electrons with a copper conductor, together with this work. The new result shows that in the planned data collection time of 3 to 4 years, the VIP-2 experiment can either set a new upper limit for the probability that the PEP is violated at the level of 10^{-31}, improving the VIP experiment result by two orders of magnitude, or find the PEP violation, which would have profound implications in science and philosophy.

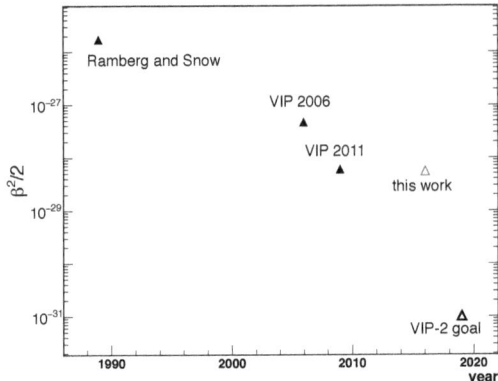

Figure 3. All the past results from Pauli exclusion principle (PEP) violation tests for electrons with a copper conductor, together with the result from this work and the anticipated goal of the VIP-2 experiment. Note that the result of this work comes from two months of data collection, and it is already compatible with the VIP result from three years of operation.

We conclude with the words of Lev Okun from his 1987 paper [18]: *"The special place enjoyed by the Pauli principle in modern theoretical physics does not mean that this principle does not require further and exhaustive experimental tests. On the contrary, it is specifically the fundamental nature of the Pauli principle which would make such tests, over the entire periodic table, of special interest"*.

Acknowledgments: We thank H. Schneider, L. Stohwasser, and D. Stückler from Stefan-Meyer-Institut for their fundamental contribution in designing and building the VIP-2 setup. We acknowledge the very important assistance of the INFN-LNGS laboratory staff during all phases of preparation, installation and data taking. We thank the Austrian Science Foundation (FWF) which supports the VIP-2 project with the grant P25529-N20. We acknowledge the support from the EU COST Action CA15220, and from Centro Fermi ("Problemi aperti nella meccania quantistica" project). Furthermore, this paper was made possible through the support of a grant from the John Templeton Foundation (ID 58158). The opinions expressed in this publication are those of the authors and do not necessarily reflect the views of the John Templeton Foundation.

Author Contributions: Sergio Bertolucci, Catalina Curceanu, Jean-Pierre Egger, Carlo Guaraldo, Edoardo Milotti, Eberhard Widmann, Johann Zmeskal conceived and designed the experiment; Massimiliano Bazzi, Carolina Berucci, Alberto Clozza, Mihail Iliescu, Andreas Pichler, Hexi Shi, Florin Sirghi, Johann Zmeskal prepared the setup; Mihail Iliescu and Hexi Shi prepared the readout and data taking system; Andreas Pichler, Hexi Shi, Johann Zmeskal performed the detctor tests; Mario Bragadireanu, Alberto Clozza, Catalina Curceanu, Mihail Iliescu, Matthias Laubenstein, Johann Marton, Marco Miliucci, Andreas Pichler, Dorel Pietreanu, Kristian Piscicchia, Alessandro Scordo, Hexi Shi, Florin Sirghi, Johann Zmeskal contributed to the installation and data taking; Hexi Shi, Andreas Pichler, Michael Cargnelli, Luca De Paolis performed the data analysis; Sergio Bartalucci, Diana Laura Sirghi, Laura Sperandio, Oton Vazquez Doce provided details of the VIP analysis; Sergio Di Matteo provided theoretical support for the data analyses; Catalina Curceanu and Hexi Shi wrote the paper.

Conflicts of Interest: The authors declare no conflict of interest.

Abbreviations

PEP	Pauli Exclusion Principle
VIP(-2) experiment	VIolation of Pauli principle (-2) experiment
CCD	Carge Coupled Device
SDD	Silicon Drift Detector
NIM	Nuclear Instrumentation Module
VME	Versa Module Europa
QDC	Charge-to-Digital Converter
LNGS	Laboratori Nazionali del Gran Sasso
FWHM	Full Width Half Maximum
ROI	Region of Interest

References

1. Messiah, A.M.L. *Quantum Mechanics, Volume II*; North-Holland: Amsterdam, The Netherlands, 1962; p. 595.
2. Messiah, A.M.L.; Greenberg, O.W. Symmetrization Postulate and Its Experimental Foundation. *Phys. Rev.* **1964**, *136*, B248.
3. Greenberg, O.W. Theories of Violation of Statistics. *AIP Conf. Proc.* **2000**, *545*, 113, doi: 10.1063/1.1337721.
4. Elliott, S.R.; LaRoque, B.H.; Gehman, V.M.; Kidd, M.F.; Chen, M. An Improved Limit on Pauli-Exclusion-Principle Forbidden Atomic Transitions. *Found. Phys.* **2012**, *42*, 1015–1030.
5. Ramberg, E.; Snow, G.A. Experimental Limit on a Small Violation of the Pauli Principle. *Phys. Lett. B* **1990**, *238*, 438–441.
6. Curceanu, C.; De Paolis, L.; Di Matteo, S.; Di Matteo, H.; Sperandio, S. Evaluation of the X-ray Transition Energies for the Pauli-Principle-Violating Atomic Transitions in Several Elements by Using the Dirac-Fock Method. Available online: http://www.lnf.infn.it/sis/preprint/detail.php?id=5330 (accessed on 23 June 2017).
7. Prange, R.; Girvin, S.M. *The Quantum Hall Effect*; Springer: New York, NY, USA, 1990.
8. Stern, A. Anyons and the quantum Hall effect—A pedagogical review. *Ann. Phys.* **2008**, *323*, 204–249.
9. Curceanu, C.; Bartalucci, S.; Bertolucci, S.; Bragadireanu, M.; Cargnelli, M.; Di Matteo, S.; Egger, J.-P.; Guaraldo, C.; Iliescu, M.; Ishiwatari, T.; et al. Experiemntal tests of quantum mechanics—Pauli exclusion principle violation (the VIP experiment) and future perspective. *J. Phys. Conf. Ser.* **2011**, *306*, 012036, doi:10.1088/1742-6596/306/1/012036.
10. Bartalucci, S.; Bertolucci, S.; Bragadireanu, M.; Cargnelli, M.; Curceanu, C.; Di Matteo, S.; Egger, J.-P.; Guaraldo, C.; Iliescu, M.; Ishiwatari, T.; et al. The VIP experimental limit on the Pauli exclusion principle violation by electrons. *Found. Phys.* **2009**, *40*, 765–775.
11. Sperandio, L. New Experimental Limit on the Pauli Exclusion Principle Violation by Electrons From the VIP Experiment. Ph.D. Thesis, Tor Vergata University, Rome, Italy, 2008.
12. Shi, H.; Bartalucci, S.; Bertolucci, S.; Berucci, C.; Bragadireanu, A.M.; Cargnelli, M.; Clozza, A.; Curceanu, C.; De Paolis, L.; Di Matteo, S.; et al. Searches for the Violation of Pauli Exclusion Principle at LNGS in VIP(-2) experiment. *J. Phys. Conf. Ser.* **2016**, *718*, 042055, doi:10.1088/1742-6596/718/4/042055.
13. Pichler, A.; Bartalucci, S.; Bazzi, M.; Bertolucci, S.; Berucci, C.; Bragadireanu, M.; Cargnelli, M.; Clozza, A.; Curceanu, C.; De Paolis, L.; et al. Application of photon detectors in the VIP-2 experiment to test the Pauli Exclusion Principle. *J. Phys. Conf. Ser.* **2016**, *718*, 052030, doi:10.1088/1742-6596/718/5/052030.
14. Shi, H.; Bartalucci, S.; Bertolucci, S.; Berucci, C.; Bragadireanu, A.M.; Cargnelli, M.; Clozza, A.; Curceanu, C.; De Paolis, L.; Di Matteo, S.; et al. Testing the Pauli Exclusion Principle for electronics at LNGS. *Phys. Procedia* **2015**, *62*, 522–559.
15. Marton, J.; Bartalucci, S.; Bertolucci, S.; Berucci, C.; Bragadireanu, M.; Cargnelli, M.; Curceanu, C.; Di Matteo, S.; Egger, J.-P.; Guaraldo, C.; et al. Testing the Pauli Exclusion Principle for Electrons. *J. Phys. Conf. Ser.* **2013**, *447*, 012060, doi:10.1088/1742-6596/335/1/012060.
16. Greenberg, O.W.; Mohapatra, R.N. Local Quantum Field Theory of Possible Violation of the Pauli Principle. *Phys. Lett.* **1987**, *59*, 2507.

17. VIP Collaboration; Bartalucci, S.; Bertolucci, S.; Bragadireanu, M.; Cargnelli, M.; Catitti, M.; Curceanu, C.; Di Matteo, S.; Egger, J.-P.; Guaraldo, C.; et al. New experimental limit on the Pauli exclusion principle violation by electrons. *Phys. Lett. B* **2006**, *641*, 18–22.
18. Okun, L. Possible violation of the Pauli principle in atoms. *JETP Lett.* **1987**, *46*, 529–532.

© 2017 by the authors. Licensee MDPI, Basel, Switzerland. This article is an open access article distributed under the terms and conditions of the Creative Commons Attribution (CC BY) license (http://creativecommons.org/licenses/by/4.0/).

Article

CSL Collapse Model Mapped with the Spontaneous Radiation

Kristian Piscicchia [1,2,*], Angelo Bassi [3,4], Catalina Curceanu [2,1], Raffaele Del Grande [2], Sandro Donadi [5], Beatrix C. Hiesmayr [6] and Andreas Pichler [7]

1. CENTRO FERMI—Museo Storico della Fisica e Centro Studi e Ricerche "Enrico Fermi", 00184 Rome, Italy
2. Istituto Nazionale di Fisica Nucleare (INFN), Laboratori Nazionali di Frascati, 00044 Frascati, Italy
3. Department of Physics, University of Trieste, 34151 Miramare-Trieste, Italy
4. Istituto Nazionale di Fisica Nucleare, Sezione di Trieste, Via Valerio 2, 34127 Trieste, Italy
5. Institute of Theoretical Physics, Ulm University, Albert-Einstein-Allee 11 D, 89069 Ulm, Germany
6. Faculty of Physics, University of Vienna, Boltzmanngasse 5, 1090 Vienna, Austria
7. Stefan-Meyer-Institut für Subatomare Physik, 1090 Vienna, Austria
* Correspondence: kristian.piscicchia@lnf.infn.it; Tel.: +39-06-9403-2654

Received: 30 April 2017; Accepted: 25 June 2017; Published: 29 June 2017

Abstract: In this paper, new upper limits on the parameters of the Continuous Spontaneous Localization (CSL) collapse model are extracted. To this end, the X-ray emission data collected by the IGEX collaboration are analyzed and compared with the spectrum of the spontaneous photon emission process predicted by collapse models. This study allows the obtainment of the most stringent limits within a relevant range of the CSL model parameters, with respect to any other method. The collapse rate λ and the correlation length r_C are mapped, thus allowing the exclusion of a broad range of the parameter space.

Keywords: quantum mechanics; the measurement problem; collapse models; X-rays

1. The CSL Collapse Model

Collapse models are phenomenological models introduced to solve the measurement problem of quantum mechanics and explain the quantum-to-classical transition [1–6]. According to these models, the linear and unitary evolution given by the Schrödinger equation is modified by adding a non-linear term and the interaction with a stochastic noise field. These modifications have two very important consequences: (i) they lead to the collapse of the wave function of the system in space (localization mechanism) and (ii) the collapse effects get amplified with the mass of the system (amplification mechanism). The combination of these two properties guarantees that macroscopic objects always have well defined positions, explaining why we do not observe quantum behaviour at the macroscopic level. On the other hand, for microscopic systems, the effect of the non-linear interaction with the noise field is very small and their dynamics is dominated by the Schrödinger evolution. Due to the presence of the non-linear interaction with the noise field, collapse models predict slight deviations from the standard quantum mechanics predictions [7].

The analysis discussed in this work sets limits on the characteristic parameters of the Continuous Spontaneous Localization (CSL) model [8–10], which is one of the most relevant and well-studied collapse models in the literature. In the CSL model, the state vector evolution is described by a modified Schrödinger equation which contains, besides the standard Hamiltonian, non-linear and stochastic terms, characterized by the interaction with a continuous set of independent noises $w(\mathbf{x}, t)$ (one for each point of the space, which is why this set is often referred to as "noise field") having zero average and white correlation in time, i.e., $E[w(\mathbf{x}, t)] = 0$ and $E[w(\mathbf{x}, t)w(\mathbf{y}, s)] = \delta(\mathbf{x} - \mathbf{y})\delta(t - s)$ where $E[...]$ denotes the average over the noises. Two phenomenological parameters (λ and r_C) are introduced in the model. The parameter λ has the dimensions of a rate and sets the strength of the collapse, while r_C is a correlation length which determines the spatial resolution of the collapse: for superposition with size much smaller than r_C, the collapse is much weaker compared to the case when the superposition has a delocalization much larger than r_C. The originally proposed values for λ and r_C are [8] $\lambda = 10^{-16}$ s^{-1}, $r_C = 10^{-7}$ m. Higher values for λ were however put forward [11], up to $\lambda = 10^{-8\pm2}$ s^{-1}.

The interaction with the noise field causes an extra emission of electromagnetic radiation for charged particles [7], which is not predicted by standard quantum mechanics. Such an effect is known as *spontaneous radiation* emission. We show that the measurement of the radiation allows for a mapping of the two relevant parameters λ and r_C (see also Ref. [12]) into a two-dimensional parameter space, i.e., we can present an exclusion plot. This gives a considerable reduction of the possible values in the parameter space of collapse models.

2. The Collapse Rate Parameter λ

The energy distribution of the spontaneous radiation, emitted as a consequence of the interaction of free electrons with the collapsing stochastic field, was first calculated by Fu [7] and later on studied in more detail in [13–15], in the framework of the non-relativistic CSL model. If the stochastic field is assumed to be a white noise, coupled to the particle mass density (mass proportional CSL model), the spontaneous emission rate is given by:

$$\frac{d\Gamma(E)}{dE} = \frac{e^2 \lambda}{4\pi^2 r_C^2 m_N^2 E}, \qquad (1)$$

where e is the charge of the proton, m_N represents the nucleon mass and E is the energy of the emitted photon. In the non-mass proportional case, the rate takes the expression:

$$\frac{d\Gamma(E)}{dE} = \frac{e^2 \lambda}{4\pi^2 r_C^2 m_e^2 E}, \qquad (2)$$

with m_e the electron mass.

Using the measured radiation emitted in an isolated slab of Germanium [16] corresponding to an energy of 11 keV, and comparing it with the predicted rate in Equations (1) and (2), Fu extracted the following upper limits on λ for the two cases:

$$\lambda \leq 2.20 \cdot 10^{-10} \text{ s}^{-1} \quad \text{mass prop.,} \qquad (3)$$

$$\lambda \leq 0.55 \cdot 10^{-16} \text{ s}^{-1} \quad \text{non-mass prop.,} \qquad (4)$$

assuming that the correlation length value is $r_C = 10^{-7}$ m. In his estimate, Fu considered the contribution to the spontaneous X-ray emission of the four valence electrons in the Germanium atoms. Such electrons can be considered as *quasi-free*, since their binding energy (of the order of ~10 eV) is much less than the emitted photons' energy. In Ref. [11], the author argues that an erroneous value for the fine structure constant is used in Ref. [7]. This correction is taken into account in the analysis described in Section 3. Further, the preliminary TWIN data set [16] used by Fu to estimate the upper limit on λ turned out to be underestimated by a factor of about 50 at 10 keV.

A new analysis was performed in Ref. [17]. Based on the improved data presented in Ref. [18], the limits corresponding to the footnote [7] in Ref. [17], for the cases of mass proportional and non-mass proportional CSL models, were:

$$\lambda \leq 8 \cdot 10^{-10} \text{ s}^{-1} \quad \text{mass prop.,} \qquad (5)$$

$$\lambda \leq 2 \cdot 10^{-16} \text{ s}^{-1} \quad \text{non-mass prop..} \qquad (6)$$

3. A New Limit on λ

In this work, the X-ray emission spectrum measured by the IGEX experiment [19] is analysed in order to set a more stringent limit on the collapse rate parameter λ. IGEX is a low-background experiment based on low-activity Germanium detectors, originally dedicated to the neutrinoless double beta decay ($\beta\beta 0\nu$) research. The published data set [20] refers to 80 kg day exposure, and was conceived to search for a dark matter WIMPs signal that originated from elastic scattering, producing Ge nuclear recoil.

For the measurement in Ref. [20], one of the IGEX detectors of 2.2 kg (active mass of about 2 kg) was used. The detector, the cryostat and the shielding were fabricated following ultra-low background techniques, in order to minimize the radionuclides emission, which represents the main background source in the measured X-ray spectrum (shown in Figure 1 as a black distribution). Moreover, a cosmic muon veto covered the top and the sides of the shield. The experiment had an overburden of 2450 m.w.e., reducing the muon flux to the value of $2 \cdot 10^{-7}$ cm^{-2} s^{-1}. The two main sources of inefficiency are represented by the muon veto anti-coincidence and the pulse shape analysis. The probability of rejecting non-coincident events with the muon veto was found to be less than 0.01. The loss of efficiency introduced by the pulse shape analysis resulted to be negligible for events above 4 keV.

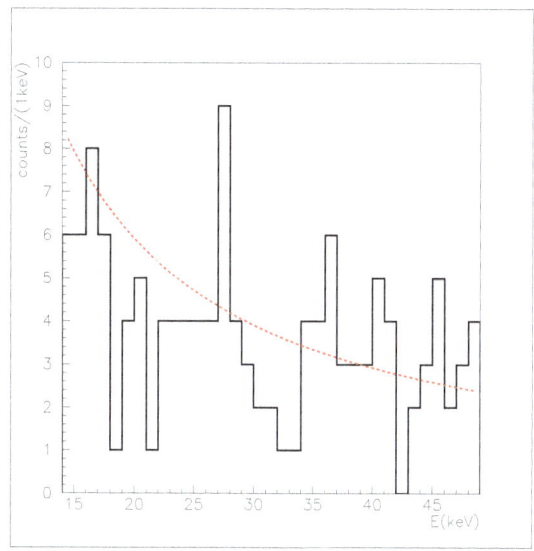

Figure 1. Fit of the X-ray emission spectrum measured by the IGEX experiment [19,20], using the theoretical fit function Equation (7). The black line corresponds to the experimental distribution; the red dashed line represents the fit. See the text for more details.

The X-ray spectrum (Figure 1) ranges in the interval (4.5 ÷ 48.5) keV, which is compatible with the non-relativistic assumption for electrons, used to derive Equations (1) and (2).

3.1. The Data Analysis: Procedure and Results

The X-ray experimental spectrum published in [20] is compared with the predicted rate Equations (1) and (2), by taking into account the spontaneous emission of the 30 outermost electrons of the Ge atoms considered as *quasi-free*. We restricted our analysis to the energy range $\Delta E = (14.5 \div 48.5)$ keV of the experimental spectrum [20], for which the binding energy of the lower lying electronic orbit (the 2s orbit) is still one order of magnitude lower than 14.5 keV, justifying the *quasi-free* hypothesis.

The X-ray spectrum is fitted in the interval ΔE by minimising a χ^2 function. The expected number of counts for each bin of 1 keV is assumed to be described by the theoretical prediction Equations (1) and (2):

$$\frac{d\Gamma(E)}{dE} = \frac{\alpha(\lambda)}{E}. \tag{7}$$

The χ^2 minimisation presumes that the bin contents y_i (number of counts in the energy bin E_i) follow Gaussian distributions. Strictly speaking, the y_is are Poissonian stochastic variables; nevertheless, the approximation is reasonable for $y_i \geq 5$; this constraint is then used for the fit. The result of the fit is shown in Figure 1 (red dashed line). For the free parameter of the fit, the minimization gives the value $\alpha(\lambda) = 115 \pm 17$, corresponding to a reduced $\chi^2/(n.d.f. - n.p.) = 0.9$. $n.d.f.$ represents the number of degrees of freedom, $n.p.$ is the number of free parameters of the fit. $\alpha(\lambda)$ is also considered to follow a Gaussian distribution with a good approximation. An upper limit can then be set as $\alpha(\lambda) \leq 143$ with a probability of 95%. Correspondingly, an upper limit on the parameter λ can be extracted using Equations (1) and (2):

$$\frac{d\Gamma(E)}{dE} = c\frac{e^2 \lambda}{4\pi^2 r_C^2 m^2 E} \leq \frac{143}{E}, \tag{8}$$

where the factor c is given by:

$$c = \left(8.29 \times 10^{24}\,\frac{\text{atoms}}{\text{kg}}\right) \cdot (80\,\text{kg day}) \cdot \left(8.64 \times 10^4\,\frac{\text{n. of seconds}}{\text{day}}\right) \cdot (30), \tag{9}$$

the first bracket accounts for the particle density of Germanium, the second represents the amount of emitting material expressed in kg day, the third term is the number of seconds in one day and 30 represents the number of spontaneously emitting electrons for each Germanium atom. Applying Equation (8), the following upper limits for the reduction rate parameter are obtained, with a probability of 95%:

$$\lambda \leq 8.1 \cdot 10^{-12}\,\text{s}^{-1}\quad \text{mass prop.,} \tag{10}$$

$$\lambda \leq 2.4 \cdot 10^{-18}\,\text{s}^{-1}\quad \text{non-mass prop..} \tag{11}$$

In order to obtain the limits in Equations (10) and (11), two implicit assumptions are made on the experimental input [20]. First, the measured spectrum is assumed to be background free, that is to say that the upper limit on λ corresponds to the case in which all the measured X-ray emission would be produced by spontaneous emission processes. This ansatz is conservative, and is imposed by our ignorance regarding the contribution from known emission processes to the measured rate. The second assumption, which is consistent with the analysis presented in Ref. [20], is that the detector efficiency, in the range ΔE, is one, and that the un-efficiencies which are introduced by the muon veto anticoincidence and the pulse shape analysis, performed to extract the experimental spectrum in Ref. [20], are very small for events above 4 keV.

Having in mind these assumptions, the measured X-ray counts in the range ΔE can be re-analysed in terms of their low-events Poissonian statistics. The number of counts y_is in each energy bin E_i can be considered as independent stochastic variables following the distributions:

$$G(y_i|P, \Lambda_i) = \frac{\Lambda_i^{y_i} e^{-\Lambda_i}}{y_i!}, \tag{12}$$

where P denotes the Poisson distribution function. The expected numbers of counts per bin Λ_i are indicated with capital letters, not to be confused with the spontaneous collapse rate λ. Let us define:

$$y = \sum_{i=1}^{n} y_i \quad , \quad \Lambda = \sum_{i=1}^{n} \Lambda_i \tag{13}$$

where n is the total number of 1 keV bins in the range ΔE, y and Λ are the total number of counts and the expected number of total counts, respectively. Here, y is distributed according to a Poissonian of parameter $\Lambda(\lambda)$, where the dependence on the collapse rate parameter, which follows the theoretical input, was explicitly indicated.

According to the Bayes theorem, the probability distribution function of $\Lambda(\lambda)$, given the measured y, assuming a uniform prior, is given by:

$$G'(\Lambda|G(y|P,\Lambda)) \propto \Lambda(\lambda)^y e^{-\Lambda(\lambda)}, \qquad (14)$$

which means that $G'(\lambda)$ is proportional to a gamma probability distribution. Due to the assumption that the background is negligible, $\Lambda(\lambda)$ also represents the expected number of total signal counts y_s, where y_s is a Poissonian variable. Thus, according to Equation (8):

$$\Lambda(\lambda) = y_s + 1 = \sum_{i=1}^{n} c \frac{e^2 \lambda}{4\pi^2 r_C^2 m^2 E_i} + 1 = \sum_{i=1}^{n} \frac{\alpha(\lambda)}{E_i} + 1. \qquad (15)$$

Substituting Equation (15) for Equation (14), the probability distribution function for the collapse rate parameter can then be obtained:

$$G'(\lambda|G(y|P,\Lambda)) \propto \left(\sum_{i=1}^{n} \frac{\alpha(\lambda)}{E_i} + 1\right)^y e^{-\left(\sum_{i=1}^{n} \frac{\alpha(\lambda)}{E_i} + 1\right)}, \qquad (16)$$

where the measured total number of counts is $y = 130$. Calculating the cumulative distribution function:

$$\int_0^{\lambda_0} G'(\lambda|G(y|P,\Lambda))\, d\lambda, \qquad (17)$$

the following upper limits can be obtained on the collapse rate parameter, setting r_C to the value 10^{-7} m, corresponding to a probability level of 95%

$$\lambda \leq 6.8 \cdot 10^{-12}\, s^{-1} \quad \text{mass prop.,} \qquad (18)$$

$$\lambda \leq 2.0 \cdot 10^{-18}\, s^{-1} \quad \text{non-mass prop..} \qquad (19)$$

4. Mapping CSL Parameters Space

In Figure 2, we present the mapping of the $\lambda - r_C$ parameters of the CSL model, where the originally proposed theoretical values are shown, together with our results. The region excluded by theoretical arguments is represented in gray. This theoretical bound (see Ref. [21]) is obtained by requiring that a single-layered graphene disk of radius ~ 0.01 mm is localized within ~ 10 ms (these are the minimum resolution and perception time of the human eye, respectively).

The region excluded by this analysis is shown in cyan for the non-mass proportional case and in magenta for the mass proportional case. Figure 2 can be compared with Figure 2 in Ref. [22], where the mapping is obtained using other measurements. It is interesting to note that, for a collapse induced by a white noise, the allowed parameter space is confined to a drastically reduced region.

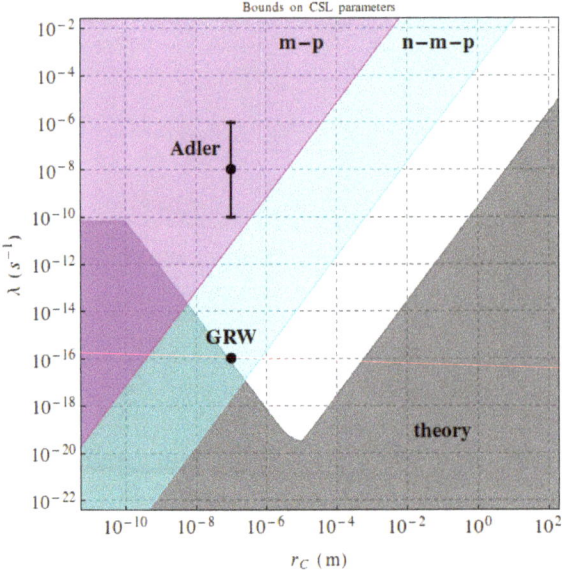

Figure 2. Mapping of the $\lambda - r_C$ Continuous Spontaneous Localization (CSL) parameters: the originally proposed theoretical values (GRW, Adler) are shown as black points; the region excluded by theory (theory) is represented in gray. The excluded region according to our analysis is shown in cyan for the non-mass proportional case (n-m-p) and in magenta for the mass proportional case (m-p).

5. Conclusions and Perspectives

We have presented an analysis of the spontaneous radiation emitted and measured by the IGEX Germanium detector, to obtain a mapping of the CSL collapse model parameters. The results shown in Figure 2 can be summarized as follows:

- the non-mass proportional model for a white noise scenario can be excluded by our analysis,
- the higher value on λ [11] can be excluded for a white noise scenario, in both mass proportional and non-mass proportional models,
- the measurement of the spontaneous radiation allows the obtainment of the most stringent limits on the CSL collapse model parameters, with respect to any other method, in a broad range of the parameter space (see also Ref. [22] for comparison).

We are presently exploring the possibility of performing a new measurement that will allow an improvement of at least one order of magnitude on the collapse rate parameter λ, exploring new regions of CSL mapping.

Acknowledgments: We acknowledge the support of the CENTRO FERMI—Museo Storico della Fisica e Centro Studi e Ricerche "Enrico Fermi" (*Open Problems in Quantum Mechanics project*), the support from the EU COST Action CA 15220 is gratefully acknowledged. Furthermore, this paper was made possible through the support of a grant from the Foundational Questions Institute, FQXi "Events" as we see them: experimental test of the collapse models as a solution of the measurement problem) and a grant from the John Templeton Foundation (ID 58158). The opinions expressed in this publication are those of the authors and do not necessarily reflect the views of the John Templeton Foundation. Beatrix C. Hiesmayr acknowledges gratefully the support by the Autrian Science Found (FWF-P26783). S. Donadi acknowledges the support by Trieste University and Istituto Nazionale di Fisica Nucleare (INFN).

Author Contributions: Kristian Piscicchia, Catalina Curceanu, Raffaele Del Grande and Andreas Pichler analyzed the data; Angelo Bassi, Sandro Donadi and Beatrix C. Hiesmayr gave the theoretical support for data analyses and interpretation; Kristian Piscicchia and Catalina Curceanu wrote the paper. All authors have read and approved the final manuscript.

Conflicts of Interest: The authors declare no conflict of interest.

References

1. Bassi, A.; Ghirardi, G.C. Dynamical reduction models. *Phys. Rep.* **2003**, *379*, 257–426.
2. Pearle, P. *Collapse Models Open Systems and Measurements in Relativistic Quantum Field Theory*; Lecture Notes in Physics; Breuer, H.-P., Petruccione, F., Eds.; Springer: Berlin/Heidelberg, Germany, 1999; Volume 526.
3. Diósi, L. Models for Universal Reduction of Macroscopic Quantum Fluctuations. *Phys. Rev. A* **1989**, *40*, 1165.
4. Bassi, A. Collapse Models: Analysis of the Free Particle Dynamics. Available online: https://arxiv.org/abs/quant-ph/0410222.pdf (accessed on 25 March 2009).
5. Adler, S.L. *Quantum Theory as an Emergent Phenomenon*; Cambridge University Press: Cambridge, UK, 2004; Chapter 6.
6. Weber, T. Quantum mechanics with spontaneous localization revisited. *Il Nuovo Cimento B* **1991**, *106*, 1111–1124.
7. Fu, Q. Spontaneous radiation of free electrons in a nonrelativistic collapse model. *Phys. Rev. A* **1997**, *56*, 1806.
8. Ghirardi, G.; Rimini, A.; Weber, T. Unified dynamics for microscopic and macroscopic systems. *Phys. Rev. D* **1986**, *34*, 470.
9. Pearle, P. Combining stochastic dynamical state-vector reduction with spontaneous localization. *Phys. Rev. A* **1989**, *39*, 2277.
10. Ghirardi, G.C.; Pearle, P.; Rimini, A. Markov processes in Hilbert space and continuous spontaneous localization of systems of identical particles. *Phys. Rev. A* **1990**, *42*, 78.
11. Adler, S.L. Lower and Upper Bounds on CSL Parameters from Latent Image Formation and IGM Heating. *J. Phys. A* **2007**, *40*, 2935–2958.
12. Curceanu, C.; Hiesmayr, B.C.; Piscicchia, K. X-rays help to unfuzzy the concept of measurement. *J. Adv. Phys.* **2015**, *4*, 263–266.
13. Adler, S.L.; Ramazanoglu, F.M. Photon emission rate from atomic systems in the CSL model. *J. Phys. A* **2007**, *40*, 13395–13406.
14. Adler, S.L.; Bassi, A.; Donadi, S. On spontaneous photon emission in collapse models. *J. Phys. A* **2013**, *46*, 245304.
15. Donadi, S.; Bassi, A.; Deckert, D.-A. On the spontaneous emission of electromagnetic radiation in the CSL model. *Ann. Phys.* **2014**, *340*, 70–86.
16. Miley, H.S.; Avignone, F.T.; Brodzinski, R.L., III; Collar, J.I.; Reeves, J.H. Suggestive evidence for the two neutrino double beta decay of Ge-76. *Phys. Rev. Lett.* **1990**, *65*, 3092.
17. Laloë, F.; Mullin, W.J.; Pearle, P. Heating of trapped ultracold atoms by collapse dynamics. *Phys. Rev. A* **2014**, *90*, 52119.
18. Collett, B.; Pearle, P.; Avignone, F.; Nussinov, S. Constraint on collapse models by limit on spontaneous X-ray emission in Ge. *Found. Phys.* **1995**, *25*, 1399–1412.
19. Aalseth, C.E.; Avignone, F.T., III; Brodzinski, R.L.; Collar, J.I.; Garcia, E.; González, D.; Hasenbalg, F.; Hensley, W.K.; Kirpichnikov, I.V.; Klimenko, A.A.; et al. Neutrinoless double-beta decay of Ge-76: First results from the International Germanium Experiment (IGEX) with six isotopically enriched detectors. *IGEX Collab. Phys. Rev. C* **1999**, *59*, 2108.
20. Morales, A.; Aalseth, C.E.; Avignone, F.T.; Brodzinski, R.L., III; Cebrian, S.; Garcia, E.; Irastorza, I.G.; Kirpichnikov, I.V.; Klimenko, A.A.; Miley, H.S.; et al. Improved constraints on WIMPs from the international Germanium experiment IGEX. *IGEX Collab. Phys. Lett. B* **2002**, *532*, 8–14.
21. Toroš, M.; Gasbarri, G.; Bassi, A. Bounds on Collapse Models from Matter-Wave Interferometry. Available online: https://arxiv.org/pdf/1601.03672.pdf (accessed on 31 May 2017).
22. Carlesso, M.; Bassi, A.; Falferi, P.; Vinante, A. Experimental bounds on collapse models from gravitational wave detectors. *Phys. Rev. D* **2016**, *94*, 124036.

© 2017 by the authors. Licensee MDPI, Basel, Switzerland. This article is an open access article distributed under the terms and conditions of the Creative Commons Attribution (CC BY) license (http://creativecommons.org/licenses/by/4.0/).

Article
Quantum Information: What Is It All About?

Robert B. Griffiths

Department of Physics, Carnegie Mellon University, Pittsburgh, PA 15213, USA; rgrif@cmu.edu

Received: 23 October 2017; Accepted: 22 November 2017; Published: 29 November 2017

Abstract: This paper answers Bell's question: What does quantum information refer to? It is about quantum properties represented by subspaces of the quantum Hilbert space, or their projectors, to which standard (Kolmogorov) probabilities can be assigned by using a projective decomposition of the identity (PDI or framework) as a quantum sample space. The single framework rule of consistent histories prevents paradoxes or contradictions. When only one framework is employed, classical (Shannon) information theory can be imported unchanged into the quantum domain. A particular case is the macroscopic world of classical physics whose quantum description needs only a single quasiclassical framework. Nontrivial issues unique to quantum information, those with no classical analog, arise when aspects of two or more incompatible frameworks are compared.

Keywords: Shannon information; quantum information; quantum measurements; consistent histories; incompatible frameworks; single framework rule

1. Introduction

A serious study of the relationship between quantum information and quantum foundations needs to address Bell's rather disparaging question, "Quantum information ... about what?" found in the third section of his polemic against the role of measurement in standard (textbook) quantum mechanics [1]. The basic issue has to do with quantum ontology, "beables" in Bell's language. I believe a satisfactory answer to Bell's question is available, indeed was already available (in a somewhat preliminary form) at the time he was writing. (If he was aware of it, Bell did not mention it in any of his publications.) Further developments have occurred since, and I have found this approach to be of some value in addressing some of the foundational issues which have come up during my own research on quantum information. So I hope the remarks which follow may assist others who find the textbook (both quantum and quantum information) presentations confusing or inadequate, and are looking for something better.

Here is a summary of the remainder of this paper. The discussion begins in Section 2 by asking Bell's question about *classical* (Shannon) information: what is *it* all about? That theory works very well in the world of macroscopic objects and properties. Hence if classical physics is fundamentally quantum mechanical, as I and many others believe, and if Shannon's approach is, as a consequence, quantum information theory applied to the domain of macroscopic phenomena, we are already half way to answering Bell's question. The other half requires extending Shannon's ideas into the microscopic domain where classical physics fails and quantum theory is essential. This is possible, Section 3, using a consistent formulation of standard (Kolmogorov) probability theory applied to the quantum domain. Current quantum textbooks do not provide this, though their discussion of measurements, Section 4, gives some useful hints. The basic approach in Section 3 follows von Neumann: Hilbert subspaces, or their projectors, represent quantum properties, and a projective decompositions of the identity (PDI) provides a quantum sample space. By not following Birkhoff and von Neumann, but instead using a simplified form of quantum logic, Section 5, one has, in the "single framework rule" of consistent histories, a means of escaping the well-known paradoxes that inhabit the quantum foundations swamp. Section 6 argues that when quantum theory is equipped with (standard!) probabilities, quantum information theory is identical to Shannon's theory in the domain of macroscopic (classical) physics, as one might have expected, since only a single quasiclassical quantum framework (PDI) is needed for a quantum mechanical description. However, classical information theory also applies, unchanged, in the microscopic quantum domain if only a *single framework* is needed. Section 7 provides a perspective on the highly nontrivial problems that are unique to quantum information and lack any simple classical analog: they arise when one wants to *compare* (not combine!) two or more *incompatible* frameworks applied to a particular situation.

2. Classical Information Theory

Let us start by asking Bell's question about *classical* information theory, the discipline which Shannon started. What is it all about? If you open any book on the subject, you will soon learn that it is all about probabilities, and information measures expressed in terms of probabilities. So we need to ask: probabilities of what? Standard (Kolmogorov) probability theory, the sort employed in classical information theory, begins with a *sample space* of mutually exclusive possibilities, like the six faces of a die. Next an *event algebra* made up of subsets of elements from the sample space, to which one assigns *probabilities*, nonnegative numbers between 0 and 1, satisfying certain *additivity* conditions.

The simplest situation, quite adequate for the following discussion, is a sample space with a finite number n of mutually exclusive possibilities, let them be labeled with an index j between 1 and n (or 0 and $n-1$ if you're a computer scientist). The event algebra consists of all 2^n subsets (including the empty set) of elements from this sample space. Then for probabilities choose a collection of n nonzero real numbers p_j lying between 0 and 1, which sum to 1. The probability of an element S in the event algebra is the sum of the p_j for j in S.

The mutually exclusive possibilities might be distinct letters of an alphabet used to send messages through a communication channel, and in the actual physical world each letter will be represented by some unique *physical property*(s) that identifies it and distinguishes it from the other letters of the alphabet. One way to visualize this is to think of a classical phase space Γ in which each point γ represents the precise state of a mechanical system, and a particular letter of the alphabet, say F, is represented by some collection of points in Γ, the set of points where the property corresponding to F is true, and where the corresponding *indicator function* $F(\gamma)$ takes the value 1, whereas for all other γ, $F(\gamma) = 0$. The different indicator functions associated with letters of the alphabet then split the phase space up into tiles, regions in which a particular indicator function for a particular letter is equal to 1, and indicators for the other letters are all equal to zero. If this tiling does not cover the entire phase space, simply add another letter to the alphabet, call it "NONE", and let its indicator be 1 on the remaining points, and 0 elsewhere. In this manner, one can map the abstract notion of an alphabet of mutually exclusive letters onto a collection of mutually exclusive physical properties, one and only one of which will be true at any given time, because the point in phase space representing the actual state of the mechanical system will be located in just one of the nonoverlapping tiles. Given the sample space of tiles and some way of assigning probabilities, we have a setup to which the ideas of classical information theory can be applied, with a fairly clear answer to the question of what the information is all about.

In summary, classical information theory is all about probabilities, and in any specific application, say to signals coming over an optical fiber, the probabilities are about, or make reference to, physical events or properties of physical systems.

3. Quantum Probabilities

If we want quantum information theory to look something like Shannon's theory, the first task is to identify a quantum sample spaces of mutually-exclusive properties to which probabilities can be assigned. The task will be simplest if these quantum probabilities obey the same rules as their classical counterparts. In particular, since Shannon's theory employs expressions like $p_j \log(p_j)$, it would be nice if the quantum probabilities were nonnegative real numbers, in contrast to the negative quasiprobabilities sometimes encountered in discussions of quantum foundations.

Can we identify a plausible sample space which relative to the quantum Hilbert space plays a similar role to a tiling of a classical phase space? (In what follows, I will assume that the quantum Hilbert space is a finite-dimensional complex vector space with an inner product. Thus, all subspaces are closed, and we can ignore certain mathematical subtleties needed for a precise discussion of infinite-dimensional spaces.) A useful beginning is suggested by the quantum textbook approach to probabilities given by the Born rule. Let A be an observable, a Hermitian operator on the quantum Hilbert space, and let

$$A = \sum_j a_j P_j \tag{1}$$

be its spectral representation: the a_j are its eigenvalues and the P_j are projectors, orthogonal projection operators, which form a *projective decomposition of the identity I* (PDI):

$$I = \sum_j P_j; \quad P_j = P_j^\dagger; \quad P_j P_k = \delta_{jk} P_j. \tag{2}$$

If the eigenvalue a_j is nondegenerate and $|\phi_j\rangle$ is the corresponding eigenvector, then

$$P_j = |\phi_j\rangle\langle\phi_j| = [\phi_j], \tag{3}$$

where $[\phi]$ is a convenient abbreviation for the Dirac dyad $|\phi\rangle\langle\phi|$.

According to the textbooks, given a normalized ket $|\psi\rangle$, the probability that when A is measured the outcome is a_j, is given by the Born rule:

$$p_j = \Pr(a_j) = \Pr(P_j) = \langle\psi|P_j|\psi\rangle = |\langle\psi|\phi_j\rangle|^2, \tag{4}$$

where the final equality applies only when P_j is the rank one projector in (3). Now a measurement of A will yield just one eigenvalue, not many, so these eigenvalues correspond to the mutually-exclusive properties P_j in the PDI used in (4). The idea that a quantum property should be associated with a subspace of the Hilbert space, or the corresponding projector, goes back at least to von Neumann, see Section III.5 of his oft-cited (but little read) book [2].

The projector P_j has eigenvalues 0 and 1, so it resembles an indicator function on the classical phase space. In fact, a PDI divides up the Hilbert space into a set of mutually exclusive subspaces—$P_j P_k = 0$ for $j \neq k$—somewhat like a tiling of the classical phase space, whereas $I = \sum_j P_j$ tells us this tiling is complete: no part of the Hilbert space has been left out. Thus, the PDI is a plausible candidate for a quantum sample space. The event algebra will then consist of the projectors in the PDI along with other projectors formed from their sums, including I, along with the zero operator. The result is a commutative Boolean algebra. We already have one scheme, (4), for assigning probabilities to elements of the PDI, and thus, by additivity, to all the projectors in the event algebra. In particular, for $j \neq k$,

$$\Pr(P_j \text{ OR } P_k) = \Pr(P_j) + \Pr(P_k) = \langle\psi|(P_j + P_k)|\psi\rangle, \tag{5}$$

and similarly for sums of three or more distinct projectors.

In summary, this looks like a plausible beginning for a theory of *quantum information*: use a PDI on the Hilbert space as a sample space; then assign probabilities to the individual projectors. Not necessarily using (4), for it is only a particular example, but by some scheme which yields nonnegative real numbers adding to 1. Indeed, this strategy works very well, and I believe it covers all legitimate uses of (standard) probability theory in quantum mechanics, at least for a Hilbert space of finite dimension.

4. Quantum Measurements

There is, of course, more to be said, and it can be motivated by noting that a carefully written quantum textbook is likely to assign the probability p_j not to the *microscopic property* of the measured system, represented by P_j, but instead to the *macroscopic measurement outcome*, the *pointer position* in the picturesque, albeit archaic, language of quantum foundations. However, in the above presentation, it looks as if the probability is assigned *directly* to the microscopic property. Was this a mistake? Not if one believes, as I do, that a properly constructed and calibrated apparatus designed to measure some quantum observable can actually do what it was designed to do. Furthermore, if there is a one-to-one correspondence between prior properties and later pointer positions, the probability p_j will be the same for both.

In support of my belief that quantum measurements measure something, I note that this is assumed by my colleagues who do experiments at accelerator laboratories. They think that when they detect a fast muon emerging from an energetic collision, there really was a fast muon that approached and triggered their detector. Are they being naive? I do not think so. In passing, I note that these colleagues do not seem to worry about the "collapse" of the muon wavefunction produced by its interaction with the detector; they are less interested in what happened to the muon after it left their measuring device, and more interested in knowing what it was doing before it arrived there.

In addition, the notion that outcome j corresponds to the earlier property P_j can in certain cases be tested by *preparing* a particle which has the property P_j (see Section IV C of [3] on the topic of preparation), sending it into the measurement apparatus, and seeing whether the result is that the pointer points to j. Given that the apparatus has been tested and calibrated in this way, is not the experimenter justified in thinking that the particle *had* the property indicated by the pointer in a run in which the particle was *not* prepared in one of the P_j states? Justified or not, this is how many of my colleagues who carry out experiments do interpret things, and if they did not it would be difficult to draw interesting conclusions from their data. Quantum physics can hardly be called an *experimental* science if experiments designed to reveal prior microscopic properties do not actually do so! For additional details on the topic of what quantum measurements measure, including POVM and weak measurements, see [3].

There is, to be sure, a conceptual difficulty lurking in the background if we assume that measurements reveal prior microscopic properties. A hint is provided by the (correct) statement in textbooks that the x and z components of spin angular momentum, S_x and S_z, of a spin-half particle *cannot be measured simultaneously*. True, but what principle lies behind this? If we assume that experimenters really do understand something about what their devices measure, their inability to carry out such a simultaneous measurement might plausibly be explained by the fact that *there is nothing there to be measured*. Even very skilled experimenters cannot measure what is not there; indeed, this could be one thing that distinguishes them from less capable colleagues.

The Hilbert space of a spin-half particle is two-dimensional, and while it contains two subspaces corresponding to $S_x = \pm 1/2$ (in units of $\hbar$), and another two corresponding to $S_z = \pm 1/2$, there is no subspace which can plausibly be associated with, to take an example, "$S_x = +1/2$ AND $S_z = -1/2$". Hence if we assume that quantum measurements measure microscopic properties represented by subspaces of the quantum Hilbert space (or their projectors), we have a ready explanation for what lies behind the assertion that S_x and S_z cannot both be measured simultaneously. This is one way in which quantum mechanics is very different from classical mechanics.

5. Incompatible Properties

5.1. Issues of Logic

The absence of a Hilbert subspace corresponding to "$S_x = +1/2$ AND $S_z = -1/2$" reflects an important difference between the logic of indicator functions on the classical phase space and quantum projectors on the Hilbert space. One analogy has already been noted: the indicator $F(\gamma)$ for a classical property F takes one of two values, 0 and 1, while a quantum projector P has eigenvalues that are either 0 or 1. In addition, the negation "NOT F" of a classical property has an indicator function $I(\gamma) - F(\gamma)$, where $I(\gamma)$ is the function which is equal to 1 everywhere on the phase space. Similarly, the negation "NOT P" of a quantum projector P is the projector $I - P$, with I the quantum identity operator. However, the analogy begins to break down when we consider the conjunction "F AND G" of two classical properties: the property which is true if and only if both F and G are true. It corresponds to the intersection of the two subsets of phase space points associated with F and G, and its indicator is the product $F(\gamma)G(\gamma)$ of the two indicators. So we might expect that the conjunction "P AND Q" of two quantum properties P and Q would be represented by the product PQ. Indeed, this is the case *if* the projectors P and Q *commute*, $PQ = QP$, in which case PQ is again a projector. However, if PQ is *not* equal to QP, then neither product is a projector, and it is not obvious how to define "P AND Q".

The point can be illustrated using S_x and S_z for a spin-half particle. The projectors representing $S_x = +1/2$ and $-1/2$ are $[x^+] = |x^+\rangle\langle x^+|$ and $[x^-]$, where $|x^+\rangle$ and $|x^-\rangle$ are the eigenvectors corresponding to $S_x = +1/2$ and $-1/2$. Since $\langle x^+|x^-\rangle = 0$ (distinct eigenvalues means the eigenvectors are orthogonal) $[x^+][x^-] = [x^-][x^+] = 0$. Thus, these projectors commute, and the property "$S_x = +1/2$ AND $S_x = -1/2$" is represented by the zero operator on the Hilbert space: the property that is always false and thus never occurs. Also $[x^+] + [x^-] = I$ so these two mutually-exclusive properties constitute a PDI, a quantum sample space. Likewise the projectors $[z^+]$ and $[z^-]$ that correspond to $S_z = +1/2$ and $-1/2$ form a PDI.

However, neither $[x^+]$ nor $[x^-]$ commutes with either $[z^+]$ or $[z^-]$, so we cannot assign a quantum property to "$S_x = +1/2$ AND $S_z = -1/2$" by taking the product of the projectors. Again, this is consistent with the idea that the reason a simultaneous measurement of S_x and S_z is impossible is that there is nothing there to be measured.

5.2. Compatible and Incompatible

Thus, one way, perhaps the most essential way, quantum physics differs from classical physics is that *projectors representing different quantum properties need not commute*. We will say that the projectors P and Q are *compatible* provided $PQ = QP$, and *incompatible* if $PQ \neq QP$. Likewise a PDI $\{P_j\}$ and another PDI $\{Q_k\}$ are *compatible* if every projector in one commutes with every projector in the other: $P_jQ_k = Q_kP_j$ for every j and k. Otherwise, they are *incompatible*. In the compatible case, there is a *common refinement* consisting of all products of the form $P_jQ_k = Q_kP_j$, and every property in the event algebra associated with $\{P_j\}$ or with $\{Q_k\}$ is also in the event algebra associated with this refinement. Hence a very central issue in quantum foundations, and also for quantum information theory if one wants to use PDI's as sample spaces, is what to do when quantum projectors do *not* commute with each other. There have been various approaches.

Von Neumann was well aware of this problem, and together with Birkhoff invented *quantum logic* [4] to deal with it. In the case of a spin-half particle, quantum logic says that "$S_x = +1/2$ AND $S_z = -1/2$" is the property represented by the zero operator; that is, it is meaningful, but it is always false. This means its negation "$S_x = -1/2$ OR $S_z = +1/2$" is always true. Think about it: is that reasonable? If you continue to try and apply ordinary logical reasoning in this situation, you will soon end up in difficulty; see Section 4.6 of [5] for details. To prevent paradoxes, Birkhoff and von Neumann modified some of the rules of ordinary logic. Alas, their quantum logic requires a revision of the rules of ordinary (propositional) logic so radical that no one (known to me) has succeeded in using it to think in a useful way about what is going on in the quantum world. Maybe we physicists are just too stupid, and will have to wait for the day when clever quantum robots with intelligence vastly superior to ours can use quantum logic to resolve the quantum mysteries. However, if they succeed, will they be able to (or even want to) explain it to us?

A second approach to the incompatibility problem is employed in quantum textbooks and is also widespread in the quantum foundations community. Instead of talking about the quantum properties revealed by measurements, discussion is limited to measurement *outcomes*, the pointer positions that are part of the macroscopic world where classical physics is an adequate approximation to quantum physics, and noncommutation can be ignored for all practical purposes. (More in Section 6 below.) I call this the "black box" approach to quantum foundations. One starts with the *preparation* of a microscopic quantum state using a *macroscopic* apparatus, and then a later *measurement* of the state using another *macroscopic* apparatus, and what lies in between—well, that is inside the black box, and we will say as little as possible about it. A quantum $|\psi\rangle$? That is just a symbolic way of representing the preparation procedure. A PDI $\{P_j\}$? That is nothing but a mathematical tool for calculating the probabilities of measurement outcomes. The black box approach has the advantage that it avoids the problem of noncommuting quantum projectors. Its disadvantage is that it provides no way of understanding in physical terms what is going on at the microscopic level inside the box.

A third approach was popularized by Bell and his followers: replace the *noncommuting* Hilbert space projectors with *commuting* hidden variables. In essence, assume that in some way classical physics applies at the microscopic level. However, if, as I believe, noncommutation of projectors and PDI's marks the frontier between classical and quantum physics, one should not be surprised that an approach which is fundamentally classical—assumes a classical sample space, as is evident from the way the mysterious symbol λ is employed in formulas—results in the famous Bell inequality that disagrees with both quantum mechanical calculations and experimental results. (Nonlocal influences can be ignored, since they do not exist; see [6].)

5.3. The Single Framework Rule

The solution to the incompatibility problem that I favor can be viewed as a lowbrow form of quantum logic, one that a physicist like me can actually make use of. Its essential idea is that as long as one is dealing with a *single* PDI the rules of classical reasoning and classical probability theory can be applied unaltered in the quantum domain. So let us do that. If two PDI's are compatible, there is a PDI which is a common refinement. So let us use it. However, if two PDI's are incompatible, combining them will lead to nonsense. So do not do it. These ideas have been worked out in considerable detail in the *consistent histories* (CH) interpretation of quantum mechanics, where the prohibition against combining incompatible PDI's is known as the *single framework rule*. Here, the term *framework* is used either for a PDI or the associated event algebra, and the single framework rule prohibits combining incompatible PDI's. The difference between CH and quantum logic can be illustrated using the example "$S_x = +1/2$ AND $S_z = -1/2$" discussed earlier. In quantum logic, this is meaningful but false, while in CH it is meaningless, neither true nor false. The negation of a false statement is a true statement, so quantum logic has to say something about it. However, the negation of a meaningless statement is equally meaningless, allowing CH to remain silent. See [7] for more details.

In order to discuss the *time development* of quantum systems, a similar approach can be used (the "histories" part of consistent histories). Once again probabilities are assigned using PDI's as sample spaces, but in this case on an extended Hilbert space of histories [8]. In addition, in order to assign probabilities to a family of histories (a PDI on the history sample space) using an extension of the Born rule, it is necessary to impose certain *consistency* conditions (the "consistent" part of consistent histories), if this family is to constitute an acceptable framework, so the single framework rule is extended to incorporate the consistency conditions. For a short introduction to the CH interpretation of quantum mechanics, see [9]. Various conceptual difficulties are discussed in [7], whereas [10] gives a fairly thorough discussion of the ontology (Hilbert subspaces as "beables"). Finally, reference [5] is a standard reference with lots of details.

One aspect of the CH approach has raised a lot of objections, so it deserves a comment. In a given situation, it may be possible to describe what is going on using various different but incompatible frameworks, so the question arises: "What is the *right* framework to use?" The right answer is that this is the wrong question to ask in the quantum domain. In classical mechanics, the state of a mechanical system at a particular instant of time can be exactly specified by a single point in its phase space, the intersection of all properties (sets of points) which are "true" at that instant. This is consistent with the idea, which I have elsewhere called *unicity* (Section 27.3 of [5]), that at every instant of time there is a single unique "state of the universe" which, even if we do not know what it is, determines all physical properties. What might be its quantum counterpart? A "wavefunction of the universe"? If there really is something of that sort, it is likely to be a horrible, uninterpretable superposition of different pointer positions at the end of a measurement, or some other form of Schrödinger cat. The corresponding projector will then not commute with properties that might resemble something in the ordinary macroscopic world, and the single framework rule will then prevent discussing the world of everyday experience. I do not see any way in which a single quantum state could plausibly represent the "true state of the world", and I believe unicity must be abandoned in the transition from classical to quantum physics.

In practice, the choice of which framework to use will depend on the problem one is interested in. Consider, for example, a situation in which a spin-half particle is prepared in an eigenstate of S_x, say $S_x = +1/2$, before being sent through a magnetic field-free region (so its spin direction will not change) into an S_z measuring device. The outcome of the measurement will be either $S_z = +1/2$ or $S_z = -1/2$; let us assume the latter. This means we can say that S_z was $-1/2$ just before the measurement took place. However, is it possible that the particle had both $S_x = +1/2$ (because it was prepared in this state) and $S_z = -1/2$ (the value measured later) *at the same time*, just before the measurement was made? This makes no sense, as the properties are incompatible. There is one framework in which at the intermediate time $S_x = +1/2$, reflecting its earlier preparation, and a different, incompatible framework in which at the intermediate time $S_z = -1/2$, reflecting the outcome of the later measurement. These frameworks cannot be combined, and each has its own uses. If we are concerned about whether S_x was perturbed (say by a stray magnetic field), then the S_x framework is helpful, while if we want to identify what the measurement measured, the S_z framework is helpful. In textbook quantum mechanics, only the S_x framework is employed. Nothing wrong with that, except that one cannot discuss in what way the measurement measures something, leaving the poor student rather confused.

This example suggests that the liberty to choose different frameworks is not as dangerous as it might at first appear. A particular choice yields some type of information, and a different choice may yield something different. By looking at a coffee cup from above you can tell if it contains some coffee, while to see if there is a crack in the bottom you need to look from below. The oddity about the quantum world is not that different views, different frameworks, are possible. Instead, it is that certain frameworks cannot be combined into a consistent quantum description, because they are incompatible. For another, less trivial, example of a case in which choosing alternative frameworks proved useful, see the end of Section 7.

6. Quantum Information Theory I

Once a proper *quantum* sample space, a PDI or framework, has been defined, standard (Kolmogorov) probability theory can be used, and this means that the whole machinery of classical (Shannon) probability theory can be imported, unchanged, into the quantum domain. However, the reasoning and the results are restricted to this *single framework*; in particular, they cannot be combined with the analysis carried out in a separate, incompatible framework. Probabilities associated with incompatible frameworks cannot be combined; paying attention to this this eliminates a lot of well-known quantum paradoxes (See Chapters 19 to 25 of [5]).

In particular, this provides a *quantum* justification for all the usual applications of classical information theory to macroscopic properties and their time development. The reason is that from a quantum perspective the classical mechanics of macroscopic objects can be discussed with quite adequate precision using a single *quasiclassical* quantum framework, in which ordinary macroscopic properties are represented by enormous subspaces—a dimension of 10 raised to the power 10^{16} should be counted as relatively small—whose projectors commute with one another for all practical purposes; and quantum dynamics, which is intrinsically stochastic, is well approximated by deterministic classical dynamics. See [11]; Chapters 7, 17, 18 of [12]; Chapter 26 of [5]; and Section 4 of [10]. Consequently, we can immediately claim that all of classical information theory, all seventeen chapters of Cover and Thomas [13], or name your favorite reference, are a valid part of *quantum* information theory when it is applied to macroscopic properties and processes. In this domain, we understand quite well what *quantum* information is all about: its probabilities refer to quasiclassical properties and processes, all the things for which classical physics provides a satisfactory approximation to a more exact quantum description.

It is worth remarking, in passing, that using a quasiclassical framework provides a solution to the infamous *measurement problem* of quantum foundations: what to do with a wavefunction which is a coherent superposition of states in which the pointer points in two (or more) directions. While in the CH approach there is nothing inherently wrong with such a thing, it can be ignored if one wants to describe the usual macroscopic outcomes of laboratory experiments. Use a quasiclassical framework, and the problems represented by Schrödinger's cat are absent—and, by the single framework rule, they are excluded from the description.

In addition, Shannon's theory can be employed, unchanged, in situations in which some or all of the properties being discussed are *microscopic, quantum properties*, provided the discussion is restricted to a *single framework*. This includes what I have elsewhere [14] referred to as the *second measurement problem*: inferring from the measurement outcome (the pointer position) something about the earlier microscopic state of the system being measured. It can be analyzed in a manner which demonstrates that my colleagues who carry out experiments at accelerator laboratories are not being foolish when they assert that a fast muon has triggered their detector. The measurement apparatus is, in effect, an information channel leading from microscopic quantum properties at the input to macroscopic quantum properties (pointer positions) at the output.

7. Quantum Information Theory II

Does this mean that *all* problems of *quantum* information can be reduced to problems of *classical* information? No, not at all, but it does provide some insight into the nature of the additional problems which are unique to quantum information, and what is needed to attack them. These problems, and there are a vast number, all have to do with *comparing* (but not *combining*!) situations involving *incompatible frameworks*. But how can this be if a strict application of the single framework rule is needed to avoid falling into nonsensical paradoxes? The answer will emerge from considering some examples, starting with that of a noisy quantum channel.

Consider a one-qubit memoryless quantum channel whose input and output is a two-dimensional Hilbert space, the quantum analog of a classical one-bit channel. The classical channel is characterized by two real parameters: the probability that a 0 entering the channel will emerge as a 1, and the probability that a 1 entering the channel will emerge as a 0. If both are zero, the channel is perfect, noiseless. I like to visualize a perfect one-qubit quantum channel as a pipe through which a spin-half particle is propelled in such a way that its spin is left unchanged. If it enters with $S_x = +1/2$ it exits with $S_x = +1/2$, if it enters with $S_z = -1/2$ it exits with $S_z = -1/2$, and so forth. Of course, on any particular run the particle can only have a well-defined spin angular momentum in a particular direction; e.g., it can be prepared in such a state, and when it comes out only one component of its spin angular momentum can be measured. So to *test* whether the channel is perfect, it is necessary to carry out many repeated measurements. This by itself is no different from a classical channel, where repeated measurements are needed to estimate the probabilities of a bit flip when a signal passes through the channel. However, in the quantum case, the probabilities that S_z gets flipped, either from $+1/2$ to $-1/2$, or from $-1/2$ to $+1/2$, can be very different from those for S_x, so repeated measurements need to be carried out using different components of the spin angular momentum. The single framework rule does not prohibit a discussion of both S_x and of S_z *provided* these refer to *different* runs of the experiment. There is no problem in supposing that in one run $S_z = -1/2$, on the next run $S_x = +1/2$, and so forth. Of course, one has to assume that the channel continues to behave in the same way, at least in a probabilistic sense, during successive runs, but the same is true for a classical channel.

Suppose Joe has built what he claims is a perfect channel, but we want to test it. This is straightforward for a 1-bit classical channel: send in a series of 0s and 1s, and see if what emerges from the channel is the same as what was sent in. A one-qubit quantum channel is more complicated. If we test it using a sequence of states in which $S_z = +1/2$ or $-1/2$, and what emerges is the same as what went in, this is not sufficient, as it could very well be the case that if one sends in $S_x = +1/2$ it will emerge with S_x either $+1/2$ or $-1/2$ in a completely random fashion, uncorrelated with the input. So we have to check something in addition to S_z. Does this mean we have to carry out experiments with $S_w = +1/2$ and $-1/2$ for *every* possible spin component w? That would take a lot of time, and is not necessary. It suffices to check both $S_z = \pm 1/2$ and $S_x = \pm 1/2$. This result is far from obvious, and to derive it one must use principles of quantum mechanics which have no classical analog. Quantum information theorists need not fear unemployment; we will be kept busy for a long time.

As another example, consider teleportation, often presented as an instance of the mysterious and almost magical way in which quantum mechanics goes beyond classical physics. A standard textbook presentation of a protocol to teleport one qubit, e.g., Section 1.3.7 of [15], consists in applying unitary time evolution to an initial quantum state, followed by a measurement which collapses it. The measurement has four possible outcomes, and the result is communicated from A to B through two uses of a perfect one-bit *classical* channel. The end result of the protocol is a quantum state transmitted unchanged from A to B; in effect, a perfect one-qubit *quantum* channel. The student will certainly learn something by working through the formulas in the textbook, but this is of limited value in developing an intuition about microscopic quantum processes. My own approach [16] to understanding teleportation employs two incompatible frameworks. One framework shows how information about S_x is transmitted from Alice to Bob with the assistance of one use of the classical channel, and the other how S_z information is transmitted with the help of the other use of the classical channel. Similar ideas (but without referring to frameworks) will be found in [17,18]. This way of "opening the black box" should, I think, assist students in gaining a better intuition for microscopic quantum processes, and I hope it will become more widespread in the quantum information community, where research, or at least its publication, is still dominated by the "shut up and calculate" mentality encouraged by textbooks.

The preceding example could be easily dismissed in that it did not lead (directly, at least) to any new results in quantum information: the original teleportation protocol [19] appeared fourteen years in advance of my analysis. Hence it may be worth mentioning another example. A student and I were trying to understand Shor's algorithm for factoring numbers, which ends with a quantum Fourier transform followed by measurements of each of the qubits in the standard basis $|0\rangle$, $|1\rangle$ basis ($|z^+\rangle, |z^-\rangle$ for a spin-half particle). We noted that if you suppose that the final measurement reveals a property that the qubit possessed *before* the measurement, there is a way of looking at the problem that leads to an alternative and simpler way to carry out the algorithm [20]. Our perspective required using a framework incompatible with that employed in the standard textbook approach: unitary time development right up to the moment when measurement "collapses" the wavefunction—which, when done properly, leads to the same final answer. I was pleased that Nielsen and Chuang mentioned our work (Exercise 4.35 on p. 188, and see p. 246 of [15]), but disappointed in that they presented it as part of one more phenomenological principle, rather than as a way of gaining insight by using measurements outcomes to infer something about what happened earlier.

In my opinion, the discipline of quantum information could benefit from paying attention to the developments in quantum foundations mentioned above. If you open your favorite book on quantum information you will discover that measurements are quite firmly embedded in the discussion, and this in the manner of other textbooks in which measurements do not actually measure something, but instead enter as a primitive concept without further definition, a rule for carrying out calculations which requires no real physical understanding of processes at the microscopic quantum level. My guess is that if quantum information texts were to provide a consistent discussion of microscopic properties and processes, it could lead to some new and interesting advances, and perhaps even some new insights into quantum foundations.

8. Conclusions

Bell's question, "Quantum information ... about what?" can be given a quite definite answer. It is about physical properties and processes, which in quantum theory are represented by subspaces of the quantum Hilbert space, and to which standard (Kolmogorov) probabilities can be assigned, using sample spaces constructed from projective decompositions of the identity operator (PDI's). The single framework rule of consistent histories forbids combining incompatible PDI's or frameworks, resulting in a consistent theory not troubled by unresolved quantum paradoxes. From a quantum perspective, classical (Shannon) information theory is the application of quantum information theory to the domain of macroscopic properties and processes, where a single quasiclassical quantum framework is sufficient for all practical purposes, and therefore quantum incompatibilities can be ignored. However, in addition, all the ideas of classical information, and in particular its probabilistic formulation, can be imported unchanged into the microscopic quantum domain, as long as one is considering only a *single* quantum framework.

That there are many distinct frameworks available in quantum theory, frameworks which cannot be combined but can be compared, represents the new frontier of information theory that is specifically *quantum*, where classical ideas no longer suffice. At this point, new, and sometimes very difficult, problems arise in the process of comparing (but not combining) different incompatible quantum frameworks. They have no analogs in classical information theory, and some of them are quite challenging. Progress in this domain might well benefit were textbooks to abandon their outdated "black box" approach to quantum theory, in which "measurement" is an undefined primitive and measurements do not actually measure anything, but are simply a calculational tool to collapse wavefunctions. It is past time to open the black box with tools that can consistently handle noncommuting projectors. Consistent histories provide one approach for doing this; if the reader can come up with something better, so much the better.

Acknowledgments: Major contributions to the consistent histories interpretation of quantum mechanics have been made over the years by Roland Omnès, Murray Gell-Mann, James Hartle, and, more recently, Richard Friedberg and Pierre Hohenberg. We may not agree about everything, but I have certainly reaped great benefit from conversations with and publications by these colleagues, and it is a pleasure to thank them. I am also grateful for comments from three anonymous referees.

Conflicts of Interest: The author declares no conflict of interest.

References

1. Bell, J.S. Against measurement. In *Sixty-Two Years of Uncertainty*; Miller, A.I., Ed.; Plenum Press: New York, NY, USA, 1990; pp. 17–31. Reprinted in *Speakable and Unspeakable in Quantum Mechanics*, 2nd ed.; Cambridge University Press: Cambridge, UK, 2004; pp. 213–231.
2. Von Neumann, J. *Mathematical Foundations of Quantum Mechanics*; Princeton University Press: Princeton, NJ, USA, 1955.
3. Griffiths, R.B. What quantum measurements measure. *Phys. Rev. A* **2017**, *96*, 32110.
4. Birkhoff, G.; von Neumann, J. The logic of quantum mechanics. *Ann. Math.* **1936**, *37*, 823–843.
5. Griffiths, R.B. *Consistent Quantum Theory*; Cambridge University Press: Cambridge, UK, 2002.
6. Griffiths, R.B. Quantum locality. *Found. Phys.* **2011**, *41*, 705–733.
7. Griffiths, R.B. The New Quantum Logic. *Found. Phys.* **2014**, *44*, 610–640.
8. Isham, C.J. Quantum logic and the histories approach to quantum theory. *J. Math. Phys.* **1994**, *35*, 2157–2185.
9. Griffiths, R.B. The Consistent Histories Approach to Quantum Mechanics. Stanford Encyclopedia of Philosophy. 2014. Available online: http://plato.stanford.edu/entries/qm-consistent-histories/ (accessed on 29 November 2017).
10. Griffiths, R.B. A consistent quantum ontology. *Stud. Hist. Philos. Mod. Phys.* **2013**, *44*, 93–114.

11. Gell-Mann, M.; Hartle, J.B. Classical equations for quantum systems. *Phys. Rev. D* **1993**, *47*, 3345–3382.
12. Omnès, R. *Understanding Quantum Mechanics*; Princeton University Press: Princeton, NJ, USA, 1999.
13. Cover, T.M.; Thomas, J.A. *Elements of Information Theory*, 2nd ed.; Wiley: New York, NY, USA, 2006.
14. Griffiths, R.B. Consistent quantum measurements. *Stud. Hist. Philos. Mod. Phys.* **2015**, *52*, 188–197.
15. Nielsen, M.A.; Chuang, I.L. *Quantum Computation and Quantum Information*; Cambridge University Press: Cambridge, UK, 2000.
16. Griffiths, R.B. Types of quantum information. *Phys. Rev. A* **2007**, *76*, 062320.
17. Renes, J.M.; Dupuis, F.; Renner, R. Efficient polar coding of quantum information. *Phys. Rev. Lett.* **2012**, *109*, 050504.
18. Coles, P.J.; Piani, M. Complementary sequential measurements generate entanglement. *Phys. Rev. A* **2014**, *89*, 010302.
19. Bennett, C.H.; Brassard, G.; Crépeau, C.; Jozsa, R.; Peres, A.; Wootters, W.K. Teleporting an unknown quantum state via dual classical and Einstein-Podolsky-Rosen channels. *Phys. Rev. Lett.* **1993**, *70*, 1895–1899.
20. Griffiths, R.B.; Niu, C.-S. Semiclassical Fourier transform for quantum computation. *Phys. Rev. Lett.* **1996**, *76*, 3228–3231.

© 2017 by the author. Licensee MDPI, Basel, Switzerland. This article is an open access article distributed under the terms and conditions of the Creative Commons Attribution (CC BY) license (http://creativecommons.org/licenses/by/4.0/).

Article

Entropic Phase Maps in Discrete Quantum Gravity

Benjamin F. Dribus

Department of Mathematics, William Carey University, 710 William Carey Parkway, Hattiesburg, MS 39401, USA; bdribus@gmail.com or bdribus@wmcarey.edu; Tel.: +1-985-285-5821

Received: 26 May 2017; Accepted: 25 June 2017; Published: 30 June 2017

Abstract: Path summation offers a flexible general approach to quantum theory, including quantum gravity. In the latter setting, summation is performed over a space of evolutionary pathways in a history configuration space. Discrete causal histories called *acyclic directed sets* offer certain advantages over similar models appearing in the literature, such as causal sets. Path summation defined in terms of these histories enables derivation of discrete Schrödinger-type equations describing quantum spacetime dynamics for any suitable choice of algebraic quantities associated with each evolutionary pathway. These quantities, called *phases*, collectively define a *phase map* from the space of evolutionary pathways to a target object, such as the unit circle $S^1 \subset \mathbb{C}$, or an analogue such as S^3 or S^7. This paper explores the problem of identifying suitable phase maps for discrete quantum gravity, focusing on a class of S^1-valued maps defined in terms of "structural increments" of histories, called *terminal states*. Invariants such as *state automorphism groups* determine multiplicities of states, and induce families of natural entropy functions. A phase map defined in terms of such a function is called an *entropic phase map*. The associated dynamical law may be viewed as an abstract combination of Schrödinger's equation and the second law of thermodynamics.

Keywords: quantum gravity; discrete spacetime; causal sets; path summation; entropic gravity

1. Introduction

1.1. Path Summation in Quantum Gravity

Feynman's path summation approach to quantum theory [1], originally developed in the non-relativistic context of four-dimensional Euclidean spacetime $\mathbb{R}^4$, has since been abstracted and generalized to apply to a wide variety of situations in which quantum effects play a significant role, including the study of fundamental spacetime structure and quantum gravity. In the latter setting, the objects over which summation is performed are no longer spaces of paths in low-dimensional real manifolds whose elements represent events, but spaces of evolutionary pathways in configuration spaces whose elements represent histories, i.e., entire spacetimes. The distinction between summing over evolutionary pathways for histories and summing over histories themselves becomes significant in the background independent context, where each pathway represents a history together with a generalized frame of reference, and where different pathways may encode identical physics. For both conceptual and computational reasons, histories incorporating a version of discreteness and a notion of causal structure are especially attractive for studying quantum gravity. Such histories include "purely causal" objects such as *causal sets* [2] and *causal networks* [3–5], "mostly causal" objects such as *causal dynamical triangulations* [6] and *quantum causal histories* [7], and objects incorporating a significant degree of additional structure, such as *spin foams* [8,9], *quantum cellular automata* [10], *causal fermion systems* [11,12], and *tensor networks* [13]. The histories studied in this paper, called *acyclic directed sets*, resemble causal sets and causal networks, but with a few important distinctions [14–16].

1.2. Path Summation Rudiments

I recall here a few basic notions regarding conventional path summation. In ordinary quantum mechanics and quantum field theory, one considers directed paths γ representing possible particle trajectories in a fixed spacetime manifold, such as Euclidean spacetime $\mathbb{R}^4$ or Minkowski spacetime $\mathbb{R}^{3+1}$. Such paths are illustrated in the left-hand diagram in Figure 1, adapted from Figure 6.2.2 of [14]. One begins with a classical theory, whose dynamics is determined by a Lagrangian $\mathcal{L}$ encoding information about motion-related or metric quantities. $\mathcal{L}$ may be regarded as an *infinitesimal path functional*, i.e., a function of the particle motion whose value depends only on instantaneous information along γ. This viewpoint generalizes naturally to more abstract settings. The classical action $\mathcal{S}(\gamma)$ is given by integrating $\mathcal{L}$ along γ with respect to time. Hamilton's principle states that the classical path γ_{CL} renders the classical action stationary. Heuristically, this means that $\mathcal{L}$ "chooses" γ_{CL} from among other alternatives by how $\mathcal{S}$ varies with γ. The classical equations of motion are the Euler–Lagrange equations for $\mathcal{L}$, derived via Hamilton's principle.

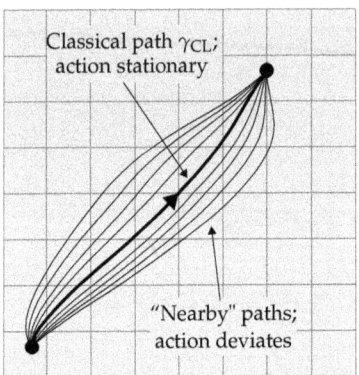
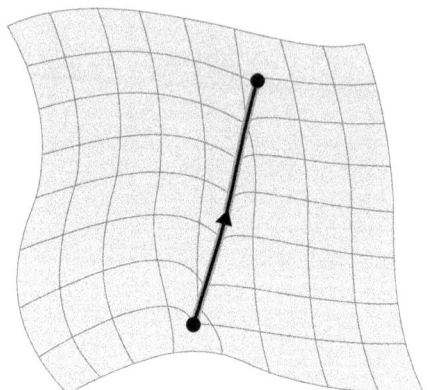

Figure 1. In a fixed spacetime background, the Lagrangian $\mathcal{L}$ "chooses" the classical path γ_{CL} via Hamilton's principle; in a background independent theory, different paths imply different spacetimes.

In the corresponding quantum theory, the behavior of the particle depends on contributions from *every possible* path. To quantify this dependence, one defines a *phase map* Θ on a space of paths in spacetime, given by Feynman's formula

$$\Theta(\gamma) = e^{\frac{i}{\hbar}\mathcal{S}(\gamma)}, \tag{1}$$

where $i = \sqrt{-1}$ and $\hbar$ is Planck's reduced constant. For convenience, I use the term "phase" for the value $e^{\frac{i}{\hbar}\mathcal{S}(\gamma)}$ itself, rather than for the "angle" $\frac{1}{\hbar}\mathcal{S}(\gamma)$ in the complex exponential. One then performs a *path integral* to "sum together" these phases. Feynman's path integral for paths in a subset R of $\mathbb{R}^4$ is the prototypical example. Its value is interpreted as a complex quantum amplitude for R, encoding the probability that the particle follows a path through R. Due to Hamilton's principle, phases for paths near the classical path γ_{CL} combine via constructive interference to yield relatively large amplitudes for neighborhoods of γ_{CL}, while phases for faraway paths destructively interfere. Schrödinger's equation for ordinary nonrelativistic quantum theory

$$i\hbar\frac{\partial \psi}{\partial t} = \mathbf{H}\psi, \tag{2}$$

may be derived from Feynman's path integral [1]. Here, ψ is the state function for the particle, and $\mathbf{H}$ is the Hamiltonian operator.

1.3. Effects of Gravity

Gravitation alters this picture by introducing interaction between spacetime and its material content. It no longer suffices to consider particle paths in a fixed spacetime manifold, because different paths induce different local responses in spacetime geometry. The right-hand diagram in Figure 1 illustrates this complication, showing a region of spacetime "warping" around a path. Absence of a fixed spacetime background in this context is called *background independence*. Einstein's equation, conventionally expressed in the form

$$R_{\mu\nu} - \frac{1}{2} R g_{\mu\nu} + \Lambda g_{\mu\nu} = \frac{8\pi G}{c^4} T_{\mu\nu}, \tag{3}$$

quantifies this coupling between geometry and matter under the framework of general relativity. Here, $R_{\mu\nu}$ is the Ricci curvature tensor, R is the scalar curvature, $g_{\mu\nu}$ is the metric tensor, Λ is the cosmological constant, G is Newton's gravitational constant, c is the speed of light, and $T_{\mu\nu}$ is the stress-energy tensor. Ultimately, one expects both geometry and matter to emerge from some deeper structural substratum, and this has been a consistent theme of fundamental physics since the early unification efforts of Einstein, Kaluza and Klein, Weyl, and a few others. Unification would offer a perfect version of background independence by eliminating all distinction between a background "arena" and foreground "objects". Discrete causal theory [14] represents one specific effort toward the goal of unification. More generally, *any* background independent adaptation of path summation associates a different copy of spacetime with each possible distribution of matter and energy, and this leads to sums involving entire configuration spaces of spacetimes. Each such spacetime is classically self-contained, in the sense that it describes its own complete version of events, and has no ordinary causal interaction with other possible spacetimes. In this context, a spacetime is often called a *history*, and a configuration space $\mathbb{S}$ of spacetimes is called a *history configuration space*.

A subset of a history configuration space $\mathbb{S}$ equipped with a total order, such as the image of a non-self-intersecting directed path γ in $\mathbb{S}$, does not represent "classical dynamics", since each history contains its own complete description of events. However, certain special totally ordered subsets of $\mathbb{S}$ may be interpreted as representing "growth" or "development" of one history into another, and such subsets are called *evolutionary pathways* in $\mathbb{S}$. Technical requirements for evolutionary pathways are discussed below. Such pathways may or may not possess initial or terminal histories, depending on the structure of $\mathbb{S}$. However, any pair of pathways in $\mathbb{S}$ sharing a common terminal history, or a common "limit" in more general settings, describe identical physics from different points of view. A familiar example is given by partitioning Minkowski spacetime $\mathbb{R}^{3+1}$ via two different integer-indexed families $\{\sigma_k\}$ and $\{\sigma'_k\}$ of spacelike sections, as illustrated in the left-hand diagram in Figure 2. This diagram follows the usual convention of suppressing two spacelike dimensions, with time running vertically up the page. Edges do not represent physical boundaries, but merely delimit the finite region shown. Discrete evolutionary pathways for $\mathbb{R}^{3+1}$ may be defined via these partitions, as shown in the middle and right-hand diagrams. One may completely foliate $\mathbb{R}^{3+1}$ by similar families, thereby defining continuous pathways in a configuration space of Lorentzian manifolds. However, the simpler discrete picture shown here, in which $\mathbb{R}^{3+1}$ is partitioned into increments of nontrivial causal extent, is more illustrative of the discrete processes studied in this paper.

Both evolutionary pathways illustrated in Figure 2 describe the same empty, flat spacetime represented by $\mathbb{R}^{3+1}$. However, they offer different perspectives regarding the evolution of this spacetime. These may be identified with different inertial frames of reference on $\mathbb{R}^{3+1}$, since $\{\sigma_k\}$ and $\{\sigma'_k\}$ are families of parallel spacelike hyperplanes. In more abstract settings, histories may not encode recognizable geometry, so the relativistic idea of frames of reference must be generalized. However, the conceptual content remains unchanged: each evolutionary pathway in a history configuration space $\mathbb{S}$ describes a history *together with* a generalized frame of reference for this history. To qualify as an evolutionary pathway, a totally ordered subset γ of $\mathbb{S}$ must satisfy the property that "later histories in γ are evolutionary descendants of earlier histories". Mathematically, this means that the total order on γ must be derived naturally from the structure of $\mathbb{S}$. The most convenient case is when $\mathbb{S}$ itself possesses natural order-theoretic structure from which evolutionary relationships may be deduced in a self-evident way. This is the case for discrete causal theory.

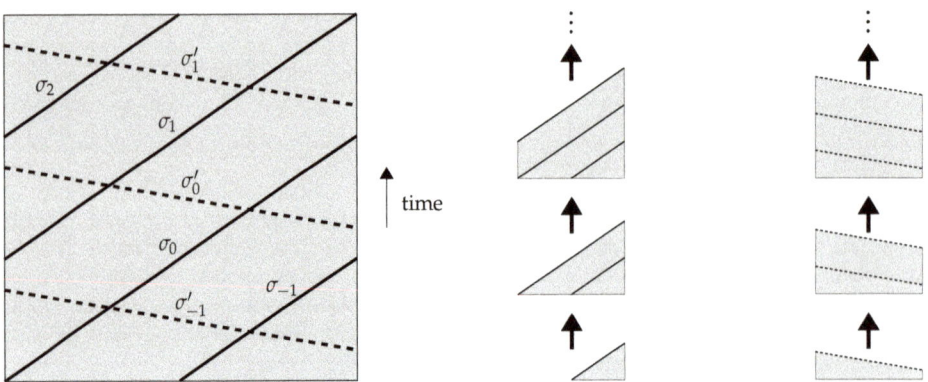

Figure 2. $\mathbb{R}^{3+1}$ partitioned via sequences of spatial sections $\{\sigma_k\}$ and $\{\sigma'_k\}$; evolutionary pathways defined by $\{\sigma_k\}$ and $\{\sigma'_k\}$. Both pathways share the same "limit history" $\mathbb{R}^{3+1}$.

1.4. Motivation for Entropic Phase Maps

Histories modeled by objects called *countable star finite acyclic directed sets* induce discrete causal history configuration spaces called *kinematic schemes,* with properties superior in some ways to those of similar spaces arising in causal set theory, causal dynamical triangulations, and related approaches. These objects are formally defined in Section 2. Path summation over a kinematic scheme $\mathbb{S}$, together with other natural machinery, enables derivation of discrete *causal Schrödinger-type equations* such as Equation (1.1.2) of [14]. This equation is reproduced here as Equation (4):

$$\psi^-_{R;\theta}(r) = \theta(r) \sum_{r^- \prec r} \psi^-_{R;\theta}(r^-). \qquad (4)$$

The meaning of this equation is explained in Section 2, and more thoroughly in [14], but I briefly describe its content here. The function $\psi^-_{R;\theta}$ is a generalized state function, called the *past state function*, while R is a set of *relations* representing natural relationships between pairs of histories in $\mathbb{S}$, called *co-relative histories*. Sequences of co-relative histories fit together to define evolutionary pathways in $\mathbb{S}$, called *co-relative kinematics*. The relations r and r^- are elements of R representing specific co-relative histories. The *precursor symbol* $\prec$ in the expression $r \prec r^-$ indicates that the evolutionary relationship represented by r is a possible sequel to the evolutionary relationship represented by r^-.

Remaining to be identified in Equation (4) is the *relation function* θ, which is the entity of principal interest in this paper. This function assigns to each element r of R a *phase* $\theta(r)$ belonging to some target object T. The most obvious choice for T is the unit circle S^1, viewed as a subobject of the complex field $\mathbb{C}$, and this is the target object focused on here. However, other choices may be studied in more general contexts. For reasons explained in [14], the unit spheres S^3 and S^7, viewed as subobjects of the quaternions $\mathbb{H}$ and octonions $\mathbb{O}$, respectively, are potentially interesting alternatives. At a finer level of detail, it may be appropriate to consider discrete subobjects of S^1, S^3, or S^7, which possess interesting algebraic properties. Alternatively, T might be an object at a higher level of algebraic hierarchy, such as a monoidal category. In any case, T must possess a "multiplicative" operation, enabling the factor $\theta(r)$ to multiply the sum $\sum_{r^- \prec r} \psi^-_{R;\theta}(r^-)$ in Equation (4). Extending θ via this operation, as described below, defines a *phase map* Θ on the space of co-relative kinematics in $\mathbb{S}$. The form of Equation (4) assumes that θ generates Θ in this way; otherwise, the equation must be generalized. Under this assumption, θ provides specific dynamical content to the equation, and thereby defines a quantum dynamical law governing fundamental spacetime structure.

The elements of the relation set R in Equation (4) encode information up to first order at the quantum level, in the sense that they represent individual stages of evolution in $\mathbb{S}$. Hence, θ is analogous to an infinitesimal path functional on $\mathbb{S}$, i.e., a generalized Lagrangian. Similarly, Θ may be regarded as a generalized action. However, to simplify the form of Equation (4), the appropriate analogue of the exponentiation appearing in Feynman's phase map (1) is "built in" to the definition of θ. Hence, the quantities I call "phases" throughout the remainder of the paper are analogous to Feynman's complex exponentials $e^{\frac{i}{\hbar}\mathcal{S}(\gamma)}$ themselves, not to the corresponding "angles" $\frac{1}{\hbar}\mathcal{S}(\gamma)$. The phase $\Theta(\gamma)$ of a co-relative kinematics γ is therefore a *product* of phases $\theta(r)$ of individual relations r along γ, rather than a sum or integral. More precisely, one may define a *concatenation product* $\sqcup$ joining co-relative kinematics "end-to-end", under which γ may be factored into a product of individual relations $\gamma = ... \sqcup r_0 \sqcup r_1 \sqcup r_2 \sqcup ...$. Extending θ multiplicatively then means that $\Theta(\gamma) = \prod_k \theta(r_k)$, where the product is in the target object T. Questions of convergence are important in general, but are not examined here, since one may go quite far under finiteness assumptions.

This paper explores the problem of identifying suitable phase maps for discrete quantum gravity, focusing on a class of S^1-valued maps defined in terms of *terminal states* Δ of histories D along evolutionary pathways γ in a history configuration space $\mathbb{S}$. Here, $\mathbb{S}$ is a kinematic scheme of star finite acyclic directed sets D, γ is a co-relative kinematics, and Δ encodes "recent" causes and effects in D. Invariants such as *state automorphism groups* $\text{Aut}(\Delta)$ determine multiplicities of states, and induce natural families of entropy functions. *Resolution entropy* is defined via a "coarse-graining" procedure called *causal atomic resolution*, analogous to conventional partitioning of state space into families of states sharing "macroscopic" properties. *Superset entropy* is defined by counting the number of ways in which a terminal state Δ may embed into a larger state Δ' called a *superset* of Δ. A large state automorphism group $\text{Aut}(\Delta)$ corresponds to a small number of such supersets, and therefore implies low entropy. *Labeled entropy* is defined by counting the number of ways to label elements of Δ; again, large $\text{Aut}(\Delta)$ implies low entropy. *Symmetry entropy*, by contrast, is defined by counting the elements of $\text{Aut}(\Delta)$ itself, so large $\text{Aut}(\Delta)$ implies high entropy in this context. A primitive version of symmetry entropy is discussed in Section 8.2 of [14]. A phase map defined in terms of such entropic quantities, or related quantities such as entropy per unit volume, is called an *entropic phase map*. The resulting version of Equation (4) may be viewed as an abstract combination of Schrödinger's equation and the second law of thermodynamics, which arises entirely from the structure of $\mathbb{S}$.

Section 2 presents the necessary background from discrete causal theory [14] to support the development and description of these ideas. Section 2.1 briefly outlines the conceptual and philosophical foundations of discrete causal theory. Section 2.2 describes the classical version of the theory, expressed in terms of countable star finite acyclic directed sets. Section 2.3 sketches the theory of *relation space*, which addresses certain technical difficulties in earlier versions of the theory such as causal set theory. Section 2.4 describes the basics of discrete quantum causal theory. Section 3 examines entropy and the second law of thermodynamics in a broad context, introduces discrete causal analogues of familiar thermodynamic ideas such as state space, and develops the specific notions of entropy mentioned above. Section 3.1 discusses entropy in general terms under a broad framework called *entropy systems*. Section 3.2 describes associated versions of the second law. Section 3.3 introduces discrete causal state spaces. Section 3.4 defines resolution, superset, labeled, and symmetry entropies. Section 4 introduces entropic phase maps, and examines some of their properties. Section 4.1 describes some simple versions of these maps explicitly. Section 4.2 discusses the problem of obtaining suitable interference effects analogous to those induced for Feynman's phase map by Hamilton's principle. Section 4.3 discusses some possible objections to the idea of entropic phase maps, and briefly examines an alternative approach involving a more conventional notion of action. Section 4.4 offers concluding remarks, and mentions some mathematical problems whose solution would enhance the study of entropic phase maps.

2. Discrete Causal Theory

2.1. Causal Metric Hypothesis

Discrete causal theory is a general approach to fundamental physics that emphasizes discrete spacetime models equipped with directed structure encoding cause-and-effect relationships between pairs of events. Included under this umbrella are causal set theory [2], causal dynamical triangulations [6], and quantum causal histories [7]. Similar ideas contribute to loop quantum gravity [8,9], information-related approaches involving causal networks or cellular automata [10,17,18], causal fermion systems [11], and the theory of tensor networks [13]. The version of discrete causal theory used in this paper is distinct from all these, but may be regarded as an enhanced version of causal set theory [14]. Clean and appealing basic structure is an asset of discrete causal theory, but its principal motivation derives from technical results called *metric recovery theorems*, discussed in Section 2.2, which demonstrate that discrete causal models can reproduce relativistic spacetime geometry at ordinary scales. Such models also avoid generic divergence problems, and offer potential explanatory advantages by allowing "pre-geometric" notions such as spacetime dimension to emerge dynamically. The reason why these models cannot yet replace relativistic geometry root and branch is because relativity explains *how* geometry evolves via Einstein's Equation (3), while discrete causal dynamics remains primitive. This paper offers a modest contribution toward rectifying this deficiency.

A radical interpretation of the aforementioned metric recovery results is the *causal metric hypothesis* [14–16], which states that *the structural properties of the universe, particularly the metric structure of spacetime, emerge from causal structure at the fundamental scale*. This general idea forms the philosophical basis for discrete causal theory, but may be accorded different weights in different versions of the theory. The *strong interpretation* of the causal metric hypothesis ascribes all of physics, including "nongravitational matter", to causal structure. In the context of entropic phase maps, the strong interpretation extends the thermodynamic hypothesis regarding gravitation [19] to treat matter and energy in similar terms. Alternatively, one may choose to restrict attention to gravity, leaving aside unification. In this context, matter and energy may be modeled by attaching auxiliary algebraic structure to causal structure. In either case, quantum theory arises via generalized path summation in a manner much simpler and more natural than conventional attempts to quantize relativistic geometry. The directed structures of individual discrete causal histories combine to induce higher-level multidirected structures on their history configuration spaces, analogous to higher-level geometric structures of moduli spaces in algebraic geometry. This *iteration of structure* enables a natural version of summation over evolutionary pathways, which leads to quantum dynamics governed by discrete causal Schrödinger-type equations such as Equation (4).

2.2. Classical Theory

The mathematical objects used to model discrete causal histories in this paper are called *countable star finite acyclic directed sets*. Before defining them formally, I make two clarifying remarks. First, these objects are conventionally called "directed graphs" rather than "directed sets", because the latter term has a more specific conventional meaning. However, graph-theoretic terminology is awkward here, and "directed set" ideally communicates the intended notion of a set D equipped with directions between distinguished pairs of elements x and y. Such a direction is called a *relation* between x and y, with *initial element* x and *terminal element* y, and is denoted by $x \prec y$. The *precursor symbol* $\prec$ generalizes the familiar *less than* symbol $<$ on a totally ordered set such as $\mathbb{Z}$. The relation $x \prec y$ is represented graphically by a directed edge between nodes representing x and y. A family of such relations is called a *binary relation* on D, denoted collectively by the same symbol $\prec$. Mathematically, $\prec$ is a subset of the Cartesian product $D \times D$. Dual usage of the word "relation" and the symbol $\prec$ for individual relations $x \prec y$ and for the set $\prec$ of all such individual relations is a standard convenience. Second, the choice to focus on *acyclic* directed sets rules out discrete causal analogues of closed causal curves, but this is a simplifying assumption that may be relaxed. It does not imply the view that quantum gravity necessarily forbids such structure. Countability and/or star finiteness may also be relaxed, though in my opinion there is limited motivation for doing so.

The following definitions are adapted from Sections 3.6 and 3.7 of [14]:

Definition 1. *A **directed set** $(D, \prec)$ is a set D equipped with a binary relation $\prec$. A **morphism** from a directed set $(D, \prec)$ to a directed set $(D', \prec')$ is a set map $f : D \to D'$ such that $f(x) \prec' f(y)$ whenever $x \prec y$. The **category of directed sets** $\mathcal{D}$ is the category whose objects are directed sets and whose morphisms are morphisms of directed sets. A **subobject** of a directed set $(D, \prec)$ is a directed set $(D', \prec')$, where D' is a subset of D, and where $\prec'$ is a subset of $\prec$ consisting of relations between pairs of elements of D'. The **causal dual** of a directed set $(D, \prec)$ is the directed set $(D, \prec^*)$, where $x \prec^* y$ if and only if $y \prec x$.*

Definition 2. *A **multidirected set** (M, R, i, t) consists of a set of elements M, a set of relations R, and **initial** and **terminal element maps** $i : R \to M$ and $t : R \to M$. A **morphism** from a multidirected set (M, R, i, t) to a multidirected set (M', R', i', t') consists of a **map of elements** $f_{\text{ELT}} : M \to M'$ and a **map of relations** $f_{\text{REL}} : R \to R'$, such that $f_{\text{ELT}}(i(r)) = i'(f_{\text{REL}}(r))$ and $f_{\text{ELT}}(t(r)) = t'(f_{\text{REL}}(r))$ for each r in R. The **category of multidirected sets** $\mathcal{M}$ is the category whose objects are multidirected sets and whose morphisms are morphisms of multidirected sets. A **subobject** of a multidirected set (M, R, i, t) is a multidirected set (M', R', i', t'), where M' and R' are subsets of M and R, respectively, and where i' and t' are the restrictions of i and t to R'. The **causal dual** of a multidirected set (M, R, i, t) is the multidirected set (M, R, t, i).*

Definition 3. *A **chain** in a multidirected set (M, R, i, t) is a sequence of relations $..., r_k, r_{k+1}, ...$ such that $t(r_k) = i(r_{k+1})$. The **past** of an element x of (M, R, i, t) is the set of all elements w in M such that there exists a chain $r_0, ..., r_N$ with $i(r_0) = w$ and $t(r_N) = x$. The **future** of x is the set of all elements y in M such that there exists a chain $r_0, ..., r_N$ with $i(r_0) = x$ and $t(r_N) = y$. An **antichain** in (M, R, i, t) is a subset σ of M with no chain connecting any pair of its elements, distinct or otherwise. The **past relation set** $R^-(x)$ of an element x in M is the set of all relations r in R such that $t(r) = x$. The **future relation set** $R^+(x)$ of x is the set of all relations r in R such that $i(r) = x$. The **relation set** $R(x)$ of x is the union $R^-(x) \cup R^+(x)$.*

For both directed sets and multidirected sets, an *isomorphism* is an invertible morphism, and an *automorphism* is a self-isomorphism. Isomorphic sets are usually considered to be equivalent. It is often convenient to denote a directed set or multidirected set by just D or M, respectively, or to write $D = (D, \prec)$ or $M = (M, R, i, t)$ to indicate that a set D or M is equipped with such structure. Similarly, the causal dual of a directed set D may be denoted by D^*, and the causal dual of a multidirected set M by M^*. A directed set $D = (D, \prec)$ may be recognized as a multidirected set whose set of relations is the binary relation $\prec$, and whose initial and terminal element maps are defined by setting $i(x \prec y) = x$ and $t(x \prec y) = y$. For multidirected sets, the notation $x \prec y$ remains useful to indicate the existence of a relation r such that $i(r) = x$ and $t(r) = y$, even though no binary relation is involved. The necessity to study multidirected sets arises at the quantum level, via iteration of structure.

A well-motivated version of discrete classical causal theory is defined by the axioms in Definition 4, adapted from Definition 4.10.1 of [14]. Symbols and terms are further discussed below.

Definition 4. *Five axioms for discrete classical causal theory are the following:*

1. **Binary axiom:** *Classical spacetime may be modeled as a directed set $D = (D, \prec)$, whose elements represent events, and whose relations represent causal relationships between pairs of events.*
2. **Generalized measure axiom:** *D is equipped with a set function μ from the power set $\mathcal{P}(D)$ of D to the extended real numbers $\mathbb{R} \cup \{\infty\}$, which assigns finite positive values to nonempty finite subsets of D, and infinite values to infinite subsets of D.*
3. **Countability:** *D is countable.*
4. **Star finiteness:** *For every element x of D, the star $\text{St}(x) = \{x\} \cup R(x)$ of x is finite.*
5. **Acyclicity:** *D possesses no cycles, i.e., sequences of relations $x_0 \prec ... \prec x_N$ with $x_0 = x_N$.*

The binary axiom specifies both a mathematical structure and a physical interpretation of this structure. The generalized measure axiom imposes no mathematical conditions on the remaining axioms, so it is allowed a range of possible versions, each specified by a choice of μ. The most attractive choices are similar to the *counting measure* used in early versions of causal set theory, which assigns to each subset of D its number of elements in fundamental units. The function μ is unrelated to the family of measures μ for an *entropy system*, introduced in Section 3.1. Since the *star* $St(x)$ of x is just $\{x\} \cup R(x)$, star finiteness is equivalent to finiteness of relation sets $R(x)$. The physical meaning of this condition is that every event has only a finite number of direct causes and effects. The reason for using $St(x)$ rather than $R(x)$ involves topological bookkeeping that plays no direct role in this paper. The meanings of countability and acyclicity are self-evident. The discreteness of D is encoded in the generalized measure axiom and the axiom of star finiteness.

Figure 3, adapted from Figure 3.6.5 of [14], illustrates different types of directed sets and multidirected sets. Elements are represented by nodes, and relations by directed edges. In the third and fourth diagrams, directions of relations are indicated by arrows, while in the first and second diagrams, directions are inferred via an "up the page" convention analogous to the convention for the direction of time in Minkowski spacetime diagrams. This convention applies only to acyclic directed sets. The first diagram illustrates a *causal set*, i.e., a countable, *irreflexive, transitive, interval finite* directed set $(C, \prec_{CS})$. Irreflexivity means that C contains no "self-relations" $x \prec_{CS} x$. Transitivity means that if $x \prec_{CS} y$ and $y \prec_{CS} z$, then $x \prec_{CS} z$. Irreflexivity and transitivity together imply acyclicity. Transitivity leads to trouble in distinguishing between direct and indirect causation in causal set theory [14,20]. Interval finiteness means that only a finite number of elements y lie between any two elements x and z of C, in the sense that $x \prec_{CS} y \prec_{CS} z$. Interval finiteness and star finiteness are incomparable, i.e., neither condition implies the other. An important class of causal sets that are generally *not* star finite are those induced by randomly "sprinkling" elements into a Lorentzian manifold. These sets are useful to illustrate metric recovery results, but they are not regarded as physically realistic, even in causal set theory. Star finite objects are preferred as the actual workhorses for quantum gravity [2,21,22]. The second diagram in Figure 3 illustrates a nontransitive acyclic directed set; in particular, the two relations $x \prec y$ and $y \prec z$ do not imply a relation $x \prec z$. The physical interpretation of this set still recognizes x as a cause of z, but not a *direct* cause. This is analogous to the relationship between a grandparent and grandchild. The third diagram illustrates a directed set D' with cycles, including the "self-relation" $t \prec' t$ and the "reciprocal relations" $u \prec' v \prec' u$. Such sets are not studied in this paper, but remain interesting in more general contexts. The fourth diagram illustrates a multidirected set M whose relation structure is more complicated than any binary relation on its set of elements. For example, there are two distinct relations in M from x to y. In discrete causal theory, multiple relations between pairs of elements arise at the quantum level, where a given pair of histories may exhibit multiple direct evolutionary relationships.

Absent from Definition 4 is any specification of classical dynamics. This reflects the philosophy that physics at the fundamental scale should be described in quantum-theoretic terms. Classical equations of motion should emerge at larger scales from underlying quantum dynamics, according to a generalized version of the correspondence principle. All histories obeying suitable axioms should contribute to this dynamics, with contributions of "well-behaved" histories reinforced via constructive interference, and contributions of "pathological" histories damped out. There should be no artificial distinction between "on-shell" histories that obey preconceived classical dynamics, and "off-shell" histories that do not. All permissible histories should begin on an equal footing, just as all permissible paths begin on equal footing in conventional path integration.

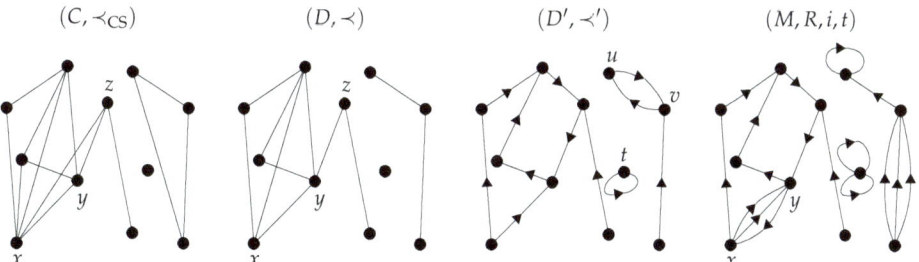

Figure 3. Causal set; acyclic directed set; directed set; multidirected set.

Structurally attractive models need not be relevant to the actual universe. Genuinely interesting models exhibit solid connections to established physics. For discrete causal theory, such connections are provided by the metric recovery theorems of Hawking [23] and Malament [24], and their generalizations [25–27]. Informally, these theorems state that *the causal structure of relativistic spacetime determines its geometric structure up to scale*. The causal metric hypothesis [14–16] strengthens and generalizes this statement by removing dependence on relativity and the caveat "up to scale". If spacetime is precisely smooth and Lorentzian to arbitrary scales, then the causal metric hypothesis is not quite true, due to this missing scale data. Hence, the hypothesis relies on the assumption that such data arises in the actual universe from some natural source *other than* a Lorentzian metric. What Finkelstein [3,4], Myrheim [28], 't Hooft [29], Sorkin [2], and others realized by around 1980 was that *discrete* causal structure supplies its own natural notion of scale via enumeration of fundamental elements. Later, it became popular to admit fluctuations in the sizes of elements to preserve systematic Lorentz invariance [30,31]. The generalized measure axiom in Definition 4 further relaxes this picture to allow the possible contribution of relation structure in determining volume. However, the basic lesson of metric recovery is unchanged by these modifications: discrete causal structure supplies natural scale data absent in continuous causal structure. Hence, Lorentzian geometry at large scales may be reasonably attributed to discrete causal structure at the fundamental scale.

2.3. Relation Space

A gem of structural philosophy from pure mathematics is Grothendieck's *relative viewpoint*, which emphasizes the study of objects *together with their natural relationships*. In discrete causal theory, the relative viewpoint is a conceptual tool of tremendous power and scope. A natural relationship between a pair of events in this setting is just a causal relationship, represented by a relation $x \prec y$ between elements x and y of a directed set $D = (D, \prec)$. The collection of all such relations is just the binary relation $\prec$. It is surprisingly useful to view $\prec$ as *a directed set in its own right*, by recognizing "relations between pairs of relations". The resulting object $\mathcal{R}(D)$ is called the *relation space* over D. Definition 5, adapted from Definition 5.1.1 of [14], generalizes this idea to multidirected sets.

Definition 5. *Let $M = (M, R, i, t)$ be a multidirected set, and let r_0 and r_1 be elements of its relation set R.*

1. *The* **induced relation** $\prec$ *on R is defined by setting $r_0 \prec r_1$ if and only if $t(r_0) = i(r_1)$.*
2. *The directed set $\mathcal{R}(M) = (R, \prec)$ is called the* **relation space over** *M.*

The induced relation involves a new use of the precursor symbol $\prec$. Figure 4, adapted from Figure 5.1.3 of [14], illustrates the relation space $\mathcal{R}(D)$ over an acyclic directed set D. The left-hand diagram shows the construction of an individual relation $r_0 \prec r_1$, while the right-hand diagram shows $\mathcal{R}(D)$ as a whole. More generally, $\mathcal{R}(M)$ may be identified with the *line digraph* [32] over the directed multigraph corresponding to M. Theorem 6 gives the essential properties of relation space.

Theorem 6. *Passage to relation space defines a functor $\mathcal{R}$ from the category $\mathcal{M}$ of multidirected sets to the category $\mathcal{D}$ of directed sets. This functor sends acyclic multidirected sets to irreducible acyclic directed sets, and preserves star finiteness.*

Proof. See [14], Theorem 5.1.4. □

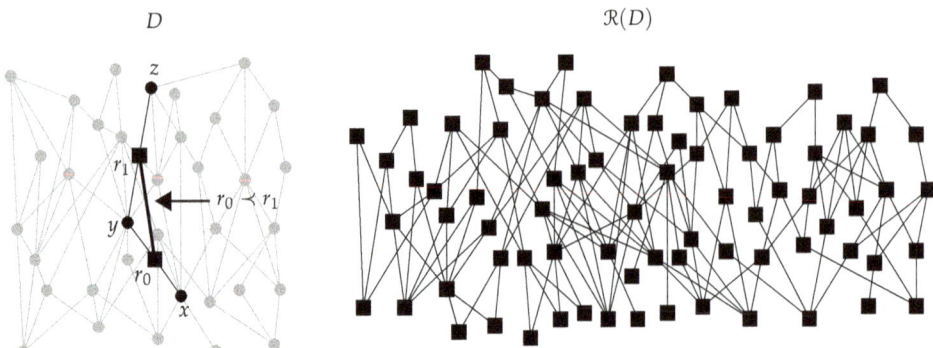

Figure 4. Induced relation between relations r_0 and r_1 in a directed set D; global view of $\mathcal{R}(D)$.

An important application of relation space in discrete causal theory is to eliminate a technical problem called *permeability* [33,34], which obstructs formulation and solution of initial value problems. In such a problem, one begins by specifying information associated with a maximal antichain σ in a directed set D, which is analogous to a spatial section of relativistic spacetime. One then attempts to solve for corresponding data throughout the future of σ. In general relativity, a *Cauchy surface* σ in a Lorentzian manifold X is an *impermeable* maximal antichain with respect to the causal structure of X, meaning that every inextensible causal curve in X intersects σ. Cauchy surfaces are useful for formulating initial value problems, because information cannot permeate a Cauchy surface σ to affect its future without being "filtered" by σ. Lorentzian manifolds containing Cauchy surfaces are called *globally hyperbolic*. The left-hand diagram in Figure 5, adapted from Figure 5.4.1 of [14], illustrates two causal curves intersecting a Cauchy surface in a globally hyperbolic manifold.

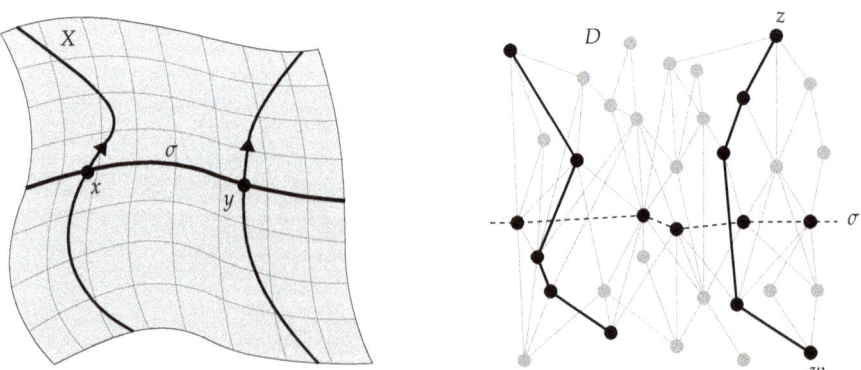

Figure 5. Cauchy surface σ in a globally hyperbolic manifold X, intersected by two causal curves; maximal antichain σ in a directed set D, permeated by two chains.

In discrete causal theory, a typical maximal antichain σ in a typical directed set D is *permeable*, meaning that chains in D may pass through σ from past to future without intersecting σ. In causal set theory [33], this phenomenon is referred to as "missing links"; the antichain σ is compared to a "sieve" [34], which is "by-passed" by a "large amount of geometric information". "Thickened antichains", obtained by adding limited quantities of past and future elements to σ, typically suffer from the same problem. Hence, maximal antichains are not good analogues of Cauchy surfaces in causal set theory, and the same statement applies to discrete causal theory in general. The right-hand diagram in Figure 5 illustrates a pair of chains permeating a maximal antichain σ in an acyclic directed set. The dashed lines connecting the elements of σ are a visual aid, not part of the structure. Permeability means that information can leak through σ, for example, from w to z. Besides posing a general obstacle to discrete causal dynamics, this problem also has as a specific bearing on the definition and analysis of entropic quantities, again typified in the causal set context [35,36]. Fortunately, however, this problem disappears upon passage to relation space.

Theorem 7. *Maximal antichains in relation space are impermeable. That is, if σ is a maximal antichain in the relation space $\mathcal{R}(M)$ over a multidirected set M, and if γ is a chain of relations in $\mathcal{R}(M)$ beginning at an element in the past of σ and terminating at an element in the future of σ, then γ intersects σ.*

Proof. See [14], Theorem 5.4.3. □

Path summation in discrete causal theory is described in terms of impermeable antichains, and therefore depends on the theory of relation space in an essential way.

2.4. Quantum Theory

Just as relations between pairs of events are central to discrete classical causal theory, so directed relationships between pairs of histories are central to discrete quantum causal theory. These relationships are called *co-relative histories*. The word "relative" refers to the relative viewpoint, while the prefix "co" derives from covariant constructions in category theory. The physical interpretation of a co-relative history is that it encodes the evolution of one history into another. The left-hand diagram in Figure 6, adapted from Figure 6.4.6 of [14], illustrates a family of four co-relative histories sharing a common initial history, called a *cobase*. The right-hand diagram illustrates how these co-relative histories are represented by morphisms of directed sets.

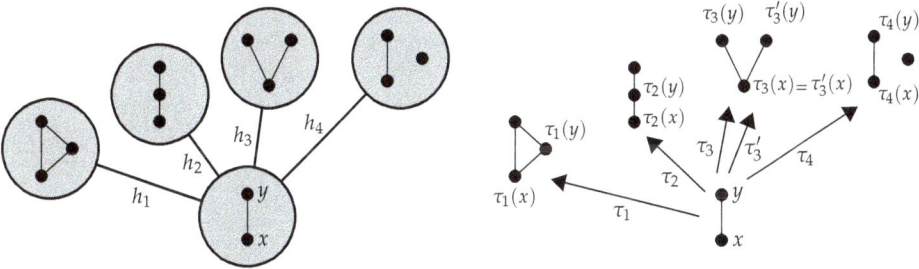

Figure 6. Four co-relative histories sharing a common cobase with two elements x and y and one relation $x \prec y$; morphisms (transitions) representing these co-relative histories.

Individual morphisms in the category $\mathcal{D}$ of directed sets do not always uniquely represent evolutionary relationships, due to symmetries. For example, the co-relative history h_3 in Figure 6 is represented by two different morphisms τ_3 and τ'_3, due to the symmetry interchanging the two maximal elements of its target history. Hence, co-relative histories are defined as equivalence classes of morphisms. It is convenient to restrict attention to special morphisms called *transitions*, which represent "growth" of directed sets. This idea is made precise in Definition 8, adapted from Definition 6.3.4 of [14]. Co-relative histories are then introduced in Definition 9, adapted from Definition 6.4.3 of [14].

Definition 8. *A* **transition** *in the category $\mathcal{D}$ of directed sets is a monomorphism $\tau : D \to D'$, embedding its* **source** *D into its* **target**, *D', as a proper, full, originary subobject. Here, "proper" means that $\tau(D)$ has nontrivial complement in D', "full" means that $\tau(x) \prec \tau(y)$ in D' if and only if $x \prec y$ in D, and "originary" means that the isomorphic image $\tau(D)$ of D in D' contains its own past.*

At a less-formal level, the condition that τ is a monomorphism means that τ does not "erase" details of the source D. The "proper" condition means that τ encodes nontrivial change. The "full" condition means that τ does not "edit" details of D. The "originary" condition means that τ does not add "prehistory" to D. These conditions support the desired evolutionary interpretation.

Definition 9. *A* **proper, full, originary co-relative history** *$h : D_i \Rightarrow D_t$ is an equivalence class of transitions $\tau : D_i \to D_t$, where two transitions τ and τ' are equivalent if and only if there exists an automorphism β of D_t mapping $\tau(D_i)$ onto $\tau'(D_i)$. The common source D_i of the transitions representing h is called the* **cobase** *of h, and the common target D_t of these transitions is called the* **target** *of h.*

The subscripts i and t in the expression $h : D_i \Rightarrow D_t$ stand for "initial" and "terminal". This notation is different from the notation for arbitrary transitions in Definition 8, since Sections 3 and 4 feature auxiliary transitions related to h that do *not* belong to the equivalence class defining h. The proper, full, and originary conditions in Definition 9 allow the unadorned term "co-relative history" to mean something more general, but co-relative histories in this paper always satisfy these conditions, except in the context of superset microstates in Definition 15, where they need not be full. Each transition in the equivalence class defining h is said to *represent* h. The "double arrow" notation $\Rightarrow$ emphasizes that h may be represented by more than one transition, but often h is uniquely represented due to the *rigidity* of typical "large" directed sets [37], which plays an important role in Sections 3 and 4. It is useful to think of h as "adding elements and relations to D_i to produce D_t", but one cannot always identify *specific* elements and relations as "the ones added" since h is an equivalence class. Multiple inequivalent transitions, and hence multiple co-relative histories, may exist between a given pair of directed sets, even a pair differing by a single element. This implies multidirected structure at the quantum level.

Choosing a suitable family $\mathcal{K}$ of directed sets, together with a suitable family $\mathcal{H}$ of co-relative histories between pairs of members of $\mathcal{K}$, one obtains a structure $\mathbb{S}$ called a *kinematic scheme,* which serves as a history configuration space. The word "kinematic" means that $\mathbb{S}$ encodes *possible* behavior, without identifying what specific behavior is determined or favored under specific conditions. The latter question involves dynamics. As an analogy, relativistic kinematics describes possible particle paths, e.g., ruling out spacelike motion, but the paths of specific particles depend on dynamical information. $\mathbb{S}$ possesses natural multidirected structure induced by $\mathcal{H}$, elaborated below. Sequences of co-relative histories in $\mathbb{S}$ define evolutionary pathways called *co-relative kinematics,* abstractly analogous to particle paths in conventional path summation. The conditions that $\mathbb{S}$ must satisfy to qualify as a kinematic scheme are that $\mathcal{H}$ must include enough co-relative histories to describe the evolution of any history in $\mathcal{K}$, and $\mathcal{K}$ must contain all "ancestors" of its members. These conditions are made precise in Definition 10, adapted from Definitions 7.4.1 and 7.4.7 of [14]. An additional desirable property, called the *generational property,* allows each co-relative history in $\mathcal{H}$ to be "factored into generations". However, this property is not studied in this paper, and it is preferable to omit it from the definition.

Definition 10. *A* **kinematic scheme** *is a pair $\mathbb{S} = (\mathcal{K}, \mathcal{H})$, where $\mathcal{K}$ is a class of directed sets, and $\mathcal{H}$ is a class of co-relative histories between pairs of members of $\mathcal{K}$ satisfying the following properties:*

1. **Accessibility**: If D is in $\mathcal{K}$, then there exists a sequence of co-relative histories in $\mathcal{H}$ terminating at D.
2. **Hereditary property**: $\mathcal{K}$ is closed under the formation of proper, full, originary subobjects.

Figure 7, adapted from Figure 7.5.2 of [14], illustrates a portion of a kinematic scheme $\mathbb{S}_{PS}$ called the *positive sequential kinematic scheme*, which serves as a source of examples throughout the remainder of the paper. $\mathbb{S}_{PS}$ is modeled after a kinematic scheme of finite causal sets appearing implicitly in Sorkin and Rideout's theory of *sequential growth dynamics* [38]. Similar structures appear elsewhere in the work of Sorkin [39], Isham [40–43], Markopoulou [7], and others. The objects illustrated inside each large open node in the figure are members of the class $\mathcal{K}$ of directed sets of $\mathbb{S}_{PS}$, which is the class of *finite* acyclic directed sets. This class is more restrictive than the class specified by Definition 4, which requires only countability. The edges connecting the large open nodes represent members of the class $\mathcal{H}$ of co-relative histories of $\mathbb{S}_{PS}$, which are those that "add a single new element to their targets". This means that if $h : D_i \Rightarrow D_t$ belongs to $\mathcal{H}$, and if $\tau : D_i \to D_t$ is a transition representing h, then the complement of $\tau(D_i)$ in D_t is a singleton. The gray-colored nodes illustrate how the set of four co-relative histories appearing in Figure 6 embeds into $\mathbb{S}_{PS}$. The thickened edges illustrate a co-relative kinematics in $\mathbb{S}_{PS}$, whereby the empty set $\varnothing$ evolves into a directed set D with four elements and three relations. The specific transition or transitions representing each co-relative history illustrated in the figure may be inferred in a straightforward manner from the directed structures of its cobase and target; for example, there is a unique transition τ representing the final co-relative history in the co-relative kinematics terminating at D. The "new element added by τ", i.e., the complement of the image of τ, is the top-right element indicated by the arrow.

Figure 7. Positive sequential kinematic scheme $\mathbb{S}_{PS}$ (first four generations); gray nodes show the four co-relative histories from Figure 6; thickened edges illustrate a co-relative kinematics.

Given a kinematic scheme $\mathbb{S} = (\mathcal{K}, \mathcal{H})$, it is useful to associate an abstract multidirected set $\mathcal{M}(\mathbb{S})$ with $\mathbb{S}$, where each member D of $\mathcal{K}$ is represented by an element $x(D)$ of $\mathcal{M}(\mathbb{S})$, and where each member $h : D_i \Rightarrow D_t$ of $\mathcal{H}$ is represented by a relation $r(h)$ from $x(D_i)$ to $x(D_t)$ in $\mathcal{M}(\mathbb{S})$.

$\mathcal{M}(\mathbb{S})$ is called the *underlying multidirected set* of $\mathbb{S}$. Chains in $\mathcal{M}(\mathbb{S})$ represent co-relative kinematics in $\mathbb{S}$. The left-hand diagram in Figure 8, adapted from Figure 7.5.4 of [14], illustrates a portion of the underlying multidirected set $\mathcal{M}(\mathbb{S}_{PS})$ of the positive sequential kinematic scheme $\mathbb{S}_{PS}$. The chain from $x(\oslash)$ to $x(D)$ represents the co-relative kinematics from $\oslash$ to D illustrated in Figure 7. This diagram illustrates the permeability problem in the context of kinematic schemes; the three nodes connected by the auxiliary dashed lines represent a maximal antichain in $\mathcal{M}(\mathbb{S}_{PS})$, which is permeated by the chain from $x(\oslash)$ to $x(D)$. It is therefore necessary to work in relation space to properly formulate the theory of path summation. The right-hand diagram in Figure 8 illustrates part of the relation space $\mathcal{R}(\mathcal{M}(\mathbb{S}_{PS}))$. The dark square nodes represent a maximal antichain, which is impermeable by Theorem 7.

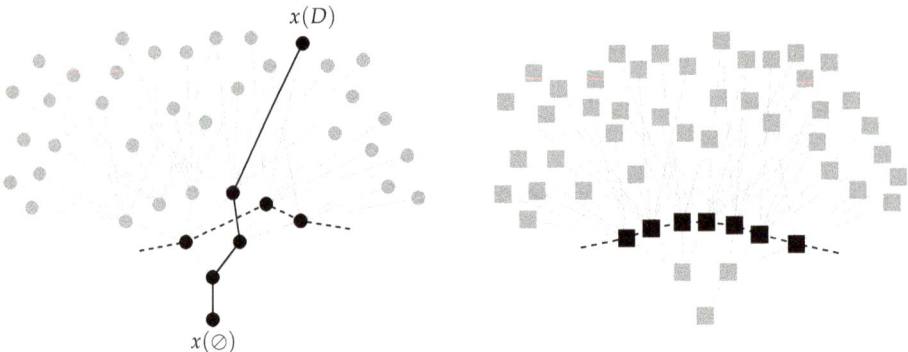

Figure 8. Portion of $\mathcal{M}(\mathbb{S}_{PS})$ illustrating the permeability problem; corresponding portion of $\mathcal{R}(\mathcal{M}(\mathbb{S}_{PS}))$ showing an impermeable maximal antichain.

While one could choose to perform path summation over a particular acyclic directed set, the resulting theory would be background dependent, and hence unsuitable for quantum gravity. Path summation in the background independent context involves summing phases $\Theta(\gamma)$ associated with co-relative kinematics γ in a kinematic scheme $\mathbb{S}$. As explained in Section 1.4, these phases are analogous to Feynman's phases $e^{\frac{i}{\hbar}S(\gamma)}$. Under modest assumptions, $\Theta(\gamma)$ is a product of phases $\theta(r)$ of individual relations representing individual co-relative histories. The relation function θ determines a specific form for Equation (4)

$$\psi^-_{R;\theta}(r) = \theta(r) \sum_{r^- \prec r} \psi^-_{R;\theta}(r^-),$$

reproduced here for convenience. The setup for deriving this equation is illustrated in Figure 9, adapted from Figure 6.9.2 of [14], where the derivation is carried out in detail. The auxiliary shading represents a finite subobject R of the relation space $\mathcal{R}(\mathcal{M}(\mathbb{S}))$. A choice of maximal antichain σ partitions R into a disjoint union $R = R^- \sqcup \sigma \sqcup R^+$, where σ represents a choice of "present", and $R^\pm$ are the corresponding past and future regions. The function $\psi^-_{R;\theta}$ is called the *past state function*, because it depends on all chains in R^-, which terminate at elements of σ. Here, one such chain γ is shown, terminating at an element $r \in \sigma$, with penultimate element r^-. This chain may be factored into a concatenation product $\gamma^- \sqcup r$, where γ^- is the subchain of γ terminating at r^-, and this factorization induces a factorization $\Theta(\gamma) = \Theta(\gamma^-)\theta(r)$ of phases. The value $\psi^-_{R;\theta}(r)$ is defined to be the sum $\sum_\gamma \Theta(\gamma)$ of the phases of all maximal chains γ in R^- terminating at r. Mathematically, Equation (4) merely organizes the factorizations $\Theta(\gamma) = \Theta(\gamma^-)\theta(r)$ for all such γ. These chains represent co-relative kinematics in the corresponding region of $\mathbb{S}$ that lead to the target history of the co-relative history represented by r. Generalizing to the case of infinite R raises questions of convergence. From an abstract perspective, the function $\psi^-_{R;\theta}$ plays a role similar to that of Feynman's "wave function" ([1], Section 5), except that no limiting process is necessary to define it, and no normalization constant is required. However, the structural context in which $\psi^-_{R;\theta}$

arises is much different than in Feynman's original non-relativistic background dependent setup, where evolutionary pathways are represented by paths in a fixed copy of $\mathbb{R}^4$. In the present discrete background independent context, each step along a chain represents a co-relative history, interpreted as the evolution of one spacetime into another. Equation (4) describes how the value of $\psi^-_{R;\theta}$ changes when the evolutionary pathways involved are extended by one additional relation r, which corresponds to multiplying the associated phases by $\theta(r)$. Abstractly, it arises in almost the same manner as the ordinary Schrödinger equation under Feynman's derivation ([1], Section 6), in which segmented paths approximating continuous evolutionary processes are extended via a time-stepping method. For Equation (4), however, no approximation is involved, so no limiting process is necessary.

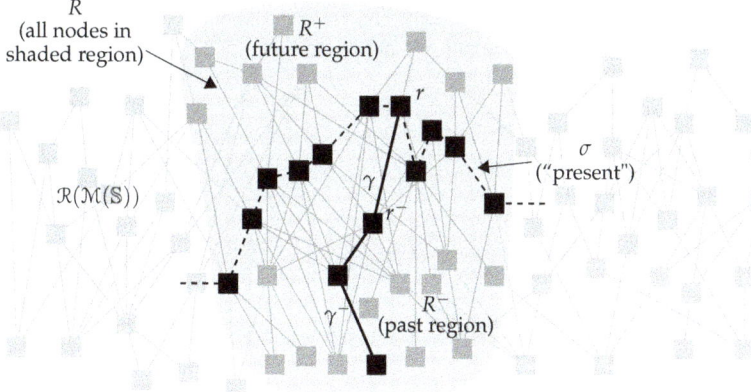

Figure 9. Setup for deriving Equation (4): $\gamma = \gamma^- \sqcup r$ and $\Theta(\gamma) = \Theta(\gamma^-)\theta(r)$.

A few further remarks regarding Equation (4) may be helpful. First, it is illuminating to spell out how the equation can describe *quantum-theoretic* behavior specifically. This depends partly on the general properties of path summation, and partly on the choice of relation function θ that determines the phase associated with each evolutionary pathway. Like virtually any formula involving path summation over a history configuration space, Equation (4) combines contributions from many distinct processes involving many distinct histories. This is a familiar feature of quantum-theoretic superposition, but is not unique to the quantum realm. For example, *classical stochastic* models such as Sorkin and Rideout's theory of sequential growth dynamics [38] organize information in a similar manner at an abstract level, but are decidedly non-quantum. The classical nature of the latter theory arises from the assignment of *real probabilities*, rather than quantum amplitudes, to evolutionary pathways. Similarly, Feynman's derivation [1] could just as easily be used to produce a continuous classical stochastic model, with real probabilities assigned to subspaces of a path space. What leads to Schrödinger's equation specifically under Feynman's setup is *Feynman's choice of phase map*, which produces the type of interference effects necessary to describe quantum-theoretic behavior. Similar considerations apply in the discrete causal context. For different choices of θ, Equation (4) could be used to describe a classical stochastic model, or a quantum-theoretic model, or neither. This highlights why the choice of phase map is so crucial to the theory. As described in Section 1.4, the most obvious choice of target object for a quantum-theoretic phase map is the choice made by Feynman, namely, S^1. Alternative choices can be interesting, but this paper focuses on S^1-valued phase maps almost exclusively. Second, due to the quantum-gravity-related focus of this paper, it is worth noting that Equation (4) shares certain similarities with the *Wheeler-Dewitt equation*, but these are not explored here. Third, allowing cycles complicates the picture, and this generalization is not considered here. Fourth, many different kinematic schemes typically share a given class $\mathcal{K}$ of directed sets, and different schemes offer different perspectives regarding the evolution of families of histories. Physical predictions must be independent of these choices, and this is expressed by saying that the theory must be *covariant*. In practical terms, this means that if one changes $\mathbb{S}$, then one generally must change θ to compensate. This paper mostly ignores covariance issues.

Figure 10 illustrates a sequential growth process in $\mathbb{S}_{PS}$, in which a history D_7 with seven elements evolves into a history D_{11} with eleven elements via a sequence of co-relative histories labeled h_7 to h_{10}. These co-relative histories are represented by relations $r(h_7)$ to $r(h_{10})$ in $\mathcal{R}(\mathcal{M}(\mathbb{S}_{PS}))$, abbreviated by r_7 to r_{10}. This growth process serves as a source of examples in Sections 3 and 4. Each pair of consecutive histories in Figure 10 encodes the same type of information associated with a single square node in Figure 9, since these nodes represent co-relative histories. Given such a process, the goal is to define phases measuring the "favorabilities" of each co-relative history. The black nodes and edges represent the *first-degree terminal states* $T^1(D_7)$ to $T^1(D_{11})$ of the histories D_7 to D_{11}, which encode the first-order information in each history, i.e., the "physically new" information, consisting of only the most recent causes and effects. First-degree terminal states are featured repeatedly in Chapters 7 and 8 of [14], where they are described via terminology such as "structural increments" or "generations". By definition, only *one* element in each history is "new" from the perspective of the sequential growth process itself; these new elements are indicated by arrows. However, this process is merely one way of describing the evolution of D_{11}, and therefore involves arbitrary extraphysical choices regarding the order of appearance of elements. Terminal states $T^n(D)$ of degree n are introduced in Definition 13. For $n > 1$, there is a distinction between degree and order; for example, second-degree terminal states may encode information of arbitrarily high order. It is convenient to use the abbreviation Δ_k for $T^1(D_k)$, which highlights the fact that Δ_k is a "structural increment" of D_k. To avoid clutter, only Δ_8 is labeled in the figure. The symbol Δ is used in later sections to denote states of arbitrary degree.

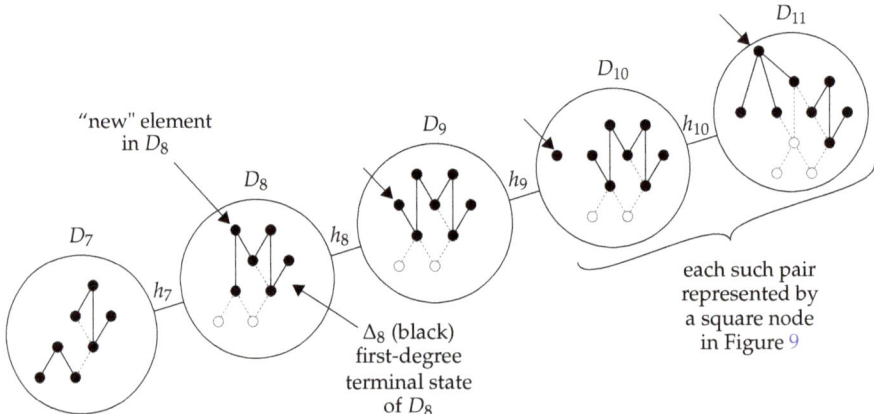

Figure 10. Sequence of co-relative histories in $\mathbb{S}_{PS}$; terminal states indicated by dark nodes and edges; "new elements" added by each co-relative history indicated by arrows.

First-degree terminal states are analogous to "present states" in conventional physics, involving data up to first order, such as position and velocity. Familiar notions of entropy are associated with such "present states", *not* with entire histories. In particular, the second law of thermodynamics compares the entropy of a "present state" to that of "previous states"; it does not involve a "higher-dimensional entropy" associated with the entire history leading up to the present state. The evolution of physical systems does not seem to be sensitive to details of the distant past; otherwise, one could not perform reliable experiments without knowing the exact history of each piece of experimental equipment. More formally, Lagrangians are typically assumed to depend on information only up to first order. The form of Equation (4) imposes an analogous assumption at the level of kinematic schemes, since the relation function θ is analogous to a Lagrangian on $\mathbb{S}$. As discussed in Section 3.3, higher-order information at the level of individual histories is not a priori irrelevant in discrete causal theory, but contributions from the distant past likely play a negligible dynamical role. Hence, the simplest "serious" entropic phase maps are defined in terms of first-degree terminal states, and more-sophisticated phase maps may be regarded as refinements of such maps.

3. Entropy and the Second Law of Thermodynamics

3.1. Entropy

Entropy, in the statistical sense pioneered by Boltzmann, may be understood very generally in terms of the distinguishability of objects described at two different levels of detail, one regarded as fine, and the other regarded as coarse. The prototypical application of this idea occurs in statistical thermodynamics, in which the fine level of detail for a system, such as a fixed quantity of ideal gas, is described in terms of microscopic data, such as the positions and momenta of individual molecules, while the coarse level of detail is described in terms of macroscopic data, such as pressure, volume, and temperature. Each possible choice of macroscopic data defines a coarse description of the system, called a *macrostate*, while each possible choice of microscopic data defines a fine description, called a *microstate*. Each macrostate generally corresponds to many different microstates, since many different choices of microscopic data may be approximated by identical macroscopic data. The *entropy* of a macrostate measures the quantity of corresponding microstates in a manner that is additive for composite systems. In more general terms, objects distinguishable at some fine level of detail may be indistinguishable at some coarser level, and a notion of entropy may be associated with the two levels to quantify this difference in distinguishability. In particular, generalizations of Boltzmann entropy such as Gibbs, Shannon, and Rényi entropies fall under the same conceptual umbrella. Measures of entropy familiar in ordinary quantum theory, such as von Neumann entropy, are less relevant, since they depend on specific algebraic apparatus less general than the path summation approach.

In statistical thermodynamics, the *state space* for a system is an abstract space parameterizing the set of possible microstates of the system for some choice of fine detail. A choice of coarse detail partitions state space into a family of subsets representing the possible macrostates of the system, where the points of each subset parameterize the microstates associated with the corresponding macrostate. Such a partition is called a *coarse-graining* of the state space. The left-hand diagram in Figure 11 illustrates such a coarse-graining, where the *cells* representing macrostates are separated by solid lines. Dotted lines and labels are explained below. Such a planar diagram could be interpreted literally as encoding the possible position and momentum of a single particle moving in one real dimension, but all such diagrams in this paper are schematic. Conventional state spaces are real manifolds, and therefore exhibit notions of proximity, volume, and other topological and metric structure. However, their dimensions are typically quite large, and this implies properties that are not well-represented by planar diagrams; for example, each region typically has very many neighbors. Even in 24-dimensional Euclidean space, each sphere in the regular packing induced by the Leech lattice is tangent to 196,560 neighbors; one may imagine the situation in 10^{24}-dimensional space. Abstract metric-related ideas remain useful for describing the properties of discrete causal state spaces, but planar diagrams only roughly represent these notions.

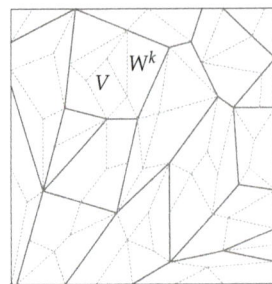 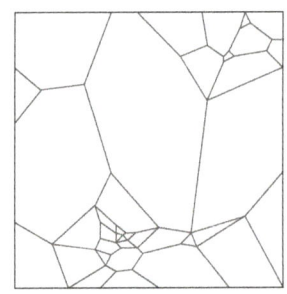 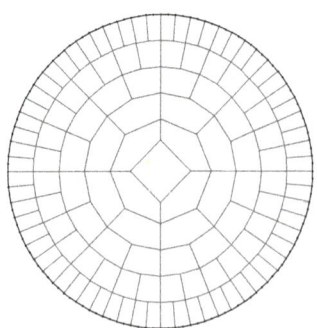

Figure 11. Partitions of state space; conventional state spaces exhibit regions of very different sizes; state space inducing an "inverse second law of thermodynamics".

Generalizing the thermodynamic picture, any set S of objects may be partitioned into a family of subsets P, where the objects belonging to each subset are regarded as equivalent at a coarse level of detail. More generally still, one may consider a strictly partially ordered family $\Pi := \{P^\alpha\}_{\alpha \in A}$ of partitions P^α of S for some index set A, where by definition $P^\alpha \prec P^\beta$ if $P^\alpha \neq P^\beta$ and if every member of P^α is a union of members of P^β. In this case, P^β is called a *refinement* of P^α. Here, $\prec$ does *not* represent causal structure, and superscript indices are used to distinguish information filtering from mere enumeration. One may define equivalence relations $\sim^\alpha$ on S for each α in A, where $s \sim^\alpha s'$ if s and s' belong to the same subset under P^α. If $P^\alpha \prec P^\beta$, then P^α induces a *quotient partition* $P^{\alpha\beta}$ of the quotient set $S^\beta := S/\sim^\beta$ in an obvious way. Any such choice of P^α and P^β may be used to define notions of coarse and fine detail. Returning to Figure 11 in this more abstract setting, the large regions bordered by solid lines in the left-hand diagram represent a choice P^α of coarse detail for a set S, while the small regions bordered by dotted lines represent a choice P^β of fine detail. Here, P^α and P^β each partition S into subsets of roughly equal size, but a typical coarse-graining in conventional thermodynamics exhibits vast differences in the sizes of regions, and correlations exist involving proximity and size. The middle diagram in Figure 11 illustrates such a coarse-graining. As emphasized by Penrose [44], such details are crucial for understanding whether a typical system can be expected to exhibit a systematic increase in entropy. For example, the right-hand diagram in Figure 11 illustrates a state space that induces an "inverse second law of thermodynamics", in the sense that a typical path in this space moves from larger to smaller cells. If $P^\alpha \prec P^\beta$, and if each member of P^α is a *finite* union of members of P^β, then one may define multiplicities and entropies via counting: if $V \subset S$ is a member of P^α, and if $V = \cup_{k=1}^{K} W^k$ for members W^k of P^β, then the multiplicity $\mu^{\alpha\beta}(V)$ of V is K, and the entropy $e^{\alpha\beta}(V)$ of V is $\log K$. The choice of notation for $\mu^{\alpha\beta}$ and $e^{\alpha\beta}$ is intended to emphasize the relative viewpoint: multiplicities and entropies are properly understood in terms of *natural relationships between levels of detail*, not in terms of any specific level of detail. For the set V shown in the left-hand diagram in Figure 11, the entropy is $e^{\alpha\beta}(V) = \log 7$, since P^β subdivides V into seven regions. In more general settings, it may be necessary to measure the sizes of members of $P^{\alpha\beta}$ via some measure $\mu^{\alpha\beta}$ other than the counting measure.

Definition 11. *An **entropy system** (S, Π, μ) consists of a set S, a set $\Pi := \{P^\alpha\}_{\alpha \in A}$ of partitions P^α of S for some index set A, strictly partially ordered by refinement, and a family μ of measures $\mu^{\alpha\beta}$ on the quotient sets S^β, one for each relation $P^\alpha \prec P^\beta$ in Π. Each such relation induces an **entropy quadruple** $(S, P^\alpha, P^\beta, \mu^{\alpha\beta})$. The **entropy** of a member V of P^α is $e^{\alpha\beta}(V) := \log \mu^{\alpha\beta}(V^\beta)$, where $V^\beta \subset S^\beta$ is the image of V under the quotient map $S \to S^\beta$, and where $\log \infty$ is understood to mean ∞.*

It is often convenient to denote an entropy quadruple by just S, or to write $S = (S, P^\alpha, P^\beta, \mu^{\alpha\beta})$ to indicate that a set S is equipped with such a structure. The functions $\mu^{\alpha\beta}$ are taken to be measures here for simplicity, but the situation could be generalized further. In particular, the target object of

$\mu^{\alpha\beta}$ need only be a totally ordered set. One may also abstain from using logarithms to "rescale" $\mu^{\alpha\beta}$. However, it suffices here to consider only the counting measure on a finite set or the Lebesgue measure on a finite-dimensional real manifold, and logarithms are useful for producing quantities that are additive for composite systems. The reason for using "e" instead of the familiar "h" for entropy is because "h" is used here to represent co-relative histories. Figure 12 illustrates a simple entropy system (S, Π, μ) whose underlying set S is the unit interval $[0,1]$ in $\mathbb{R}$. The set Π of partitions of S has members P^0, P^1, P^2, and P^3, which subdivide S into segments of equal lengths $1, 1/2, 1/3$, and $1/6$, respectively. P^0 is the trivial partition, under which S represents a single macrostate. The strict partial order $\prec$ on Π consists of five individual relations $P^0 \prec P^1$, $P^0 \prec P^2$, $P^0 \prec P^3$, $P^1 \prec P^3$, and $P^2 \prec P^3$, each of which induces an entropy quadruple. The quotient sets S^0, S^1, S^2, and S^3 have 1, 2, 3 and 6 elements, respectively. There are two nontrivial quotient partitions, P^{13} and P^{23}, which subdivide the quotient set S^3 into equal-sized subsets with 3 and 2 elements, respectively. Multiplicities and entropies of some representative subsets of S with respect to different entropy quadruples are also listed. For example, the subset $U = (\frac{1}{2}, 1]$ of S has measure $\mu^{13}(U) = 3$ and entropy $e^{13}(U) = \log 3$ with respect to the entropy quadruple (S, P^1, P^3, μ^{13}).

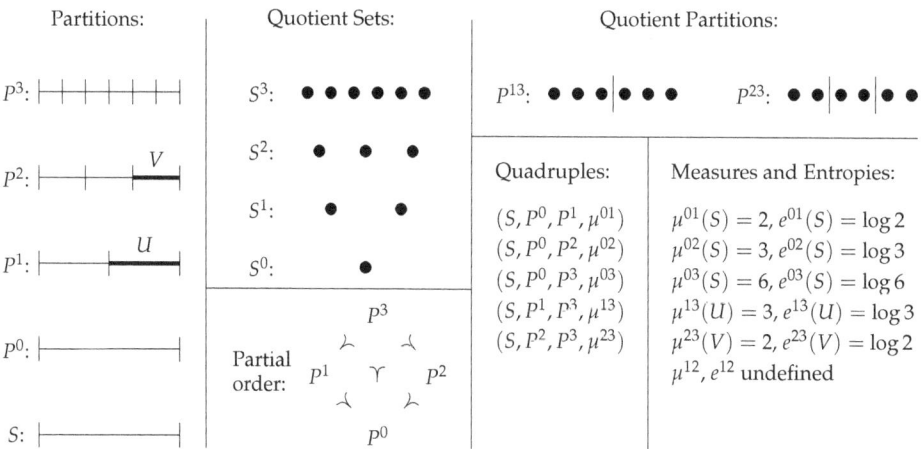

Figure 12. A simple entropy system on the unit interval $S = [0,1] \subset \mathbb{R}$.

The motivation for adopting such a general viewpoint is that multiple "levels" of entropy are evident in discrete causal theory. An important example involves the nth-degree terminal states $T^n(D)$ mentioned in Section 2.4 and formally introduced in Definition 13. Given two directed sets D and D', it may be the case that $T^n(D)$ and $T^n(D')$ are isomorphic, while $T^{n+1}(D)$ and $T^{n+1}(D')$ differ. In this case, D and D' are indistinguishable at the level of detail specified by the index value n, but become distinguishable at the finer level of detail specified by the index value $n + 1$. On the level of individual elements, two elements x and y belonging to a subobject Δ of a directed set D may be "locally indistinguishable", in the sense that they are interchanged by an automorphism of Δ, but may be "globally distinguishable", in the sense that no such automorphism extends to an automorphism of D. More generally, one may consider chains of subobjects $\Delta = \Delta^1 \subset \Delta^2 \subset ... \subset \Delta^n \subset D$ containing x and y, some of which possess automorphism groups interchanging x and y, and some of which do not. Of obvious interest is the case in which Δ^1 is a low-order terminal state of a history, and Δ^n for $n > 1$ are progressive "thickenings" of Δ.

While entropy is defined by associating entire families of "fine" states with individual "coarse" states, it is sometimes interesting to compare the amount of detail encoded by specific pairs of states. It is then natural to relate such "local comparisons" to the "global comparisons" leading to entropy systems. In this context, one need not distinguish a priori between macrostates and microstates; states are defined individually by specifying varying degrees and types information about an object or system, and are then compared and categorized. Given two such states Δ and Δ', it is sometimes possible to unambiguously identify Δ' as more detailed than Δ, or vice versa. In other cases, Δ and Δ' are incomparable, in the sense that Δ contains more of one type of information, while Δ' contains more of another. In this setting, one may recognize a natural partial order $\prec$ on the family of states under consideration, where $\Delta \prec \Delta'$ if and only if Δ' is unambiguously more detailed than Δ. This type of partial order is different from the partial orders on sets of partitions in Definition 11, but the two types of structure are related. For example, given an entropy quadruple $(S, P^\alpha, P^\beta, \mu^{\alpha\beta})$, the set $P^\alpha \cup P^\beta$ is a subset of the power set $\mathcal{P}(S)$ of all subsets of S. The relation $P^\alpha \prec P^\beta$ means that every member V of P^α is a union of members W of P^β. One may define an induced relation on $P^\alpha \cup P^\beta$, also denoted by $\prec$, where $V \prec W$ if and only if V is a proper superset of W. Hence, a single relation between two partitions induces a partial order on a corresponding family of subsets. This partial order is of a special type, with maximal chain length 1, because its only relations are those of the form $V \prec W$ for $V \in P^\alpha$ and $W \in P^\beta$ such that $W \subset V$. However, one may easily define partially ordered sets with longer chains by considering sequences of partitions $\ldots \prec P^n \prec P^{n+1} \prec \ldots$.

Working in the opposite direction, one may begin with a partial order $\prec$ on an arbitrary set Σ. Here, Σ is viewed as an abstract analogue of a family of states encoding various types and quantities of detail, while $\prec$ is viewed as an abstract analogue of the partial order relating pairs of states Δ and Δ' whenever Δ' is unambiguously more detailed than Δ. One may partition Σ into a family of antichains σ with respect to $\prec$. There are generally many different choices of partition, each analogous to a frame of reference in relativity. In the entropic setting, elements of a given antichain σ are viewed as abstract analogues of states sharing an equal level of detail. In the simplest case, the antichains σ "foliate" Σ, in the sense that each nonextremal antichain σ_k has an unambiguous maximal predecessor σ_{k-1} and minimal successor σ_{k+1}. More generally, the antichains σ form a partially ordered family. In either case, the partition defines an *atomic decomposition* of Σ with respect to $\prec$, an idea revisited in a different context in Section 3.3. In many cases, detail may be quantified in a variety of different ways, and this leads to the consideration of families $\{\prec^\alpha\}_{\alpha \in A}$ of partial orders on Σ. Such families are themselves partially ordered via the order-theoretic version of refinement, under which $\prec^\alpha$ precedes $\prec^\beta$ if and only if $\Delta \prec^\beta \Delta'$ whenever $\Delta \prec^\alpha \Delta'$. An antichain with respect to $\prec^\beta$ is then automatically an antichain with respect to $\prec^\alpha$, so any partition of Σ induced by $\prec^\beta$ refines at least one such partition induced by $\prec^\alpha$. In this manner, the partial ordering by refinement of the family of partitions induced by $\{\prec^\alpha\}_{\alpha \in A}$ respects the partial ordering on $\{\prec^\alpha\}_{\alpha \in A}$ itself. Hence, entropy systems defined in terms of such partitions automatically respect the order-theoretic structure of Σ.

3.2. The Second Law

The familiar intuition regarding the second law of thermodynamics is that "entropy increases with time". Generalizing this idea to apply to the broad framework of entropy systems introduced in Section 3.1 requires suitable analogues of "time" and "increase". Time evolution is conventionally represented by a directed curve in state space, and in this context the second law says that motion along such a curve tends to pass from smaller to larger cells in a specified coarse-graining. The left-hand diagram in Figure 13 illustrates such a curve γ. A typical curve originating in one of the two shaded areas is likely to exhibit a systematic increase in entropy, at least for early times, since such curves begin in small cells whose borders are dominated by larger cells. A typical curve originating elsewhere in the state space does not exhibit such an increase in entropy. This illustrates the fact that both the structure of state space and the region of origin of the curve describing the system of interest are relevant to the existence of a recognizable second law. In the cosmology of the early universe, for example, the question of why specific measures of entropy were initially relatively low is just as important as the question of why entropy increased thereafter [44].

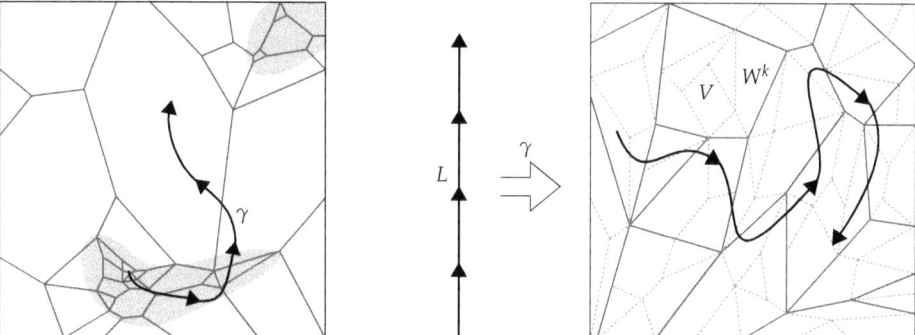

Figure 13. Curve in state space along which entropy increases; map from a linearly ordered set into an entropy quadruple, showing no discernible second law.

The abstract analogue of a directed curve in state space is a map γ from a linearly ordered set L into an entropy quadruple $S = (S, P^\alpha, P^\beta, \mu^{\alpha\beta})$. Such a map is illustrated in the right-hand diagram in Figure 13. Here, L is drawn to suggest an interval in $\mathbb{R}$, but in more general settings L may be a non-continuous object such as an interval in $\mathbb{Q}$, a discrete object such as an interval in $\mathbb{Z}$, a finite object such as the set $\{0, ..., N\}$, or even a transfinite object, such as the long line. The notion of an increasing function requires similar generalization beyond the familiar setting of real analysis. Even in conventional thermodynamics, strict definition of an increasing function must be relaxed, since the second law is understood not as a prescription that entropy *must* increase over any time interval, but as a description of the fact that entropy *does* increase with overwhelming likelihood over sufficiently long time intervals. The map γ in the figure passes through cells of multiplicities 5, 2, 3, 7, 6, 6, 7 (again), 4, 2, 4, and 6 (again). Hence, the associated system does not obey a discernible version of the second law. In the general case, it seems preferable to describe a variety of ways to define a version of the second law for such a system than to isolate a particular choice via formal definition. An individual map γ from a totally ordered set L into an entropy quadruple $S = (S, P^\alpha, P^\beta, \mu^{\alpha\beta})$, obeys a strict version of the second law if for every pair of subsets V and V' of S belonging to P^α, and for every pair of elements ℓ and ℓ' in L such that $\gamma(\ell) \in V$ and $\gamma(\ell') \in V'$, it is true that $\mu^{\alpha\beta}(V) \leq \mu^{\alpha\beta}(V')$. Intuitively, this means that γ never passes from a large cell into a smaller cell. There are various ways to relax this strict description. If L possesses a metric, then one may specify a rule relating the size of the interval (ℓ, ℓ') to the probability that $\mu^{\alpha\beta}(V) \leq \mu^{\alpha\beta}(V')$. If the target object of $\mu^{\alpha\beta}$ also possesses a metric, then one may define something like a derivative, i.e., a rule relating the sizes of the intervals $(\mu^{\alpha\beta}(V), \mu^{\alpha\beta}(V'))$ to the sizes of the corresponding intervals (ℓ, ℓ'). More generally, a region U of S obeys a version of the second law if a typical map $\gamma : L \to S$ originating in U obeys an individual version of the second law. The word "typical" may be made precise in terms of a generalized measure on the space of maps γ. It is sometimes necessary to restrict attention to special maps to obtain a clear pattern; for example, some entropy quadruples exhibit entropy increases along typical "short curves", but not along typical "long curves". In particular, some cosmological models posit a reversal of the second law in the distant past and/or future.

3.3. Discrete Causal State Spaces

In statistical thermodynamics, microstates are determined by information up to first order, e.g., by positions and momenta of individual molecules. Such information, together with the dynamical laws of classical mechanics, is sufficient to recover higher-order information; one may uniquely evolve a given state "backward in time". Hence, if two states are indistinguishable up to first order, then they are *absolutely* indistinguishable. In discrete causal theory, the situation is different. The analogue of information up to first order in a finite acyclic directed set D is its first-degree terminal state $T^1(D)$, which consists of all maximal elements of D, all relations terminating at these elements, and all initial elements of these relations. Knowledge of $T^1(D)$ generally does not enable recovery of D. One may propose a choice of classical dynamics implying such a relationship for very special classes of directed sets, for example, by abstracting the Einstein–Hilbert action from general relativity, which takes the form

$$S_{EH} = \frac{c^4}{16\pi G} \int_X R\sqrt{-\det(g)} d^4x, \tag{5}$$

in the simple vacuum case with zero cosmological constant. Here, g is a Lorentzian metric on a 4-dimensional manifold X, R is the curvature scalar arising from the metric connection, G is Newton's gravitational constant, and c is the speed of light. Yet despite interesting efforts in this direction, for example, in causal set theory [45–47], such a strategy is dubious due to the amount of geometric structure taken for granted in relativity. Geometric data such as metrics and curvature, and even "pre-geometric" data such as dimension and topology, are emergent notions in discrete causal theory. Action functionals in this context must be defined more fundamentally, and cannot be expected to produce straightforward analogues of deterministic, time-symmetric Euler–Lagrange-type equations that uniquely determine classical dynamics via information up to first order. In particular, elements of a directed set D that are indistinguishable up to first order, i.e., permuted by an automorphism of $T^1(D)$, may be *distinguishable* when one considers higher-order information. It is therefore necessary to consider higher-degree terminal states in what follows. The form of Equation (4) does assume that first-order information suffices at the level of kinematic schemes, in the sense that the phase of an arbitrary co-relative kinematics is the product of the phases of its individual co-relative histories. This picture may be generalized without leaving the general framework of path summation, but such generalization is not undertaken here. In any case, the latter phases do generally depend nontrivially on information above first order in the corresponding cobases and targets.

The simplest discrete causal analogues of familiar thermodynamic state spaces are *nth-order state spaces* $\mathbb{D}^n$, whose elements represent isomorphism classes of countable star finite acyclic directed sets Δ with maximal chain length n. Equivalently, $\mathcal{R}^n(\Delta)$ is a nonempty antichain. It is useful to preface formal definitions involving $\mathbb{D}^n$ with some informal remarks. First, while the notion of order identifying a state Δ as a member of $\mathbb{D}^n$ is intrinsic to Δ itself, the desired interpretation of Δ is as a terminal state of a history D, containing information encoded by chains of length at most n terminating at maximal elements of D. Second, it is usually impossible to choose a member of $\mathbb{D}^n$ that includes *all* such information for $n > 1$, because chains of length at most n terminating at different maximal elements of D may intersect to produce longer chains, thereby defining a higher-order state. One might consider re-defining $\mathbb{D}^n$ to include such states, requiring only that each element be connected to a maximal element by at least one chain of length at most n. In physical terms, such states are still composed of elements exerting "recent influence", but may contain chains of arbitrary length. However, such a definition would not be ideal for the desired applications. For example, it would allow any countable star finite acyclic directed set in which all chains are bounded above to be converted to a member of $\mathbb{D}^1$ or $\mathbb{D}^2$ by adding new relations terminating at new maximal elements, thereby flouting the intuition that low-order states should be "causally simple". It is preferable to define a separate notion called *degree*, which facilitates the definition of terminal states containing all information up to a given order in a particular history. Following this idea, Definition 13 introduces special states $T^n(D)$, called *nth-degree terminal states*, which include all information encoded in chains of length at most n terminating at a maximal element in D. Third, as mentioned in Section 2.4, the distinction between order and degree does not arise for $n = 1$; the first-degree terminal state $T^1(D)$ of D automatically belongs to $\mathbb{D}^1$. Fourth, the *nth superset microstates* introduced in Definition 18 are constructed by adding n "prehistorical" elements to a state, which may not increase its maximal chain length at all. These subtleties reflect the fact that more than one natural-number grading is useful in studying discrete causal state spaces.

It is useful to define terminal states in terms of transitions between *pairs* of histories, using the relative viewpoint. Though the ultimate goal is to use information encoded in terminal states to assign phases to sequences of co-relative histories, i.e., co-relative kinematics, the states of principal interest in studying a given co-relative history $h : D_i \Rightarrow D_t$ are typically *not* those induced by transitions representing h. This is because the "physically new" structure associated with D_i and D_t is more meaningful than whatever structure h "adds to" D_i to produce D_t. For example, each co-relative history $h : D_i \Rightarrow D_t$ in $\mathbb{S}_{PS}$ adds only one element to D_i, so most of the physically new structure in D_t is typically already present in D_i. Yet what one is really interested in is whether or not the physically new structure in D_t is "more favorable" than the physically new structure in D_i; i.e., one wishes to compare terminal states of D_i and D_t. These may be defined in terms of auxiliary transitions that are determined by h, but do not represent h under Definition 9. First, however, one must define terminal states associated with arbitrary transitions.

Definition 12. *Let $\tau : D \to D'$ be a transition of acyclic directed sets. The subobject Δ^τ of D' consisting of all elements of $D' - \tau(D)$, all relations terminating at such elements, and all initial elements of such relations, is called the **terminal state** of τ. If $\mathcal{R}^n(\Delta^\tau)$ is a nonempty antichain, then the **order** $\mathrm{ord}(\Delta^\tau)$ of Δ^τ is n.*

Despite the relative nature of Definition 12, it is convenient to refer to Δ^τ as a terminal state of the target set D' in many cases. Δ^τ does *not* include relations between elements of $\tau(D)$; it includes only relations that are "new" with respect to τ. If the context is expanded to include cycles, a different definition of order is necessary. For example, one may define $\mathrm{ord}(\Delta^\tau)$ to be the maximal length of non-self-intersecting chains in Δ^τ. Here, however, I focus almost exclusively on the acyclic case. Any directed set D' is itself the terminal state of the unique transition $\oslash \to D'$. This transition may be denoted by $\tau_\oslash$ when the choice of target set D' is obvious. As mentioned above, is useful to define special terminal states that encode *all* information up to order n in a given history.

Definition 13. *Let D be an acyclic directed set in which every chain is bounded above.*

1. *The **nth-degree terminal state** $T^n(D)$ of D is the subobject of D consisting of all elements connected to a maximal element of D by a chain of length at most n, together with all relations in such chains.*
2. *The **nth-degree initial state** $I^n(D)$ of D is the subobject of D constructed by deleting all non-minimal elements of $T^n(D)$ from D, together with all relations in D terminating at such elements.*
3. *The **nth-degree transition** $\tau_D^n : I^n(D) \to D$ associated with D is the inclusion map $I^n(D) \to D$.*

The boundedness hypothesis in Definition 13 is included to rule out situations in which D has maximal elements but also has chains "extending to infinity", since it is awkward to exclude such chains from consideration when studying terminal behavior. Such histories are not considered here.

Definition 14. *The nth-order state space $\mathbb{D}^n$ is the set of all isomorphism classes of countable star finite acyclic directed sets Δ such that $\mathcal{R}^n(\Delta)$ is a nonempty antichain. The **finite-order state space** $\mathbb{D}$ is the disjoint union $\coprod_{n=0}^\infty \mathbb{D}^n$, and the **(total, countable, acyclic) state space** $\overline{\mathbb{D}}$ is the set of all isomorphism classes of countable acyclic directed sets, which may be viewed as limits of sequences in $\mathbb{D}$.*

Since the elements and relations in a member Δ of $\mathbb{D}^n$ are assumed to possess no internal structure, one might expect Δ to be treated as a microstate. However, since discrete causal theory does not rule out the dynamical relevance of information above order n at the level of individual histories, data describing how Δ might fit into a larger history can be important in determining future behavior influenced by Δ. Such data defines an even finer level of detail than Δ itself, permitting Δ to be viewed as a macrostate. Ambiguity regarding the status of Δ is not surprising, due to the relative nature of entropy. Figure 14 illustrates four different methods of defining coarse and fine levels of detail using $\mathbb{D}^n$. Informal discussion of these methods then precedes formal treatment in Definition 15. The first diagram shows a third-order state Δ embedded in a history D. In this case, Δ does not contain all the third-order information in D; in particular, it is *not* the third-degree terminal state $T^3(D)$ of D. The second diagram illustrates one way to treat Δ as a microstate, called a *resolution microstate*, by approximating its structure via the method of *causal atomic resolution*, introduced in [14]. This method involves choosing special subsets of Δ, called *causal atoms*, which serve as individual elements of a coarser directed set. Such a choice defines a *causal atomic decomposition* of Δ. A sequence of such decompositions is a causal atomic resolution, with each subsequence defining "initial" and "terminal" levels of detail, and hence a notion of entropy. More generally, one may define partially ordered families of decompositions, also called resolutions, which induce entropy systems. The resolution in the figure involves a single decomposition, and hence just two levels of detail. Causal atomic resolution provides perhaps the most obvious discrete causal analogue of conventional coarse-graining. In particular, it involves actual approximation, meaning that the information contained in a causal atomic decomposition is not only incomplete, but also imprecise. However, there is generally no canonical choice of resolution for a given state, and different resolutions may be very dissimilar. Further, resolutions reaching far above the fundamental scale can produce objects that are obviously "too granular" to resemble physical spacetime. Members of $\mathbb{D}^n$ are usually treated as macrostates in this paper, but methods such as causal atomic resolution remain worthy of further study in more general entropic settings.

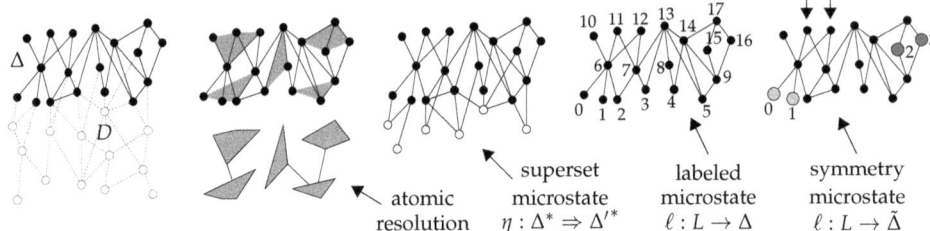

Figure 14. History D and terminal state Δ; causal atomic resolution of Δ; superset microstate of Δ; labeled microstate of Δ; symmetry microstate of Δ.

The third diagram in Figure 14 illustrates the most obvious way to treat a member Δ of $\mathbb{D}^n$ as a macrostate, by adding "prehistory" to define larger states called *superset microstates*. Different superset microstates of Δ impose different constraints on the family of histories of which Δ could be a terminal state. In particular, the superset Δ' of Δ shown in the diagram is induced by the history D. At a higher level of detail, Δ' may itself be viewed as a macrostate, with its own superset microstates adding more prehistory. One may imagine "flipping over" this diagram to obtain a co-relative history $\eta : \Delta^* \Rightarrow \Delta'^*$ between the causal duals Δ^* and Δ'^* of Δ and Δ', and this is how superset microstates are formalized in Definition 15. Hence, the convenient term "superset" is not quite precise, because co-relative histories involve equivalence classes. Naïve amalgamation of superset microstates produces a state space with an infinite number of elements in each cell, since one may always add more prehistory to a directed set. This leads a priori to infinite multiplicities and entropies for finite states. However, supersets adding "recent" data are expected to dominate dynamically, and families of superset microstates may be filtered to reflect this expectation. In the case of finite states, one may work with finite families of microstates defined in terms of numbers of elements and relations, lengths of chains, sizes of antichains, and similar quantities. Here, I focus on families defined via the number of prehistorical elements added to Δ. The quantity of superset microstates of a given type is decreased by symmetries of Δ, which render equivalent different subsets of Δ. This meshes with the intuition that high-entropy states should be "disordered". For example, if Δ is an antichain of cardinality K with automorphism group $\mathrm{Aut}(\Delta) \cong S_K$, then there is only one way to add a single prehistorical element and k relations to Δ for any $k \leq K$, since the terminal elements of these relations in Δ may be exchanged for any other k elements of Δ under $\text{Aut}(\Delta)$. By contrast, there are $\binom{K}{k}$ ways to add such an element and relations to Δ if $\text{Aut}(\Delta)$ is trivial.

The fourth and fifth diagrams in Figure 14 illustrate contrasting ways to treat a member Δ of $\mathbb{D}^n$ as a macrostate by focusing on its symmetries directly. Under the method illustrated in the fourth diagram, a microstate of Δ is simply a copy of Δ labeled via a map $\ell : L \to \Delta$, where L is a set of consecutive natural numbers starting with zero, and where two labelings are regarded as equivalent if they are related by an automorphism of Δ. Such a microstate is called a *labeled microstate*. The number of labeled microstates associated with a state Δ of cardinality K ranges from 1 if $\text{Aut}(\Delta) \cong S_K$ to $K!$ if $\text{Aut}(\Delta)$ is trivial. This method agrees qualitatively with the superset approach in the sense that high-entropy states are those for which $\text{Aut}(\Delta)$ is small. The method illustrated in the fifth diagram essentially reverses this relationship. Here, one begins with an arbitrary labeling $\ell : L \to \tilde{\Delta}$, where $\tilde{\Delta}$ is the subset of Δ not fixed by $\text{Aut}(\Delta)$. Automorphisms of Δ convert ℓ to other labelings, each of which represents a *symmetry microstate*. Such a microstate may be viewed as a "mode of symmetry breaking", since it breaks the symmetries of Δ in a specific way. For a finite state Δ, the number of symmetry microstates is just $|\text{Aut}(\Delta)|$, so high-entropy states are those for which $\text{Aut}(\Delta)$ is large. More generally, one may work with non-surjective *partial labelings* $\ell : L \to \tilde{\Delta}$ that leave a subgroup of $\text{Aut}(\Delta)$ unbroken. The labeling in the figure is of this type, since there remains an automorphism of Δ interchanging the elements indicated by arrows. The set of such partial labelings is partially ordered by extension, which is interesting from the perspective of state-specific detail discussed at the end of Section 3.1. While it is counterintuitive to associate high entropy with symmetry, there are arguments for entertaining such possibilities. Symmetry is central to the theory of "elementary" particles, so certain special structures that are locally symmetric, at least at measurable scales, are favored by the actual dynamics of the physical universe. Such structures may be "attached" to underlying causal structure via auxiliary algebraic information, but the strong interpretation of the causal metric hypothesis demands an emergent description of both spacetime symmetries and internal symmetries. The most obvious way to satisfy this demand is to incorporate some type of symmetry data directly into Equation (4). Notions of entropy associated with superset microstates and/or labeled microstates might accomplish a similar purpose, since their enumeration depends largely on symmetry considerations. Regardless of the type of entropy chosen, an attractive though speculative idea is that elementary particles might arise via *local entropic traps*, whereby certain regular structures that are small by conventional measures but large compared to the fundamental scale might be very stable from an entropic perspective.

A mathematical result important in the study of superset microstates, labeled microstates, and symmetry microstates is Bender and Robinson's proof [37] that a typical acyclic directed set D has trivial automorphism group, i.e., is *rigid*. This result applies asymptotically under modest assumptions about the number of relations in D. However, these assumptions fail to hold for a typical low-order terminal state Δ, since such a state has unusually large "spatial size" and small "causal size", and typically lacks enough relations to "bind elements in place". Hence, $\text{Aut}(\Delta)$ is often nontrivial for such a state. The extreme case is a zeroth-order state, whose automorphism group is the entire symmetric group permuting its elements transitively. However, states tend to become increasingly rigid as their order increases. Bender and Robinson's result enables rough enumerations of the number of high-order superset microstates and labeled microstates for a state Δ of a given cardinality. It also suggests a novel explanation for *why* the details of the distant past seem to be irrelevant to future dynamics, namely, because relatively few additional generations of elements must be added to a typical low-order state to break most of its symmetries.

Definition 15. $\mathbb{D}^n$, $\mathbb{D}$, and $\overline{\mathbb{D}}$ *may be used to define finer state spaces, for which their members are macrostates.*

1. *The* **nth-order superset state space** $\mathbb{D}^n_{\text{SUP}}$ *is the set of full, originary co-relative histories* $\eta : \Delta^* \Rightarrow \Delta'^*$, *where Δ is a member of $\mathbb{D}^n$ and Δ' is a member of $\mathbb{D}$. Its elements are called* **superset microstates**. *The corresponding* **finite-order superset state space** $\mathbb{D}_{\text{SUP}}$ *and* **(total, countable, acyclic) superset state space** $\overline{\mathbb{D}}_{\text{SUP}}$ *are defined in the obvious ways.*

2. The **nth-order labeled state space** $\mathbb{D}^n_{\text{LAB}}$ is the set of complete labelings of members Δ of $\mathbb{D}^n$, where two labelings of Δ are considered to be equivalent if they are related by an element of $\text{Aut}(\Delta)$. Its elements are called **labeled microstates**. The corresponding **finite-order labeled state space** $\mathbb{D}_{\text{LAB}}$ and **(total, countable, acyclic) labeled state space** $\overline{\mathbb{D}_{\text{LAB}}}$ are defined in the obvious ways.
3. The **nth-order symmetry state space** $\mathbb{D}^n_{\text{SYM}}$ is the set of partial labelings of members Δ of $\mathbb{D}^n$ induced by applying elements of $\text{Aut}(\Delta)$ to arbitrary initial labelings of the subsets $\tilde{\Delta}$ of Δ not fixed by $\text{Aut}(\Delta)$. Its elements are called **symmetry microstates**. The corresponding **finite-order symmetry state space** $\mathbb{D}_{\text{SYM}}$ and **(total, countable, acyclic) symmetry state space** $\overline{\mathbb{D}_{\text{SYM}}}$ are defined in the obvious ways.

The spaces $\mathbb{D}^n_{\text{SUP}}$, $\mathbb{D}^n_{\text{LAB}}$, and $\mathbb{D}^n_{\text{SYM}}$, together with their larger counterparts, offer many alternative notions of states at many different levels of detail, and induce a variety of entropy systems. The reason why the co-relative history η in the definition of $\mathbb{D}^n_{\text{SUP}}$ is not assumed to be proper is because it is sometimes convenient to view a state Δ as a superset microstate of itself, i.e., to take η to be the co-relative history represented by the identity morphism $\Delta \to \Delta$. The "full" and "originary" conditions on η merely formalize the idea that η adds "prehistory" to Δ. It is sometimes convenient to refer to a superset Δ' of Δ as a superset microstate of Δ if the choice of co-relative history $\eta : \Delta^* \Rightarrow \Delta'^*$ is clear from context, for example, if there is only one such co-relative history. Using this convention, Figure 15 illustrates some of the superset microstates of the first-degree terminal state Δ_7 appearing in the sequential growth process in Figure 10. Each of these microstates is constructed by adding a single prehistorical element to Δ_7, along with a family of prehistorical relations. The 22 microstates shown in the figure each involve one or two extra relations. Overall, there are 96 such microstates, with between zero and seven extra relations.

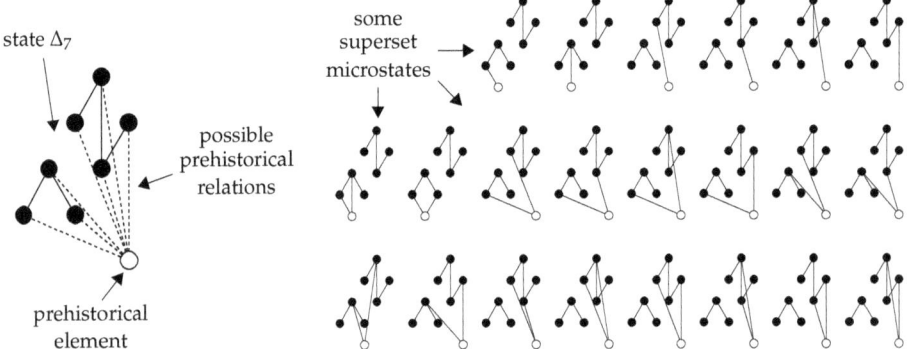

Figure 15. 22 of the 96 superset microstates of Δ_7 given by adding one prehistorical element.

For a state Δ^τ of cardinality K, the number of superset microstates adding a single element is "roughly" 2^K, if one ignores the contribution of symmetries. This reflects the idea that one may choose any family of elements in Δ^τ to be in the direct future of the single prehistorical element, since 2^K is the sum of the binomial coefficients $\binom{K}{k}$ for $0 \leq k \leq K$. Nontrivial symmetries of Δ^τ reduce this number; in particular, the number of superset microstates of the first-degree terminal states Δ_7 to Δ_{11} in Figure 10 are 96, 64, 72, 144, and 132. Ignoring symmetries need not yield exactly 2^K microstates, due to a curious graph-theoretic phenomenon called *pseudosimilarity*, whereby one directed set may be a terminal state of another in multiple distinct ways, even if the two sets differ by only a single element. Figure 16 illustrates this subtlety via an example provided by Brendan McKay, in which augmenting two copies of a state Δ^τ by a single prehistorical element in two different ways produces isomorphic supersets. The drawing emphasizes the latter isomorphism; the fact that the black nodes and edges represent two copies of the same state Δ^τ may be seen by matching up the elements labeled x and y.

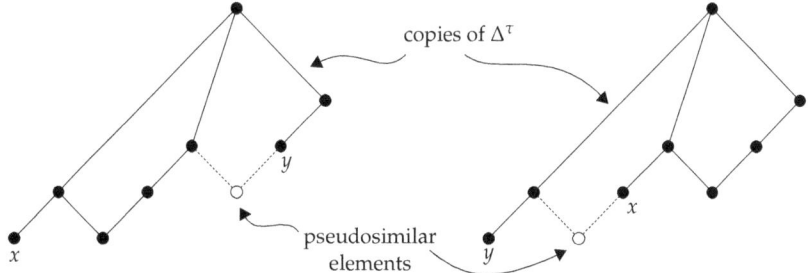

Figure 16. McKay's example: a superset may induce multiple microstates via pseudosimilarity.

Figure 17 illustrates a small region of $\mathbb{D}^1_{\text{SYM}}$ whose macrostates are the first-degree terminal states Δ_7 to Δ_{11} appearing in the sequential growth process from Figure 10. The left-hand diagram reproduces this process. In the middle diagram, Δ_7 to Δ_{11} are represented by large cells labeled 7 to 11, subdivided into smaller cells representing symmetry microstates. Because the histories D_7 to D_{11} are rigid, $\mathbb{D}^1_{\text{SYM}}$ accurately reflects relative distinguishability properties between terminal states and their histories in this case, since every state symmetry is broken by its ambient history. The figure highlights the fact that symmetry microstates of a given terminal state are isomorphic as partially labeled directed sets, which raises the question of how they are distinct. The answer is that there are multiple ways to break the automorphisms of the original states involved, even though the resulting objects remain *isomorphic*. $\mathbb{D}^1_{\text{SYM}}$ generally has "too many microstates" for terminal states of nonrigid histories, since it includes symmetry breaking information for symmetries that remain unbroken. This issue may be addressed by restricting the class of permissible labelings. The right-hand diagram represents the sequential growth process abstractly via a "curve" in $\mathbb{D}^1_{\text{SYM}}$. Since $\mathbb{D}^1_{\text{SYM}}$ encodes information only up to first order at the level of individual histories, the entire curve is necessary to reconstruct the evolution of D_{11}. The corresponding regions of $\mathbb{D}^1_{\text{SUP}}$ and $\mathbb{D}^1_{\text{LAB}}$ are much too large and cluttered to illustrate here, but the basic structural aspects are similar.

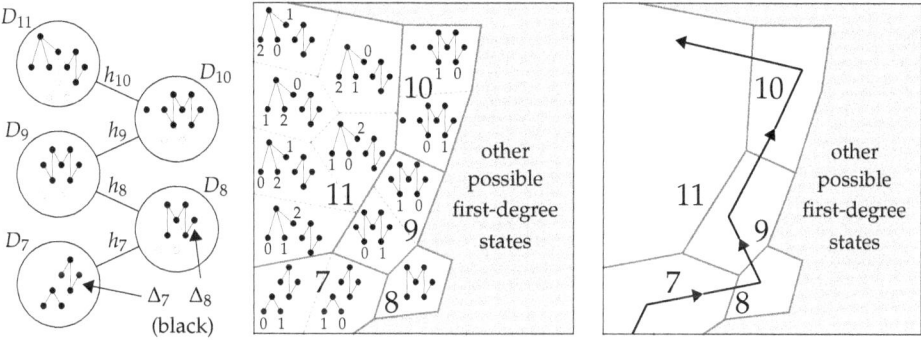

Figure 17. Sequential growth process from Figure 10; region of $\mathbb{D}^1_{\text{SYM}}$ through which this process moves; abstract view of the process.

Definitions 14 and 15 identify discrete causal state spaces as sets, but one may recognize additional "geometric" structure on these spaces defined in terms of discrete operations that convert one state to another. It is useful to define such operations for multidirected sets in general.

Definition 16. *Let M and M' be multidirected sets. Elementary operations on such sets are defined as follows:*

1. *Add or delete an isolated element.*

2. Add or delete a relation between two elements.

The **absolute distance** $d(M, M')$ between M and M' is the minimal number of elementary operations required to convert M to M', if this number is finite. Otherwise, $d(M, M') = \infty$.

Notions of distance between pairs of states facilitate useful analogues of familiar evolutionary ideas. For example, in conventional thermodynamics, one may ask why every system does not immediately transition to the cell in state space representing thermal equilibrium. The answer is that curves in state space are continuous in this context, so a typical system beginning far from thermal equilibrium must pass through a sequence of intervening macrostates before reaching it. Although literal continuity does not apply in the discrete causal context, similar ideas may be invoked whenever one can define notions of distance and neighbors. In particular, even if a given co-relative history is "favored" from a purely entropic perspective, it may be "costly" in the sense that it entails direct passage between widely separated regions of a discrete causal state space. Similarly, "short" paths between a given pair of states might be favored over "long" paths that involve drastic changes in structure. These ideas are revisited in Section 4.2 in the context of spacetime expansion, and again in Section 4.3 in the context of discrete causal action principles.

Alternative, relative notions of distance between pairs of directed or multidirected sets may be defined in terms of "ambient" structure from a configuration space. In the case of directed sets, such structure may originate from a kinematic scheme.

Definition 17. *Let $\mathbb{S} = (\mathcal{K}, \mathcal{H})$ be a kinematic scheme, and let D be a member of $\mathcal{K}$ in which every chain is bounded above. Let $T^n(D)$ be the nth-degree terminal state of D, and let Δ be any other element of $\overline{\mathbb{D}}$.*

1. *The **directed distance** $d_{\mathbb{S},D}(T^n(D), \Delta)$ between $T^n(D)$ and Δ in $\mathbb{S}$ with respect to D is the minimal length of chains $x(D) \prec x(D_1) \prec ... \prec x(D_N)$ in $\mathcal{M}(\mathbb{S})$, where $T^n(D_N) = \Delta$.*
2. *The **undirected distance** $\ell_{\mathbb{S},D}(T^n(D), \Delta)$ between $T^n(D)$ and Δ in $\mathbb{S}$ with respect to D is the minimal length of undirected paths $x(D), x(D_1), ..., x(D_N)$ in $\mathcal{M}(\mathbb{S})$ with initial element $x(D)$ and terminal element $x(D_N)$, where $T^n(D_N) = \Delta$.*

The reason why $d_{\mathbb{S},D}$ and $\ell_{\mathbb{S},D}$ depend on a choice of D is because $T^n(D)$ and Δ may appear as terminal states of many different histories in $\mathbb{S}$. If $T^n(D) = T^n(D_1) = T^n(D_2)$, then it may be easier to reach a history with nth-degree terminal state Δ from D_1 than from D_2. The distinction between a chain $x(D) \prec x(D_1) \prec ... \prec x(D_N)$ and an undirected path $x(D), x(D_1), ..., x(D_N)$ is that chains respect the directions of relations in $\mathcal{M}(\mathbb{S})$, while undirected paths generally do not. States close together in an undirected sense may be far apart in a directed sense, since undirected paths are more general than chains. Dependence on D implies that $d_{\mathbb{S},D}$ and $\ell_{\mathbb{S},D}$ are inherently asymmetric. It is reasonable to expect that $d_{\mathbb{S},D}$ and $\ell_{\mathbb{S},D}$ may closely approximate more conventional notions of distance for suitable classes of "large" directed sets, but this topic is not further explored here.

3.4. Multiplicities and Entropies

Four approaches to defining discrete causal microstates via terminal states of transitions were introduced in Section 3.3. A preliminary step, given in Definition 14, was to define spaces $\mathbb{D}^n$ of nth-order states, along with larger spaces $\mathbb{D}$ and $\overline{\mathbb{D}}$ including states of arbitrary order. The first approach was to treat the states making up these spaces as individual microstates, called resolution microstates, and apply a discrete causal analogue of conventional coarse-graining, called causal atomic resolution, to partition these spaces into cells. The remaining approaches treated such states as macrostates, with finer state spaces of microstates introduced in Definition 15. The second approach was to add detail to terminal states by specifying prehistorical information, leading to the spaces $\mathbb{D}^n_{\text{SUP}}$, $\mathbb{D}_{\text{SUP}}$, and $\overline{\mathbb{D}_{\text{SUP}}}$ of superset microstates. The third approach was to add detail to terminal states by labeling their elements, leading to the spaces $\mathbb{D}^n_{\text{LAB}}$, $\mathbb{D}_{\text{LAB}}$, and $\overline{\mathbb{D}_{\text{LAB}}}$ of labeled microstates. The fourth approach was to add detail to terminal states via partial labelings specifying symmetry breaking information, leading to the spaces $\mathbb{D}_{\text{SYM}}^n$, $\mathbb{D}_{\text{SYM}}$, and $\overline{\mathbb{D}_{\text{SYM}}}$ of symmetry microstates.

Before explaining how discrete causal entropies may be defined via these four approaches, I mention progress in the study of causal set entropy by Sorkin and collaborators [35,36]. This work exhibits interesting relationships with analogous continuum-based notions, is supported by numerical simulations involving "low-dimensional" causal sets, and incorporates covariance considerations. However, it is very different in its assumptions and emphasis from the approaches examined in this paper. First, the entropies involved are defined in terms of auxiliary fields on causal sets, and are therefore not completely background independent quantities. Sorkin does consider causal set "vacuum solutions", whose entropies may be attributed solely to causal structure, but entropies associated with nontrivial interactions typically involve large quantities of extra-causal data. Second, pre-packaged quantum-theoretic machinery such as Hilbert spaces, operator algebras, density matrices, and von Neumann-type entropy are applied to individual causal sets under this approach, rather than emerging naturally from a history configuration space. Third, the permeability problem and other technical obstructions arising in the absence of relation space methods render it difficult to define terminal states or associated entropic data in this setting. The resulting measures of entropy are a priori "higher-dimensional", and can be associated only indirectly with conventional notions of time-dependent entropy and the second law of thermodynamics. Fourth, many of the cases considered under this approach involve special causal sets of the type mentioned in Section 2.2, induced by sprinkling elements into relativistic spacetime manifolds. Such causal sets are naturally limited in their potential to reveal structural features beyond the scope of general relativity.

I give only a brief sketch of how one may construct entropy systems via resolution microstates. For simplicity, I describe this construction in terms of an individual nth-order state space $\mathbb{D}^n$. The first step is to choose a resolution of each state Δ in this space. In the simplest case, these resolutions may be chosen to consist of single causal atomic decompositions. A choice of such decompositions defines a coarse-graining of $\mathbb{D}^n$, which induces an entropy quadruple, while a choice of resolutions involving longer sequences of decompositions, or partially ordered families of decompositions, defines an entropy system. In the general case, one may define a partially ordered family of equivalence relations on $\mathbb{D}^n$, specified by treating states as equivalent if their resolutions agree beyond a certain level of detail. The associated equivalence classes then define partitions of $\mathbb{D}^n$, and their cardinalities define multiplicities. The resulting notion of entropy is called *resolution entropy*. One may choose to define resolutions in such a way that each decomposition reduces the maximal length of chains in each state by a specified quantity. For example, the decomposition illustrated in the second diagram in Figure 14 converts a "fine" third-order state to a "rough" first-order state. An analogue of resolution entropy appears in Sorkin's approach to causal set entropy [35,36], but involves a random "decimation" version of coarse-graining that does not incorporate causal structure in the same way that causal atomic resolution does. It also involves "higher-dimensional" entropy, rather than entropy associated with terminal states. However, numerical examples do hint at interesting universal behavior for this type of entropy, and this evidence provides motivation for studying resolution entropy in more detail.

Numerous questions must be answered, however, before one may have confidence in the resolution approach. The most basic is how sensitive resolution entropy is to changes of resolution, since resolutions generally involve arbitrary extraphysical choices regarding the organization of information. Another question, already mentioned in Section 3.3, is how one may reconcile the increasing "granularity" produced by multi-level resolutions with the basic philosophy of metric recovery, under which discrete causal structure at the fundamental scale should produce effectively smooth structure at sufficiently large scales. A third issue arises from the empirical dynamical irrelevance of details of the distant past. If only very low-order terminal states play a substantial dynamical role in the future evolution of histories, then repeated causal atomic decompositions of dynamically relevant states will produce antichains at relatively fine levels of detail. Antichains possess no internal structure besides cardinality, which seems much too crude to determine meaningful dynamics, especially locally. Therefore, the utility of resolution entropy seems to be limited by the "causal depth" of relevant information. This issue does not necessarily disqualify the resolution approach, however, due to the scales involved. In particular, the difference in magnitude between the Planck scale and presently-measurable scales suggests than information up to order 10^{10} or 10^{15} could be relevant without producing noticeable deviations from the empirical obsolescence of high-order information. A resolution involving decompositions similar to the one illustrated in Figure 14 would require perhaps 30 decompositions to cover 10–15 orders of magnitude, and could therefore contain a large quantity of information. However, such illustrations involving small histories can be misleading; for example, it would not be surprising if each element in a typical physically realistic history were directed related to 10^{10} or more other elements. Such large numbers of relations would affect the qualitative properties of realistic resolutions.

Superset microstates offer a variety of different ways to define entropy systems via the state spaces $\mathbb{D}_{\text{SUP}}^n$, $\mathbb{D}_{\text{SUP}}$, and $\overline{\mathbb{D}_{\text{SUP}}}$. I begin by discussing simple notions of entropy involving individual partitions of these spaces. For simplicity, I focus on the case of finite states. Let Δ be such a state, and consider all superset microstates $\eta : \Delta^* \Rightarrow \Delta'^*$ adding a single prehistorical element to Δ. The number of such microstates is the cardinality of the future relation set $R^+(x(\Delta^*))$ in $\mathcal{M}(\mathbb{S}_{\text{PS}})$, since the number of different ways in which Δ can be the terminal state of a history with one additional element is the same as the number of ways in which Δ^* can evolve into a history with one additional element. As a reminder, $x(\Delta^*)$ is the element in the underlying multidirected set $\mathcal{M}(\mathbb{S}_{\text{PS}})$ of $\mathbb{S}_{\text{PS}}$ representing Δ^*, and $R^+(x(\Delta^*))$ is the set of relations in $\mathcal{M}(\mathbb{S}_{\text{PS}})$ beginning at $x(\Delta^*)$, each of which represent a co-relative history with cobase Δ^*. The *first superset multiplicity* $\mu_{\text{SUP}}^1(\Delta)$ of Δ is then defined to be the number $|R^+(x(\Delta^*))|$ of such microstates η, and the *first superset entropy* $e_{\text{SUP}}^1(\Delta)$ is defined to be $\log \mu_{\text{SUP}}^1(\Delta)$. Following essentially the same reasoning, nth superset multiplicities and entropies may be defined.

Definition 18. *The nth* **superset multiplicity** $\mu_{\text{SUP}}^n(\Delta)$ *of a finite state Δ is the number of co-relative histories $\eta : \Delta^* \Rightarrow \Delta'^*$, where the complement of the image of Δ^* under any transition representing η has cardinality n. The nth* **superset entropy** $e_{\text{SUP}}^n(\Delta)$ *of Δ is $\log \mu_{\text{SUP}}^n(\Delta)$.*

An interesting entropy system on $\mathbb{D}_{\text{SUP}}$ is given by filtering superset microstates $\eta : \Delta^* \Rightarrow \Delta'^*$ by both the number of prehistorical elements added to Δ by η, and the order of the resulting supersets Δ'. $\mathbb{D}_{\text{SUP}}$ has a natural partition whose members are the infinite sets $C_{\text{SUP}}(\Delta)$ parameterizing *all* full, originary co-relative histories η with cobase Δ^* and target belonging to $\mathbb{D}$. One may partition each set $C_{\text{SUP}}(\Delta)$ by numbers of elements added to Δ, or by orders of supersets Δ', or by both. A general way to formalize the idea that two superset microstates $\eta_1 : \Delta^* \Rightarrow \Delta_1'^*$ and $\eta_2 : \Delta^* \Rightarrow \Delta_2'^*$ of Δ are equivalent up a given level of detail is to specify a *common interpolating microstate* $\eta_3 : \Delta^* \Rightarrow \Delta_3'^*$, characterized by the property that η_1 and η_2 both factor through η_3. This means that there exist pairs of transitions $\Delta^* \xrightarrow{\tau_3} \Delta_3'^* \xrightarrow{\tau_1} \Delta_1'^*$ and $\Delta^* \xrightarrow{\tau_3'} \Delta_3'^* \xrightarrow{\tau_2} \Delta_2'^*$, where τ_3 and τ_3' both represent η_3, and where the compositions $\tau_1 \circ \tau_3$ and $\tau_2 \circ \tau_3'$ represent η_1 and η_2, respectively. Informally, this means that besides being supersets of Δ, the states Δ_1' and Δ_2' also share common prehistorical elements. One may then define equivalence relations $\sim^m$ and $\sim^n$ on $\mathbb{D}_{\text{SUP}}$, for each $m, n \in \mathbb{N}$, where $\eta_1 \sim^m \eta_2$ if η_1 and η_2 factor through a common interpolating microstate η_3 adding m prehistorical elements to Δ, and where $\eta_1 \sim^n \eta_2$ if η_1 and η_2 factor through a common interpolating microstate η_3 whose superset has order n. Equivalence relations $\sim^{(m,n)}$ combine these two requirements. The corresponding partitions $P^{(m,n)}$ are partially ordered lexicographically; i.e., $P^{(m,n)} \prec P^{(m',n')}$ if and only if $m < m'$ or $m = m'$ and $n < n'$. It is convenient to denote the pair (m, n) by the single symbol α, regarded as an element of $\mathbb{N}^2 = \mathbb{N} \times \mathbb{N}$. Informally, the partition P^α groups together superset microstates that agree both up to a given number of prehistorical elements and a given order.

Definition 19. *Let $\alpha = (m, n) \in \mathbb{N}^2$, and let $\Pi_{\text{LEX}} := \{P^\alpha\}_{\alpha \in \mathbb{N}^2}$ be the set of partitions P^α of $\mathbb{D}_{\text{SUP}}$ defined by taking superset microstates η_1 and η_2 of Δ to be equivalent if they factor through a common interpolating* microstate $\eta_3 : \Delta^* \Rightarrow {\Delta_3}'^*$ of Δ represented by a transition $\tau_3 : \Delta^* \to {\Delta_3}'^*$ such that $|{\Delta_3}'^* - \tau_3(\Delta^*)| = m$ and $\text{ord}({\Delta_3}') = n$. Let $\sim^\alpha$ be the corresponding equivalence relation, and for any subset $V \subset \mathbb{D}_{\text{SUP}}$, let V^α be the corresponding quotient set. For any relation $P^\alpha \prec P^\beta$ under the lexicographic order induced by $\mathbb{N}^2$, and for any subset V belonging to P^α, let $\mu^{\alpha\beta}(V^\beta)$ be the cardinality of V^β. Let μ_{LEX} be the family of measures $\mu^{\alpha\beta}$. Then the triple $(\mathbb{D}_{\text{SUP}}, \Pi_{\text{LEX}}, \mu_{\text{LEX}})$ is called the **lexicographic superset entropy system**.

The measures $\mu^{\alpha\beta}(V^\beta)$ may take on infinite values; for example, there are infinitely many ways to add a single prehistorical element to $\mathbb{N}$. Definition 19 does not specify the number of relations added to Δ by each microstate, or the maximal sizes of antichains in the corresponding supersets, or any of a variety of other basic combinatorial data that may be used to partition $\mathbb{D}_{\text{SUP}}$ in different ways. Using such quantities, one may define alternative entropy systems, involving, for example, "higher-dimensional" lexicographic orders. This particular entropy system merely formalizes some of the simpler properties that may be used to organize families of superset microstates.

Labeled microstates also induce a variety of entropic notions. The most obvious is given by simply counting the number of equivalence classes of labelings of a state Δ. If Δ has cardinality K, then its total number of labelings is $K!$. These labelings are partitioned by the action of $\text{Aut}(\Delta)$ into equivalence classes of cardinality $|\text{Aut}(\Delta)|$, so the number of such classes is $K!/|\text{Aut}(\Delta)|$.

Definition 20. *The **labeled multiplicity** $\mu_{\text{LAB}}(\Delta)$ of a state Δ of cardinality K is $K!/|\text{Aut}(\Delta)|$. The **labeled entropy** $e_{\text{LAB}}(\Delta)$ of Δ is $\log \mu_{\text{LAB}}(\Delta) = \log K! - \log |\text{Aut}(\Delta)|$.*

It is sometimes desirable to decompose the subset $C_{\text{LAB}}(\Delta)$ of $\mathbb{D}_{\text{LAB}}$ consisting of all equivalence classes of labelings of Δ. This may be accomplished via equivalence classes of partial labelings of Δ, i.e., labelings of special subsets U of Δ. To yield a suitable version of equivalence, U must be a union of orbits under $\text{Aut}(\Delta)$, and the labeling must be by consecutive natural numbers beginning with zero. The set of equivalence classes of such partial labelings is partially ordered by extension of class representatives. A labeling ℓ of U corresponds to a subset $C_{\text{LAB}}(\ell)$ of $C_{\text{LAB}}(\Delta)$ defined by labelings of Δ extending ℓ. Letting U and ℓ vary, one obtains a family of sets $\{C_{\text{LAB}}(\ell)\}$ that cover $C_{\text{LAB}}(\Delta)$, generally in a highly redundant fashion. A *partition of $C_{\text{LAB}}(\Delta)$ induced by partial labelings of Δ* is defined to be a partition whose members are open sets in the topology on $C_{\text{LAB}}(\Delta)$ generated by $\{C_{\text{LAB}}(\ell)\}$, i.e., unions of finite intersections of members of $\{C_{\text{LAB}}(\ell)\}$. Choosing such a partition for each Δ defines a partition of $\mathbb{D}_{\text{LAB}}$, and the collection of all such partitions forms a "large" entropy system. Smaller subsystems may be more convenient to work with in practice.

Definition 21. *Let Δ be a member of $\mathbb{D}$, and let $C_{\text{LAB}}(\Delta)$ be the subset of $\mathbb{D}_{\text{LAB}}$ consisting of all equivalence classes of labelings of Δ. Let $\Pi_{\text{LAB}}(\Delta)$ be the set of partitions of $C_{\text{LAB}}(\Delta)$ induced by partial labelings of Δ, and let Π_{LAB} be the set of partitions of $\mathbb{D}_{\text{LAB}}$ constructed from the partitions $\Pi_{\text{LAB}}(\Delta)$, partially ordered by refinement. For any relation $P^\alpha \prec P^\beta$ in Π_{LAB}, and for any subset V belonging to P^α, let $\mu^{\alpha\beta}(V^\beta)$ be the cardinality of the quotient set V^β of V under the equivalence relation $\sim^\beta$ induced by P^β. Let μ_{LAB} be the family of measures $\mu^{\alpha\beta}$. Then the triple $(\mathbb{D}_{\text{LAB}}, \Pi_{\text{LAB}}, \mu_{\text{LAB}})$ is called the **labeled entropy system**.*

Symmetry microstates share entropic similarities with labeling microstates, since both approaches involve labelings. The principal differences are that symmetry microstates label only elements of a state Δ that are not fixed by its automorphisms, and labelings related by automorphisms are *not* considered to be equivalent. It is convenient to fix an arbitrary "initial" labeling on the set $\tilde{\Delta}$ of elements of Δ not fixed by $\text{Aut}(\Delta)$, i.e., the union of nonsingleton orbits under $\text{Aut}(\Delta)$. A labeling of $\tilde{\Delta}$ is then considered *permissible* if it is generated by applying an element of $\text{Aut}(\Delta)$ to this initial labeling. The number of such labelings is just the order $|\text{Aut}(\Delta)|$ of $\text{Aut}(\Delta)$.

Definition 22. *The **symmetry multiplicity** $\mu_{\text{SYM}}(\Delta)$ of a finite state Δ is $|\text{Aut}(\Delta)|$. The **symmetry entropy** $e_{\text{SYM}}(\Delta)$ of Δ is $\log \mu_{\text{SYM}}(\Delta) = \log |\text{Aut}(\Delta)|$.*

By Definitions 20 and 22, $\mu_{\text{LAB}}(\Delta)\mu_{\text{SYM}}(\Delta) = K!$ for a state Δ of cardinality K. Processes exhibiting an increase in e_{LAB} therefore exhibit a decrease in e_{SYM} for a fixed state cardinality, and vice versa, although "expanding universes" may exhibit simultaneous increases in both types of entropy. As in the case of labeled microstates, it is sometimes desirable to decompose the subset $C_{\text{SYM}}(\Delta)$ of $\mathbb{D}_{\text{SYM}}$ consisting of all permissible labelings of $\tilde{\Delta}$. This may be accomplished by partially labeling $\tilde{\Delta}$ in a suitable manner; in particular, the set U of elements labeled must be a union of nonsingleton orbits under $\text{Aut}(\Delta)$. Such a labeling ℓ defines a subset $C_{\text{SYM}}(\ell)$ of $C_{\text{SYM}}(\Delta)$ consisting of all labelings of $\tilde{\Delta}$ extending ℓ. The set of all such labelings for all such U is partially ordered by extension. The collection of sets $\{C_{\text{SYM}}(\ell)\}$ define a family of partitions of $\mathbb{D}_{\text{SYM}}$, and hence an entropy system.

Definition 23. *Let Δ be a member of $\mathbb{D}$, and let $C_{\text{SYM}}(\Delta)$ be the subset of $\mathbb{D}_{\text{SYM}}$ consisting of all permissible labelings of the set $\tilde{\Delta}$ of elements of Δ not fixed by $\text{Aut}(\Delta)$, with respect to an arbitrary initial labeling. Let $\Pi_{\text{SYM}}(\Delta)$ be the set of partitions of $C_{\text{SYM}}(\Delta)$ induced by partial labelings of $\tilde{\Delta}$, and let Π_{SYM} be the set of partitions of $\mathbb{D}_{\text{SYM}}$ constructed from the partitions $\Pi_{\text{SYM}}(\Delta)$, partially ordered by refinement. For any relation $P^\alpha \prec P^\beta$ in Π_{SYM}, and for any subset V belonging to P^α, let $\mu^{\alpha\beta}(V^\beta)$ be the cardinality of the quotient set V^β of V under the equivalence relation $\sim^\beta$ induced by P^β. Let μ_{SYM} be the family of measures $\mu^{\alpha\beta}$. Then the triple $(\mathbb{D}_{\text{SYM}}, \Pi_{\text{SYM}}, \mu_{\text{SYM}})$ is called the **symmetry entropy system**.*

It may often suffice on physical grounds to restrict attention to notions of entropy more specific than those associated with the entropy systems of Definitions 19, 21 and 23, although it may be necessary to supersede the simplistic notions of Definitions 18, 20 and 22. For superset microstates, *weighted sums* of entropies can be useful to naturally distill finite entropic values from infinite families of microstates. Abstractly, such sums are analogous to Gibbs or Shannon entropies. A practical reason to study such sums is to quantify the degree to which prehistorical data of various orders is dynamically relevant. A simple example of such a weighted sum is

$$e(\Delta) = \sum_{n=1}^{\infty} \frac{e_{\text{SUP}}^n(\Delta)}{n^4}, \tag{6}$$

where the denominator n^4 dominates the rapid growth of $e_{\text{SUP}}^n(\Delta)$ as n increases. For both labeled microstates and symmetry microstates, symmetry considerations are paramount. Interesting generalizations of Definitions 20 and 22 include those involving the study of symmetries that are broken or preserved by specific prehistorical information. This leads to the concept of *extension groups,* which measure how many automorphisms of a terminal state extend to automorphisms of a specified superset. One may formalize this idea in terms of pairs of transitions (τ_1, τ_2), where τ_1 specifies a terminal state Δ^{τ_1}, and τ_2 specifies a superset Δ^{τ_2} of Δ^{τ_1} that breaks some of the symmetries of Δ^{τ_1}. Finiteness assumptions may be added as necessary.

Definition 24. *Let τ, τ_1 and τ_2 be transitions of directed sets with sources D, D_1 and D_2, and common target D'. Assume that $\tau_2(D_2) \subset \tau_1(D_1)$ in D'. Let Δ^τ, Δ^{τ_1} and Δ^{τ_2} be the terminal states of τ, τ_1, and τ_2.*

1. *The **state automorphism group** of τ is $\text{Aut}(\Delta^\tau)$.*
2. *The **relative extension group** $E^{\tau_1 \tau_2}$ of (τ_1, τ_2) is the subgroup of $\text{Aut}(\Delta^{\tau_1})$ of automorphisms of Δ^{τ_1} that extend to automorphisms of Δ^{τ_2}.*
3. *The **relative symmetry multiplicity** $\mu_{\text{SYM}}^{\tau_1 \tau_2}$ of (τ_1, τ_2) is $|\text{Aut}(\Delta^{\tau_1})| - |E^{\tau_1 \tau_2}|$.*
4. *The **relative symmetry entropy** $e_{\text{SYM}}^{\tau_1 \tau_2}$ of (τ_1, τ_2) is $\log \mu_{\text{SYM}}^{\tau_1 \tau_2}$.*

The *generational automorphism groups* discussed in Section 8.2 of [14] are special cases of state automorphism groups. The quantities $\mu_{\text{SYM}}^{\tau_1 \tau_2}$ and $e_{\text{SYM}}^{\tau_1 \tau_2}$ may be derived from the symmetry entropy system, if desired. $E^{\tau_1 \tau_2}$ is generally not a normal subgroup of $\text{Aut}(\Delta^{\tau_1})$. The superset Δ^{τ_2} may acquire "new" symmetries that do not extend nontrivial symmetries of Δ^{τ_1}, but this is atypical due to rigidity. Since the purpose of studying entropic phase maps is to assign quantum-theoretic phases to co-relative kinematics, it is necessary to adapt the preceding notions to apply to co-relative histories $h : D_i \Rightarrow D_t$ in a kinematic scheme $\mathbb{S}$. The states of principal interest in this context are terminal states of the cobase D_i and target D_t of h. For generality, it is convenient to work with an unspecified entropy function on a subset of $\overline{\mathbb{D}}$. Again, finiteness assumptions may be added as necessary.

Definition 25. *Let $h : D_i \Rightarrow D_t$ be a co-relative history. Let Δ^{τ_i} and Δ^{τ_t} be terminal states of D_i and D_t, respectively. Let e be an entropy function on a subset of $\overline{\mathbb{D}}$.*

1. *The **initial entropy** $e_i^{\tau_i}(h)$ of h with respect to τ_i is $e(\Delta^{\tau_i})$.*
2. *The **terminal entropy** $e_t^{\tau_t}(h)$ of h with respect to τ_t is $e(\Delta^{\tau_t})$.*
3. *The **relative entropy** $e^{\tau_i \tau_t}(h)$ of h with respect to the pair (τ_i, τ_t) is $e(\Delta^{\tau_t}) - e(\Delta^{\tau_i})$.*

It is useful to specialize Definition 25 to the case where τ_i and τ_t are transitions of specific degrees, as specified in Definition 13.

Definition 26. *Let $h : D_i \Rightarrow D_t$ be a co-relative history, and let e be an entropy function on a subset of $\overline{\mathbb{D}}$.*

1. *The nth **initial entropy** $e_i^n(h)$ of h is $e(T^n(D_i))$.*
2. *The nth **terminal entropy** $e_t^n(h)$ of h is $e(T^n(D_t))$.*
3. *The nth **relative entropy** $e^n(h)$ of h is $e(T^n(D_t)) - e(T^n(D_i))$.*

4. Entropic Phase Maps

4.1. Examples of Phase Maps

Given an entropy function e on a subset U of the state space $\overline{\mathbb{D}}$, one may assign relative entropies $e^{\tau_i \tau_t}(h) = e(\Delta^{\tau_t}) - e(\Delta^{\tau_i})$ to each co-relative history $h : D_i \Rightarrow D_t$ in a kinematic scheme $\mathbb{S}$ whose histories have terminal states in U, where Δ^{τ_i} and Δ^{τ_t} are terminal states of D_i and D_t with respect to transitions τ_i and τ_t. Abstracting Feynman's approach, one may then associate a quantum-theoretic phase $\theta_e(r(h)) = \exp(i e^{\tau_i \tau_t}(h))$ with the relation $r(h)$ representing h in $\mathcal{R}(\mathcal{M}(\mathbb{S}))$. As explained in Section 1.4, this approach may be generalized to allow for target objects other than the unit circle S^1, but such generalization is not carried out here. The subscript e in the expression θ_e indicates that this function is defined directly in terms of entropy, rather than multiplicity, entropy per unit volume, or some other variant of entropic information. Of course, θ_e also depends on the choices of transitions τ_i and τ_t, but this dependence is suppressed to avoid notational clutter. For a co-relative kinematics in $\mathbb{S}$, represented by a chain $\gamma = r(h_0) \prec ... \prec r(h_N)$ of relations $r(h_k)$ in $\mathcal{R}(\mathcal{M}(\mathbb{S}))$ representing co-relative histories $h_k : D_{ik} \Rightarrow D_{tk}$ for $0 \leq k \leq N$, one may extend θ_e multiplicatively to define a phase map

$$\Theta_e(\gamma) = \prod_{k=0}^{N} \exp(i e^{\tau_{ik} \tau_{tk}}(h_k)), \tag{7}$$

where $\Delta^{\tau_{ik}}$ and $\Delta^{\tau_{tk}}$ are terminal states of D_{ik} and D_{tk} with respect to transitions τ_{ik} and τ_{tk}. This approach restricts attention to causal Schrödinger-type equations of the form given in Equation (4), since this equation is defined in terms of a relation function θ, rather than a possibly nonmultiplicative phase map. Since the target of h_k coincides with the cobase of h_{k+1}, it is often reasonable to choose $\tau_{i(k+1)} = \tau_{tk}$. With these choices, the product in Equation (7) telescopes to yield the simpler expression

$$\Theta_e(\gamma) = \exp\Big(i\big(e(\Delta^{\tau_{tN}}) - e(\Delta^{\tau_{i0}})\big)\Big). \tag{8}$$

This telescoping property implies that the value of Θ_e is independent of the choice of chain γ in $\mathcal{R}(\mathcal{M}(\mathbb{S}))$ between $r(h_0)$ and $r(h_N)$, a feature revisited in Section 4.2. It is sometimes convenient to use the shorthand $e^{\tau_{i0} \tau_{tN}}(\gamma)$ for the entropic quantity $e(\Delta^{\tau_{tN}}) - e(\Delta^{\tau_{i0}})$ multiplying i in the exponential in Equation (8), which generalizes the expression $e^{\tau_i \tau_t}(h) = e(\Delta^{\tau_t}) - e(\Delta^{\tau_i})$ appearing in Definition 25 for a single co-relative history $h : D_i \Rightarrow D_t$. The simplest such phase maps Θ_e are given by choosing $\Delta^{\tau_{ik}}$ and $\Delta^{\tau_{tk}}$ to be the mth-degree terminal states $T^m(D_{ik})$ and $T^m(D_{tk})$ defined via the mth-degree transitions $\tau_{ik} = \tau_{D_{ik}}^m$ and $\tau_{tk} = \tau_{D_{tk}}^m$ under Definition 13, for some natural number m. I focus principally on phase maps of this form in what follows. The primitive phase maps discussed in Section 8.2 of [14] are defined exclusively in terms of terminal states of transitions representing the co-relative histories $h_0, ..., h_N$. The approach described here is more general.

Referring to Section 3.4, there are many possible ways to define an entropy function e to determine specific content for Equation (7) or Equation (8). No specific examples involving resolution entropy are computed here, since the details of this approach are outside the scope of this paper. In rough terms, however, the multiplicities assigned to terminal states in this context are the numbers of such states sharing common resolutions, and the corresponding entropies are the logarithms of these multiplicities. An obvious qualitative conclusion that may be drawn in this context is that maximizing the entropic quantity $e^{\tau_{i0}\tau_{tN}}(\gamma) = e(\Delta^{\tau_{tN}}) - e(\Delta^{\tau_{i0}})$ tends to favor "expanding universe" scenarios, in which the cardinality of $\Delta^{\tau_{tN}}$ exceeds that of $\Delta^{\tau_{i0}}$, provided that the sizes of causal atoms are roughly equal in decompositions of states of different sizes. This qualitative relationship may be understood by "inverting" the decomposition process, replacing each element in a directed set with a causal atom; there are clearly more ways to do this for larger sets. Qualitative entropic preference for expanding universe scenarios is in fact a generic feature of discrete causal notions of entropy; this is a posteriori obvious on basic enumerative grounds. Cosmological observations *do* favor accelerating expansion of spacetime, but the correspondence between large universes and high overall entropy is much too general to favor discrete causal theory specifically. Conventional thermodynamic systems exhibit increasing entropy without acquiring new degrees of freedom, and this suggests examining the notion of *entropy per unit volume* to "correct" for differences in the sizes of states. This idea is revisited in more detail below. It should also be emphasized that the quantity $e^{\tau_{i0}\tau_{tN}}(\gamma)$ appears here in a role analogous to that of the classical action $\mathcal{S}$ in Feynman's phase map, which is typically *minimized* for favored trajectories under Hamilton's principle of stationary action. This suggests the possibility of adding a minus sign to the exponents in Equations (7) and (8), thus treating $e^{\tau_{i0}\tau_{tN}}(\gamma)$ as a "negative action". Regardless of this choice, the quantity $e^{\tau_{i0}\tau_{tN}}(\gamma)$ must obey some analogue of stationary action to produce suitable interference effects, for example, by exhibiting similar values for similar states of high entropy. This nontrivial requirement is elaborated in Section 4.2.

A simple specific choice for the entropy function e in Equations (7) and (8) is the nth superset entropy function e_{SUP}^n of Definition 18. Choosing $\Delta^{\tau_{i0}} = T^m(D_{i0})$ and $\Delta^{\tau_{tN}} = T^m(D_{tN})$ in Equation (8) yields the phase map

$$\Theta_e(\gamma) = \exp\left(i\left(e_{\text{SUP}}^n(T^m(D_{tN})) - e_{\text{SUP}}^n(T^m(D_{i0}))\right)\right). \quad (9)$$

Even this simple phase map is difficult to compute exactly for arbitrary values of m and n, since it requires calculating all possible ways to add n prehistorical elements and an unspecified number of relations to $T^m(D_{i0})$ and $T^m(D_{tN})$. However, a few special cases may be computed, and rough qualitative conclusions may be drawn. Beginning with $m = 0$, $T^0(D_{i0})$ and $T^0(D_{tN})$ are just antichains consisting of the maximal elements of D_{i0} and D_{tN}, respectively. In the finite case, their cardinalities are natural numbers K_{i0} and K_{tN}. If also $n = 0$, then

$$\Theta_e(\gamma) = \exp\left(i\left(e_{\text{SUP}}^0(T^0(D_{tN})) - e_{\text{SUP}}^0(T^0(D_{i0}))\right)\right) = \exp\left(i\left(\log 1 - \log 1\right)\right) = e^0 = 1,$$

for any choice of γ, since there is exactly one way to add zero elements to each of the directed sets $T^0(D_{i0})$ and $T^0(D_{tN})$. More generally, trivial supersets produce trivial superset entropies. Taking $m = 0$ and $n = 1$ in Equation (9) still involves zeroth-degree terminal states, but adds nontrivial information to these states. The first superset multiplicity $\mu_{\text{SUP}}^1(T^0(D_{i0}))$ of $T^0(D_{i0})$ under Definition 18 is $K_{i0} + 1$, because a superset of an antichain given by adding a single prehistorical element is determined up to isomorphism by its number of relations, which may range from 0 to K_{i0} in this case. Similarly, the multiplicity $\mu^1_{\text{SUP}}(T^0(D_{tN}))$ is $K_{tN} + 1$, so with these choices

$$\Theta_e(\gamma) = \exp\Big(i\big(\log(K_{tN} + 1) - \log(K_{i0} + 1)\big)\Big).$$

Here, the entropic preference for "expanding universe" scenarios is quantitatively obvious, and the same effect clearly extends to higher-order states and higher-index superset entropy functions, since there are typically more ways to add families of prehistorical elements to large directed sets than to small ones. Conventional thermodynamics suggests that working with zeroth-degree terminal states is likely inadequate to determine relevant entropic quantities, so a more serious treatment involves states of higher degree. Substituting first-degree terminal states $T^1(D_{i0})$ and $T^1(D_{tN})$ into Equation (9) yields the most obvious discrete causal analogue of conventional thermodynamic entropy in the superset context. Zeroth superset entropies offer no useful information, so the first interesting case is given by setting $m = n = 1$. This requires computing the number of ways to add a single prehistorical element to a first-degree terminal state of cardinality K, an interesting enumerative problem. Referring to the discussion following Figure 15, a very rough estimate of this number is 2^K, assuming that the state is nearly rigid. This produces an estimate of

$$\Theta_e(\gamma) \approx \exp\Big(i(K_{tN} - K_{i0})\log 2\Big)$$

for the resulting phase map, which again suggests an entropic preference for "expanding universe" scenarios. Applying higher-index entropy maps e^n_{SUP} in this context leads to further intricate enumerations, but rough estimates may again be formulated. Ignoring symmetries, overcounting, and multidirected structure of the type illustrated by McKay's example in Figure 16, the nth superset multiplicity $\mu^n_{\text{SUP}}(\Delta)$ of a state Δ of cardinality K and arbitrary order is roughly

$$\mu^n_{\text{SUP}}(\Delta) \approx \prod_{k=1}^n 2^{K+k} = 2^{\binom{n}{2}+Kn} = 2^{\frac{n^2}{2}+O(n)}, \tag{10}$$

which corresponds to superset entropies of roughly $n^2 \log \sqrt{2} + O(n)$. This estimate is derived by adding prehistorical elements sequentially, and naïvely multiplying together the estimated multiplicities at each step. The factor n^2 explains the choice of denominators n^4 in the summands in Equation (6), which offers a simple way to ensure convergence of the series. Equation (10) yields better estimates for higher-order states, which are typically more rigid. For zeroth-order states, it is a very poor estimate, particularly for low-index superset entropies. For first-degree terminal states, its overall accuracy depends on the asymptotic behavior of automorphism groups of states of increasing size. The mathematical interest of terminal states of low but nonzero degree arises largely from the fact that their behavior is *balanced* between the rigidity of high-order states and the transitivity of zeroth-order states in a group-theoretic sense. Estimates assuming rigidity, such as Equation (10), are naturally rough in this context, but can nonetheless provide useful upper bounds. As in the case of resolution entropy, conventional thermodynamic analogies suggest studying entropies per unit volume in the superset context. The necessity of demonstrating suitable interference effects under path summation also remains central. Since there is generally no natural limit to "how far back in time" one may extend supersets, filtering methods associated with the lexicographic superset entropy system of Definition 19, such as such the weighted sum of entropies in Equation (6), are of interest for organizing relevant information, while respecting the relative insignificance of the distant past, and producing finite values for physically meaningful quantities.

The labeled entropy function e_{LAB} of Definition 20 offers another choice for the entropy function e in Equations (7) and (8). A trivial case is when $\Delta^{T_{i0}} = T^0(D_{i0})$ and $\Delta^{T_{tN}} = T^0(D_{tN})$. Since these states are antichains, they are transitive under their automorphism groups; i.e., each consists of a single orbit.

Hence, all labelings of these states are equivalent, so their labeled multiplicities are equal to 1, and their labeled entropies are equal to zero. Thus, $\Theta_e(\gamma) = e^0 = 1$ for any choice of γ. For higher-degree states, the situation is more interesting. Referring again to Definition 20, the labeled multiplicity $\mu_{\text{LAB}}(\Delta)$ of an arbitrary state Δ of cardinality K is $K!/|\text{Aut}(\Delta)|$. In particular, the multiplicity of 1 for a zeroth-order state may be interpreted as the ratio $K!/K!$. This ratio typically increases toward $K!$ for a sequence of states of increasing order, since such states tend to become increasingly rigid. For such a sequence constructed by adding new levels of structure to an initial state, the state cardinality K in the ratio $K!/|\text{Aut}(\Delta)|$ is itself an increasing function, but this ratio is particularly interesting in the study of entropy per unit volume, which corrects for increasing K. Low-order states often possess nontrivial automorphism groups, and the computation of labeled entropies for such states leads to interesting enumerative problems. The dynamical insignificance of the distant past suggests that these states are also the most interesting from an evolutionary perspective. For high-degree states $T^m(D_{i0})$ and $T^m(D_{tN})$ of cardinalities K_{i0} and K_{tN}, abbreviated to K and K' for legibility, typical labeled multiplicities are approximately $K!$ and $K'!$ by rigidity, and the corresponding entropies are approximately

$$e_{\text{LAB}}(T^m(D_{i0})) \approx \log K! = K \log K - K + O(\log K)$$

and

$$e_{\text{LAB}}(T^m(D_{tN})) \approx \log K'! = K' \log K' - K' + O(\log K'),$$

by Stirling's approximation. These estimates lead to a phase map with values of roughly

$$\Theta_e(\gamma) \approx \exp\left(i \log(K'!/K!)\right) \approx \exp\left(i(K' \log K' - K \log K)\right), \tag{11}$$

where the last expression omits the linear and logarithmic terms in Stirling's approximation, since rigidity is only generic and asymptotic. As in previous examples, maximizing the entropic quantity $e^{T_{i0}T_{tN}}(\gamma) \approx K' \log K' - K \log K$ in this context favors "expanding universe" scenarios. More sophisticated phase maps involving filtering methods such as weighted sums associated with the labeled entropy system of Definition 21 are also of interest in this context.

Phase maps derived from symmetry entropies may be treated in a similar manner, although high labeled entropies correspond to low symmetry entropies, and vice versa, after accounting for the cardinalities of the states under consideration. If $e = e_{\text{SYM}}$, then the symmetry multiplicities of the zeroth-degree states $T^0(D_{i0})$ and $T^0(D_{tN})$ of cardinalities K and K' are $K!$ and $K'!$, so the corresponding phase $\Theta_e(\gamma) = \exp\left(i \log(K'!/K!)\right)$ is the same as the estimate given in Equation (11) for the phase induced by labeled entropies of nearly-rigid states $T^m(D_{i0})$ and $T^m(D_{tN})$ of the same cardinalities. Conversely, for nearly-rigid states, phase values induced by symmetry entropies are near $e^0 = 1$. Again, the most interesting behavior occurs for terminal states of relatively low but nonzero degree, which possess limited but nontrivial causal structure, and have limited but nontrivial symmetries. More sophisticated phase maps may be constructed in terms of the symmetry entropy system of Definition 23. For example, it is interesting to compare entropies associated with terminal states of different degrees for the same history, using the relative notions introduced in Definition 24.

4.2. Interference Effects

Feynman's path integral reinforces the contributions of paths near the classical path γ_{CL} of a particle, via constructive interference, while faraway paths are damped out via destructive interference. Mathematically, this means that the phases assigned to paths near γ_{CL} tend to cluster near each other on the unit circle S^1, inducing large amplitudes for neighborhoods of γ_{CL}, while the phases assigned to faraway paths tend to scatter around S^1, leading to cancellation. To produce this type of behavior, paths near γ_{CL} must possess similar phases. As explained in Section 1.2, Feynman's phase map $\Theta(\gamma) = e^{\frac{i}{\hbar}\mathcal{S}(\gamma)}$ satisfies this condition due to Hamilton's principle, i.e., because γ_{CL} renders the classical action $\mathcal{S}$ stationary. In the discrete causal context, analogous relationships must be identified and exploited for the path summation approach to succeed. Much of the appeal of entropic phase maps in this setting arises from the fact that the idea of entropy is sufficiently general to produce a variety of discrete causal quantities with interesting interference-related behavior that may resemble that of $\mathcal{S}$, while remaining sufficiently specific to offer meaningful physical interpretations. This is not to suggest that $\mathcal{S}$ is similar to conventional entropy in other ways; indeed, $\mathcal{S}$ is a cumulative quantity that is typically minimized by favored processes, which are typically time-symmetric, while entropy is conventionally understood as an instantaneous quantity whose increase is observed to follow, and in some settings is believed to possibly generate, the arrow of time. It is the *role* of discrete causal entropy in producing desirable interference effects that must be "action-like" in the context of entropic phase maps. This is one reason why it is reasonable to simultaneously entertain essentially opposite versions of entropy in this setting, such as labeled entropy and symmetry entropy. In a similar manner, discrete causal action principles need not closely resemble conventional motion-related or metric-related action principles in general, provided that they play an analogous abstract role. The action principles discussed in Section 4.3 are chosen with conventional definitions in mind, but many other choices are possible.

It is therefore interesting to explore which, if any, discrete causal notions of entropy can produce "clustering effects" for phases that mimic stationary action in a suitable manner. I begin with a simple "very early universe scenario" in $\mathbb{S}_{PS}$, involving a toy co-relative kinematics represented by a chain $\gamma = r(h_0) \prec \dots \prec r(h_N)$ of relations $r(h_k)$ in $\mathcal{R}(\mathcal{M}(\mathbb{S}_{PS}))$ representing co-relative histories $h_k: D_{ik} \Rightarrow D_{tk}$ for $0 \leq k \leq N$. In the general telescoping entropic phase map

$$\Theta_e(\gamma) = \exp\left(i\left(e(\Delta^{\tau_{iN}}) - e(\Delta^{\tau_{i0}})\right)\right)$$

of Equation (8), I choose e to be the symmetry entropy function e_{SYM} of Definition 22, and $\Delta^{\tau_{i0}}$ and $\Delta^{\tau_{iN}}$ to be zeroth-degree terminal states $T^0(D_{i0})$ and $T^0(D_{tN})$ of cardinalities 5 and 10, respectively. With these choices, $\Theta_e(\gamma) = \exp(i(\log 10! - \log 5!)) = e^{i(10.3169\dots)}$. Phases determined by this particular map are very unstable for small changes in the sizes of $T^0(D_{i0})$ and $T^0(D_{tN})$. For example, adding one additional element to $T^0(D_{tN})$ yields a phase of $e^{i(12.7148\dots)}$, which is separated from $\Theta_e(\gamma)$ by an angle of about $3\pi/4$ on S^1. More generally, since $\log(K+1)! - \log K! = \log(K+1)$, adding even a single additional maximal element to an arbitrary zeroth-order terminal state produces a much different symmetry multiplicity, and this behavior only increases for large histories. Working with entropy per unit volume, instead of raw entropy, trades this instability for a profound, and perhaps excessive, stability. By Stirling's approximation, the entropy per unit volume of $T^0(D_{tN})$ is roughly $\log |T^0(D_{tN})|$ in this example, a quantity which is very stable under small changes in the size of $T^0(D_{tN})$. Using ballpark figures for fundamental units, the observable universe may possess a spatial volume of about 10^{180} in a suitable frame of reference, and treating Hubble's "constant" as actually constant gives a doubling time of about 10^{60}. Depending on the choice of kinematic scheme, one may therefore imagine a chain of perhaps 10^{60} to 10^{180} co-relative histories leading to a change in entropy per unit volume of about $\log 2$. Hence, this simplistic notion of entropy per unit volume does not seem to change very rapidly in the actual universe.

The chain independence property for the general telescoping entropic phase map Θ_e of Equation (8) is at least superficially attractive in the path summation context, since it suggests large amplitudes for processes possessing large numbers of evolutionary pathways. What is really needed, however, is a stronger property that produces "nearly identical phases" for "nearly identical physics", rather than merely producing identical phases for alternative descriptions of identical physics. A class of maps that often exhibits this type of behavior is the class of *telescoping multiplicity phase maps*

$$\Theta_\mu(\gamma) = \exp\left(i\mu(\Delta^{\tau_{iN}})/\mu(\Delta^{\tau_{i0}})\right). \tag{12}$$

Even a modest increase in entropy between $\Delta^{\tau_{i0}}$ and $\Delta^{\tau_{iN}}$ corresponds to a ratio $\mu(\Delta^{\tau_{iN}})/\mu(\Delta^{\tau_{i0}})$ that is near zero. Phases $\Theta_\mu(\gamma)$ for chains γ exhibiting large increases in entropy therefore constructively interfere, clustering near the complex number $e^{i0} = 1$. Similar behavior is not evident in

Equation (8), because the entropic quantity $e^{\tau_{i0}\tau_{tN}}(\gamma) = e(\Delta^{\tau_{tN}}) - e(\Delta^{\tau_{i0}})$ in the exponent of Θ_e typically has nonnegligible magnitude compared to the circumference 2π of S^1. Hence, two chains γ and γ' with "similar" final co-relative histories exhibiting large but distinct entropies may possess phases $\Theta_e(\gamma)$ and $\Theta_e(\gamma')$ far apart on S^1, which does not suggest encouraging interference properties for Θ_e. For example, suppose that $\Delta^{\tau_{i0}}$ is rigid, and compare two different chains γ and γ' with final co-relative histories h_N and $h'_{N'}$ exhibiting symmetry multiplicities $\mu_{SYM}(\Delta^{\tau_{tN}}) = K$ and $\mu_{SYM}(\Delta^{\tau_{tN'}}) = 6K$. Here, $\Delta^{\tau_{tN}}$ and $\Delta^{\tau_{tN'}}$ may be nearly-identical first-degree terminal states, differing, for example, by a single "trident-shaped" component contributing a symmetry factor of S_3. However, the difference between the entropic quantities $e^{\tau_{i0}\tau_{tN}}(\gamma)$ and $e^{\tau_{i0}\tau_{tN'}}(\gamma')$ in $\Theta_e(\gamma)$ and $\Theta^1_e(\gamma')$ is $\log 6$, which translates to an angular separation exceeding $\pi/2$. This example suggests that very similar processes can destructively interfere under Θ_e. In contrast, the angular separation between $\Theta_\mu(\gamma')$ and $\Theta_\mu(\gamma)$ in this example is $1/6K$, so that both phases are very near $e^{i0} = 1$ for large K. Unfortunately, the map Θ_μ in Equation (12) seems to exhibit *too much* constructive interference, in the sense that it assigns a phase near 1 to *every* chain involving a modest increase in entropy. The precedent of Feynman's phase map $\Theta(\gamma) = e^{\frac{i}{\hbar}S(\gamma)}$ suggests that the entropic quantities multiplying i in a phase map should not be uniformly small for "physically reasonable" chains. Indeed, by scaling the classical action S by Planck's reduced constant $\hbar$, Feynman's map allows these multipliers to differ appreciably for modestly different paths describing the behavior of systems for which quantum effects are noticeable, such as the motion of individual electrons.

It seems, then, that the "additive recipe" of Equation (8) may produce too little constructive interference, while the "multiplicative recipe" of Equation (12) may produce too much. There are many possible ways to address this issue. It should be noted that the problem with Equation (12) seems to be much more serious, producing an obviously wrong answer, whereas for Equation (8) it is merely unclear what the interference behavior looks like for physically realistic histories. If one chooses, then, to study modifications of Equation (8), there are at least two obvious methods to explore. First, one may adjust Θ_e via a positive real-valued scale factor s, analogous to $\hbar$. The resulting phase map is of the form

$$\Theta_s(\gamma) = \exp\left(\frac{i}{s}\left(e(\Delta^{\tau_{tN}}) - e(\Delta^{\tau_{i0}})\right)\right). \tag{13}$$

Choosing $s > 1$ produces more tightly-clustered phases, thereby increasing constructive interference for similar processes. The obvious question then becomes how to choose s in a non-arbitrary manner. This immediately suggests a second method of modifying Θ_e, by adjusting the entropies $e(\Delta^{\tau_{i0}})$ and $e(\Delta^{\tau_{tN}})$ individually, via information derived in a natural manner from the co-relative histories h_0 and h_N. An interesting variant of this approach, foreshadowed above, is to focus on *entropy per unit volume*, rather than raw entropy. This involves completely different considerations than does the conventional thermodynamic study of a variable-volume system, such as a quantity of gas in a chamber compressed by a piston. Such a system is background dependent and does not involve spacetime expansion. In the present more-fundamental setting, the study of entropy per unit volume is partly motivated by the idea that the production of "new spacetime" ought to involve some "cost", or obey some analogue of continuity. In particular, one does not observe immediate runaway expansion of spacetime, even though this tends to produce a large increase in entropy. A general phase map for finite states defined in terms of entropy per unit volume is the telescoping map

$$\Theta_{e/V}(\gamma) = \exp\left(i\left(e(\Delta^{\tau_{tN}})/|\Delta^{\tau_{tN}}| - e(\Delta^{\tau_{i0}})/|\Delta^{\tau_{i0}}|\right)\right). \tag{14}$$

For an "early universe scenario" involving a version of this map, let $\Delta^{\tau_{i0}}$ and $\Delta^{\tau_{tN}}$ be first-degree terminal states $T^1(D_{i0})$ and $T^1(D_{tN})$ of cardinalities 10 and 20, respectively, and suppose that $|\text{Aut}(\Delta^{\tau_{i0}})| = 10^2$ and $|\text{Aut}(\Delta^{\tau_{tN}})| = 10^4$. Then using $e = e_{SYM}$ in Equation (14) yields

$$\Theta_{e/V}(\gamma) = \exp\left(i\left(\log(10^4)/20 - \log(10^2)/10\right)\right) = e^{i0} = 1.$$

A similar process represented by a chain γ' whose final co-relative history has the same size for its first-degree terminal state but twice the symmetry multiplicity produces a phase of $\Theta^1_{e/V}(\gamma') \approx e^{i(0.0346...)}$. The angular difference of 0.0346... between these two values is much smaller than the corresponding difference of $\log 2 = 0.6931...$ produced by Θ^1_e. Hence, $\Theta_{e/V}$ offers an example of how one may increase constructive interference effects via natural information associated with evolutionary processes. Precise characterization of these effects in physically realistic scenarios depends on asymptotic behavior of large states. For example, working with symmetry entropy, states that are "too rigid" will typically produce values near $e^{i0} = 1$ under Equation (14), regardless of the process involved. On the other hand, states that are "too free" will produce phases for similar processes insufficiently close to generate adequate constructive interference. Other state-specific modifications of Equation (8) are also worth considering. For example, natural data associated with states may be used to determine weights in more sophisticated phase maps involving weighted sums, such as generalizations of the map given by Equation (6). This is analogous to assigning density functions to state spaces or weights to individual outcomes in Gibbs or Shannon entropy.

4.3. Objections and Alternatives

Entropic phase maps may be criticized in various ways, and alternative approaches are possible under the general framework of path summation. Given a choice of dynamics favoring an increase in a specified type of entropy, it is prudent to ask whether this dynamics obviously contradicts established physics. If so, then it can be at best a toy model. Figure 18 illustrates one type of scenario that may be considered in this context, involving a sequence of co-relative histories h'_7 to h'_{11} beginning with the initial history D_7 from the evolutionary process illustrated in Figure 10. Subsequent histories in the present process are much different; each is constructed by adding a new element related to *all* previously-existing elements. New elements are illustrated by large black nodes. This process is visually suggestive of *gravitational collapse*, leading to a "black hole" represented by the chain of new elements. This analogy is motivated by the fact that causal influence flows exclusively toward the "back hole". The automorphism groups $\text{Aut}(T^1(D'_k))$ are large symmetric groups; in fact, they are the largest possible automorphism groups for states of cardinality $|T^1(D'_k)|$ that are not antichains. In particular, they are much larger than the corresponding groups associated with the process illustrated in Figure 10. Hence, the present process maximizes symmetry entropy for first-degree terminal states.

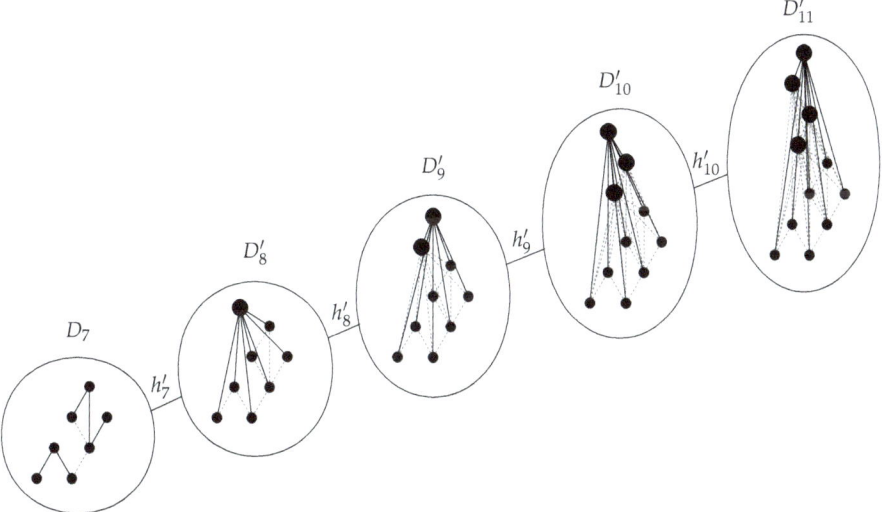

Figure 18. Sequence of co-relative histories h'_k suggestive of gravitational collapse.

Since gravitational collapse is an important feature of general relativity, one should expect such processes to be favored for certain histories that are large in ordinary terms but small on cosmological scales. Similarly, one should expect "expanding universe" scenarios such as those discussed in Section 4.1 to be favored in an appropriate cosmological sense. However, one should *not* expect extreme versions of such processes to dominate all others in every situation, and such behavior would disqualify any choice of dynamics producing it. Generalizing the present example, it would discredit the entire idea of entropic phase maps if gravitational collapse scenarios were found to entropically dominate all other evolutionary pathways combined. Rough computations suggest that this is not the case. For example, beginning with a history D, one may estimate its number of direct descendants in $\mathbb{S}_{PS}$, along with the possible sizes of their first-degree terminal state automorphism groups. If D has cardinality K, then there exists one direct descendant D' of D in $\mathbb{S}_{PS}$ for which $\text{Aut}(T^1(D'))$ is isomorphic to S_K, with cardinality $K!$, namely, the directed set D' with one new element related to all elements of D. The co-relative history $D \Rightarrow D'$ represents the beginning of the global gravitational collapse scenario for D. Similarly, there are typically about K direct descendants of D constructed by adding one new element connected to $K - 1$ elements of D. There may be fewer such descendants, due to symmetries, but this is atypical due to rigidity. The first-degree terminal state automorphism groups of these direct descendants may be as large as S_{K-1}, with cardinalities as large as $(K - 1)!$, though they may be smaller due to symmetry breaking by the "excluded element". Next, there are typically about $\binom{K}{2}$ direct descendants of D in $\mathbb{S}_{PS}$ constructed by adding one new element connected to $K - 2$ elements of D, with first-degree terminal state automorphism groups as large as $(K - 2)!$. Continuing this rough enumeration leads to an overestimate of the sum of the symmetry multiplicities for first-degree terminal states over all direct descendants of D in $\mathbb{S}_{PS}$:

$$\text{multiplicity sum} \approx \sum_{k=0}^{K} \binom{K}{k}(K-k)! = \sum_{k=0}^{K} \frac{K!}{k!}.$$

The ratio of the individual multiplicity associated with the beginning of gravitational collapse to the overall multiplicity sum is therefore roughly

$$K! / \sum_{k=0}^{K} \frac{K!}{k!} = 1 / \left(\frac{1}{K!} \sum_{k=0}^{n} \frac{K!}{k!} \right) = 1 / \sum_{k=0}^{K} \frac{1}{k!} \approx \frac{1}{e} = 0.3678...$$

Though this ratio is actually somewhat larger due to symmetry considerations, as well as the tiny effect of truncating the rapidly convergent series for e, this computation suggests that the gravitational collapse scenario does not always entropically dominate all other evolutionary pathways in the case of symmetry entropy.

A much more general objection to the idea of entropic phase maps, already mentioned in Section 4.2, is that it forces together notions that are only distantly related in conventional situations where the path summation approach to quantum theory is known to succeed and where the second law of thermodynamics is known to hold. In particular, the interference behavior of Feynman's phase map for paths in $\mathbb{R}^4$ is not closely related to conventional entropic data. As explained in Section 1.2, Feynman's map $\Theta(\gamma) = e^{\frac{i}{\hbar} \mathcal{S}(\gamma)}$ is determined by the classical action $\mathcal{S}(\gamma) = \int_\gamma \mathcal{L}\, dt$, where $\mathcal{L}$ is the Lagrangian. Hamilton's principle states that the classical path γ_{CL} renders $\mathcal{S}(\gamma)$ stationary, and for "sufficiently short" paths, $\mathcal{S}(\gamma)$ is generally minimized by γ_{CL}. In this context, the Lagrangian $\mathcal{L}$ is symmetric under time reversal, so Hamilton's principle certainly does not imply the second law. While paths favored by Hamilton's principle typically do exhibit increases in entropy in realistic scenarios, this behavior may be attributed to auxiliary details such as where these paths originate in state space. However, time reversal of a classical system, which generally involves a systematic *decrease* in entropy, obeys the equations of motion determined by $\mathcal{L}$ just as well as does the original system. Hence, an analogy between "high entropy" and "stationary action" is not necessarily motivated by established physics in any compelling way. From this viewpoint, it is not at all obvious that discrete causal analogues of Feynman's phase map should depend directly on entropy.

The answer to this objection, already summarized in Section 4.2, is that discrete causal entropy is neither expected, nor required, to play an "action-like" role in every sense. Nor must it resemble conventional thermodynamic entropy in the sense of approximation, under which macrostates are defined via imprecise, rather than merely incomplete, data. Indeed, the only version of entropy introduced in Section 3 that fits this description is resolution entropy. The remaining versions all differ from conventional thermodynamic entropy in at least two important respects: first, they do not involve actual approximation; second, they depend nontrivially on information above first order at the level of individual histories. More generally, discrete causal entropy must be "action-like" only in that it produces desirable interference effects, and it must be "entropic" only in that it arises via comparison of levels of detail under the basic framework of entropy systems. Regardless of such conventional analogies, combinatorial data encoded in terminal states is likely, on basic structural grounds, to determine discrete causal dynamics in the background independent setting. The entropic notions introduced in Section 3.4 enjoy the additional benefits of possessing clear physical meaning and suggesting effects that are known to be among the most universal in physics. Hence, these notions stand out from among a relatively limited assortment of reasonable alternatives for determining specific data for path summation.

Nevertheless, it is illuminating to briefly examine an alternative approach to path summation in the discrete causal context, expressed via discrete causal action principles related more directly to conventional motion-related or metric-related ideas. This involves defining discrete causal "Lagrangians" and "actions" that mimic their conventional counterparts as closely as possible, in the sense that they are defined in terms of specific "alterations" of individual histories. This is a much narrower prescription than that of the relation function θ in Equation (4), which is "Lagrangian-like" in an abstract sense regardless of its actual information content. An immediate difficulty with this strategy is that notions such as energy, metric structure, and curvature, which are central to conventional definitions of $\mathcal{L}$ and $\mathcal{S}$, are themselves emergent in discrete causal theory. The same is true of related quantities such as mass and momentum, which are often used to determine these notions. In partially-background-dependent versions of discrete causal theory, such as quantum causal set theory, "nongravitational matter" is ascribed to auxiliary fields and particles existing on directed sets, and it is not too difficult to define reasonable analogues of $\mathcal{L}$ and $\mathcal{S}$ in this setting. However, the situation is subtler in the perfectly-background-independent context under the strong version of the causal metric hypothesis. As explained in Section 3.3, a popular problem in the study of discrete gravity is how to abstract and generalize the Einstein–Hilbert action $\mathcal{S}_{EH}$ [45–47]. However, the metric g and the scalar curvature R used to define $\mathcal{S}_{EH}$ are unlikely to possess meaningful direct analogues at the fundamental scale, where even primitive notions such as dimension and topological structure are relatively obscure. Success in abstracting such quantities would accomplish only part of the desired objective in any case, since a genuinely fundamental theory of spacetime should explain the origins of more basic geometric and pre-geometric properties.

For these reasons, it seems preferable to work at a more conceptual level in defining discrete causal analogues of $\mathcal{L}$ and $\mathcal{S}$. The conceptual content of Hamilton's principle is that nature is basically conservative; it favors as little overall alteration as possible in evolving from one state to another. Setting aside conventional ideas involving the conversion of one type of energy into another, or the overall motion represented by a path between two points in a manifold, one may formulate discrete causal action principles embodying this basic concept, hypothesizing that the resulting dynamics will faithfully preserve the desired physical *meaning* as one works up from the fundamental scale. In this context, the most natural discrete causal analogues of $\mathcal{L}$ and $\mathcal{S}$ are functionals that describe the extent to which a given history or terminal state is altered in a process leading to another history or terminal state. One way of describing such alteration is in terms of the elementary operations introduced in Definition 16, which define the absolute distance between pairs of directed or multidirected sets. There are at least two possible choices for how to quantify such an action: one may either count the number of elementary operations necessary to convert one state Δ to another state Δ', ignoring ambient histories, or one may count the number of operations involved in converting a history with terminal state Δ to a history with terminal state Δ'. The difference between these two notions of action is analogous to the difference between absolute distance in Definition 16 and scheme-dependent distances in Definition 17.

Definition 27. *Let $h : D_i \Rightarrow D_t$ be a co-relative history in a kinematic scheme $\mathbb{S}$. Let Δ^{τ_i} and Δ^{τ_t} be terminal states of D_i and D_t with respect to transitions τ_i and τ_t, respectively.*

1. *The* **state-level Lagrangian quantity** *$\mathcal{L}^{\tau_i \tau_t}(h)$ of h with respect to the pair (τ_i, τ_t) is the number of elementary operations necessary to convert Δ^{τ_i} to Δ^{τ_t}.*
2. *The* **history-level Lagrangian** *$\mathcal{L}$ is the functional assigning to each co-relative history h the number of elementary operations involved in converting D_i to D_t, i.e., the number of elements and relations added to D_i by h.*

Both $\mathcal{L}^{\tau_i \tau_t}(h)$ and $\mathcal{L}$ may take on either finite or infinite values in this general setting, though it is often useful and appropriate to impose finiteness conditions. $\mathcal{L}^{\tau_i \tau_t}(h)$ is called a "Lagrangian quantity" rather than a "Lagrangian" because it depends on choices of transitions τ_i and τ_t. One may specialize this definition to define standard Lagrangian functionals. For example, one might define the *first-degree state-level Lagrangian* $\mathcal{L}^1$ to be the functional assigning the state-level Lagrangian quantity $\mathcal{L}^{\tau_{D_i}^1 \tau_{D_t}^1}(h)$ to each co-relative history $h : D_i \Rightarrow D_t$. The history-level quantity $\mathcal{L}$ seems much more natural than the state-level quantity $\mathcal{L}^{\tau_i \tau_t}(h)$ in a structural sense. An unattractive aspect of $\mathcal{L}^{\tau_i \tau_t}(h)$ is that a sequence of elementary operations converting Δ^{τ_i} to Δ^{τ_t} typically identifies structural components of these two sets that arise from different parts of their corresponding histories. For example, the first-degree terminal state Δ_7 of the history D_7 appearing in the evolutionary process illustrated in Figure 10 may be converted into the first-degree terminal state Δ_8 by a sequence of three elementary operations, but only at the expense of identifying "early" structure in D_7 with "later" structure in D_8.

A good motivation to study state-level quantities such as $\mathcal{L}^{\tau_i \tau_t}(h)$ despite this awkwardness is that they are related to conventional evolutionary ideas in certain important ways. For example, one may imagine a history in which "nothing changes", in the sense that each terminal state of a given degree "exactly replicates itself". The simplest example is given by sequential growth of a chain; at each stage of evolution, the first-degree terminal state of this chain consists of a single relation connecting its penultimate element to its terminal element. Such a "frozen" or "static" history exhibits a value of zero at every stage of evolution for an appropriate uniform choice of state-level Lagrangian quantities $\mathcal{L}^{\tau_i \tau_t}(h)$, such as those induced by the first-degree state-level Lagrangian $\mathcal{L}^1$. This agrees with the naïve idea of dynamical stasis for this history. By contrast, the value $\mathcal{L}(h)$ of the history-level Lagrangian $\mathcal{L}$ at every stage h of the evolution of such a history is a nonzero constant, and a similar average value for $\mathcal{L}(h)$ occurs in "non-static" histories adding roughly the same number of elements and relations at each evolutionary stage. Such histories may exhibit extreme structural differences among generations, which may be essentially invisible to $\mathcal{L}$. More generally, state-level quantities may often detect interesting changes that are invisible to history-level quantities. A closely-related issue is the problem of how to obtain suitable analogues of conventional evolutionary continuity. As explained in Section 3.3, the conventional entropic preference for thermal equilibrium is balanced by the continuity of evolution curves in state space and the fact that such curves may not originate near the cell representing thermal equilibrium. The same topic was revisited in Section 4.2 in the context of entropy per unit volume and spacetime expansion. Dynamics that explicitly resists drastic changes in state-level quantities seems a priori more likely to avoid serious pathologies along these lines than dynamics defined in terms of history-level quantities.

Each discrete causal Lagrangian induces a corresponding discrete causal action by summing Lagrangian quantities over sequences of co-relative histories.

Definition 28. Let $\mathbb{S}$ be a kinematic scheme, and let $\gamma = r(h_0) \prec ... \prec r(h_N)$ be a chain in $\mathcal{R}(\mathcal{M}(\mathbb{S}))$ representing a co-relative kinematics in $\mathbb{S}$, where each relation $r(h_k)$ represents a co-relative history $h_k : D_{ik} \to D_{tk}$. Let $\Delta^{\tau_{ik}}$ and $\Delta^{\tau_{tk}}$ be terminal states of D_{ik} and D_{tk} with respect to transitions τ_{ik} and τ_{tk}.

1. The **state-level action quantity** $\mathcal{S}^{\{\tau_{ik}\},\{\tau_{tk}\}}(\gamma)$ along γ with respect to the pair of sequences of transitions $\{\tau_{ik}\} = \{\tau_{i0},...,\tau_{iN}\}$ and $\{\tau_{tk}\} = \{\tau_{t0},...,\tau_{tN}\}$ is the sum

$$\mathcal{S}^{\{\tau_{ik}\},\{\tau_{tk}\}}(\gamma) = \sum_{k=0}^{N} \mathcal{L}^{\tau_{ik}\tau_{tk}}(h_k)$$

2. The **history-level action** $\mathcal{S}$ is the functional assigning to each chain γ the number of elementary operations involved in converting D_{i0} to D_{tN}, i.e., the number of elements and relations added to D_{i0} by the sequence of co-relative histories $h_0, ..., h_N$.

As in the case of Lagrangians, the history-level action $\mathcal{S}$ seems to be much more natural in a basic structural sense than the state-level action quantity $\mathcal{S}^{\{\tau_{ik}\},\{\tau_{tk}\}}(\gamma)$. One obvious complication involving the latter quantity is that fewer elementary operations are typically required to convert a state Δ directly to a state Δ'' than to first convert Δ to an "interpolating state" Δ', then convert Δ' to Δ''. However, the awkwardness of $\mathcal{S}^{\{\tau_{ik}\},\{\tau_{tk}\}}(\gamma)$ may be ameliorated to some extent by specifying a uniform choice of transitions $\{\tau_{ik}\}$ and $\{\tau_{tk}\}$, for example, first-degree transitions. The resulting *first-degree state-level action* functional may be denoted by $\mathcal{S}^1$. Again, a good motivation for considering state-level functionals is that they are more closely related to conventional evolutionary ideas in certain respects than are history-level functionals. In particular, the history-level functional $\mathcal{S}$ does not distinguish between co-relative kinematics involving state-replicating "static histories" and co-relative kinematics involving histories in which considerable state-level change occurs, provided that the same total number of elements and relations are added over the course of each process.

Discrete causal Lagrangians and actions defined in terms of elementary operations on directed sets supply dynamical alternatives to entropic phase maps under the path summation approach to quantum theory. For example, one might define an *action-induced phase map* $\Theta(\gamma) = e^{i\mathcal{S}^1(\gamma)}$ using the first-degree state-level action functional $\mathcal{S}^1$ introduced above. This raises the obvious question of how these two general types of dynamics compare. For example, one may consider the gravitational collapse scenario illustrated in Figure 18. The value of the first-degree state-level Lagrangian $\mathcal{L}^1$ at the kth stage of evolution is 2, because the kth first-degree terminal state Δ_k differs from the $(k+1)$st first-degree terminal state Δ_{k+1} by a single element and a single relation, up to isomorphism. However, the elements and relations that are identified under such a comparison are completely different from the perspective of the entire terminal history D_{k+1}. The value of the history-level Lagrangian $\mathcal{L}$ at the kth stage of evolution is $(k+1)$, because one new element and k new relations are added to the initial history D_k. The state automorphism group $\text{Aut}(\Delta_k)$ of Δ_k, meanwhile, is typically isomorphic to S_{k-1}, of cardinality $(k-1)!$, and the state automorphism group $\text{Aut}(\Delta_{k+1})$ of Δ_{k+1} is typically isomorphic to S_k, of cardinality $k!$. The ratio of the symmetry multiplicities $\mu_{\text{SYM}}(\Delta_{k+1})/\mu_{\text{SYM}}(\Delta_k)$ is therefore typically k, and the corresponding increase in symmetry entropy is typically $\log k$.

Interesting structural relationships exist between the Lagrangians and actions introduced in this section and the entropic notions developed in Section 3. Here, I can only offer vague sketches of a few of these relationships. For example, the construction of superset microstates may be expressed via "elementary operations" at the level of kinematic schemes. In particular, the first superset multiplicity $\mu_{\text{SUP}}^1(\Delta)$ in Definition 18 is the number $|R^+(x(\Delta^*))|$ of relations in $\mathcal{M}(\mathbb{S}_{\text{PS}})$ beginning at the element $x(\Delta^*)$ representing the causal dual Δ^* of a state Δ. If this multiplicity is N, then one may imagine a "growth process" for $\mathbb{S}_{\text{PS}}$ that adds the N co-relative histories represented by the elements of $R^+(x(\Delta^*))$ at some stage of growth. This corresponds to a "history-level action" of roughly $2N$ for the corresponding stage of growth of $\mathcal{M}(\mathbb{S}_{\text{PS}})$, ignoring multidirected structure, so in this case large entropy corresponds to large action. However, since supersets encode "growth into the past", one might argue for associating a minus sign with this "action", reversing this relationship. Relative notions of symmetry entropy such as those introduced in Definition 24 also involve supersets, and may therefore be related to such higher-level "action". However, the most basic question in comparing a "non-entropic" discrete causal action principle to a choice of discrete causal entropy is whether or not such a principle, together with the structure of an appropriate discrete causal state space, at least favors *increasing* entropy, regardless of whether or not it favors the *maximal* possible increase at each evolutionary stage. In this context, an action principle applied to a state space may lead indirectly to a version of the second law of thermodynamics, even if it is not derived from, or equivalent to, such a law. This is certainly the case for conventional thermodynamics based on Newtonian physics applied to ordinary state spaces. Corresponding relationships between discrete causal action principles and discrete causal entropy remain mostly unexplored.

4.4. Summary and Conclusions

Entropic phase maps offer one possible method of supplying specific dynamical content for the path summation approach to discrete quantum causal theory developed in [14]. Background and basics of this approach are reviewed in Sections 1 and 2 of this paper. Such maps assign phases to evolutionary pathways called co-relative kinematics in a discrete causal history configuration space called a kinematic scheme. Their role is analogous to the role of Feynman's phase map in the path summation approach to ordinary quantum theory [1], which assigns phases to particle paths in a background spacetime manifold. Each co-relative kinematics consists of a sequence of individual evolutionary relationships between pairs of histories, called co-relative histories, mathematically represented by equivalence classes of transitions between pairs of directed sets. A phase map whose values are multiplicative for concatenation of co-relative kinematics is generated by a relation function θ, which assigns phases to relations representing individual co-relative histories. Such a phase map determines a specific version of the causal Schrödinger-type equation

$$\psi^-_{R;\theta}(r) = \theta(r) \sum_{r^- \prec r} \psi^-_{R;\theta}(r^-),$$

reproduced here from Equation (4). In physical terms, a suitable phase map must produce interference effects that reinforce "reasonable" evolutionary processes, while damping out pathological processes. In the case of entropic phase maps, this means that the entropic quantities defining these maps should satisfy a property analogous to Hamilton's principle of stationary action. In other respects, these quantities need not resemble the classical action that determines Feynman's phase map. In particular, they need not be directly associated with familiar motion-related concepts such as potential and kinetic energy, which define classical Lagrangians and actions in Newtonian mechanics, or with metric structure, which determines the Einstein–Hilbert action in general relativity.

Entropy systems, introduced in Section 3.1, offer a general approach to entropy and the second law of thermodynamics. Conventional versions of the second law involve notions of entropy associated with "present states", not with entire histories. In the discrete causal context, this suggests defining entropies for terminal states of histories, which encode "recent" causes and effects. Such states are defined in Section 3.3 in terms of transitions between pairs of directed sets. Aside from their evident physical importance, such states are mathematically interesting due to their symmetry properties, which exhibit a balance between the typical rigidity of general acyclic directed sets demonstrated by Bender and Robinson [37], and the transitivity of antichains under their automorphism groups. There are a variety of ways to define entropies for such states, all of which involve comparing distinguishability properties of states at different levels of detail. Since multiple such levels merit simultaneous consideration in discrete causal theory, a sufficiently general approach to discrete causal entropy requires the use of entropy systems, which organize such levels in a systematic way. Given two levels of detail, descriptions of a system at the coarser level are called macrostates, while descriptions at the finer level are called microstates. The corresponding notion of entropy measures the quantity of microstates corresponding to each macrostate in a manner that is additive for composite systems. An important distinction between conventional thermodynamics and discrete causal theory is that precise information up to first order typically suffices to determine future evolution in the former setting, while higher-order information at the level of individual histories is a priori relevant in the latter setting. In both cases, however, empirical evidence suggests that details of the distant past should exert negligible influence on future events.

Four general methods of defining discrete causal macrostates and microstates, along with their associated notions of entropy, and the resulting entropic phase maps, are examined in this paper. Spaces of states are studied in Section 3.3, entropies in Section 3.4, and phase maps in Section 4.1. The first method uses the theory of causal atomic resolution, whereby causal structure at the fundamental scale is approximated by families of coarser causal structures constructed from special subsets of directed sets, called causal atoms. This leads to the notion of resolution entropy. This approach is very similar to coarse-graining of state space in conventional thermodynamics; in particular, it involves actual approximation. The second method supplements the information encoded in terminal states by describing how they may embed into larger states called supersets. This leads to the notion of superset entropy. The level of detail in the original states is regarded as "coarse" because it is incomplete, not because it is approximate. Supersets offer finer detail in the sense that they encode more complete information. The third method measures distinguishability properties intrinsic to states by counting the number of distinct ways in which they may be labeled. This leads to the notion of labeled entropy. Labeled entropy is maximal for states lacking nontrivial symmetries, which meshes with the intuition that high-entropy states should be "disordered". The fourth method follows essentially the opposite approach, by counting symmetries. This leads to the notion of symmetry entropy. Like superset entropy, both labeled entropy and symmetry entropy involve organizing precise but incomplete information, rather than actual approximation.

Computation of entropic phase maps in physically realistic situations is analytically involved, and most of the results in this paper involve toy examples or qualitative results. Many of these appear in Sections 4.1, 4.2 and 4.3. Discrete causal versions of the second law of thermodynamics favor expanding universe scenarios, but this conclusion is obvious on basic enumerative grounds, and does not favor discrete causal theory over other theories in any specific way. There is some evidence that raw measures of entropy may be too sensitive to minor changes in structure to produce desirable interference effects. The notion of entropy per unit volume seems more stable in this regard, and is also attractive in other respects. Since the theory of entropic phase maps is almost completely unexplored, many versions of the approach can likely be eliminated without serious effort. Symmetry entropy is doubtful on conventional grounds, and also seems to be vulnerable to pathological instabilities such as universal gravitational collapse scenarios. However, the idea is not obviously unworkable, and the desire to model symmetric structures in nature, such as "elementary" particles, renders such notions worth entertaining. Discrete causal action principles involving elementary operations on directed sets offer an alternative to entropic phase maps in the path summation context. Relationships exist between these two approaches, but the details of these connections are unclear at present.

Problems that must be solved to further develop the theory of entropic phase maps include the enumeration of certain classes of acyclic directed sets, and the computations of their automorphism groups. These problems may be approached from a mathematical perspective via the theory of random graphs, and interesting and important results of this nature may be found in the graph-theoretic literature. However, most of these results are developed from a perspective very different than the study of fundamental spacetime structure, and the perception of what problems are interesting is different in this setting as well. Hence, it is not easy to mine the existing body of graph theory for such results, and many physically relevant topics remain underdeveloped. This is likely due both to difficulty of problems and differences in emphasis. Particularly useful in this context would be a thorough analysis of families of directed graphs corresponding to nth-order states.

For example, how would one compute the average number of superset microstates adding 10^3 elements to a first-order state of cardinality 10^4? What is the average size of the automorphism group of a first-order state with 10^9 elements and 10^{12} relations? For a fixed degree n, how does the average size of $\text{Aut}(T^n(D))$ scale with the cardinality of D? For a fixed ratio of order to cardinality for states Δ, how does the average size of $\text{Aut}(\Delta)$ scale with the cardinality of Δ? Going beyond average quantities, how are the numbers of superset microstates, or the sizes of state automorphism groups, distributed for certain classes of states? Are they randomly scattered, or do they tend to cluster around certain values? Many questions of this nature must be answered before the physical implications of entropic phase maps can be understood in any detail. Computational resources may also be used to compile numerical evidence about the behavior of various entropic phase maps for relatively small histories. For example, it would be very interesting to compute some of the entropic quantities examined in this paper for the first few generations of the positive sequential kinematic scheme $\mathbb{S}_{\text{PS}}$.

Acknowledgments: The author thanks Brendan McKay, Johnny Feng, Jessica Garriga, Kiran Bist, and Stephanie Dribus for useful discussions.

Conflicts of Interest: The author declares no conflict of interest.

References

1. Feynman, R. Space-Time Approach to Non-Relativistic Quantum Mechanics. *Rev. Mod. Phys.* **1948**, *20*, 367.
2. Bombelli, L.; Lee, J.; Meyer, D.; Sorkin, R. Space-Time as a Causal Set. *Phys. Rev. Lett.* **1987**, *59*, 521.
3. Finkelstein, D. Space-Time Code. *Phys. Rev.* **1969**, *184*, 1261.
4. Finkelstein, D. "Superconducting" Causal Nets. *Int. J. Theor. Phys.* **1988**, *27*, 473–519.
5. Knuth, K.H.; Bahreyni, N. A potential foundation for emergent space-time. *J. Math. Phys.* **2014**, *55*, 112501.
6. Ambjorn, J.; Dasgupta, A.; Jurkiewicz, J.; Loll, R. A Lorentzian cure for Euclidean troubles. *Nucl. Phys. B Proc. Suppl.* **2002**, *106*, 977–979.
7. Markopoulou, F. Quantum Causal Histories. *Class. Quantum Gravity* **2000**, *17*, 2059.
8. Rovelli, C. Quantum Gravity. In *Cambridge Monographs on Mathematical Physics*; Cambridge University Press: Cambridge, UK, 2004.
9. Thiemann, T. Modern Canonical Quantum General Relativity. In *Cambridge Monographs on Mathematical Physics*; Cambridge University Press: Cambridge, UK, 2007.
10. D'Ariano, G.M.; Perinotti, P. Derivation of the Dirac Equation from Principles of Information Processing. *Phys. Rev. A* **2014**, *90*, 062106.
11. Finster, F. Causal Fermion Systems: An Overview. In *Quantum Mathematical Physics*; Springer: Berlin, Germany, 2016.
12. Finster, F. The Continuum Limit of Causal Fermion Systems: From Planck Scale Structures to Macroscopic Physics. In *Fundamental Theories of Physics*; Springer: Berlin, Germany, 2016.
13. Chen, H.; Sasakura, N.; Sato, Y. Emergent Classical Geometries on Boundaries of Randomly Connected Tensor Networks. *arXiv* **2016**, arXiv:1601.04232.
14. Dribus, B.F. *Discrete Causal Theory: Emergent Spacetime and the Causal Metric Hypothesis*; Springer: Berlin, Germany, 2017.
15. Dribus, B.F. On the Foundational Assumptions of Modern Physics. In *Questioning the Foundations, the Frontiers Collection*; Springer: Berlin, Germany, 2015; pp. 45–60.
16. Dribus, B.F. On the Axioms of Causal Set Theory. *arXiv* **2013**, arXiv:1311.2148.
17. D'Ariano, G.M.; Chiribella, G.; Perinotti, P. *Quantum Theory From First Principles*; Cambridge University Press: Cambridge, UK, 2017.
18. Knuth, K.H. Information-based Physics: An observer-centric foundation. *Contemp. Phys.* **2014**, *55*, 12–32.
19. Verlinde, E. On the origin of gravity and the laws of Newton. *J. High Energy Phys.* **2011**, *4*, 29.
20. Kleitman, D.J.; Rothschild, B.L. Asymptotic Enumeration of Partial Orders on a Finite Set. *Trans. Am. Math. Soc.* **1975**, *205*, 205–220.
21. Moore, C. Comment on "Space-Time as a Causal Set". *Phys. Rev. Lett.* **1988**, *60*, 655.
22. Bombelli, L.; Lee, J.; Meyer, D.; Sorkin, R. Bombelli et al. Reply to Comment on "Space-Time as a Causal Set". *Phys. Rev. Lett.* **1988**, *60*, 656.

23. Hawking, S.W.; King, A.R.; McCarthy, P.J. A new topology for curved space-time which incorporates the causal, differential, and conformal structures. *J. Math. Phys.* **1976**, *17*, 174–181.
24. Malament, D.B. The class of continuous timelike curves determines the topology of spacetime. *J. Math. Phys.* **1977**, *18*, 1399–1404.
25. Martin, K.; Panangaden, P. A Domain of Spacetime Intervals in General Relativity. *Commun. Math. Phys.* **2006**, *267*, 563–586.
26. Bombelli, L.; Meyer, D. Origin of Lorentzian geometry. *Phys. Lett. A* **1989**, *141*, 226–228.
27. Parrikar, O.; Surya, S. Causal topology in future and past distinguishing spacetimes. *Class. Quantum Gravity* **2011**, *28*, 155020.
28. Myrheim, J. Statistical Geometry. Available online: https://cds.cern.ch/record/293594/files/197808143.pdf (accessed on 30 June 2017).
29. Hooft, G. Quantum Gravity: A Fundamental Problem and some Radical Ideas. In *Recent Developments in Gravitation*; Springer: New York, NY, USA, 1978; pp. 323–345.
30. Ahmed, M.; Dodelson, S.; Greene, P.B.; Sorkin, R. Everpresent Λ. *Phys. Rev. D* **2004**, *69*, 103523.
31. Bombelli, L.; Henson, J.; Sorkin, R. Discreteness without symmetry breaking: A theorem. *Mod. Phys. Lett. A* **2009**, *24*, 2579–2587.
32. Harary, F.; Norman, R.Z. Some Properties of Line Digraphs. *Rediconti del Circolo Matematico di Palermo* **1960**, *9*, 161–168.
33. Major, S.A.; Rideout, D.; Surya, S. Spatial Hypersurfaces in Causal Set Cosmology. *Class. Quantum Gravity* **2006**, *23*, 4743–4751.
34. Surya, S. Directions in Causal Set Quantum Gravity. In *Recent Research in Quantum Gravity*; Dasgupta, A., Ed.; Nova Science Publishing Incorporated: Hauppauge, NY, USA, 2012.
35. Sorkin, R. Expressing entropy globally in terms of (4D) field-correlations. *J. Phys. Conf. Ser.* **2014**, *484*, 012004.
36. Sorkin, R.; Yazdi, Y. Entanglement Entropy in Causal Set Theory. *arXiv* **2016**, arXiv:1611.10281v1.
37. Bender, E.A.; Robinson, R.W. The Asymptotic Number of Acyclic Digraphs II. *J. Comb. Theory Ser. B* **1988**, *44*, 363–369.
38. Rideout, D.; Sorkin, R. Classical sequential growth dynamics for causal sets. *Phys. Rev. D* **2000**, *61*, 024002.
39. Sorkin, R. Toward a Fundamental Theorem of Quantal Measure Theory. *Math. Struct. Comput. Sci.* **2012**, *22*, 816–852.
40. Isham, C. Quantum Logic and the Histories Approach to Quantum Theory. *J. Math. Phys.* **1994**, *35*, 2157.
41. Isham, C. Topos Theory and Consistent Histories: The Internal Logic of the Set of all Consistent Sets. *Int. J. Theor. Phys.* **1997**, *36*, 785.
42. Isham, C. Quantising on a Category. *Found. Phys.* **2005**, *35*, 271–297.
43. Isham, C. Topos Methods in the Foundations of Physics. In *Deep Beauty: Understanding the Quantum World through Mathematical Innovation*; Halvorson, H., Ed.; Cambridge University Press: Cambridge, UK, 2011.
44. Penrose, R. *Cycles of Time*; Vintage Books: New York, NY, USA, 2010.
45. Benincasa, D.M.T.; Dowker, F. Scalar Curvature of a Causal Set. *Phys. Rev. Lett.* **2010**, *104*, 181301.
46. Glaser, L. A closed form expression for the causal set D'Alembertian. *Class. Quantum Gravity* **2014**, *31*, 5007.
47. Aslanbeigi, S.; Saravani, M.; Sorkin, R. Generalized Causal Set d'Alembertians. *arXiv* **2014**, arXiv:1403.1622.

© 2017 by the author. Licensee MDPI, Basel, Switzerland. This article is an open access article distributed under the terms and conditions of the Creative Commons Attribution (CC BY) license (http://creativecommons.org/licenses/by/4.0/).

Article

Nonclassicality by Local Gaussian Unitary Operations for Gaussian States

Yangyang Wang [1,†], Xiaofei Qi [1,2,*,†] and Jinchuan Hou [1,3,†]

1 Department of Mathematics, Shanxi University, Taiyuan 030006, China; wyy19860927@163.com (Y.W.); houjinchuan@tyut.edu.cn (J.H.)
2 Institute of Big Data Science and Industry, Shanxi University, Taiyuan 030006, China
3 Department of Mathematics, Taiyuan University of Technology, Taiyuan 030024, China
* Correspondence: qixf1981@sxu.edu.cn; Tel.:+86-351-7010555
† These authors contributed equally to this work.

Received: 19 January 2018; Accepted: 6 April 2018; Published: 11 April 2018

Abstract: A measure of nonclassicality $\mathcal{N}$ in terms of local Gaussian unitary operations for bipartite Gaussian states is introduced. $\mathcal{N}$ is a faithful quantum correlation measure for Gaussian states as product states have no such correlation and every non product Gaussian state contains it. For any bipartite Gaussian state ρ_{AB}, we always have $0 \leq \mathcal{N}(\rho_{AB}) < 1$, where the upper bound 1 is sharp. An explicit formula of $\mathcal{N}$ for $(1+1)$-mode Gaussian states and an estimate of $\mathcal{N}$ for $(n+m)$-mode Gaussian states are presented. A criterion of entanglement is established in terms of this correlation. The quantum correlation $\mathcal{N}$ is also compared with entanglement, Gaussian discord and Gaussian geometric discord.

Keywords: quantum correlations; Gaussian states; Gaussian unitary operations; continuous-variable systems

1. Introduction

The presence of correlations in bipartite quantum systems is one of the main features of quantum mechanics. The most important one among such correlations is entanglement [1]. However, recently much attention has been devoted to the study and the characterization of quantum correlations that go beyond the paradigm of entanglement, being necessary but not sufficient for its presence. Non-entangled quantum correlations also play important roles in various quantum communications and quantum computing tasks [2–5].

For the last two decades, various methods have been proposed to quantify quantum correlations, such as quantum discord (QD) [6,7], geometric quantum discord [8,9], measurement-induced nonlocality (MIN) [10] and measurement-induced disturbance (MID) [11] for discrete-variable systems. It is also important to develop new simple criteria for witnessing correlations beyond entanglement for continuous-variable systems. In this direction, Giorda, Paris [12] and Adesso, Datta [13] independently introduced the definition of Gaussian QD for Gaussian states and discussed its properties. Adesso and Girolami in [14] proposed the concept of Gaussian geometric discord (GD) for Gaussian states. Measurement-induced disturbance of Gaussian states was studied in [15], while MIN for Gaussian states was discussed in [16]. For other related results, see [17,18] and the references therein. Note that not every quantum correlation defined for discrete-variable systems has a Gaussian analogy for continuous-variable systems [16]. On the other hand, the values of Gaussian QD and Gaussian GD are very difficult to be computed and the known formulas are only for some $(1+1)$-mode Gaussian states. Little information is revealed by Gaussian QD and GD. The purpose of this paper is to introduce a new measure of nonclassicality for $(n+m)$-mode quantum states in continuous-variable systems, which is simpler to be computed and can be used with any $(n+m)$-mode Gaussian states.

Given a bipartite quantum state ρ acting on Hilbert space $H_A \otimes H_B$, denote by $\rho_A = \text{Tr}_B(\rho)$ the reduced density operator in subsystem A. For the case of finite dimensional systems, the author of [19] proposed a quantity $d_{U_A}(\rho)$ defined by $d_{U_A}(\rho) = \frac{1}{\sqrt{2}}\|\rho - (U_A \otimes I)\rho(U_A \otimes I)^\dagger\|_F$, where $\|A\|_F = \sqrt{\text{Tr}(A^\dagger A)}$ denotes the Frobenius norm and U_A is any unitary operator satisfying $[\rho_A, U_A] = 0$. This quantity demands that the reduced density matrix of the subsystem A is invariant under this unitary transformation. However, the global density matrix may be changed after such local unitary operation, and therefore $d_{U_A}(\rho)$ may be non-zero for some U_A. Then, Datta, Gharibian, et al. discussed respectively in [20,21] the properties of $d_{U_A}(\rho)$ and revealed that $\max_{U_A} d_{U_A}(\rho)$ can be used to investigate the nonclassical effect.

Motivated by the works in [19–21], we can consider an analogy for continuous-varible systems. In the present paper, we introduce a quantity $\mathcal{N}$ in terms of local Gaussian unitary operations for $(n+m)$-mode quantum states in Gaussian systems. Different from the finite dimensional case, besides the local Gaussian unitary invariance property for quantum states, we also show that $\mathcal{N}(\rho_{AB}) = 0$ if and only if ρ_{AB} is a Gaussian product state. This reveals that the quantity $\mathcal{N}$ is a kind of faithful measure of the nonclassicality for Gaussian states that a state has this nonclassicality if and only if it is not a product state. In addition, we show that $0 \leq \mathcal{N}(\rho_{AB}) < 1$ for each $(n+m)$-mode Gaussian state ρ_{AB} and the upper bound 1 is sharp. An estimate of $\mathcal{N}$ for any $(n+m)$-mode Gaussian states is provided and an explicit formula of $\mathcal{N}$ for any $(1+1)$-mode Gaussian states is obtained. As an application, a criterion of entanglement for $(1+1)$-mode Gaussian states is established in terms of $\mathcal{N}$ by numerical approaches. Finally, we compare $\mathcal{N}$ with Gaussian QD and Gaussian GD to illustrate that it is a better measure of the nonclassicality.

2. Gaussian States and Gaussian Unitary Operations

Recall that, for arbitrary state ρ in an n-mode continuous-variable system, its characteristic function χ_ρ is defined as

$$\chi_\rho(z) = \text{Tr}(\rho W(z)),$$

where $z = (x_1, y_1, \cdots, x_n, y_n)^T \in \mathbb{R}^{2n}$ with $\mathbb{R}$ the field of real numbers and $(\cdot)^T$ the transposition, and $W(z) = \exp(iR^T z)$ is the Weyl operator. Let $R = (R_1, R_2, \cdots, R_{2n})^T = (\hat{Q}_1, \hat{P}_1, \cdots, \hat{Q}_n, \hat{P}_n)^T$. As usual, $\hat{Q}_i$ and $\hat{P}_i$ stand respectively for the position and momentum operators for each $i \in \{1, 2, \cdots, n\}$. They satisfy the Canonical Commutation Relation (CCR) in natural units ($\hbar = 1$)

$$[\hat{Q}_i, \hat{P}_j] = \delta_{ij} iI \text{ and } [\hat{Q}_i, \hat{Q}_j] = [\hat{P}_i, \hat{P}_j] = 0,$$

$i, j = 1, 2, \ldots, n$.

Gaussian states: ρ is called a Gaussian state if $\chi_\rho(z)$ is of the form

$$\chi_\rho(z) = \exp[-\frac{1}{4}z^T \Gamma z + i\mathbf{d}^T z],$$

where

$$\begin{aligned}\mathbf{d} &= (\langle \hat{R}_1 \rangle, \langle \hat{R}_2 \rangle, \ldots, \langle \hat{R}_{2n} \rangle)^T \\ &= (\text{Tr}(\rho R_1), \text{Tr}(\rho R_2), \ldots, \text{Tr}(\rho R_{2n}))^T \in \mathbb{R}^{2n}\end{aligned}$$

is called the mean or the displacement vector of ρ and $\Gamma = (\gamma_{kl}) \in M_{2n}(\mathbb{R})$ is the covariance matrix (CM) of ρ defined by $\gamma_{kl} = \text{Tr}[\rho(\Delta \hat{R}_k \Delta \hat{R}_l + \Delta \hat{R}_l \Delta \hat{R}_k)]$ with $\Delta \hat{R}_k = \hat{R}_k - \langle \hat{R}_k \rangle$ ([22–24]). Here, $M_{l \times k}(\mathbb{R})$ stands for the set of all l-by-k real matrices and, when $l = k$, we write $M_{l \times k}(\mathbb{R})$ as $M_l(\mathbb{R})$. Note that the CM Γ of a state is symmetric and must satisfy the uncertainty principle $\Gamma + i\Delta \geq 0$, where $\Delta = \oplus_{i=1}^n \Delta_i$ with $\Delta_i = \begin{pmatrix} 0 & 1 \\ -1 & 0 \end{pmatrix}$ for each i. From the diagonal terms of the above inequality, one can easily derive the usual Heisenberg uncertainty relation for position and momentum $V(\hat{Q}_i)V(\hat{P}_i) \geq 1$ with $V(\hat{R}_i) = \langle (\Delta \hat{R}_i)^2 \rangle$ [25].

Now assume that ρ_{AB} is any $(n+m)$-mode Gaussian state. Then, the CM Γ of ρ_{AB} can be written as

$$\Gamma = \begin{pmatrix} A & C \\ C^T & B \end{pmatrix}, \tag{1}$$

where $A \in M_{2n}(\mathbb{R})$, $B \in M_{2m}(\mathbb{R})$ and $C \in M_{2n \times 2m}(\mathbb{R})$. Particularly, if $n = m = 1$, by means of local Gaussian unitary (symplectic at the CM level) operations, Γ has a standard form:

$$\Gamma_0 = \begin{pmatrix} A_0 & C_0 \\ C_0^T & B_0 \end{pmatrix}, \tag{2}$$

where $A_0 = \begin{pmatrix} a & 0 \\ 0 & a \end{pmatrix}$, $B_0 = \begin{pmatrix} b & 0 \\ 0 & b \end{pmatrix}$, $C_0 = \begin{pmatrix} c & 0 \\ 0 & d \end{pmatrix}$, $\Gamma_0 > 0$, $\det \Gamma_0 \geq 1$ and $\det \Gamma_0 + 1 \geq \det A_0 + \det B_0 + 2 \det C_0$ ([26–29]).

Gaussian unitary operations. Let us consider an n-mode continuous-variable system with $R = (\hat{Q}_1, \hat{P}_1, \cdots, \hat{Q}_n, \hat{P}_n)^T$. For a unitary operator U, the unitary operation $\rho \mapsto U\rho U^\dagger$ is said to be Gaussian if its output is a Gaussian state whenever its input is a Gaussian state, and such U is called a Gaussian unitary operator. It is known that a unitary operator U is Gaussian if and only if

$$U^\dagger R U = SR + \mathbf{m},$$

for some vector $\mathbf{m}$ in $\mathbb{R}^{2n}$ and some $S \in \mathrm{Sp}(2n, \mathbb{R})$, the symplectic group of all $2n \times 2n$ real matrices S that satisfy

$$S \in \mathrm{Sp}(2n, \mathbb{R}) \Leftrightarrow S\Delta S^T = \Delta.$$

Thus, every Gaussian unitary operator U is determined by some affine symplectic map $(S, \mathbf{m})$ acting on the phase space, and can be denoted by $U = U_{S,\mathbf{m}}$ ([23,24]).

The following well-known facts for Gaussian states and Gaussian unitary operations are useful for our purpose.

Lemma 1 ([23]). *For any $(n+m)$-mode Gaussian state ρ_{AB}, write its CM Γ as in Equation (1). Then, the CMs of the reduced states $\rho_A = \mathrm{Tr}_B \rho_{AB}$ and $\rho_B = \mathrm{Tr}_A \rho_{AB}$ are matrices A and B, respectively.*

Denote by $S(H_A \otimes H_B)$ the set of all quantum states of $H_A \otimes H_B$, where H_A and H_B are respectively the state space for n-mode and m-mode continuous-variable systems.

Lemma 2 ([30]). *If $\rho_{AB} \in S(H_A \otimes H_B)$ is an $(n+m)$-mode Gaussian state, then ρ_{AB} is a product state, that is, $\rho_{AB} = \sigma_A \otimes \sigma_B$ for some $\sigma_A \in S(H_A)$ and $\sigma_B \in S(H_B)$, if and only if $\Gamma = \Gamma_A \oplus \Gamma_B$, where Γ, Γ_A and Γ_B are the CMs of ρ_{AB}, σ_A and σ_B, respectively.*

Lemma 3 ([23,24]). *Assume that ρ is any n-mode Gaussian state with CM Γ and displacement vector $\mathbf{d}$, and $U_{S,\mathbf{m}}$ is a Gaussian unitary operator. Then, the characteristic function of the Gaussian state $\sigma = U\rho U^\dagger$ is of the form $\exp(-\frac{1}{4}z^T \Gamma_\sigma z + i \mathbf{d}_\sigma^T z)$, where $\Gamma_\sigma = S\Gamma S^T$ and $\mathbf{d}_\sigma = \mathbf{m} + S\mathbf{d}$.*

3. Quantum Correlation Introduced by Gaussian Unitary Operations

Now, we introduce a quantum correlation $\mathcal{N}$ by local Gaussian unitary operations in the continuous-variable system.

Definition 1. For any $(n+m)$-mode quantum state $\rho_{AB} \in \mathcal{S}(H_A \otimes H_B)$, the quantum correlation $\mathcal{N}(\rho_{AB})$ of ρ_{AB} by Gaussian unitary operations is defined by

$$\mathcal{N}(\rho_{AB}) = \frac{1}{2} \sup_{U} \|\rho_{AB} - (I \otimes U)\rho_{AB}(I \otimes U^\dagger)\|_2^2, \qquad (3)$$

where the supremum is taken over all Gaussian unitary operators $U \in \mathcal{B}(H_B)$ satisfying $U\rho_B U^\dagger = \rho_B$, and $\rho_B = \mathrm{Tr}_A(\rho_{AB})$ is the reduced state. Here, $\mathcal{B}(H_B)$ is the set of all bounded linear operators acting on H_B.

Observe that $\mathcal{N}(\rho_{AB}) = 0$ holds for every product state. Thus, the product state contains no such correlation.

Remark 1. For any Gaussian state ρ_{AB}, there exist many Gaussian unitary U so that $U\rho_B U^\dagger = \rho_B$. This ensures that the definition of the quantity $\mathcal{N}(\rho_{AB})$ makes sense for each Gaussian state ρ_{AB}.

To see this, we need Williamson Theorem ([31]), which states that, for any n-mode Gaussian state $\rho \in \mathcal{S}(H)$ with CM Γ_ρ, there exists a $2n \times 2n$ symplectic matrix $\mathbf{S}$ such that $\mathbf{S}\Gamma_\rho\mathbf{S}^T = \oplus_{i=1}^n v_i I_2$ with $v_i \geq 1$. The diagonal matrix $\oplus_{i=1}^n v_i I_2$ and v_is are called respectively the Williamson form and the symplectic eigenvalues of Γ_ρ. By the Williamson Theorem, there exists a Gaussian unitary operator $U = U_{\mathbf{S},\mathbf{m}} = U_{\mathbf{S},-\mathbf{S}d}$ such that $U\rho U^\dagger = \otimes_{i=1}^n \rho_i$, where ρ_i are thermal states. Let $\mathbf{S}_\theta = \oplus_{i=1}^n \mathbf{S}_{\theta_i}$ with

$$\mathbf{S}_{\theta_i} = \begin{pmatrix} \cos\theta_i & \sin\theta_i \\ -\sin\theta_i & \cos\theta_i \end{pmatrix}, \theta_i \in [0, \frac{\pi}{2}].$$ Then, $\mathbf{S}_\theta$ is a symplectic matrix, and the corresponding Gaussian unitary operator $U_{\mathbf{S}_\theta,0} = U_{\mathbf{S}_\theta}$ has the form $U_{\mathbf{S}_\theta} = \otimes_{i=1}^n U_{\mathbf{S}_{\theta_i}} = \otimes_{i=1}^n \exp(\theta_i \hat{a}_i^\dagger \hat{a}_i)$. It is easily checked that $\mathbf{S}_\theta(\oplus_{i=1}^n v_i I)\mathbf{S}_\theta^T = \oplus_{i=1}^n v_i I$, and so $U_{\mathbf{S}_\theta}(\otimes_{i=1}^n \rho_i)U_{\mathbf{S}_\theta}^\dagger = \otimes_{i=1}^n \rho_i$. Now, write $W = U^\dagger U_{\mathbf{S}_\theta} U$. Obviously, W is Gaussian unitary and satisfies $W\rho W^\dagger = U^\dagger U_{\mathbf{S}_\theta} U\rho U^\dagger U_{\mathbf{S}_\theta}^\dagger U = \rho$.

We first prove that $\mathcal{N}$ is local Gaussian unitary invariant for all quantum states.

Proposition 1 (Local Gaussian unitary invariance). *If $\rho_{AB} \in \mathcal{S}(H_A \otimes H_B)$ is an $(n+m)$-mode quantum state, then $\mathcal{N}((U \otimes V)\rho_{AB}(U^\dagger \otimes V^\dagger)) = \mathcal{N}(\rho_{AB})$ holds for any Gaussian unitary operators $U \in \mathcal{B}(H_A)$ and $V \in \mathcal{B}(H_B)$.*

Proof of Proposition 1. Let $\rho_{AB} \in \mathcal{S}(H_A \otimes H_B)$ be an $(n+m)$-mode Gaussian state. For any Gaussian unitary operators $U \in \mathcal{B}(H_A)$ and $V \in \mathcal{B}(H_B)$, denote $\sigma_{AB} = (U \otimes V)\rho_{AB}(U^\dagger \otimes V^\dagger)$. Then, $\sigma_B = V\rho_B V^\dagger$. For any Gaussian unitary operator $W \in \mathcal{B}(H_B)$ satisfying $W\sigma_B W^\dagger = \sigma_B$, we have $WV\rho_B V^\dagger W^\dagger = V\rho_B V^\dagger$. Let $W' = V^\dagger WV$. Then, W' is also a Gaussian unitary operator and satisfies $W'\rho_B W'^\dagger = V^\dagger WV\rho_B V^\dagger W^\dagger V = \rho_B$. It is clear that W' runs over all Gaussian unitary operators that commutes with ρ_B when W runs over all Gaussian unitary operators commuting with σ_B. Hence, by Equation (3), we have

$$\begin{aligned}
&\mathcal{N}(\sigma_{AB})\\
&=\frac{1}{2}\sup_W \|\sigma_{AB} - (I\otimes W)\sigma_{AB}(I\otimes W)\|_2^2\\
&=\frac{1}{2}\sup_W \|(U\otimes V)\rho_{AB}(U^\dagger\otimes V^\dagger) - (I\otimes W)(U\otimes V)\rho_{AB}(U^\dagger\otimes V^\dagger)(I\otimes W)\|_2^2\\
&=\sup_W \{\mathrm{Tr}(\rho_{AB}^2) - \mathrm{Tr}(\rho_{AB}(I\otimes V^\dagger W V)\rho_{AB}(I\otimes V^\dagger W^\dagger V))\}\\
&=\sup_{W'} \{\mathrm{Tr}(\rho_{AB}^2) - \mathrm{Tr}(\rho_{AB}(I\otimes W')\rho_{AB}(I\otimes W'^\dagger))\}\\
&=\frac{1}{2}\sup_{W'} \|\rho_{AB} - (I\otimes W')\rho_{AB}(I\otimes W'^\dagger)\|_2^2\\
&=\mathcal{N}(\rho_{AB})
\end{aligned}$$

as desired. □

The next theorem shows that $\mathcal{N}(\rho_{AB})$ is a faithful nonclassicality measure for Gaussian states.

Theorem 1. *For any $(n+m)$-mode Gaussian state $\rho_{AB} \in \mathcal{S}(H_A \otimes H_B)$, $\mathcal{N}(\rho_{AB}) = 0$ if and only if ρ_{AB} is a product state.*

Proof of Theorem 1. By Definition 1, the "if" part is apparent. Let us check the "only if" part. Since the mean of any Gaussian state can be transformed to zero under some local Gaussian unitary operation, it is sufficient to consider those Gaussian states whose means are zero by Proposition 1. In the sequel, assume that ρ_{AB} is an $(n+m)$-mode Gaussian state with zero mean vector and CM $\Gamma = \begin{pmatrix} A & C \\ C^T & B \end{pmatrix}$ as in Equation (1), so that $\mathcal{N}(\rho_{AB}) = 0$.

By Lemma 1, the CM of ρ_B is B. According to the Williamson Theorem, there exists a symplectic matrix S_0 such that $S_0 B S_0^T = \oplus_{i=1}^m v_i I$ and $U_0 \rho_B U_0^\dagger = \otimes_{i=1}^m \rho_i$, where $U_0 = U_{S_0,0}$ and ρ_i are of the thermal states. Write $\sigma_{AB} = (I\otimes U_0)\rho_{AB}(I\otimes U_0^\dagger)$. It follows from Proposition 1 that $\mathcal{N}(\sigma_{AB}) = \mathcal{N}(\rho_{AB}) = 0$. Obviously, σ_{AB} has the CM of form:

$$\Gamma' = \begin{pmatrix} A' & C' \\ C'^T & \oplus_i^m v_i I \end{pmatrix}$$

and the mean 0.

For any $\theta_i \in [0, \frac{\pi}{2}]$ for $i = 1, 2, \cdots, m$, let S_θ be the symplectic matrix as in Remark 1. Then, $S_\theta(\oplus_{i=1}^m v_i I)S_\theta^T = \oplus_{i=1}^m v_i I$ and $U_{S_\theta,0}\sigma_B U_{S_\theta,0}^\dagger = \sigma_B = \mathrm{Tr}_A(\sigma_{AB})$. As $\mathcal{N}(\sigma_{AB}) = 0$, by Equation (3), $\sigma_{AB} = (I\otimes U_{S_\theta,0})\sigma_{AB}(I\otimes U_{S_\theta,0}^\dagger)$, and hence they must have the same CMs, that is,

$$\begin{pmatrix} A' & C' \\ C'^T & \oplus_{i=1}^m v_i I \end{pmatrix} = \begin{pmatrix} A' & C'S_\theta^T \\ S_\theta C'^T & \oplus_{i=1}^m v_i I \end{pmatrix}.$$

Note that $I - S_\theta^T$ is an invertible matrix if we take $\theta_i \in (0, \frac{\pi}{2})$ for each i. Then, it follows from $C' = C'S_\theta^T$ that we must have $C' = 0$. Thus, σ_{AB} is a product state by Lemma 2, and, consequently, $\rho_{AB} = (I\otimes U_0^\dagger)\sigma_{AB}(I\otimes U_0)$ is also a product state. □

We can give an analytic formula of $\mathcal{N}(\rho_{AB})$ for $(1+1)$-mode Gaussian state ρ_{AB}. Since $\mathcal{N}$ is locally Gaussian unitary invariant, it is enough to assume that the mean vector of ρ_{AB} is zero and the CM is standard.

Theorem 2. For any $(1+1)$-mode Gaussian state ρ_{AB} with CM Γ whose standard form is $\Gamma_0 = \begin{pmatrix} A_0 & C_0 \\ C_0^T & B_0 \end{pmatrix}$ as in Equation (2), we have

$$\mathcal{N}(\rho_{AB}) = \frac{1}{\sqrt{(ab-c^2)(ab-d^2)}} - \frac{1}{\sqrt{(ab-\frac{c^2}{2})(ab-\frac{d^2}{2})}}. \qquad (4)$$

Particularly, $\mathcal{N}(\rho_{AB}) = 1 - \sqrt{\frac{2}{2-c^2d^2+ab(c^2+d^2)}}$ whenever ρ_{AB} is pure.

Proof of Theorem 2. By Proposition 1, we may assume that the mean vector of ρ_{AB} is zero. Let $U_{S,m}$ be a Gaussian unitary operator such that $U_{S,m}\rho_B U_{S,m}^\dagger = \rho_B$. Then, S and m meet the conditions $SB_0S^T = B_0$ and $Sd_B + m = d_B = 0$. It follows that $m = 0$. Thus, we can denote $U_{S,m}$ by U_S. As $S\Delta S^T = \Delta$, there exists some $\theta \in [0, \frac{\pi}{2}]$ such that $S = S_\theta = \begin{pmatrix} \cos\theta & \sin\theta \\ -\sin\theta & \cos\theta \end{pmatrix}$. Thus, the CM of Gaussian state $(I \otimes U_S)\rho_{AB}(I \otimes U_S^\dagger)$ is

$$\Gamma_\theta = \begin{pmatrix} a & 0 & c\cos\theta & -c\sin\theta \\ 0 & a & d\sin\theta & d\cos\theta \\ c\cos\theta & d\sin\theta & b & 0 \\ -c\sin\theta & d\cos\theta & 0 & b \end{pmatrix},$$

and the mean of $(I \otimes U_S)\rho_{AB}(I \otimes U_S^\dagger)$ is $(I \oplus S)d + 0 \oplus 0 = 0$ as $d = 0$. Hence, by Equations (3) and (4), one gets

$$\mathcal{N}(\rho_{AB})$$
$$= \frac{1}{2} \sup_{U_{S,m}} \|\rho_{AB} - (I \otimes U)\rho_{AB}(I \otimes U_{S,m}^\dagger)\|_2^2$$
$$= \sup_{U_{S,m}} \{\operatorname{Tr}(\rho_{AB}^2) - \operatorname{Tr}(\rho_{AB}(I \otimes U_{S,m})\rho_{AB}(I \otimes U_{S,m}^\dagger))\}$$
$$= \sup_{\theta \in [0, \frac{\pi}{2}]} \{\frac{1}{\sqrt{\det\Gamma}} - \frac{1}{\sqrt{\det[(\Gamma+\Gamma_\theta)/2]}}\}$$
$$= \max_{\theta \in [0, \frac{\pi}{2}]} \{\frac{1}{\sqrt{a^2b^2+c^2d^2-ab(c^2+d^2)}}$$
$$\quad - \frac{1}{\sqrt{[ab-c^2(1+\cos\theta)/2][ab-d^2(1+\cos\theta)/2]}}\}$$
$$= \frac{1}{\sqrt{(ab-c^2)(ab-d^2)}} - \frac{1}{\sqrt{(ab-c^2/2)(ab-d^2/2)}}.$$

Hence, Equation (4) is true.

Particularly, if ρ_{AB} is a pure state, then, by [29], we have $1 = \operatorname{Tr}(\rho^2) = \frac{1}{\sqrt{\det\Gamma}} = \frac{1}{\sqrt{(ab-c^2)(ab-d^2)}}$. This entails that $\mathcal{N}(\rho_{AB}) = 1 - \sqrt{\frac{2}{2-c^2d^2+ab(c^2+d^2)}}$. □

For the general $(n+m)$-mode case, it is difficult to give an analytic formula of $\mathcal{N}(\rho_{AB})$ for all $(n+m)$-mode Gaussian states ρ_{AB}. However, we are able to give an estimate of $\mathcal{N}(\rho_{AB})$.

Theorem 3. *For any $(n+m)$-mode Gaussian state ρ_{AB} with CM $\Gamma = \begin{pmatrix} A & C \\ C^T & B \end{pmatrix}$ as in Equation (1), we have*

$$0 \leq \mathcal{N}(\rho_{AB}) \leq \frac{1}{\sqrt{\det \Gamma}} - \frac{1}{\sqrt{(\det A)(\det B)}} < 1. \tag{5}$$

Particularly, when ρ_{AB} is pure, $\mathcal{N}(\rho_{AB}) \leq 1 - \frac{1}{\sqrt{(\det A)(\det B)}}$. Moreover, the upper bound 1 in the inequality (5) is sharp, that is, we have

$$\sup_{\rho_{AB}} \mathcal{N}(\rho_{AB}) = 1.$$

Proof of Theorem 3. By Proposition 1, without loss of generality, we may assume that the mean of ρ_{AB} is 0. Let $U_{\mathbf{S},\mathbf{m}}$ be a Gaussian unitary operator such that $U_{\mathbf{S},\mathbf{m}}\rho_B U_{\mathbf{S},\mathbf{m}}^\dagger = \rho_B$. Then, the CM and the mean of the Gaussian state $(I \otimes U_{\mathbf{S},\mathbf{m}})\rho_{AB}(I \otimes U_{\mathbf{S},\mathbf{m}}^\dagger)$ are $\Gamma_U = \begin{pmatrix} A & CS^T \\ SC^T & B \end{pmatrix}$ and $\mathbf{0}$, respectively. Note that, for any n-mode Gaussian states ρ, σ with CMs V_ρ, V_σ and means $\mathbf{d}_\rho, \mathbf{d}_\sigma$, respectively, it is shown in [32] that

$$\mathrm{Tr}(\rho\sigma) = \frac{1}{\sqrt{\det[(V_\rho + V_\sigma)/2]}} \exp[-\frac{1}{2}\delta\langle d\rangle^T \det[(V_\rho + V_\sigma)/2]^{-1}\delta\langle d\rangle], \text{ where } \delta\langle d\rangle = \mathbf{d}_\rho - \mathbf{d}_\sigma. \tag{6}$$

Hence,

$$\mathcal{N}(\rho_{AB}) = \frac{1}{2}\sup_U \|\rho_{AB} - (I\otimes U)\rho_{AB}(I\otimes U^\dagger)\|_2^2$$
$$= \sup_U \{\mathrm{Tr}(\rho_{AB}^2) - \mathrm{Tr}(\rho_{AB}(I\otimes U)\rho_{AB}(I\otimes U^\dagger))\}$$
$$= \sup_S \{\frac{1}{\sqrt{\det \Gamma}} - \frac{1}{\sqrt{\det[(\Gamma + \Gamma_U)/2]}}\}.$$

Since $A > 0$, $B > 0$ and $\frac{\Gamma + \Gamma_U}{2} = \begin{pmatrix} A & \frac{C + CS^T}{2} \\ \frac{C^T + SC^T}{2} & B \end{pmatrix}$, by Fischer's inequality (p. 506, [33]), we have $\det \frac{\Gamma + \Gamma_U}{2} \leq (\det A)(\det B)$. Thus, we get $\mathcal{N}(\rho_{AB}) \leq \frac{1}{\sqrt{\det \Gamma}} - \frac{1}{\sqrt{(\det A)(\det B)}}$. If ρ_{AB} is a pure state, then $1 = \mathrm{Tr}(\rho_{AB}^2) = \frac{1}{\sqrt{\det \Gamma}}$, which gives $\mathcal{N}(\rho_{AB}) \leq 1 - \frac{1}{\sqrt{(\det A)(\det B)}}$.

Notice that, by Equation (6), we have $\frac{1}{\det \Gamma} = \mathrm{Tr}(\rho_{AB}^2)^2 \leq 1$. This implies that $\mathcal{N}(\rho_{AB}) \leq \frac{1}{\sqrt{\det \Gamma}} - \frac{1}{\sqrt{(\det A)(\det B)}} < 1$ since $\det A > 0$ and $\det B > 0$, that is, the inequality (5) is true.

To see that the upper bound 1 is sharp, consider the two-mode squeezed vacuum state $\rho(r) = S(r)|00\rangle\langle 00|S^\dagger(r)$, where $S(r) = \exp(-r\hat{a}_1\hat{a}_2 + r\hat{a}_1^\dagger\hat{a}_2^\dagger)$ is the two-mode squeezing operator with squeezed number $r \geq 0$ and $|00\rangle$ is the vacuum state ([24]). The CM of $\rho(r)$ is $\frac{1}{2}\begin{pmatrix} A_0 & B_0 \\ B_0 & A_0 \end{pmatrix}$, where $A_0 = \begin{pmatrix} \exp(-2r) + \exp(2r) & 0 \\ 0 & \exp(-2r) + \exp(2r) \end{pmatrix}$ and $B_0 = \begin{pmatrix} -\exp(-2r) + \exp(2r) & 0 \\ 0 & \exp(-2r) - \exp(2r) \end{pmatrix}$. By Theorem 2, it is easily calculated that

$$\mathcal{N}(\rho(r)) = 1 - \frac{8}{6 + \exp(-4r) + \exp(4r)}.$$

Clearly, $\mathcal{N}(\rho(r)) \to 1$ as $r \to \infty$, thus

$$\sup_r \mathcal{N}(\rho(r)) = 1,$$

completeing the proof. □

4. Comparison with Other Quantum Correlations

Entanglement is one of the most important quantum correlations, being central in most quantum information protocols [1]. However, it is an extremely difficult task to verify whether a given quantum state is entangled or not. Recall that a quantum state $\rho_{AB} \in S(H_A \otimes H_B)$ is said to be separable if it belongs to the closed convex hull of the set of all product states $\rho_A \otimes \rho_B \in S(H_A \otimes H_B)$. Note that a state ρ_{AB} is separable if and only if it admits a representation $\rho_{AB} = \int_{\mathcal{X}} \rho_A(x) \otimes \rho_B(x) \pi(dx)$, where $\pi(dx)$ is a Borel probability measure and $\rho_{A(B)}(x)$ is a Borel $S(H_{A(B)})$-valued function on some complete, separable metric space $\mathcal{X}$ [34]. One of the most useful separability criteria is the positive partial transpose (PPT) criterion, which can be found in [35,36]. The PPT criterion states that if a state is separable, then its partial transposition is positive. For discrete systems, the positivity of the partial transposition of a state is necessary and sufficient for its separability in the $2 \otimes 2$ and $2 \otimes 3$ cases. However, it is not true for higher dimensional systems [36]. For continuous systems, in [27,37], the authors extended the PPT criterion to $(n + m)$-mode continuous systems. It is remarkable that, for any $(1 + n)$-mode Gaussian state, it has PPT if and only if it is separable. Furthermore, for the $(1 + 1)$-mode case, it is shown that a $(1 + 1)$-mode Gaussian state ρ_{AB} is separable if and only if $\tilde{v}_- \geq 1$, where $\tilde{v}_-$ is the smallest symplectic eigenvalue of the CM of the partial transpose $\rho_{AB}^{T_B}$ [24,29].

Comparing $\mathcal{N}$ with the entanglement, we conjecture that there exists some positive number $d < 1$ such that $\mathcal{N}(\rho_{AB}) \leq d$ for any $(n + m)$-mode separable Gaussian state ρ_{AB}, that is,

$$\sup_{\rho_{AB} \text{ is separable}} \mathcal{N}(\rho_{AB}) \leq d < 1.$$

If this is true, then ρ_{AB} is entangled when $\mathcal{N}(\rho_{AB}) > d$. This will give a criterion of entanglement for $(n + m)$-mode Gaussian states in terms of correlation $\mathcal{N}$. Though we can not give a mathematical proof, we show that this is true for $(1 + 1)$-mode separable Gaussian states with $d \leq \frac{1}{10}$ by a numerical approach (Firstly, we randomly generated one million, five million, ten million, fifty million, one hundred million, five hundred million separable Gaussian states with $a, b, |c|, |d|$ ranging from 1 to 2, respectively. We found that the maximum of $\mathcal{N}$ is smaller than 0.09. Secondly, we used the same method and extended the range to 5. Then, the maximum of $\mathcal{N}$ is smaller than 0.1. Thirdly, using the same method and extending the range to 10, 100, 1000, 10000, respectively, we found that the maximum of $\mathcal{N}$ is still smaller than 0.1. We repeated the above computations ten times, and the result is just the same).

Proposition 2. $\mathcal{N}(\rho_{AB}) \leq 0.1$ for any $(1 + 1)$-mode separable Gaussian state ρ_{AB}.

It is followed from Theorem 1 that the quantum correlation $\mathcal{N}$ exists in all entangled Gaussian states and almost all separable Gaussian states except product states. In addition, Proposition 2 can be viewed as a sufficient condition for the entanglement of two-mode Gaussian states: if $\mathcal{N}(\rho_{AB}) > 0.1$, then ρ_{AB} is entangled.

To have an insight into the behavior of this quantum correlation by $\mathcal{N}$ and to compare it with the entanglement and the discords, we consider a class of physically relevant states–squeezed thermal state (STS). This kind of Gaussian state is used by many authors to illustrate the behavior of several interesting quantum correlations [12,13]. Recall that a two-mode Gaussian state ρ_{AB} is an STS if $\rho_{AB} = S(r)v_1(\bar{n}_1) \otimes v_2(\bar{n}_2)S(r)^\dagger$, where $v_i(\bar{n}_i) = \sum_k \frac{\bar{n}_i^k}{(1+\bar{n}_i)^{k+1}}|k\rangle\langle k|$ is the thermal state with thermal photon number $\bar{n}_i$ ($i = 1, 2$) and $S(r) = \exp\{r(\hat{a}_1^\dagger \hat{a}_2^\dagger - \hat{a}_1 \hat{a}_2)\}$ is the two-mode squeezing operator. Particularly, when $\bar{n}_1 = \bar{n}_2 = 0$, ρ_{AB} is a pure two-mode squeezed vacuum state, also known as an Einstein–Podolski–Rosen (EPR) state [24]. When $\bar{n}_1 > 0$ or $\bar{n}_2 > 0$, ρ_{AB} is a mixed Gaussian state.

For fixed r, ρ_{AB} is separable (not in product form) for large enough $\bar{n}_1, \bar{n}_2$. Notice that if ρ is a STS with the CM Γ_0 in the standard form in Equation (2), then $c = -d$. In this case, by Theorem 2, we have

$$\mathcal{N}(\rho_{AB}) = \frac{1}{ab - c^2} - \frac{1}{ab - c^2/2}. \quad (7)$$

Using this parametrization, one can get $a = 2\bar{n}_r + 1 + 2\bar{n}_1(1 + \bar{n}_r) + 2\bar{n}_2\bar{n}_r$, $b = 2\bar{n}_r + 1 + 2\bar{n}_2(1 + \bar{n}_r) + 2\bar{n}_1\bar{n}_r$ and $c = -d = 2(1 + \bar{n}_1 + \bar{n}_2)\sqrt{\bar{n}_r(1 + \bar{n}_r)}$, where $\bar{n}_r = \sinh^2 r$ ([12]). Especially, if $\bar{n}_1 = \bar{n}_2 = \bar{n}$, then ρ_{AB} is called a symmetric squeezed thermal state (SSTS). Now assume that ρ_{AB} is a SSTS. Then, ρ_{AB} is a mixed state if and only if $\bar{n} > 0$. The global purity of ρ_{AB} is $\mu = \text{Tr}(\rho_{AB}^2) = \frac{1}{(1+2\bar{n})^2}$ and the smallest symplectic eigenvalue $\tilde{v}_-$ of CM of $\rho_{AB}^{T_B}$ is $\tilde{v}_- = \frac{1+2\bar{n}}{\exp(2r)}$. Moreover, ρ_{AB} is entangled if and only if $\tilde{v}_- < 1$.

We first discuss the relation between $\mathcal{N}$ and the entanglement by considering SSTS. Regard $\mathcal{N}(\rho_{AB})$ as a function of μ and $\tilde{v}_-$. From Figure 1a, for separable states, we see that the value $\mathcal{N}$ at the separable SSTS is always smaller than 0.06, which supports positively Proposition 2. From Figure 1b, for fixed purity μ, $\mathcal{N}$ turns out to be a decreasing function of $\tilde{v}_-$. However, for fixed $\tilde{v}_-$, $\mathcal{N}$ tends to 0 when μ increases.

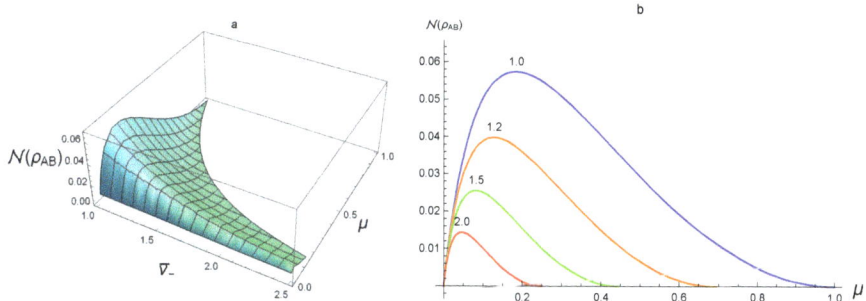

Figure 1. (a) $\mathcal{N}(\rho_{AB})$ for separable SSTSs as a function of μ and $\tilde{v}_-$; (b) from top to bottom, $\tilde{v}_-$ = 1.0, 1.2, 1.5, 2.0.

For the entangled SSTS, one sees from Figure 2a,b that the value of $\mathcal{N}$ is from 0 to 1. This reveals that, for some entangled SSTSs, $\mathcal{N}$ can be smaller than $\frac{1}{10}$. Thus, Proposition 2 is only a necessary condition for a Gaussian state to be separable. For fixed purity μ, from Figure 1b and 2b, $\mathcal{N}(\rho_{AB})$ increases when entanglement increases (that is, $\tilde{v}_- \to 0$) and $\lim_{\mu \to 1, \tilde{v}_- \to 0} \mathcal{N} = 1$. However, for fixed $\tilde{v}_-$, the behavior of $\mathcal{N}$ on μ is more complex.

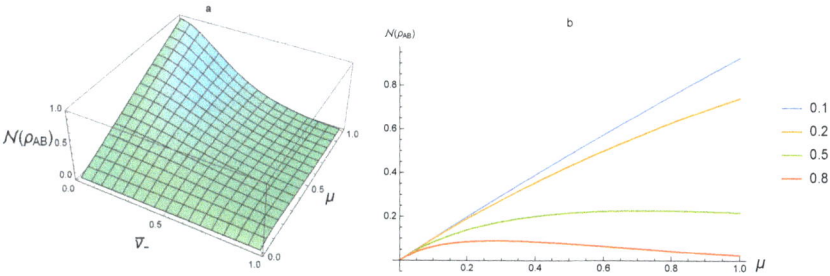

Figure 2. (a) $\mathcal{N}(\rho_{AB})$ for entangled SSTS as a function of μ and $\tilde{v}_-$; (b) from top to bottom, $\tilde{v}_-$ = 0.1, 0.2, 0.5, 0.8.

Regarding $\mathcal{N}$ as a function of r and $\bar{n}$, Figure 3 shows that $\mathcal{N}(\rho_{AB})$ is an increasing function of r and a decreasing function of $\bar{n}$, respectively. The value of $\mathcal{N}(\rho_{AB})$ always gains the maximum at $\bar{n} = 0$, that is, at pure states. Figure 3b also shows that $\mathcal{N}(\rho_{AB})$ almost depends only on $\bar{n}$ when r is large enough because the curves for $r = 5, 10, 20$ are almost the same.

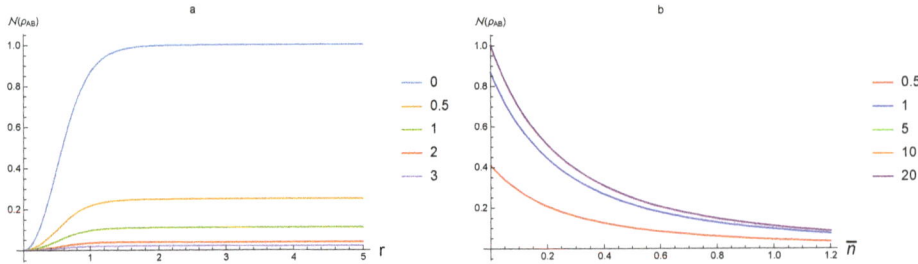

Figure 3. $\mathcal{N}(\rho_{AB})$ for SSTS as a function of $\bar{n}$ and r. (**a**) from top to bottom $\bar{n} = 0, 0.5, 1, 2, 3$; (**b**) from top to bottom $r = 0.5, 1, 5, 10, 20$.

Recall that an n-mode Gaussian positive operator-valued measure (GPOVM) is a collection of positive operators $\Pi = \{\Pi(z)\}$ satisfying $\int_z \Pi(z)dz = I$, where $\Pi(z) = W(z)\omega W^\dagger(z), z \in \mathbb{R}^{2n}$ with $W(z)$ the Weyl operators and ω an n-mode Gaussian state, which is called the seed of the GPOVM Π [38,39]. Let ρ_{AB} be a $(n+m)$-mode Gaussian state and $\Pi = \{\Pi(z)\}$ be a GPOVM of the subsystem B. Denote by $\rho_A(z) = \frac{1}{p(z)}\mathrm{Tr}_B(\rho_{AB}I \otimes \Pi(z))$ the reduced state of the system A after the GPOVM Π performed on the system B, where $p(z) = \mathrm{Tr}(\rho_{AB}I \otimes \Pi(z))$. Write the von Neumann entropy of a state ρ as $S(\rho)$, that is, $S(\rho) = -\mathrm{Tr}(\rho \log \rho)$. Then, the Gaussian QD of ρ_{AB} is defined as $D(\rho_{AB}) = S(\rho_B) - S(\rho_{AB}) + \inf_\Pi \int dzp(z)S(\rho_A(z))$ [12,13], where the infimum takes over all GPOVMs Π performed on the system B. It is known that a $(1+1)$-mode Gaussian state has zero Gaussian QD if and only if it is a product state; in addition, for all separable $(1+1)$-mode Gaussian states, $D(\rho_{AB}) \leq 1$; if the standard form of the CM of a $(1+1)$-mode Gaussian state ρ_{AB} is as in Equation (2), then

$$D(\rho_{AB}) = f(\sqrt{\det B_0}) + f(v_-) + f(v_+) + f(\sqrt{\inf_\omega \det E_\omega}), \tag{8}$$

where the infimum takes over all one-mode Gaussian states ω, $f(x) = \frac{x+1}{2}\log\frac{x+1}{2} - \frac{x-1}{2}\log\frac{x-1}{2}$, v_- and v_+ are the symplectic eigenvalues of the CM of ρ_{AB}, $E_\omega = A_0 - C_0(B_0 + \Gamma_\omega)^{-1}C_0^T$ with Γ_ω the CM of ω. Let $\alpha = \det A_0, \beta = \det B_0, \gamma = \det C_0, \delta = \det \Gamma_0$, then we have [13]

$$\inf_\omega \det E_\omega = \begin{cases} \frac{2\gamma^2 + (\beta-1)(\delta-\alpha) + 2|\gamma|\sqrt{\gamma^2 + (\beta-1)(\delta-\alpha)}}{(\beta-1)^2} & \text{if } (\delta - \alpha\beta)^2 \leq (1+\beta)\gamma^2(\alpha+\delta), \\ \frac{\alpha\beta - \gamma^2 + \delta - \sqrt{\gamma^4 + (\delta-\alpha\beta)^2 - 2\gamma^2(\alpha\beta+\delta)}}{2\beta} & \text{otherwise.} \end{cases} \tag{9}$$

In [14], the quantum GD D_G is proposed. Consider an $(n+m)$-mode Gaussian state ρ_{AB}, its Gaussian GD is defined by $D_G(\rho_{AB}) = \inf_\Pi ||\rho_{AB} - \Pi(\rho_{AB})||_2^2$, where the infimum takes over all GPOVM Π performed on system B, $||\cdot||_2$ stands for the Hilbert–Schmidt norm and $\Pi(\rho_{AB}) = \int dz(I \otimes \sqrt{\Pi(z)})\rho_{AB}(I \otimes \sqrt{\Pi(z)})$. If ρ_{AB} is a $(1+1)$-mode Gaussian state with the CM Γ as in Equation (1) and Π is an one-mode Gaussian POVM performed on mode B with seed ω_B, then $\Pi(\rho_{AB}) = \omega_A \otimes \omega_B$, where ω_A is a Gaussian state of which the CM $\Gamma_{\omega_A} = A + C(B + \Gamma_B)^{-1}C^T$ with Γ_{ω_B} the CM of ω_B. It is known from [14] that

$$D_G(\rho) = \inf_{\omega_B} ||\rho_{AB} - \omega_A \otimes \omega_B||_2^2. \tag{10}$$

Now it is clear that, for $(1+1)$-mode Gaussian state ρ_{AB}, $D_G(\rho_{AB}) = 0$ if and only if ρ_{AB} is a product state.

By Theorem 1 and the results mentioned above, D, D_G and $\mathcal{N}$ describe the same quantum correlation for $(1+1)$-mode Gaussian states. However, from the definitions, D, D_G use all GPOVMs, while $\mathcal{N}$ only employs Gaussian unitary operations, which is simpler and may consume less physical resources. Moreover, though an analytical formula of D is given for two-mode Gaussian states, the expression is more complex and more difficult to calculate (Equations (8) and (9)). D_G is not handled in general and there is no analytical formula for all $(1+1)$-mode Gaussian states (Equation (10)). As far as we know, there are no results obtained on D, D_G for general $(n+m)$-mode case.

To have a better insight into the behavior of $\mathcal{N}$ and D_G, we compare them in scale with the help of two-mode STS. Note that D_G of any two-mode STS ρ_{AB} is given by [14]

$$D_G(\rho_{AB}) = \frac{1}{ab - c^2} - \frac{9}{(\sqrt{4ab - 3c^2} + \sqrt{ab})^2}. \tag{11}$$

Clearly, our formula (7) for $\mathcal{N}$ is simpler then formula (11) for D_G.

Figures 4 and 5 are plotted in terms of photo number $\bar{n}$ and squeezing parameter r. Figure 4 shows that, for the case of SSTS and for $0 < r \leq 2.5$, we have $D_G(\rho_{AB}) < \mathcal{N}(\rho_{AB})$. This means that $\mathcal{N}$ is better than D_G when they are used to detect the correlation that they describe in the SSTS with $r < 2.5$. Figure 5a reveals that, for the case of nonsymmetric STS and for $r = 0.5$, we have $D_G(\rho_{AB}) < \mathcal{N}(\rho_{AB})$; that is, $\mathcal{N}$ is better in this situation too. However, for $r = 5$, $\mathcal{N}$ and D_G can not be compared with each other globally, which suggests that one may use $\max\{\mathcal{N}(\rho_{AB}), D_G(\rho_{AB})\}$ to detect the correlation.

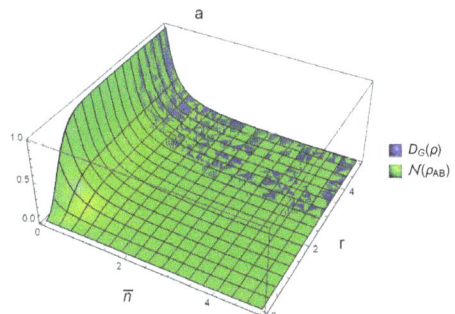

Figure 4. Comparison with $D_G(\rho_{AB})$ for SSTS.

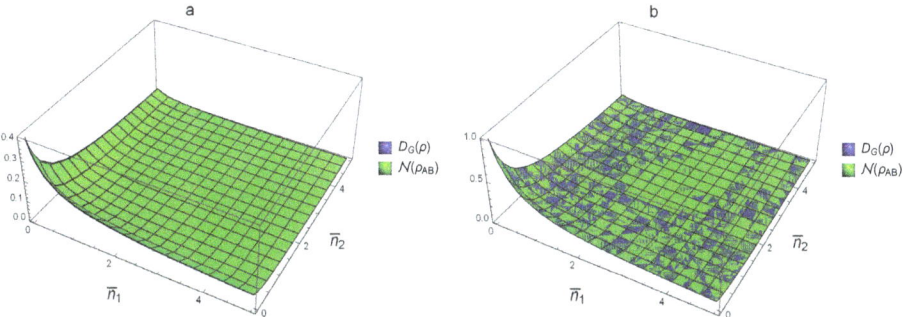

Figure 5. Comparison with $D_G(\rho_{AB})$ for nonsymmetric STS. (**a**) and (**b**) are correspond to nonsymmetric STS with $r = 0.5, 5$, respectively.

5. Conclusions

In conclusion, we introduce a measure of quantum correlation by $\mathcal{N}$ for bipartite quantum states in continuous-variable systems. This measure is introduced by performing Gaussian unitary operations to a subsystem and the value of it is invariant for all quantum states under local Gaussian unitary operations. $\mathcal{N}$ exists in all $(n+m)$-mode Gaussian states except product ones. In addition, $\mathcal{N}$ takes values in $[0,1)$ and the upper bound 1 is sharp. An analytical formula of $\mathcal{N}$ for any $(1+1)$-mode Gaussian states is obtained. Moreover, for any $(n+m)$-mode Gaussian states, an estimate of $\mathcal{N}$ is established in terms of its covariance matrix. Numerical evidence shows that the inequality $\mathcal{N}(\rho_{AB}) \leq 0.1$ holds for any $(1+1)$-mode separable Gaussian states ρ_{AB}, which can be viewed as a criterion of entanglement. It is worth noting that Gaussian QD, Gaussian GD and $\mathcal{N}$ measure the same quantum correlation for $(1+1)$-mode Gaussian states. However, $\mathcal{N}$ is easier to calculate and can be applied to any $(n+m)$-mode Gaussian states.

Acknowledgments: The authors would like to thank the anonymous referees for helpful comments and suggestions that improved the original paper. This work is partially supported by the Natural Science Foundation of China (11671006, 11671294) and the Outstanding Youth Foundation of Shanxi Province (201701D211001).

Author Contributions: Yangyang Wang completed the proofs of main theorems. The rest work of this paper was accomplished by Xiaofei Qi and Jinchuan Hou.

Conflicts of Interest: The authors declare no conflict of interest.

References

1. Horodecki, R.; Horodecki, P.; Horodecki, M.; Horodecki, K. Quantum entanglement. *Rev. Mod. Phys.* **2009**, *81*, 865.
2. Dakić, B.; Lipp, Y.O.; Ma, X.; Ringbauer, M.; Kropatschek, S.; Barz, S.; Paterek, T.; Vedral, V.; Zeilinger, A.; Brukner, Č.; et al. Quantum discord as resource for remote state preparation. *Nat. Phys.* **2012**, *8*, 666–670.
3. Madhok, V.; Datta, A. Interpreting quantum discord through quantum state merging. *Phys. Rev. A* **2011**, *83*, 032323.
4. Cavalcanti, D.; Aolita, L.; Boixo, S.; Modi, K.; Piani, M.; Winter, A. Operational interpretations of quantum discord. *Phys. Rev. A* **2011**, *83*, 032324.
5. Datta, A.; Shaji, A.; Caves, C.M. Quantum discord and the power of one qubit. *Phys. Rev. Lett.* **2008**, *100*, 050502.
6. Ollivier, H.; Zurek, W.H. Quantum Discord: A Measure of the Quantumness of Correlations. *Phys. Rev. Lett.* **2001**, *88*, 017901.
7. Dakić, B.; Vedral, V.; Brukner, Č. Necessary and Sufficient Condition for Nonzero Quantum Discord. *Phys. Rev. Lett.* **2010**, *105*, 190502.
8. Luo, S.; Fu, S. Geometric measure of quantum discord. *Phys. Rev. A* **2010**, *82*, 034302.
9. Miranowicz, A.; Horodecki, P.; Chhajlany, R.W.; Tuziemski, J.; Sperling, J. Analytical progress on symmetric geometric discord: Measurement-based upper bounds. *Phys. Rev. A* **2012**, *86*, 042123.
10. Luo, S.; Fu, S. Measurement-induced nonlocality. *Phys. Rev. Lett.* **2011**, *82*, 120401.
11. Luo, S. Using measurement-induced disturbance to characterize correlations as classical or quantum. *Phys. Rev. A* **2008**, *77*, 022301.
12. Giorda, P.; Paris, M.G.A. Gaussian Quantum Discord. *Phys. Rev. Lett.* **2010**, *105*, 020503.
13. Adesso, G.; Datta, A. Quantum versus Classical Correlations in Gaussian States. *Phys. Rev. Lett.* **2010**, *105*, 030501.
14. Adesso, G.; Girolami, D. Gaussian geometric discord. *Int. J. Quantum Inf.* **2011**, *9*, 1773–1786.
15. Mišta, L.; Tatham, R., Jr.; Girolami, D.; Korolkova, N.; Adesso, G. Measurement-induced disturbances and nonclassical correlations of Gaussian states. *Phys. Rev. A* **2011**, *83*, 042325.
16. Ma, R.F.; Hou, J.C.; Qi, X.F. Measurement-induced nonlocality for Gaussian states. *Int. J. Theor. Phys.* **2017**, *56*, 1132–1140.
17. Farace, A.; de Pasquale, A.; Rigovacca, L.; Giovannetti, V. Discriminating strength: A bona fide measure of non-classical correlations. *New J. Phys.* **2014**, *16*, 073010.

18. Rigovacca, L.; Farace, A.; de Pasquale, A.; Giovannetti, V. Gaussian discriminating strength. *Phys. Rev. A* **2015**, *92*, 042331.
19. Fu, L. Nonlocal effect of a bipartite system induced by local cyclic operation. *Europhys. Lett.* **2006**, *75*, 1.
20. Datta, A.; Gharibian, S. Signatures of nonclassicality in mixed-state quantum computation. *Phys. Rev. A* **2009**, *79*, 042325.
21. Gharibian, S. Quantifying nonclassicality with local unitary operations. *Phys. Rev. A* **2012**, *86*, 042106.
22. Braunstein, S.L.; van Loock, P. Quantum information with continuous variables. *Rev. Mod. Phys.* **2005**, *77*, 513.
23. Wang, X.B.; Hiroshimab, T.; Tomitab, A.; Hayashi, M. Quantum information with Gaussian states. *Phys. Rep.* **2007**, *448*, 1–111.
24. Weedbrook, C.; Pirandola, S.; García-Patrón, R.; Cerf, N.J.; Ralph, T.C.; Shapiro, J.H.; Lloyd, S. Gaussian quantum information. *Rev. Mod. Phys.* **2012**, *84*, 621.
25. Simon, R.; Mukunda, N.; Dutta, B. Quantum-noise matrix for multimode systems: U(n) invariance, squeezing, and normal forms. *Phys. Rev. A* **1994**, *49*, 1567.
26. Duan, L.M.; Giedke, G.; Cirac, J.I.; Zoller, P. Inseparability Criterion for Continuous Variable Systems. *Phys. Rev. Lett.* **2000**, *84*, 2722.
27. Simon, R. Peres-Horodecki Separability Criterion for Continuous Variable Systems. *Phys. Rev. Lett.* **2000**, *84*, 2726.
28. Serafini, A. Multimode Uncertainty Relations and Separability of Continuous Variable States. *Phys. Rev. Lett.* **2006**, *96*, 110402.
29. Pirandola, S.; Serafini, A.; Lloyd, S. Correlation matrices of two-mode bosonic systems. *Phys. Rev. A* **2009**, *79*, 052327.
30. Anders, J. Estimating the degree of entanglement of unknown Gaussian states. *arXiv* **2012**, arXiv:quant-ph/0610263v1.
31. Williamson, J. On the algebraic problem concerning the normal forms of linear dynamical systems. *Am. J. Math.* **1936**, *58*, 141–163.
32. Marian, P.; Marian, T.A. Uhlmann fidelity between two-mode Gaussian states. *Phys. Rev. A* **2012**, *86*, 022340.
33. Horn, R.A.; Johnson, C.R. *Matrix Analysis*; Cambridge University Press: Cambridge, UK, 2012.
34. Holevo, A.S. *Quantum Systems, Channels, Information: A Mathematical Introduction*; De Gruyter: Berlin, Germany, 2012.
35. Peres, A. Separability Criterion for Density Matrices. *Phys. Rev. Lett.* **1997**, *77*, 1413.
36. Horodecki, M.; Horodecki, P.; Horodecki, R. Separability of mixed states: necessary and sufficient conditions. *Phys. Lett. A* **1996**, *1*, 223.
37. Werner, R.F.; Wolf, M.M. Bound Entangled Gaussian States. *Phys. Rev. Lett.* **2001**, *86*, 3658.
38. Giedke, G.; Cirac, J.I. Characterization of Gaussian operations and distillation of Gaussian states. *Phys. Rev. A* **2002**, *66*, 032316.
39. Fiurášek, J.; Mišta, L., Jr. Gaussian localizable entanglement. *Phys. Rev. A* **2007**, *75*, 060302.

 © 2018 by the authors. Licensee MDPI, Basel, Switzerland. This article is an open access article distributed under the terms and conditions of the Creative Commons Attribution (CC BY) license (http://creativecommons.org/licenses/by/4.0/).

Article

Entropic Updating of Probabilities and Density Matrices

Kevin Vanslette

Department of Physics, University at Albany (SUNY), Albany, NY 12222, USA; kvanslette@albany.edu

Received: 2 November 2017; Accepted: 2 December 2017; Published: 4 December 2017

Abstract: We find that the standard relative entropy and the Umegaki entropy are designed for the purpose of inferentially updating probabilities and density matrices, respectively. From the same set of inferentially guided design criteria, both of the previously stated entropies are derived in parallel. This formulates a quantum maximum entropy method for the purpose of inferring density matrices in the absence of complete information.

Keywords: probability theory; entropy; quantum relative entropy; quantum information; quantum mechanics; inference

1. Introduction

We *design* an inferential updating procedure for probability distributions and density matrices such that inductive inferences may be made. The inferential updating tools found in this derivation take the form of the standard and quantum relative entropy functionals, and thus we find the functionals are *designed* for the purpose of updating probability distributions and density matrices, respectively. Previously formulated design derivations which found the entropy to be a tool for inference originally required five *design criteria* (DC) [1–3], this was reduced to four in [4–6], and then down to three in [7]. We reduced the number of required DC down to two while also providing the first *design* derivation of the quantum relative entropy—*using the same design criteria and inferential principles in both instances*.

The designed quantum relative entropy takes the form of Umegaki's quantum relative entropy, and thus it has the "proper asymptotic form of the relative entropy in quantum (mechanics)" [8–10]. Recently, Wilming, etc. [11] gave an axiomatic characterization of the quantum relative entropy that "uniquely determines the quantum relative entropy". Our derivation differs from their's, again in that we *design* the quantum relative entropy for a purpose, but also that our DCs are imposed on what turns out to be the functional derivative of the quantum relative entropy rather than on the quantum relative entropy itself. The use of a quantum entropy for the purpose of inference has a large history: Jaynes [12,13] invented the notion of the quantum maximum entropy method [14], while it was perpetuated by [15–22] and many others. However, we find the quantum *relative* entropy to be the suitable entropy for updating density matrices, rather than the von Neuman entropy [23], as is suggested in [24]. I believe the present article provides the desired motivation for why the appropriate quantum relative entropy for updating density matrices, from prior to posterior, should be logarithmic in form while also providing a solution for updating non-uniform prior density matrices [24]. The relevant results of these papers may be found using the quantum relative entropy with suitably chosen prior density matrices.

It should be noted that because the relative entropies were reached by design, they may be interpreted as such, "the relative entropies are tools for updating", which means we no longer need to attach an interpretation *ex post facto*—as a measure of disorder or amount of missing information. In this sense, the relative entropies were built for the purpose of saturating their own interpretation [4,7], and, therefore, the quantum relative entropy *is the tool designed for updating density matrices*.

This article takes an inferential approach to probabilities and density matrices that is expected to be notionally consistent with the Bayesian derivations of Quantum Mechanics, such as Entropic Dynamics [7,25–27], as well as Bayesian interpretations of Quantum Mechanics, such as QBism [28]. The quantum maximum entropy method is, however, expected to be useful independent of one's interpretation of Quantum Mechanics because the entropy is designed at the level of density matrices rather than being formulated from arguments about the "inner workings" of Quantum Mechanics. This inferential approach is, at the very least, *verbally convenient* so we will continue writing in this language.

A few applications of the quantum maximum entropy method are given in an another article [29]. By maximizing the quantum relative entropy with respect to a "data constraint" and the appropriate prior density matrix, the Quantum Bayes Rule [30–34] (a positive-operator valued measure (POVM) measurement and collapse) is derived. The quantum maximum entropy method can reproduce the density matrices in [35,36] that are cited as "Quantum Bayes Rules", but the required constraints are difficult to motivate; however, it is expected that the results of this paper may be useful for further understanding Machine Learning techniques that involve the quantum relative entropy [37]. The Quantum Bayes Rule derivation in [29] is analogous to the standard Bayes Rule derivation from the relative entropy given in [38], as was suggested to be possible in [24]. This article provides the foundation for [29], and thus, the quantum maximum entropy method unifies a few topics in Quantum Information and Quantum Measurement through entropic inference.

As is described in this article and in [29], the quantum maximum entropy method is able to provide solutions even if the constraints and prior density matrix in question do not all mutually commute. This might be useful for subjects as far reaching as [39], which seeks to use Quantum Theory as a basis for building models for cognition. The immediate correspondence is that the quantum maximum entropy method might provide a solution toward addressing the empirical evidence for noncommutative cognition, which is how one's cognition changes when addressing questions in permuted order [39]. A simpler model for noncommutative cognition may also be possible by applying sequential updates via the standard maximum entropy method with their order permuted. Sequential updating does not, in general, give the same resultant probability distribution when the updating order is permuted—this is argued to be a feature of the standard maximum entropy method [40]. Similarly, sequential updating in the quantum maximum entropy method also has this feature, but it should be noted that the noncommutativity of sequential updating is different in principle than simultaneously updating with respect to expectation values of noncommuting operators.

The remainder of the paper is organized as follows: first, we will discuss some universally applicable principles of inference and motivate the design of an entropy function able to rank probability distributions. This entropy function will be designed such that it is consistent with inference by applying a few reasonable design criteria, which are guided by the aforementioned principles of inference. Using the same principles of inference and design criteria, we find the form of the quantum relative entropy suitable for inference. The solution to an example of updating 2×2 prior density matrices with respect to expectation values over spin matrices that do not commute with the prior via the quantum maximum entropy method is given in the Appendix B. We end with concluding remarks (I thank the reviewers for providing several useful references in this section).

2. The Design of Entropic Inference

Inference is the appropriate updating of probability distributions when new information is received. Bayes rule and Jeffrey's rule are both equipped to handle information in the form of data; however, the updating of a probability distribution due to the knowledge of an expectation value was realized by Jaynes [12–14] through the method of maximum entropy. The two methods for inference were thought to be devoid of one another until the work of [38,40], which showed Bayes Rule and Jeffrey's Rule to be consistent with the method of maximum entropy when the expectation values were in the form of data [38,40]. In the spirit of the derivation we will carry on as if the maximum entropy method were not known and show how it may be derived as an application of inference.

Given a probability distribution $\varphi(x)$ over a general set of propositions $x \in X$, it is self evident that if new information is learned, we are entitled to assign a new probability distribution $\rho(x)$ that somehow reflects this new information while also respecting our prior probability distribution $\varphi(x)$. The main question we must address is: "Given some information, to what posterior probability distribution $\rho(x)$ should we update our prior probability distribution $\varphi(x)$?", that is,

$$\varphi(x) \xrightarrow{*} \rho(x)?$$

This specifies the problem of inductive inference. Since "information" has many colloquial, yet potentially conflicting, definitions, we remove potential confusion by defining **information** operationally ($*$) as the *rationale* that causes a probability distribution to change (inspired by and adapted from [7]). Directly from [7]:

> Our goal is to design a method that allows a systematic search for the preferred posterior distribution. The central idea, first proposed in [4], is disarmingly simple: to select the posterior, first rank all candidate distributions in increasing *order of preference* and then pick the distribution that ranks the highest. Irrespective of what it is that makes one distribution preferable over another (we will get to that soon enough), it is clear that any ranking according to preference must be transitive: if distribution ρ_1 is preferred over distribution ρ_2, and ρ_2 is preferred over ρ_3, then ρ_1 is preferred over ρ_3. Such transitive rankings are implemented by assigning to each $\rho(x)$ a real number $S[\rho]$, which is called the entropy of ρ, in such a way that if ρ_1 is preferred over ρ_2, then $S[\rho_1] > S[\rho_2]$. The selected distribution (one or possibly many, for there may be several equally preferred distributions) is that which maximizes the entropy functional.

Because we wish to update from prior distributions φ to posterior distributions ρ by ranking, the entropy functional $S[\rho, \varphi]$ is a real function of both φ and ρ. In the absence of new information, there is no available *rationale* to prefer any ρ to the original φ, and thereby the relative entropy should be designed such that the selected posterior is equal to the prior φ (in the absence of new information). The prior information encoded in $\varphi(x)$ is valuable and we should not change it unless we are informed otherwise. Due to our definition of information, and our desire for objectivity, we state the predominate guiding principle for inductive inference:

> The Principle of Minimal Updating (PMU):
> *A probability distribution should only be updated to the extent required by the new information.*

This simple statement provides the foundation for inference [7]. If the updating of probability distributions is to be done objectively, then possibilities should not be needlessly ruled out or suppressed. Being informationally stingy, that we should only update probability distributions when the information requires it, pushes inductive inference toward objectivity. Thus, using the PMU helps formulate a pragmatic (and objective) procedure for making inferences using (informationally) subjective probability distributions [41].

This method of inference is only as universal and general as its ability to apply *equally well* to *any* specific inference problem. The notion of "specificity" is the notion of statistical independence; a special case is only special in that it is separable from other special cases. The notion that systems may be "sufficiently independent" plays a central and deep-seated role in science and the idea that some things can be neglected and that not everything matters, is implemented by imposing criteria that tells us how to handle independent systems [7]. Ironically, the universally *shared* property by all specific inference problems is their ability to be *independent* of one another—they share independence. Thus, a universal inference scheme based on the PMU permits:

Properties of Independence (PI):

Subdomain Independence: When information is received about one set of propositions, it should not affect or change the state of knowledge (probability distribution) of the other propositions (else information was also received about them too);

And,

Subsystem Independence: When two systems are a priori believed to be independent and we only receive information about one, then the state of knowledge of the other system remains unchanged.

The PIs are special cases of the PMU that ultimately take the form of *design criteria* in this design derivation. The process of constraining the form of $S[\rho, \varphi]$ by imposing design criteria may be viewed as the process of *eliminative induction*, and after sufficient constraining, a single form for the entropy remains. Thus, the justification behind the surviving entropy is not that it leads to demonstrably correct inferences, but, rather, that all other candidate entropies demonstrably fail to perform as desired [7]. Rather than the *design criteria* instructing one how to update, they instruct in what instances one should *not* update. That is, rather than justifying one way to skin a cat over another, we tell you when *not* to skin it, which is operationally unique—namely you don't do it—luckily enough for the cat.

The Design Criteria and the Standard Relative Entropy

The following *design criteria* (DC), guided by the PMU, are imposed and formulate the standard relative entropy as a tool for inference. The form of this presentation is inspired by [7].

DC1: Subdomain Independence

We keep DC1 from [7] and review it below. DC1 imposes the first instance of when one should not update—the Subdomain PI. Suppose the information to be processed does *not* refer to a particular subdomain $\mathcal{D}$ of the space $\mathcal{X}$ of xs. In the absence of new information about $\mathcal{D}$, the PMU insists we do not change our minds about probabilities that are conditional on $\mathcal{D}$. Thus, we design the inference method so that $\varphi(x|\mathcal{D})$, the prior probability of x conditional on $x \in \mathcal{D}$, is not updated and therefore the selected conditional posterior is

$$P(x|\mathcal{D}) = \varphi(x|\mathcal{D}). \tag{1}$$

(The notation will be as follows: we denote priors by φ, candidate posteriors by lower case ρ, and the selected posterior by upper case P.) We emphasize the point is not that we make the unwarranted assumption that keeping $\varphi(x|\mathcal{D})$ unchanged is guaranteed to lead to correct inferences. It need not; induction is risky. The point is, rather, that, in the absence of any evidence to the contrary, there is no reason to change our minds and the prior information takes priority.

DC1 Implementation

Consider the set of microstates $x_i \in \mathcal{X}$ belonging to either of two non-overlapping domains $\mathcal{D}$ or its compliment $\mathcal{D}'$, such that $\mathcal{X} = \mathcal{D} \cup \mathcal{D}'$ and $\emptyset = \mathcal{D} \cap \mathcal{D}'$. For convenience, let $\rho(x_i) = \rho_i$. Consider the following constraints:

$$\rho(\mathcal{D}) = \sum_{i \in \mathcal{D}} \rho_i \quad \text{and} \quad \rho(\mathcal{D}') = \sum_{i \in \mathcal{D}'} \rho_i, \tag{2}$$

such that $\rho(\mathcal{D}) + \rho(\mathcal{D}') = 1$, and the following "local" expectation value constraints over $\mathcal{D}$ and $\mathcal{D}'$,

$$\langle A \rangle = \sum_{i \in \mathcal{D}} \rho_i A_i \quad \text{and} \quad \langle A' \rangle = \sum_{i \in \mathcal{D}'} \rho_i A'_i, \tag{3}$$

where $A = A(x)$ is a scalar function of x and $A_i \equiv A(x_i)$. As we are searching for the candidate distribution which maximizes S while obeying (2) and (3), we maximize the entropy $S \equiv S[\rho, \varphi]$ with respect to these expectation value constraints using the Lagrange multiplier method,

$$0 = \delta\Big(S - \lambda[\rho(\mathcal{D}) - \sum_{i \in \mathcal{D}} \rho_i] - \mu[\langle A \rangle - \sum_{i \in \mathcal{D}} \rho_i A_i]$$
$$-\lambda'[\rho(\mathcal{D}') - \sum_{i \in \mathcal{D}'} \rho_i] - \mu'[\langle A' \rangle - \sum_{i \in \mathcal{D}'} \rho_i A_i]\Big),$$

and, thus, the entropy is maximized when the following differential relationships hold:

$$\frac{\delta S}{\delta \rho_i} = \lambda + \mu A_i \quad \forall i \in \mathcal{D}, \tag{4}$$

$$\frac{\delta S}{\delta \rho_i} = \lambda' + \mu' A_i' \quad \forall i \in \mathcal{D}'. \tag{5}$$

Equations (2)–(5), are $n+4$ equations we must solve to find the four Lagrange multipliers $\{\lambda, \lambda', \mu, \mu'\}$ and the n probability values $\{\rho_i\}$ associated to the n microstates $\{x_i\}$. If the subdomain constraint DC1 is imposed in the most restrictive case, then it will hold in general. The most restrictive case requires splitting $\mathcal{X}$ into a set of $\{\mathcal{D}_i\}$ domains such that each $\mathcal{D}_i$ singularly includes one microstate x_i. This gives,

$$\frac{\delta S}{\delta \rho_i} = \lambda_i + \mu_i A_i \quad \text{in each } \mathcal{D}_i. \tag{6}$$

Because the entropy $S = S[\rho_1, \rho_2, \ldots; \varphi_1, \varphi_2, \ldots]$ is a functional over the probability of each microstate's posterior and prior distribution, its variational derivative is also a function of said probabilities in general,

$$\frac{\delta S}{\delta \rho_i} \equiv \phi_i(\rho_1, \rho_2, \ldots; \varphi_1, \varphi_2, \ldots) = \lambda_i + \mu_i A_i \quad \text{for each } (i, \mathcal{D}_i). \tag{7}$$

DC1 is imposed by constraining the form of $\phi_i(\rho_1, \rho_2, \ldots; \varphi_1, \varphi_2, \ldots) = \phi_i(\rho_i; \varphi_1, \varphi_2, \ldots)$ to ensure that changes in $A_i \to A_i + \delta A_i$ have no influence over the value of ρ_j in domain $\mathcal{D}_j$, through ϕ_i, for $i \neq j$. If there is no new information about propositions in $\mathcal{D}_j$, its distribution should remain equal to φ_j by the PMU. We further restrict ϕ_i such that an arbitrary variation of $\varphi_j \to \varphi_j + \delta\varphi_j$ (a change in the prior state of knowledge of the microstate j) has no effect on ρ_i for $i \neq j$ and therefore DC1 imposes $\phi_i = \phi_i(\rho_i, \varphi_i)$, as is guided by the PMU. At this point, it is easy to generalize the analysis to continuous microstates such that the indices become continuous $i \to x$, sums become integrals, and discrete probabilities become probability densities $\rho_i \to \rho(x)$.

Remark

We are designing the entropy for the purpose of ranking posterior probability distributions (for the purpose of inference); however, the highest ranked distribution is found by setting the variational derivative of $S[\rho, \varphi]$ equal to the variations of the expectation value constraints by the Lagrange multiplier method,

$$\frac{\delta S}{\delta \rho(x)} = \lambda + \sum_i \mu_i A_i(x). \tag{8}$$

Therefore, the real quantity of interest is $\frac{\delta S}{\delta \rho(x)}$ rather than the specific form of $S[\rho, \varphi]$. All forms of $S[\rho, \varphi]$ that give the correct form of $\frac{\delta S}{\delta \rho(x)}$ are *equally valid* for the purpose of inference. Thus, every design criteria may be made on the variational derivative of the entropy rather than the entropy itself, which we do. When maximizing the entropy, for convenience, we will let,

$$\frac{\delta S}{\delta \rho(x)} \equiv \phi_x(\rho(x), \varphi(x)), \tag{9}$$

and further use the shorthand $\phi_x(\rho, \varphi) \equiv \phi_x(\rho(x), \varphi(x))$, in all cases.

DC1': *In the absence of new information, our new state of knowledge $\rho(x)$ is equal to the old state of knowledge $\varphi(x)$.*

This is a special case of DC1, and is implemented differently than in [7]. The PMU is in principle a statement about informational honestly—that is, one should not "jump to conclusions" in light of new information and in the absence of new information, one should not change their state of knowledge. If no new information is given, the prior probability distribution $\varphi(x)$ does not change, that is, the posterior probability distribution $\rho(x) = \varphi(x)$ is equal to the prior probability. If we maximizing the entropy without applying constraints,

$$\frac{\delta S}{\delta \rho(x)} = 0, \tag{10}$$

then DC1' imposes the following condition:

$$\frac{\delta S}{\delta \rho(x)} = \phi_x(\rho, \varphi) = \phi_x(\varphi, \varphi) = 0, \tag{11}$$

for all x in this case. This special case of the DC1 and the PMU turns out to be incredibly constraining as we will see over the course of DC2.

Comment

If the variable x is continuous, DC1 requires that when information refers to points infinitely close but just outside the domain $\mathcal{D}$, that it will have no influence on probabilities conditional on $\mathcal{D}$ [7]. This may seem surprising as it may lead to updated probability distributions that are discontinuous. Is this a problem? No.

In certain situations (e.g., physics) we might have explicit reasons to believe that conditions of continuity or differentiability should be imposed and this information might be given to us in a variety of ways. The crucial point, however—and this is a point that we keep and will keep reiterating—is that unless such information is explicitly given, we should not assume it. If the new information leads to discontinuities, so be it.

DC2: Subsystem Independence

DC2 imposes the second instance of when one should not update—the Subsystem PI. We emphasize that DC2 *is not a consistency requirement.* The argument we deploy is *not* that both the prior *and* the new information tells us the systems are independent, in which case consistency requires that it should not matter whether the systems are treated jointly or separately. Rather, DC2 refers to a situation where the new information does not say whether the systems are independent or not, but information is given about each subsystem. The updating is being *designed* so that the independence reflected in the prior is maintained in the posterior by default via the PMU and the second clause of the PIs [7].

The point is not that when we have no evidence for correlations we draw the firm conclusion that the systems must necessarily be independent. They could indeed have turned out to be correlated and then our inferences would be wrong. Again, induction involves risk. The point is rather that if the joint prior reflects independence and the new evidence is silent on the matter of correlations, then the prior independence takes precedence. As before, in this case subdomain independence, the probability distribution should not be updated unless the information requires it [7].

DC2 Implementation

Consider a composite system, $x = (x_1, x_2) \in \mathcal{X} = \mathcal{X}_1 \times \mathcal{X}_2$. Assume that all prior evidence led us to believe the subsystems are independent. This belief is reflected in the prior distribution: if the individual system priors are $\varphi_1(x_1)$ and $\varphi_2(x_2)$, then the prior for the whole system is their product

$\varphi_1(x_1)\varphi_2(x_2)$. Further suppose that new information is acquired such that $\varphi_1(x_1)$ would by itself be updated to $P_1(x_1)$ and that $\varphi_2(x_2)$ would be itself be updated to $P_2(x_2)$. By design, the implementation of DC2 constrains the entropy functional such that, in this case, the joint product prior $\varphi_1(x_1)\varphi_2(x_2)$ updates to the selected product posterior $P_1(x_1)P_2(x_2)$ [7].

The argument below is considerably simplified if we expand the space of probabilities to include distributions that are not necessarily normalized. This does not represent any limitation because a normalization constraint may always be applied. We consider a few special cases below:

Case 1: We receive the extremely constraining information that the posterior distribution for system 1 is completely specified to be $P_1(x_1)$ while we receive no information at all about system 2. We treat the two systems jointly. Maximize the joint entropy $S[\rho(x_1, x_2), \varphi(x_1)\varphi(x_2)]$ subject to the following constraints on the $\rho(x_1, x_2)$:

$$\int dx_2 \rho(x_1, x_2) = P_1(x_1). \tag{12}$$

Notice that the probability of each $x_1 \in \mathcal{X}_1$ within $\rho(x_1, x_2)$ is being constrained to $P_1(x_1)$ in the marginal. We therefore need a one Lagrange multiplier $\lambda_1(x_1)$ for each $x_1 \in \mathcal{X}_1$ to tie each value of $\int dx_2 \rho(x_1, x_2)$ to $P_1(x_1)$. Maximizing the entropy with respect to this constraint is,

$$\delta\left[S - \int dx_1 \lambda_1(x_1) \left(\int dx_2 \rho(x_1, x_2) - P_1(x_1)\right)\right] = 0, \tag{13}$$

which requires that

$$\lambda_1(x_1) = \phi_{x_1 x_2}\left(\rho(x_1, x_2), \varphi_1(x_1)\varphi_2(x_2)\right), \tag{14}$$

for arbitrary variations of $\rho(x_1, x_2)$. By design, DC2 is implemented by requiring $\varphi_1\varphi_2 \to P_1\varphi_2$ in this case, therefore,

$$\lambda_1(x_1) = \phi_{x_1 x_2}\left(P_1(x_1)\varphi_2(x_2), \varphi_1(x_1)\varphi_2(x_2)\right). \tag{15}$$

This equation must hold for all choices of x_2 and all choices of the prior $\varphi_2(x_2)$ as $\lambda_1(x_1)$ is independent of x_2. Suppose we had chosen a different prior $\varphi_2'(x_2) = \varphi_2(x_2) + \delta\varphi_2(x_2)$ that disagrees with $\varphi_2(x_2)$. For all x_2 and $\delta\varphi_2(x_2)$, the multiplier $\lambda_1(x_1)$ remains unchanged as it constrains the independent $\rho(x_1) \to P_1(x_1)$. This means that any dependence that the right-hand side might potentially have had on x_2 and on the prior $\varphi_2(x_2)$ *must cancel out*. This means that

$$\phi_{x_1 x_2}\left(P_1(x_1)\varphi_2(x_2), \varphi_1(x_1)\varphi_2(x_2)\right) = f_{x_1}(P_1(x_1), \varphi_1(x_1)). \tag{16}$$

Since φ_2 is arbitrary in f, suppose further that we choose a constant prior set equal to one, $\varphi_2(x_2) = 1$, therefore

$$f_{x_1}(P_1(x_1), \varphi_1(x_1)) = \phi_{x_1 x_2}\left(P_1(x_1) * 1, \varphi_1(x_1) * 1\right) = \phi_{x_1}\left(P_1(x_1), \varphi_1(x_1)\right) \tag{17}$$

in general. This gives

$$\lambda_1(x_1) = \phi_{x_1}(P_1(x_1), \varphi_1(x_1)). \tag{18}$$

The left-hand side does not depend on x_2, and therefore neither does the right-hand side. An argument exchanging systems 1 and 2 gives a similar result.

Case 1—Conclusion: When the system 2 is not updated the dependence on φ_2 and x_2 drops out,

$$\phi_{x_1 x_2}\left(P_1(x_1)\varphi_2(x_2), \varphi_1(x_1)\varphi_2(x_2)\right) = \phi_{x_1}\left(P_1(x_1), \varphi_1(x_1)\right), \tag{19}$$

and vice-versa when system 1 is not updated,

$$\phi_{x_1 x_2}\left(\varphi_1(x_1)P_2(x_2), \varphi_1(x_1)\varphi_2(x_2)\right) = \phi_{x_2}\left(P_2(x_2), \varphi_2(x_2)\right). \tag{20}$$

As we seek the general functional form of $\phi_{x_1 x_2}$, and because the x_2 dependence drops out of (19) and the x_1 dependence drops out of (20) for arbitrary φ_1, φ_2 and $\varphi_{12} = \varphi_1 \varphi_2$, the explicit coordinate dependence in ϕ consequently drops out of both such that,

$$\phi_{x_1 x_2} \to \phi, \tag{21}$$

as $\phi = \phi(\rho(x), \varphi(x))$ must only depend on coordinates through the probability distributions themselves. (As a double check, explicit coordinate dependence was included in the following computations but inevitably dropped out due to the form the functional equations and DC1'. By the argument above, and for simplicity, we drop the explicit coordinate dependence in ϕ here.)

Case 2: Now consider a different special case in which the marginal posterior distributions for systems 1 and 2 are both completely specified to be $P_1(x_1)$ and $P_2(x_2)$, respectively. Maximize the joint entropy $S[\rho(x_1, x_2), \varphi(x_1)\varphi(x_2)]$ subject to the following constraints on the $\rho(x_1, x_2)$,

$$\int dx_2 \, \rho(x_1, x_2) = P_1(x_1) \quad \text{and} \quad \int dx_1 \, \rho(x_1, x_2) = P_2(x_2). \tag{22}$$

Again, this is one constraint for each value of x_1 and one constraint for each value of x_2, which, therefore, require the separate multipliers $\mu_1(x_1)$ and $\mu_2(x_2)$. Maximizing S with respect to these constraints is then,

$$0 = \delta \left[S - \int dx_1 \mu_1(x_1) \left(\int dx_2 \, \rho(x_1, x_2) - P_1(x_1) \right) \right.$$
$$\left. - \int dx_2 \mu_2(x_2) \left(\int dx_1 \, \rho(x_1, x_2) - P_2(x_2) \right) \right], \tag{23}$$

leading to

$$\mu_1(x_1) + \mu_2(x_2) = \phi\left(\rho(x_1, x_2), \varphi_1(x_1)\varphi_2(x_2)\right). \tag{24}$$

The updating is being designed so that $\varphi_1 \varphi_2 \to P_1 P_2$, as the independent subsystems are being updated based on expectation values which are silent about correlations. DC2 thus imposes,

$$\mu_1(x_1) + \mu_2(x_2) = \phi\left(P_1(x_1) P_2(x_2), \varphi_1(x_1)\varphi_2(x_2)\right). \tag{25}$$

Write (25) as,

$$\mu_1(x_1) = \phi\left(P_1(x_1) P_2(x_2), \varphi_1(x_1)\varphi_2(x_2)\right) - \mu_2(x_2). \tag{26}$$

The left-hand side is independent of x_2 so we can perform a trick similar to that we used before. Suppose we had chosen a different *constraint* $P_2'(x_2)$ that differs from $P_2(x_2)$ and a new prior $\varphi_2'(x_2)$ that differs from $\varphi_2(x_2)$ except at the value $\tilde{x}_2$. At the value $\tilde{x}_2$, the multiplier $\mu_1(x_1)$ remains unchanged for all $P_2'(x_2)$, $\varphi_2'(x_2)$, and thus x_2. This means that any dependence that the right-hand side might potentially have had on x_2 and on the choice of $P_2(x_2)$, $\varphi_2'(x_2)$ must cancel out, leaving $\mu_1(x_1)$ unchanged. That is, the Lagrange multiplier $\mu(x_2)$ "pushes out" these dependences such that

$$\phi\left(P_1(x_1) P_2(x_2), \varphi_1(x_1)\varphi_2(x_2)\right) - \mu_2(x_2) = g(P_1(x_1), \varphi_1(x_1)). \tag{27}$$

Because $g(P_1(x_1), \varphi_1(x_1))$ is independent of arbitrary variations of $P_2(x_2)$ and $\varphi_2(x_2)$ on the left hand side (LHS) above—it is satisfied equally well for all choices. The form of $g = \phi(P_1(x_1), q_1(x_1))$ is apparent if $P_2(x_2) = \varphi_2(x_2) = 1$ as $\mu_2(x_2) = 0$ similar to Case 1 as well as DC1'. Therefore, the Lagrange multiplier is

$$\mu_1(x_1) = \phi\left(P_1(x_1), \varphi_1(x_1)\right). \tag{28}$$

A similar analysis carried out for $\mu_2(x_2)$ leads to

$$\mu_2(x_2) = \phi\left(P_2(x_2), \varphi_2(x_2)\right). \tag{29}$$

Case 2—Conclusion: Substituting back into (25) gives us a functional equation for ϕ,

$$\phi\left(P_1 P_2, \varphi_1 \varphi_2\right) = \phi\left(P_1, \varphi_1\right) + \phi\left(P_2, \varphi_2\right). \tag{30}$$

The general solution for this functional equation is derived in the Appendix A.3, and is

$$\phi(\rho, \varphi) = a_1 \ln(\rho(x)) + a_2 \ln(\varphi(x)), \tag{31}$$

where a_1, a_2 are constants. The constants are fixed by using DC1'. Letting $\rho_1(x_1) = \varphi_1(x_1) = \varphi_1$ gives $\phi(\varphi, \varphi) = 0$ by DC1', and, therefore,

$$\phi(\varphi, \varphi) = (a_1 + a_2) \ln(\varphi) = 0, \tag{32}$$

so we are forced to conclude $a_1 = -a_2$ for arbitrary φ. Letting $a_1 \equiv A = -|A|$ such that we are really maximizing the entropy (although this is purely aesthetic) gives the general form of ϕ to be

$$\phi(\rho, \varphi) = -|A| \ln\left(\frac{\rho(x)}{\varphi(x)}\right). \tag{33}$$

As long as $A \neq 0$, the value of A is arbitrary as it always can be absorbed into the Lagrange multipliers. The general form of the entropy designed for the purpose of inference of ρ is found by integrating ϕ, and, therefore,

$$S(\rho(x), \varphi(x)) = -|A| \int dx \left(\rho(x) \ln\left(\frac{\rho(x)}{\varphi(x)}\right) - \rho(x)\right) + C[\varphi]. \tag{34}$$

The constant in ρ, $C[\varphi]$, will always drop out when varying ρ. The apparent extra term $(|A| \int \rho(x)dx)$ from integration cannot be dropped while simultaneously satisfying DC1', which requires $\rho(x) = \varphi(x)$ in the absence of constraints or when there is no change to one's information. In previous versions where the integration term $(|A| \int \rho(x)dx)$ is dropped, one obtains solutions like $\rho(x) = e^{-1}\varphi(x)$ (independent of whether $\varphi(x)$ was previously normalized or not) in the absence of new information. Obviously, this factor can be taken care of by normalization, and, in this way, both forms of the entropy are equally valid; however, this form of the entropy better adheres to the PMU through DC1'. Given that we may regularly impose normalization, we may drop the extra $\int \rho(x)dx$ term and $C[\varphi]$. For convenience then, (34) becomes

$$S(\rho(x), \varphi(x)) \to S^*(\rho(x), \varphi(x)) = -|A| \int dx\, \rho(x) \ln\left(\frac{\rho(x)}{\varphi(x)}\right), \tag{35}$$

which is a special case when the normalization constraint is being applied. Given normalization is applied, the same selected posterior $\rho(x)$ maximizes both $S(\rho(x), \varphi(x))$ and $S^*(\rho(x), \varphi(x))$, and the star notation may be dropped.

Remarks

It can be seen that the relative entropy is invariant under coordinate transformations. This implies that a system of coordinates carry no information and it is the "character" of the probability distributions that are being ranked against one another rather than the specific set of propositions or microstates they describe.

The general solution to the maximum entropy procedure with respect to N linear constraints in ρ, $\langle A_i(x)\rangle$, and normalization gives a canonical-like selected posterior probability distribution,

$$\rho(x) = \varphi(x) \exp\left(\sum_i \alpha_i A_i(x)\right). \tag{36}$$

The positive constant $|A|$ may always be absorbed into the Lagrange multipliers so we may let it equal unity without loss of generality. DC1' is fully realized when we maximize with respect to a constraint on $\rho(x)$ that is already held by $\varphi(x)$, such as $\langle x^2\rangle = \int x^2 \rho(x)\,dx$, which happens to have the same value as $\langle x^2\rangle_\varphi = \int x^2 \varphi(x)\,dx$, then its Lagrange multiplier is forcibly zero $\alpha_1 = 0$ (as can be seen in (36) using (34)), in agreement with Jaynes. This gives the expected result $\rho(x) = \varphi(x)$ as there is no new information. Our design has arrived at a refined maximum entropy method [12] as a universal probability updating procedure [38].

3. The Design of the Quantum Relative Entropy

In the last section, we assumed that the universe of discourse (the set of relevant propositions or microstates) $\mathcal{X} = \mathcal{A} \times \mathcal{B} \times \ldots$ was known. In quantum physics, things are a bit more ambiguous because many probability distributions, or many experiments, can be associated with a given density matrix. In this sense, it is helpful to think of density matrices as "placeholders" for probability distributions rather than a probability distributions themselves. As any probability distribution from a given density matrix, $\rho(\cdot) = \mathrm{Tr}(|\cdot\rangle\langle\cdot|\hat{\rho})$, may be ranked using the standard relative entropy, it is unclear why we would chose one universe of discourse over another. In lieu of this, such that one universe of discourse is not given preferential treatment, we consider ranking entire density matrices against one another. Probability distributions of interest may be found from the selected posterior density matrix. This moves our universe of discourse from sets of propositions $\mathcal{X} \to \mathcal{H}$ to Hilbert space(s).

When the objects of study are quantum systems, we desire an objective procedure to update from a prior density matrix $\hat{\varphi}$ to a posterior density matrix $\hat{\rho}$. We will apply the same intuition for ranking probability distributions (Section 2) and implement the PMU, PI, and design criteria to the ranking of density matrices. We therefore find the quantum relative entropy $S(\hat{\rho}, \hat{\varphi})$ to be designed for the purpose of inferentially updating density matrices.

3.1. Designing the Quantum Relative Entropy

In this section, we design the quantum relative entropy using the same inferentially guided *design criteria* as were used in the standard relative entropy.

DC1: Subdomain Independence

The goal is to design a function $S(\hat{\rho}, \hat{\varphi})$ that is able to rank density matrices. This insists that $S(\hat{\rho}, \hat{\varphi})$ be a real scalar valued function of the posterior $\hat{\rho}$, and prior $\hat{\varphi}$ density matrices, which we will call the quantum relative entropy or simply the entropy. An arbitrary variation of the entropy with respect to $\hat{\rho}$ is,

$$\delta S(\hat{\rho},\hat{\varphi}) = \sum_{ij}\frac{\delta S(\hat{\rho},\hat{\varphi})}{\delta \rho_{ij}}\delta\rho_{ij} = \sum_{ij}\left(\frac{\delta S(\hat{\rho},\hat{\varphi})}{\delta\hat{\rho}}\right)_{ij}\delta(\hat{\rho})_{ij} = \sum_{ij}\left(\frac{\delta S(\hat{\rho},\hat{\varphi})}{\delta\hat{\rho}^T}\right)_{ji}\delta(\hat{\rho})_{ij} = \mathrm{Tr}\left(\frac{\delta S(\hat{\rho},\hat{\varphi})}{\delta\hat{\rho}^T}\delta\hat{\rho}\right), \tag{37}$$

where $\mathrm{Tr}(\ldots)$ is the trace. We wish to maximize this entropy with respect to expectation value constraints, such as $\langle A\rangle = \mathrm{Tr}(\hat{A}\hat{\rho})$ on $\hat{\rho}$. Using the Lagrange multiplier method to maximize the entropy with respect to $\langle A\rangle$ and normalization, and setting the variation equal to zero,

$$\delta\left(S(\hat{\rho},\hat{\varphi}) - \lambda[\mathrm{Tr}(\hat{\rho}) - 1] - \alpha[\mathrm{Tr}(\hat{A}\hat{\rho}) - \langle A\rangle]\right) = 0, \tag{38}$$

where λ and α are the Lagrange multipliers for the respective constraints. Because $S(\hat{\rho}, \hat{\varphi})$ is a real number, we inevitably require δS to be real, but without imposing this directly, we find that requiring δS to be real requires $\hat{\rho}, \hat{A}$ to be Hermitian. At this point, it is simpler to allow for arbitrary variations of $\hat{\rho}$ such that,

$$\text{Tr}\left(\left(\frac{\delta S(\hat{\rho}, \hat{\varphi})}{\delta \hat{\rho}^T} - \lambda \hat{1} - \alpha \hat{A}\right)\delta \hat{\rho}\right) = 0. \tag{39}$$

For these arbitrary variations, the variational derivative of S must satisfy,

$$\frac{\delta S(\hat{\rho}, \hat{\varphi})}{\delta \hat{\rho}^T} = \lambda \hat{1} + \alpha \hat{A} \tag{40}$$

at the maximum. As in the remark earlier, *all* forms of S that give the correct form of $\frac{\delta S(\hat{\rho}, \hat{\varphi})}{\delta \hat{\rho}^T}$ under variation are *equally valid* for the purpose of inference. For notational convenience, we let

$$\frac{\delta S(\hat{\rho}, \hat{\varphi})}{\delta \hat{\rho}^T} \equiv \phi(\hat{\rho}, \hat{\varphi}), \tag{41}$$

which is a matrix valued function of the posterior and prior density matrices. The form of $\phi(\hat{\rho}, \hat{\varphi})$ is already "local" in $\hat{\rho}$ (the variational derivative is with respect to the whole density matrix), so we don't need to constrain it further as we did in the original DC1.

DC1': *In the absence of new information, the new state $\hat{\rho}$ is equal to the old state $\hat{\varphi}$*

Applied to the ranking of density matrices, in the absence of new information, the density matrix $\hat{\varphi}$ should not change, that is, the posterior density matrix $\hat{\rho} = \hat{\varphi}$ is equal to the prior density matrix. Maximizing the entropy without applying any constraints gives,

$$\frac{\delta S(\hat{\rho}, \hat{\varphi})}{\delta \hat{\rho}^T} = \hat{0}, \tag{42}$$

and, therefore, DC1' imposes the following condition in this case:

$$\frac{\delta S(\hat{\rho}, \hat{\varphi})}{\delta \hat{\rho}^T} = \phi(\hat{\rho}, \hat{\varphi}) = \phi(\hat{\varphi}, \hat{\varphi}) = \hat{0}. \tag{43}$$

As in the original DC1', if $\hat{\varphi}$ is known to obey some expectation value $\langle \hat{A} \rangle$, and then if one goes out of their way to constrain $\hat{\rho}$ to that expectation value and nothing else, it follows from the PMU that $\hat{\rho} = \hat{\varphi}$, as no information has been gained. This is not imposed directly but can be verified later.

DC2: Subsystem Independence

The discussion of DC2 is the same as the standard relative entropy DC2—it is not a consistency requirement, and the updating is *designed* so that the independence reflected in the prior is maintained in the posterior by default via the PMU when the information provided is silent about correlations.

DC2 Implementation

Consider a composite system living in the Hilbert space $\mathcal{H} = \mathcal{H}_1 \otimes \mathcal{H}_2$. Assume that all prior evidence led us to believe the systems were independent. This is reflected in the prior density matrix: if the individual system priors are $\hat{\varphi}_1$ and $\hat{\varphi}_2$, then the joint prior for the whole system is $\hat{\varphi}_1 \otimes \hat{\varphi}_2$. Further suppose that new information is acquired such that $\hat{\varphi}_1$ would itself be updated to $\hat{\rho}_1$ and that $\hat{\varphi}_2$ would be itself be updated to $\hat{\rho}_2$. By design, the implementation of DC2 constrains the entropy functional such that in this case, the joint product prior density matrix $\hat{\varphi}_1 \otimes \hat{\varphi}_2$ updates to the product posterior $\hat{\rho}_1 \otimes \hat{\rho}_2$ so that inferences about one do not affect inferences about the other.

The argument below is considerably simplified if we expand the space of density matrices to include density matrices that are not necessarily normalized. This does not represent any limitation because normalization can always be easily achieved as one additional constraint. We consider a few special cases below:

Case 1: We receive the extremely constraining information that the posterior distribution for system 1 is completely specified to be $\hat{\rho}_1$ while we receive no information about system 2 at all. We treat the two systems jointly. Maximize the joint entropy $S[\hat{\rho}_{12}, \hat{\varphi}_1 \otimes \hat{\varphi}_2]$, subject to the following constraints on the $\hat{\rho}_{12}$,

$$\text{Tr}_2(\hat{\rho}_{12}) = \hat{\rho}_1. \tag{44}$$

Notice all of the N^2 elements in $\mathcal{H}_1$ of $\hat{\rho}_{12}$ are being constrained. We therefore need a Lagrange multiplier which spans $\mathcal{H}_1$ and therefore it is a square matrix $\hat{\lambda}_1$. This is readily seen by observing the component form expressions of the Lagrange multipliers $(\hat{\lambda}_1)_{ij} = \lambda_{ij}$. Maximizing the entropy with respect to this $\mathcal{H}_2$ independent constraint is

$$0 = \delta\left(S - \sum_{ij} \lambda_{ij} \left(\text{Tr}_2(\hat{\rho}_{1,2}) - \hat{\rho}_1\right)_{ij}\right), \tag{45}$$

but reexpressing this with its transpose $(\hat{\lambda}_1)_{ij} = (\hat{\lambda}_1^T)_{ji}$, gives

$$0 = \delta\left(S - \text{Tr}_1(\hat{\lambda}_1[\text{Tr}_2(\hat{\rho}_{1,2}) - \hat{\rho}_1])\right), \tag{46}$$

where we have relabeled $\hat{\lambda}_1^T \to \hat{\lambda}_1$, for convenience, as the name of the Lagrange multipliers are arbitrary. For arbitrary variations of $\hat{\rho}_{12}$, we therefore have

$$\hat{\lambda}_1 \otimes \hat{1}_2 = \phi\left(\hat{\rho}_{12}, \hat{\varphi}_1 \otimes \hat{\varphi}_2\right). \tag{47}$$

DC2 is implemented by requiring $\hat{\varphi}_1 \otimes \hat{\varphi}_2 \to \hat{\rho}_1 \otimes \hat{\varphi}_2$, such that the function ϕ is designed to reflect subsystem independence in this case; therefore, we have

$$\hat{\lambda}_1 \otimes \hat{1}_2 = \phi\left(\hat{\rho}_1 \otimes \hat{\varphi}_2, \hat{\varphi}_1 \otimes \hat{\varphi}_2\right). \tag{48}$$

Had we chosen a different prior $\hat{\varphi}_2' = \hat{\varphi}_2 + \delta\hat{\varphi}_2$, for all $\delta\hat{\varphi}_2$ the LHS $\hat{\lambda}_1 \otimes \hat{1}_2$ remains unchanged given that ϕ is independent of scalar functions (I would like to thank M. Krumm for pointing this out.) of $\hat{\varphi}_2$, as those could be lumped into $\hat{\lambda}_1$ while keeping $\hat{\rho}_1$ fixed. The potential dependence on scalar functions of $\hat{\varphi}_2$ can be removed by imposing DC2 in a subsystem independent situation where $\hat{\rho}_1'$ in ϕ need not be fixed under variations of $\hat{\varphi}_2$. The resulting equation in such a situation, for instance maximizing the entropy of an independent joint prior with respect to $\text{Tr}(\hat{A}_1 \otimes \hat{1}_2 \cdot \hat{\rho}_{12}) = \langle A \rangle$, facilitated by a scalar Lagrange multiplier λ, and after imposing DC2,

$$\lambda \hat{A}_1 \otimes \hat{1}_2 = \phi\left(\hat{\rho}_1' \otimes \hat{\varphi}_2, \hat{\varphi}_1 \otimes \hat{\varphi}_2\right). \tag{49}$$

For subsystem independence to be imposed here, $\hat{\rho}_1'$ must be independent of variations in $\hat{\varphi}_2$, and, therefore, in a general subsystem independent case, ϕ is independent of scalar functions of $\hat{\varphi}_2$. This means that any dependence that the right-hand side of (48) might potentially have had on $\hat{\varphi}_2$ must drop out, meaning,

$$\phi\left(\hat{\rho}_1 \otimes \hat{\varphi}_2, \hat{\varphi}_1 \otimes \hat{\varphi}_2\right) = f(\hat{\rho}_1, \hat{\varphi}_1) \otimes \hat{1}_2. \tag{50}$$

Since $\hat{\varphi}_2$ is arbitrary, suppose further that we choose a unit prior, $\hat{\varphi}_2 = \hat{1}_2$, and note that $\hat{\rho}_1 \otimes \hat{1}_2$ and $\hat{\varphi}_1 \otimes \hat{1}_2$ are block diagonal in $\mathcal{H}_2$. Because the LHS is block diagonal in $\mathcal{H}_2$,

$$f(\hat{\rho}_1, \hat{\varphi}_1) \otimes \hat{1}_2 = \phi\left(\hat{\rho}_1 \otimes \hat{1}_2, \hat{\varphi}_1 \otimes \hat{1}_2\right). \tag{51}$$

The RHS is block diagonal in $\mathcal{H}_2$ and, because the function ϕ is understood to be a power series expansion in its arguments,

$$f(\hat{\rho}_1, \hat{\varphi}_1) \otimes \hat{1}_2 = \phi\left(\hat{\rho}_1 \otimes \hat{1}_2, \hat{\varphi}_1 \otimes \hat{1}_2\right) = \phi\left(\hat{\rho}_1, \hat{\varphi}_1\right) \otimes \hat{1}_2. \tag{52}$$

This gives

$$\hat{\lambda}_1 \otimes \hat{1}_2 = \phi\left(\hat{\rho}_1, \hat{\varphi}_1\right) \otimes \hat{1}_2, \tag{53}$$

and, therefore, the $\hat{1}_2$ factors out and $\hat{\lambda}_1 = \phi\left(\hat{\rho}_1, \hat{\varphi}_1\right)$. A similar argument exchanging systems 1 and 2 shows $\hat{\lambda}_2 = \phi\left(\hat{\rho}_2, \hat{\varphi}_2\right)$.

Case 1—Conclusion: The analysis leads us to conclude that when the system 2 is not updated, the dependence on $\hat{\varphi}_2$ drops out,

$$\phi\left(\hat{\rho}_1 \otimes \hat{\varphi}_2, \hat{\varphi}_1 \otimes \hat{\varphi}_2\right) = \phi\left(\hat{\rho}_1, \hat{\varphi}_1\right) \otimes \hat{1}_2, \tag{54}$$

and, similarly,

$$\phi\left(\hat{\varphi}_1 \otimes \hat{\rho}_2, \hat{\varphi}_1 \otimes \hat{\varphi}_2\right) = \hat{1}_1 \otimes \phi\left(\hat{\rho}_2, \hat{\varphi}_2\right). \tag{55}$$

Case 2: Now consider a different special case in which the marginal posterior distributions for systems 1 and 2 are both completely specified to be $\hat{\rho}_1$ and $\hat{\rho}_2$, respectively. Maximize the joint entropy, $S[\hat{\rho}_{12}, \hat{\varphi}_1 \otimes \hat{\varphi}_2]$, subject to the following constraints on the $\hat{\rho}_{12}$,

$$\text{Tr}_2(\hat{\rho}_{12}) = \hat{\rho}_1 \quad \text{and} \quad \text{Tr}_1(\hat{\rho}_{12}) = \hat{\rho}_2, \tag{56}$$

where $\text{Tr}_i(\ldots)$ is the partial trace function, which a trace over the vectors in over $\mathcal{H}_i$. Here, each expectation value constrains the entire space $\mathcal{H}_i$, where $\hat{\rho}_i$ lives. The Lagrange multipliers must span their respective spaces, so we implement the constraint with the Lagrange multiplier operator $\hat{\mu}_i$, then,

$$0 = \delta\left(S - \text{Tr}_1(\hat{\mu}_1[\text{Tr}_2(\hat{\rho}_{12}) - \hat{\rho}_1]) - \text{Tr}_2(\hat{\mu}_2[\text{Tr}_1(\hat{\rho}_{12}) - \hat{\rho}_2])\right). \tag{57}$$

For arbitrary variations of $\hat{\rho}_{12}$, we have

$$\hat{\mu}_1 \otimes \hat{1}_2 + \hat{1}_1 \otimes \hat{\mu}_2 = \phi\left(\hat{\rho}_{12}, \hat{\varphi}_1 \otimes \hat{\varphi}_2\right). \tag{58}$$

By design, DC2 is implemented by requiring $\hat{\varphi}_1 \otimes \hat{\varphi}_2 \to \hat{\rho}_1 \otimes \hat{\rho}_2$ in this case; therefore, we have

$$\hat{\mu}_1 \otimes \hat{1}_2 + \hat{1}_1 \otimes \hat{\mu}_2 = \phi\left(\hat{\rho}_1 \otimes \hat{\rho}_2, \hat{\varphi}_1 \otimes \hat{\varphi}_2\right). \tag{59}$$

Write (59) as

$$\hat{\mu}_1 \otimes \hat{1}_2 = \phi\left(\hat{\rho}_1 \otimes \hat{\rho}_2, \hat{\varphi}_1 \otimes \hat{\varphi}_2\right) - \hat{1}_1 \otimes \hat{\mu}_2. \tag{60}$$

The LHS is independent of changes that might occur in $\mathcal{H}_2$ on the RHS of (60). This means that any variation of $\hat{\rho}_2$ and $\hat{\varphi}_2$ must be "pushed out" by $\hat{\mu}_2$—it removes the dependence of $\hat{\rho}_2$ and $\hat{\varphi}_2$ in ϕ. Any dependence that the RHS might potentially have had on $\hat{\rho}_2$, $\hat{\varphi}_2$ must cancel out in a general subsystem independent case, leaving $\hat{\mu}_1$ unchanged. Consequently,

$$\phi\left(\hat{\rho}_1 \otimes \hat{\rho}_2, \hat{\varphi}_1 \otimes \hat{\varphi}_2\right) - \hat{1}_1 \otimes \hat{\mu}_2 = g(\hat{\rho}_1, \hat{\varphi}_1) \otimes \hat{1}_2. \tag{61}$$

Because $g(\hat{\rho}_1, \hat{\varphi}_1)$ is independent of arbitrary variations of $\hat{\rho}_2$ and $\hat{\varphi}_2$ on the LHS above—it is satisfied equally well for all choices. The form of $g(\hat{\rho}_1, \hat{\varphi}_1)$ reduces to the form of $f(\hat{\rho}_1, \hat{\varphi}_1)$ from Case 1 when $\hat{\rho}_2 = \hat{\varphi}_2 = \hat{1}_2$ and, similarly, DC1' gives $\hat{\mu}_2 = 0$. Therefore, the Lagrange multiplier is

$$\hat{\mu}_1 \otimes \hat{1}_2 = \phi(\hat{\rho}_1, \hat{\varphi}_1) \otimes \hat{1}_2. \tag{62}$$

A similar analysis is carried out for $\hat{\mu}_2$ leading to

$$\hat{1}_1 \otimes \hat{\mu}_2 = \hat{1}_1 \otimes \phi(\hat{\rho}_2, \hat{\varphi}_2). \tag{63}$$

Case 2—Conclusion: Substituting back into (59) gives us a functional equation for ϕ,

$$\phi(\hat{\rho}_1 \otimes \hat{\rho}_2, \hat{\varphi}_1 \otimes \hat{\varphi}_2) = \phi(\hat{\rho}_1, \hat{\varphi}_1) \otimes \hat{1}_2 + \hat{1}_1 \otimes \phi(\hat{\rho}_2, \hat{\varphi}_2), \tag{64}$$

which is

$$\phi(\hat{\rho}_1 \otimes \hat{\rho}_2, \hat{\varphi}_1 \otimes \hat{\varphi}_2) = \phi(\hat{\rho}_1 \otimes \hat{1}_2, \hat{\varphi}_1 \otimes \hat{1}_2) + \phi(\hat{1}_1 \otimes \hat{\rho}_2, \hat{1}_1 \otimes \hat{\varphi}_2). \tag{65}$$

The general solution to this matrix valued functional equation is derived in Appendix A.5 and is

$$\phi(\hat{\rho}, \hat{\varphi}) = \widetilde{A} \ln(\hat{\rho}) + \widetilde{B} \ln(\hat{\varphi}), \tag{66}$$

where tilde $\widetilde{A}$ is a "super-operator" having constant coefficients and twice the number of indicies as $\hat{\rho}$ and $\hat{\varphi}$ as discussed in the Appendix (i.e., $\left(\widetilde{A} \ln(\hat{\rho})\right)_{ij} = \sum_{k\ell} A_{ijk\ell} (\log(\hat{\rho}))_{k\ell}$ and similarly for $\widetilde{B} \ln(\hat{\varphi})$). DC1' imposes

$$\phi(\hat{\varphi}, \hat{\varphi}) = \widetilde{A} \ln(\hat{\varphi}) + \widetilde{B} \ln(\hat{\varphi}) = \hat{0}, \tag{67}$$

which is satisfied in general when $\widetilde{A} = -\widetilde{B}$, and, now,

$$\phi(\hat{\rho}, \hat{\varphi}) = \widetilde{A} \left(\ln(\hat{\rho}) - \ln(\hat{\varphi}) \right). \tag{68}$$

We may fix the constant $\widetilde{A}$ by substituting our solution into the RHS of Equation (64), which is equal to the RHS of Equation (65),

$$\left(\widetilde{A}_1 \left(\ln(\hat{\rho}_1) - \ln(\hat{\varphi}_1) \right) \right) \otimes \hat{1}_2 + \hat{1}_1 \otimes \left(\widetilde{A}_2 \left(\ln(\hat{\rho}_2) - \ln(\hat{\varphi}_2) \right) \right)$$

$$= \widetilde{A}_{12} \left(\ln(\hat{\rho}_1 \otimes \hat{1}_2) - \ln(\hat{\varphi}_1 \otimes \hat{1}_2) \right) + \widetilde{A}_{12} \left(\ln(\hat{1}_1 \otimes \hat{\rho}_2) - \ln(\hat{1}_1 \otimes \hat{\varphi}_2) \right), \tag{69}$$

where $\widetilde{A}_{12}$ acts on the joint space of 1 and 2 and $\widetilde{A}_1$, $\widetilde{A}_2$ acts on single subspaces 1 or 2, respectively. Using the well known log tensor product identity in this case (The proof is demonstrated by taking the log of $\hat{\rho}_1 \otimes \hat{1}_2 \equiv \exp(\hat{\rho}_1') \otimes \hat{1}_2 = \exp(\hat{\rho}_1' \otimes \hat{1}_2)$ and substituting $\hat{\rho}_1' = \log(\hat{\rho}_1)$.), $\ln(\hat{\rho}_1 \otimes \hat{1}_2) = \ln(\hat{\rho}_1) \otimes \hat{1}_2$, the RHS of Equation (69) becomes

$$= \widetilde{A}_{12} \left(\ln(\hat{\rho}_1) \otimes \hat{1}_2 - \ln(\hat{\varphi}_1) \otimes \hat{1}_2 \right) + \widetilde{A}_{12} \left(\hat{1}_1 \otimes \ln(\hat{\rho}_2) - \hat{1}_1 \otimes \ln(\hat{\varphi}_2) \right). \tag{70}$$

Note that arbitrarily letting $\hat{\rho}_2 = \hat{\varphi}_2$ gives

$$\left(\widetilde{A}_1 \left(\ln(\hat{\rho}_1) - \ln(\hat{\varphi}_1) \right) \right) \otimes \hat{1}_2 = \widetilde{A}_{12} \left(\ln(\hat{\rho}_1) \otimes \hat{1}_2 - \ln(\hat{\varphi}_1) \otimes \hat{1}_2 \right), \tag{71}$$

or arbitrarily letting $\hat{\rho}_1 = \hat{\varphi}_1$ gives

$$\hat{1}_1 \otimes \left(\tilde{A}_2 \left(\ln(\hat{\rho}_2) - \ln(\hat{\varphi}_2) \right) \right) = \tilde{A}_{12} \left(\hat{1}_1 \otimes \ln(\hat{\rho}_2) - \hat{1}_1 \otimes \ln(\hat{\varphi}_2) \right). \tag{72}$$

As $\tilde{A}_{12}$, $\tilde{A}_1$, and $\tilde{A}_2$ are constant tensors, inspecting the above equalities determines the form of the tensor to be $\tilde{A} = A \tilde{1}$ where A is a scalar constant and $\tilde{1}$ is the super-operator identity over the appropriate (joint) Hilbert space.

Because our goal is to maximize the entropy function, we let the arbitrary constant $A = -|A|$ and distribute $\tilde{1}$ identically, which gives the final functional form,

$$\phi(\hat{\rho}, \hat{\varphi}) = -|A| \left(\ln(\hat{\rho}) - \ln(\hat{\varphi}) \right). \tag{73}$$

"Integrating" ϕ gives a general form for the quantum relative entropy,

$$S(\hat{\rho}, \hat{\varphi}) = -|A| \text{Tr}(\hat{\rho} \log \hat{\rho} - \hat{\rho} \log \hat{\varphi} - \hat{\rho}) + C[\hat{\varphi}] = -|A| S_U(\hat{\rho}, \hat{\varphi}) + |A| \text{Tr}(\hat{\rho}) + C[\hat{\varphi}], \tag{74}$$

where $S_U(\hat{\rho}, \hat{\varphi})$ is Umegaki's form of the relative entropy [42–44], the extra $|A|\text{Tr}(\hat{\rho})$ from integration is an artifact present for the preservation of DC1', and $C[\hat{\varphi}]$ is a constant in the sense that it drops out under arbitrary variations of $\hat{\rho}$. This entropy leads to the same inferences as Umegaki's form of the entropy with an added bonus that $\hat{\rho} = \hat{\varphi}$ in the absence of constraints or changes in information—rather than $\hat{\rho} = e^{-1}\hat{\varphi}$, which would be given by maximizing Umegaki's form of the entropy. In this sense, the extra $|A|\text{Tr}(\hat{\rho})$ only improves the inference process as it more readily adheres to the PMU though DC1'; however, now, because $S_U \geq 0$, we have $S(\hat{\rho}, \hat{\varphi}) \leq \text{Tr}(\hat{\rho}) + C[\hat{\varphi}]$, which provides little nuisance. In the spirit of this derivation, we will keep the $\text{Tr}(\hat{\rho})$ term there, but, for all practical purposes of inference, as long as there is a normalization constraint, it plays no role, and we find (letting $|A| = 1$ and $C[\hat{\varphi}] = 0$),

$$S(\hat{\rho}, \hat{\varphi}) \to S^*(\hat{\rho}, \hat{\varphi}) = -S_U(\hat{\rho}, \hat{\varphi}) = -\text{Tr}(\hat{\rho} \log \hat{\rho} - \hat{\rho} \log \hat{\varphi}), \tag{75}$$

Umegaki's form of the relative entropy. $S^*(\hat{\rho}, \hat{\varphi})$ is an equally valid entropy because, given normalization is applied, the same selected posterior $\hat{\rho}$ maximizes both $S(\hat{\rho}, \hat{\varphi})$ and $S^*(\hat{\rho}, \hat{\varphi})$.

3.2. Remarks

Due to the universality and the equal application of the PMU by using the same design criteria for both the standard and quantum case, the quantum relative entropy reduces to the standard relative entropy when $[\hat{\rho}, \hat{\varphi}] = 0$ or when the experiment being preformed $\hat{\rho} \to \rho(a) = \text{Tr}(\hat{\rho}|a\rangle\langle a|)$ is known. The quantum relative entropy we derive has the correct asymptotic form of the standard relative entropy in the sense of [8–10]. Further connections will be illustrated in a follow up article that is concerned with direct applications of the quantum relative entropy. Because two entropies are derived in parallel, we expect the well-known inferential results and consequences of the relative entropy to have a quantum relative entropy representation.

Maximizing the quantum relative entropy with respect to some constraints $\langle \hat{A}_i \rangle$, where $\{\hat{A}_i\}$ are a set of arbitrary Hermitian operators, and normalization $\langle \hat{1} \rangle = 1$, gives the following general solution for the posterior density matrix:

$$\hat{\rho} = \exp \left(\alpha_0 \hat{1} + \sum_i \alpha_i \hat{A}_i + \ln(\hat{\varphi}) \right) = \frac{1}{Z} \exp \left(\sum_i \alpha_i \hat{A}_i + \ln(\hat{\varphi}) \right) \equiv \frac{1}{Z} \exp \left(\hat{C} \right), \tag{76}$$

where α_i are the Lagrange multipliers of the respective constraints and normalization may be factored out of the exponential in general because the identity commutes universally. If $\hat{\varphi} \propto \hat{1}$, it is well known that the analysis arrives at the same expression for $\hat{\rho}$ after normalization, as it would if the von Neumann entropy were used, and thus one can find expressions for thermalized quantum states $\hat{\rho} = \frac{1}{Z} e^{-\beta \hat{H}}$. The remaining problem is to solve for the N Lagrange multipliers using their N associated expectation value constraints. In principle, their solution is found by computing Z and using standard methods from Statistical Mechanics,

$$\langle \hat{A}_i \rangle = -\frac{\partial}{\partial \alpha_i} \ln(Z), \tag{77}$$

and inverting to find $\alpha_i = \alpha_i(\langle \hat{A}_i \rangle)$, which has a unique solution due to the joint concavity (convexity depending on the sign convention) of the quantum relative entropy [8,9] when the constraints are linear in $\hat{\rho}$. The simple proof that (77) is monotonic in α, and therefore invertible, is that its derivative $\frac{\partial}{\partial \alpha} \langle \hat{A}_i \rangle = \langle \hat{A}_i^2 \rangle - \langle \hat{A}_i \rangle^2 \geq 0$. Between the Zassenhaus formula [45]

$$e^{t(\hat{A}+\hat{B})} = e^{t\hat{A}} e^{t\hat{B}} e^{-\frac{t^2}{2}[\hat{A},\hat{B}]} e^{\frac{t^3}{6}(2[\hat{B},[\hat{A},\hat{B}]]+[\hat{A},[\hat{A},\hat{B}]])} \ldots, \tag{78}$$

and Horn's inequality [46–48], the solutions to (77) lack a certain calculational elegance because it is difficult to express the eigenvalues of $\hat{C} = \log(\hat{\rho}) + \sum \alpha_i \hat{A}_i$ (in the exponential) in simple terms of the eigenvalues of the $\hat{A}_i$'s and $\hat{\rho}$, in general, when the matrices do not commute. The solution requires solving the eigenvalue problem for $\hat{C}$, such the the exponential of $\hat{C}$ may be taken and evaluated in terms of the eigenvalues of the $\alpha_i \hat{A}_i$s and the prior density matrix $\hat{\rho}$. A pedagogical exercise is starting with a prior that is a mixture of spin-z up and down $\hat{\rho} = a|+\rangle\langle+| + b|-\rangle\langle-|$ ($a, b \neq 0$), maximizing the quantum relative entropy with respect to an expectation of a general Hermitian operator with which the prior density matrix does not commute. This example for spin is given in the Appendix B.

4. Conclusions

This approach emphasizes the notion that entropy is a tool for performing inference and downplays counter-notional issues that arise if one interprets entropy as a measure of disorder, a measure of distinguishability, or an amount of missing information [7]. Because the same design criteria, guided by the PMU, are applied equally well to the design of a relative and quantum relative entropy, we find that both the relative and quantum relative entropy are designed for the purpose of inference. Because the quantum relative entropy is the functional that fits the requirements of a tool designed for the inference of density matrices, we now know what it is and how to use it—formulating an inferential quantum maximum entropy method. This article provides the foundation for [29], which, in particular, derives the Quantum Bayes Rule and collapse as special cases of the quantum maximum entropy method, as was craved in [24], analogous to [38,40]'s treatment for deriving Bayes Rule using the standard maximum entropy method. The quantum maximum entropy method thereby unifies a few topics in Quantum Information and Quantum Measurement through entropic inference.

Acknowledgments: I must give ample acknowledgment to Ariel Caticha who suggested the problem of justifying the form of the quantum relative entropy as a criterion for ranking of density matrices. He cleared up several difficulties by suggesting that design constraints be applied to the variational derivative of the entropy rather than the entropy itself. In addition, he provided substantial improvements to the method for imposing DC2 that led to the functional equations for the variational derivatives ($\phi_{12} = \phi_1 + \phi_2$)—with more rigor than in earlier versions of this article. His time and guidance are all greatly appreciated—thanks, Ariel. I would also like to thank M. Krumm, the reviewers, as well as our information physics group at UAlbany for our many intriguing discussions about probability, inference, and quantum mechanics.

Conflicts of Interest: The author declares no conflict of interest.

Appendix A

The Appendix loosely follows the relevant sections in [49], and then uses the methods reviewed to solve the relevant functional equations for ϕ. The last section is an example of the quantum maximum entropy method applied to a mixed spin state.

Appendix A.1. Simple Functional Equations

From [49] pages 31–44.

Theorem A1. *If Cauchy's functional equation*

$$f(x+y) = f(x) + f(y) \tag{A1}$$

is satisfied for all real x, y, and if the function $f(x)$ is (a) continuous at a point, (b) nonegative for small positive x's, or (c) bounded in an interval, then,

$$f(x) = cx \tag{A2}$$

is the solution to (A1) for all real x. If (A1) is assumed only over all positive x, y, then under the same conditions, (A2) holds for all positive x.

Proof. The most natural assumption for our purposes is that $f(x)$ is continuous at a point (which later extends to continuity all points as given by Darboux [50]). Cauchy solved the functional equation by induction. In particular, Equation (A1) implies,

$$f(\sum_i x_i) = \sum_i f(x_i), \tag{A3}$$

and if we let each $x_i = x$ as a special case to determine f, we find

$$f(nx) = nf(x). \tag{A4}$$

We may let $nx = mt$ such that

$$f(x) = f(\frac{m}{n}t) = \frac{m}{n}f(t). \tag{A5}$$

Letting $\lim_{t \to 1} f(t) = f(1) = c$ gives

$$f(\frac{m}{n}) = \frac{m}{n}f(1) = \frac{m}{n}c, \tag{A6}$$

and, because for $t = 1$, $x = \frac{m}{n}$ above, we have

$$f(x) = cx, \tag{A7}$$

which is the general solution of the linear functional equation. In principle, c can be complex. The importance of Cauchy's solution is that it can be used to give general solutions to the following Cauchy equations:

$$\begin{aligned} f(x+y) &= f(x)f(y), & \text{(A8)} \\ f(xy) &= f(x) + f(y), & \text{(A9)} \\ f(xy) &= f(x)f(y), & \text{(A10)} \end{aligned}$$

by preforming consistent substitution until they are the same form as (A1), as given by Cauchy. We will briefly discuss the first two. □

Theorem A2. *The general solution of $f(x+y) = f(x)f(y)$ is $f(x) = e^{cx}$ for all real or for all positive x, y that are continuous at one point and, in addition to the exponential solution, the solution $f(0) = 1$ and $f(x) = 0$ for $(x > 0)$ are in these classes of functions.*

The first functional $f(x+y) = f(x)f(y)$ is solved by first noting that it is strictly positive for real x, y, $f(x)$, which can be shown by considering $x = y$,

$$f(2x) = f(x)^2 > 0. \tag{A11}$$

If there exists $f(x_0) = 0$, then it follows that $f(x) = f((x - x_0) + x_0) = 0$, a trivial solution, hence the reason why the possibility of being equal to zero is excluded above. Given $f(x)$ is nowhere zero, we are justified in taking the natural logarithm $\ln(x)$, due to its positivity $f(x) > 0$. This gives,

$$\ln(f(x+y)) = \ln(f(x)) + \ln(f(y)), \tag{A12}$$

and letting $g(x) = \ln(f(x))$ gives,

$$g(x+y) = g(x) + g(y), \tag{A13}$$

which is Cauchy's linear equation, and thus has the solution $g(x) = cx$. Because $g(x) = \ln(f(x))$, one finds in general that $f(x) = e^{cx}$.

Theorem A3. *If the functional equation $f(xy) = f(x) + f(y)$ is valid for all positive x, y then its general solution is $f(x) = c \ln(x)$ given it is continuous at a point. If $x = 0$ (or $y = 0$) are valid, then the general solution is $f(x) = 0$. If all real x, y are valid except 0, then the general solution is $f(x) = c \ln(|x|)$.*

In particular, we are interested in the functional equation $f(xy) = f(x) + f(y)$ when x, y are positive. In this case, we can again follow Cauchy and substitute $x = e^u$ and $y = e^v$ to get,

$$f(e^u e^v) = f(e^u) + f(e^v), \tag{A14}$$

and letting $g(u) = f(e^u)$ gives $g(u + v) = g(u) + g(v)$. Again, the solution is $g(u) = cu$ and, therefore, the general solution is $f(x) = c \ln(x)$ when we substitute for u. If x could equal 0, then $f(0) = f(x) + f(0)$, which has the trivial solution $f(x) = 0$. The general solution for $x \neq 0$, $y \neq 0$ and x, y positive is therefore $f(x) = c \ln(x)$.

Appendix A.2. Functional Equations with Multiple Arguments

From [49] pages 213–217. Consider the functional equation,

$$F(x_1 + y_1, x_2 + y_2, ..., x_n + y_n) = F(x_1, x_2, ..., x_n) + F(y_1, y_2, ..., y_n), \tag{A15}$$

which is a generalization of Cauchy's linear functional Equation (A1) to several arguments. Letting $x_2 = x_3 = ... = x_n = y_2 = y_3 = ... = y_n = 0$ gives

$$F(x_1 + y_1, 0, ..., 0) = F(x_1, 0, ..., 0) + F(y_1, 0, ..., 0), \tag{A16}$$

which is the Cauchy linear functional equation having solution $F(x_1, 0, ..., 0) = c_1 x_1$, where $F(x_1, 0, ..., 0)$ is assumed to be continuous or at least measurable majorant. Similarly,

$$F(0, ..., 0, x_k, 0, ..., 0) = c_k x_k, \tag{A17}$$

and if you consider

$$F(x_1 + 0, 0 + y_2, 0, ..., 0) = F(x_1, 0, ..., 0) + F(0, y_2, 0, ..., 0) = c_1 x_1 + c_2 y_2, \tag{A18}$$

and, as y_2 is arbitrary, we could have let $y_2 = x_2$ such that in general

$$F(x_1, x_2, ..., x_n) = \sum c_i x_i, \tag{A19}$$

formulating the general solution.

Appendix A.3. Relative Entropy

We are interested in the following functional equation:

$$\phi(\rho_1\rho_2, \varphi_1\varphi_2) = \phi(\rho_1, \varphi_1) + \phi(\rho_2, \varphi_2). \tag{A20}$$

This is an equation of the form,

$$F(x_1y_1, x_2y_2) = F(x_1, x_2) + F(y_1, y_2), \tag{A21}$$

where $x_1 = \rho(x_1)$, $y_1 = \rho(x_2)$, $x_2 = \varphi(x_1)$, and $y_2 = \varphi(x_2)$. First, assume all q and p are greater than zero. Then, substitute: $x_i = e^{x'_i}$ and $y_i = e^{y'_i}$ and let $F'(x'_1, x'_2) = F(e^{x'_1}, e^{x'_2})$ and so on such that

$$F'(x'_1 + y'_1, x'_2 + y'_2) = F'(x'_1, x'_2) + F'(y'_1, y'_2), \tag{A22}$$

which is of the form of (A15). The general solution for F is therefore

$$F'(x'_1 + y'_1, x'_2 + y'_2) = a_1(x'_1 + y'_1) + a_2(x'_2 + y'_2) = a_1 \ln(x_1y_1) + a_2 \ln(x_2y_2) = F(x_1y_1, x_2y_2), \tag{A23}$$

which means the general solution for ϕ is

$$\phi(\rho_1, \varphi_1) = a_1 \ln(\rho(x_1)) + a_2 \ln(\varphi(x_1)). \tag{A24}$$

In such a case, when $\varphi(x_0) = 0$ for some value $x_0 \in \mathcal{X}$, we may let $\varphi(x_0) = \epsilon$, where ϵ is as close to zero as we could possibly want—the trivial general solution $\phi = 0$ is saturated by the special case when $\rho = \varphi$ from DC1'. Here, we return to the text.

Appendix A.4. Matrix Functional Equations

(This derivation is implied in [49] pages 347–349). First, consider a Cauchy matrix functional equation,

$$f(\hat{X} + \hat{Y}) = f(\hat{X}) + f(\hat{Y}), \tag{A25}$$

where $\hat{X}$ and $\hat{Y}$ are $n \times n$ square matrices. Rewriting the matrix functional equation in terms of its components gives

$$f_{ij}(x_{11} + y_{11}, x_{12} + y_{12}, ..., x_{nn} + y_{nn}) = f_{ij}(x_{11}, x_{12}, ..., x_{nn}) + f_{ij}(y_{11}, y_{12}, ..., y_{nn}) \tag{A26}$$

and is now in the form of (A15), and, therefore, the solution is

$$f_{ij}(x_{11}, x_{12}, ..., x_{nn}) = \sum_{\ell,k=0}^{n} c_{ij\ell k} x_{\ell k} \tag{A27}$$

for $i, j = 1, ..., n$. We find it convenient to introduce super indices, $A = (i, j)$ and $B = (\ell, k)$ such that the component equation becomes

$$f_A = \sum_B c_{AB} x_B, \tag{A28}$$

and resembles the solution for the linear transformation of a vector from [49]. In general, we will be discussing matrices $\hat{X} = \hat{X}_1 \otimes \hat{X}_2 \otimes ... \otimes \hat{X}_N$ which stem from tensor products of density matrices. In this situation, $\hat{X}$ can be thought of as $2N$ index tensor or a $z \times z$ matrix where $z = \prod_i^N n_i$ is the product of the ranks of the matrices in the tensor product or even as a vector of length z^2. In such a case, we may abuse the super index notation where A and B lump together the appropriate number of indices such that (A28) is the form of the solution for the components in general. The matrix form of the general solution is

$$f(\hat{X}) = \tilde{C}\hat{X},\tag{A29}$$

where $\tilde{C}$ is a constant super-operator having components c_{AB}.

Appendix A.5. Quantum Relative Entropy

The functional equation of interest is

$$\phi\left(\hat{\rho}_1 \otimes \hat{\rho}_2, \hat{\varphi}_1 \otimes \hat{\varphi}_2\right) = \phi\left(\hat{\rho}_1 \otimes \hat{1}_2, \hat{\varphi}_1 \otimes \hat{1}_2\right) + \phi\left(\hat{1}_1 \otimes \hat{\rho}_2, \hat{1}_1 \otimes \hat{\varphi}_2\right).\tag{A30}$$

These density matrices are Hermitian, positive semi-definite, have positive eigenvalues, and are not equal to $\hat{0}$. Because every invertible matrix can be expressed as the exponential of some other matrix, we can substitute $\hat{\rho}_1 = e^{\hat{\rho}_1'}$, and so on for all four density matrices giving,

$$\phi\left(e^{\hat{\rho}_1'} \otimes e^{\hat{\rho}_2'}, e^{\hat{\varphi}_1'} \otimes e^{\hat{\varphi}_2'}\right) = \phi\left(e^{\hat{\rho}_1'} \otimes \hat{1}_2, e^{\hat{\varphi}_1'} \otimes \hat{1}_2\right) + \phi\left(\hat{1}_1 \otimes e^{\hat{\rho}_2'}, \hat{1}_1 \otimes e^{\hat{\varphi}_2'}\right).\tag{A31}$$

Now, we use the following identities for Hermitian matrices:

$$e^{\hat{\rho}_1'} \otimes e^{\hat{\rho}_2'} = e^{\hat{\rho}_1' \otimes \hat{1}_2 + \hat{1}_1 \otimes \hat{\rho}_2'}\tag{A32}$$

and

$$e^{\hat{\rho}_1'} \otimes \hat{1}_2 = e^{\hat{\rho}_1' \otimes \hat{1}_2},\tag{A33}$$

to recast the functional equation as,

$$\phi\left(e^{\hat{\rho}_1' \otimes \hat{1}_2 + \hat{1}_1 \otimes \hat{\rho}_2'}, e^{\hat{\varphi}_1' \otimes \hat{1}_2 + \hat{1}_1 \otimes \hat{\varphi}_2'}\right) = \phi\left(e^{\hat{\rho}_1' \otimes \hat{1}_2}, e^{\hat{\varphi}_1' \otimes \hat{1}_2}\right) + \phi\left(e^{\hat{1}_1 \otimes \hat{\rho}_2'}, e^{\hat{1}_1 \otimes \hat{\varphi}_2'}\right).\tag{A34}$$

Letting $G(\hat{\rho}_1' \otimes \hat{1}_2, \hat{\varphi}_1' \otimes \hat{1}_2) = \phi\left(e^{\hat{\rho}_1' \otimes \hat{1}_2}, e^{\hat{\varphi}_1' \otimes \hat{1}_2}\right)$, and the like, gives

$$G(\hat{\rho}_1' \otimes \hat{1}_2 + \hat{1}_1 \otimes \hat{\rho}_2', \hat{\varphi}_1' \otimes \hat{1}_2 + \hat{1}_1 \otimes \hat{\varphi}_2') = G(\hat{\rho}_1' \otimes \hat{1}_2, \hat{\varphi}_1' \otimes \hat{1}_2) + G(\hat{1}_1 \otimes \hat{\rho}_2', \hat{1}_1 \otimes \hat{\varphi}_2').\tag{A35}$$

This functional equation is of the form

$$G(\hat{X}_1' + \hat{Y}_1', \hat{X}_2' + \hat{Y}_2') = G(\hat{X}_1', \hat{X}_2') + G(\hat{Y}_1', \hat{Y}_2'),\tag{A36}$$

which has the general solution

$$G(\hat{X}', \hat{Y}') = \tilde{A}\,\hat{X}' + \tilde{B}\hat{Y}',\tag{A37}$$

analogous to (A19), and finally, in general,

$$\phi(\hat{\rho}, \hat{\varphi}) = \tilde{A}\,\ln(\hat{\rho}) + \tilde{B}\ln(\hat{\varphi}),\tag{A38}$$

where $\tilde{A}, \tilde{B}$ are super-operators having constant coefficients. Here, we return to the text.

Appendix B. Spin Example

Consider an arbitrarily mixed prior (in the spin-z basis for convenience) with $a, b \neq 0$,

$$\hat{\varphi} = a|+\rangle\langle+| + b|-\rangle\langle-| \tag{A39}$$

and a general Hermitian matrix in the spin-1/2 Hilbert space,

$$c_\mu \hat{\sigma}^\mu = c_1 \hat{1} + c_x \hat{\sigma}_x + c_y \hat{\sigma}_x + c_z \hat{\sigma}_z \tag{A40}$$

$$= (c_1 + c_z)|+\rangle\langle+| + (c_x - ic_y)|+\rangle\langle-| + (c_x + ic_y)|-\rangle\langle+| + (c_1 - c_z)|-\rangle\langle-|, \tag{A41}$$

having a known expectation value,

$$\text{Tr}(\hat{\rho} c_\mu \hat{\sigma}^\mu) = c. \tag{A42}$$

Maximizing the entropy with respect to this general expectation value and normalization is:

$$0 = \left(\delta S - \lambda [\text{Tr}(\hat{\rho}) - 1] - \alpha (\text{Tr}(\hat{\rho} c_\mu \hat{\sigma}^\mu) - c) \right), \tag{A43}$$

which after varying gives the solution,

$$\hat{\rho} = \frac{1}{Z} \exp(\alpha c_\mu \hat{\sigma}^\mu + \log(\hat{\varphi})). \tag{A44}$$

Letting

$$\hat{C} = \alpha c_\mu \hat{\sigma}^\mu + \log(\hat{\varphi}) \tag{A45}$$

gives

$$\hat{\rho} = \frac{1}{Z} e^{\hat{C}} = U e^{U^{-1} \hat{C} U} U^{-1} = \frac{1}{Z} U e^{\hat{\lambda}} U^{-1}$$

$$= \frac{e^{\lambda_+}}{Z} U |\lambda_+\rangle\langle\lambda_+| U^{-1} + \frac{e^{\lambda_-}}{Z} U |\lambda_-\rangle\langle\lambda_-| U^{-1}, \tag{A46}$$

where $\hat{\lambda}$ is the diagonalized matrix of $\hat{C}$ having real eigenvalues. They are

$$\lambda_\pm = \lambda \pm \delta\lambda, \tag{A47}$$

due to the quadratic formula, where explicitly:

$$\lambda = \alpha c_1 + \frac{1}{2} \log(ab), \tag{A48}$$

and

$$\delta\lambda = \frac{1}{2} \sqrt{\left(2\alpha c_z + \log(\frac{a}{b})\right)^2 + 4\alpha^2 (c_x^2 + c_y^2)}. \tag{A49}$$

Because $\lambda_\pm$ and a, b, c_1, c_x, c_y, c_z are real, $\delta\lambda$ is real and ≥ 0. The normalization constraint specifies the Lagrange multiplier Z,

$$1 = \text{Tr}(\hat{\rho}) = \frac{e^{\lambda_+} + e^{\lambda_-}}{Z}, \tag{A50}$$

so $Z = e^{\lambda_+} + e^{\lambda_-} = 2e^{\lambda}\cosh(\delta\lambda)$. The expectation value constraint specifies the Lagrange multiplier α,

$$c = \text{Tr}(\hat{\rho}c_\mu\sigma^\mu) = \frac{\partial}{\partial\alpha}\log(Z) = c_1 + \tanh(\delta\lambda)\frac{\partial}{\partial\alpha}\delta\lambda, \tag{A51}$$

which becomes

$$c = c_1 + \frac{\tanh(\delta\lambda)}{2\delta\lambda}\left(2\alpha(c_x^2 + c_y^2 + c_z^2) + c_z\log(\frac{a}{b})\right),$$

or

$$c = c_1 + \tanh\left(\frac{1}{2}\sqrt{\left(2\alpha c_z + \log(\frac{a}{b})\right)^2 + 4\alpha^2(c_x^2 + c_y^2)}\right)\frac{2\alpha(c_x^2 + c_y^2 + c_z^2) + c_z\log(\frac{a}{b})}{\sqrt{\left(2\alpha c_z + \log(\frac{a}{b})\right)^2 + 4\alpha^2(c_x^2 + c_y^2)}}. \tag{A52}$$

This equation is monotonic in α and therefore it is uniquely specified by the value of c. Ultimately, this is a consequence from the concavity of the entropy. The specific proof of (A52)'s monotonicity is below:

Proof. For $\hat{\rho}$ to be Hermitian, $\hat{C}$ is Hermitian and $\delta\lambda = \frac{1}{2}\sqrt{f(\alpha)}$ is real—furthermore, because $\delta\lambda$ is real $f(\alpha) \geq 0$ and thus $\delta\lambda \geq 0$. Because $f(\alpha)$ is quadratic in α and positive, it may be written in vertex form,

$$f(\alpha) = a(\alpha - h)^2 + k, \tag{A53}$$

where $a > 0, k \geq 0$, and (h, k) are the (x, y) coordinates of the minimum of $f(\alpha)$. Notice that the form of (A52) is

$$F(\alpha) = \frac{\tanh(\frac{1}{2}\sqrt{f(\alpha)})}{\sqrt{f(\alpha)}} \times \frac{\partial f(\alpha)}{\partial\alpha}. \tag{A54}$$

Making the change of variables $\alpha' = \alpha - h$ centers the function such that $f(\alpha') = f(-\alpha')$ is symmetric about $\alpha' = 0$. We can then write

$$F(\alpha') = \frac{\tanh(\frac{1}{2}\sqrt{f(\alpha')})}{\sqrt{f(\alpha')}} \times 2a\alpha', \tag{A55}$$

where the derivative has been computed. Because $f(\alpha')$ is a positive, symmetric, and monotonically increasing on the (symmetric) half-plane (for α' greater than or less that zero), $S(\alpha') \equiv \frac{\tanh(\frac{1}{2}\sqrt{f(\alpha')})}{\sqrt{f(\alpha')}}$ is also positive and symmetric, but it is unclear whether $S(\alpha)$ is strictly monotonic in the half-plane or not. We may restate

$$F(\alpha') = S(\alpha') \times 2a\alpha'. \tag{A56}$$

We are now in a convenient position to preform the derivate test for monotonic functions:

$$\begin{aligned}\frac{\partial}{\partial\alpha'}F(\alpha') &= 2aS(\alpha') + 2a\alpha'\frac{\partial}{\partial\alpha'}S(\alpha') \\ &= 2aS(\alpha')\left(1 - \frac{a\alpha'^2}{a\alpha'^2 + k}\right) + a\frac{a\alpha'^2}{a\alpha'^2 + k}\left(1 - \tanh^2(\frac{1}{2}\sqrt{a\alpha'^2 + k})\right) \\ &\geq 2aS(\alpha')\left(1 - \frac{a(\alpha')^2}{a\alpha'^2 + k}\right) \geq 0\end{aligned} \tag{A57}$$

because $a, k, S(\alpha')$, and therefore $\frac{a\alpha'^2}{a\alpha'^2+k}$ are all > 0. The function of interest $F(\alpha')$ is therefore monotonic for all α', and therefore it is monotonic for all α, completing the proof that there exists a unique real Lagrange multiplier α in (A52).

Although (A52) is monotonic in α, it is seemingly a transcendental equation. This can be solved graphically for the given values c, c_1, c_x, c_y, c_z, i.e., given the Hermitian matrix and its expectation value are specified. Equation (A52) and the eigenvalues take a simpler form when $a = b = \frac{1}{2}$ because, in this instance, $\hat{\varphi} \propto \hat{1}$ and commutes universally so it may be factored out of the exponential in (A44). □

References

1. Shore, J.E.; Johnson, R.W. Axiomatic derivation of the Principle of Maximum Entropy and the Principle of Minimum Cross-Entropy. *IEEE Trans. Inf. Theory* **1980**, *26*, 26–37.
2. Shore, J.E.; Johnson, R.W. Properties of Cross-Entropy Minimization. *IEEE Trans. Inf. Theory* **1981**, *27*, 472–482.
3. Csiszár, I. Why least squares and maximum entropy: An axiomatic approach to inference for linear inverse problems. *Ann. Stat.* **1991**, *19*, 2032.
4. Skilling, J. The Axioms of Maximum Entropy. In *Maximum-Entropy and Bayesian Methods in Science and Engineering*; Erickson, G.J., Smith, C.R., Eds.; Kluwer Academic Publishers: Dordrecht, The Netherlands, 1988.
5. Skilling, J. Classic Maximum Entropy. In *Maximum-Entropy and Bayesian Methods in Science and Engineering*; Kluwer Academic Publishers: Dordrecht, The Netherlands, 1988.
6. Skilling, J. Quantified Maximum Entropy. In *Maximum-Entropy and Bayesian Methods in Science and Engineering*; Fougére, P.F., Ed.; Kluwer Academic Publishers: Dordrecht, The Netherlands, 1990.
7. Caticha, A. Entropic Inference and the Foundations of Physics (Monograph Commissioned by the 11th Brazilian Meeting on Bayesian Statistics—EBEB-2012). Available online: http://www.albany.edu/physics/ACaticha-EIFP-book.pdf (accessed on 30 November 2017).
8. Hiai, F.; Petz, D. The Proper Formula for Relative Entropy and its Asymptotics in Quantum Probability. *Commun. Math. Phys.* **1991**, *143*, 99–114.
9. Petz, D. Characterization of the Relative Entropy of States of Matrix Algebras. *Acta Math. Hung.* **1992**, *59*, 449–455.
10. Ohya, M.; Petz, D. *Quantum Entropy and Its Use*; Springer: New York, NY, USA, 1993; ISBN 0-387-54881-5.
11. Wilming, H.; Gallego, R.; Eisert, J. Axiomatic Characterization of the Quantum Relative Entropy and Free Energy. *Entropy* **2017**, *19*, 241.
12. Jaynes, E.T. Information Theory and Statistical Mechanics. *Phys. Rev.* **1957**, *106*, 620–630.
13. Jaynes, E.T. *Probability Theory: The Logic of Science*; Cambridge University Press: Cambridge, UK, 2003.
14. Jaynes, E.T. Information Theory and Statistical Mechanics II. *Phys. Rev.* **1957**, *108*, 171–190.
15. Balian, R.; Vénéroni, M. Incomplete descriptions, relevant information, and entropy production in collision processes. *Ann. Phys.* **1987**, *174*, 229–224.
16. Balian, R.; Balazs, N.L. Equiprobability, inference and entropy in quantum theory. *Ann. Phys.* **1987**, *179*, 97–144.
17. Balian, R. Justification of the Maximum Entropy Criterion in Quantum Mechanics. In *Maximum Entropy and Bayesian Methods*; Skilling, J., Ed.; Kluwer Academic Publishers: Dordrecht, The Netherlands, 1989; pp. 123–129.
18. Balian, R. On the principles of quantum mechanics. *Am. J. Phys.* **1989**, *57*, 1019–1027.
19. Balian, R. Gain of information in a quantum measurement. *Eur. J. Phys.* **1989**, *10*, 208–213
20. Balian, R. Incomplete descriptions and relevant entropies. *Am. J. Phys.* **1999**, *67*, 1078–1090.
21. Blankenbecler, R.; Partovi, H. Uncertainty, Entropy, and the Statistical Mechanics of Microscopic Systems. *Phys. Rev. Lett.* **1985**, *54*, 373–376.
22. Blankenbecler, R.; Partovi, H. Quantum Density Matrix and Entropic Uncertainty. In Proceedings of the Fifth Workshop on Maximum Entropy and Bayesian Methods in Applied Statistics, Laramie, WY, USA, 5–8 August 1985.
23. Von Neumann, J. *Mathematische Grundlagen der Quantenmechanik*; Springer: Berlin, Germany, 1932. English Translation: *Mathematical Foundations of Quantum Mechanics*; Princeton University Press: Princeton, NY, USA, 1983.

24. Ali, S.A.; Cafaro, C.; Giffin, A.; Lupo, C.; Mancini, S. On a Differential Geometric Viewpoint of Jaynes' Maxent Method and its Quantum Extension. *AIP Conf. Proc.* **2012**, *1443*, 120–128.
25. Caticha, A. Entropic Dynamics: Quantum Mechanics from Entropy and Information Geometry. Available online: https://arxiv.org/abs/1711.02538 (accessed on 30 November 2017).
26. Reginatto, M.; Hall, M.J.W. Quantum-classical interactions and measurement: A consistent description using statistical ensembles on configuration space. *J. Phys. Conf. Ser.* **2009**, *174*, 012038.
27. Reginatto, M.; Hall, M.J.W. Information geometry, dynamics and discrete quantum mechanics. *AIP Conf. Proc.* **2013**, *1553*, 246–253.
28. Caves, C.; Fuchs, C.; Schack, R. Quantum probabilities as Bayesian probabilities. *Phys. Rev. A* **2002**, *65*, 022305.
29. Vanslette, K. The Quantum Bayes Rule and Generalizations from the Quantum Maximum Entropy Method. Available online: https://arxiv.org/abs/1710.10949 (accessed on 30 November 2017).
30. Schack, R.; Brun, T.; Caves, C. Quantum Bayes rule. *Phys. Rev. A* **2001**, *64*, 014305.
31. Korotkov, A. Continuous quantum measurement of a double dot. *Phys. Rev. B* **1999**, *60*, 5737–5742.
32. Korotkov, A. Selective quantum evolution of a qubit state due to continuous measurement. *Phys. Rev. B* **2000**, *63*, 115403.
33. Jordan, A.; Korotkov, A. Qubit feedback and control with kicked quantum nondemolition measurements: A quantum Bayesian analysis. *Phys. Rev. B* **2006**, *74*, 085307.
34. Hellmann, F.; Kamiński, W.; Kostecki, P. Quantum collapse rules from the maximum relative entropy principle. *New J. Phys.* **2016**, *18*, 013022.
35. Warmuth, M. A Bayes Rule for Density Matrices. In *Advances in Neural Information Processing Systems 18, Proceedings of the Neural Information Processing Systems Conference, Montréal, QC, Canada, 7–12 December 2005*; Neural Information Processing Systems Foundation, Inc.: La Jolla, CA, USA, 2015.
36. Warmuth, M.; Kuzmin, D. A Bayesian Probability Calculus for Density Matrices. *Mach. Learn.* **2010**, *78*, 63–101.
37. Tsuda, K. Machine learning with quantum relative entropy. *J. Phys. Conf. Ser.* **2009**, *143*, 012021.
38. Giffin, A.; Caticha, A. Updating Probabilities. Presented at the 26th International Workshop on Bayesian Inference and Maximum Entropy Methods (MaxEnt 2006), Paris, France, 8–13 July 2006.
39. Wang, Z.; Busemeyer, J.; Atmanspacher, H.; Pothos, E. The Potential of Using Quantum Theory to Build Models of Cognition. *Top. Cogn. Sci.* **2013**, *5*, 672–688.
40. Giffin, A. Maximum Entropy: The Universal Method for Inference. Ph.D. Thesis, University at Albany (SUNY), Albany, NY, USA, 2008.
41. Caticha, A. Toward an Informational Pragmatic Realism. *Minds Mach.* **2014**, *24*, 37–70.
42. Umegaki, H. Conditional expectation in an operator algebra, IV (entropy and information). *Kōdai Math. Sem. Rep.* **1962**, *14*, 59–85.
43. Uhlmann, A. Relative entropy and the Wigner-Yanase-Dyson-Lieb concavity in an interpolation theory. *Commun. Math. Phys.* **1997**, *54*, 21–32.
44. Schumacher, B.; Westmoreland, M. Relative entropy in quantum information theory. In Proceedings of the AMS Special Session on Quantum Information and Computation, Washington, DC, USA, 19–21 January 2000.
45. Suzuki, M. On the Convergence of Exponential Operators—The Zassenhaus Formula, BCH Formula and Systematic Approximants. *Commun. Math. Phys.* **1977**, *57*, 193–200.
46. Horn, A. Eigenvalues of sums of Hermitian matrices. *Pac. J. Math.* **1962**, *12*, 225–241.
47. Bhatia, R. Linear Algebra to Quantum Cohomology: The Story of Alfred Horn's Inequalities. *Am. Math. Mon.* **2001**, *108*, 289–318.
48. Knutson, A.; Tao, T. Honeycombs and Sums of Hermitian Matrices. *Not. AMS* **2001**, *48*, 175–186.
49. Aczél, J. *Lectures on Functional Equations and Their Applications*; Academic Press Inc.: New York, NY, USA, 1966; Volume 19, pp. 31–44, 141–145, 213–217, 301–302, 347–349.
50. Darboux, G. Sur le théorème fondamental de la géométrie projective. *Math. Ann.* **1880**, *17*, 55–61.

© 2017 by the author. Licensee MDPI, Basel, Switzerland. This article is an open access article distributed under the terms and conditions of the Creative Commons Attribution (CC BY) license (http://creativecommons.org/licenses/by/4.0/).

Article

Finding a Hadamard Matrix by Simulated Quantum Annealing

Andriyan Bayu Suksmono

Telecommunication Engineering Scientific and Research Group (TESRG), School of Electrical Engineering and Informatics and The Research Center on Information and Communication Technology (PPTIK-ITB), Institut Teknologi Bandung, Jl. Ganesha No.10, Bandung 40132, Indonesia; suksmono@stei.itb.ac.id

Received: 2 January 2018; Accepted: 16 February 2018; Published: 22 February 2018

Abstract: Hard problems have recently become an important issue in computing. Various methods, including a heuristic approach that is inspired by physical phenomena, are being explored. In this paper, we propose the use of simulated quantum annealing (SQA) to find a Hadamard matrix, which is itself a hard problem. We reformulate the problem as an energy minimization of spin vectors connected by a complete graph. The computation is conducted based on a path-integral Monte-Carlo (PIMC) SQA of the spin vector system, with an applied transverse magnetic field whose strength is decreased over time. In the numerical experiments, the proposed method is employed to find low-order Hadamard matrices, including the ones that cannot be constructed trivially by the Sylvester method. The scaling property of the method and the measurement of residual energy after a sufficiently large number of iterations show that SQA outperforms simulated annealing (SA) in solving this hard problem.

Keywords: quantum annealing; adiabatic quantum computing; hard problems; Hadamard matrix; binary optimization

1. Introduction

1.1. Background

Finding a solution to a hard problem is a challenging task in computing. Such a problem is characterized by its complexity, as it grows beyond the polynomial against the size of the input. A class of particularly important ones are NP (non-deterministic polynomial) problems, in which verifying a solution can be conducted in polynomial time, whereas finding the solution is of exponential order. Examples of such problems are, among others, the TSP (traveling salesman problem), SAT (Boolean satisfiability), graph coloring, graph isomorphism, and subset sums.

An interesting approach to the hard problems is a method inspired by physical phenomena, such as classical annealing (CA) or quantum annealing (QA). Both CA and QA are physical processes that obtain an ordered (physical) system from an unordered one, which can be done either thermally (as is the case in CA) or quantum-mechanically (as is the case in QA). To simulate the physical processes on a (classical/non-quantum) computer, numerical methods, such as MC (Monte Carlo) for CA and PIMC (path-integral Monte Carlo) for QA, have been developed. The algorithm or computational method inspired by classical/thermal annealing is called simulated annealing (SA), whereas the one based on quantum annealing is called simulated quantum annealing (SQA). Both of these methods make use of the methods in numerical CA or numerical QA. They encode the problem into a Hamiltonian of a spin system [1] and then evolve the system from a high energy state down to the ground state. The annealing process enables the system to avoid local minima trapping and therefore is capable of achieving a global optimum, which represents the best solution of the problem. The main difference between SA and SQA is in the evolution of the systems; whereas SA uses classical/thermal annealing, SQA employs quantum mechanism.

In SA [2–4], one starts the system in total randomness with regard to a high temperature state. The temperature is then lowered and the system is evolved, which causes the energy to decrease so that the system becomes increasingly ordered. To avoid local-optima trapping, a particular updating rule, such as the Metropolis [2], is applied. The rule allows the system to (sometimes) move to a higher energy state. Upon completion of the algorithm, the system achieves the ground state, at which point a solution is found.

In [5], Kadowaki and Nishimori introduced quantum fluctuations to replace the thermal fluctuations in SA to accelerate the convergence. They applied the method on an Ising model, where a transverse field plays the role of temperature in classical SA, enabling the system to achieve the ground state with greater probability. Santoro et al. [6] compared classical and quantum Monte Carlo annealing protocols on a two-dimensional Ising model. They found that the quantum Monte Carlo annealing is superior to classical annealing. In [7], Boixo et al. show experimental results on a 108 qubit *D-Wave One*, which is a kind of hardware implementation of QA. A strong correlation between D-Wave and SQA, compared to the device with classical annealing, was found, which indicates that the D-Wave performs quantum annealing. This result raised the important issue of whether QA actually outperforms SA [8]. Rønnow et al. [9] showed how quantum speedup should be defined and measured. In an experiment with random spin glass instances on 503 qubits of *D-Wave Two*, they did not find any evidence of such speedup.

Regardless of these issues, different results have been achieved via SQA. Isakov [10] performed quantum Monte Carlo (QMC) simulations and found that the QMC tunneling rate displayed scaled according to system size. He also found quadratic speedup in QMC simulations when, instead of periodic conditions, open boundary conditions were employed. In [11], Mazzola et al. demonstrated that QMC simulations can recover the scaling of ground-state tunneling rates, which validates QA in terms of solving combinatorial problems.

Some classes of hard problems, including ones with exponential or combinatorial complexity, have been a subject of interest in SQA research. Martonak et al. [12] introduced an application of SQA to solve the TSP problem. They found that a PIMC algorithm was more efficient than SA in terms of finding an approximately minimal tour in a given graph. SQA has also been used to successfully address other hard problems related to graphs, such as graph coloring [13] and graph isomorphism [14].

In this paper, we propose SQA as a mean to find a Hadamard matrix (H-matrix). Previously, in [15], we successfully employed SA to perform a similar task, in which low-order H-matrices were found. Compared to existing H-matrix construction methods, an SA-based method is more general in terms of its capability of finding (or constructing probabilistically) an $m = 4k$ order H-matrix, without any restriction on the property of the order m, whereas the Sylvester method requires $m = 2^n$, where k and n are positive integers. This paper extends this classical SA method to its quantum version, where PIMC based on Suzuki–Trotter formulation [16,17] is employed to simulate the quantum process.

1.2. Finding A Hadamard Matrix

A Hadamard matrix, or H-matrix, is an orthogonal binary $\{\pm 1\}$ matrix of size $4k \times 4k$, where k is a positive integer. This matrix was discovered by J. J. Sylvester [18] in 1867 and then studied more extensively by J. Hadamard [19] during his investigation of the maximal determinant problem. The orthogonal property makes the H-matrix popular in applied areas, such as information coding and signal transform. In the 1960s and 1970s, Hadamard code was used in space exploration for information transmission [20,21]. In a recent technological case, CDMA (code-division multiple access), which is widely used in cellular mobile phone systems, employs Walsh–Hadamard signals to reduce interference between its users [22,23].

One of the most important issues in the theory of H-matrix is its existence. Any 2^l order H-matrix with l a positive integer can be constructed using Sylvester's method. Furthermore, if there is an m order H-matrix, $m = 4k$ can be shown for a positive integer k. On the other hand, no one yet knows if there is always a $4k$ order H-matrix [20,21]. The latter case is formulated as the Hadamard matrix conjecture. Up to this writing, the smallest unknown $4k$ order H-matrix is 668.

Various reconstruction methods have been proposed [24–29]. Nevertheless, these methods force the order m to follow a particular rule. In [15], a general $m = 4k$ order algorithm employing SA is proposed. The method works on a special H-matrix called a seminormalized Hadamard (SH) matrix, in which the first column is a $4k$ order unity vector $\vec{v}_0 = (1, \cdots, 1)^T$, and the rest are $4k$ order SH vectors $\vec{v}_i \in V$.

A brute-force method needs to verify all N_B of the $4k$ order binary matrix to find an H-matrix, where $N_B(4k) = 2^{16k^2}$ [15]. Let all matrices constructed where $\vec{v}_0$ is the first column and a combination of $\vec{v}_i \in V$ constitutes the remaining $(4k-1)$ columns be called quasi-SH (QSH) matrices. Since there are $N_V = C(4k, 2k)$ SH vectors, there are about $N_{QU}(4k) \approx \left(\frac{2^{4k}}{8k^{3/2}}\right)^{4k}$ unique QSH-matrices. Although the number has been greatly reduced compared to N_B, exhaustive checking still requires a great amount of computational resources. The SA method proposed in [15] is capable of finding a few low-order SH matrices in a more reasonable time.

Following the convention in our previous paper [15], the role of the spin, i.e., its ± 1 eigenvalues, is replaced by SH spin vectors $\vec{v}_i \in V$. To find a $4k$ order SH-matrix, one needs $(4k-1)$ fully connected SH spin vectors, which initially are set randomly. With a defined energy $E(\vec{Q})$, the SH spin vectors are randomly changed in accordance with conditions whereby a transition into another SH spin vector is allowed but a transition into a non-SH-spin-vector is forbidden.

2. Methods

2.1. Simulated Quantum Annealing

The Hamiltonian of an Ising system with spin configuration $\{\hat{\sigma}_k\}$, where $k \in K = \{1, 2, \cdots, i, j, \cdots\}$ is the set of the lattice's indices, can be expressed as

$$\hat{H} = -\sum_{i \neq j} J_{ij} \hat{\sigma}_i^z \hat{\sigma}_j^z - \sum_i h_i \hat{\sigma}_i^z \quad (1)$$

where J_{ij} is a coupling constant/strength between a spin at site i with a spin at site j, h_j is the magnetic strength at site j, and $\{\hat{\sigma}_i^z, \hat{\sigma}_i^x\}$ are Pauli's matrices at site i. In SQA, quantum fluctuation is elaborated by introducing a transverse magnetic field Γ. The Hamiltonian of the system takes the following form [5]:

$$\hat{H}_{QA} = -\sum_{i \neq j} J_{ij} \hat{\sigma}_i^z \hat{\sigma}_j^z - \sum_i h_i \hat{\sigma}_i^z - \Gamma \sum_i \hat{\sigma}_i^x. \quad (2)$$

In Equation (2), the transverse field is changed (reduced) over time, i.e., $\Gamma \equiv \Gamma(t)$. On the right hand side of the equation, the first two terms corresponds to potential energy $\hat{H}_{pot}$, while the third one is the Hamiltonian introduced by the transverse field, which is related to kinetic energy $\hat{H}_{kin}$; i.e, we can define

$$\hat{H}_{pot} \equiv -\sum_{i \neq j} J_{ij} \hat{\sigma}_i^z \hat{\sigma}_j^z - \sum_i h_i \hat{\sigma}_i^z \quad (3)$$

$$\hat{H}_{kin} \equiv -\Gamma \sum_i \hat{\sigma}_i^x. \quad (4)$$

In general, $\hat{H}_{pot}$ and $\hat{H}_{kin}$ do not commute, so $[\hat{H}_{pot}, \hat{H}_{kin}] \neq 0$. Denoting the Hamiltonian of the potential as a function of spin configurations $\hat{H}_{pot} \equiv \hat{H}(\{\hat{\sigma}_i^z\})$, we can also express Equation (2) in a more general form as follows:

$$\hat{H}_{QA} = -\hat{H}(\{\hat{\sigma}_i^z\}) - \Gamma \sum_i \hat{\sigma}_i^x. \tag{5}$$

To simulate a quantum system described by Equation (5) using the classical method, we have to formulate PIMC by introducing imaginary time. It can be then approximated by the Suzuki–Trotter transform by adding one dimension in the imaginary time direction, which, for $(P \times N)$ degrees of freedom, takes the following form [13,30]:

$$H_{ST} = \frac{1}{P} \sum_{p=1}^{P} H_{pot}(\{S_{i,p}\}) - J_{\Gamma} \left(\sum_{p=1}^{P-1} \sum_i^N S_{i,p} S_{i,p+1} + \sum_j^N S_{j,1} S_{j,P} \right) \tag{6}$$

where N is the number of spins in the lattice, P is the number of Trotter's replicas, $S_i = \pm 1$ are the eigenvalues of the spin matrices, and

$$J_{\Gamma} = -\frac{PT}{2} \ln \tanh\left(\frac{\Gamma}{PT}\right) > 0 \tag{7}$$

is the nearest-neighbor coupling of the transverse magnetic field [30].

2.2. SQA Formulation of the SH Spin Vector

Similar to the previous paper [15], we employ a seminormalized Hadamard spin vector, abbreviated here as an SH spin vector, instead of an ordinary spin. In a 4k order SH spin vector, for a given positive integer k, 2k spins are -1 and another 2k spins are $+1$. Therefore, an SH spin vector transition is allowed only if these balance numbers are conserved; otherwise, such a transition is forbidden. We also treat the SH spin vector as a single entity, even though it consists of 4k spins, and is denoted as $\vec{v}_i \in V$, where V is the set of all 4k-order SH vectors. We formulate the energy of a particular configuration of spin vectors $\{\vec{v}_i\}$ as follows:

$$E(\{\vec{v}_i\}) = \left| \sum_{i \neq j} \vec{v}_i \cdot \vec{v}_j + \sum_i \vec{1} \cdot \vec{v}_i \right| - 16k^2 \tag{8}$$

where $\vec{v}_i \cdot \vec{v}_j$ denotes the inner product of the vector $\vec{v}_i$ with $\vec{v}_j$.

Figure 1 shows an Ising system with four SH spin vectors with an additional Trotter's dimension. In the lower part of Figure 1a, each circle represents a binary spin, whereas the solid line represents the connection among the spins. Interacting spin i with binary variable S_i and spin j with binary variable S_j contributes the term $J_{ij} S_i S_j$ to the Hamiltonian. For a 4k order case, every 4k non-connected spins are grouped into one SH vector $\vec{v}_i$, which is illustrated as a dashed line. To simplify the diagram, each SH vector is represented by a filled circle; thus, we obtain the upper part of Figure 1a, which is called a slice or a replica. In the PIMC, the slice is replicated P-times, and these slices are arranged as layers in imaginary time. Each neighboring SH vector in a replica, i.e., $\vec{v}_{i,p}$ with $\vec{v}_{i,p-1}$ and $\vec{v}_{i,p}$ with $\vec{v}_{i,p+1}$, interacts. The extension (in imaginary time) is illustrated in Figure 1b. The Hamiltonian in Equation (6) becomes a Hamiltonian of an SH vector spin system H_{QV} that can be rewritten as follows:

$$H_{QV} = \frac{1}{P} \sum_{p=1}^{P} H_{pot}(\{\vec{v}_{i,p}\}) - J_{\Gamma} \left(\sum_{p=1}^{P-1} \sum_i \vec{v}_{i,p} \cdot \vec{v}_{i,p+1} + \sum_i \vec{v}_{i,1} \cdot \vec{v}_{i,P} \right) \tag{9}$$

where $J_\Gamma \equiv J_\Gamma(t)$ and $H_{pot}\left(\{\vec{v}_{i,p}\}\right)$ represent complete-graph connections among the SH spin vectors, similar to Equation (8), which is given by

$$H_{pot}\left(\{\vec{v}_{i,p}\}\right) = \left|\sum_{i \neq j} \vec{v}_{i,p} \cdot \vec{v}_{j,p} + \sum_i \vec{1} \cdot \vec{v}_{i,p}\right| - 16k^2. \qquad (10)$$

The evolution of H_{QV} in Equation (9) leads to the solution to the H-matrix search problem.

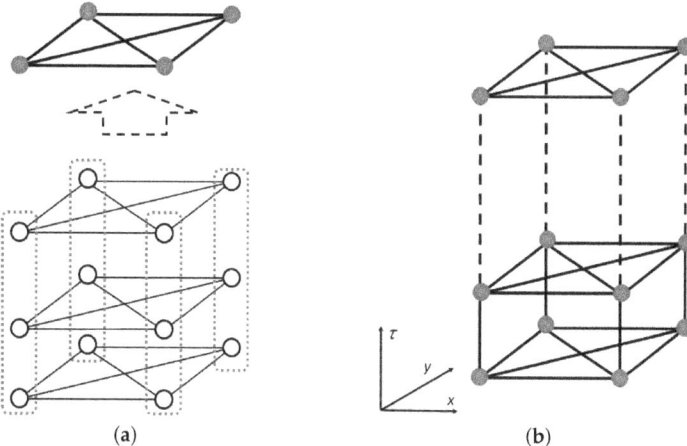

Figure 1. Connection diagrams of the spins and spin vectors. We consider a four-order SH vector in this example: (**a**) four SH spins are connected by a complete graph K_4, and each column is then grouped into a single SH spin vector; (**b**) an extension of fully connected SH spin vectors into a Trotter dimension (imaginary time) τ.

We will now formulate the SQA method for finding the H-matrix into an algorithm, which is displayed as pseudo-code in Algorithm 1. It takes the matrix order, the number of replicas, the initial temperature, the initial value of Γ, and the amount of iterations and sub-iterations as inputs. This algorithm yields either an SH-matrix or a QSH-matrix that has more orthogonal column vectors than the initial one. The algorithm starts with a random initialization of replicas with QSH-matrices, which are $(4k-1)$ sets of SH vectors, and then calculates its initial energy. Following the schedule of a linear transverse field, a trial transition is performed for each replica. The acceptance and rejection of the transition is based on the Metropolis criterion. The iteration will be stopped when either the number of maximum iterations is reached or an SH-matrix is found.

Algorithm 1 Finding an H-Matrix via Simulated Quantum Annealing

1: **Input:** Order of SH-matrix $4k$, number of replicas P, T_0, Γ_0, $MaxIter$, $SubIter$.
2: **Output:** A $4k$-order SH-matrix $\vec{H}_F$ or a partially orthogonal matrix $\vec{Q}$.
3: Initialize $T = T_0, \Gamma = \Gamma_0$
4: Initialize all-replicas R with randomly generated QSH-matrix: $R \leftarrow \{\vec{Q}_1, ..., \vec{Q}_P\}$
5: $idx \leftarrow 0$
6: $\vec{H}_F \leftarrow \vec{0}$
7: $FLAG \leftarrow 0$
8: **while** ($idx < MaxIter$) or ($FLAG == 0$) **do**
9:     Calculate $J_\Gamma(idx; \Gamma, P, T)$
10:     Calculate current all-replicas energy: $E_{rep} = H_{QV}(R, J_\Gamma)$
11:     $r \leftarrow 0$
12:     **while** $r < P$ **do**
13:         Select a replica at position r: $\vec{Q}_r$
14:         Calculate potential energy of the replica: $E_{pot} = H_{pot}(\vec{Q}_r)$
15:         **if** $E_{pot} > 0$ **then**
16:             $m \leftarrow 0$
17:             **while** ($m < SubIter$) and ($FLAG == 0$) **do**
18:                 Flip SH spin vector randomly: $\vec{Q}_r \to \vec{Q}'_r$
19:                 Calculate energy of the updated replica: $E_{pot1} = H_{pot}(\vec{Q}'_r)$
20:                 **if** $E_{pot1} == 0$ **then**
21:                     $E_{pot} \leftarrow E_{pot1}$
22:                     $\vec{H}_F \leftarrow \vec{Q}'_r$
23:                     $FLAG \leftarrow 1$
24:                     $r \leftarrow P$
25:                 **else**
26:                     Update all-replicas: $R \to R'$
27:                     Calculate energy of updated all-replicas $E_{rep1} \leftarrow H_{QV}(R', J_\Gamma)$
28:                     $\Delta E_{rep} \leftarrow E_{rep1} - E_{rep}$
29:                     $\Delta E_{pot} \leftarrow E_{pot1} - E_{pot}$
30:                     Perform a transition if allowed (Metropolis update rule):
31:                     **if** $(\Delta E_{pot} < 0)$ or $(\Delta E_{rep} < 0)$ or $(e^{-\frac{\Delta E_{rep}}{T}} > rand)$ **then**
32:                         Accept the transition: $R \leftarrow R'$, $E_{rep} \leftarrow E_{rep1}$
33:                     **end if**
34:                 **end if**
35:                 $m \leftarrow m + 1$
36:             **end while**
37:         **else**
38:             $\vec{H}_F \leftarrow \vec{Q}_r$
39:             $FLAG \leftarrow 1$
40:             $r \leftarrow P$
41:         **end if**
42:         $r \leftarrow r + 1$
43:     **end while**
44: **end while**

3. Numerical Experiments and Analysis

3.1. Finding a 12-Order SH-Matrix Using SQA

We have performed numerical experiments to find low-order H-matrices. Here we present results for the H-matrix of order 12 for detailed analysis, since it is the lowest-order H-matrix that cannot be constructed by the Sylvester method. Initially, all of the slices (replica) were filled with randomly generated $\vec{v}_i \in V$. Note that there are two nested iterations in Algorithm 1. The first one is an iteration of all replicas with the maximum number set to $k \cdot M \times M$, where $M = 12$ is the H-order. The second one is an iteration of flipping within a slice of a replica, whose number is $c \cdot M$, c can be any small number.

The energy evolution during the iteration is shown in Figure 2. The figure shows curves of replica energy E_{rep}, mean potential energy Ep_{mean}, and minimum potential energy Ep_{min}. The replica energy is defined similarly to Equation (9), i.e., $E_{rep} \equiv H_{QV}$, whereas the potential energy is given by Equation (10) $E_{pot} \equiv H_{pot}$. The mean and minimum values have been taken across the replicas. Based on the figure, both Ep_{mean} and E_{rep} fluctuate over time, but they tend to decrease. The minimum energy of a lattice in the replica Ep_{min} also tends to decrease. When $Ep_{min} = 0$, the H-matrix is found.

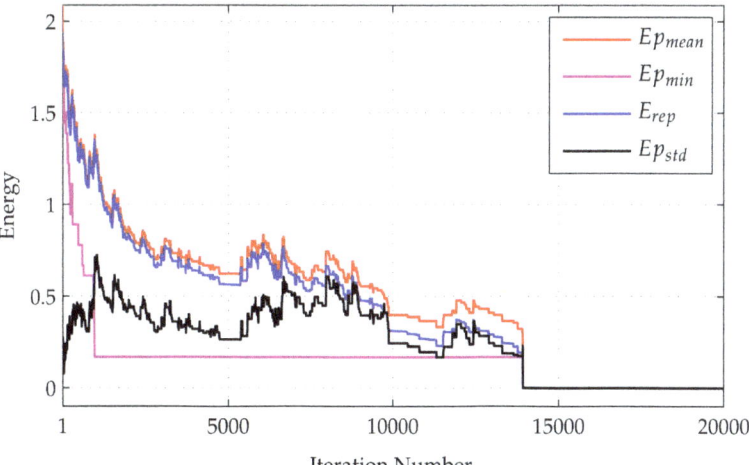

Figure 2. Energy evolution during the SQA algorithm runs to find an SH-matrix of order 12. Four curves are drawn in the graph, which are the mean potential energy Ep_{mean}, the minimum potential energy Ep_{min}, the replica energy E_{rep}, and the deviation standard of the potential energy Ep_{std}. When Ep_{min} equals zero, the iteration is stopped since an SH-matrix has been found. The Ep_{std} curve indicates high variation in the configuration of replicas at the initial stage, which is then reduced in later stages.

The degree of orthogonality of the matrix $\vec{Q}$ is displayed by the indicator matrix $\vec{D} \equiv \vec{Q}^T \vec{Q}$. Figure 3 shows the initial QSH-matrix and its related indicator matrix. We also show the initial and final indicators for the first and last slices of the replica in Figure 4. It is expected that all of the QSH-matrices become more orthogonal, indicated by a lower number of zeros in off-diagonal entries. The last figure showing the last slice of the replica condition after the iterations are completed clearly show this case. The found H-matrix is shown in the left part of Figure 5, with its corresponding indicator shown on the right, which is a diagonal matrix.

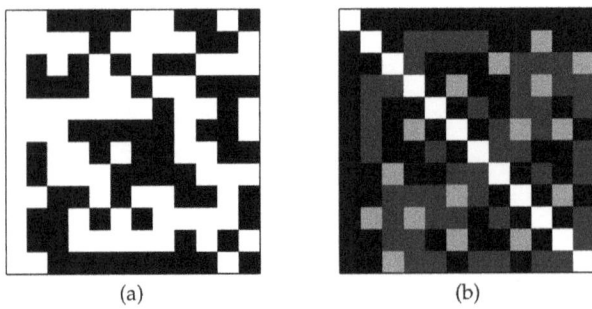

Figure 3. The initial state of the found H-matrix: (**a**) The QSH-matrix, white squares indicate +1, black squares indicate −1. (**b**) Orthogonality indicator, gray squares show the non-orthogonality condition of related pair of vectors.

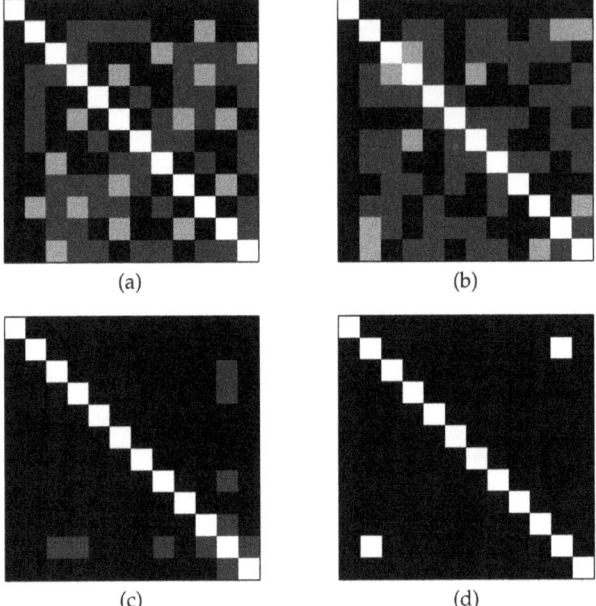

Figure 4. Indicator matrices of the replica content: (**a**) the first replica at the initial stage; (**b**) the last replica at the initial stage; (**c**) the first replica at the final stage, and (**d**) the last replica at the final stage. The matrices at the initial stages show most of the vectors as non-orthogonal, whereas those at the final stages show most of the vectors as orthogonal.

 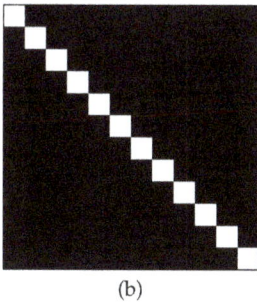

Figure 5. Final results: (**a**) the found H-matrix and (**b**) its orthogonality indicator. The diagonal form of the indicator matrix indicates that all of the column vectors are now orthogonal.

3.2. The Number of The Replicas and Convergence Issue

In theory, the number of Trotter's replicas P should be as large as possible. However, in practice, we should also consider the convergence issue when a running time restriction (iteration number) is given. As explained in [13,30], replicas provide diversity of solutions; a greater P selects the best solution with minimum energy. On the other hand, the replicas are not merely running Monte Carlo on several replicas; the interactions between replicas $J_\Gamma(t)$ also define their behavior, i.e., a large value of Γ at the initial stage implies a low value of J_Γ, which loosens the connections, and the interactions then become independent. A low Γ value at the end of an iteration implies a high J_Γ value, which tightens the replica connections such that they become similar. To measure these variations, we used a simple deviation standard of energy across the replicas. Figure 6 shows the curves of variation of the energy evolution for $P = 5, 10, 15$, and 20 in finding a 12-order H-matrix.

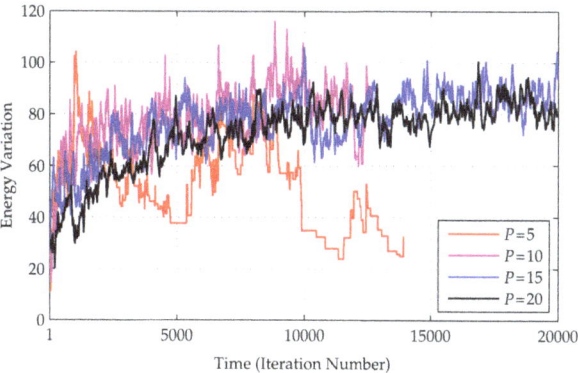

Figure 6. The effect of the replica number P in the algorithm: although ideally a large P is desired, it also needs to be adjusted to the problem. Variation in replica energy (in terms of deviation standards of the energy across the replicas) when searching for an H-matrix is shown. The numbers of replicas $P = 10, 15, 20$ yield large variations up to the end of the iteration, whereas $P = 5$ yields a better result with steady values at the end. In all of these cases, for the construction of a 12-order H-matrix, the total maximum iteration is set to 20,000, consisting of a global iteration count of 20,000 for each P.

Since initially the replicas were set randomly, they will have almost identical energy, so variation in the energy will be very low. In later iterations, the value will increase as a new configuration is explored, and this will be followed by a decrease, which indicates that the replicas have become homogeneous. This cycle of increasing–decreasing energy should be observed if P is chosen properly with respect to the dimension of the problem (H-order) and a sufficient number of iterations. When P is too small, the system will perform akin to classical SA, whereas a P that is too large will cause the system to fail. The figure shows that, for a given number of maximum iterations 20,000, the number of replicas $P = 5$ is the most suitable; anything higher is too high. This also shows that frequent updates on a limited number of replicas, compared to less frequent updates on a larger number of replicas, better achieve convergence.

3.3. Performance Comparison: SQA vs. SA

To compare performances, in the first experiment, we measured the residual error of both algorithms. Since the ground state is achieved when the matrix becomes orthogonal, in which case Equation (10) will equal zero, the residual error ϵ will be defined as the minimum H_{pot} over all of the replicas, i.e., $\epsilon = \min(H_{pot})$. We have chosen the order of the H-matrix to be sufficiently large so that we will still have a residual error at the end of the execution of the algorithm, i.e., so that the H-matrix is not found. We considered order $M = 28$ to be sufficient for this purpose, where we actually have $28^3 = 21,952$ spins. We also chose a Trotter slice of $P = 5$ and plotted the curve for iterations 50 up to 5,000,000.

Following [30], the annealing schedule was linear; i.e., the temperature T was reduced linearly in SA, and was the transverse magnetic strength Γ. Even though T is reduced linearly, the threshold probability P_{thresh} will change exponentially. By using the function

$$P_{thresh}(t) = 1 - \frac{1}{2} e^{\frac{-1}{T(t)}} \quad (11)$$

the threshold will start a bit higher than 0.5, which asymptotically approaches 1.0 at the end of iteration time t. Figure 7 shows the curve of $T(t)$, $P_{thresh}(t)$, $\Gamma(t)$, and $J_\Gamma(t)$.

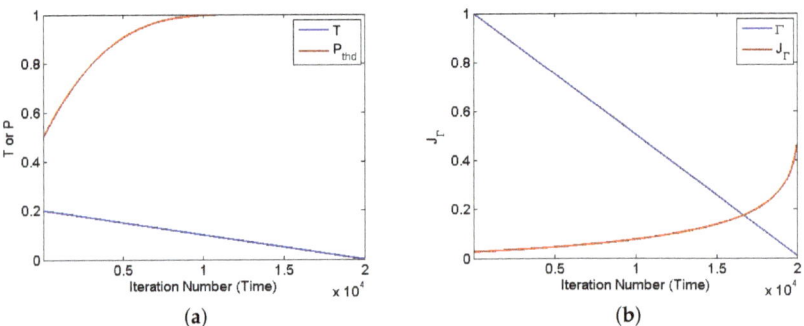

Figure 7. The annealing schedules in SA and SQA: (**a**) Linear temperature schedule and corresponding threshold schedule in SA. (**b**) Linear transverse-field $\Gamma(t)$ and corresponding $J_\Gamma(t)$ in SQA.

The experiments were repeated 10 times for each case. The averages of residual errors for each iteration numbers are plotted in Figure 8 for both SA and SQA.

The figure shows that, although initially the residual error of SQA is larger than SA, the slope is steeper. With a higher number of iterations, which in this case is around 100,000, SQA is superior. Considering that SQA shows the least amount of error among the replica slices, it seems that variation in the replica is an ideal solution. In SA, once a solution is selected, the change in spin configuration will be less significant by the time the system reaches a lower energy state. Therefore, in terms of finding an H-matrix, SQA is superior to SA.

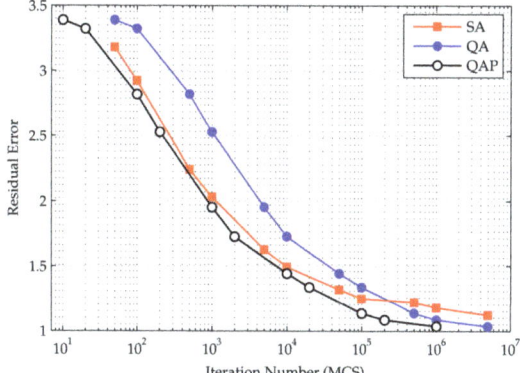

Figure 8. Residual energy left by the SA and SQA algorithms. The QAP curve shows when the horizontal axis accounts for the MCS (the Monte Carlo step); i.e., the number of iterations in the SQA curve is divided by the number of slices P. The figure shows that SQA outperform SA in finding an H-matrix. Even when the number of steps is counted without the MCS, SQA eventually outperforms SA at higher iterations, demonstrated by the steeper slope of the SQA performance curve, compared to SA.

In the second experiment, both SA and SQA were applied to matrices with an increasing size (order). Figure 9 shows a graph of computational gain, which is defined as the ratio of the number of SA iterations to the number of SQA iterations needed to achieve 50 percent of the residual energy of the initial mean energy of all replicas. The horizontal axis shows the order of the H-matrix, from 4 to 20, whereas the vertical axis shows the computational gain. The gain grows with the order of the H-matrix, which shows that speedup increases with problem size. Based on this curve, we observe that SQA outperforms SA for the Hadamard search problem.

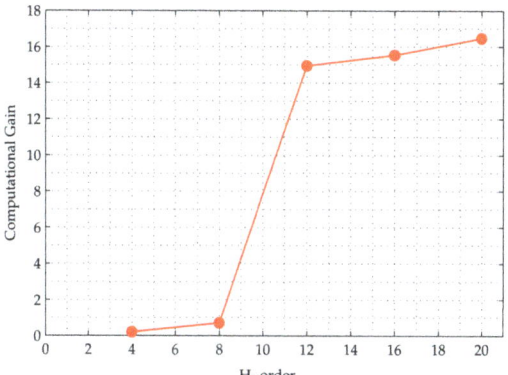

Figure 9. Curve of computational gain, which is the ratio of the number of SA iterations to the number of SQA iterations needed for the algorithm to achieve 50 percent of its initial residual error. The horizontal axis represent the problem size, which is the order of the H-matrix. The figure shows that the gain grows non-linearly with problem size, indicating that SQA outperforms SA.

4. Conclusions

We here propose a new method of finding an H-matrix based on SQA. We have formulated the method into an algorithm, which has been implemented, tested, and analyzed. Low-order H-matrices, including one of order 12 that cannot be constructed via the Sylvester method, were found. We have also discussed the advantages of the method over classical SA. Measurements of the residual error and the relative running time on an increasing order of H-matrices indicate that SQA is superior to SA in solving the Hadamard search problem.

Acknowledgments: This research was funded by ITB Grant of Research *P3MI* 2017. The author would like to thank ITB (Institut Teknologi Bandung) for their continuous support to his research. He also thanks Donny Danudirjdo and Andika Triwidada for their assistance in the manuscript layout and English editing.

Conflicts of Interest: The author declares no conflict of interest.

References

1. Lucas, A. Ising formulations of many NP problem. *Front. Phys.* **2014**, *2*, 5, doi:10.3389/fphy.2014.00005.
2. Metropolis, N.; Rosenbluth, A.W.; Rosenbluth, M.N.; Teller, A.H. Equation of state calculations by fast computing machines. *J. Chem. Phys.* **1953**, *21*, 1087, doi:10.1063/1.1699114.
3. Kirkpatrick, S.; Gelatt, C.D., Jr.; Vecchi, M.P. Optimization by simulated annealing. *Science* **1983**, *220*, 671–680, doi:10.1126/science.220.4598.671.
4. Cerny, V. Thermodynamical approach to the traveling salesman problem: An efficient simulation algorithm. *J. Optim. Theory Appl.* **1985**, *45*, 41–51, doi:10.1007/BF00940812.
5. Kadowaki, T.; Nishimori, H. Quantum annealing in the transverse Ising model. *Phys. Rev. E* **1988**, *58*, 5355, doi:10.1103/PhysRevE.58.5355.
6. Santoro, G.E.; Martonak, R.; Tosatti, E.; Car, E. Theory of quantum annealing of an Ising spin glass. *Science* **2002**, *295*, 2427–2730, doi:10.1126/science.1068774.
7. Boixo, S.; Rønnow, T.F.; Isakov, S.V.; Wang, Z.; Wecker, D.; Lidar, D.A.; Martinis, J.M.; Troyer, M. Evidence for quantum annealing with more than one hundred qubits. *Nat. Phys.* **2014**, *10*, 218–224, doi:10.1038/nphys2900.
8. Heim, B.; Rønnow, T.F.; Isakov, S.V.; Troyer, M. Quantum versus classical annealing of Ising spin glasses. *Science* **2015**, *348*, 215–217, doi:10.1126/science.aaa4170.
9. Rønnow, T.F.; Wang, Z.; Job, J.; Boixo, S.; Isakov, S.V.; Wecker, D.; Martinis, J.M.; Lidar, D.A.; Troyer, M. Defining and detecting quantum speedup. *Science* **2014**, *345*, 420–424, doi:10.1126/science.1252319.
10. Isakov, S.V.; Mazzola, G.; Smelyanskiy, V.N.; Jiang, Z.; Boixo, S.; Neven, H.; Troyer, M. Understanding Quantum Tunneling through Quantum Monte Carlo Simulation. *Phys. Rev. Lett.* **2016**, *117*, 180402, doi:10.1103/PhysRevLett.117.180402.
11. Mazzola, G.; Smelyanskiy, V.N.; Troyer, M. Quantum Monte Carlo Tunneling from quantum chemistry to quantum annealing. *Phys. Rev. B* **2017**, *96*, 134305, doi:10.1103/PhysRevB.96.134305.
12. Martonak, R.; Santoro, G.E.; Tosatti, E. Quantum annealing of the traveling-salesman problem. *Phys. Rev. E* **2004**, *70*, doi:10.1103/PhysRevE.70.057701.
13. Titiloye, O.; Crispin, A. Quantum annealing of the graph coloring problem. *Discret. Optim.* **2011**, *8*, 376–384, doi:10.1016/j.disopt.2010.12.001.
14. Zick, K.M.; Shehab, O.; French, M. Experimental quantum annealing: Case study involving the graph isomorphism problem. *Sci. Rep.* **2015**, *5*, 11168, doi:10.1038/srep11168.
15. Suksmono, A.B. Finding a Hadamard matrix by simulated annealing of spin-vectors. *J. Phys. Conf. Ser.* **2012**, *856*, 012012, doi:10.1088/1742-6596/856/1/012012.
16. Suzuki, M. Relationship between d-dimensional quantal spin systems and (d+1)-dimensional Ising systems: Equivalence, critical exponents and systematic approximants of the partition function and spin correlations. *Prog. Theor. Phys.* **1976**, *56*, 1454–1469, doi:10.1143/PTP.56.1454.
17. Trotter, H.F. On the product of semi-groups of operators. *Proc. Am. Math. Soc.* **1959**, *10*, 545–551, doi:10.1090/S0002-9939-1959-0108732-6.
18. Sylvester, J.J. Thoughts on inverse orthogonal matrices, simultaneous sign successions, and tessellated pavements in two or more colours, with applications to Newton's Rule, ornamental tile-work, and the theory of numbers. *Lond. Edinb. Dublin Philos. Mag. J. Sci.* **1867**, *34*, 461–475.
19. Hadamard, J. Resolution d'une question relative aux determinants. *Bull. Sci. Math.* **1893**, *17*, 240–246.
20. Hedayat, A.; Wallis, W.D. Hadamard Matrices and Their Applications. *Ann. Stat.* **1978**, *6*, 1184–1238.
21. Horadam, K.J. *Hadamard Matrices and Their Applications*; Princeton University Press: Princeton, NJ, USA, 2007; ISBN 978-1-40-084290-2.

22. Garg, V. *Wireless Communications and Networking*; Morgan-Kaufman: San Francisco, CA, USA, 2007; ISBN 978-0-12-373580-5.
23. Seberry, J.; Wysocki, B.J.; Wysocki, T.A. On some applications of Hadamard matrices. *Metrika* **2005**, *62*, 221–239, doi:10.1007/s00184-005-0415-y.
24. Paley, R.E.A.C. On Orthogonal Matrices. *J. Math. Phys.* **1933**, *12*, 311–320, doi:10.1002/sapm1933121311.
25. Dade, E.C.; Goldberg, K. The construction of Hadamard matrices. *Mich. Math. J.* **1959**, *6*, 247–250, doi:10.1307/mmj/1028998229.
26. Williamson, J. Hadamard's determinant theorem and the sum of four squares. *Duke Math. J.* **1944**, *11*, 65–81, doi:10.1215/S0012-7094-44-01108-7.
27. Bush, K.A. Unbalanced Hadamard matrices and finite projective planes of even order. *J. Comb. Theory Ser. A* **1971**, *11*, 38–44, doi:10.1016/0097-3165(71)90005-7.
28. Bush, K.A. *Atti del Convegno di Geometria Combinatoria e sue Applicazioni*; University Perugia: Perugia, Italy, 1971, Volume 131.
29. Wallis, J.S. On the existence of Hadamard matrices. *J. Comb. Theory A* **1976**, *21*, 188–195, doi:10.1016/0097-3165(76)90062-5.
30. Battaglia, D.A.; Santoro, G.E.; Tosatti, E. Optimization by quantum annealing: Lessons from hard satisfiability problems. *Phys. Rev. E* **2005**, *71*, 066707, doi:10.1103/PhysRevE.71.066707.

© 2018 by the author. Licensee MDPI, Basel, Switzerland. This article is an open access article distributed under the terms and conditions of the Creative Commons Attribution (CC BY) license (http://creativecommons.org/licenses/by/4.0/).

Article

Quantum Genetic Learning Control of Quantum Ensembles with Hamiltonian Uncertainties

Ameneh Arjmandzadeh and Majid Yarahmadi *

Department of Mathematics and Computer sciences, Lorestan University, Khorramabad, Lorestan 465, Iran; arjmandzadeh.am@fs.lu.ac.ir
* Correspondence: yarahmadi.m@lu.ac.ir; Tel.: +98-916-665-3079

Received: 9 April 2017; Accepted: 19 July 2017; Published: 1 August 2017

Abstract: In this paper, a new method for controlling a quantum ensemble that its members have uncertainties in Hamiltonian parameters is designed. Based on combining the sampling-based learning control (SLC) and a new quantum genetic algorithm (QGA) method, the control of an ensemble of a two-level quantum system with Hamiltonian uncertainties is achieved. To simultaneously transfer the ensemble members to a desired state, an SLC algorithm is designed. For reducing the transfer error significantly, an optimization problem is defined. Considering the advantages of QGA and the nature of the problem, the optimization problem by using the QGA method is solved. For this purpose, N samples through sampling of the uncertainty parameters via uniform distribution are generated and an augmented system is also created. By using QGA in the training step, the best control signal is obtained. To test the performance and validation of the method, the obtained control is implemented for some random selected samples. A couple of examples are simulated for investigating the proposed model. The results of the simulations indicate the effectiveness and the advantages of the proposed method.

Keywords: quantum control; quantum genetic algorithm; sampling-based learning control (SLC)

---

1. Introduction

In quantum phenomena, as in the classical systems, the existence of uncertainties and noises are unavoidable. For example, in superconducting qubits, the coupling energy of a Josephson junction may have fluctuations [1,2]. Noises and fluctuations may exist in magnetic fields and electric fields in cavity quantum electrodynamics (QED) [3,4]. The spins of an ensemble in nuclear magnetic resonance (NMR) experiments may not be exactly known with respect to the strength of the applied radio frequency field [5].

The classification of inhomogeneous quantum ensembles is a significant issue which has many applications in the discrimination of atoms (or molecules), the separation of isotopic molecules, and quantum information extraction. Thus, treating the quantum systems with uncertainties is an important and applicable subject which needs to be considered.

A quantum ensemble consists of a large number of single quantum systems. In the practical world, some of the quantum systems exist in the form of quantum ensembles. Each single quantum system in a quantum ensemble is referred to as a member of the ensemble [6]. Quantum ensembles have wide applications in emerging quantum technology, including long-distance quantum communication [7], quantum computation [8], and magnetic resonance imaging [9].

Control of inhomogeneous quantum ensembles is an important issue in practical applications. Control of inhomogeneous quantum systems for discrimination between two or more similar systems, for instance, is an attractive field of study [10]. In practical applications, the members of quantum ensembles could have variations in some parameters of dynamic systems. These situations are referred to as inhomogeneous quantum ensembles [6].

There are many approaches which can be used for solving quantum control problems with uncertainties. For instance, an optimal control for NMR pulse sequences is designed by applying gradient algorithms [11]. Additionally, a sequential convex programming method is proposed for designing robust quantum manipulations [12]. Dong and his collogues have designed a development of the variable structure control approach with sliding modes to improve the robustness of quantum systems in which a sliding mode control method is presented for two-level quantum systems to treat bounded uncertainties in the system Hamiltonian [13]. In addition to these works, a Lyapunov control method is presented to attain a universal quantum control [14]. For the first time a sampling-based learning control (SLC) of inhomogeneous quantum ensembles is presented for overcoming the compensation for parameter dispersion [6]. As an important application, the sampling-based learning controller is used for designing of a superconducting quantum control of systems [15]. Construction of universal quantum gates by using a sampling-based learning control are presented in order to find robust optimal control fields in the presence of different fluctuations and uncertainties [16]. Furthermore, an extended sampling-based learning control for designing a robust quantum unitary transformation in quantum information processing is presented and implemented [17]. In other applications, to prevent a control field failing in laser-assisted collisions, a sampling-based robust control is used [18].

In [19], a systematic sampling-based learning control method with gradient-based learning algorithms for steering the components of inhomogeneous quantum ensembles with uncertainties to the same ideal state is investigated by Dong and coworkers. There are some challenges in gradient algorithms. For instance, they may fall into a local optimum depending on the initial choices of problem variables or, in complex situations, function derivatives may not be easily found.

Genetic-type algorithms (GAs) have being used in optimization problem-solving. For this purpose, by applying cross-over and mutation operators on current solutions, new solutions are generated and, statistically, they are moving toward optimal solutions in the search space. The set of solutions, however, converges to an optimum solution according to the principle of the Darwinian theory of evolution.

The quantum genetic algorithm (QGA) was identified by Narayanan and Moore [20]. The QGA, with even a smaller population, presents a great ability of global optimization and good robustness. Therefore, as compared with the common genetic algorithm, QGA has greater effectiveness [21,22]. QGAs are mostly constructed based on qubits (or quantum bits) and state superposition in quantum mechanics. In contrast to classical representations of chromosomes (a binary string, for instance), here they are represented by vectors of qubits (quantum registers).

In this paper, for controlling the quantum systems with uncertainties, a hybrid method based on the SLC method and QGA is used. Specially, artificial samples are generated by sampling the uncertainty parameters in the system model and an augmented system is constructed by using these samples in the training step. Then, to train a control law with the desired performance for the augmented system, QG (quantum genetic) learning and optimization algorithms are used. In the process of testing, a set of selected uncertainty samples is tested to evaluate the control performance. Additionally, an improvement of QGA is conducted to attain better results. In [22] an adding quantum mutation operation in the conventional quantum genetic algorithm is used as an improving device. Quantum mutation, by swapping the value of the probability amplitude of qubits (α, β), can completely reverse the individual's evolutionary direction. In this paper the mutation operation is implemented on measured qubits (bit strings), which is more effective than adding quantum mutation. Reduction of learning iterations, test error and training error, and also increasing the fidelity index are advantages of the proposed method.

This paper is organized as follows: Section 2 represents the quantum control model and formulates the control problem; A quantum genetic learning ensemble control algorithm is designed in Section 3; Simulation results and control performance are illustrated in Section 4; Conclusions are presented in Section 5.

2. Problem Formulation

In this paper, a finite-dimensional (N-level) closed quantum system with a state in an underlying Hilbert space is considered. The states can be written as a superposition of eigenstates as follows:

$$|\psi(t)\rangle = \sum_{i=1}^{N} c_i(t)|\phi_i\rangle \tag{1}$$

where complex numbers $c_i(t)$ satisfy $\sum_{i=1}^{N}|c_i(t)|^2 = 1$ and $\{|\phi_i\rangle\}_{i=1}^{N}$ are the eigenstates of the N-level quantum system [23]. Usually, the states of two-level quantum systems are considered as arrows from the origin to points on the Bloch sphere [24].

The dynamical equation can be described as the following Schrödinger equation:

$$\begin{array}{l} i\hbar \frac{d}{dt}|\psi(t)\rangle = H(t)|\psi(t)\rangle \\ |\psi(t=0)\rangle = |\psi_0\rangle \end{array} \tag{2}$$

where $\hbar$ is Plank constant (assume $\hbar = 1$ in this paper), $H(t)$ is the system Hamiltonian and $i = \sqrt{-1}$. The dynamics of the system are governed under the following Hamiltonian:

$$H(t) = H_0 + H_c(t) = H_0 + \sum_{m=1}^{M} u_m(t) H_m \tag{3}$$

where H_0 is the free Hamiltonian of the system and $H_c(t)$ is the time-dependent control Hamiltonian that represents the interaction of the system with the external control fields $u_m(t)$, $m = 1, 2, \ldots, M$ (scalar functions). Additionally, H_m for $m = 1, 2, \ldots, M$ are Hermitian operators.

In practical applications, there exist external disturbances affecting the control fields. Assume that the system Hamiltonian is disturbed as follows:

$$H_\Theta(t) = f_0(\theta_0) H_0 + \sum_{m=1}^{M} f_m(\theta_m) u_m(t) H_m \tag{4}$$

where functions $f_m(\theta_m)$, $(m = 0, 1, \ldots, M)$ characterize uncertainty functions and $\Theta = (\theta_0, \theta_1, \ldots, \theta_M)$. To compare and indicate the advantages of the proposed method, it is assumed that the situations and assumptions are similar to the system described in [19]. Therefore, let $f_m(\theta_m)$, for $m = 1, 2, \ldots, M$, be continuous functions and the parameters $\theta_m \in [1 - E_m, 1 + E_m]$ could be time-dependent. For simplicity, one can assume that the uncertainty bounds $E_0 =, \ldots, = E_m =, \ldots, = E_M = E$ are all equal in this paper. Additionally, let the nominal values of θ_m are 1 and the fluctuations of the uncertainty parameters θ_m be $2E$ (where $E \in [0, 1]$).

The objective is to design the controls $\{u_m(t), m = 1, 2, \ldots, M\}$ to steer the quantum system with uncertainties from an initial state $|\psi_0\rangle$ to a target state $|\psi_{target}\rangle$ with high fidelity. The fidelity between two pure quantum states $|\psi_1\rangle$ and $|\psi_2\rangle$ is defined as [25]:

$$F(|\psi_1\rangle, |\psi_2\rangle) = |\langle \psi_1|\psi_2\rangle|. \tag{5}$$

Suppose that a similar ensemble's members with different Hamiltonians are given. The main objective is to drive the members from an initial state to a desired state. To control the ensemble, one can select a set of samples instead of all ensemble members and create an augmented system to be controlled. Let $\{H_{\Theta_n}, n = 1, 2, \ldots, N\}$ be the Hamiltonian of the selected samples, where N is the number of the training samples. The augmented system is constructed as follows:

$$\frac{d}{dt}\begin{pmatrix} |\psi_1(t)\rangle \\ |\psi_2(t)\rangle \\ \vdots \\ |\psi_N(t)\rangle \end{pmatrix} = -i \begin{pmatrix} H_{\Theta_1}(t)|\psi_1(t)\rangle \\ H_{\Theta_2}(t)|\psi_2(t)\rangle \\ \vdots \\ H_{\Theta_N}(t)|\psi_N(t)\rangle \end{pmatrix}, \tag{6}$$

where $\Theta_n \in \{(\theta_{0n_0}, \theta_{1n_1}, \ldots, \theta_{Mn_M}), n_0 = 1, 2, \ldots, N_0, \ldots, n_M = 1, 2, \ldots, N_M\}$ and $N = \prod_{j=0}^{M} N_j$ is number of the training samples. The task is to find the best control u^* such that the performance function

$$J(u) = \frac{1}{N} \sum_{n=1}^{N} \left| \left\langle \psi_n(t) \middle| \psi_{n_{target}} \right\rangle \right|^2 \tag{7}$$

for each control strategy in $u = \{u_m(t), m = 1, 2, \ldots, M\}$, is maximized. Thus, the control problem can be formulated as a maximization problem as follows:

$$\max J(u) = \frac{1}{N} \sum_{n=1}^{N} \left| \left\langle \psi_n(T) \middle| \psi_{n_{target}} \right\rangle \right|^2$$

$$s.t. \quad \frac{d}{dt} \begin{pmatrix} |\psi_1(t)\rangle \\ |\psi_2(t)\rangle \\ \vdots \\ |\psi_N(t)\rangle \end{pmatrix} = -i \begin{pmatrix} H_{\Theta_1}(t)|\psi_1(t)\rangle \\ H_{\Theta_2}(t)|\psi_2(t)\rangle \\ \vdots \\ H_{\Theta_N}(t)|\psi_N(t)\rangle \end{pmatrix}, \tag{8}$$

$$\psi(t=0)\rangle = |\psi_0\rangle, H_{\Theta_n}(t) = f_{0,n}(\theta_{0,n})H_0 + \sum_{m=1}^{M} f_{m,n}(\theta_{m,n})u_m(t)H_m, \ n = 1, 2, \ldots, N$$

when $\theta_{m,n} \in [1-E, 1+E]$, $t \in [0, T]$ and $n = 1, 2, \ldots, N$.

Note that $J(u)$ depends on the control signal u, implicitly, subject to Schrödinger equation, be satisfied.

3. Quantum Genetic Learning Ensemble Control Algorithm

In this section a systematic methodology for control design of a quantum ensemble is presented during two training and testing steps. Solving Equation (8) by using QGA a quantum learning controller is designed.

3.1. Solving Process

If $u_m(t) = u_m$, $m = 1, 2, \ldots, M$ for $t \in [0, \Delta t]$ where u_m is a constant, then according to the Schrodinger equation and time-evolution equation for each sample, from Equation (6) we have:

$$|\psi_n(\Delta t)\rangle = e^{-i(H_0 f_0(\theta_{0n_0}) + \sum_{m=1}^{M} u_m f_m(\theta_{mn_m}) H_m) \Delta t} |\psi_n(0)\rangle, n = 1, 2, \ldots, N. \tag{9}$$

So, for $t \in [0, \Delta t]$ considering Equation (9), the objective function of Equation (8), changes to:

$$\text{Max } J(u) = \frac{1}{N} \sum_{n=1}^{N} \left| \left\langle \psi_n(0) e^{-i(H_0 f_0(\theta_{0n_0}) + \sum_{m=1}^{M} u_m f_m(\theta_{mn_m}) H_m) \Delta t} \middle| \psi_{n_{target}} \right\rangle \right|^2. \tag{10}$$

Hence, $[0, T]$ is divided into Q subintervals and suppose that $u_m(t), m = 1, 2, \ldots, M$ are constants in any subinterval with the same length $\Delta t = T/Q$. Let $|\psi_n^{j-1}(0)\rangle$ be the initial state of the control system in the j-th subinterval, then for j-th subinterval the following problem must be solved:

$$\text{Max } J^j(u) = \|\langle|e^{-i(H_0 f_0(\theta_{0n_0}) + \sum_{m=1}^{M} u_m f_m(\theta_{mn_m}) H_m) \Delta t} \psi_n^{j-1}(0) \middle| \psi_{n_{target}}^j \rangle \|^2, \tag{11}$$

where

$$\left| \psi_{n_{target}}^j \right\rangle = \frac{|\psi_n(0)\rangle + j \cdot \frac{|\psi_{n_{target}}\rangle - |\psi_n(0)\rangle}{Q}}{\| |\psi_n(0)\rangle + j \cdot \frac{|\psi_{n_{target}}\rangle - |\psi_n(0)\rangle}{Q} \|} \tag{12}$$

is the target state of *j*-th subinterval for *n*-th sample. In each subinterval, Equation (11), by QGA is solved and the best control u_m^{*j}, $m = 1, \ldots, M$ is obtained. Then, for $j = 1, \ldots, Q$,

$$\left|\psi_n^j\right> = e^{-i(H_0 f_0(\theta_{0n_0}) + \sum_{m=1}^{M} u_m^{*j} f_m(\theta_{mn_m}) H_m) \Delta t} \left|\psi_n^{j-1}(0)\right> \tag{13}$$

is the state transferred by optimal control u_m^{*j}, $m = 1, 2, \ldots, M$ in the *j*-th subinterval, which is considered as the initial state of the next subinterval, that is, $\left|\psi_n^j(0)\right> = \left|\psi_n^j\right>$ is the initial state of the (*j* + 1)-th subinterval, and the process continues.

3.2. Structure of Quantum Chromosomes

The smallest unit of information stored in a two-state quantum unit is called a quantum bit or qubit, which can be in a superposition of states. QGA is an algorithm based on the concepts of qubit and superposition of the states in quantum mechanics theory [22]. A chromosome is made as a string of *m* qubits that forms a quantum register. Additionally, the *j*-th individual chromosome of the *t*-th generation can be indicated as

$$u_j^t = \begin{bmatrix} \alpha_{11}^{jt} & \alpha_{12}^{jt} & \cdots & \alpha_{1k}^{jt} & \alpha_{21}^{jt} & \alpha_{22}^{jt} & \cdots & \alpha_{2k}^{jt} & \cdots & \cdots & \alpha_{m1}^{jt} & \alpha_{m2}^{jt} & \cdots & \alpha_{mk}^{jt} \\ \beta_{11}^{jt} & \beta_{12}^{jt} & \cdots & \beta_{1k}^{jt} & \beta_{21}^{jt} & \beta_{22}^{jt} & \cdots & \beta_{2k}^{jt} & \cdots & \cdots & \beta_{m1}^{jt} & \beta_{m2}^{jt} & \cdots & \beta_{mk}^{jt} \end{bmatrix} \tag{14}$$

where *m* indicates the number of genes in any chromosomes and *k* represents the number of qubits encoding each gene. In the initial generation (when $t = 0$), quantum encoding (α, β) of each individual in the population is initialized with ($\frac{1}{\sqrt{2}}, \frac{1}{\sqrt{2}}$), which denotes that the probability of collapsing the superposed state into each basic states is equal.

3.3. Quantum Rotating Gates

Unlike the conventional genetic algorithm that uses a crossover operation, the quantum genetic algorithm applies the probability amplitude of qubits to encode chromosomes and uses quantum rotating gates to update generations. The genetic utilization of the quantum genetic algorithm is mainly through acting on the superposition state or entanglement state by the quantum rotating gates to change the probability amplitude. Accordingly, the construction of quantum rotating gates is the key issue of the quantum genetic algorithm, and it directly affects the performance of the algorithm. Quantum rotating gates can be organized according to the practical problems and usually can be defined as [26]

$$R(\xi_i) = \begin{bmatrix} \cos(\xi_i) & -\sin(\xi_i) \\ \sin(\xi_i) & \cos(\xi_i) \end{bmatrix}. \tag{15}$$

Therefore, the updating process is defined as follows:

$$\begin{bmatrix} \alpha_i' \\ \beta_i' \end{bmatrix} = R(\xi_i) \begin{bmatrix} \alpha_i \\ \beta_i \end{bmatrix} = \begin{bmatrix} \cos(\xi_i) & -\sin(\xi_i) \\ \sin(\xi_i) & \cos(\xi_i) \end{bmatrix} \begin{bmatrix} \alpha_i \\ \beta_i \end{bmatrix} \tag{16}$$

where $(\alpha_i, \beta_i)^T$ and $(\alpha_i', \beta_i')^T$ are the probability amplitudes of the *i*-th qubit in a chromosome before and after the quantum rotating gates update, respectively. Additionally, θ_i is the rotating angle. In Table 1, the updating strategies, for the chromosomes, are presented. The value and the sign of θ_i are determined by the adjustment strategy. Here, x_i is the *i*-th bit of the current chromosome; Ref_i is the *i*-th bit of the current optimal binary solution, named the reference binary solution, that all quantum chromosomes should be steered toward its corresponding chromosome; $f(x)$ is the fitness function; $s(\alpha_i, \beta_i)$ is the rotate direction of the rotating angle and $\Delta\theta_i$ is the increment value of the *i*-th rotating angle. The value of $\Delta\theta_i$ is a constant and is usually around 0.01π. The overall process in QGA is similar to the GAs but with some differences in changing from one generation to the next one. In fact, a new generation $P(t)$ is achieved by operating quantum rotating gates on any individuals.

Table 1. Adjustment strategy of rotating angle.

x_i	Ref_i	$f(x) > f(Ref)$	$\Delta\theta_i$	$s(ff_i, fi_i)$			
				$ff_i fi_i > 0$	$ff_i fi_i < 0$	$ff_i = 0$	$fi_i = 0$
0	0	FALSE	0	0	0	0	0
0	0	TRUE	0	0	0	0	0
0	1	FALSE	$\Delta\theta_i$	+1	−1	0	±1
0	1	TRUE	$\Delta\theta_i$	−1	+1	±1	0
1	0	FALSE	$\Delta\theta_i$	−1	+1	±1	0
1	0	TRUE	$\Delta\theta_i$	+1	−1	0	±1
1	1	FALSE	0	0	0	0	0
1	1	TRUE	0	0	0	0	0

A genetic type-based iterative learning algorithm is shown in Algorithm 1. The algorithm is written according to Section 3.1.

Algorithm 1. Genetic Type Based Iterative Learning Algorithm

set $j = 1$ (counter of subintervals)
set $\left|\psi^{j-1}(0)\right\rangle = |\psi(0)\rangle$
$\left|\psi_{n_{target}}^{j}\right\rangle = \dfrac{|\psi_n(0)\rangle + j\frac{|\psi_{n_{target}}\rangle - |\psi_n(0)\rangle}{Q}}{\| |\psi_n(0)\rangle + j\frac{|\psi_{n_{target}}\rangle - |\psi_n(0)\rangle}{Q} \|}$ (target state)
Repeat (for each subinterval)
Choose a set of arbitrary controls u_m^{j0}, $m = 1, 2, \ldots, m$
Solve problem (11) by using QGA and find u_m^{j*}, $m = 1, 2, \ldots, m$
$\left|\psi_n^{j}\right\rangle = e^{-i(H_0 f_0(\theta_{0n_0}) + \sum_{k=1}^{M} u_k^{*j} f_k(\theta_{kn_k}) H_k)\Delta t} \left|\psi_n^{j-1}(0)\right\rangle$
$\left|\psi_n^{j}(0)\right\rangle := \left|\psi_n^{j}\right\rangle$
$j = j + 1$
Until $j = Q$
The optimal control $u_m^* = \{u_{m}^{j*}, j = 1, 2, \ldots, Q\}$, $m = 1, 2, \ldots, M$

Additionally, in Figure 1, a schematic diagram of the proposed method is given. In this diagram, first, a random population of quantum chromosomes $P(t)$ is generated. A binary population $P_b(t)$ by measuring the present population is obtained. After evaluating $P_b(t)$ and specifying the best solution Ref, the whole of the quantum chromosomes are rotated toward the corresponding chromosome of Ref, according to Table 1. This process generates a new population with better fitness. As indicated in Figure 1, the above processes are repeated until the stop criterion is satisfied for all $j = 1, 2, \ldots, Q$.

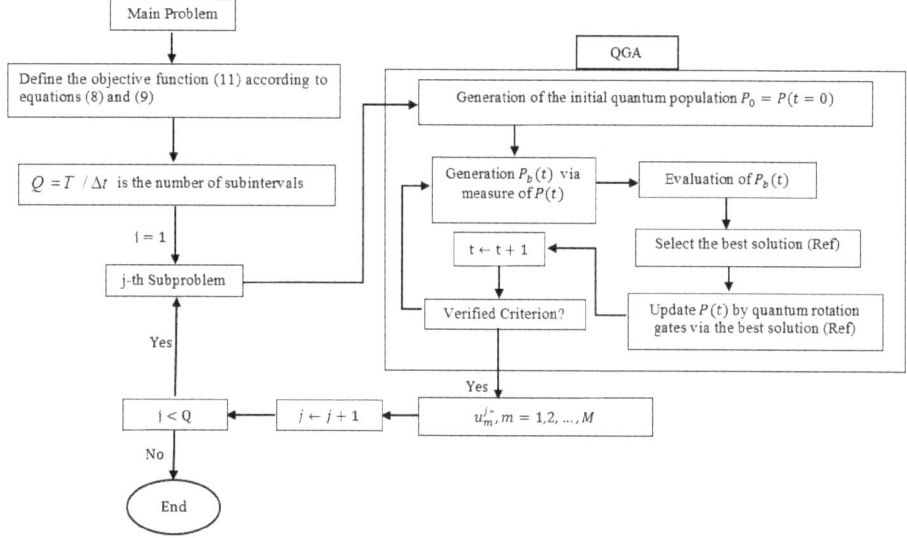

Figure 1. Diagram description of finding the signal control process.

4. Simulation Results

In this section two examples are simulated. Assume that all of the control signals are bounded in a known interval $[u_{min}, u_{max}]$.

Objective and protocols of simulation are explained as follows:

Let $[u_{min}, u_{max}] = [-4, 6]$, the initial state $|\psi(0)\rangle = (0, 0, 1)$ in real coordinates (i.e., $|\psi_0\rangle = [1\ 0]^t$), the time interval $[0, 5]$ ($T = 5$) is divided by $Q = 20$ and time slices $\Delta t = 0.25$. Additionally, the quantum genetic populations are the input control signals. The evolution generation number, the size of the population and the length of the each quantum chromosome are 200, 100, and 24, respectively. The mutation rate is 0.05 and the selection percentage of individuals is 50%. The stop condition for the iterative algorithm is considered as $|1 - J(u)| < \varepsilon$ ($\varepsilon = 0.001$). The objective of the problem-solving is transferring all of the initial states to the target state $|\psi(T)\rangle = (0, 0, -1)$ (i.e., $|\psi_T\rangle = [0\ 1]^t$), with maximum fidelity. The value of $\Delta\theta_i$ is set 0.01π.

Example 1: Consider the following two level quantum systems:

$$i\tfrac{d}{dt}|\psi(t)\rangle = (f_0(\theta_0)H_0 + \sum_{m=1}^{2} f_m(\theta_m)u_m(t)H_m)|\psi(t)\rangle \tag{17}$$
$$|\psi(t=0)\rangle = |\psi_0\rangle$$

where $H_0 = \tfrac{1}{2}\sigma_z = \tfrac{1}{2}\begin{pmatrix}1 & 0\\ 0 & -1\end{pmatrix}$ is the free Hamiltonian and $H_1 = \tfrac{1}{2}\sigma_x = \tfrac{1}{2}\begin{pmatrix}0 & 1\\ 1 & 0\end{pmatrix}$, $H_2 = \tfrac{1}{2}\sigma_y = \tfrac{1}{2}\begin{pmatrix}0 & -i\\ i & 0\end{pmatrix}$. Additionally, $\sigma_x = \begin{pmatrix}0 & 1\\ 1 & 0\end{pmatrix}$, $\sigma_y = \begin{pmatrix}0 & -i\\ i & 0\end{pmatrix}$, and $\sigma_z = \begin{pmatrix}1 & 0\\ 0 & -1\end{pmatrix}$ are Pauli matrices. Assume that the system's state is written as

$$|\psi(t)\rangle = c_1(t)|1\rangle + c_2(t)|2\rangle \tag{18}$$

where $B = \{|1\rangle, |2\rangle\}$ is the orthonormal basis of the corresponding Hilbert space. Let $C(t) = (c_1(t), c_2(t))$, where $c_i(t)$ are complex time depended coefficients. Therefore, Equation (17) is equivalent to

$$i\dot{C}(t) = (f_0(\theta_0)H_0 + \sum_{m=1}^{2} f_m(\theta_m)u_m(t)H_m)C(t). \tag{19}$$

In this example, let $f_m(\theta_m) = (1 - 2\theta_m^2)\exp(-\theta_m^2/2)$ be the Mexican hat wavelet functions for $m = 1, 2$ and $f_0(\theta_0) = 1$ on $[1 - E, 1 + E]$ for $E = 0.21$. After sampling the uncertainty parameters, every sample can be described as follows:

$$\begin{pmatrix} \dot{c}_1(t) \\ \dot{c}_2(t) \end{pmatrix} = -i \begin{pmatrix} 0.5f_0(\theta_0) & G(\theta_1, \theta_2) \\ G^*(\theta_1, \theta_2) & -f_0(\theta_0) \end{pmatrix} \begin{pmatrix} c_1(t) \\ c_2(t) \end{pmatrix} \tag{20}$$

where $G(\theta_1, \theta_2) = 0.5(f_1(\theta_1)u_1(t) - f_2(\theta_2)u_2(t)i)$ and $\theta_i \in [1 - E, 1 + E]$. Additionally, G^* is the complex conjugate of G. To construct an augmented system for the training step of the SLC method, consider N training samples that are selected through sampling the uncertainties, as follows:

$$\begin{pmatrix} \dot{c}_{1,n}(t) \\ \dot{c}_{2,n}(t) \end{pmatrix} = -i \begin{pmatrix} 0.5f_0(\theta_{0,n}) & G(\theta_{1,n}, \theta_{2,n}) \\ G^*(\theta_{1,n}, \theta_{2,n}) & -f_0(\theta_{0,n}) \end{pmatrix} \begin{pmatrix} c_{1,n}(t) \\ c_{2,n}(t) \end{pmatrix}, n = 1, 2, \ldots, N. \tag{21}$$

The results of simulation are illustrated in Figure 2. Figure 2a illustrates the control signals $u_m(t)$, $m = 1, 2$ obtained in the training step.

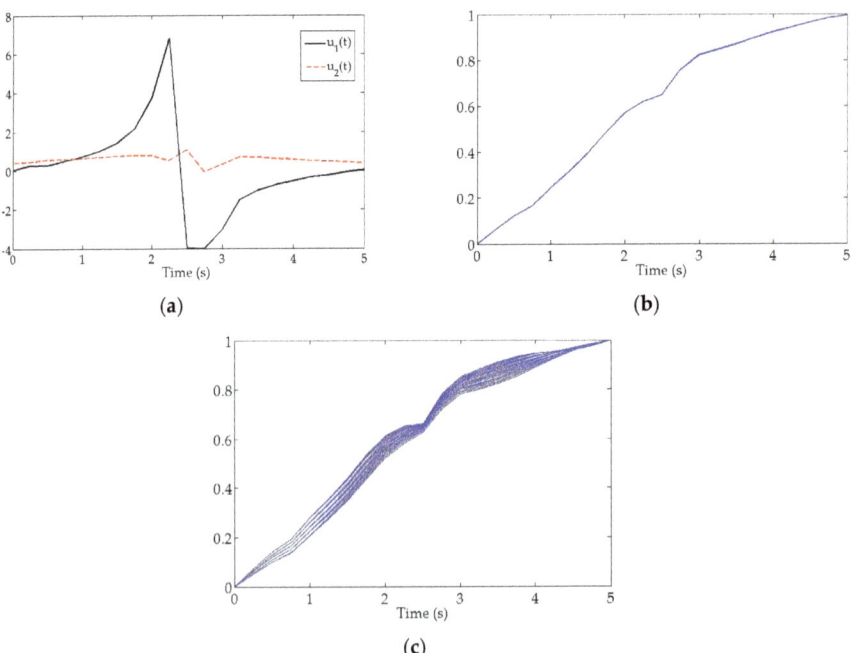

Figure 2. Control of an ensemble of two level quantum system with uncertainties: (**a**) control signals $u_m(t)$, $m = 1, 2$; (**b**) fidelity function (Fitness function performance); (**c**) simultaneously steering ensemble members to the desired state.

Figure 2b illustrates the mean of fidelity function of any states as a fitness function of the QGA. Finally, Figure 2c illustrates simultaneously steering ensemble members to the desired state. As simulation results indicate, 25 training samples are steered to the target state with a fidelity amplitude 0.9986 and error = 0.001. After running the control system, with founded control signals of the training step, for 200 test samples the fidelity amount is 0.9968 and the corresponding error is 0.003.

Example 2: *The second example is a three-level quantum system with uncertainties in Hamiltonian parameters that are found widely in natural and artificial atoms. Some atoms can be explained by a V-type three-level quantum system model. It is important to reach a robust preparation of this class of states for practical applications of quantum technology. The SLC, contributed with QGA, is used for a V-type quantum control system.* Assume the initial state is:

$$|\psi(t)\rangle = c_1(t)|1\rangle + c_2(t)|2\rangle + c_3(t)|3\rangle \tag{22}$$

with $B = \{|1\rangle, |2\rangle, |3\rangle\}$, the orthonormal basis of the corresponding Hilbert space. Let $C(t) = (c_1(t), c_2(t), c_3(t))$, where $c_i(t)$ are complex numbers. Then we have

$$i\dot{C}(t) = (f_0(\theta_0)H_0 + \sum_{m=1}^{4} f_m(\theta_m)u_m(t)H_m)C(t). \tag{23}$$

We take $H_0 = \text{diag}(1.5, 1, 0)$ as the free Hamiltonian and choose H_1, H_2, H_3, and H_4 as follows:

$$H_1 = \begin{pmatrix} 0 & 1 & 0 \\ 1 & 0 & 0 \\ 0 & 0 & 0 \end{pmatrix}, H_2 = \begin{pmatrix} 0 & -i & 0 \\ i & 0 & 0 \\ 0 & 0 & 0 \end{pmatrix}, H_3 = \begin{pmatrix} 0 & 0 & 1 \\ 0 & 0 & 0 \\ 1 & 0 & 0 \end{pmatrix}, H_4 = \begin{pmatrix} 0 & 0 & -i \\ 0 & 0 & 0 \\ i & 0 & 0 \end{pmatrix}. \tag{24}$$

After sampling the uncertainty parameters, every sample can be described as follows:

$$\begin{pmatrix} \dot{c}_1(t) \\ \dot{c}_2(t) \\ \dot{c}_3(t) \end{pmatrix} = -i \begin{pmatrix} 1.5 f_0(\theta_0) & G(\theta_1, \theta_2) & G(\theta_3, \theta_4) \\ G^*(\theta_1, \theta_2) & f_0(\theta_0) & 0 \\ G^*(\theta_3, \theta_4) & 0 & 0 \end{pmatrix} \begin{pmatrix} c_1(t) \\ c_2(t) \\ c_3(t) \end{pmatrix}, \tag{25}$$

where $G(\theta_1, \theta_2) = f_1(\theta_1)u_1(t) - f_2(\theta_2)u_2(t)i$, $G(\theta_3, \theta_4) = f_3(\theta_3)u_3(t) - f_4(\theta_4)u_4(t)i$, and $\theta_i \in [1 - E, 1 + E]$. $E \in [0,1]$ is a given constant and G^* is the complex conjugate of G. Comparing the results with previous works, uncertainty coefficients are chosen the same as what is given in [19], that is, $f_m(\theta_m) = \theta_m$ and $f_0(\theta_0) = \theta_0$ have uniform distributions over $[0.79, 1.21]$. To construct an augmented system for the training step of the SLC design, we choose N training samples (denoted as $n = 1, 2, \ldots, N$) through sampling the uncertainties as follows:

$$\begin{pmatrix} \dot{c}_{1,n}(t) \\ \dot{c}_{2,n}(t) \\ \dot{c}_{3,n}(t) \end{pmatrix} = -i \begin{pmatrix} 1.5 f_0(\theta_{0,n}) & G(\theta_{1,n}, \theta_{2,n}) & G(\theta_{3,n}, \theta_{4,n}) \\ G^*(\theta_{1,n}, \theta_{2,n}) & f_0(\theta_{0,n}) & 0 \\ G^*(\theta_{3,n}, \theta_{4,n}) & 0 & 0 \end{pmatrix} \begin{pmatrix} c_{1,n}(t) \\ c_{2,n}(t) \\ c_{3,n}(t) \end{pmatrix} \tag{26}$$

where $G(\theta_{1,n}, \theta_{2,n}) = f_1(\theta_{1,n})u_1(t) - f_2(\theta_{2,n})u_2(t)i$ and $G(\theta_3, \theta_4) = f_3(\theta_{3,n})u_3(t) - f_4(\theta_{4,n})u_4(t)i$. Now, the objective is to find a robust control strategy $u(t) = \{u_m(t), m = 1, 2, 3, 4\}$ to drive the quantum system from $|\psi_0\rangle = |1\rangle$ (i.e., $C_0 = (1, 0, 0)$) to $|\psi_{target}\rangle = \left(1/\sqrt{2}\right)(|2\rangle + |3\rangle)$ (i.e., $C_{target} = \left(0, 1/\sqrt{2}, 1/\sqrt{2}\right)$). The general conditions here are similar to ones mentioned in previous example but $Q = 10$. Apart from the initial values, the results are always converged and it is more precise than the gradient method as shown in Table 2.

Table 2. Comparison between the results of QGA and gradient algorithm.

Method	Training Error	Test Error
Gradient based learning control	0.004	0.08
Quantum Genetic algorithm	0.002	0.005

The training error is computed as $|1 - J(u^*(T))|$ in which $J(u^*(T))$ is the fidelity function for training samples. For calculating the test error, optimal control u^* is implemented to the test samples, which are selected randomly. Additionally, the amount of $|1 - J(u^*(T))|$ is computed for test samples, as a test error index. The method presented in this paper always converges and does not depend on initial choices of $u = \{u_m(t), m = 1, 2, \ldots, M\}$. Figure 3a–c demonstrate the control signals $u_m(t)$, $m = 1, 2, 3, 4$ and fidelity function for steering training samples simultaneously to the target state.

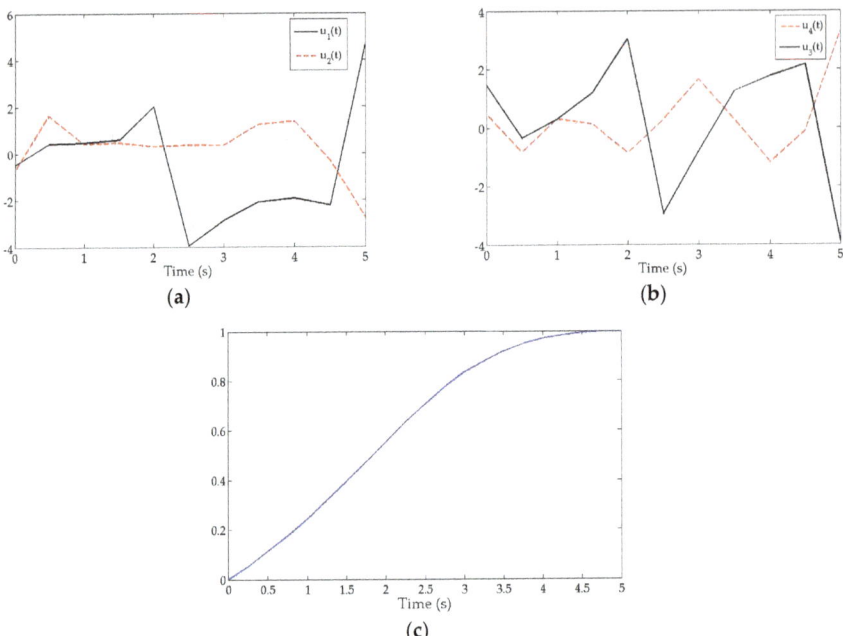

Figure 3. Control of an ensemble of a two-level quantum system with uncertainties: (a) control signals $u_m(t)$, $m = 1, 2$; (b) control signals $u_m(t)$, $m = 3, 4$; (c) fidelity function.

The training samples are steered to the target state with a fidelity of 0.9982. The control values found in training step are applied on 200 testing samples. The fidelity amount of 0.9954 is achieved with a test error equal to 0.005. Figure 3a,b show the control signals $u_1(t)$, $u_2(t)$, $u_3(t)$, and $u_4(t)$ through the time interval $[0, 5]$ and Figure 3c illustrates the mean of the fidelity function of all of the states as a fitness function of QGA.

5. Conclusions

In this paper a new quantum genetic sampling-based learning controller is designed. For this purpose an unconstrained nonlinear optimization problem is designed and is solved by a new quantum genetic algorithm. All of the members of an inhomogeneous quantum ensemble transfers to a known target state, simultaneously. In this method controller performance is independent of the initial input values and this is an important advantage of the proposed method as compared with gradient-based learning methods. Additionally, transfer process errors and learning iteration numbers are reduced, significantly. A couple of examples for two- and three-level quantum systems are simulated by using the proposed method. The simulation results indicate the advantages and efficiency of the presented method.

Acknowledgments: The authors are very grateful to the editor and anonymous reviewers for their suggestions in improving the quality of the paper.

Author Contributions: They conceived of the presented idea and developed the theory of the presented paper. Both authors discussed the results and contributed to the final manuscript. Both authors have read and approved the final manuscript.

Conflicts of Interest: The authors declare no conflict of interest.

References

1. Shnirman, A.; Schön, G.; Hermon, Z. Quantum manipulations of small Josephson Junctions. *Phys. Rev. Lett.* **1997**, *79*, 2371–2374. [CrossRef]
2. Makhlin, Y.; Schön, G.; Shnirman, A. Josephson junction quantum logic gates. *Comput. Phys. Commun.* **2000**, *127*, 156–164. [CrossRef]
3. Giovannetti, V.; Vitali, D.; Tombesi, P.; Ekert, A. Scalable quantum computation with cavity QED systems. *Phys. Rev. A* **2000**, *62*, 032306. [CrossRef]
4. Shu, J.; Zou, X.; Xiao, Y.; Guo, G. Quantum phase gate of photonic qubits in a cavity QED system. *Phys. Rev. A* **2007**, *74*, 044302. [CrossRef]
5. Li, J.S.; Khaneja, N. Control of inhomogeneous quantum ensembles. *Phys. Rev. A* **2006**, *73*, 030302. [CrossRef]
6. Chen, C.; Dong, D.; Long, R.; Petersen, I.R.; Rabitz, H.A. Sampling-based learning control of inhomogeneous quantum ensembles. *Phys. Rev. A* **2014**, *89*, 023402. [CrossRef]
7. Duan, L.M.; Lukin, M.D.; Cirac, J.I.; Zoller, P. Long-distance quantum communication with atomic ensembles and linear optics. *Nature* **2001**, *414*, 413–418. [CrossRef] [PubMed]
8. Cory, D.G.; Fahmy, A.F.; Havel, T.F. Ensemble quantum computing by NMR spectroscopy. *Proc. Natl. Acad. Sci. USA* **1997**, *94*, 1634–1639. [CrossRef] [PubMed]
9. Li, J.S.; Ruths, J.; Yu, T.Y.; Arthanari, H.; Wagner, G. Optimal pulse design in quantum control: A unified computational method. *Proc. Natl. Acad. Sci. USA* **2011**, *108*, 1879–1884. [CrossRef] [PubMed]
10. Mitra, A.; Rabitz, H. Mechanistic Analysis of Optimal Dynamic Discrimination of Similar Quantum Systems. *J. Phys. Chem. A* **2004**, *108*, 4778–4785. [CrossRef]
11. Khanejia, N.; Reiss, T.; Kehlet, C.; Schulte-Herbrüggen, T.; Glaser, S.J. Optimal control of coupled spin dynamics: Design of NMR pulse sequences by gradient ascent algorithm. *J. Magn. Reson.* **2005**, *172*, 296–305. [CrossRef] [PubMed]
12. Kosut, R.L.; Grace, M.D.; Brif, C. Robust control of quantum gates via sequential convex programming. *Phys. Rev. A* **2013**, *88*, 1–12. [CrossRef]
13. Dong, D.; Petersen, I.R. Sliding mode control of two-level quantum systems. *Automatica* **2012**, *48*, 725–735. [CrossRef]
14. Hou, S.C.; Wang, L.C.; Yi, X.X. Realization of quantum gates by Lyapunov control. *Phys. Lett. A* **2014**, *378*, 699–704. [CrossRef]
15. Dong, D.; Chen, C.; Qi, B.; Petersen, I.R.; Nori, F. Robust manipulation of superconducting qubits in the presence of fluctuations. *Sci. Rep.* **2015**, *5*, 7873. [CrossRef] [PubMed]
16. Dong, D.; Wu, C.; Chen, C.; Qi, B.; Petersen, I.R.; Nori, F. Learning robust pulses for generating universal quantum gates. *Sci. Rep.* **2015**, *6*, 36090. [CrossRef] [PubMed]
17. Wu, C.; Qi, B.; Chen, C. Robust learning control design for quantum unitary transformations. *IEEE Trans. Cybern.* **2016**, *99*, 1–13. [CrossRef] [PubMed]
18. Zhang, W.; Dong, D.; Petersen, I.R.; Rabitz, H.A. Sampling-based robust control in synchronizing collision with shaped laser pulses: An application. *RSC Adv.* **2016**, *6*, 92962–92969. [CrossRef]
19. Dong, D.; Mabrok, M.A.; Petersen, I.R.; Qi, B.; Chen, C.; Rabitz, H. Sampling-Based Learning Control for Quantum Systems with Uncertainties. *IEEE Trans. Control Syst. Technol.* **2015**, *23*, 2155–2166. [CrossRef]

20. Narayanan, A.; Moore, M. Quantum-inspired genetic algorithm. In Proceedings of the IEEE International Conference on Evolutionary Computation, Nagoya, Japan, 20–22 May 1996.
21. Laboudi, Z.; Chikhi, S. Comparison of Genetic Algorithm and Quantum Genetic Algorithm. *Int. Arab J. Inf. Technol.* **2012**, *9*, 243–249.
22. Wang, H.; Liu, J.; Zhi, J.; Fu, C. The Improvement of Quantum Genetic Algorithm and Its Application on Function Optimization. *Math. Probl. Eng.* **2013**, *2013*, 1–10. [CrossRef]
23. Wu, C.; Chen, C.; Qi, B.; Dong, D. Robust quantum operation for two-level systems using sampling-based learning control. In Proceedings of the IEEE International Conference on Systems, Man, and Cybernetics, Hongkong, China, 9–12 October 2015.
24. Wang, L.C.; Hou, S.C.; Yi, X.X.; Dong, D.; Petersen, I.R. Optimal Lyapunov quantum control of two-level systems: Convergence and extended techniques. *Phys. Lett. A* **2014**, *378*, 1074–1080. [CrossRef]
25. Nielsen, M.A.; Chuang, I.L. *Distance Measures for Quantum Information*; Cambridge University Press: Cambridge, UK, 2000.
26. Lahoz-Beltra, R. Quantum Genetic Algorithms for Computer Scientists. *Computers* **2016**, *5*, 24. [CrossRef]

© 2017 by the authors. Licensee MDPI, Basel, Switzerland. This article is an open access article distributed under the terms and conditions of the Creative Commons Attribution (CC BY) license (http://creativecommons.org/licenses/by/4.0/).

Article

Discrete Wigner Function Derivation of the Aaronson—Gottesman Tableau Algorithm

Lucas Kocia, Yifei Huang and Peter Love *

Department of Physics, Tufts University, Medford, MA 02155, USA; lucas.kocia@tufts.edu (L.K.); yifei.huang@tufts.edu (Y.H.)
* Correspondence: peter.love@tufts.edu; Tel.: +1-617-627-3029 (ext. 7-1065)

Received: 3 May 2017; Accepted: 4 July 2017; Published: 11 July 2017

Abstract: The Gottesman–Knill theorem established that stabilizer states and Clifford operations can be efficiently simulated classically. For qudits with odd dimension three and greater, stabilizer states and Clifford operations have been found to correspond to positive discrete Wigner functions and dynamics. We present a discrete Wigner function-based simulation algorithm for odd-d qudits that has the same time and space complexity as the Aaronson–Gottesman algorithm for qubits. We show that the efficiency of both algorithms is due to harmonic evolution in the symplectic structure of discrete phase space. The differences between the Wigner function algorithm for odd-d and the Aaronson–Gottesman algorithm for qubits are likely due only to the fact that the Weyl–Heisenberg group is not in $SU(d)$ for $d = 2$ and that qubits exhibit state-independent contextuality. This may provide a guide for extending the discrete Wigner function approach to qubits.

Keywords: quantum information; quantum computation; semiclassical physics

1. Introduction

The cost of brute-force classical simulation of the time evolution of n-qubit states grows exponentially with n. An important exception to this involves the set of Clifford operators acting on stabilizer states. This set of states plays an important role in quantum error correction [1] and is closed under action by Clifford gates. Efficient simulation of such systems was demonstrated with the tableau algorithm of Aaronson and Gottesman [1,2] for qubits ($d = 2$). Finding the underlying reason for why such an efficient algorithm is possible for Clifford circuit simulation has since been the subject of much study [3–5].

Recent progress has been the result of work by Wootters [6], Gross [7], Veitch et al. [8,9], Mari et al. [4], and Howard et al. [5], who have formulated a new perspective based on the discrete phase spaces of states and operators in finite Hilbert spaces using discrete Wigner functions. In odd-dimensional systems, they have shown that stabilizer states have positive-definite discrete Wigner functions and that Clifford operators are positive-definite maps. This implies that Clifford circuits are non-contextual and are efficiently simulatable on classical computers. In odd-dimensional systems, stabilizer states have been shown to be the discrete analogue to Gaussian states in continuous systems [7] and Clifford group gates have been shown to have underlying harmonic Hamiltonians that preserve the discrete Weyl phase space points [10]. This means Clifford circuits are expressible by path integrals truncated at order $\hbar^0$ and are thus manifestly classical [10,11].

This poses the question: what is the relationship between past efficient algorithms for Clifford circuits and the propagation of discrete Wigner functions of stabilizer states under Clifford operators? In the present paper, we show that the original Aaronson–Gottesman tableau algorithm for qubit stabilizer states is actually equivalent to such a discrete Wigner function propagation and that the tableau matrix coincides with the discrete Wigner function of a stabilizer state. We accomplish this by first developing a Wigner function-based algorithm that classically simulates stabilizer state evolution under Clifford gates and measurements in the $\hat{Z}$ Pauli basis for odd d. We then show its equivalence to the well-known Aaronson–Gottesman tableau algorithm [2] for qubits ($d = 2$). Both algorithms require $\mathcal{O}(n^2)$ dits to represent n stabilizer states, $\mathcal{O}(n)$ operations per Clifford operator, and both deterministic and random measurements require $\mathcal{O}(n^2)$ operations.

The Aaronson–Gottesman tableau algorithm makes use of the Heisenberg representation. This means that time evolution is accomplished by updating an associated tableau or matrix representation of the Clifford operators instead of the stabilizer states themselves. The algorithm we present is framed in the Schrödinger picture and involves evolving the Wigner function of stabilizer states. By demonstrating that the two algorithms are equivalent, we show that the formulation of Clifford simulation in the Heisenberg picture is a choice and not a necessity for its efficient simulation. Furthermore, by instead working in the Schrödinger picture we are able to more easily reveal the purely classical basis of both algorithms and the physically intuitive phase space structures and symplectic properties on which they rely.

2. Discrete Wigner Function for Odd d Qudits

Before we discuss the discrete Wigner function, we introduce a basic framework that defines how a phase space behaves for odd d-dimensional Hilbert spaces. To begin, we associate the computational basis with the position basis, such that the Pauli $\hat{Z}_j$ operator on the jth qudit for n qudits acts as a "boost" operator:

$$\hat{Z}_j |k_1, \ldots, k_j, \ldots, k_n\rangle = e^{\frac{2\pi i}{d} k_j} |k_1, \ldots, k_j, \ldots, k_n\rangle, \tag{1}$$

where $k_j \in \mathbb{Z}/d\mathbb{Z}$ for $1 \leq j \leq n$.

The discrete Fourier transform operator is defined by:

$$\hat{F}_j = \frac{1}{\sqrt{d}} \sum_{\substack{k_j, l_j \in \\ \mathbb{Z}/d\mathbb{Z}}} e^{-\frac{2\pi i}{d} k_j l_j} |k_1, \ldots, k_j, \ldots, k_n\rangle \langle l_1, \ldots, l_j, \ldots, l_n|.$$

This is the d-dimensional equivalent of the Hadamard gate and allows us to define the Pauli $\hat{X}_j$ operator as follows:

$$\hat{X}_j \equiv \hat{F}_j \hat{Z}_j \hat{F}_j^\dagger. \tag{2}$$

While $\hat{Z}_j$ is a boost, $\hat{X}_j$ is a shift operator because

$$\hat{X}_j^{\delta q} |k_1, \ldots, k_j, \ldots, k_n\rangle \equiv |k_1, \ldots, k_j \oplus \delta q, \ldots, k_n\rangle, \tag{3}$$

where $\oplus$ denotes integer addition mod d.

We can reexpress the boost $\hat{Z}_j$ and shift $\hat{X}_j$ operators in terms of their generators, which are the conjugate $\hat{q}_j$ and $\hat{p}_j$ operators, respectively:

$$\hat{Z}_j = e^{\frac{2\pi i}{d} \hat{q}_j} \tag{4}$$

and

$$\hat{X}_j = e^{-\frac{2\pi i}{d} \hat{p}_j}. \tag{5}$$

Thus, we can refer to the $\hat{X}_j$ basis as the momentum (p_j) basis, which is equivalent to the Fourier transform of the q_j basis:

$$\hat{p}_j = \hat{F}_j \hat{q}_j \hat{F}_j^\dagger. \tag{6}$$

These bases form the discrete Weyl phase space (p, q).

The Wigner function $W_\Psi(p,q)$ of a pure state $|\Psi\rangle$ is defined on this discrete Weyl phase space:

$$W_\Psi(p,q) = d^{-n} \sum_{\substack{\xi_q \in \\ (\mathbb{Z}/d\mathbb{Z})^n}} e^{-\frac{2\pi i}{d}\xi_q \cdot p} \Psi\left(q + \frac{(d+1)\xi_q}{2}\right) \Psi^*\left(q - \frac{(d+1)\xi_q}{2}\right). \tag{7}$$

This is equivalent to the discrete Wigner function introduced by Gross [7]. We will shortly be interested in the discrete Wigner function of stabilizer states. However, first, we introduce the effect that the Clifford gates have in this discrete Weyl phase space.

2.1. Clifford Gates

A Clifford group gate $\hat{V}$ is related to a symplectic transformation on the discrete Weyl phase space, governed by a symplectic matrix $\mathcal{M}_{\hat{V}}$ and vector $\alpha_{\hat{V}}$ [7]:

$$\begin{pmatrix} p' \\ q' \end{pmatrix} = \mathcal{M}_{\hat{V}}\left[\begin{pmatrix} p \\ q \end{pmatrix} + \frac{1}{2}\alpha_{\hat{V}}\right] + \frac{1}{2}\alpha_{\hat{V}}. \tag{8}$$

Wigner functions $W_\Psi(x)$ of states evolve under Clifford operators $\hat{V}$ by

$$W_\Psi\left(\mathcal{M}_{\hat{V}}(x + \alpha_{\hat{V}}/2) + \alpha_{\hat{V}}/2\right), \tag{9}$$

where $x \equiv (p,q)$. When considering Clifford gate propagation, we can restrict to a set of gates which are generators of the Clifford group. One such set of generators is made up of the phase-shift gate $\hat{P}_i$, the Hadamard gate $\hat{F}_i$, and the controlled-not (CNOT) $\hat{C}_{ij}$ (which act on the ith and jth qudits).

The phase shift $\hat{P}_i$ is a one-qudit gate with the underlying Hamiltonian $H_{\hat{P}_i} = -\frac{d+1}{2}q_i^2 + \frac{d+1}{2}q_i$ [10]. Without loss of generality, we will instead consider

$$\hat{P}'_i = \hat{P}_i \hat{P}_i \hat{Z}_i, \tag{10}$$

which we will refer to as the phase-shift gate in this paper. We note that the usual phase-shift can be obtained from the new one within the Clifford group:

$$\hat{P}_i = \hat{P}'_i \hat{P}'_i \hat{Z}_i, \tag{11}$$

where $[\hat{P}_i, \hat{Z}_i] = [\hat{P}'_i, \hat{Z}_i] = 0$. Hence, $\hat{P}'_i$ is an adequate replacement generator for $\hat{P}_i$, and we will use it instead of $\hat{P}_i$ from now on. Since its Hamiltonian has no linear term ($H_{\hat{P}'_i} = -q_i^2$), this leads to an easier presentation ahead since $\alpha_{\hat{P}'_i} = 0$. The corresponding equations of motion for $\hat{P}'_i$ are $\dot{p}_i = 2q_i$ and $\dot{q}_i = 0$. Hence, for $\Delta t = 1$,

$$\left(\mathcal{M}_{\hat{P}'_i}\right)_{j,k} = \delta_{j,k} + 2\delta_{i,j}\delta_{n+i,k}. \tag{12}$$

The Hadamard gate $\hat{F}_i$ is a one-qudit gate and has the underlying Hamiltonian $H_{\hat{F}_i} = -\frac{\pi}{4}(p_i^2 + q_i^2)$ [10]. The corresponding equations of motion are $\dot{p}_i = \frac{\pi}{2}q_i$ and $\dot{q}_i = -\frac{\pi}{2}p_i$. Hence, for $\Delta t = 1$,

$$\left(\mathcal{M}_{\hat{F}_i}\right)_{j,k} = \delta_{j,k} - \delta_{i,j}\delta_{i,k} - \delta_{n+i,j}\delta_{n+i,k} \tag{13}$$
$$+ \delta_{i,j}\delta_{n+i,k} - \delta_{n+i,j}\delta_{i,k},$$

and $\alpha_{\hat{F}_i} = 0$.

Finally, the two-qudit CNOT $\hat{C}_{ij}$ on control qudit i and second qudit j has the corresponding Hamiltonian $H_{\hat{C}_{ij}} = p_i q_j$ [10]. The corresponding equations of motion are $(\dot{p}_i, \dot{p}_j) = -(0, p_i)$ and $(\dot{q}_i, \dot{q}_j) = (q_j, 0)$. Hence, for $\Delta t = 1$,

$$\left(\mathcal{M}_{\hat{C}_{ij}}\right)_{k,l} = \delta_{k,l} - \delta_{i,k}\delta_{j,l} + \delta_{n+j,k}\delta_{n+i,l}, \tag{14}$$

and $\alpha_{\hat{C}_{ij}} = 0$.

2.2. Wigner Functions of Stabilizer States

A discrete Wigner function for stabilizer states associated with the boost and shift operators defined in Equations (4) and (5) is given by the following theorem [10]:

Theorem 1. *The discrete Wigner function $W_\Psi(x)$ of a stabilizer state Ψ for any odd d and n qudits is $\delta_{\Phi \times x, r}$ for $2n \times 2n$ matrix Φ and $2n$ vector r with entries in $\mathbb{Z}/d\mathbb{Z}$.*

An equivalent form was proven by Gross [7] who also showed that these discrete Wigner functions of stabilizer states are non-negative. In particular, if we begin with a stabilizer state defined as $|\Psi_0\rangle = |q_0\rangle$, then $W_{\Psi_0}(x) = \delta_{\Phi_0 \times x, r_0}$, where $\Phi_0 = \begin{pmatrix} 0 & 0 \\ 0 & \mathbb{I}_n \end{pmatrix}$ for $\mathbb{I}_n$ the $n \times n$ identity matrix, and $r_0 = (0, q_0)$.

3. Wigner Stabilizer Algorithm for Odd d Qudits

With the discrete Wigner function of a stabilizer state defined in Theorem 1 and the effect of the Clifford group generators on discrete Wigner functions defined in Equation (9), we can now examine the effect Clifford operators have on stabilizer states. We note that since the discrete Wigner functions of stabilizer states are non-negative and Clifford operations take stabilizer states to stabilizer states, it follows that Clifford operations (if associated positive-operator valued measures (POVMs) also have non-negative Wigner functions) can always be efficiently classically simulated by sampling from these Wigner functions as probability distributions [4]. However, here we pursue a description that is not dependent on classical sampling.

3.1. Stabilizer Representation

From Theorem 1, propagation of the stabilizer state Ψ can be represented by considering the state's Wigner function: $W_\Psi(x) = \delta_{\Phi_t \cdot x, r_t}$. In this way, Φ_t and r_t specify a linear system of equations in terms of p_t and q_t. The first n rows of Φ_t are the coefficients of $(p_t, q_t)^T$ in $p_0(p_t, q_t)$ and the last n rows of Φ_t are the coefficients of $(p_t, q_t)^T$ in $q_0(p_t, q_t)$:

$$\begin{pmatrix} p_0 \\ q_0 \end{pmatrix} = \Phi_t \begin{pmatrix} p_t \\ q_t \end{pmatrix}. \tag{15}$$

The Kronecker delta function sets this linear system of equations equal to r_t. In this way, an affine map—a linear transformation displaced from the origin by r_t—is defined. This system of equations must be updated after every unitary propagation and measurement.

Since the Wigner functions $W_\Psi(x)$ of stabilizer states propagate under $\mathcal{M}$ as $W_\Psi(\mathcal{M}x)$, it follows that

$$\Phi_t \to \Phi_t \mathcal{M}_t^{-1}. \tag{16}$$

(The importance of vector r_t and when it must be updated will become evident when we consider random measurements.) Hence, after n operations $\mathcal{M}_1, \mathcal{M}_2, \ldots, \mathcal{M}_n$,

$$\mathcal{M}_t^{-1} = \mathcal{M}_1^{-1} \mathcal{M}_2^{-1} \ldots \mathcal{M}_n^{-1}. \tag{17}$$

The matrices are ordered chronologically left-to-right instead of right-to-left.

Since $\mathcal{M}$ is symplectic, $\mathcal{M}_t^{-1} = -\mathcal{J}\mathcal{M}_t^T\mathcal{J}$ where

$$\mathcal{J} = \begin{pmatrix} 0 & -\mathbb{I}_n \\ \mathbb{I}_n & 0 \end{pmatrix}. \tag{18}$$

Thus, the the stability matrices $\mathcal{M}$ for $\hat{F}_i$, $\hat{P}'_i$ and $\hat{C}_{ij}$ given in Equations (12)–(14) differ from their inverses only by sign changes in their off-diagonal elements:

$$\left(\mathcal{M}_{\hat{P}'_i}^{-1}\right)_{j,k} = \delta_{j,k} - 2\delta_{i,j}\delta_{n+i,k}, \tag{19}$$

$$\left(\mathcal{M}_{\hat{F}_i}^{-1}\right)_{j,k} = \delta_{j,k} - \delta_{i,j}\delta_{i,k} - \delta_{n+i,j}\delta_{n+i,k} \tag{20}$$
$$- \delta_{i,j}\delta_{n+i,k} + \delta_{n+i,j}\delta_{i,k},$$

and

$$\left(\mathcal{M}_{\hat{C}_{ij}}^{-1}\right)_{k,l} = \delta_{k,l} + \delta_{i,k}\delta_{j,l} - \delta_{n+j,k}\delta_{n+i,l}. \tag{21}$$

We assume the quantum state is initialized in the computational basis state $\Psi_0 = \underbrace{|0\rangle \otimes \cdots \otimes |0\rangle}_{n}$ and so initially we should set $\Phi_0 = \begin{pmatrix} 0 & 0 \\ 0 & \mathbb{I}_n \end{pmatrix}$ and $r_0 = 0$. The initial stabilizer state is $W_{\Psi_0} = \delta_{q_t,0}$. However, it will become clear when we discuss measurements that it is practically useful to instead set

$$\Phi_0 = \begin{pmatrix} \mathbb{I}_n & 0 \\ 0 & \mathbb{I}_n \end{pmatrix}, \tag{22}$$

thereby setting $W_{\Psi_0} = \delta_{(p_t,q_t),(0,0)}$—not a true Wigner function. This new matrix Φ_0 is equivalent to the last matrix if the first n rows in $\Phi_t x$ and r_t are ignored—the same as ignoring $p_0(p_t, q_t)$. In fact, we have two Wigner functions here: one defined by the first n rows and another by the last n rows. We proceed in this manner, ignoring the first n rows, until their usefulness becomes apparent to us.

For n qudits unitary propagation requires $\mathcal{O}(n^2)$ dits of storage to track Φ_t and r_t. More precisely, since Φ_t is a $2n \times 2n$ matrix and r_t is an $2n$-vector, $2n(2n+1)$ dits of storage are necessary.

3.2. Unitary Propagation

Φ_t contains the coefficients of the linear equations relating x_0 to x_t. Each row is one equation relating q_{0_i} or p_{0_i} to x_t. When manipulating rows of Φ_t we shall refer to the linear equations that these rows define.

Examining Equations (19)–(21), we see that the inverse stability matrices of the generator gates $\hat{F}_i$, $\hat{P}_i$ and $\hat{C}_{ij}$ are the sum of an identity matrix and a matrix with a finite number of non-zero off-diagonal elements. The number of these off-diagonal elements is independent of the number of qudits, n. Hence, multiplying Φ_t with a new stability matrix in Equation (16) and evaluating the matrix multiplication is equivalent to performing a finite number of n-vector dot products and so requires $\mathcal{O}(n)$ operations. Therefore, keeping track of propagation of stabilizer states by Clifford gates can be simulated with $\mathcal{O}(n)$ operations.

Let us examine these unitary operations more closely. Defining $\oplus$ and $\ominus$ to be mod d addition and subtraction respectively, we find:

Phase gate on qudit i ($\hat{P}'_i$). For all $j \in \{1, \ldots, 2n\}$, set $\Phi_{j,n+i} \mapsto \Phi_{j,n+i} \ominus 2\Phi_{j,n}$.

Hadamard gate on qudit i ($\hat{F}_i$). For all $j \in \{1, \ldots, 2n\}$, negate $\Phi_{j,i}$ mod d, and then swap $\ominus\Phi_{j,i}$ and $\Phi_{j,n+i}$.

CNOT from control i to target j ($\hat{C}_{ij}$). For all $j \in \{1, \ldots, 2n\}$, set $\Phi_{k,j} \mapsto \Phi_{k,j} \oplus \Phi_{k,i}$ and $\Phi_{k,n+i} \mapsto \Phi_{k,n+i} \ominus \Phi_{k,n+j}$.

This confirms that unitary propagation in this scheme requires $\mathcal{O}(n)$ operations.

3.3. Measurement

The outcome of a measurement $\hat{Z}_i$ on a stabilizer state can be either random or deterministic. As described above, the bottom half of Φ_t defines q_{0j} for $j \in \{1, \ldots, n\}$, each of which is a linear combination of q_{t_j} and p_{t_j}. The entries in the $(n+j)$th row of Φ_t give the coefficient of p_{t_i} and q_{t_i} in q_{0j} for $j \in \{1, \ldots, n\}$. If the coefficient of p_{t_i} in any q_{0j} is non-zero then the measurement $\hat{Z}_i$ will be random. If all coefficients of p_{t_i} are zero for q_{0j} $\forall j$, then the measurement of $\hat{Z}_i$ will be deterministic. This can be seen from the fact that if our stabilizer state $|\Psi\rangle$ is an eigenstate of $\hat{Z}_i$, then $\hat{Z}_i |\Psi\rangle = e^{i\phi} |\Psi\rangle$ for some $\phi \in \mathbb{R}$ and (discrete) Wigner functions do not change under a global phase. Thus, measuring $\hat{Z}_i$ leaves the Wigner function of $|\Psi\rangle$ invariant if the measurement is deterministic. Since $\hat{Z}_i$ is a boost operator that increments the momentum of a state by one, its effect on the linear system of equations specified by the Wigner function is:

$$\Phi_t \begin{pmatrix} p_{t_1} \\ \vdots \\ p_{t_i} \\ \vdots \\ p_{t_n} \\ q_t \end{pmatrix} = \begin{pmatrix} r_{tp} \\ r_{tq} \end{pmatrix} \xmapsto{\hat{Z}_i} \Phi_t \begin{pmatrix} p_{t_1} \\ \vdots \\ p_{t_i}+1 \\ \vdots \\ p_{t_n} \\ q_t \end{pmatrix} = \begin{pmatrix} r_{tp} \\ r_{tq} \end{pmatrix}. \quad (23)$$

Thus, if the lower half of the ith column of Φ_t is zero, then $\hat{Z}_i$ leaves the Wigner function invariant (and so the measurement is deterministic). Verifying that these coefficients are all zero takes $\mathcal{O}(n)$ operations for each $\hat{Z}_i$.

In other words, to see if a given measurement of $\hat{Z}_i$ is random or deterministic, a search must be performed for non-zero $\Phi_{tn+j,i}$ elements. If such a non-zero element exists, then the measurement is random since it means that the final momentum of qudit i affects the state of the stabilizer and so its position must be undetermined (by Heisenberg's uncertainty principle). If no such finite $\Phi_{tn+j,i}$ element exists, then the measurement $\hat{Z}_i$ is deterministic. We now describe the algorithm in detail for these two cases:

Case 1: Random Measurement

Let the $(n+j)$th row in the bottom half of Φ_t have a non-zero entry in the ith column, $\Phi_{tn+j,i} \neq 0$. Since the random measurement $\hat{Z}_i$ will project qudit i onto a position state, we will replace the $(n+j)$th row with $q_{0i} = q'_i$ (the uniformly random outcome of this measurement). After this projection onto a position state, none of the other qudits' positions should depend on qudit i's momentum, p_{t_i}. To accomplish this, before we replace row $(n+j)$, we solve its equation for p_{t_i} and substitute every instance of p_{t_i} in the linear system of equations with this solution. As a result, every equation will no longer depend on p_{t_i} and we can go ahead and replace the $(n+j)$th row with $q_{0i} = q'_i$.

There is one more thing to do, which will be important for deterministic measurements: replace the jth row with the old $(n+j)$th row. This sets $p_{0i} = q_{0j}(\boldsymbol{p_t}, \boldsymbol{q_t})$, which becomes the only remaining equation explicitly dependent on p_{t_i}. In other words, $p_{0i} \propto p_{t_i}$, similar to the beginning when we set $p_{0i} = p_{t_i}$ by setting $\Phi = \mathbb{I}_{2n}$. However, now we also preserve any dependence p_{0i} has on the other qudits incurred during unitary propagation. In other words, we preserve p_{t_i}'s dependence upon the other qudits, but only in the Wigner function specified by the top n rows, which we ignore otherwise.

After replacing the equation specified by row $(n+j)$ of Φ_t and r_t with a randomly chosen measurement outcome q'_i (i.e., $q_{0i} = q'_i$), the identification of rows $(n+i)$ and $(n+j)$ are exchanged, so that the former now specifies $q_{0j}(p_t, q_t)$ while the latter specifies $q_{0i}(p_t, q_t)$. p_{0i} has also been updated by replacing the jth row in the first half of Φ_t, with the $(n+j)$th row we just changed. Again, this row now describes $p_{0i}(p_t, q_t)$ while the ith row now specifies $p_{0j}(p_t, q_t)$. Overall, this takes $\mathcal{O}(n^2)$ operations since we are replacing $\mathcal{O}(n)$ rows with $\mathcal{O}(n)$ entries.

Case 2: Deterministic Measurement

Since the measurement is deterministic, Φ_t and r_t do not change. The n equations specified by the bottom half of $\Phi_t x_t = r_t$ can be used to solve for q_{t_i}—the deterministic measurement outcome. In general, this can also be done by inverting Ψ_t and evaluating $x_t = \Phi_t^{-1} \cdot r_t$ for q_i. Aaronson and Gottesman themselves noted that such a matrix inversion is possible, but practically takes $\mathcal{O}(n^3)$ operations. (However, we are not certain if Aaronson and Gottesman were referring to the Φ_t matrix corresponding to the $2n \times 2n$ part of their tableau when they discuss matrix inversion in [2].)

Fortunately, there is another method that scales as $\mathcal{O}(n^2)$ and requires use of the n equations represented by the top n rows of Φ_t, which were included in our description by setting $\Phi_0 = \mathbb{I}_{2n}$. The linear system of n equations represented by $\Phi_t x_t = r_t$ can be written as

$$\Phi_t x_t = r_t, \tag{24}$$

$$\begin{pmatrix} p_0(p_t, q_t) \\ q_0(p_t, q_t) \end{pmatrix} = \begin{pmatrix} r_{tp} \\ r_{tq} \end{pmatrix}, \tag{25}$$

where we are interested in linear combinations of the bottom half, $q_0(p_t, q_t)$, to solve for the measurement outcome q_{t_i}:

$$\sum_{j=1}^{n} c_{ij} q_{0j} = q_{t_i}, \tag{26}$$

where $c_{ij} \in \mathbb{Z}/d\mathbb{Z}$.

Lemma 1. *The coefficient in front of p_{t_i} in the row of Φ_t that specifies $p_{0j}(p_t, q_t)$, Φ_{tji}, is equal to the coefficient c_{ij} in front of q_{0j} that makes up q_{t_i} in Equation (26). Equivalently,*

$$c_{ij} = q_{0j} \cdot q_{t_i}(p_0, q_0) = p_{0j}(p_t, q_t) \cdot p_{t_i} = \Phi_{tji}. \tag{27}$$

Proof. Under evolution under the Clifford group operators,

$$\begin{pmatrix} p_t \\ q_t \end{pmatrix} = \mathcal{M}_t \begin{pmatrix} p_0 \\ q_0 \end{pmatrix}. \tag{28}$$

$\mathcal{M}_t^{-1} = -\mathcal{J} \mathcal{M}_t^T \mathcal{J}$ since $\mathcal{M}_t$ is symplectic. This means that we can express the matrix inversion as follows:

$$\begin{pmatrix} p_0 \\ q_0 \end{pmatrix} = \mathcal{M}_t^{-1} \begin{pmatrix} p_t \\ q_t \end{pmatrix} \tag{29}$$

$$= -\mathcal{J} \mathcal{M}_t^T \mathcal{J} \begin{pmatrix} p_t \\ q_t \end{pmatrix} \tag{30}$$

$$= -\mathcal{J} \begin{pmatrix} (\mathcal{M}_t)_{11} & (\mathcal{M}_t)_{12} \\ (\mathcal{M}_t)_{21} & (\mathcal{M}_t)_{22} \end{pmatrix}^T \mathcal{J} \begin{pmatrix} p_t \\ q_t \end{pmatrix} \tag{31}$$

$$= \begin{pmatrix} (\mathcal{M}_t)_{22} & (-\mathcal{M}_t)_{12} \\ (-\mathcal{M}_t)_{21} & (\mathcal{M}_t)_{11} \end{pmatrix} \begin{pmatrix} p_t \\ q_t \end{pmatrix}. \tag{32}$$

Therefore, $\left(\mathcal{M}_t^{-1}\right)_{11i,j} = (\mathcal{M}_t)_{22i,j}$, and so

$$c_{ij} = q_{0j} \cdot q_{ti}(\boldsymbol{p}_0, \boldsymbol{q}_0) = p_{0j}(\boldsymbol{p}_t, \boldsymbol{q}_t) \cdot p_{ti} = \Phi_{tji}. \tag{33}$$

This property can also be seen in the drawing of phase space shown in Figure 1. There, initial perpendicular p_{0j} and q_{0j} manifolds are drawn along with harmonically evolved p_{ti} and q_{ti} manifolds, which remain perpendicular to each other and make an angle α to the first p_{0j} and q_{0j} manifolds, respectively. The projection of $q_{ti}(\boldsymbol{p}_0, \boldsymbol{q}_0)$ onto q_{0j} can be represented as the length b of a right triangle's adjacent side to the angle α, with an opposite side set to some length a. The projection of $p_{0j}(\boldsymbol{p}_t, \boldsymbol{q}_t)$ onto p_{ti} is similarly represented by the length b' of a right triangle's adjacent side to the angle α, with an opposite side also set to length a. It follows that the third angle β in both triangles must be the same, and so by the law of sines

$$\frac{a}{\sin \alpha} = \frac{b}{\sin \beta} = \frac{b'}{\sin \beta}. \tag{34}$$

Therefore, $b = b'$ and so these two projections are equal to one another. In the discrete Weyl phase space such manifolds must lie along grid phase points and obey the periodicity in x_p and x_q, but the premise is the same. □

Overall, the procedure outlined in Lemma 1 for deterministic measurements takes $\mathcal{O}(n^2)$ operations since Equation (27) is a sum of $\mathcal{O}(n)$ vectors made up of $\mathcal{O}(n)$ components. Therefore, the overall measurement protocol takes $\mathcal{O}(n^2)$ operations. Note that this formulation of the algorithm shows that it is the symplectic structure on phase space and the linear transformation under harmonic evolution that allows the inversion (Equation (32)) to be performed efficiently.

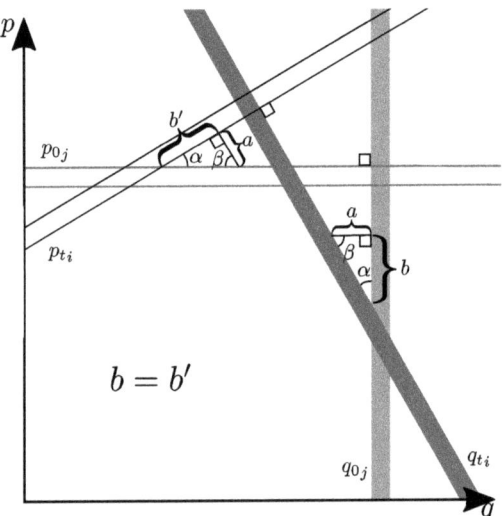

Figure 1. The initial perpendicular manifolds p_{0j} and q_{0j} and the harmonically evolved perpendicular manifolds p_{ti} and q_{ti}. Description of the various lengths and angles are given in the text in the proof of Lemma 1.

4. Aaronson–Gottesman Tableau Algorithm for Qubits ($d = 2$)

The Aaronson–Gottesman tableau algorithm was originally defined for qubits ($d = 2$) [2]. Like the algorithm we presented in the previous section, it only requires overall $\mathcal{O}(n^2)$ operations for propagation and measurement for n qubits. The algorithm has been proven to be extendable to $d > 2$ [12] and similar algorithms have been formulated in $d > 2$ [13]. Alternatives have also been developed to the tableau formalism, though they prove to be equally efficient in worst-case scenarios [14]. However, we are not aware of any direct extension of the Aaronson–Gottesman tableau algorithm to dimensions greater than two. In this and the next section, we will show that the Wigner algorithm presented in Section 3 is equivalent to the Aaronson–Gottesman tableau algorithm extended to odd d.

4.1. Stabilizer Representation

The Aaronson–Gottesman algorithm is defined in the stabilizer formalism. It keeps track of the evolution of a stabilizer state by updating the generators of the stabilizer group, elements of which are defined as follows:

Definition 1. *A set of operators that satisfies $S = \{\hat{g} \in \mathcal{P} \text{ such that } \hat{g}|\psi\rangle = |\psi\rangle\}$ are called the stabilizers of state $|\psi\rangle$, where $\mathcal{P}$ is the set of Pauli operators, each of which has the form $e^{\frac{\pi i}{2}\alpha}\hat{P}_1 \otimes \cdots \otimes \hat{P}_n$ where $\alpha \in \{0, 1, 2, 3\}$ for n qubits with $\hat{P}_i \in \{\hat{I}_i, \hat{Z}_i, \hat{X}_i, \hat{Y}_i\}$.*

For the sake of completeness, we present here a summary of the qubit Aaronson–Gottesman algorithm, in order to compare it to our odd d qudit algorithm. For more details, see [1,2].

Each n-qubit stabilizer state is uniquely determined by 2^n Pauli operators. There are only n generators of this Abelian group of 2^n operators. Therefore, an n-qubit stabilizer state is defined by the n generators of its stabilizer state. Every element in this set of generators, $\{\hat{g}_1, \hat{g}_2, \ldots, \hat{g}_n\}$, is in the Pauli group, and each generator has the form:

$$\hat{g}_i = \pm \hat{P}_{i1} \ldots \hat{P}_{in}. \tag{35}$$

Any unitary propagation by Clifford operators or measurement of the stabilizer state changes at least some of the $\hat{P}_{ij}$ elements of the n generators of the state's stabilizer. This includes the ± 1 phase in Equation (35), which must also be kept track of in Aaronson–Gottesman's algorithm.

4.2. Unitary Propagation

For each Clifford operation, Aaronson and Gottesman showed that only $\mathcal{O}(n)$ operations are necessary to update all generators [2]. Specifically, according to the update rules in Table 1, each generator can be updated with a constant number of operators for every single Clifford gate, therefore $\mathcal{O}(n)$ in total. However, it is a little more complicated to update the generators after each measurement. To do this efficiently, Aaronson introduced "*destabilizers*":

Definition 2. *Destabilizers $\{\hat{g}'_1, \ldots, \hat{g}'_n\}$ are the operators that generate the full Pauli group with the stabilizers $\{\hat{g}_1, \ldots, \hat{g}_n\}$. They have the following properties:*

(i) $\hat{g}'_1, \hat{g}'_2, \ldots, \hat{g}'_n$ *commute.*
(ii) *Each destabilizer $\hat{g}'_h$ anti-commutes with the corresponding stabilizer $\hat{g}_h$, and commutes with all other stabilizers.*

Table 1. Transformation of stabilizer generators under Clifford operations.

Gates	Input	Output
Hadamard	$\hat{X}$	$\hat{Z}$
	$\hat{Z}$	$\hat{X}$
phase	$\hat{X}$	$\hat{Y}$
	$\hat{Z}$	$\hat{Z}$
CNOT	$\hat{X} \otimes \hat{I}$	$\hat{X} \otimes \hat{X}$
	$\hat{I} \otimes \hat{X}$	$\hat{I} \otimes \hat{X}$
	$\hat{Z} \otimes \hat{I}$	$\hat{Z} \otimes \hat{I}$
	$\hat{I} \otimes \hat{Z}$	$\hat{Z} \otimes \hat{Z}$

To incorporate the destabilizers, a *tableau* becomes useful to see how they play a role in updating the stabilizer generators during measurement [2].

Aaronson–Gottesman defined such a $2n \times (2n+1)$ binary tableau matrix as:

$$\begin{pmatrix} x_{11} & \cdots & x_{1n} & z_{11} & \cdots & z_{1n} & r_1 \\ \vdots & \ddots & \vdots & \vdots & \ddots & \vdots & \vdots \\ x_{n1} & \cdots & x_{nn} & z_{n1} & \cdots & z_{nn} & r_n \\ x_{(n+1)1} & \cdots & x_{(n+1)n} & z_{(n+1)1} & \cdots & z_{(n+1)n} & r_{n+1} \\ \vdots & \ddots & \vdots & \vdots & \ddots & \vdots & \vdots \\ x_{(2n)1} & \cdots & x_{(2n)n} & z_{(2n)1} & \cdots & z_{(2n)n} & r_{2n} \end{pmatrix}.$$

This matrix contains $2n$ rows. The first n rows denote the destabilizers $\hat{g}'_1$ to $\hat{g}'_n$ while rows $(n+1)$ to $2n$ represent the stabilizers $\hat{g}_1$ to $\hat{g}_n$. The $(n+1)$th bit in each row denotes the phase $(-1)^{r_i}$ for each generator. We encode the jth Pauli operator in the ith row as shown in Table 2.

Table 2. Binary representation of the Pauli operators and the Pauli group phase used in their tableau representation.

x_{ij}	z_{ij}	$\hat{P}_j$
0	0	$\hat{I}_j$
0	1	$\hat{Z}_j$
1	0	$\hat{X}_j$
1	1	$\hat{Y}_j$

r_i	Phase
0	+1
1	−1

We can update the stabilizers and destabilizers as follows:

Hadamard gate on qubit i For all $j \in \{1, 2, ..., 2n\}$, $r_j \mapsto r_j \oplus x_{ji} z_{ji}$, then swap x_{ji} with z_{ji}.

Phase gate on qubit i For all $j \in \{1, 2, ..., 2n\}$, $r_j \mapsto r_j \oplus x_{ji} z_{ji}$, $z_{ji} \mapsto z_{ji} \oplus x_{ji}$.

CNOT gate on control qubit i and target qubit j For all $k \in \{1, 2, ..., 2n\}$, $r_k \mapsto r_k \oplus x_{ki} z_{kj}(x_{kj} \oplus z_{ki} \oplus 1)$, $x_{kj} \mapsto x_{kj} \oplus x_{ki}$, $z_{ki} \mapsto z_{ki} \oplus z_{kj}$.

These actions correspond to those given in Table 1.

Notice the striking similarity of these tableau transformation rules under unitary propagation to the Φ transformation rules in Section 4. The most notable difference is that the Aaronson–Gottesman algorithm involves updates of the vector r. We will discuss this and its connection to the dimension $d = 2$ of the system in Section 5. It is clear that these transformations also take $O(n)$ operations each.

4.3. Measurement

To describe the measurement part of the algorithm, we need to first define a rowsum operation in the tableau that corresponds to multiplying two Pauli operators together. As defined in [2]:

Rowsum: To sum row i and j, first update the bits that represent operators by $x_{ik} \oplus x_{jk}$ and $z_{ik} \oplus z_{jk}$ for $k = 1, \ldots, n$. To calculate the resultant phase, Aaronson and Gottesman first defined the following function:

$$f(x_{ik}, x_{jk}, z_{ik}, z_{jk}) = \begin{cases} 0 & \text{if } x_{ik} = z_{ik} = 0, \\ z_{jk} - x_{jk} & \text{if } x_{ik} = z_{ik} = 1, \\ z_{jk}(2x_{jk} - 1) & \text{if } x_{ik} = 1, z_{ik} = 0, \\ x_{jk}(1 - 2z_{jk}) & \text{if } x_{ik} = 0, z_{ik} = 1. \end{cases} \quad (36)$$

Since each stabilizer generator is the tensor product of n single qubit Pauli operators (see Equation (35)), they must be multiplied together to obtain the phase:

$$\begin{cases} 0 & \text{if } r_i + r_j + \sum_{k=1}^{n} f(x_{ik}, x_{jk}, z_{ik}, z_{jk}) \equiv 0 \pmod{4}, \\ 1 & \text{if } r_i + r_j + \sum_{k=1}^{n} f(x_{ik}, x_{jk}, z_{ik}, z_{jk}) \equiv 2 \pmod{4}. \end{cases} \quad (37)$$

Having defined the rowsum function, let us now consider a measurement of $\hat{Z}_i$ on qubit i. For $d = 2$, Pauli group operators can only commute or anti-commute with each other. If $\hat{Z}_i$ anti-commutes with one or more of the generators, then the measurement is random. If $\hat{Z}_i$ commutes with all of the generators, then the measurement is deterministic. We consider these two cases:

Case 1: Random Measurement

$\hat{Z}_i$ anti-commutes with one or more of the generators. If there is more than one, we can always pick a single anti-commuting generator, $\hat{g}_j$, and update the rest by replacing them with their product with $\hat{g}_j$ (i.e., taking the rowsum of their corresponding rows) such that they commute with $\hat{Z}_i$. These updates take $\mathcal{O}(n^2)$ operations. Finally, we only need to replace $\hat{g}_j$ by $\hat{Z}_i$.

In other words, with respect to the tableau, there should exist at least one $j \in \{n+1, n+2, \ldots, 2n\}$ such that $x_{ji} = 1$. Replacing all rows where $x_{ki} = 1$ for $k \neq j$ with the sum of the jth and kth row (using the rowsum function) sets all $x_{ki} = 0$ for $k \neq j$.

Finally, we replace the $(j - n)$th row with the jth row and update the jth row by setting $z_{ji} = 1$ and all other x_{jk}s and z_{jk}s to 0 for all k. We output $r_j = 0$ or $r_j = 1$ with equal probability for the measurement result. This procedure takes $O(n^2)$ operations because each rowsum operation takes $O(n)$ operations and up to $n - 1$ rowsums may be necessary.

Case 2: Deterministic Measurement

$\hat{Z}_i$ commutes with all generators. In this case, there is no $j \in \{n+1, n+2, \ldots, 2n\}$ such that $x_{ji} = 1$ and we don't need to update any of the generators. However, we do need to do some work to retrieve the measurement outcome.

Measurement $\hat{Z}_i$ commutes with all of the stabilizers; therefore, either $+\hat{Z}_i$ or $-\hat{Z}_i$ is a stabilizer of the state. Therefore, it must be generated by the generators. The sign ± 1 is the measurement outcome we are looking for. This means that

$$\prod_{j=1}^{n} \hat{g}_j^{c_j} = \pm \hat{Z}_i, \quad (38)$$

where $c_j = 1$ or 0.

For those destabilizers g'_k that satisfy

$$\{g'_k, \pm \hat{Z}_i\} = 0, \tag{39}$$

$c_k = 1$. Otherwise, $c_k = 0$. This can be seen from

$$\{g'_k, \pm \hat{Z}_i\} = \{g'_k, \prod_{j=1}^{n} \hat{g}_j^{c_j}\} = \prod_{\substack{j=1 \\ j \neq k}}^{n} \hat{g}_j^{c_j} \{g'_k, \hat{g}_k^{c_k}\} = 0, \tag{40}$$

where we used part (ii) of Definition 2 of the destabilizers and Equation (39). The last equality requires $c_k = 1$.

Therefore, to find the deterministic measurement outcome, the stabilizers whose corresponding destabilizer anti-commutes with the measurement operation $\hat{Z}_i$ must be multiplied together. Every row $(n + j)$ in the bottom half of the tableau, such that $x_{ji} = 1$ (for $j \in \{1, \ldots, n\}$), can be added up together and stored in a temporary register. The resultant phase ± 1 of this sum is the measurement result we are looking for.

Checking if each destabilizer commutes or anti-commutes with $\hat{Z}_i$ takes a constant number of operations. One multiplication takes $O(n)$ operations, and there are $O(n)$ multiplications needed. Therefore, a measurement takes $O(n^2)$ operations overall.

5. Discussion

As we made clear throughout Section 4, the scaling of the number of required operations with respect to number of qudits n is exactly the same in the ($d = 2$) Aaronson–Gottesman algorithm as in the (odd d) Wigner algorithm presented in Section 3. The two algorithms also require the same number of dits of temporary storage for performing the deterministic measurement. Moreover, there is a correspondence between the tableau employed by Aaronson–Gottesman and the matrix Φ_t and vector r_t we use. In particular, the tableau is equal to $\begin{pmatrix} \Phi_t & | & r_t \end{pmatrix}$:

$$\Phi_t = \begin{pmatrix} \frac{\partial p_0}{\partial p_t} & \frac{\partial p_0}{\partial q_t} \\ \frac{\partial q_0}{\partial p_t} & \frac{\partial q_0}{\partial q_t} \end{pmatrix} \equiv \begin{pmatrix} x_{11} & \cdots & x_{1n} & z_{11} & \cdots & z_{1n} \\ \vdots & \ddots & \vdots & \vdots & \ddots & \vdots \\ x_{n1} & \cdots & x_{nn} & z_{n1} & \cdots & z_{nn} \\ x_{(n+1)1} & \cdots & x_{(n+1)n} & z_{(n+1)1} & \cdots & z_{(n+1)n} \\ \vdots & \ddots & \vdots & \vdots & \ddots & \vdots \\ x_{(2n)1} & \cdots & x_{(2n)n} & z_{(2n)1} & \cdots & z_{(2n)n} \end{pmatrix} \tag{41}$$

and

$$r_t = \begin{pmatrix} r_p \\ r_q \end{pmatrix} \equiv \begin{pmatrix} r_1 \\ \vdots \\ r_n \\ r_{n+1} \\ \vdots \\ r_{2n} \end{pmatrix}. \tag{42}$$

This can be seen through the following equation:

$$\exp\left(\frac{2\pi i}{d} \sum_{j=1}^{2n} \Phi_{tn+i,j} \hat{x}_j\right) |\Psi_t\rangle = \prod_{j=1}^{2n} \exp\left(\frac{2\pi i}{d} \Phi_{tn+i,j} \hat{x}_j\right) |\Psi_t\rangle$$

$$= \exp\left(\frac{2\pi i}{d} r_{ti}\right) |\Psi_t\rangle, \tag{43}$$

where $\hat{x} \equiv (\hat{p}, \hat{q})$. Multiplying the right-hand side of the first equation and the second equation by $\exp\left(-\frac{2\pi i}{d} r_{ti}\right)$, it follows that

$$\exp\left(-\frac{2\pi i}{d} r_{ti}\right) \prod_{j=1}^{2n} \exp\left(\frac{2\pi i}{d} \Phi_{tn+i,j} \hat{x}_j\right) \Psi_t = \hat{g}_i |\Psi_t\rangle = |\Psi_t\rangle. \tag{44}$$

In other words, r_{ti} specifies the phase $\exp\left(-\frac{2\pi i}{d} r_{ti}\right)$ of the ith stabilizer, which is itself specified by $\Phi_{tn+i,j}$ for $j \in \{0,\ldots,2n\}$. These are the same roles for r and the tableau in the Aaronson–Gottesman tableau algorithm [2].

Indeed, both algorithms check the bottom half of their matrices for finite elements of $\Phi_{n+j,i}$ to determine if a measurement on the ith qudit will be random or not. They also use a very similar protocol to determine the outcome of deterministic measurements. The Wigner-based algorithm motivates these manipulations in terms of the symplectic structure of Weyl phase space and the relationship between the two Wigner functions specified by the top and bottom of Φ, providing a strong physical intuition for their effects. Aaronson and Gottesman motivate these manipulations using the anti-commutation relations between the stabilizer and destabilizer generators. In addition, the latter half of both the Wigner function's r_t and Aaronson–Gottesman's r are used to determine measurement outcomes. The only fundamental algorithmic difference between the approaches is that the Wigner-based algorithm does not require updates of r_t during unitary propagation. The reason for this lies in the fact that Aaronson–Gottesman's algorithm deals with systems with $d = 2$ while the Wigner-based algorithm is restricted to odd d.

In particular, for the one-qubit Clifford group gate operator $\hat{A} = \{\hat{P}_i, \hat{F}_i\} \, \forall i = \{1,\ldots,n\}$, the Aaronson–Gottesman algorithm specifies that for a q- or p-state, its Wigner function evolves by:

$$W_\Psi(\mathcal{M}_{\hat{A}} x). \tag{45}$$

However, for $|r\rangle = \frac{1}{\sqrt{2}}(|0\rangle \pm i|1\rangle)$, a Y-state which is diagonal in the pq plane, its Wigner function must first be translated:

$$W_\Psi(\mathcal{M}_{\hat{A}} x + \beta), \tag{46}$$

where the translation β can be $(1,0)$ or $(0,1)$ equivalently. There is a similar state-dependence for the two-qubit CNOT gate $\hat{C}_{ij}$.

This demonstrates that the Aaronson–Gottesman algorithm is state-dependent on the qubit stabilizer state it is acting on. On the other hand, the Wigner function algorithm on odd d qudit stabilizer states is state-independent. This likely is a consequence of the fact that the Weyl–Heisenberg group, which is made up of the boost and shift operators defined in Equations (4) and (5) that underlie the discrete Wigner formulation, are a subgroup of $U(d)$ instead of $SU(d)$ for $d = 2$ [15]. Furthermore, qubits exhibit state-independent contextuality while odd d qudits do not [16]. Recent progress on this subject relating non-contextuality to classical simulatability for qubits can be found here [17,18].

6. Example of Stabilizer Evolution

As a demonstration of what stabilizer state propagation looks like in the Wigner formalism, we proceed to go through an example of Bell state preparation and measurement starting from the state $|0\rangle \otimes |0\rangle$. To illustrate this process we decompose the two qutrit Wigner function of this state into nine 3×3 grids, as shown in Figure 2. The prepared Wigner function is denoted in Figure 3 with the color black, and the Wigner function represented by setting $\Phi_0 = \begin{pmatrix} 1 & 0 \\ 0 & 0 \end{pmatrix}$ (i.e., considering the top n rows of Φ to be a separate Wigner function, as discussed at the end of Section 3.1) is denoted with the color gray.

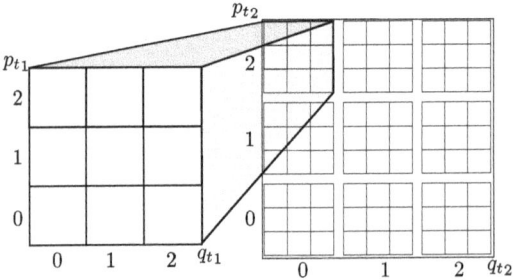

Figure 2. A decomposition of the two qutrit Wigner function into nine 3 × 3 grids, where each 3 × 3 grid denotes the value of the Wigner function at all p_{t_1} and q_{t_1} for a fixed value of p_{t_2} and q_{t_2} denoted by the external axes. This organization is used in Figure 3 below.

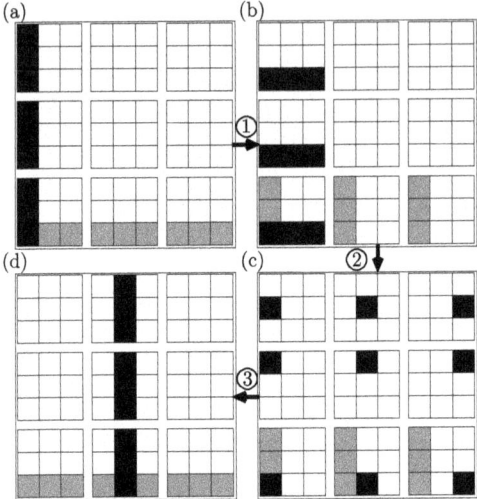

Figure 3. The Wigner function of two qutrits initially prepared in (**a**) the state $|0\rangle \otimes |0\rangle$. (1) This is evolved under $\hat{F}_1$ to produce (**b**) $\frac{1}{\sqrt{3}}(|0\rangle + |1\rangle + |2\rangle) \otimes |0\rangle$. (2) Subsequently, this state is evolved under $\hat{C}_{12}$ producing (**c**) the Bell state $\frac{1}{\sqrt{3}}(|00\rangle + |11\rangle + |22\rangle)$. (3) Qutrit 1 is then measured producing the random outcome 1, which collapses qutrit 2 into the same state, so that (**d**) $|1\rangle \otimes |1\rangle$ results. The black color indicates the Wigner function specified by the lowest n rows of $\delta_{\Phi_i x, r_i}$, and the gray color indicates the Wigner function specified by the highest n rows ($q_0(p_t, q_t)$ and $p_0(p_t, q_t)$, respectively). The evolution and algorithmic implementation are explained in the text.

We begin with

$$W_\Psi(x) = \delta \begin{pmatrix} 1 & 0 & 0 & 0 \\ 0 & 1 & 0 & 0 \\ 0 & 0 & 1 & 0 \\ 0 & 0 & 0 & 1 \end{pmatrix} \begin{pmatrix} p_{t_1} \\ p_{t_2} \\ q_{t_1} \\ q_{t_2} \end{pmatrix} \begin{pmatrix} 0 \\ 0 \\ 0 \\ 0 \end{pmatrix} = \delta \begin{pmatrix} p_{t_1} \\ p_{t_2} \\ q_{t_1} \\ q_{t_2} \end{pmatrix} \begin{pmatrix} 0 \\ 0 \\ 0 \\ 0 \end{pmatrix},$$

(47)

denoting an initially prepared state of $|0\rangle \otimes |0\rangle$. This is clear in Figure 3a by the black band that lies along all Weyl phase space points with $q_{t_1} = 0$ and $q_{t_2} = 0$. On the other hand, the gray manifold is perpendicular to the black one, and lies along Weyl phase space points with $p_{t_1} = 0$ and $p_{t_2} = 0$.

Acting on this state with $\hat{F}_1$ produces $\frac{1}{\sqrt{3}}\left(e^{\frac{2\pi i}{3}0\times 0}|0\rangle + e^{\frac{2\pi i}{3}1\times 0}|1\rangle + e^{\frac{2\pi i}{3}2\times 0}|2\rangle\right) \otimes |0\rangle$. Applying the algorithm specified at the end of Section 3.2, we find:

$$W_\Psi(x) = \tag{48}$$

$$\delta\begin{pmatrix} 0 & 0 & -1 & 0 \\ 0 & 1 & 0 & 0 \\ 1 & 0 & 0 & 0 \\ 0 & 0 & 0 & 1 \end{pmatrix}\begin{pmatrix} p_{t_1} \\ p_{t_2} \\ q_{t_1} \\ q_{t_2} \end{pmatrix}\begin{pmatrix} 0 \\ 0 \\ 0 \\ 0 \end{pmatrix} = \delta\begin{pmatrix} -q_{t_1} \\ p_{t_2} \\ p_{t_1} \\ q_{t_2} \end{pmatrix}\begin{pmatrix} 0 \\ 0 \\ 0 \\ 0 \end{pmatrix}.$$

Thus, the momentum of qutrit 1 is now determined and is 0 while the second qutrit is unchanged. This can be seen in Figure 3b, where the q_{t_2} values of the non-zero Weyl phase space points are the same, while the state has rotated by $-\pi/2$ in (p_{t_1}, q_{t_1})-space. A similar transformation has occurred for the perpendicular gray manifold.

Acting next with $\hat{C}_{12}$ produces the Bell state $\frac{1}{\sqrt{3}}(|00\rangle + |11\rangle + |22\rangle)$, which is represented by the following Wigner function:

$$W_\Psi(x) = \tag{49}$$

$$\delta\begin{pmatrix} 0 & 0 & -1 & 0 \\ 0 & 1 & 0 & 0 \\ 1 & 1 & 0 & 0 \\ 0 & 0 & -1 & 1 \end{pmatrix}\begin{pmatrix} p_{t_1} \\ p_{t_2} \\ q_{t_1} \\ q_{t_2} \end{pmatrix}\begin{pmatrix} 0 \\ 0 \\ 0 \\ 0 \end{pmatrix} = \delta\begin{pmatrix} -q_{t_1} \\ p_{t_2} \\ p_{t_1} + p_{t_2} \\ -q_{t_1} + q_{t_2} \end{pmatrix}\begin{pmatrix} 0 \\ 0 \\ 0 \\ 0 \end{pmatrix}.$$

The entanglement between the two qutrits is evident in both of their dependence on each other's momenta and positions, $p_{t_1} = -p_{t_2}$ and $q_{t_1} = q_{t_2}$, specified by the last two rows. Figure 3c shows that the state is still representable as lines in Weyl phase space, except they now traverse through the different planes of (q_{t_1}, p_{t_1}) associated with each value of (q_{t_2}, p_{t_2}). However, if you consider the left column in Figure 3c corresponding to $q_{t_2} = 0$, you can see that the only black Weyl phase points are at $q_{t_1} = 0$. Similarly, the middle column corresponding to $q_{t_2} = 1$ shows that $q_{t_1} = 1$, and the right column corresponding to $q_{t_2} = 2$ shows that $q_{t_1} = 2$ too, confirming that $|\Phi\rangle = \frac{1}{\sqrt{3}}(|00\rangle + |11\rangle + |22\rangle)$. Thus, the entanglement of the two qutrits' positions is clearly evident in this Figure of the Wigner function.

We then proceed to measure qutrit 1. Since the lower two equations involve p_{t_1}, we know that this is a random measurement. Let us pick the outcome to be 1 and set the third row as such, replacing the first row with the old third row. This collapses qutrit 2 into the same state:

$$W_\Psi(x) = \tag{50}$$

$$\delta\begin{pmatrix} 1 & 1 & 0 & 0 \\ 0 & 1 & 0 & 0 \\ 0 & 0 & 1 & 0 \\ 0 & 0 & -1 & 1 \end{pmatrix}\begin{pmatrix} p_{t_1} \\ p_{t_2} \\ q_{t_1} \\ q_{t_2} \end{pmatrix}\begin{pmatrix} 0 \\ 0 \\ 1 \\ 0 \end{pmatrix} = \delta\begin{pmatrix} p_{t_1} + p_{t_2} \\ p_{t_2} \\ q_{t_1} \\ -q_{t_1} + q_{t_2} \end{pmatrix}\begin{pmatrix} 0 \\ 0 \\ 1 \\ 0 \end{pmatrix}.$$

The lower two rows show that now $q_{t_1} = 1$, as we chose, and $q_{t_2} = q_{t_1} = 1$. The collapse of qutrit 2 into $|1\rangle$ can also been seen in Figure 3c by the fact that $q_{t_1} = 1$ only in the 3 × 3 grids that correspond to $q_{t_2} = 1$ too.

Finally, the fact that a measurement of q_{t_2} would be deterministic at this point can be seen in the fact that p_{t_2} is not present in the last two rows of Φ_t. Furthermore, it is clear, since the first row has a coefficient of 1 in front of p_{t_1}, that the corresponding third row must be added with weight 1 to the fourth row to obtain this deterministic measurement outcome of $q_{t_2} = 1$. This can also be seen in

Figure 3 by finding the projection of p_{0_1} onto p_{t_2}, which are shown by the gray manifolds in panels (a) and (d), respectively. They are collinear and so the projection is equal to 1. (Perpendicular manifolds corresponds to a projection of 0, and those that lie $\pi/4$ diagonally with respect to each other have a projection equal to 2 in this discrete geometry.)

7. Conclusions

In summary, we introduced an algorithm that efficiently simulates stabilizer state evolution under Clifford gates and measurements in the $\hat{Z}$ Pauli basis for odd d qudits. We accomplished this by relying on the phase-space perspective of stabilizer states as discrete Gaussians and Clifford operators as having underlying harmonic Hamiltonians. We showed the equivalence of our algorithm, through Equations (43) and (44), to the well-known Aaronson–Gottesman tableau algorithm [2] for qubits, revealing that Aaronson–Gottesman's tableau corresponds to a discrete Wigner function. As a consequence, we revealed the physically intuitive phase space perspective of Aaronson–Gottesman's algorithm, as well as its extension to higher odd d.

This work illustrates that no efficiency advantage is gained by using the Heisenberg representation for stabilizer propagation. Equation (44) indicates that the Heisenberg representation is equivalent to the Schrödinger representation in this context; evolving the operators is just as efficient as evolving the states, as perhaps expected.

Lastly, the correspondence between the Wigner-based algorithm and the Aaronson–Gottesman tableau algorithm may point the direction on how to resolve the long-standing issue of describing the Wigner–Weyl–Moyal and center-chord formalism for $d = 2$ systems. We have shown that the Aaronson–Gottesman algorithm is essentially a $d = 2$ treatment of the Wigner approach. The salient difference appears to be the state-dependence of this evolution, and likely is related to the state-independent contextuality that qubits exhibit, which odd d qudits do not. Exploring the details of this state-dependence is a promising subject of future study.

Acknowledgments: This work was supported by the Air Force Office of Scientific Research (AFOSR) award No. FA9550-12-1-0046.

Author Contributions: All authors contributed to the work presented here.

Conflicts of Interest: The authors declare no conflict of interest.

References

1. Gottesman, D. The Heisenberg Representation of Quantum Computers. *arXiv* **1998**, arXiv:quant-ph/9807006.
2. Aaronson, S.; Gottesman, D. Improved simulation of stabilizer circuits. *Phys. Rev. A* **2004**, *70*, 052328.
3. Gottesman, D. Fault-tolerant quantum computation with higher-dimensional systems. In *Quantum Computing and Quantum Communications*; Springer: Heidelberg, Germany, 1999; pp. 302–313.
4. Mari, A.; Eisert, J. Positive Wigner functions render classical simulation of quantum computation efficient. *Phys. Rev. Lett.* **2012**, *109*, 230503.
5. Howard, M.; Wallman, J.; Veitch, V.; Emerson, J. Contextuality supplies the 'magic' for quantum computation. *Nature* **2014**, *510*, 351–355.
6. Wootters, W.K. A Wigner-function formulation of finite-state quantum mechanics. *Ann. Phys.* **1987**, *176*, 1–21.
7. Gross, D. Hudson's theorem for finite-dimensional quantum systems. *J. Math. Phys.* **2006**, *47*, 122107.
8. Veitch, V.; Ferrie, C.; Gross, D.; Emerson, J. Negative quasi-probability as a resource for quantum computation. *New J. Phys.* **2012**, *14*, 113011.
9. Veitch, V.; Wiebe, N.; Ferrie, C.; Emerson, J. Efficient simulation scheme for a class of quantum optics experiments with non-negative Wigner representation. *New J. Phys.* **2013**, *15*, 013037.
10. Kocia, L.; Love, P. Semiclassical Formulation of Gottesman–Knill and Universal Quantum Computation. *arXiv* **2016**, arXiv:1612.05649.
11. Koh, D.E.; Penney, M.D.; Spekkens, R.W. Computing quopit Clifford circuit amplitudes by the sum-over-paths technique. *arXiv* **2017**, arXiv:1702.03316.
12. De Beaudrap, N. A linearized stabilizer formalism for systems of finite dimension. *arXiv* **2011**, arXiv:1102.3354.

13. Yoder, T.J. A Generalization of the Stabilizer Formalism for Simulating Arbitrary Quantum Circuits. 2012. Available online: https://pdfs.semanticscholar.org/b200/efe1709d07ffc1b5b7bd90e61c09e2729bdf.pdf (accessed on 6 July 2017).
14. Anders, S.; Briegel, H.J. Fast simulation of stabilizer circuits using a graph-state representation. *Phys. Rev. A* **2006**, *73*, 022334.
15. Bengtsson, I.; Zyczkowski, K. On discrete structures in finite Hilbert spaces. *arXiv* **2017**, arXiv:1701.07902.
16. Mermin, N.D. Hidden variables and the two theorems of John Bell. *Rev. Mod. Phys.* **1993**, *65*, 803.
17. Raussendorf, R.; Browne, D.E.; Delfosse, N.; Okay, C.; Bermejo-Vega, J. Contextuality as a resource for qubit quantum computation. *arXiv* **2015**, arXiv:1511.08506.
18. Kocia, L.; Love, P. Discrete Wigner Formalism for Qubits and the Non-Contextuality of Clifford Operations on Qubit Stabilizer States. *arXiv* **2017**, arXiv:1705.08869.

© 2017 by the authors. Licensee MDPI, Basel, Switzerland. This article is an open access article distributed under the terms and conditions of the Creative Commons Attribution (CC BY) license (http://creativecommons.org/licenses/by/4.0/).

Article

Concepts and Criteria for Blind Quantum Source Separation and Blind Quantum Process Tomography

Alain Deville [1],* and Yannick Deville [2]

[1] Institut Matériaux Microélectronique et Nanosciences de Provence (IM2NP), Aix-Marseille Université, 13397 Marseille, France
[2] Institut de Recherche en Astrophysique et Planétologie (IRAP), Université de Toulouse, 31400 Toulouse, France; yannick.deville@irap.omp.eu
* Correspondence: alain.deville@univ-amu.fr; Tel.: +33-5-61-33-28-24

Received: 6 April 2017; Accepted: 23 June 2017; Published: 6 July 2017

Abstract: Blind Source Separation (BSS) is an active domain of Classical Information Processing, with well-identified methods and applications. The development of Quantum Information Processing has made possible the appearance of Blind Quantum Source Separation (BQSS), with a recent extension towards Blind Quantum Process Tomography (BQPT). This article investigates the use of several fundamental quantum concepts in the BQSS context and establishes properties already used without justification in that context. It mainly considers a pair of electron spins initially separately prepared in a pure state and then submitted to an undesired exchange coupling between these spins. Some consequences of the existence of the entanglement phenomenon, and of the probabilistic aspect of quantum measurements, upon BQSS solutions, are discussed. An unentanglement criterion is established for the state of an arbitrary qubit pair, expressed first with probability amplitudes and secondly with probabilities. The interest of using the concept of a random quantum state in the BQSS context is presented. It is stressed that the concept of statistical independence of the sources, widely used in classical BSS, should be used with care in BQSS, and possibly replaced by some disentanglement principle. It is shown that the coefficients of the development of any qubit pair pure state over the states of an orthonormal basis can be expressed with the probabilities of results in the measurements of well-chosen spin components.

Keywords: blind source separation (BSS); qubit pair; exchange coupling; entangled pure state; unentanglement criterion; probabilities in quantum measurements; independence of random quantum sources

1. Introduction

The book entitled "Do we really understand quantum mechanics?" [1] was published five years ago. Some forty years earlier, its author, Laloë, had co-authored a treatise on quantum mechanics, together with Cohen-Tannoudji, later a Nobel laureate, and Diu [2]. While this recent book illustrates the present strong interest for the foundations of Quantum Theory (QT), already in 1929, Dirac could claim: *"The general theory of quantum mechanics is now almost complete"* and *"The underlying physical laws necessary for the mathematical theory of a large part of physics and the whole of chemistry are thus completely known"* [3]. Since that time, the development of both telecommunications through electromagnetic waves and solid state electronics favoured the appearance first of classical Information Theory, and then of Quantum Information Theory and Processing (QIT, QIP).

This special issue, *Quantum Information and Foundations*, in the Quantum Information Section of Entropy, reflects the existence of links between QIP/QIT and the foundations of QT. An instance of such links is given by the approach adopted e.g., in Timpson's Thesis [4]. This methodology, in the framework of Philosophy of Science, is difficult because of its rather general character. For the last decade, we have been following another approach. Starting from a problem in the domain of classical information processing, namely Source Separation (SS) with its more difficult so-called Blind version (BSS), introduced around 1985 and now a mature field [5,6], we are developing its quantum counterpart, which we proposed to call Blind Quantum Source Separation (BQSS). Each step of this more pedestrian approach may be controlled, presently e.g., through simulations. This approach has been achieved in our 2007 paper introducing BQSS [7], and in those describing the solutions which we have built since then (see e.g., [6,8–14]), and which led to our recent introduction of Blind Quantum Process Tomography (cf. [12,14] and more explanations at the end of this section and in Part A.2 of the Appendix).

A short presentation of the problem of classical (i.e., non quantum) or conventional BSS, and of its interest, is needed here. In BSS, typically, at first, a set of users (the Writer) presents a set of simultaneous signals (input signals, or sources) at the input of a multi-user communication system (the Mixer). The sources, constrained to possess some general properties (e.g., mutual statistical independence), are combined (mixed, in the SS sense) in the Mixer, often specified through a model, e.g., the linear memoryless one (cf. Chapter 11 from [15]). Another set of users (the Reader) receives the signals arriving at the Mixer output. The Writer possibly knows the sources, but the Reader does not know them, and cannot access the inputs of the Mixer. That Mixer uses one or several parameter values, unknown to the Reader, who only knows some of its general properties. The Reader's final task is the restoration of the sources (possibly up to some so-called acceptable indeterminacies) from the signals at the Mixer output, during the inversion phase. An intermediate task is the determination of the unknown parameters of the Mixer, or of its inverse. Before receiving the signals to be separated at the Mixer output, derived from the sources sent by the Writer, the Reader therefore enters an "adaptation phase", during which he knows that the Writer is sending one (or possibly a limited number of) signal(s) submitted to some definite, and known by the Reader, constraints. The particular signal sent is not known by the Reader (blind separation problem), who knows the class of the input signal(s) and the signal(s) at the Mixer output in the adaptation phase, and, of course, the mixed signals to be separated in the inversion phase.

Conventional BSS is already used to extract some or all source signals in various application fields, e.g., in some audio systems, or when using radio-frequency signals to transmit digital data, or in the biomedical field, in the processing of signals such as electrocardiograms, electroencephalograms or magnetoencephalograms, as explained in Part A.1 of the Appendix. More information on the applications of conventional BSS may be found in our previous papers [11,14], in [6], and in the papers or books they cite.

BSS is moreover closely linked to a well-known domain of signal processing technology called system identification. More precisely, BSS is linked to Blind Mixture Identification (BMI), as briefly explained in Part A.1 of the Appendix and developed in [6], and BSS may be used in the corresponding applications.

Conventional (B)SS has favoured the introduction of concepts and the development of specific methods [5,6]. Its extension to the quantum domain seems suitable for at least three reasons. First, the source concept may be extended from a classical to a quantum context. Secondly, as any classical phenomenon, conventional (B)SS may be seen as the limit of a quantum phenomenon. When developing solutions to the BQSS problem, it seems legitimate to try and import concepts and methods from the classical to the quantum SS domain. However, the presence of entanglement in a quantum approach should be clearly identified and the consequences of its existence should not be underestimated. In addition, the concepts of quantum sources and of their statistical independence deserve some discussion, and consequences of the probabilistic aspect of the results of measurements in the quantum domain must be drawn. Furthermore, last but not least, since some of the basic concepts of QT are still open to discussion, when e.g., using measurements, even in an abstract process, the adopted point of view should once be made explicit, in order to minimize confusion. The nature of this special issue gave us the opportunity to clarify concepts and justify properties already used in our previous papers upon BQSS, a task postponed up to now, and which should be of use in the

BQSS domain, and maybe in other fields. These two motivations stimulate a third natural one, namely the hope of extending the field of BSS applications toward the quantum world. In the following sections, in order to illustrate our methods and help reading, some aspects or results of our previous papers will be occasionally presented, but the building of any specific BQSS solution is outside their scope. The reader interested in the results from simulations may consult [8,11], obtained through BQSS methods with classical processing, and [14], with quantum processing in the forward path. This recent paper moreover contains a table with a detailed comparison of the key features and performance from the existing methods.

In all of our previous papers, we considered two distinguishable qubits numbered 1 and 2, and we presently keep this situation. When it is meaningful to speak of the state of a quantum system, and specifically if this system is a qubit, this state may be either pure or mixed. In order to avoid any confusion with the meaning of a mixture in the SS context, if it is needed to speak of a (quantum) mixed state in the following, we will systematically speak of a statistical mixture. A typical situation is the following one: at an initial time t_0, the Writer prepares both qubits, each in a given pure state, described by some ket. This ket carries information, an idea contained in the expression "quantum source". The initial state $|\Psi(t_0)>$ of the qubit pair is then the tensor product of the corresponding kets. The time between t_0 (writing) and t_1 (reading) is supposed to be short enough for the qubit pair to be treated as isolated, a choice already made by Feynman [16,17] in the context of the quantum computer, and presently refined at the beginning of Section 4.1 for qubits physically realized with spins. At any time t between t_0 and t_1, the state of the qubit pair may then be described by a ket $|\Psi(t)>$. In the Schrödinger picture, this time evolution of the pair is described by a time-dependent unitary operator $U(t_0, t_1)$. It is assumed that an undesired coupling exists between these qubits. Because of this undesired coupling, as time goes on the state of the pair generally becomes entangled. Coupling is then interpreted as a mixing (in the SS sense), realized by an abstract Mixer depending upon one or several parameter values, unknown to the Reader, who only knows some general properties of that Mixer. It is said that the input of the Mixer receives state $|\Psi(t_0)>$, and that its output provides state $|\Psi(t)>$. It should be well appreciated that *inverting* $U(t_0, t_1)$ in order to get $|\Psi(t_0)>$ from $|\Psi(t_1)>$ *is not that easy, because* $U(t_0, t_1)$ *is unknown* (blind QSS). In Section 2, it is first explained why both state and process quantum tomography are unable to solve this BQSS problem, and secondly why the Schmidt criterion is ill-suited for following the degree of entanglement of $|\Psi(t_1)>$ during the adaptation phase. The Peres–Horodecki criterion [18,19] is valid for separable statistical mixtures of bipartite systems, and not specifically for unentangled pure states. A better suited unentanglement criterion is therefore established in Section 2.

In Section 3, a model situation, for a single spin and then for a pair of spins, in inhomogeneous magnetic fields with random directions, allows us to speak of random and possibly independent variables, in that quantum context. We explain why, although this random quantum state corresponds to a statistical mixture, it is simpler, in the BQSS context, to speak of a random pure state than to introduce a density operator. In Section 4, we first make brief comments about the description of quantum states (including the existence of statistical mixtures as source states, in a more general context), about the act of measurement and about the physical realization of qubits with electron spins. We then discuss questions related to the probabilities of the possible results obtained in measurements of spin components, in the context of spins 1/2 as qubits. We first present their use when the Reader makes measurements at the Mixer output in order to restore the sources (cf. Figure 1). These measurements establish a link between the output of the Mixer and the classical world. It is stressed that while the macroscopic support of the results of measurements has a classical behaviour, the probabilities of these results obey quantum laws. We then establish an unentanglement criterion using probabilities, equivalent to the one established in Section 2 for the probability amplitudes c_i. It is shown that the c_i coefficients can be expressed as functions of the probabilities of results in the measurements of well-chosen spin components. In Section 5, we derive the expression of the above unentanglement criterion for all possible source states, at the output of the so-called separating system, with respect to the parameters of both the cylindrical Heisenberg coupling, an abstract Mixer largely used in our previous papers, and that separating system.

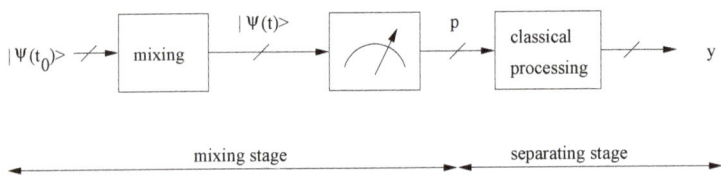

Figure 1. Block diagram of a system using classical-processing BQSS.

In Part A.2 of the Appendix, the question of the applications of BQSS is addressed. Partly because the appearance of BQSS is recent, the subject of its applications is presently largely speculative. Two main subdomains should be distinguished. The first one is BQSS in a strict sense. It aims at recovering the source states and is the quantum counterpart of conventional BSS. The second subdomain focuses on an intermediate step possibly found in methods developed for BQSS and aiming at the knowledge of the mixer function or of its inverse. The corresponding classical problem is known as Blind Mixture Identification (BMI), a subfield of System Identification. The non-blind quantum version of System Identification is that already mentioned and well-established field of QIP called Quantum Process Tomography (as opposed to Quantum State Tomography). We recently introduced the quantum version of BMI, which we proposed to call Blind Quantum Process Tomography (BQPT).

2. An Unentanglement Criterion for a Qubit Pair

A superficial look may suggest that it is possible to restore the initial product state through State or Process Tomography (ST, PT). ST aims at determining a quantum state if a lot of copies of that state are available [20]. However, in BQSS, the Reader is unable to access the input of the Mixer, and ST is therefore obviously presently strictly useless. PT would presently consist of placing (preparing) successive well-defined and known quantum states at the input of the Mixer, thus operating in the non-blind mode (cf. [15], p. 202) and observing the corresponding signals at its output. However, in the BQSS problem, the Reader is strictly unable to operate that way, as he is unable to ask the Writer to prepare him the quite specific input states asked for by PT. Therefore, quantum tomography is unable to solve the BQSS problem, which needs dedicated methods (for more details, see [8]).

Up to now, in the BQSS problem, we developed two main approaches for both determination of the unknown parameter(s) of the mixing or separating system and source separation. In the first approach [7,8,11], the Reader measures observables, using the signals at the Mixer output (cf. Figure 1). The results, and properties associated with them, e.g., the probabilities of their occurrences, are kept upon a macroscopic device, e.g., the memory of a classical computer, and then used in a separating system. Since this macroscopic device and the separating system have a classical behaviour, we called this processing aimed at restoring the sources "classical-processing BQSS". In the second, quite different, and more recently introduced approach [9,10,14], the quantum state at the Mixer output is sent to the input of a quantum-processing subsystem (cf. Figure 2), the inverting block of the separating system. This block is so designed that its output provides a quantum pure state equal to $|\Psi(t_0)>$ (possibly up to some acceptable indeterminacies), after the adaptation phase.

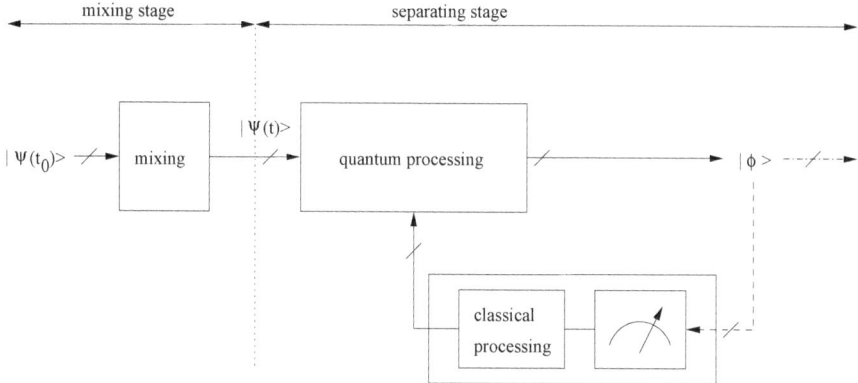

Figure 2. Block diagram of a system using BQSS, with quantum processing in the forward path (no cloning [14], with permision from Elsevier).

From now on, the state spaces of two arbitrary qubits, called qubits 1 and 2, are denoted as $\mathcal{E}_1$ and $\mathcal{E}_2$, respectively. The possible (pure) states of the pair are the kets in $\mathcal{E}_1 \otimes \mathcal{E}_2$. We assume that the qubits are physically realized with spins 1/2, which, e.g., allows us to speak of the spin component s_{1z} or s_{2z}, but many results established hereafter keep true without this assumption. We introduce the orthonormal basis $\mathcal{B}_+$, $\{|++>, |+->, |-+>, |-->\}$, where e.g., $|+->$ means $|1+> \otimes |2->$ and $|i,+>, |i,->$ are normed eigenkets of the s_{iz} component of (reduced) spin $\vec{s_i}$ (with i = 1, 2), for the eigenvalues +1/2 and $-1/2$, respectively. Any pure pair state, entangled or not, may be expanded in $\mathcal{B}_+$ as

$$|\Psi> = c_1 |++> + c_2 |+-> + c_3 |-+> + c_4 |-->, \quad (1)$$

where the complex coefficients c_j (j = 1 to 4) respect $\sum_j |c_j|^2 = 1$. If a pure state or a statistical mixture of a bipartite system S_{12} (parts S_1 and S_2) is described by a density operator ρ, the corresponding reduced traces $\rho_1 = Tr_2 \rho$ and $\rho_2 = Tr_1 \rho$ have all the mathematical properties of a density operator [2]. In addition, if S_{12} is in a pure state, ρ_1 and ρ_2 have the same eigenvalues [21]. This pure state is unentangled if and only if its Schmidt number N_S (the number of non-zero eigenvalues of ρ_1 and ρ_2) is equal to 1 [21]. We are particularly interested in the case when $|\Psi>$ is the state found at the output of the inverting block. Then, any pure state may be expanded in the standard basis $\mathcal{B}+$ as in Equation (1), where the values of the c_i coefficients are affected by both the coupling between the qubits and, during the adaptation phase, by the adaptation procedure. This adaptation phase typically consists of an iterative numerical algorithm, which aims at optimizing a continuous-valued function, traditionally called the "cost function". For any given values of the adjustable parameters of the inverting block, the cost function measures a kind of "distance" between $|\Psi>$ at the output of the inverting block and an unentangled pure state. The Schmidt unentanglement criterion cannot be used in our problem because the considered state remains (at least slightly) entangled throughout the adaptation procedure, and the Schmidt number thus remains higher than one. The Schmidt criterion provides a binary-valued unentanglement detector, with a Schmidt number equal to one or not and, if taking into account all possible integer values of N_S beyond unentanglement detection, the Schmidt criterion provides a discrete-valued quantity. What we eventually need instead is a quantitative, continuous-valued, measure of that "distance" of the considered state with respect to unentanglement, in order to keep the adjustable parameter values of the inverting block, yielding the state which is the closest to unentanglement. Moreover, even if the Schmidt approach could be modified to this end, it would yield high computational complexity, as it would require one to diagonalize ρ_1 or ρ_2 for each of the quite numerous steps of the iterative adaptation algorithm. We avoid these issues as follows. Since the qubit pair is in a pure state, its partial traces ρ_1 and ρ_2 satisfy

$$\mathrm{Tr}\rho_1^2 = \mathrm{Tr}\rho_2^2 \leq 1, \tag{2}$$

and the common value for $\mathrm{Tr}\rho_1^2$ and $\mathrm{Tr}\rho_2^2$ is 1 if and only if the pure state is unentangled (cf. [21]). One could think of using $\mathrm{Tr}\rho_1^2 - 1$ as a cost function. However, $\mathrm{Tr}\rho_1^2$ depends upon the c_i, which suggests one to try and establish an unentanglement criterion using the c_i explicitly. To this end, we consider state $|\Psi\rangle$ defined through Equation (1). When it is assumed that $|\Psi\rangle$ is unentangled, i.e., that it can be written as

$$|\Psi\rangle = (a|+\rangle + b|-\rangle) \otimes (c|+\rangle + d|-\rangle), \tag{3}$$

then, in Equation (1), $c_1 = ac$, $c_2 = ad$, $c_3 = bc$, $c_4 = bd$, so $c_1 c_4$ and $c_2 c_3$ are both equal to $abcd$:

$$c_1 c_4 = c_2 c_3. \tag{4}$$

Conversely, when it is assumed that Equation (4) is satisfied, if $c_1 \neq 0$ then $|\Psi\rangle$ may be written as

$$|\Psi\rangle = c_1(|+\rangle + \frac{c_3}{c_1}|-\rangle) \otimes (|+\rangle + \frac{c_2}{c_1}|-\rangle), \tag{5}$$

which means that $|\Psi\rangle$ is then unentangled. If Equation (4) is satisfied and $c_1 = 0$, then $c_2 = 0$ and $c_3 \neq 0$, or $c_3 = 0$ and $c_2 \neq 0$, or $c_2 = c_3 = 0$, and in each case $|\Psi\rangle$ is unentangled. Therefore, if the qubit pair is in a pure state $|\Psi\rangle$ written as in Equation (1), then:

$$|\Psi\rangle \text{ is unentangled} \iff c_1 c_4 = c_2 c_3. \tag{6}$$

This unentanglement criterion for a qubit pair pure state was used without justification in [9,10]. In Equation (1), $|\Psi\rangle$ was expanded in the standard basis. It is possible instead to introduce e.g., the normed eigenvectors of s_{1x} and s_{2x}, or more generally those of s_{1u} and s_{2v}, the components of the spins along respective arbitrary directions $\vec{u}(\theta_{1E}, \varphi_{1E})$ and $\vec{v}(\theta_{2E}, \varphi_{2E})$, defined through their Euler angles. For each component, the possible results are again $\pm 1/2$. The possible results for the pair may be symbolically written as $(+u+v)$, $(+u-v)$, $(-u+v)$ and $(-u-v)$, and the corresponding probabilities as P_{1uv}, P_{2uv}, P_{3uv}, P_{4uv}. Equation (1) is replaced by

$$|\Psi\rangle = c_{1uv}|+u+v\rangle + c_{2uv}|+u-v\rangle + c_{3uv}|-u+v\rangle + c_{4uv}|-u-v\rangle. \tag{7}$$

With the same reasoning within the new basis, (6) is replaced by

$$|\Psi\rangle \text{ is unentangled} \iff c_{1uv} c_{4uv} = c_{2uv} c_{3uv}. \tag{8}$$

3. Random Quantum Sources and Their Independence

The qubits are again supposed to be physically realized with spins 1/2. Standard Electron Spin and Nuclear Magnetic Resonance (ESR, NMR) use a non-microscopic number of resonant spins, but methods have been proposed for more than twenty years in order to detect a single spin, particularly with Optically Detected Magnetic Resonance (ODMR [22,23]) or with Magnetic Resonance Force Microscopy (MRFM [24]), and more recently at low temperature (0.5 K) with Spin Excitation Spectroscopy [25], or even with ESR, in extreme conditions [26]. These approaches are still under development. Here, anticipating upon advances in spintronics, we rather consider a pair of spins, or even a single spin, submitted to a static magnetic field.

When speaking e.g., of a microwave source for satellite television, one speaks of the device emitting the microwave carrier. Similarly, the expression "laser source" generally refers to the device creating the coherent radiation. In conventional SS, "source" is an abbreviation for "source signal".

Furthermore, in Quantum SS with abstract qubits corresponding to physical spins 1/2, the word "source" does not refer to some atomic beam delivering atoms carrying an electron or nuclear magnetic moment, but still means "source signal", then referring to some information from the quantum states of these qubits.

In conventional SS, an important concept is that of statistical independence of the sources, at the root of the frequent use of Independent Component Analysis (ICA) [27]. In [7,8,11], we postulated the existence of statistically independent quantum sources when using the classical-processing SS defined at the beginning of Section 2. Hereafter, we show that statistical independence may exist in that context. Quantum Mechanics (QM) does e.g., consider random *operators*, the matrix elements of which are random quantities (see the random lattice operators $F^{(q)}$ in the quantum description of the motions of nuclear moments in liquids, in the study of Spin-Lattice Relaxation (SLR), in [28]). As a simple model situation, a magnetic moment $\vec{\mu}$ associated with a single electron spin 1/2, with $\vec{\mu} = -G\vec{s}$ (isotropic $\bar{\bar{g}}$ tensor), placed in a Stern–Gerlach device, is now introduced. The static field is $\vec{B_0} = B_0 \vec{Z}$, with amplitude B_0. The system of interest consists of this spin and the magnet. Writing the Zeeman Hamiltonian as $h = -\vec{\mu}\vec{B}_0 = GB_0 s_Z$ indicates that while the spin is a quantum object, the magnetic field is treated classically. The Writer first prepares the spin in the $|+Z\rangle$ eigenstate of s_Z (eigenvalue $+1/2$). The moment is then received by the Reader, supposed to ignore the direction of $\vec{B_0}$, and who chooses some direction attached to the Laboratory as the quantization direction, called z (unit vector $\vec{u_z}$) and introduces a Laboratory-tied cartesian reference frame xyz, used to define θ_E and φ_E, the Euler angles of $\vec{Z}$. Since the field is treated classically, θ_E and φ_E behave as classical variables, while s_Z is an operator. The Reader measures $s_z = \vec{s}\vec{u_z}$ (eigenstates: $|+\rangle$ and $|-\rangle$), and is interested in the probability p_{+z} of getting $+1/2$. An elementary calculation indicates that

$$|+Z\rangle = r|+\rangle + \sqrt{1-r^2}e^{i\varphi}|-\rangle, \tag{9}$$

with

$$r = \cos\frac{\theta_{2E}}{2}, \qquad \varphi = \varphi_E, \tag{10}$$

and therefore $p_{+z} = \cos^2\theta_E/2$. Once the direction of the magnetic field has been chosen, state $|+Z\rangle$ is then unambiguously defined. If this direction has a deterministic nature, r and φ are deterministic variables, and $|+Z\rangle$ may then be called a deterministic quantum state. If θ_E and φ_E, defining the direction of $\vec{B_0}$ chosen by the Writer, obey probabilistic laws, one may consider that the quantum quantities r and φ, which depend upon the classical Random Variables (RV) θ_E and φ_E, do possess the properties of conventional, i.e., classical, RV. It may e.g., happen that they be uncorrelated, or even independent (which happens if θ_E and φ_E are independent). In addition, if θ_E and φ_E depend on time in a random way, r and φ are then random time functions. We are not strictly facing the quantum equivalent of a classical situation here. Rather, the stochastic character of the field direction, with classical nature, is reflected in the random behaviour of the quantum state expressed through Equation (9). Therefore, rather than a random operator, we meet here a random quantum state. The concept of a random state, if not the expression, was already used e.g., in the early and canonical books [29,30]. The probability p_{+z}, presently a function of the RV θ_E, is itself an RV. This results from both the randomness of the field direction and the standard probabilistic interpretation of QM. Probabilities of results of measurements for a qubit pair were treated as RV, without the present justification, in most of our previous papers, including [7,8,11].

If one measures the scalar observable O when the spin is in the state $|\Psi\rangle = \alpha|+\rangle + \beta|-\rangle = \Sigma_k f_k|\varphi_k\rangle$ (where k is associated with $+$ and $-$), had the f_k been deterministic the mean value would have been:

$$\langle\Psi|O|\Psi\rangle = \sum_{k,l} f_k^* f_l O_{kl}, \qquad O_{kl} = \langle\varphi_k|O|\varphi_l\rangle. \tag{11}$$

Since the f_k are random, one must moreover calculate the statistical mean, denoted as $\overline{\langle \Psi|O|\Psi\rangle}$:

$$\overline{\langle \Psi|O|\Psi\rangle} = \sum_{k,l} \overline{f_k^* f_l} O_{kl} = Tr\rho O, \tag{12}$$

where ρ is the density operator, the matrix elements of which, in the $(|+\rangle, |-\rangle)$ basis, are $\rho_{l,k} = \overline{f_k^* f_l}$. Therefore, it is in principle possible to presently introduce a density operator, which is a non-random operator (its matrix elements are not random quantities, but statistical averages). However, this does not present any interest, since in the BQSS problem examined up to now, the Reader knows that e.g., qubit 1 has been prepared in a pure state, but does not know the values of the ρ_{ij} coefficients in any basis, and is consequently unable to choose a basis in which ρ would be diagonal. It is simpler to keep speaking of a random pure state.

As a model situation, we now consider two spins 1/2 numbered 1 and 2, each with conditions similar to the previous ones, with fields along directions with respective unit vectors $\vec{Z_1}(\theta_{1E}, \varphi_{1E})$ and $\vec{Z_2}(\theta_{2E}, \varphi_{2E})$, and each spin initially prepared in the state

$$|\psi_i(t_0)\rangle = r_i|i+\rangle + \sqrt{1 - r_i^2} e^{i\varphi_i}|i-\rangle, \quad i = 1, 2, \tag{13}$$

where $|i+\rangle$ and $|i-\rangle$ are the eigenkets of s_{iz}, the component of $\vec{s_i}$ along the quantization direction, for the eigenvalues 1/2 and $-1/2$, respectively. For the same reason, if the field directions are random, r_1, φ_1, r_2 and φ_2 have the properties of conventional RV. If $(\theta_{1E}, \varphi_{1E})$ and $(\theta_{2E}, \varphi_{2E})$ are mutually statistically independent, the same is then true for the couples of RV (r_1, φ_1) and (r_2, φ_2). In addition, if e.g., θ_{1E} and φ_{1E} are independent, the same is true for r_1 and φ_1 (cf. Equation (10)). These properties are of major importance for our quantum-source independent component analysis (QSICA) methods described in [11]. We may then say that the initial state of each qubit is random, i.e., that in Equation (13) r_i and φ_i are RV. When considering the preparation of a pair of qubits each in a pure state, one may assume either a deterministic or a random direction for each magnetic field. This discussion shows that the relevant concept, in the latter case, is that of random quantum states, rather than that of random quantum operators mentioned earlier in this section.

Keeping our assumption of a pair of qubits each prepared in a pure state, we now consider the second approach for the adaptation and inversion phases (cf. the beginning of Section 2 and Figure 2), with a quantum state $|\Phi\rangle$ present at the output of the inverting block. The presence of $|\Phi\rangle$ and the Reader's final aim, the recovery of the initial pure state, prompts the Reader: (1) to speak of a deterministic or random pure state, rather than to use a density operator; (2) to consider that the first constraint to be respected in BQSS is then the very existence of an unentangled state at the output of this inverting block. If unentanglement has first been achieved, then and only then is it possible to speak of a deterministic or random state for each part of that product state. While entanglement has no classical counterpart, the following point may be noted here: if a bipartite system is in a pure (deterministic) state $|\Phi\rangle$, to which a density operator $\rho = |\Phi\rangle\langle\Phi|$ corresponds, $|\Phi\rangle$ is unentangled if and only if the partial traces ρ_1 and ρ_2 satisfy the equality $\rho = \rho_1 \otimes \rho_2$ [31]. This unentanglement condition is reminiscent of the relation $\rho = \rho_1 \cdot \rho_2$ between ρ, the joint probability density function of independent classical RV X_1 and X_2, and ρ_1 and ρ_2, the respective marginal probability density functions. Presently, operators replace functions, a tensor product replaces the ordinary product, and this reminiscence reflects the existence of a classical analogue to unentangled states. Condition (4) for unentanglement was established using spins 1/2, but is valid for any pair of two-level systems. This discussion suggests that, in the BQSS problem, when considering a pair of qubits prepared in a pure state, and moreover using the second approach of Section 2 for adaptation and inversion, instead of trying to directly import ICA methods into the BQSS context, one should focus on disentanglement at the output of the inverting block, which recently led us to introduce a disentanglement-based separation principle [9,10].

In the next section, use will be made of the number of real independent parameters necessary to define an arbitrary normed ket $|\Psi\rangle$ in $\mathcal{E}_1 \otimes \mathcal{E}_2$, written as in Equation (1), and a ket in $\mathcal{E}_1 \otimes \mathcal{E}_2$ forced to be unentangled. These numbers are specified hereafter. An arbitrary normed ket $|\Psi\rangle$ in $\mathcal{E}_1 \otimes \mathcal{E}_2$ depends upon the four complex quantities c_1 to c_4 linked through two relations between real numbers ($\sum_i |c_i|^2$ is equal to 1, and $|\Psi\rangle$ and $e^{i\varphi}|\Psi\rangle$, with φ an arbitrary real quantity, should be considered identical). An *arbitrary normed* ket $|\Psi\rangle$ in $\mathcal{E}_1 \otimes \mathcal{E}_2$ therefore depends upon *six* real independent parameters. If it is forced to be unentangled, it has to satisfy the equality $c_1 c_4 = c_2 c_3$ between complex quantities. An *unentangled* normed ket $|\Psi\rangle$ therefore depends upon *four* real parameters. This corresponds to the fact that $|\Psi\rangle$ is then restricted to the form $|\Psi\rangle = |\psi_1\rangle \otimes |\psi_2\rangle$, where the normed kets $|\psi_1\rangle$ and $|\psi_2\rangle$, describing the state of qubits 1 and 2, respectively, each depend upon two real parameters (r_1, φ_1), (r_2, φ_2) (cf. Equation (13)).

4. BQSS and Probabilities in Spin Component Measurements

4.1. Some General Considerations

Faced with the variety of existing interpretations of QM, Fuchs and Peres have argued that "quantum theory needs no interpretation" [32]. Concerning the question of interpreting QM, one may distinguish between claims that can be experimentally tested (i.e., confirmed or refuted) through experience, and those which cannot. This may be illustrated by an instance from the early days of QM, related to the measurement act. At first, Bohr apparently introduced some dichotomy between the quantum system of interest and the classical behaviour of the apparatus. Chapter VI of Von Neumann's 1932 book [30] was perhaps the first attempt to treat the system of interest and the apparatus (with a so-called pointer) as a single system obeying the laws of QM. However, in his book, Von Neumann also introduced a postulate (wave-function reduction) specifiying the state of the system of interest at the end of the measurement. Since then, this postulate has been criticized, first by Margenau, who introduced the concept of preparation, to be distinguished from the one of measurement, and who insisted that e.g., when a photon is absorbed, the measurement act does not bring the photon into a new state, but destroys it [33,34]. The measurement act has been largely debated, including recent discussions through the concept of decoherence (see e.g., [1,21]). When trying to develop the domain of BQSS, we got some control of the proposed separation methods, through simulations, but we moreover tried to avoid using ideas linked with some specific "interpretation" of QM. In [8], we did mention Von Neumann's book and the irreversible behaviour of the system during measurements, but, after getting a result through some measurement upon a qubit pair, we never used the state of that qubit pair at the end of that measurement. On the contrary, after such a measurement, the qubit pair was often (in an abstract process) submitted to a new preparation, which is not linked to any specific interpretation of QM.

In the previous sections, the concepts of a pure state and a statistical mixture were both used. The concept of a statistical mixture may be introduced through a different and more general situation [35] than the one used in Section 3. The system of interest S and its environment E are viewed as a global quantum system Σ. If S and E are uncoupled, and isolated from the rest of the world, and have been separately prepared in a pure state at time t_a, then they evolve separetely, each in a (time-dependent) pure state. If, after t_a, a coupling between S and E exists between some times t_b and t_c, then from t_b on their state generally becomes entangled. In addition, if, starting from t_c, one focuses upon the behaviour of S, use of the partial trace tool shows that everything then occurs as if S were in a state of statistical mixture described by a well-chosen density operator, obeying the Von Neumann equation. If one takes the qubit pair as S, up to now we did not discuss the BQSS problem found when the Writer proposes the qubit pair in a state described by a statistical mixture resulting from some past interaction with its environment.

In recent discussions about the measurement problem, the concept of decoherence [21] was used for discussing the effect of a transfer of energy from the system to its environment, an irreversible phenomenon corresponding to SLR in the ESR/NMR context (with, in the simplest situations, a characteristic time called T_1) [28,36]. In our previous papers and in the present one, starting from time t_0 when the Writer operates, then, at the chosen time scale, the qubit pair is assumed to be isolated from its environment.

In the ESR/NMR domain, a well-known situation exists when a collection of identical (nuclear or electron) spins placed in a fixed resonant magnetic field are transiently submitted to an intense, oscillating magnetic field with a frequency equal to (or near) its resonant value, and with well-chosen polarization. If each spin is coupled to the magnetic fields only, at the end of the pulse the density matrix (written in the basis in which the static Zeeman Hamiltonian is diagonal) describing the state of these spins possesses non-diagonal elements, called coherences. If a weak internal coupling (spin-spin coupling) such as the dipolar magnetic coupling exists between the spins, and if it is able to manifest itself at a time scale allowing one to neglect SLR, it progressively induces a decrease of the coherences, a reversible phenomenon allowing spin echo techniques.

There is presently a second reason for referring to these behaviours in the MR domain, namely the fact that DiVincenzo suggested the use of *electron* spins for the physical realization of qubits more than twenty years ago [37]. Between two neighbouring electron spins, there may exist a strong exchange interaction, a strictly quantum phenomenon historically first identified by Heisenberg in magnetically ordered materials. This is the first reason for our choice of a Heisenberg coupling in the BQSS problem. The second one is that, on the formal side, the version of the Heisenberg Hamiltonian with spherical or cylindrical symmetry, simple enough to be used in theoretical works, may serve as a benchmark in that BQSS problem. It should be recalled that an Ising coupling, simpler to manipulate theoretically than the Heisenberg one, was present in the DiVincenzo 1995 paper, where it helped in the operating process, while the presence of the Heisenberg coupling is undesired and should be compensated for in the BQSS context.

It is well-known that the ESR lines of transition ions in insulators at moderate concentrations are broadened by the dipolar magnetic coupling between the electron spins, the exchange interaction being negligible then. In concentrated samples, exchange is stronger than dipolar coupling and produces a narrowing of the lines [36]. Dipolar coupling is long ranged and anisotropic, which should lead to heavy theoretical treatments if considering a three-dimensional configuration in the BQSS context. Future technological developments could possibly make e.g., the consideration of a planar square lattice of dipolar coupled spins meaningful in that context.

4.2. Probabilities in Measurements, Classical versus Quantum World

In this subsection, we are interested in our first approach as defined in Section 2, with measurements at the Mixer output (cf. Figure 1). We specifically consider the solutions to BQSS discussed in [7,8,11], with two spins 1/2, each prepared in a pure state at t_0, then submitted to an undesired Heisenberg cylindrical coupling [28,38] (axial component: J_z, normal component: J_{xy}, cf. Equation (4) and Appendix E of [8], and [36]), and measurements of s_{1z} and s_{2z} at the output of the formal Mixer at t_1. The probabilities of obtaining $(+1/2,+1/2)$, $(+1/2,-1/2)$, $(-1/2,+1/2)$ and $(-1/2,-1/2)$ are denoted, respectively, as p_1, p_2, p_3 and p_4 (as in [8], while in [7] e.g., our present p_4 was denoted as p_2). We keep Equation (13) for both qubits, with the choice $\varphi_1 = 0$. One then gets [8]:

$$p_1 = r_1^2 r_2^2, \qquad p_4 = (1 - r_1^2)(1 - r_2^2). \qquad (14)$$

p_2 depends upon a mixing parameter $v = \mathrm{sgn}(\cos \Delta_E) \sin \Delta_E$, with [8] $\Delta_E = -J_{xy}(t_1 - t_0)/\hbar$. This expression for Δ_E may be vizualized as the opposite of the phase rotation $\Delta \phi = \omega(t_1 - t_0)$ between states coupled by a Hamiltonian term with energy J_{xy}, during the time interval $(t_1 - t_0)$, with ω given by the Planck–Einstein relation $\omega = J_{xy}/\hbar$. Probability p_2 satisfies

$$p_2 = r_1^2(1 - r_2^2)(1 - v^2) + (1 - r_1^2)r_2^2 v^2 - 2r_1 r_2 \sqrt{1 - r_1^2}\sqrt{1 - r_2^2}\sqrt{1 - v^2}\, v \sin \Delta_I \qquad (15)$$

and, with our choice for φ_1, $\Delta_I = \varphi_2$.

In Equation (13), which describes the initial state of the qubit pair, r_1, r_2, φ_1 and φ_2, are used to define probability amplitudes, i.e., quantum quantities. Expressions (14) and (15) show that p_1, p_4 and p_2 depend upon both r_1 and r_2, and that p_2 moreover depends upon Δ_I and therefore the probabilities clearly follow quantum laws. This instance illustrates the distinction to be made between the quantum status of these probabilities and the validity of the classical approximation for the physical supports that store them. In [7,8,11], once r_1, r_2 and Δ_I were known, the initially prepared qubit states were completely known, and in the context of classical-processing BQSS, we called r_1, r_2 and Δ_I the sources (cf. Section 3) in order to focus on the quantities used in the SS process.

The concept of RV is often used in a classical context. Since on the contrary probabilities p_1, p_4 and p_2 follow quantum laws, treating them as RV does not go without saying. However, Equations (14) and (15) establish that when r_1, r_2, φ_2 are RV (cf. Section 3) the same is true for p_1, p_4 and p_2. They also indicate that p_1, p_4 and p_2 depend upon both r_1 and r_2, and that p_2 also depends upon Δ_I. When $J_{xy} = 0$ (Ising Hamiltonian $-2Js_{1z}s_{2z}$), then $v = 0$ and, for the state at the Mixer output, $p_1p_4 = p_2p_3$, which can be interpreted as follows. The four states defining the $\mathcal{B}_+$ basis are then eigenstates of the Hamiltonian, but time evolution introduces phase differences, and it can be verified that the state at the Mixer output is *entangled* (except if, accidentally, $J(t_1 - t_0)/\hbar = k\pi$, k being an integer). However, when measuring s_{1z} and s_{2z}, the probability of getting $(1/2, 1/2)$ is then time-independent, which is also true for the probabilities of getting $(1/2, -1/2)$, $(-1/2, 1/2)$ or $(-1/2, -1/2)$. Therefore, both products p_1p_4 and p_2p_3 are time-independent, and since $p_1p_4 = p_2p_3$ at t_0, because the qubit pair is then in a product state, this equality is preserved as time goes on, although the state has become entangled.

In the end, these measurements made at the output of the Mixer establish a bridge between the classical and the quantum worlds, the results being kept on macroscopic devices for which the classical approximation is valid, while the probabilities of their occurrences follow quantum laws.

4.3. An Unentanglement Criterion Using Probabilities

The unentanglement criterion expressed through Equation (4) uses the c_i coefficients, i.e., probability amplitudes. However, measurements give access to probabilities, not to probability amplitudes, and the question of establishing whether this unentanglement criterion could be formulated with probabilities (of the results from spin component measurements) therefore seems relevant. State $|\Phi\rangle$ being present at the ouput of the inverting block, and the components s_{1u} and s_{2u} being then measured, we denote the probabilities of obtaining $(1/2, 1/2)$, $(1/2, -1/2)$, $(-1/2, 1/2)$ and $(-1/2, -1/2)$ as P_{1u}, P_{2u}, P_{3u}, P_{4u}, respectively, and the corresponding eigenstates of $s_{1u}.s_{2u}$ as $|+u,+u\rangle$, $|+u,-u\rangle$, $|-u,+u\rangle$ and $|-u,-u\rangle$. If e.g., s_{1x} and s_{2x} are measured, the probabilities are denoted as P_{ix}, with $i = 1$ to 4. In Section 3, it was said that an unentangled normed ket $|\Psi\rangle$ in $\mathcal{E}_1 \otimes \mathcal{E}_2$ possesses four degrees of freedom. Taking the squared modulus of each member of the equality $c_1c_4 = c_2c_3$ leads to

$$P_{1z}P_{4z} = P_{2z}P_{3z}. \tag{16}$$

Then, taking $\vec{u}$ and $\vec{v}$ of Section 2 both along direction x, we know that $c_{1x}c_{4x} = c_{2x}c_{3x}$ for an unentangled state (cf. Equation (8)), and therefore that

$$P_{1x}P_{4x} = P_{2x}P_{3x}. \tag{17}$$

Equation (16) together with (17) is however weaker than condition $c_1c_4 = c_2c_3$, as can be tested by considering the following state:

$$|\Psi_{i-i11}\rangle = \frac{1}{2}(i|++\rangle - i|+-\rangle + |-+\rangle + |--\rangle). \tag{18}$$

$|\Psi_{i-i11}\rangle$ is entangled since $c_1 c_4 = -c_2 c_3$. It can be written

$$|\Psi_{i-i11}\rangle = \frac{1}{2}(|+x,+x\rangle + i|+x,-x\rangle - |-x,+x\rangle + i|-x,-x\rangle). \tag{19}$$

Equation (19) shows that the four probabilities P_{ix} attached to $|\Psi_{i-i11}\rangle$ are all equal to 1/4. Therefore, $|\Psi_{i-i11}\rangle$ satisfies (16) and (17), while being entangled.

The two qubits being in the state $|\Psi\rangle$ expressed through (1), one may decide to treat the three orthogonal directions on the same footing, measuring successively s_x for both spins, then, in a new set of preparations/measurements, s_y for both spins, and finally s_z for both spins. The probabilities of obtaining $(1/2, 1/2))$, $(1/2, -1/2)$, $(-1/2, 1/2)$, $(-1/2, -1/2)$, respectively, when measuring s_{1k} and s_{2k} (with k successively equal to x, y, and z), will be denoted as P_{1k}, P_{2k}, P_{3k} and P_{4k}. For e.g., the entangled state $|\Psi_{i-i11}\rangle$, as $P_{1z}P_{4z} = P_{2z}P_{3z}$ and $P_{1x}P_{4x} = P_{2x}P_{3x}$, the hope is that entanglement can be detected thanks to $P_{1y}P_{4y} \neq P_{2y}P_{3y}$, but, in fact, the four P_{iy} are equal to 1/4. Therefore, measuring the same spin component for both qubits, successively for x, y and z, fails to allow us to build up an unentanglement criterion.

However, since two spins are present, there is still the possibility of not systematically measuring the same spin component for both spins. One chooses to measure successively s_z for both spins, then s_{1z} and s_{2x} in a new set of preparations/measurements, and finally s_{1z} and s_{2y}. The presence of the s_{1z} measurement in each of these sets corresponds to recognizing that (1) uses the standard basis. The probabilities of obtaining $(1/2, 1/2)$, $(1/2, -1/2)$, $(-1/2, 1/2)$, $(-1/2, -1/2)$, respectively, when measuring s_{1i} and s_{2j} (with $i = z$, x, or y, and $j = z$, x, or y) will be denoted as P_{1ij}, P_{2ij}, P_{3ij} and P_{4ij}. Denoting the c_i introduced in Equation (1) as $c_i = \rho_i e^{i\psi_i}$, then from Equation (4) it is known that $|\Psi\rangle$ is unentangled if and only if

$$\{\rho_1 \rho_4 = \rho_2 \rho_3 \quad \text{and} \quad \psi_1 + \psi_4 = \psi_2 + \psi_3 \quad \mod 2\pi\}. \tag{20}$$

Measuring $\{s_{1z}, s_{2z}\}$ allows us to know the moduli $|c_i|^2 = \rho_i^2$ in (1), and to express the first equality in Equation (20) as

$$P_{1zz}P_{4zz} = P_{2zz}P_{3zz}. \tag{21}$$

The P_{kzx} and P_{kzy} (with $k = 1$ to 4), when expressed as functions of the moduli ρ_l and angles ψ_m, depend upon trigonometric functions of the ψ_m angles. For instance, for any state $|\Psi\rangle$ entangled or not

$$2P_{1zx} = (\rho_1^2 + \rho_2^2) + 2\rho_1 \rho_2 \cos(\psi_1 - \psi_2). \tag{22}$$

When expressing unentanglement through probabilities, one then has to try and respect both $\cos \alpha = \cos \beta$ and $\sin \alpha = \sin \beta$ with α and β values compatible with the equality $\psi_1 + \psi_4 = \psi_2 + \psi_3$, rather than to respect the equality $\psi_1 + \psi_4 = \psi_2 + \psi_3$ (mod 2π) itself. If it is first known that simultaneously $P_{1zz}P_{4zz} = P_{2zz}P_{3zz}$ and $P_{1zx}P_{4zx} = P_{2zx}P_{3zx}$ are true, then one immediately deduces that $\cos(\psi_1 - \psi_2) = \cos(\psi_3 - \psi_4)$. In addition, if $P_{1zy}P_{4zy} = P_{2zy}P_{3zy}$ replaces the second equality, one deduces that $\sin(\psi_1 - \psi_2) = \sin(\psi_3 - \psi_4)$. Therefore, when the three equalities between probability products are satisfied, then $\rho_1 \rho_4 = \rho_2 \rho_3$ and $\psi_1 + \psi_4 = \psi_2 + \psi_3$ (mod 2π). Conversely, if $|\psi\rangle$ is unentangled, then Equation (8) implies that $P_{1zj}P_{4zj} = P_{2zj}P_{3zj}$, with $j = z, x, y$ respectively. Finally,

$$c_1 c_4 = c_2 c_3 \iff \{P_{1zj}P_{4zj} = P_{2zj}P_{3zj}, \quad \text{with} \quad j = x, y, z\}. \tag{23}$$

The equivalence therefore is between a single relation between probability amplitudes and a triplet of relations between probabilities. This criterion, although established in the context of BQSS, has the same general validity as Equation (4).

Use of criterion (23) necessitates successive measurements first of s_{1z} and s_{2z}, then (after new preparations) of s_{1z} and s_{2x}, and finally (again after new preparations) of s_{1z} and s_{2y}, in order to successively estimate first the P_{izz} probabilities, then the P_{izx} and finally the P_{izy}. One must measure

s_{1z} each time, because (1) getting e.g., $(+1/2, -1/2)$ when measuring s_{1z} and s_{2z} is an event to be distinguished from the one realized when measuring s_{1z} and s_{2x} and getting $(+1/2, -1/2)$, (2) results of measurements of s_{1z} and s_{2x} are independent only if $|\Psi\rangle$ is unentangled, which precisely can't be assumed when Equation (23) is to be used.

The two distinguishable spins were made to play different roles in the process, which led to Equation (23) (systematic measurement of s_{1z}). This dissymmetry is only partial, as Equation (23) can be replaced by a version obtained by exchanging the spin numbers. The next subsection makes a symmetrical use of measurements of spin components, allowing one to get the *values* of both the ρ_i moduli and the ψ_i angles for the c_i coefficients in Equation (1).

4.4. Knowing 2-Qubit Pure States from s_{ij} Measurements

If a qubit pair physically realized with spins 1/2 is known to be in an arbitrary pure state described by $|\Psi\rangle$ written as in Equation (1), with $c_i = \rho_i e^{i\psi_i}$ and $i = 1$ to 4, then in order to know $|\Psi\rangle$, one should know three moduli ρ_i and three angles ψ_i. Accessing these six real quantities is more demanding than testing $|\Psi\rangle$ unentanglement, since once these quantities are known, it is always possible to know whether $|\Psi\rangle$ is unentangled, by testing whether both equalities $\rho_1\rho_4 = \rho_2\rho_3$ and $\psi_1 + \psi_4 = \psi_2 + \psi_3$ are satisfied. On the contrary, when one focuses upon entanglement, these two equalities may be found to be satisfied, while the values of the ρ_i and ψ_i are unknown. In the previous subsection, an unentanglement criterion using only probabilities in the measurements of the s_{ij} components, equivalent to the $c_1 c_4 = c_2 c_3$ criterion, was given. Its existence suggests the following question: is it possible to access these six real quantities using only probabilities of results in the measurements of the spin components? We are going to show that the answer is yes. It is already known that measurements of both s_{1z} and s_{2z} give access to the moduli ρ_i, through the probabilities P_{izz} introduced in Section 4.3. One is left with e.g., determining the three angle differences $(\psi_1 - \psi_3)$, $(\psi_2 - \psi_3)$ and $(\psi_4 - \psi_3)$ from well-chosen probabilities. We first consider measurements of s_{1z} and s_{2i}, with $i = x$ or y, as in Section 4.3. When measuring s_{1z} and s_{2x}, the probabilities of getting $(1/2, 1/2)$ and $(-1/2, 1/2)$ are, respectively,

$$P_{1zx} = \frac{1}{2}|c_1 + c_2|^2, \qquad P_{3zx} = \frac{1}{2}|c_3 + c_4|^2, \tag{24}$$

which leads to

$$\cos(\psi_1 - \psi_2) = \frac{2P_{1zx} - P_{1zz} - P_{2zz}}{2\sqrt{P_{1zz}P_{2zz}}}, \quad \cos(\psi_3 - \psi_4) = \frac{2P_{3zx} - P_{3zz} - P_{4zz}}{2\sqrt{P_{3zz}P_{4zz}}}. \tag{25}$$

Similarly, when measuring s_{1z} and s_{2y}, the probabilities of getting $(1/2, 1/2)$ and $(-1/2, 1/2)$ are, respectively,

$$P_{1zy} = \frac{1}{2}|c_1 - ic_2|^2, \qquad P_{3zy} = \frac{1}{2}|c_3 - ic_4|^2, \tag{26}$$

which leads to

$$\sin(\psi_1 - \psi_2) = -\frac{2P_{1zy} - P_{1zz} - P_{2zz}}{2\sqrt{P_{1zz}P_{2zz}}}, \quad \sin(\psi_3 - \psi_4) = -\frac{2P_{3zy} - P_{3zz} - P_{4zz}}{2\sqrt{P_{3zz}P_{4zz}}}. \tag{27}$$

Expressions (25) and (27) allow us to know both $(\psi_1 - \psi_2)$ and $(\psi_3 - \psi_4)$ (mod 2π).

Now, exchanging the roles of spins 1 and 2, we successively measure $\{s_{1x}, s_{2z}\}$ and (after new preparations) $\{s_{1y}, s_{2z}\}$. The probabilities of getting $(1/2, 1/2)$ in these measurements are, respectively,

$$P_{1xz} = \frac{1}{2}|c_1 + c_3|^2, \qquad P_{1yz} = \frac{1}{2}|c_1 - ic_3|^2, \tag{28}$$

which leads to

$$\cos(\psi_1 - \psi_3) = \frac{2P_{1xz} - P_{1zz} - P_{3zz}}{2\sqrt{P_{1zz}P_{3zz}}}, \quad \sin(\psi_1 - \psi_3) = -\frac{2P_{1yz} - P_{1zz} - P_{3zz}}{2\sqrt{P_{1zz}P_{3zz}}}. \tag{29}$$

$(\psi_1 - \psi_3)$ is therefore known (mod 2π).

If one wants to identify not the state at the Mixer input but a pure state at the Inverter output, State Tomography (ST) may in principle be used. However, it is far simpler to make measurements for the five $\{s_{1i}, s_{2j}\}$ pairs just considered and to access the corresponding probabilities, than to use ST. The reason is that ST claims to be valid for any quantum state, and therefore does not take advantage of the fact that the qubit pair is presently known to be in a pure state. The dimension of the state space of the qubit pair being four, then, for ST, one has to introduce sixteen operators, namely the Identity, the six operators s_{1i} and s_{2j} (with $i = x, y, z$, and $j = x, y, z$), and the nine products $s_{1i}s_{2j}$ [20]. One should determine experimentally fifteen mean values, giving access to fifteen independent real values together defining the density operator describing the qubit pair state (three diagonal real elements, and six non-diagonal complex elements).

The simpler state estimation procedure proposed in this section therefore opens the way to new classes of BQSS methods, that we just started to explore in [12,13], and then applying this procedure to the Mixer output.

5. Disentanglement and Cylindrical-Symmetry Heisenberg Coupling

In Section 4.2, we considered measurements made at the Mixer output. We now come to the method for BQSS used, e.g., in [9], with classical processing in the adapting block of the separating system, using the notations of [9]. $|\Psi(t_0)\rangle$, the initial product state of the qubit pair, is given by Equation (1), with the values of the coefficients c_i (in the $\mathcal{B}_+$ basis) taken at t_0 and denoted as $c_i(t_0)$. These components form the source vector

$$C_+(t_0) = [c_1(t_0), c_2(t_0), c_3(t_0), c_4(t_0)]^T, \quad T: \text{transpose}. \tag{30}$$

Similarly, the state at the Mixer output at time t, here denoted as $|\Psi(t)>$, is given by Equation (1), with the values of the coefficients c_i (in the $\mathcal{B}_+$ basis) taken at t and denoted as $c_i(t)$. The coupling-induced transition from state $|\Psi(t_0)\rangle$ to $|\Psi(t)\rangle$ is interpreted as the transformation induced by the Mixer, leading to the appearance of $|\Psi(t)\rangle$ at its output. In the same basis, $|\Psi(t)\rangle$ is described by the column vector $C_+(t)$ given by (30), with t replacing t_0. In the matrix formalism, the relation between $C_+(t_0)$ and $C_+(t)$ is written as

$$C_+(t) = MC_+(t_0), \tag{31}$$

where the square fourth-order matrix M describes the effect of the coupling. In [8], it was shown that when the coupling may be described by a Heisenberg cylindrical Hamiltonian, then $M = QDQ^{-1}$, where $Q = Q^{-1}$ is a square matrix with the following non-zero matrix elements:

$$Q_{11} = Q_{44} = 1, \quad Q_{22} = -Q_{33} = Q_{23} = Q_{32} = \frac{1}{\sqrt{2}}, \tag{32}$$

and D is a Diagonal square matrix with its diagonal elements equal to $D_{ii} = e^{-i\omega_i(t-t_0)}$ ($i = 1...4$), the ω_i being real quantities depending upon J_z and J_{xy}, with generally unknown numerical values. The input of the inverting block then receives this state $|\Psi(t)\rangle$. Its output provides a state $|\Phi\rangle$ described in the $\mathcal{B}_+$ basis by a column vector C, with

$$C = UC_+(t) = UMC_+(t_0), \tag{33}$$

where the square matrix U (Unmixing matrix) describes the effect of the inverting block of the separating system. If it is possible to choose U in the form $U = M^{-1}$, then $|\Phi\rangle$ will be equal to $|\Psi(t_0)\rangle$. However, strictly speaking, operating this way is impossible because $M = QDQ$, and D is unknown. In [9], the inverting block was formally built using a chain of quantum gates globally realizing matrix U in the form $U = Q\widetilde{D}Q$, where $\widetilde{D}$ is a diagonal matrix with its four diagonal elements $\widetilde{D}_{ii}$ ($i = 1...4$) equal to

$$\widetilde{D}_{ii} = e^{i\gamma_i}, \quad \gamma_i : \text{free real parameters.} \tag{34}$$

$\widetilde{D}D = \Delta$ is therefore a diagonal matrix with diagonal elements $\Delta_{ii} = e^{i\delta_i}$, where

$$\delta_i = \gamma_i - \omega_i(t - t_0). \tag{35}$$

The $\widetilde{D}$ matrix and the adaptation phase were introduced because it is not possible to modify the values of the D matrix. In the following discussion, it is assumed that the ω_i are time-independent and that the adaptation phase has been successful with respect to unentanglement, i.e., that it has been possible to adjust the γ_i in such a way that, in the inversion phase, if the Writer has prepared each qubit of the qubit pair in an arbitrary pure state at time t_0, we are then sure that state $|\Phi\rangle$ at the output of the inverting block is unentangled. The column vectors $C_+(t_0)$ and C are associated with $|\Psi(t_0)\rangle$ and $|\Phi\rangle$ respectively, and $C = Q\Delta QC_+(t_0)$ is therefore the column vector

$$\begin{pmatrix} e^{i\delta_1} c_1(t_0) \\ [e^{i\delta_2}(c_2(t_0) + c_3(t_0)) + e^{i\delta_3}(c_2(t_0) - c_3(t_0))]/2 \\ [e^{i\delta_2}(c_2(t_0) + c_3(t_0)) - e^{i\delta_3}(c_2(t_0) - c_3(t_0))]/2 \\ e^{i\delta_4} c_4(t_0) \end{pmatrix}. \tag{36}$$

State $|\Phi\rangle$ is unentangled if and only if Equation (4) is fulfilled, i.e., if

$$e^{i(\delta_1+\delta_4)} c_1 c_4 = \frac{1}{4}[2 c_2 c_3 (e^{i2\delta_2} + e^{i2\delta_3}) + (c_2^2 + c_3^2)(e^{i2\delta_2} - e^{i2\delta_3})] \tag{37}$$

(c_i meaning $c_i(t_0)$, for $i = 1$ to 4). We want this relation to be satisfied for any unentangled $|\Psi(t_0)\rangle$. Starting with a $|\Psi(t_0)\rangle$ state with $c_2(t_0)c_3(t_0) \neq 0$ and remembering that $c_1(t_0)c_4(t_0) = c_2(t_0)c_3(t_0)$, Equation (37) may then be written

$$e^{i(\delta_1+\delta_4)} - \frac{1}{2}(e^{i2\delta_2} + e^{i2\delta_3}) = \frac{c_2^2(t_0) + c_3^2(t_0)}{4 c_2(t_0) c_3(t_0)}(e^{i2\delta_2} - e^{i2\delta_3}). \tag{38}$$

Equation (38) is required to be fulfilled for all possible states $|\Psi(t_0)\rangle$ with $c_2(t_0)c_3(t_0) \neq 0$, and for fixed δ_i values (defined once for all during the adaptation phase). The left-hand term does not depend upon the $c_i(t_0)$, whereas its right-hand term does depend upon them. Therefore, Equation (38) is satisfied only if

$$e^{i2\delta_2} - e^{i2\delta_3} = 0, \quad \text{i.e.,} \quad \delta_3 - \delta_2 = m\pi, \; m : \text{integer,} \tag{39}$$

and then Equation (38) moreover imposes that

$$\delta_1 + \delta_4 = 2\delta_2 + 2k\pi, \quad k : \text{integer.} \tag{40}$$

If Equations (39) and (40) and relation $c_1(t_0)c_4(t_0) = c_2(t_0)c_3(t_0)$ are inserted into Equation (36), it is easy to write $|\Phi\rangle$ as a product state, which confirms that if Equations (39) and (40) are fulfilled, and then $|\Phi\rangle$ is unentangled indeed.

If one now supposes e.g., a $|\Psi(t_0)\rangle$ with $c_3(t_0) = 0$, $c_2(t_0) \neq 0$, $c_4(t_0) \neq 0$, and therefore $c_1(t_0) = 0$, then in order for $|\Phi\rangle$ to be unentangled Equation (37) has to be fulfilled. Putting $c_1(t_0) = c_3(t_0) = 0$ into Equation (37) leads to Equation (39), and the δ_i are then not submitted to another constraint.

The same behaviour is found if $c_4(t_0) = c_3(t_0) = 0$, and $c_1(t_0) \neq 0$, $c_2(t_0) \neq 0$, and this remains true if $c_1(t_0) = c_2(t_0) = c_4(t_0) = 0$, $c_3(t_0) \neq 0$.

When one starts with an *arbitrary* initial unentangled state $|\Psi(t_0)\rangle$, the following property is a consequence of the results of the previous discussion. If during the adaptation phase it has been possible to rightly fix the γ_i values, one may claim that the corresponding $|\Phi\rangle$ is unentangled if and only if during that adaptation phase the choice of the γ_i has allowed conditions (39) and (40) to be both fulfilled. This, however, does not guarantee that $|\Phi\rangle$ is identical to $|\Psi(t_0)\rangle$. The latter identification corresponds to source restoration itself, outside the scope of this article.

6. Conclusions

Conventional BSS is a mature field of Signal Processing, with various applications. Its extension into a quantum context has been developing for a decade, first through the creation of theoretical methods for Blind Quantum Source Separation (BQSS), with classical and/or quantum processing, and recently through the use of BQSS in the exploration of Blind Quantum Process Tomography (BQPT). The present paper examined in detail concepts (e.g., those of quantum sources and of their independence) and established properties (e.g., an unentanglement criterion) introduced in our previous papers. In the BQSS context, with qubits supposed to be realized with spins 1/2, one has to face two major consequences of the quantum behaviour. First, if each qubit of a spin qubit pair is initially prepared in a pure state, and the time evolution of the pair state is governed by some undesired coupling between the spins, the Reader at the Mixer output accesses an unknown generally entangled qubit pair quantum state. This entangled state may be sent to a quantum processing system in order to restore the initially prepared state. Writing the output state of this processing system as e.g., $|\Phi\rangle = \sum_i c_i |i\rangle$ in the standard basis, with well-ordered basis states, we showed that this state is unentangled if and only if $c_1 c_4 = c_2 c_3$, a constraint between probability amplitudes. Secondly, results of measurements of the qubit spin components have a probabilistic nature, and the corresponding probabilities follow quantum properties even when processed with classical means. This article shows precautions to be taken when trying to extend to Blind Quantum SS the concept of source statistical independence used in conventional BSS. Using the probabilities P_{izj} of getting the different possible results when measuring s_{1z} and s_{2j}, successively with $j = z$, x and y, it is shown that the above unentanglement criterion may be written as $\{P_{1zj}P_{4zj} = P_{2zj}P_{3zj}\}$, a set of three constraints between probabilities. This unentanglement criterion has already been used in the adaptation phase of Blind Quantum SS, through a disentanglement-based separation principle, before restoration of the initial unentangled state. The already developed BQSS/BQPT methods do not depend on some specific interpretation of Quantum Theory, while respecting its general postulates.

Acknowledgments: This theoretical study was performed without financial support. The costs to publish in open access were handled by Yannick Deville, in the framework of the research activities and projects that he is heading in his lab (Institut de Recherche en Astrophysique et Planétologie).

Author Contributions: This theoretical study was performed by Alain Deville and Yannick Deville, in connection with the research activities about related topics that they also performed together (see above-mentioned papers). Both authors participated in writing this paper.

Conflicts of Interest: The authors declare no conflict of interest.

Appendix A. About Applications of Blind Conventional and Quantum Source Separation

Appendix A.1. Conventional BSS

Some audio systems aim at automatic recognition of speech by a processing unit, e.g., in order to control actuators (for instance, a car driver can thus control various car functions by speech). When a speech signal is recorded by a set of microphones situated in a noisy environment, each recorded signal is a mixture of speech and of various noise signals. In order to avoid a degraded recognition performance in case these plain recordings were directly provided to an automatic speech recognition (ASR) system, these recordings may be first pre-processed by means of a BSS system, so as to extract the speech signal. The denoised speech output of this BSS system is then provided to the ASR system (see [11] and references therein).

When using radio-frequency signals to transmit digital data, reception antennas may simultaneously receive several mixed data streams. BSS is then applied to first unmix these signals. Each extracted signal may then be separately used as required in the considered application. Its use in the radio-frequency identification (RFID) system instance is briefly presented in [11].

The biomedical field makes a systematic use of signals such as electrocardiograms (ECGs) or electroencephalograms (EEGs), processed by human experts or computers. This "main task" is often difficult because each signal in the recorded set is a mixture of various contributions, and the information of interest thus cannot be easily extracted from any such mixed signal. Again, a solution to this problem consists of pre-processing the original recordings by means of BSS methods, so as to extract each signal component of interest separately on each output of this BSS system. In [11], information is given about the extraction of foetus's heartbeats from ECG recordings which were mixtures of large-magnitude mother's heartbeats, low-magnitude foetus's heartbeats and noise components. These foetus's heartbeats were hardly visible in the original recordings.

BSS is closely related to the so-called Blind System Identification (BSI). The problem of describing an unknown classical (i.e., non quantum) system through a realistic model is called system identification. When e.g., this system may be described by a matrix, the task is the determination of its matrix elements. In Blind System Identification, some properties of the input signals are known, but the input signals themselves are unknown. Methods for BSS often include the determination of the unknown mixer function or of its inverse. This is a kind of BSI problem, called Blind Mixture Identification (BMI).

Appendix A.2. Blind Quantum Source Separation

The acronym BQSS describes the operations aimed at recovering the source state(s) (possibly up to some accepted indeterminacies), in a context already described in this paper. BQSS with classical processing can already be used, e.g., by physicists, in possible experiments requiring methods for retrieving information about individual quantum states from measurements performed after undesired coupling between these states, e.g., when dealing with quantum phenomena involving electron spins 1/2. BQSS with quantum processing keeps the quantum form of the available mixed data and processes them by means of quantum circuits in order to retrieve the quantum sources. This version of our QSS methods could be of interest for the core of future quantum computers, where both the data to be processed and the processing means will have a quantum form. Quantum-processing BQSS may then be used as a pre-processing stage, to remove undesired alterations (e.g., due to Heisenberg coupling between physical qubits made with electron spins) of the data to be provided to the input of the main processing stage, which then applies the final quantum algorithm to these pre-processed data. It was explained in Part A.1 of this Appendix that such a two-stage system architecture is already used in conventional BSS.

Independently from BQSS, the QIP community has already developed what is called Quantum Process Tomography (QPT), the quantum version of system identification, and which operates in a non-blind way. It turns out that BQSS, by estimating the inverse of the mixing function, is also able to estimate this function itself, i.e., the parameters of the considered coupling operator (possibly up to some residual transforms, called indeterminacies as in classical BSS). BQSS therefore opens the way to introducing the blind version of QPT (called BQPT), i.e., performing QPT essentially without knowing the values of the input quantum states of the considered process (but e.g., requesting them to be unentangled). The applications related to BQSS thus include applications of BQPT, as a spin-off. In [14], it was recalled that QPT is considered the gold standard for fully characterising quantum systems, and in particular for characterising the quantum logic gates that form the basic elements of a quantum computer. Extending the standard QPT tool to BQPT, its blind version, should be of interest, e.g., when the input states of the considered process indeed cannot be known, or when it is important to benefit from the fact that BQSS avoids the intrisic complexity of standard QPT methods. For more details about the applications of BQSS and BQPT, the interested reader may refer to [11,14], and to references therein.

References

1. Laloë, F. *Comprenons-Nous Vraiment la MéCanique Quantique*; EDP Sciences Les Ulis: Les Ulis, France, 2011; English version: *Do We Really Understand Quantum Mechanics?* Cambridge University Press: Cambridge, UK, 2012.
2. Cohen-Tannoudji, C.; Diu, B.; Laloë, F. *Mécanique Quantique*; Hermann: Paris, France, 1973; English version: *Quantum Mechanics*; John Wiley: New York, NY, USA, 1977.
3. Dirac, P. Quantum Mechanics of Many-Electron Systems. *Proc. R. Soc. A* **1929**, *123*, 714–733.
4. Timpson, C.G. Quantum Information Theory and the Foundations of Quantum Mechanics. Ph.D. Thesis, University of Oxford, Oxford, UK, 2004.
5. Comon, P.; Jutten, C. (Eds.) *Handbook of Blind Source Separation: Independent Component Analysis and Applications*; Academic Press: Oxford, UK, 2010.
6. Deville, Y. Blind Source Separation and Blind Mixture Identification Methods. In *Wiley Encyclopedia of Electrical and Electronics Engineering*; Webster, J., Ed.; Wiley: Hoboken, NJ, USA, 2016; pp. 1–33.
7. Deville, Y.; Deville, A. Blind separation of quantum states: Estimating two qubits from an isotropic Heisenberg spin coupling model. In Proceedings of the 7th International Conference on Independent Component Analysis and Signal Separation, London, UK, 9–12 September 2007; Davies, M.E., James, C.J., Abdallah, S.A., Plumbley, M.D., Eds.; Springer: Berlin, Germany, 2007; pp. 706–713.
8. Deville, Y.; Deville, A. Classical-processing and quantum-processing signal separation methods for qubit uncoupling. *Quantum Inf. Process.* **2012**, *11*, 1311–1347.
9. Deville, Y.; Deville, A. A quantum-feedforward and classical-feedback separating structure adapted with monodirectional measurements; blind qubit uncoupling capability and links with ICA. In Proceedings of the 23rd IEEE International Workshop on Machine Learning for Signal Processing, Southampton, UK, 22–25 September 2013.
10. Deville, Y.; Deville, A. Blind qubit state disentanglement with quantum processing: Principle, criterion and algorithm using measurements along two directions. In Proceedings of the 2014 IEEE International Conference on Acoustics, Speech and Signal Processing, Florence, Italy, 4–9 May 2014; pp. 6262–6266.
11. Deville, Y.; Deville, A. Quantum-Source Independent Component Analysis and Related Statistical Blind Qubit Uncoupling Methods. In *Blind Source Separation: Advances in Theory, Algorithms and Applications*; Naik, G.R., Wang, W., Eds; Springer: Berlin, Germany, 2014; pp. 3–37.
12. Deville, Y.; Deville, A. From blind quantum source separation to blind quantum process tomography. In Proceedings of the 12th International Conference on Latent Variable Analysis and Signal Separation, Liberec, Czech Republic, 25–28 August 2015; Vincent, E., Yeredor, A., Koldovský, Z., Tichavský, P., Eds.; Springer: Berlin, Germany, 2015; pp. 184–192.
13. Deville, Y.; Deville, A. Blind quantum computation: Blind quantum source separation and blind quantum process tomography. In Proceedings of the 19th Conference on Quantum Information Processing, Banff, AB, Canada, 10–15 January 2016.
14. Deville, Y.; Deville, A. Blind quantum source separation: Quantum-processing qubit uncoupling systems based on disentanglement. *Digit. Signal Process.* **2017**, *67*, 30–51.
15. Deville, Y. *Traitement du Signal: Signaux Temporels et Spatiotemporels—Analyse des Signaux, Théorie de L'information, Traitement D'antenne, Séparation Aveugle de Sources*; Ellipses Editions Marketing: Paris, France, 2011. (In French)
16. Feynman, R.P. Quantum Mechanical Computers. *Opt. News* **1985**, *11*, 11–20.
17. Feynman, R.P. *Feynman Lectures on Computation*; Perseus Publishing: Cambridge, MA, USA, 1996.
18. Peres, A. Separability Criterion for Density Matrices. *Phys. Rev. Lett.* **1996**, *77*, 1413–1415.
19. Horodecki, M.; Horodecki, P.; Horodecki, R. Separability of mixed states: Necessary and sufficient conditions. *Phys. Lett. A* **1996**, *223*, 1–8.
20. Nielsen, M.A.; Chuang, I.L. *Quantum Computation and Quantum Information*; Cambridge University Press: Cambridge, UK, 2000.

21. Buchleitner, A.; Viviescas, C.; Tiersch, M. (Eds.) *Entanglement and Decoherence (Lectures Notes in Physics)*; Springer: Berlin, Germany, 2009.
22. Köhler, J.; Disselhorst, J.A.J.M.; Donckers, M.C.J.M.; Groenen, E.J.J.; Schmidt, J.; Moerner, W.E. Magnetic resonance of a single molecular spin. *Nature* **1993**, *363*, 242–244.
23. Gruber, A.; Dräbenstedt, A.; Tietz, C.; Fleury, L.; Wrachtrup, J.; von Borczyskowski, C. Scanning Confocal Optical Microscopy and Magnetic Resonance on Single Defect Centers. *Science* **1997**, *276*, 2012–2014.
24. Rugar, D.; Budakian, R.; Mamin, H.J.; Chui, B.W. Single spin detection by magnetic resonance force microscopy. *Nature* **2004**, *430*, 329–332.
25. Otte, A.F. Can data be stored in a single magnetic atom? *Europhys. News* **2008**, *38*, 31–34.
26. Bienfait, A.; Pla, J.J.; Kubo, Y.; Stern, M.; Zhou, X.; Lo, C.C.; Weis, C.D.; Schenkel, T.; Thewalt, M.L.W.; Vion, D.; et al. Reaching the quantum limit of sensitivity in electron spin resonance. *arXiv* **2015**, arXiv:1507.06831.
27. Hyvärinen, A.; Karhunen, J.; Oja, E. *Independent Component Analysis*; Wiley: New York, NY, USA, 2001.
28. Abragam, A. *The Principles of Nuclear Magnetism*; Oxford University Press: Oxford, UK, 1961.
29. Tolman, R.C. *The Principles of Statistical Mechanics*; Oxford University Press: Oxford, UK, 1938; p. 327.
30. Von Neumann, J. *Les Fondements Mathématiques de la Mécanique Quantique*; Alcan: Paris, France, 1946; Editions Jacques Gabay: Paris, France, 1988. (In French)
31. Barnett, S.M. *Quantum Information*; Oxford University Press: Oxford, UK, 2009.
32. Fuchs, C.A.; Peres, A. Quantum theory needs no "interpretation". *Phys. Today* **2000**, *53*, 70–71.
33. Margenau, H. Quantum-Mechanical description. *Phys. Rev.* **1936**, *49*, 240–242.
34. Margenau, H. Critical Points in Modern Physical Theory. *Philos. Sci.* **1937**, *4*, 337–370.
35. Feynman, R.P. *Statistical Mechanics*; Basic Books: New York, NY, USA, 1972.
36. Abragam, A.; Bleaney, B. *Electron Paramagnetic Resonance of Transition Ions*; Oxford University Press: Oxford, UK, 1970.
37. DiVincenzo, D.P. Quantum Computation. *Science* **1995**, *270*, 255–261.
38. Fazekas, P. *Electron Correlation and Magnetism*; World Scientific: Hackensack, NJ, USA, 1999.

© 2017 by the authors. Licensee MDPI, Basel, Switzerland. This article is an open access article distributed under the terms and conditions of the Creative Commons Attribution (CC BY) license (http://creativecommons.org/licenses/by/4.0/).

MDPI
St. Alban-Anlage 66
4052 Basel
Switzerland
Tel. +41 61 683 77 34
Fax +41 61 302 89 18
www.mdpi.com

Entropy Editorial Office
E-mail: entropy@mdpi.com
www.mdpi.com/journal/entropy

www.ingramcontent.com/pod-product-compliance
Lightning Source LLC
LaVergne TN
LVHW071933080526
838202LV00064B/6604